Mendocino F.Z.

Murray F.Z.

Molokai F.Z.

Clarion F.Z.

Clipperton F.Z.

Galapagos Rift

Hawaiian Is.

Line Is.

Marshall-Gilbert Is.

Marquesas Is.

Tuamotu Is.

Society Is.

Cook Is.

Easter Island F.Z.

Pacific Rise

Chile Rise

Peru-Chile Trench

Puerto Rico Trench

Eltanin F.Z.

East

Ridge

Aleutian Trench

Kuril Trench

Juan de Fuca Ridge

ALASKA

NORTH AMERICA

ROCKY MOUNTAINS

SOUTH AMERICA

AUSTRALIA

ANTARCTICA

BORNEO

NEW GUINEA

TASMANIA

LAPTEV SEA

NEW SIBERIAN ISLANDS

WRANGEL ISLAND

CHUKCHI SEA

CHUKCHI CAP

CANADA ABYSSAL PLAIN

CONTINENTAL RISE

MELVILLE ISLAND

BANKS I.

VICTORIA ISLAND

BAFFIN ISLAND

ELLESMERE ISLAND

BAFFIN BAY

BROOKS RANGE

BERING STRAIT

BERING SEA

HUDSON BAY

OKHOTSK SEA

BERING ABYSSAL PLAIN

ALEUTIAN ABYSSAL PLAIN

JAPAN SEA

YELLOW SEA

PEKING

SHANGHAI

TOKYO

STANOVOY MTS.

VERKHOYANSK MTS.

KOLYMA

GIBSON DESERT

SIMPSON DESERT

NULLARBOR PLAIN

Perth

Sydney

NEW ZEALAND

WILKES ABYSSAL PLAIN

WILKES LAND

MT. COOK

Chicago

Montreal

New York

Washington

New Orleans

México

GRAND BANKS

Buenos Aires

MATO GROSSO

FALKLAND ISLANDS

BELLINGSHAUSEN ABYSSAL PLAIN

Essentials of
Oceanography

SEVENTH EDITION

Essentials of
Oceanography

Harold V. Thurman

Professor Emeritus
Mt. San Antonio College

Alan P. Trujillo

Associate Professor
Palomar College

Upper Saddle River, New Jersey 07458

Library of Congress Cataloging-in-Publication Data

Thurman, Harold V.
  Essentials of oceanography / Harold V. Thurman, Alan P. Trujillo.—7th ed.
    p. cm.
  Includes bibliographical references and index.
  ISBN 0-13-065235-0
  1. Oceanography.   I. Trujillo, Alan P.   II. Title.
  GC11.2.T49 2002
  551.46—dc21   2001040922

Senior Editor: Patrick Lynch
Editor-in-Chief: Sheri L. Snavely
Editor-in-Chief, Development: Carol Trueheart
Development Editor: John Murdzek
Vice President of Production and Manufacturing: David W. Riccardi
Executive Managing Editor: Kathleen Schiaparelli
Assistant Managing Editor: Beth Sturla
Senior Marketing Manager: Christine Henry
Assistant Editor: Amanda Griffith
Director of Creative Services: Paul Belfanti
Director of Design: Carole Anson
Managing Editor, Audio/Video Assets: Grace Hazeldine
Art Editor: Adam Velthaus
Art Director: Jonathan Boylan
Interior/Cover Design: Maureen Eide
Manufacturing Manager: Trudy Pisciotti
Assistant Manufacturing Manager: Michael Bell
Photo Researcher: Christine Pullo
Photo Editor: Reynold Reiger
Art Studio: Artworks
    Senior Manager: Patty Burns
    Production Manager: Rhonda Whitson
    Manager, Production Technologies: Matt Haas
    Project Coordinator: Connie Long
    Illustrators: Kathryn Anderson, Mark Landis
    Art Quality Assurance: Tim Nguyen, Stacy Smith, Pamela Taylor
Editorial Assistants: Nancy Bauer; Sean Hale
Production Supervision/Composition: Lithokraft II
Cover / Title Page Photo: Full view of iceberg, Ralph A. Clevenger/Corbis, National Maritime Museum, © National Maritime Museum Picture Library, London, England.

Prentice Hall ©2002, 1999, 1996 by Prentice-Hall, Inc.
Upper Saddle River, New Jersey 07458

Earlier editions ©1993, 1990, 1987, 1983 by Macmillan Publishing Company

ISBN 0-13-065235-0

Pearson Education Ltd., *London*
Pearson Education Australia Pty. Limited, *Sydney*
Pearson Education Singapore Pte. Ltd.
Pearson Education North Asia Ltd., *Hong Kong*
Pearson Education Canada, Ltd., *Toronto*
Pearson Educación de Mexico, S.A. de C.V.
Pearson Education—Japan, *Tokyo*
Pearson Education Malaysia, Pte. Ltd.

To my parents Anthony and Ann Trujillo
and for my daughter Eva Ryan Trujillo

About the Authors

HAL THURMAN retired in May 1994 after 24 years of teaching in the Earth Sciences Department of Mt. San Antonio College in Walnut, California. Interest in geology led to a bachelor's degree from Oklahoma A&M University, followed by seven years working as a petroleum geologist, mainly in the Gulf of Mexico. Here his interest in oceans developed, and he earned a master's degree from California State University at Los Angeles, then joined the Earth Sciences faculty at Mt. San Antonio College. Hal Thurman is also the author of *Introductory Oceanography* and has co-authored a marine biology textbook, written articles on the Pacific, Atlantic, Indian, and Arctic Oceans for the 1994 edition of *World Book Encyclopedia,* and served as a consultant on the National Geographic publication, *Realms of the Sea.* He enjoys going to sea on vacations with his wife Iantha.

AL TRUJILLO teaches at Palomar Community College in San Marcos, California, where he is co-Director of the Oceanography Program. He received his bachelor's degree in geology from the University of California at Davis and his master's degree in geology from Northern Arizona University, afterwards working for several years in industry as a developmental geologist, hydrogeologist, and computer specialist. Al began teaching in the Earth Sciences Department at Palomar in 1990 and in 1997 was awarded Palomar's Distinguished Faculty Award for Excellence in Teaching. He has contributed to other Prentice Hall Earth science textbooks, including *Earth,* 7th Edition and *Earth Science,* 10th Edition. He is currently working with Hal Thurman as a co-author on *Introductory Oceanography,* 10th Edition. In addition to writing and teaching, Al works as a naturalist and lecturer aboard natural history expedition vessels in Alaska and the Sea of Cortez/Baja California. His research interests include beach processes, sea cliff erosion, and computer applications in oceanography. Al and his wife, Sandy, have two children, Karl and Eva.

Brief Contents

Introduction   1

1 Introduction to Planet "Earth"   6

2 Plate Tectonics and the Ocean Floor   34

3 Marine Provinces   74

4 Marine Sediments   97

5 Water and Seawater   130

6 Air-Sea Interaction   162

7 Ocean Circulation   197

8 Waves and Water Dynamics   236

9 Tides   265

10 The Coast: Beaches and Shoreline Processes   288

11 The Coastal Ocean   316

12 The Marine Habitat   346

13 Biological Productivity and Energy Transfer   372

14 Animals of the Pelagic Environment   406

15 Animals of the Benthic Environment   438

Afterword   471

Appendices   473
Appendix I   473
Appendix II   476
Appendix III   478
Appendix IV   481
Appendix V   485

Glossary   489

Credits and Acknowledgments   510

Index   514

Contents

Preface   xv

To the Student   xv
To the Instructor   xv
What's New in this Edition?   xvi
The New Instructional Package   xvii
Acknowledgments   xviii
Environmental Issues in Oceanography   xix

Introduction   1

What is Oceanography?   1
Earth's Oceans   2
Rational Use of Technology?   3

1

Introduction to Planet "Earth"   6

Key Questions   7
Feature: Diving into the Marine Environment   7
Geography of the Oceans   7
   The Four Principal Oceans, Plus One   7
   The Seven Seas?   9   Comparing the Oceans to the
   Continents   10
Explorations of the Oceans: Some Historical Notes
   About Oceanography   10
   Early History   10   The Middle Ages   13
   The Age of Discovery in Europe   13
Box 1–1   How do Sailors Know Where They Are at Sea?:
   From Stick Charts to Satellites   14
   The Beginning of Voyaging for Science   17
   History of Oceanography...To Be Continued   17
The Nature of Scientific Inquiry   17
   Observations   18   Hypothesis   18   Testing   18
   Theory   18   Theories and the Truth   18
Origins   19
   Origin of the Solar System and Earth   19
   Origin of the Atmosphere and the Oceans   21
Life Begins in the Oceans   22
   The Importance of Oxygen to Life   22
   Plants and Animals Evolve   23
Radiometric Dating and the Geologic Time Scale   25
Students Sometimes Ask... 26
Box 1–2   "Deep" Time   28
Chapter in Review   31
Key Terms   31

Questions and Exercises   31
References   32
   Suggested Reading in Scientific American   32
Oceanography on the Web   33

2

Plate Tectonics and the Ocean Floor   34

Key Questions   34
Feature: Voyages to Inner Space: Visiting the Deep
  Ocean Floor in Submersibles   34
Evidence for Continental Drift   35
   Fit of the Continents   35   Matching Sequences of
   Rocks and Mountain Chains   36   Glacial Ages
   and Other Climate Evidence   37   Distribution of
   Organisms   38   Objections to the Continental
   Drift Model   39
Evidence for Plate Tectonics   40
   Earth's Magnetic Field and Paleomagnetism   40
   Sea Floor Spreading and Features of the Ocean
   Basins   43
Box 2–1   Do Sea Turtles (and Other Animals) Use Earth's
   Magnetic Field for Navigation?   44
   Other Evidence from the Ocean Basins   45
   The Acceptance of a Theory   47
Earth Structure   47
   Chemical Composition Versus Physical
   Properties   47   Near the Surface   50   Isostatic
   Adjustment   50
Plate Boundaries   52
   Divergent Boundaries   53   Convergent
   Boundaries   55   Transform Boundaries   59
Testing the Model: Some Applications
   of Plate Tectonics   60
   Mantle Plumes and Hotspots   60   Seamounts and
   Tablemounts   62   Coral Reef Development   63
   Detecting Plate Motion with Satellites   65
   The Past: Paleoceanography   65   The Future:
   Some Bold Predictions   66
Students Sometimes Ask... 67
Chapter in Review   70
Key Terms   70
Questions and Exercises   71
References   72
   Suggested Reading in Scientific American   73
Oceanography on the Web   73

3

Marine Provinces   74

Key Questions 74
Feature: Experiments in Underwater Living 74
Bathymetry 75
Bathymetric Techniques 75   Provinces of the Ocean Floor 77   Features of Continental Margins 77
Box 3–1   Sea Floor Mapping from Space 80
Features of the Deep-Ocean Basin 84
Box 3–2   A Grand "Break": Evidence for Turbidity Currents 86
Features of the Mid-Ocean Ridge 87
Students Sometimes Ask... 91
Chapter in Review 94
Key Terms 94
Questions and Exercises 95
References 95
Suggested Reading in Scientific American 95
Oceanography on the Web 96

4

Marine Sediments   97

Key Questions 97
Feature: Collecting the Historical Record of the Deep-Ocean Floor 97
Lithogenous Sediment 100
Origin 100   Composition 101   Sediment Texture 101   Distribution 103
Biogenous Sediment 105
Origin 105   Composition 106
Hydrogenous Sediment 110
Origin 110   Composition and Distribution 110
Box 4–1   Diatoms: The Most Important Things You Have (Probably) Never Heard Of 111
Distribution 112
Cosmogenous Sediment 114
Origin, Composition, and Distribution 114
Box 4–2   When a Sea Was Dry: Clues from the Mediterranean 115
Distribution of Neritic and Pelagic Deposits: A Summary 116
Box 4–3   When the Dinosaurs Died: The Cretaceous–Tertiary (K–T) Event 117
Mixtures 118
Ocean Sediments as a Resource 120
Petroleum 120   Gas Hydrates 121   Sand and Gravel 121   Evaporative Salts 122   Phosphorite (Phosphate Minerals) 123   Manganese Nodules and Crusts 123
Students Sometimes Ask... 124

Chapter in Review 125
Key Terms 126
Questions and Exercise 127
References 128
Suggested Reading in Scientific American 129
Oceanography on the Web 129

5

Water and Seawater   130

Key Questions 130
Feature: The HMS *Challenger* Expedition: Birth of Oceanography 130
Atomic Structure 131
The Water Molecule 132
Geometry 132   Polarity 132   Interconnections of Molecules 133   Water: The Universal Solvent 133
Water's Thermal Properties 134
Heat, Temperature, and Changes of State 134   Water's Freezing and Boiling Points 135   Water's Heat Capacity 135   Water's Latent Heats 136
Water Density 139
Seawater 140
Salinity 140   Salinity Variations 142
Box 5–1   How to Avoid Goiters 143
Determining Salinity 144
Dissolved Components Added and Removed from Seawater 144
Acidity and Alkalinity of Seawater 146
The pH Scale 146   The Carbonate Buffering System 146
Processes Affecting Seawater Salinity 147
Processes That Decrease Seawater Salinity 147   Processes That Increase Seawater Salinity 148   The Hydrologic Cycle 148
Surface and Depth Salinity Variation 148
Surface Salinity Variation 148   Depth Salinity Variation 149
Seawater Density 150
Pycnocline and Thermocline 152
Box 5–2   The Hot and Cold About OTEC Systems 154
Comparing Pure Water and Seawater 155
Desalination 155
Distillation 155   Membrane Processes 156   Other Methods of Desalination 157
Students Sometimes Ask... 157
Chapter in Review 158
Key Terms 159
Questions and Exercises 160
References 160
Suggested Reading in Scientific American 161
Oceanography on the Web 161

6

Air–Sea Interaction   162

Key Questions 162
Feature: RMS *Titanic*: Lost (1912) and Found (1985) 162
Uneven Solar Heating on Earth 164
Distribution of Solar Energy 164   Earth's Seasons 164   Oceanic Heat Flow 165 The Atmosphere: Physical Properties 166 An Example: A Non-Spinning Earth 168
The Coriolis Effect 168
Example 1: Perspectives and Frames of Reference on a Merry-Go-Round 169   Example 2: A Tale of Two Missiles 170   Changes in the Coriolis Effect with Latitude 171
Atmospheric Circulation Cells on a Spinning Earth 172
Circulation Cells 172   Pressure 172   Wind Belts 173   Boundaries 173   Circulation Cells: Idealized or Real? 173
Box 6–1   Why Christopher Columbus Never Set Foot on North America 174
The Oceans, Weather, and Climate 175
Winds 175   Storms 175   Tropical Cyclones (Hurricanes) 178   Climate Patterns in the Oceans 182
Box 6–2   The Storm of the Century: Galveston, Texas (1900) 184
The Atmosphere's Greenhouse Effect 185
Which Gases Contribute to the Greenhouse Effect? 186   What Changes Will Occur as a Result of Increased Global Warming? 186 The IPCC and the Kyoto Protocol 189   The Ocean's Role in Reducing the Greenhouse Effect 189 What Should We Do About the Increasing Greenhouse Gases? 190
Students Sometimes Ask... 190
Chapter in Review 191
Box 6–3   The ATOC Experiment: SOFAR so Good? 192
Key Terms 194
Questions and Exercises 194
References 195
Suggested Reading in Scientific American 195
Oceanography on the Web 196

7

Ocean Circulation   197

Key Questions 197
Feature: Benjamin Franklin: The World's Most Famous Physical Oceanographer 197
Measuring Ocean Currents 198

Surface Currents 199
Equatorial Currents, Boundary Currents, and Gyres 200   Ekman Spiral and Ekman Transport 202
Box 7–1   Running Shoes as Drift Meters: Just Do It 203
Geostrophic Currents 205 Western Intensification 205
Box 7–2   The Voyage of the *Fram*: A 1000-Mile Journey Locked in Ice 206
Equatorial Countercurrents 207   Ocean Currents and Climate 208
Upwelling and Downwelling 209
Diverging Surface Water 209   Converging Surface Water 209   Coastal Upwelling and Downwelling 209   Other Upwelling 211
Surface Currents of the Oceans 211
Antarctic Circulation 211   Atlantic Ocean Circulation 213   Pacific Ocean Circulation 217 Indian Ocean Circulation 223
Box 7–3   El Niño and the Incredible Shrinking Marine Iguanas of the Galápagos Islands 224
Deep Currents 225
Origin of Thermohaline Circulation 225   Sources of Deep Water 226   Worldwide Deep-Water Circulation 227
Students Sometimes Ask... 230
Chapter in Review 231
Key Terms 233
Questions and Exercises 233
References 233
Suggested Reading in Scientific American 234
Oceanography on the Web 235

8

Waves and Water Dynamics   236

Key Questions 236
Feature: The Biggest Wave in Recorded History: Lituya Bay, Alaska (1958) 236
What Causes Waves? 238
How Waves Move 240
Wave Characteristics 240
Circular Orbital Motion 241   Deep-Water Waves 241   Shallow Water Waves 242 Transitional Waves 243
Wind-Generated Waves 243
"Sea" 244   Swell 245   Surf 247
Box 8–1   Rogue Waves: Ships Beware! 249
Wave Refraction 251   Wave Reflection 252
Tsunami 253
Coastal Effects 254   Historic Tsunami 254 Tsunami Warning System 256
Box 8–2   The Big Shake: A Tsunami from the Cascadia Subduction Zone Hits Japan 257

Power from Waves 259
Students Sometimes Ask... 259
Chapter in Review 261
Key Terms 262
Questions and Exercises 263
References 264
    Suggested Reading in Scientific American *264*
Oceanography on the Web 264

9

Tides   265

Key Questions 265
Feature: A Brief History of Some Successful Tidal
Power Plants 265
Generating Tides 266
    Tide-Generating Forces 266   Tidal Bulges: The
    Moon's Effect 270   Tidal Bulges: The Sun's
    Effect 270   Earth's Rotation 270   The
    Monthly Tidal Cycle 271   Other Factors 273
    Idealized Tide Prediction 274
Tides in the Ocean 274
    Tidal Patterns 276   An Example of Tidal
    Extremes: The Bay of Fundy 277   Coastal Tidal
    Currents 277
Box 9–1   Tidal Bores: Boring Waves these Are Not! 278
    Some Considerations of Tidal Power 280
Students Sometimes Ask... 282
Box 9–2   Grunions: Doing What Comes Naturally on the
Beach 282
Chapter in Review 285
Key Terms 286
Questions and Exercises 286
References 287
    Suggested Reading in Scientific American *287*
Oceanography on the Web 287

10

The Coast: Beaches and Shoreline
Processes   288

Key Questions 288
Feature: The Ultimate Protection: The National
Flood Insurance Program (NFIP) 288
The Coastal Region 289
    Beach Terminology 289   Beach Composition 290
    Movement of Sand on the Beach 290
Erosional- and Depositional-Type Shores 292
    Features of Erosional-Types Shores 293
Box 10–1   Warning: Rip Currents ... Do You Know What
to Do? 294

    Features of Depositional-Types Shores 295
Emerging and Submerging Shorelines 300
    Tectonic and Isostatic Movements of Earth's
    Crust 301   Eustatic Changes in Sea Level 302
    Sea Level and the Greenhouse Effect 303
Characteristics of U.S. Coasts 304
    The Atlantic Coast 304   The Gulf Coast 305
    The Pacific Coast 305
Hard Stabilization 305
    Groins and Groin Fields 307   Jetties 307
    Breakwaters 308   Seawalls 309   Alternatives
    to Hard Stabilization 310
Students Sometimes Ask... 310
Box 10–2   The Move of the Century: Relocating the Cape
Hatteras Lighthouse 312
Chapter in Review 313
Key Terms 314
Questions and Exercises 314
References 315
    Suggested Reading in Scientific American *315*
Oceanography on the Web 315

11

The Coastal Ocean   316

Key Questions 316
Feature: The Law of the Sea 316
Coastal Waters 318
    Salinity 318   Temperature 318   Coastal
    Geostrophic Currents 318
Estuaries 320
    Origin of Estuaries 320   Water Mixing in
    Estuaries 321   Estuaries and Human
    Activities 322
Coastal Wetlands 323
    Serious Loss of Valuable Wetlands 324
Lagoons 326
    Laguna Madre 326
A Case Study: The Mediterranean Sea 327
    Mediterranean Circulation 327
Pollution in Coastal Waters 327
    What is Pollution 327   Petroleum 329
Box 11–1   The *Exxon Valdez* Oil Spill: Not the Worst Spill
Ever 330
    Sewage Sludge 335   DDT and PCBs 336
    Mercury and Minamata Disease 338
    Non-Point-Source Pollution and Trash 340
Students Sometimes Ask... 340
Box 11–2   From A to Z in Plastics: The Miracle
Substance? 342
Chapter in Review 344
Key Terms 345
Questions and Exercises 345
References 345

Suggested Reading in Scientific American *345*
Oceanography on the Web *345*

12

The Marine Habitat   346

Key Questions 346
Feature: Charles Darwin and the Voyage of HMS
  ***Beagle* 346**
Classification of Living Things *347*
Classification of Marine Organisms *349*
  Plankton (Floaters) 349   Nekton
  (Swimmers) 350   Benthos (Bottom
  Dwellers) 350
Distribution of Life in the Oceans *352*
  Why Are There So Few Marine Species? 352
Adaptations of Organisms to the Marine
  Environment *352*
  Need for Physical Support 353   Water's
  Viscosity 355   Temperature 356   Salinity 358
  Dissolved Gases 360   Water's High
  Transparency 361   Pressure 361
Divisions of the Marine Environment *363*
  Pelagic (Open Sea) Environment 363
Box 12–1   A False Bottom: The Deep Scattering Layer
  (DSL) 364
  Benthic (Sea Bottom) Environment 365
Students Sometimes Ask... 367
Chapter in Review *369*
Key Terms *369*
Questions and Exercises *370*
References *371*
  Suggested Reading in Scientific American *371*
Oceanography on the Web *371*

13

Biological Productivity and Energy
Transfer   372

Key Questions 372
Feature: Baseline Studies in the California Current:
  The CalCOFI Program 372
Primary Productivity *373*
  Photosynthetic Productivity 373   Availability of
  Nutrients 374   Availability of Solar
  Radiation 375   Margins of the Oceans 375
  Light Transmission in Ocean Water 376
Photosynthetic Marine Organisms *379*
  Seed-Bearing Plants (Spermatophyta) 379
  Macroscopic (Large) Algae 380   Microscopic
  (Small) Algae 382

Regional Productivity *382*
  Productivity in Polar Oceans 383
Box 13–1   Red Tides: Was Alfred Hitchcock's *The Birds*
  Based on Fact? 384
  Productivity in Tropical Oceans 384
Box 13–2   *Pfiesteria:* A Morphing Peril to Fish and
  Humans 386
  Productivity in Temperate Oceans 388
Energy Flow *389*
  Energy Flow in Marine Ecosystems 389
  Symbiosis 390
Biogeochemical Cycling *391*
Trophic Levels and Biomass Pyramids *392*
  Trophic Levels 392   Transfer Efficiency 393
  Biomass Pyramid 393
Ecosystems and Fisheries *394*
  Incidental Catch 397   Fisheries Management 398
Students Sometimes Ask... 398
Box 13–3   A Case Study in Fisheries Mismanagement: The
  Peruvian Anchoveta Fishery 399
Chapter in Review *401*
Key Terms *402*
Questions and Exercises *403*
References *404*
  Suggested Reading in Scientific American *404*
Oceanography on the Web *405*

14

Animals of the Pelagic
Environment   406

Key Questions 406
Feature: Alexander Agassiz: Advancements in
  Ocean Sampling 406
Staying above the Ocean Floor *407*
  Gas Containers 407   Floating Organisms
  (Zooplankton) 408   Swimming Organisms
  (Nekton) 412
Adaptations for Seeking Prey *415*
  Lungers versus Cruisers 415   Speed and Body
  Size 416
Box 14–1   Some Myths (and Facts) About Sharks 417
  Cold-Blooded Versus Warm-Blooded 418
  Circulatory System Modifications 418
Adaptations to Avoid Being Prey *418*
  Schooling 418
Marine Mammals *420*
  Order Carnivora 420   Order Sirenia 421
  Order Cetacea 422   An Example of Migration:
  Gray Whales 430
Box 14–2   Killer Whales: A Reputation Deserved? 431
Students Sometimes Ask... 432
Chapter in Review *434*

Key Terms 435
Questions and Exercises 435
References 436
    Suggested Reading in Scientific American 437
Oceanography on the Web 437

15

Animals of the Benthic Environment 438

Key Questions 438
Feature: The Great Debate on Life in the Deep Ocean: The Rosses and Edward Forbes 438
Rocky Shores 439
    Spray (Supralittoral) Zone 439   High Tide
    Zone 440   Middle Tide Zone 440   Low Tide
    Zone 441
Sediment-Covered Shores 446
    The Sediment 446   Intertidal Zonation 446
    Life in the Sediment 446   Sandy Beaches 447
    Mud Flats 449
Shallow Offshore Ocean Floor 450
    Rocky Bottoms (Sublittoral) 450   Coral
    Reefs 452
Box 15–1   How White I Am: Coral Bleaching and Other
Diseases 454
    The Deep-Ocean Floor 458   The Physical
    Environment 458   Food Sources and Species
    Diversity 459   Deep-Sea Hydrothermal Vent
    Biocommunities 459
Box 15–2   How Long Would Your Remains Remain on the
Sea Floor? 460

Low Temperature Seep Biocommunities 464
Students Sometimes Ask... 466
Chapter in Review 468
Key Terms 469
Questions and Exercises 469
References 470
    Suggested Reading in Scientific American 470
Oceanography on the Web 470

Afterword 471

Marine Sanctuaries and Marine Reserves 471
What Can I Do? 471

Appendices 473

    I Metric and English Units Compared 473
    II Geographic Locations 476
    III Latitude and Longitude on Earth 478
    IV A Chemical Background: Why Water Has
        2 H's and 1 O 481
    V Careers in Oceanography 485

Glossary 489

Credits and Acknowledgments 510

Index 514

Preface

To the Student

Welcome! You're about to embark on a journey that is far from ordinary. Over the course of this term, you will discover the central role the oceans play in the vast global system of which you are a part.

The book's content was carefully developed to provide a foundation in science by examining the vast body of oceanic knowledge. This knowledge includes information from a variety of scientific disciplines—geology, chemistry, physics, and biology—as they relate to the oceans. However, no formal background in any of these disciplines is required to successfully master the subject matter contained within this book. Our desire is to have you take away from your oceanography course much more than just a collection of facts. Instead, we want you to develop a fundamental understanding of *how the oceans work.*

This book is intended to help you in your quest to know more about the oceans. Taken as a whole, the components of the ocean—its sea floor, chemical constituents, physical components, and life forms—comprise one of Earth's largest interacting, interrelated, and interdependent systems. Because humans are beginning to impact Earth systems, it is important to understand not only how the oceans operate, but also how the oceans interact with Earth's other systems (such as its atmosphere, biosphere, and hydrosphere) as part of a larger picture. Thus, this book uses a systems approach to highlight the interdisciplinary relationship between oceanographic phenomena and how those phenomena affect other Earth systems.

To that end—and to help you make the most of your study time—we focused the presentation in this book by organizing the material around three essential components:

1. **Concepts:** General ideas derived or inferred from specific instances or occurrences (for instance, the concept of density can be used to explain why the oceans are layered).

2. **Processes:** Actions or occurrences that bring about a result (for instance, the process of waves breaking at an angle to the shore results in the movement of sediment along the shoreline).

3. **Principles:** Rules or laws concerning the functioning of natural phenomena or mechanical processes (for instance, the principle of sea floor spreading suggests that the geographic positions of the continents have changed through time).

Interwoven within these concepts, processes, and principles are hundreds of photographs, illustrations, real-world examples, and applications that make the material relevant and accessible (and maybe sometimes even *entertaining*) by bringing the science to life.

New in this edition are web-based **Environmental Issues in Oceanography (EIO)** features that are designated within selected chapters by a special notation. Each EIO calls attention to a content-related environmental concern, many of which are the result of human interaction with the marine environment. The EIO features are web-based activities with worksheets in an accompanying workbook. The EIO exercises have been specially designed by two dedicated environmental scientists (Dan Abel of Coastal Carolina University and Robert McConnell of Mary Washington College). Page xix of this Preface explains the EIO feature in more detail.

Ultimately, it is our hope that by understanding how the oceans work, you will develop a new awareness and appreciation of all aspects of the marine environment and its role in Earth systems. To this end, the book has been written for you, the student of the oceans. So enjoy and immerse yourself! You're in for an exciting ride.

Alan Trujillo
Harold Thurman

To the Instructor

The seventh edition of *Essentials of Oceanography* is designed to accompany an introductory college-level course in general or physical oceanography taught to students with no formal background in mathematics or science. Like previous editions, the goal of this edition of the textbook is to clearly present the relationships of scientific principles to ocean phenomena in an engaging and meaningful way. Further, this edition has benefited from being thoroughly student reviewed and edited by hundreds of students.

This edition has also been reviewed by a dozen instructors from leading institutions across the country. As one reviewer of the sixth edition stated, *"The material is presented in a nicely organized fashion, with clarity and in a readable style. The diagrams are well done and the tables neatly delineate passages of text into a simple, presentable form. The boxes provide interesting diversions to the text, and are nicely linked in terms of their general interest and importance to the chapter material. The 'Students Sometimes Ask…' sections are a neat trick for engaging the reader in additional topics, and these follow nicely from the level of the chapters."*

The 15-chapter format of this textbook is designed for easy coverage of the material in a 15- or 16-week semester. For courses taught on a 10-week quarter system, the instructor may need to select those chapters that cover the topic and concepts of primary relevance to their course. Chapters are self-contained and thus can be covered in any order. Following the introductory chapter (Chapter 1, which covers the general geography of the oceans, a historical perspective of oceanography, the reasoning behind the scientific method, and a discussion of the origin of Earth, the atmosphere, the oceans, and life itself), the four major academic disciplines of oceanography are represented in these chapters:

- Geological oceanography (Chapters 2, 3, 4, and parts of Chapters 10 and 11)

- Chemical oceanography (Chapter 5 and part of Chapter 11)

- Physical oceanography (Chapters 6–9 and parts of Chapters 10 and 11)

- Biological oceanography (Chapters 12–15)

However, we believe that oceanography is at its best when it links together several scientific disciplines and shows how they are interrelated in the oceans. Therefore, this interdisciplinary approach is a key element of every chapter.

One of the major additions to this edition of the text is the inclusion of web-based "Environmental Issues in Oceanography" (EIO) features within selected chapters. EIO features are on-line projects designed to: add relevance to the material by tying the concepts in the text to environmental issues, involve students in critical-thinking exercises, and help students better evaluate and understand the ecological significance of problems that have been in the forefront of recent news stories. Each EIO can be found on-line at http://www.prenhall.com/oceanissues and the projects can be completed in the accompanying workbook for submission. The EIO projects are authored by two environmental scientists (Dan Abel of Coastal Carolina University and Robert McConnell of Mary Washington College) with a special interest in critical thinking regarding environmental issues. Page xix of this Preface explains the EIO feature in more detail.

What's New in this Edition?

Changes in this edition are designed to increase the readability, relevance, and appeal of this book. The major changes include:

- Incorporation of comments from hundreds of students who thoroughly reviewed and edited the previous edition in small focus group discussions and one-on-one meetings with author Al Trujillo

- Inclusion of new chapter-related web-based "Environmental Issues in Oceanography" features that are designed to increase environmental awareness and critical-thinking activities

- Addition of "Key Questions" at the beginning of each chapter and a revised "Chapter in Review" summary feature at the end of each chapter

- Use of yellow highlighted "Concept Statements" imbedded within the text to focus attention on key concepts

- Inclusion of eight new feature boxes that present some of the most recent discoveries in oceanography

- Feature boxes are organized around three new themes:

 Research Methods in Oceanography, which highlights how oceanographic knowledge is obtained

 People and the Ocean Environment, which illustrates human interaction with the ocean

 Historical Feature, which focuses on historical developments in oceanography

- Addition or modification of over a dozen tables, which organize and summarize important data

- Inclusion of over 100 new photos and illustrations, and modification or redrawing of over 150 existing figures to add clarity and improve the illustration package

- Major reorganization, revision, and/or additions to Chapters 1, 2, 4, 5, 7, 11, 12, and 13

- Addition of a new appendix, "A Chemical Background: Why Water has 2 H's and 1 O"

- All text in the chapters has been thoroughly reviewed and edited in a continued effort to refine the style and clarity of the writing

Additionally, this edition continues to offer some of the previous edition's most popular features, including:

- Chapter-opening features that highlight an important aspect of the history of oceanography related to each chapter

- Use of the international metric system (Système International or SI units) with comparable English system units in parenthesis

- Explanation of word etymons (*etumon* = the true sense of a word) as new terms are introduced in an effort to demystify scientific terms by showing what the terms actually mean

- Notation of key terms with **bold print,** which are defined when they are introduced and are included in the glossary

- The "Students Sometimes Ask..." section at the end of each chapter that contains actual student questions, along with the authors' answers

- The end-of-chapter questions and exercises

- A dedicated Web site (http://www.prenhall.com/thurman) that features chapter-specific learning objectives, on-line quizzes, critical-thinking exercises, links to the EIO Web site, and relevant Internet links

The New Instructional Package

For the Student:

- A **Companion Web site** (http://www.prenhall.com/thurman), which is designed to function both as an on-line study guide and as a launching pad for further exploration. The site is designed using the latest technology, written by co-author Al Trujillo and Molly Trecker, and tied chapter-by-chapter to the text.

  - **To aid in reviewing the text material,** the site contains several self-testing modules, including multiple choice and true/false, fill-in-the-blanks, and image-labeling exercises. The answers can be submitted via the Internet to Prentice Hall's server for a grade, which can then be submitted to your instructor via e-mail.

  - **To foster critical thinking,** the site contains *Web Essay* questions—short-answer questions that require the student to use the Internet to research an issue, evaluate the information, and formulate a response. Responses to the Web Essays can also be e-mailed to your instructor for grading.

  - **To encourage and enable further exploration** using the incredible array of resources now available on the Internet, every chapter contains both general and chapter-specific annotated *Destinations* links to some of the best oceanography sites on the World Wide Web. These sites are researched and annotated by oceanography instructors to insure quality and relevancy and continually checked by Prentice Hall to insure validity.

- An **Environmental Issues in Oceanography workbook,** which is packaged free with *Essentials of Oceanography.* The workbook contains all the questions from the web-based EIO projects in a format suitable for submission to your instructor.

- The **New York Times** *Themes of the Times*—**Oceanography** is a unique newspaper-format supplement featuring recent articles about oceanography culled from the pages of the *New York Times.* This supplement, available for wrapping with the text at no charge, encourages you to make connections between what you're learning in the classroom and recent news events as reported in the media.

- *Science on the Internet: A Student's Guide* is a "guidebook" resource that helps science students locate and explore myriad science resources on the World Wide Web. It also provides an over-view of the Web itself, general navigational strategies, and brief student activities. *Science on the Internet* can be packaged free with *Essentials of Oceanography.*

For the Instructor:

- A **Transparency Set** of all line illustrations, some tables, and selected photographs from the text, enlarged for excellent classroom visibility. (Note: All line illustrations are available on the Digital Image Gallery CD-ROM.)

- A **Slide Set** of 180 slides from the text including key line illustrations plus most photographs not available with the transparency set. (Note: All photographs for which permission could be acquired are available on the Digital Image Gallery CD-ROM.)

- The **Digital Image Gallery (DIGIT) CD-ROM,** which includes all illustrations, tables, and photographs (for which permission could be acquired) from the text in high-resolution, 16-bit JPEG files. The JPEG files are organized by chapter and can be easily imported into lecture presentation software (such as Microsoft® PowerPoint®). The CD-ROM comes with a comprehensive reference list of all figures and their captions.

- A customizable **PowerPoint® presentation** for each chapter, written by Al Trujillo. All PowerPoint presentations are included on the Digital Image Gallery CD-ROM and can be easily modified to suit each instructor's needs.

- A new **Instructor's Manual,** written by Al Trujillo and containing answers to end-of-chapter questions, a selection of various types of exam questions, and a wealth of instructional resources for each chapter.

- **Computerized Test Manager CD-ROM,** with which you can easily create and tailor exams to your own needs: produce multiple versions of your exam, with answer keys; select questions by number, type, level of difficulty, or random distribution; export your exam to a rich text file for import into your word processing program; and more. The software comes with a comprehensive reference guide and includes the toll-free technical support line.

Acknowledgments

The authors are indebted to many individuals for their helpful comments and suggestions during the revision of this book. Al Trujillo is particularly indebted to his colleagues at Palomar Community College, Patty Deen and Lisa DuBois, for their keen interest in the project and for allowing him to use some of their creative ideas in the book. They are simply the finest colleagues imaginable.

Many individuals at Scripps Institution of Oceanography have been particularly helpful, especially the staff at *Explorations* magazine. Thanks also go to the many people around the world who helped locate images or willingly donated photographs for use in the text. Without the kind help of these people, the job of tracking down experts and finding images would have been a much larger task.

Many people were instrumental in helping the text evolve from its manuscript stage. Patrick Lynch, Geoscience Editor at Prentice Hall, expertly guided the project and offered many insightful suggestions to improve the text. Developmental Editor John Murdzek read the entire manuscript and made vast improvements in the writing style. The editorial staff at Lithokraft II also made many thoughtful suggestions to improve the manuscript.

Al Trujillo would also like to thank his students, whose questions provided the material for the "Students Sometimes Ask…" sections and whose continued input has proved invaluable for improving the text. Since scientists (and good teachers) are always experimenting, thanks also for allowing yourselves to be a captive audience with which to conduct my experiments.

Al Trujillo also thanks his patient and understanding family for putting up with his absence during the long hours of preparing "The Book." Lastly, appreciation is extended to the chocolate manufacturers Hershey, See's, and Ghiradelli for providing inspiration. A heartfelt thanks to all of you!

Many other individuals (including several anonymous reviewers) have provided valuable technical reviews for this and previous works. The following reviewers are gratefully acknowledged:

William Balsam, *University of Texas at Arlington*
Steven Benham, *Pacific Lutheran University*
Lori Bettison-Varga, *College of Wooster*
Thomas Bianchi, *Tulane University*
Mark Boryta, *Consumnes River College*
Laurie Brown, *University of Massachusetts*
Nancy Bushell, *Kauai Community College*
G. Kent Colbath, *Cerritos Community College*
Thomas Cramer, *Brookdale Community College*

Hans G. Dam, *University of Connecticut*
Wallace W. Drexler, *Shippensburg University*
Walter C. Dudley, *University of Hawaii*
Charles H. V. Ebert, *SUNY Buffalo*
Kenneth L. Finger, *Irvine Valley College*
Dave Gosse, *University of Virginia*
Joseph Holliday, *El Camino Community College*
Mary Anne Holmes, *University of Nebraska, Lincoln*
Timothy C. Horner, *California State University, Sacramento*
Ron Johnson, *Old Dominion University*
M. John Kocurko, *Midwestern State University*
Lawrence Krissek, *Ohio State University*
Richard A. Laws, *University of North Carolina*
Richard D. Little, *Greenfield Community College*
Stephen A. Macko, *University of Virginia, Charlottesville*
Matthew McMackin, *San Jose University*
James M. McWhorter, *Miami-Dade Community College*
Gregory Mead, *University of Florida*
Johnnie N. Moore, *University of Montana*
B. L. Oostdam, *Millersville University*
William W. Orr, *University of Oregon*
Curt Peterson, *Portland State University*
Edward Ponto, *Onondaga Community College*
Donald L. Reed, *San Jose State University*
Cathryn L. Rhodes, *University of California, Davis*
James Rine, *University of South Carolina*
Jill K. Singer, *SUNY College, Buffalo*
Arthur Snoke, *Virginia Polytechnic Institute*
Lenore Tedesco, *Indiana University Purdue University at Indianapolis*
Bess Ward, *Princeton University*
Jackie L. Watkins, *Midwestern State University*
Arthur Wegweiser, *Edinboro University of Pennsylvania*

Although this book has benefited from careful review by many individuals, the accuracy of the information rests with the authors. If you find errors or have comments about the text, please contact us at:

Al Trujillo
Department of Earth Sciences
Palomar College
1140 W. Mission Rd.
San Marcos, CA 92069
atrujillo@palomar.edu
Web: http://daphne.palomar.edu/atrujillo

Hal Thurman
8031 Stonewyck Rd.
Germantown, TN 38138
ann_harold_thurman@email.msn.com

▬EIO▬

Environmental Issues in Oceanography

by Daniel C. Abel, Coastal Carolina University and Robert L. McConnell, Mary Washington College

What are the EIO projects?

The **Environmental Issues in Oceanography (EIO)** are web-based, critical-thinking projects that guide you in applying the concepts of oceanography and scientific reasoning in analyzing a range of immediate, relevant environmental issues. The goal of these projects is not only to draw your attention to some pressing ecological issues, but to help your instructor in training you to gather, evaluate, and analyze information as a scientist would in order to draw your own conclusions—a skill that will benefit you for the rest of your life.

> **▬EIO▬**
>
> For more information and on-line exercises about the Environmental Issue in Oceanography (EIO) "Coastal Population Growth," visit the EIO Web site at **http://www.prenhall.com/oceanissues** and select Issue #1.

Where do you find them?

Twelve EIO boxes (pictured above) in selected chapters alert you that the concept you've just been studying in theory will be applied in practice to one of the eight EIO projects. You will see that each of these projects integrate many concepts across chapters, but to avoid cluttering the text they are called out only in discussions of the most critical concepts. The EIO workbook comes free with every *Essentials of Oceanography,* seventh edition text purchased from Prentice Hall.

How do you use them?

The EIO projects integrate the concepts and theory from the text, news and data sources from the Web, and the critical-thinking questions found in the EIO workbook. To use the resources most effectively:

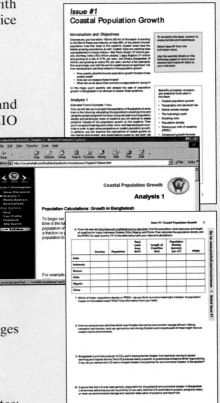

- First, review the project in your EIO workbook. The first page of each project outlines the project, alerts you to the concepts and tools that you will be applying, and anticipates the time needed to complete each part of the project.

- Next, connect to the EIO Web site at www.prenhall.com/oceanissues and access the project. The Web site will lead you on a structured exploration of the issue, linking you to outside sources for data and other information necessary to complete your analysis.

- Lastly, use the workbook to record your responses, calculations, and conclusions. The questions on the Web site are recreated here, with space for your answers; perforated pages make it easy to tear out these pages and submit them to your instructor.

The Issues

1. Coastal Population Growth • 2. Sonar and Whales • 3. Methane Hydrates: Energy Boom or Climate Bust? • 4. Toxic Chemicals in Seawater • 5. The Lasting Influences of Hurricanes • 6. Beaches or Bedrooms?: The Dynamic Coastal Environment • 7. Illegal Immigration: Ballast Water and Exotic Species • 8. Lifestyles of the Large and Blubbery: How to Grow a Blue Whale

Introduction

Welcome to a book about the oceans. As you read this book, we hope that it elicits a sense of wonder and a spirit of curiosity about our watery planet. The oceans represent many different things to different people. To some, it is a wilderness of beauty and tranquility, a refuge from hectic civilized lives. Others see it as a vast recreational area that inspires either rest or physical challenge. To others, it is a mysterious place that is full of unknown wonders. And to others, it is a place of employment unmatched by any on land. To be sure, its splendor has inspired artists, writers, and poets for centuries (Figure I–1). Whatever your view, we hope that understanding the way the oceans work will increase your appreciation of the marine environment. Above all, take time to admire the oceans.

Essentials of Oceanography was first written in the late 1970s to help students develop an *awareness about the marine environment*—that is, develop an appreciation for the oceans by learning about oceanic processes (how the oceans behave) and their interrelationships (how physical entities are related to one another in the oceans). In this seventh edition, our goal is the same: to give the reader the scientific background to understand the basic principles underlying oceanic phenomena. In this way, one can then make informed decisions about the oceans in the years to come. We hope that some of you will be inspired so much by the oceans that you will continue to study them formally or informally in the future. (For those who may be considering a life-long career in oceanography, see Appendix V, "Careers in Oceanography.")

What Is Oceanography?

Oceanography (*ocean* = the marine environment, *graphy* = the name of a descriptive science) is quite literally the description of the marine environment. Unfortunately, this definition does not fully portray the extent of what oceanography encompasses: oceanography is much more than just *describing* marine phenomena. Oceanography could be more accurately called the scientific study of all aspects of the marine environment. Hence, the field of study called oceanography could (and maybe *should*) be called oceanology (*ocean* = the marine environment, *ology* = the study of). However, the science of studying the oceans has traditionally been called oceanography. It is also called *marine science* and includes the study of the water of the ocean, the life within it, and the (not so) solid earth beneath it.

Since prehistoric time, people have used the oceans as a means of transportation and as a source of food. However, the importance of ocean processes has been studied technically only since the 1930s. The impetus for this study began with the search for petroleum, continued with the emphasis on ocean warfare during World War II, and more recently has been expressed in the concern for the well-being of the ocean environment. Historically, those who make their living fishing in the ocean go where the

- What is oceanography?
- How have the oceans influenced conditions on Earth?
- What is rational use of technology?

key questions

> Oceanography is not so much a science as a collection of scientists who find common cause in trying to understand the complex nature of the ocean. In the vast salty seas that encompass the earth, there is plenty of room for persons trained in physics, chemistry, biology, and engineering to practice their specialties. Thus, an oceanographer is any scientifically trained person who spends much of his [or her] career on ocean problems.
> — *Willard Bascom(1980)*

Figure I–I The ocean environment at Jalama Beach, California.

physical processes of the oceans offer good fishing. But how marine life interrelates with ocean geology, chemistry, and physics to create good fishing grounds has been more or less a mystery until only recently when scientists in these disciplines began to investigate the oceans with new technology.

Oceanography is typically divided into different academic disciplines (or subfields) of study. The four main disciplines of oceanography that are covered in this book are:

- *Geological oceanography,* which is the study of the structure of the sea floor and how the sea floor has changed through time; the creation of sea floor features; and the history of sediments deposited on it.

- *Chemical oceanography,* which is the study of the chemical composition and properties of seawater; how to extract certain chemicals from seawater; and the effects of pollutants.

- *Physical oceanography,* which is the study of waves, tides, and currents; the ocean-atmosphere relationship that influences weather and climate; and the transmission of light and sound in the oceans.

- *Biological oceanography,* which is the study of the various oceanic life forms and their relationships to one another; adaptations to the marine environment; and developing ecologically sound methods of harvesting seafood.

Other disciplines include ocean engineering, marine archaeology, and marine policy. Since the study of oceanography often examines in detail all the different disciplines of oceanography, it is frequently described as being an *interdisciplinary* science, or one covering all the disciplines of science as they apply to the oceans (Figure I–2). The content of this book includes the broad range of interdisciplinary science topics that comprises the field of oceanography. In essence, this is a book about *all* aspects of the oceans.

Earth's Ocean

The oceans are the largest and most prominent feature on Earth. In fact, they are the single most defining feature of our planet. As viewed from space, our planet is a beautiful blue, white, and brown globe (Figure I–3). It is our oceans of liquid water that sets us apart in the solar system. No other planet has an ocean; however, a recent discovery of fluid-filled cracks on some of Jupiter's moons—most notably Europa—has lead to speculation that there may be an ocean of liquid water beneath the ice. The fact that our planet has so much water, *and in the liquid form,* is unique in the solar system.

The oceans determine where our continents end, and thus have shaped political boundaries and human history many times. The oceans conceal many features; in fact, the majority of Earth's geographic features are on the ocean floor. Remarkably, there was once more known about the surface of the moon than about the floor of the oceans! Fortunately, over the past 30 years, our knowledge of both has increased dramatically.

The oceans influence weather all over the globe, even in continental areas far from any ocean. The oceans are also the lungs of the planet, taking carbon dioxide gas (CO_2) out of the atmosphere and replacing it with oxygen gas (O_2). Some scientists have estimated that the oceans supply as much as 70% of the oxygen humans breathe.

The oceans are in large part responsible for the development of life on Earth, providing a stable environment in which life could evolve over millions of years. Today, the oceans contain the greatest number of living things on the planet, from microscopic bacteria and algae to the largest life form alive today (the blue whale). Interestingly, water is the major component of nearly every life form on Earth, and our own body fluid chemistry is remarkably similar to the chemistry of seawater.

The oceans hold many secrets waiting to be discovered, and new discoveries about the oceans are made nearly every day. The oceans are a source of food, minerals, and energy that remain largely untapped. Over half of the world population lives in coastal areas near the oceans, taking advantage of the mild climate, an inexpensive form of transportation, and vast recreational opportunities. And unfortunately, the oceans are also the dumping ground for many of society's wastes.

Rational Use of Technology?

Many stresses have been put on the oceans by an ever-increasing human population. For instance, population studies reveal that over 50% of world population—some 3.2 billion people—live along the coastline and over 80% of all Americans live within an hour's drive from an ocean or the Great Lakes. In the future, these figures are expected to increase (in the United States, eight of the 10 largest cities are in coastal environments

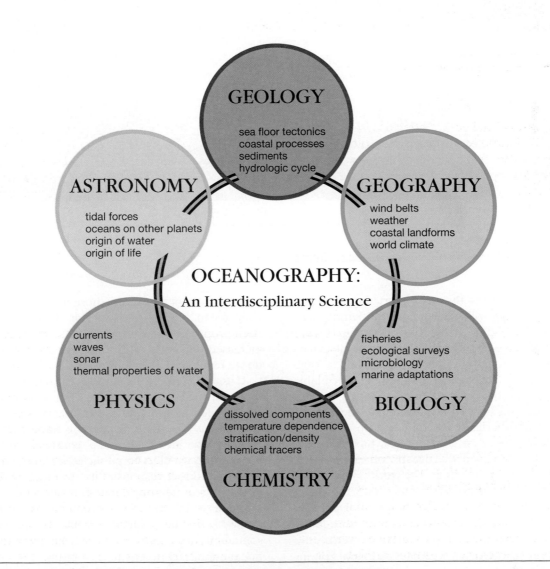

Figure I–2 A Venn diagram showing the interdisciplinary nature of oceanography.

Figure I–3 Earth from space.

and 3600 people move to the coast every day). This migration to the coasts will further mar the delicate natural balance that exists in the coastal ocean (Figure I–4). Specifically, the migration is resulting in more harbor and channel dredging, industrial waste and sewage disposal at sea, chemical spills, cooling of power plants with seawater, and the destruction of wetlands that are vital to the cleansing of runoff waters and to the maintenance of coastal fisheries.

Although it may seem as if humans have severely and irreversibly damaged the oceans, our impact has been felt mostly in coastal areas. The world's oceans are a vast resource that has not yet been lethally damaged. Humans have been able to inflict only minor damage here and there along the margins of the oceans. However, as our technology makes us more powerful, the threat of irreversible harm becomes greater. For in-

stance, in the open ocean (those areas far from shore), deep-ocean mining and nuclear waste disposal have been proposed. How do we as a society deal with these increased demands on the marine environment? How do we regulate the ocean's use?

If used wisely, our technology can actually reduce the threat of irreversible harm. Which path will we take? We all need to evaluate carefully our own actions and the effects those actions have on the environment. In addition, we need to make conscientious decisions about those we elect to public office. Some of you may even have direct responsibility for initiating legislature that affects our environment. It is our hope that you, as a student of the marine environment, will gain enough knowledge while studying oceanography to help your community (and perhaps even your nation) make rational use of technology in the oceans.

Figure I–4 Heavily developed coastline of Miami Beach, Florida.

Our environment, our health, our economic prospects, our national defense, the foods we eat, and the air that we breathe—even our genetic future—will depend upon how wisely we apply the technologies that become available. And to do this we need a population of scientists, but also of citizens, of workers, of administrators, of policy makers…who can grasp the scientific way of thinking.
                                        —Leon Lederman (2000)

◼EIO◼

For more information and on-line exercises about the Environmental Issue in Oceanography (EIO) "Coastal Population Growth," visit the EIO Web site at **http://www. prenhall.com/oceanissues** and select Issue #1.

CHAPTER
1
Introduction to Planet "Earth"

- What are the four principal oceans on Earth?

- Where is the deepest part of the oceans, and have humans ever visited there?

- How was early exploration of the oceans achieved?

- What is the nature of scientific inquiry?

- Where did Earth's oceans come from?

- Did life on Earth begin in the oceans?

- How old is Earth?

key questions

"When you have eliminated the impossible, whatever remains, however improbable, must be the truth."

—*Sir Arthur Conan Doyle,*
The Sign of the Four *(1890)*

DIVING INTO THE MARINE ENVIRONMENT

Throughout history, humans have submerged themselves in the marine environment to observe it directly for scientific exploration, profit, or adventure (Figure 1A). As early as 4500 B.C., brave and skillful divers reached depths of 30 meters[1] (98 feet) on one breath of air to retrieve red coral and mother-of-pearl shells. Later, diving bells (bell-shaped structures full of trapped air) were lowered into the sea to provide passengers or underwater divers with an air supply. In 360 B.C., Aristotle, in his *Problematum*, recorded the use by Greek sponge divers of kettles full of air lowered into the sea. However, technology to move around freely while breathing underwater was not developed until 1943, when Jacques-Yves Cousteau and

Figure 1A Oceanographer and explorer Willard Bascom.

---

[1]Throughout this book, metric measurements are used (and the corresponding English measurements are in parenthesis). See Appendix I, "Metric and English Units Compared," for conversion factors between the two systems of units.

Émile Gagnan invented the fully automatic, compressed-air Aqualung. The equipment was later dubbed "**scuba**," an acronym for <u>s</u>elf-<u>c</u>ontained <u>u</u>nderwater <u>b</u>reathing <u>a</u>pparatus, and is used by millions of recreational divers today. By using scuba, divers can experience the ocean first hand, leading to a fuller appreciation of the wonder and beauty of the marine environment.

Those who venture underwater must contend with many obstacles inherent in ocean diving, such as low temperatures, darkness, and the effects of greatly increased pressure. To combat low temperatures, specially designed clothing is worn. Waterproof, high-intensity diving lights are used to combat darkness. To combat the deleterious effects of pressure, depth and duration of dives must be limited. As a result, most scuba divers rarely venture below a depth of 30 meters (98 feet)—where the pressure is three times that at the surface—and they stay there less than 30 minutes.

It is relatively dangerous for humans to enter the marine environment because our bodies are adapted to living in the relatively low pressure of the atmosphere. In water, pressure increases rapidly with depth, to which anyone who has been to the bottom of the deep end of a swimming pool can attest. The increased pressure at depth in the ocean can cause problems for divers. For instance, higher pressure causes more nitrogen to be dissolved in a diver's body and may cause a disorienting condition known as *nitrogen narcosis*, or "rapture of the deep." Further, if a diver surfaces too rapidly, expanding gases within the body can catastrophically rupture cell membranes (a condition called *barotrauma*).

In addition, when divers return to the surface, they may experience *decompression illness* (the "bends"). The "bends" affects divers that ascend to the lower pressure at the surface too rapidly, causing nitrogen bubbles to form in the bloodstream and other tissues (analogous to the bubbles that form in a carbonated beverage when the container is opened). Various symptoms can result, from nosebleed and joint pain (which causes divers to stoop over, hence the term the "bends") to permanent neurological injury and even fatal paralysis. To avoid it, divers must ascend slowly, allowing time for excess dissolved nitrogen to be eliminated from the blood via the lungs.

Despite these risks, divers venture to greater and greater depths in the ocean. In 1962, Hannes Keller and Peter Small made an open-ocean dive from a diving bell to a then record-breaking depth of 304 meters (1000 feet). Although they used a special gas mixture, Small died once they returned to the surface. Presently, the record ocean dive is 534 meters (1752 feet), but researchers who study the physiology of deep divers have simulated a dive to 701 meters (2300 feet) in a pressure chamber using a special mix of oxygen, hydrogen, and helium gases. Researchers believe that humans will eventually be able to stay under water for extended periods of time at depths below 600 meters (1970 feet).

———————————●———————————

It seems perplexing that our planet is called "Earth" when 70.8% of its surface is covered by oceans. Many early human cultures that lived in the Mediterranean (*medi* = middle, *terra* = land) Sea area envisioned the world as composed of large land masses surrounded by marginal bodies of water (Figure 1–1). How surprised they must have been when they ventured into the larger oceans of the world. Our planet is misnamed "Earth" because we live on the land portion of the planet. If we were marine animals, our planet would probably be called "Ocean," "Water," "Hydro," "Aqua," or even "Oceanus," to indicate the prominence of Earth's oceans.

Geography of the Oceans

A world map (Figure 1–2) shows how extensive our oceans really are. Notice that *the oceans dominate the surface area of the globe*. If you have traveled by boat across an ocean (or flown across one in an airplane) you soon realize that the oceans are enormous. Notice, also, that *the oceans are interconnected* and form a single continuous body of seawater, which is why it is commonly referred to as a "world ocean" (singular, not plural). For instance, a vessel at sea can travel from one ocean to another, whereas it is impossible to travel on land from one continent to most others without crossing an ocean. In addition, the oceans contain 97.2% of all the water on or near Earth's surface, so *the volume of the oceans is immense*.

The Four Principal Oceans, Plus One

Our world ocean can be divided into four principal oceans (plus an additional ocean), based on the shape of the ocean basins and the positions of the continents (Figure 1–2).

Pacific Ocean The **Pacific Ocean** is the world's largest, covering over half of the ocean surface area on Earth (Figure 1–3). The Pacific Ocean is the single largest geographic feature on the planet, covering over one-third of Earth's entire surface. The Pacific Ocean is so large that *all* of the continents could fit into the space

Figure 1–1 An early map of the world. The world according to the Greek Herodotus in 450 B.C., showing the prominence of the Mediterranean Sea surrounded by the continents of Europe, Libya (Africa), and Asia. Seas are labeled "mare" and are shown as a band of water encircling the land.

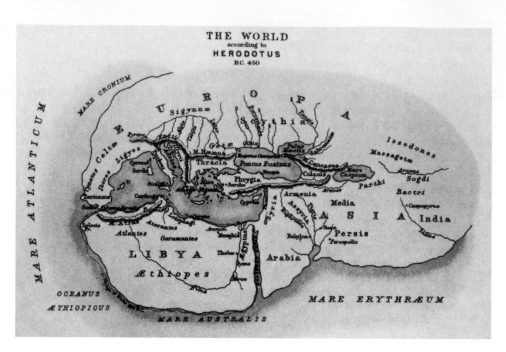

occupied by it—with room left over! Although the Pacific Ocean is also the deepest ocean in the world, it contains many small tropical islands. It was named in 1520 by explorer Ferdinand Magellan's party, in honor of the fine weather they encountered while crossing into the Pacific (*paci* = peace) Ocean.

Atlantic Ocean   The **Atlantic Ocean** is about half the size of the Pacific Ocean and is not quite as deep (Figure 1–3). It separates the Old World (Europe, Asia, and

Africa) from the New World (North and South America). The Atlantic Ocean was named after the Atlas Mountains in northwest Africa.

Indian Ocean   The **Indian Ocean** is slightly smaller than the Atlantic Ocean and has about the same average depth (Figure 1–3). It is mostly in the Southern Hemisphere (south of the Equator, or below 0 degrees latitude in Figure 1–2). The Indian Ocean was named for its proximity to the subcontinent of India.

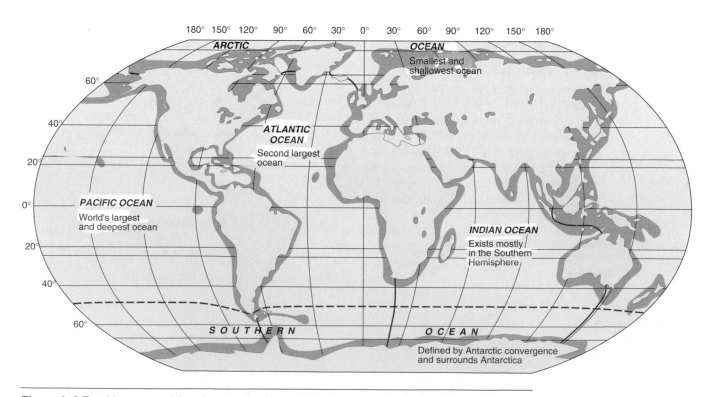

Figure 1–2 Earth's oceans. Map showing the four principal oceans, plus the Southern or Antarctic Ocean. Dark blue shading represents shallow areas.

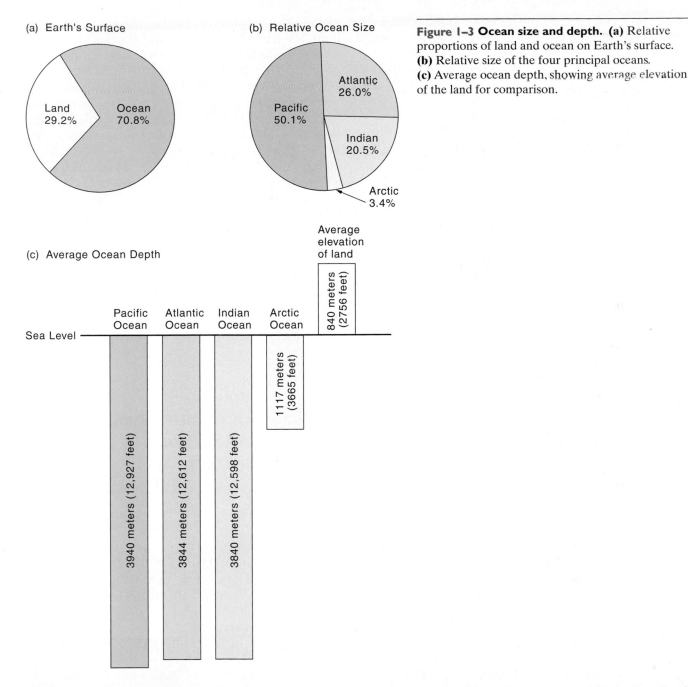

(a) Earth's Surface

Land 29.2%
Ocean 70.8%

(b) Relative Ocean Size

Atlantic 26.0%
Pacific 50.1%
Indian 20.5%
Arctic 3.4%

(c) Average Ocean Depth

Average elevation of land

840 meters (2756 feet)

Sea Level

Pacific Ocean — 3940 meters (12,927 feet)
Atlantic Ocean — 3844 meters (12,612 feet)
Indian Ocean — 3840 meters (12,598 feet)
Arctic Ocean — 1117 meters (3665 feet)

Figure 1–3 Ocean size and depth. (a) Relative proportions of land and ocean on Earth's surface. **(b)** Relative size of the four principal oceans. **(c)** Average ocean depth, showing average elevation of the land for comparison.

Arctic Ocean The **Arctic Ocean** is about 7% the size of the Pacific Ocean and is only a little more than one-quarter as deep as the rest of the oceans (Figure 1–3). Although it has a permanent layer of sea ice at the surface, the ice is only a few meters thick. The Arctic Ocean was named after the northern constellation Ursa Major, otherwise known as the Big Dipper, or the Bear (*arktos* = bear).

Southern Ocean or Antarctic Ocean Oceanographers recognize an additional ocean near the continent of Antarctica in the Southern Hemisphere (Figure 1–2). Defined by the meeting of currents near Antarctica called the Antarctic Convergence, the **Southern Ocean** or **Antarctic Ocean** is really the portions of the Pacific, Atlantic, and Indian Oceans south of about 50 degrees

south latitude. This ocean was named for its location in the Southern Hemisphere.

The four principal oceans are the Pacific, Atlantic, Indian, and Arctic Oceans. An additional ocean, the Southern or Antarctic Ocean, is also recognized.

The Seven Seas?

"Sailing the seven seas" is a familiar phrase in literature and song, but are there really seven seas? If there are four principal oceans (plus one), what is the difference between a sea and an ocean? In common use, the terms "sea" and "ocean" are often used interchangeably. For instance, a *sea* star lives in the *ocean*; the *ocean* is full of *sea*water; *sea* ice forms in the *ocean*; and one might stroll the *sea*shore while living in *ocean*-front property. Technically, however, seas are defined as follows:

- Smaller and shallower than an ocean (this is why the Arctic Ocean might be more appropriately called a sea)
- Composed of salt water (many "seas," such as the Caspian Sea in Asia, are actually large freshwater lakes)
- Somewhat enclosed by land (but some seas, such as the Sargasso Sea in the Atlantic Ocean, are defined by strong ocean currents rather than by land)

Although the definition of a sea leaves some room for interpretation, it does allow us to answer the question about the seven seas. If we count the oceans as seas and split the Pacific Ocean and the Atlantic Ocean arbitrarily at the Equator, then we have seven seas: the North Pacific, the South Pacific, the North Atlantic, the South Atlantic, the Indian, the Arctic, and the Southern or Antarctic.

Comparing the Oceans to the Continents

Figure 1–4 shows that the average depth of the world's oceans is 3729 meters (12,234 feet), so there must be some extremely deep areas in the ocean to offset the shallow areas close to shore. Figure 1–4 also shows that the deepest depth in the oceans (the Challenger Deep region of the Mariana Trench near Guam) is a staggering 11,022 meters (36,161 feet) below sea level.

Could anything live at these depths, where conditions include crushing high pressure, absolutely no light, and water temperature just above freezing? Could humans ever visit this region? Remarkably, the answer to both questions is yes. In January 1960, United States Navy Lt. Don Walsh and explorer Jacques Piccard descended to the bottom of the Challenger Deep in the **Trieste**, a deep-diving **bathyscaphe** (*bathos* = depth, *scaphe* = a small ship) (Figure 1–5). During the descent at 9906 meters (32,500 feet), the men heard a loud cracking sound that shook the cabin. They were unable to see that the 7.6-centimeter (3-inch) thick Plexiglas viewing port on the entranceway had cracked. Miraculously, though, it held for the rest of the dive. Over five hours after leaving the surface, they reached the bottom at 10,912 meters (35,800 feet)—a record depth of human descent that has not been broken since. Not surprisingly, they did see some life forms there: a small flatfish from the sole family, a shrimp, and some jellyfish, all of which are adapted to life in the deep.

By comparison, Figure 1–4 shows that the average height of the continents is only 840 meters (2756 feet).

The average height of the land is not that far above sea level. The highest mountain in the world (the mountain with the greatest height above sea level) is Mount Everest in the Himalaya Mountains of Asia at 8850 meters (29,035 feet)Everest is a full 2172 meters (7126 feet) shorter than the Mariana Trench is deep. The mountain with the *greatest total height* from base to top is Mauna Kea on the island of Hawaii in the United States. It measures 4206 meters (13,800 feet) above sea level and 5426 meters (17,800 feet) from sea level down to its base, for a total height of 9632 meters (31,601 feet). The total height of Mauna Kea is 782 meters (2566 feet) higher than Mount Everest but it is still 1390 meters (4560 feet) shorter than the Mariana Trench is deep. No mountain on Earth is taller than the Mariana Trench is deep.

The deepest part of the ocean is the Mariana Trench in the Pacific Ocean. It is 11,022 meters (36,161 feet) deep and was visited by humans in 1960 in a specially designed bathysphere.

Explorations of the Oceans: Some Historical Notes About Oceanography

Early History

Humankind probably first viewed the oceans as a source of food. Later, vessels were built to move upon the ocean's surface and transport ocean-going people to new fishing grounds. The oceans also provided an inexpensive and efficient way to move large and heavy objects, facilitating trade and interaction between cultures.

Pacific Navigators   It is not known who first developed navigation, but it may have been ancestors of Pacific Islanders. The peopling of the Pacific Islands (Oceania) is somewhat perplexing because there is no evidence that people actually evolved on these islands. Their presence required travel over hundreds or even thousands of miles of open ocean from the continents, probably in small vessels of that time (double canoes, outrigger canoes, or balsa rafts). The islands in the Pacific Ocean are widely scattered, so it is likely that only a fortunate few of the voyagers made landfall and that many others perished during the voyage. Figure 1–6 shows the three major island regions in the Pacific Ocean: Micronesia (*micro* = small, *nesia* = islands), Melanesia (*melan* = black, *nesia* = islands), and Polynesia (*poly* = many, *nesia* = islands), which covers the largest area.

No written records of Pacific human history exist before the arrival of Europeans in the 16th century.

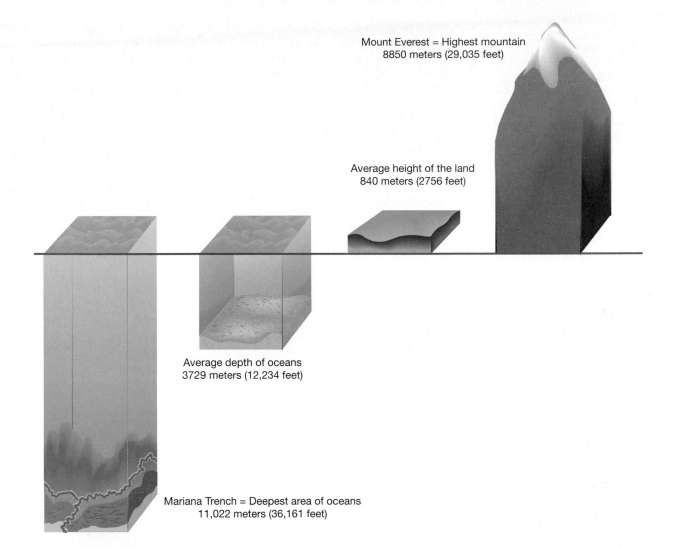

Figure 1–4 Comparing depth of the oceans to elevation of the land. The deepest and average depth of the oceans compared to the highest and average elevation on land.

Nevertheless, the movement of Asian peoples into Micronesia and Melanesia is easy to imagine because distances between islands are relatively short. In Polynesia, however, large distances separate island groups, which must have presented great challenges to ocean voyagers. Easter Island, for example, at the southeastern corner of the triangular-shaped Polynesian islands region, is over 1600 kilometers (1000 miles) from Pitcairn Island, the next nearest island. Clearly, a voyage to the Hawaiian Islands must have been one of the most difficult because they are 3000 kilometers (1860 miles) from the nearest inhabited islands, the Marquesas Islands (Figure 1–6).

Archeological evidence suggests that humans from New Guinea may have occupied New Ireland as early as 4000 or 5000 B.C. However, there is little evidence of human travel farther into the Pacific Ocean before 1100 B.C. By then, pottery makers called the "Lapita

people" had traveled on to Fiji, Tonga, and Samoa. From there, Polynesians sailed on to the Marquesas (A.D. 300). The Marquesas appear to have been the starting point for voyages to Easter Island (A.D. 400), the Hawaiian Islands (A.D. 500), Society Islands (A.D. 600), and New Zealand (A.D. 800).

The first archeological evidence from the eastern Pacific indicates human occupation of the Marquesas Islands by A.D. 300. Although the Maori of New Zealand, the Hawaiians, and the Easter Islanders of the 16th century were Polynesian, there is no clear evidence to explain how these peoples arrived at these destinations.

Thor Heyerdahl, an adventurous biologist/anthropologist, proposed that voyagers from South America may have reached islands of the South Pacific before the coming of the Polynesians. In 1947, he sailed the *Kon Tiki*—a balsa raft designed like those that were used by South American navigators at the time of

Figure 1–5 The U.S. Navy's bathyscaphe *Trieste*. The *Trieste* suspended on a crane before its record-setting deep dive. The 1.8-meter (6-foot) diameter diving chamber (round ball below the float) accommodated two people and had steel walls 7.6 centimeters (3 inches) thick.

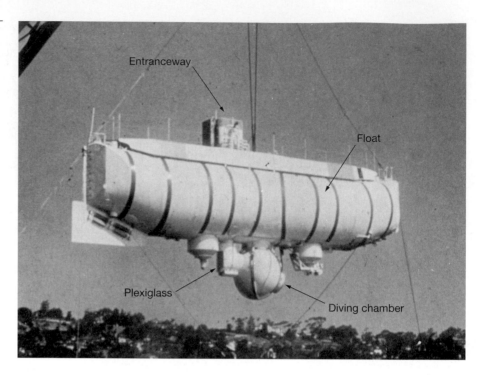

Figure 1–6 The peopling of the Pacific islands. The major island groups of the Pacific Ocean are Micronesia (*brown shading*), Melanesia (*red shading*), and Polynesia (*green shading*). The "Lapita people" present in New Ireland 5000–4000 B.C. can be traced to Fiji, Tonga, and Samoa by 1100 B.C. The route of Thor Heyerdahl's balsa raft *Kon Tiki* is also shown.

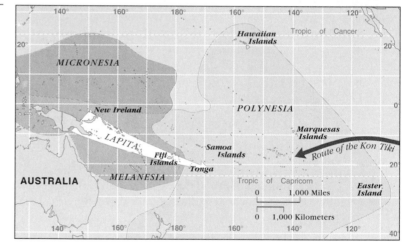

European discovery—from South America to the Tuamotu Islands, a journey of over 11,300 kilometers (7000 miles) (Figure 1–6). Although the voyage of the *Kon Tiki* demonstrates that early South Americans could have traveled to Polynesia just as easily as early Asian cultures, the anthropological community can find no evidence of such a migration.

European Navigators The first humans from the Western Hemisphere known to have developed the art of navigation were the **Phoenicians**, who lived at the eastern end of the Mediterranean Sea, in the present-day area of Egypt, Syria, Lebanon, and Israel. As early as 2000 B.C., they investigated the Mediterranean Sea, the Red Sea, and the Indian Ocean. The first recorded circumnavigation of Africa, in 590 B.C., was made by the Phoenicians, who had also sailed as far north as the British Isles.

The Greek astronomer-geographer **Pytheas** sailed northward to Iceland in 325 B.C. using a simple yet elegant method for determining latitude (one's position north or south) in the Northern Hemisphere. His method involved measuring the angle between an observer's line of sight to the North Star and line of sight to the northern horizon (see Appendix III, "Latitude and Longitude on Earth"). Despite Pytheas's method for determining latitude, it was still impossible to accurately determine longitude (one's position east or west).

Like many Egyptians who considered Earth to be a sphere, **Eratosthenes** (276-192 B.C.), a Greek librarian in the Egyptian city of Alexandria, cleverly used the shadow of a stick in a hole in the ground and elementary geometry to determine Earth's circumference. His value of 40,000 kilometers (24,840 miles) compares well with the value of 40,032 kilometers (24,875 miles) accepted today.

In about 150 A.D., **Ptolemy** produced a map of the world that represented Roman knowledge at that time. The map included the continents of Europe, Asia, and Africa, as did earlier Greek maps, but it also included vertical lines of longitude and horizontal lines of latitude. Moreover, Ptolemy showed the known seas to be surrounded by land, much of which was as yet unknown.

The Middle Ages

After the fall of the Roman Empire in the fifth century A.D., the achievements of the Phoenicians, Greeks, and Romans were retained by the Arabs, who controlled northern Africa and Spain, but lost or suppressed by those who controlled Europe. The Arabs used this knowledge to become the dominant navigators in the Mediterranean area and to trade extensively with East Africa, India, and southeast Asia. The Arabs were able to trade across the Indian Ocean because they had learned how to take advantage of the seasonal patterns of monsoon winds. During the summer, when monsoon winds blow from the southwest, ships laden with goods would leave the Arabian ports and sail eastward across the Indian Ocean. During the winter, when the trade winds blow from the northeast, ships would return west. In the rest of southern and eastern Europe, the Western concept of world geography degenerated considerably. For example, one notion envisioned the world as a disk with Jerusalem at the center.

In northern Europe, the **Vikings** of Scandinavia, who had excellent ships and good navigation skills, actively explored the Atlantic Ocean (Figure 1–7). Late in the ninth century, aided by a period of worldwide climatic warming, the Vikings colonized Iceland. In about 981, **Erik "the Red" Thorvaldson** sailed westward from Iceland and discovered Greenland. He may also have traveled further westward to Baffin Island. He returned to Iceland and led the first wave of Viking colonists to Greenland in 985. **Bjarni Herjolfsson** sailed from Iceland to join the colonists, but he sailed too far southwest and is thought to be the first Viking to have seen what is now called Newfoundland. Bjarni did not land but instead returned to the new colony at Greenland. **Leif Eriksson**, son of Erik the Red, became intrigued by Bjarni's stories about the new land Bjarni had seen. In 995, Lief bought Bjarni's ship and set out from Greenland for the land that Bjarni had seen to the southwest. Leif spent the winter in that portion of North America and named the land Vinland (now Newfoundland) after the grapes that were found there. Climatic cooling and inappropriate farming practices for the region caused these Viking colonies in Greenland and Vinland to struggle and die out by about 1450.

The Age of Discovery in Europe

The 30-year period from 1492 to 1522 is known as the **Age of Discovery**. During this time, Europeans explored the continents of North and South America and the globe was circumnavigated for the first time. As a result, Europeans learned the true extent of the world's oceans and that human populations existed elsewhere on newly "discovered" continents and islands with cultures vastly different from those familiar to European voyagers.

Why was there such an increase in ocean exploration during the Age of Discovery? One reason was that Sultan Mohammed II had captured Constantinople (the capital of eastern Christendom) in 1453, which isolated Mediterranean port cities from the riches of the India, Asia, and the East Indies (modern-day Indonesia). The Western world had to search for new Eastern trade routes by sea.

The Portuguese, under the leadership of Prince **Henry the Navigator** (1392-1460), led a renewed effort to explore outside Europe. The prince established a marine institution at Sagres to improve Portuguese sailing skills. The treacherous journey around the tip of Africa was a great obstacle to an alternative trade route. Cape Agulhas (at the southern tip of Africa) was first rounded by **Bartholomeu Diaz** in 1486. He was followed in 1498 by **Vasco da Gama**, who continued around the tip of Africa to India, thus establishing a new Eastern trade route to Asia.

Christopher Columbus sailed west from Spain in 1492, hoping to find a new route to the East Indies across the Atlantic Ocean. The true size of Earth's

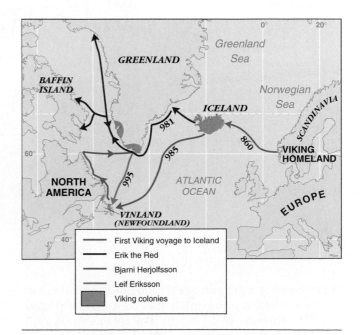

Figure 1–7 Viking colonies in the north Atlantic. Map showing the routes and dates of Viking explorations and the locations of the colonies that were established in Iceland, Greenland, and parts of North America.

Box 1–1 How do Sailors Know Where They Are at Sea?: From Stick Charts to Satellites

How do you know where you are in the ocean without roads, signposts, or any land in sight? How do you determine the distance to a destination? How do you find your way back to a good fishing spot or where you have discovered sunken treasure? Sailors have relied on a variety of navigation tools to help answer questions such as these by locating where they are at sea.

Some of the first navigators were the Polynesians. Despite the enormity of the Pacific Ocean, the Polynesians were able to inhabit distant islands. These early navigators must have been very aware of the marine environment and been able to read subtle differences in the ocean and sky. The tools they used to help them navigate between islands included the Sun and Moon, the nighttime stars, the behavior of marine organisms, various ocean properties, and an ingenious device called a **stick chart** (Figure 1B). The stick charts accurately depicted the consistent pattern of ocean waves. By orienting their vessels relative to this regular ocean wave direction, sailors could successfully navigate at sea. The bent wave directions let them know when they were getting close to an island—even one that was located beyond the horizon.

The importance of knowing where you are at sea is illustrated by a tragic incident in 1707, when a British battle fleet was over 160 kilometers (100 miles) off course and ran aground in the Isles of Scilly near England, with the loss of four ships and 2000 men. **Latitude** (location north or south) was relatively easy to determine at sea by measuring the position of the Sun and stars using a device called a **sextant** (*sextant* = sixth, in reference to the instrument's arc, which is one-sixth of a circle) (Figure 1C). The accident occurred because the ship's crew had no way of keeping track of their **longitude** (location east or west; see Appendix III, "Latitude and Longitude on Earth"). To determine longitude, which is a function of time, it was necessary to know the time difference between a reference meridian and when the Sun was directly overhead of a ship at sea (noon local time). However, clocks in the early 1700s were driven by pendulums and would not work for long on a rocking ship at sea. In 1714, the British Parliament offered a £20,000 prize (about $12 million today) for developing a device that would work well enough at sea to determine longitude within 0.5 degree after a voyage to the West Indies.

A cabinetmaker in Lincolnshire named **John Harrison** began working on such a timepiece in 1728, which was dubbed the **chronometer** (*chrono* = time, *meter* = measure). Harrison's first chronometer, H1 (Figure 1D) was successfully tested in 1736, but he received only £500 of the prize

Figure 1B Navigational stick chart. This bamboo stick chart of Micronesia's Marshall Islands shows islands (represented by shells at the junctions of the sticks), regular ocean wave direction (represented by the straight strips), and waves that bend around islands (represented by the curved strips). Similar stick charts were used by early Polynesian navigators.

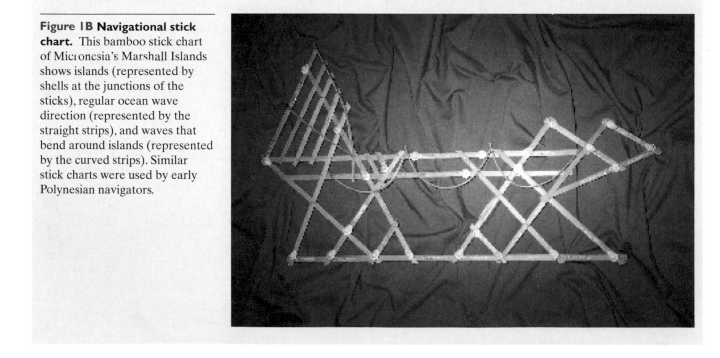

circumference had been underestimated and instead he reached what is now called the West Indies (Figure 1–8).[2] Europeans first saw the Pacific Ocean in 1513, when **Vasco Núñez de Balboa** attempted a land crossing of the Isthmus of Panama and sighted a large ocean to the west from atop a mountain.

The culmination of the Age of Discovery was a remarkable circumnavigation of the globe initiated by **Ferdinand Magellan** (Figure 1–8). Magellan left Spain in September 1519 with five ships and 280 sailors. He crossed the Atlantic Ocean, sailed down the eastern

[2]For more information about Columbus's voyage, see Box 6-1 in Chapter 6, "Air-Sea Interaction."

because the device was deemed too complex, costly, and fragile. Eventually, his more compact fourth version, (H4)—which resembles an oversized pocket watch—was tested during a trans-Atlantic voyage in 1761. Upon reaching Jamaica, it was so accurate that it had lost only *five seconds* of time, a longitude error of only *0.02 degree*! Although Harrison's chronometer greatly exceeded the requirements of the government, the committee in charge of the prize withheld payment because most of them were astronomers who wanted the solution to come from measurement of the stars. Additionally, Harrison was a commoner with little political influence and the committee members were unconvinced that reliable versions of the chronometer could be produced in large numbers. For his life's work, Harrison was finally awarded the balance of the prize in 1773, when he was 80 years old—and only after the committee was persuaded to do so by King George III.

Figure 1C Sextant. This hand-held sextant is similar to the ones used by early navigators to determine latitude.

Figure 1D John Harrison's chronometer H1. The first of Harrison's chronometers was driven by a helical balance spring that remained horizontal and independent of ship motion.

coast of South America, and traveled through a passage to the Pacific Ocean at 52 degrees south latitude, now named the Strait of Magellan in his honor. After landing in the Philippines on March 15, 1521, Magellan was killed in a fight with the inhabitants of these islands. Juan Sebastian del Caño completed the circumnavigation by taking the last of the ships, the *Victoria*, across the Indian Ocean, around Africa, and back to Spain in 1522. After three years, just one ship and 18 men completed the voyage.

Following these voyages, the Spanish initiated many others to take gold from the Aztec and Inca cultures in Mexico and South America. The English and Dutch, meanwhile, used smaller, more maneuverable ships to

Today, navigating at sea relies on the **Global Positioning System (GPS)**, which was initiated in the 1970s by the United States Department of Defense when they began launching a $13 billion system of satellites (Figure 1E). Initially designed for military purposes but now available for a variety of civilian uses, GPS relies on 24 satellites that send continuous radio signals to the surface. Position is determined by very accurate measurement of the time of travel of radio signals from the satellites to receivers on board a ship (or on land). Thus, a vessel can determine its exact latitude and longitude to within a few meters—a small fraction of the length of most ships. Navigators from days gone by would be amazed at how quickly and accurately a vessel's location can be determined, but they might say that it has taken all the adventure out of navigating at sea.

Figure 1E A Global Positioning System (GPS) satellite. Currently, 24 of these satellites orbit about 20,000 kilometers (12,400 miles) above Earth, each taking 12 hours to go around Earth once. The satellite measures 5.2 meters (17 feet) across with its solar panels unfurled.

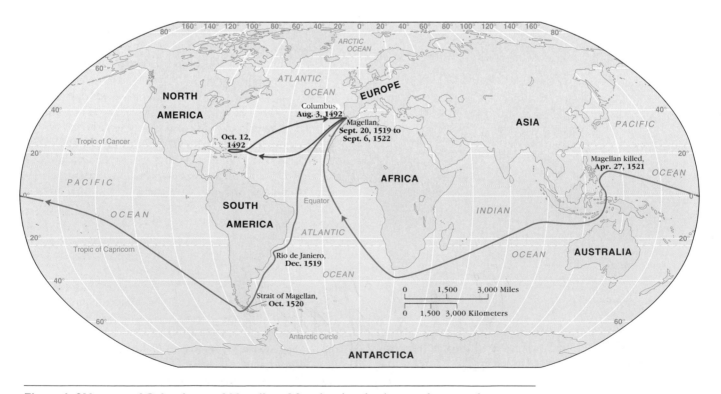

Figure 1–8 Voyages of Columbus and Magellan. Map showing the dates and routes of Columbus's first voyage and the first circumnavigation of the globe by Magellan's party.

rob the gold from bulky Spanish galleons, which resulted in many confrontations at sea. The maritime dominance of Spain ended when the English defeated the Spanish Armada in 1588. With control of the seas, the English thus became the dominant world power and remained so until early in the twentieth century.

The Beginning of Voyaging for Science

The English realized that increasing their scientific knowledge of the oceans would help maintain their maritime superiority. For this reason, Captain **James Cook** (1728-1779), an English navigator and prolific explorer (Figure 1–9), undertook three voyages of scientific discovery with the ships *Endeavour, Resolution,* and *Adventure* between 1768 and 1779. He searched for the continent Terra Australis ("Southern Land," or Antarctica) and concluded that it lay beneath or beyond the extensive ice fields of the southern oceans, if it existed at all. Cook also mapped many previously unknown islands, including the South Georgia, South Sandwich, and the Hawaiian Islands. During his last voyage, Cook searched for the fabled "northwest passage" from the Pacific Ocean to the Atlantic Ocean and stopped in Hawaii, where he was killed in a skirmish with native Hawaiians.

Cook's expeditions added greatly to the scientific knowledge of the oceans. He determined the outline of the Pacific Ocean and was the first person known to cross the Antarctic Circle in his search for Antarctica. Cook initiated systematic sampling of subsurface water temperatures, measuring winds and currents, taking depth measurements (called *soundings*), and collecting data on coral reefs. Cook also discovered that a ship-board diet containing the German staple sauerkraut prevented his crew from contracting scurvy, a disease that incapacitated sailors. Scurvy is caused by a vitamin C deficiency and the cabbage used to make sauerkraut contains large quantities of vitamin C. In addition, by proving the value of John Harrison's chronometer as a means of determining longitude (see Box 1–1), Cook made possible the first accurate maps of Earth's surface, some of which are still in use today.

History of Oceanography . . . To Be Continued

Additional events in the history of oceanography can be found at the beginning of subsequent chapters. Each chapter-opening feature introduces an important historical event that is related to the subject of the chapter.

> The ocean's large size did not prohibit early explorers from venturing into all parts of the ocean for discovery, trade, or conquest. Voyaging for science began relatively recently.

The Nature of Scientific Inquiry

In modern society, scientific studies are increasingly used to substantiate the need for action. However, there is often little understanding of how science operates. For instance, how certain are we about a particular scientific theory? How are facts different from theories?

The overall goal of science is to discover underlying patterns in the natural world. This knowledge is then used to make predictions about what should or should not be expected to happen given a certain set of circumstances. Scientists develop explanations about the causes and effects of various natural phenomena (such as why Earth has seasons or what the structure of matter is). Scientific analyses are based on an assumption that all natural phenomena are controlled by understandable physical processes and the same physical processes operating today have been operating throughout time. Consequently, science has demonstrated remarkable power in allowing scientists to describe the natural world accurately, to identify the underlying causes of natural phenomena, and to better predict future events that rely on natural processes.

Science supports the explanation of the natural world that best explains all available observations. Scientific inquiry is formalized into what is called the **scientific method**, which is used to formulate scientific theories (Figure 1–10).

Figure 1–9 Captain James Cook of the British Royal Navy.

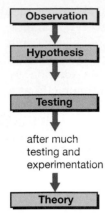

Observation — Collection of scientific facts through observation and measurement

Hypothesis — A tentative, testable statement about the natural world that can be used to build more complex inferences and explanations

Testing — Development of observations, experiments, and models to test (and, if necessary, revise) the hypothesis

after much testing and experimentation

Theory — In science, a well-substantiated explanation of some aspect of the natural world that can incorporate facts, laws, logical inferences, and tested hypotheses

Figure 1–10 The scientific method.

Observations

The scientific method begins with **observations**, which are occurrences we can measure with our senses. They are things we can manipulate, see, feel, hear, touch, or smell. We often sense them directly or by using sophisticated tools such as a microscope or telescope. If an observation is repeatedly confirmed—that is, made so many times that it is assumed to be completely valid—then it can be called a scientific **fact**.

Hypothesis

As observations are being made, the human mind attempts to sort out the observations in a way that reveals some underlying order or pattern in the objects or phenomena being observed. This sorting process—which involves a lot of trial and error—seems to be driven by a fundamental human urge to make sense of our world. This is how **hypotheses** (*hypo* = under, *thesis* = an arranging) are made. A hypothesis is sometimes labeled as an informed or educated guess, but it is more than that. A hypothesis is a tentative, testable statement about the general nature of the phenomena observed. In other words, a hypothesis is an initial idea of how or why things happen in nature.

Suppose we wanted to understand why whales breach (that is, why whales sometimes throw their entire bodies out of the water). After scientists observe breaching many times, they can organize their observations into a hypothesis. For instance, one hypothesis is that a breaching whale is trying to dislodge parasites from its body. Scientists often have multiple working hypotheses (for example, whales may use breaching to communicate with other whales). If a hypothesis cannot be tested, it is not scientifically useful no matter how interesting it might seem.

Testing

Hypotheses are used to predict certain occurrences that lead to further research and the refinement of those hypotheses. For instance, the hypothesis that a breaching whale is trying to dislodge its parasites suggests that breaching whales have more parasites than whales that don't breach. Analyzing the number of parasites on breaching versus nonbreaching whales would either support that hypothesis or cause it to be recycled and modified.

In science, the validity of any explanation is determined by its coherence with observations in the natural world and its ability to predict further observations. Only after much testing and experimentation—usually done by many experimenters using a wide variety of repeatable tests—does a hypothesis gain validity where it can be advanced to the next step.

Theory

If a hypothesis has been strengthened by additional observations and if it is successful in predicting additional phenomena, then it can be advanced to what is called a **theory** (*theoria* = a looking at). A theory is a well-substantiated explanation of some aspect of the natural world that can incorporate facts, laws (descriptive generalizations about the behavior of an aspect of the natural world), logical inferences, and tested hypotheses. A theory is not a guess or a hunch. Rather, it is an understanding that develops from extensive observation, experimentation, and creative reflection.

In science, theories are formalized only after many years of testing and verifying predictions. Thus, scientific theories are those that have been rigorously scrutinized to the point where most scientists agree that they are the best explanation of certain observable facts. Examples of prominent, well-accepted theories that are held with a very high degree of confidence include geology's theory of plate tectonics (which will be covered in the next chapter) and biology's theory of evolution.

Theories and the Truth

We've seen how the scientific method is used to develop theories, but does science ever arrive at the undisputed "truth?" Science never reaches an absolute truth because we can never be certain that we have all the observations, especially considering that new technology will be available in the future to examine phenomena in different ways. New observations are always possible, so the nature of scientific truth is subject to change. It is accurate to say that science arrives at that which is *probably* true, based on the available observations.

It is not science's downfall that scientific ideas are modified as more observations are collected. In fact, the opposite is true. Science is a process that depends on reexamining ideas as new observations are made. Thus, science progresses when new observations yield new hypotheses and modification of theories. As a result, science is littered with hypotheses that have been abandoned in favor of later explanations that fit new observations. One of the best known is the idea that Earth was at the center of the universe, a proposal that was supported by the apparent daily motion of the Sun, Moon, and stars around Earth.

The statements of science should never be accepted as the "final truth." Over time, however, they generally form a sequence of increasingly more accurate statements. Theories are the end-points in science and do not turn into facts through accumulation of evidence. Nevertheless, the data can become so convincing that the accuracy of a theory is no longer questioned. For instance, the *heliocentric* (*helios* = sun, *centric* = center) *theory* of our solar system states that Earth revolves around the Sun rather than vise versa. Such concepts are supported by such abundant observational and experimental evidence that they are no longer questioned in science.

Is there really such a formal method to science as the scientific method suggests? Actually, the work of scientists is much less formal and is not always done in a clearly logical and systematic manner. Like a detective analyzing a crime scene, a scientist uses ingenuity and serendipity, visualizes models, and sometimes follows hunches in order to unravel the mysteries of nature.

> Science supports the explanation of the natural world that best explains all available observations. Because new observations can modify existing theories, science is always developing.

Origins

Where did Earth come from? Where did the atmosphere and the oceans come from? When were all of these created? In this section, we'll present scientific answers to these kinds of questions using the fragmentary evidence that remains from several billion years ago.

Origin of the Solar System and Earth

Earth is the third of nine planets in our **solar system** that revolve around the Sun (Figure 1–11). Evidence suggests that the Sun and solar system formed at the same time from a huge cloud of debris (gases and dust) called a **nebula** (*nebula* = a cloud). Astronomers base this hypothesis on the orderly nature of our solar system and the consistent age of meteorites (pieces of the early solar system).

Figure 1–11 The solar system. The Sun and planets of the solar system drawn to scale.

The Nebular Hypothesis According to the **nebular hypothesis** (Figure 1–12), all bodies in the solar system formed from an enormous cloud composed mostly of hydrogen and helium, with only a small percentage of heavier elements. As this huge accumulation of gas and

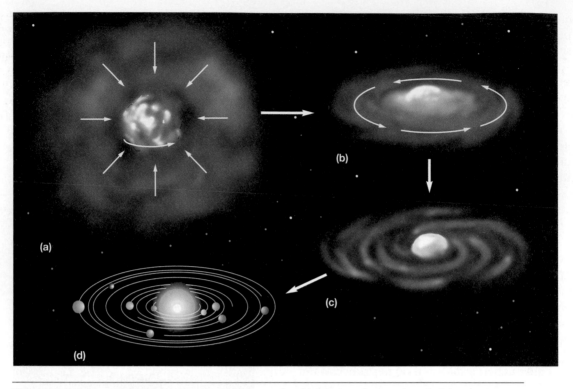

Figure 1–12 The nebular hypothesis. (a) A huge cloud of dust and gases (a nebula) contracts. **(b)** Most of the material is gravitationally swept toward the center, producing the sun, while the remainder flattens into a disk. **(c)** Small eddies are created by the circular motion. **(d)** In time, most of the remaining debris forms the planets and their moons.

dust revolved around its center, the Sun began to form as the force of gravity concentrated particles. In its early stages, the diameter of the Sun may have equaled or exceeded the diameter of our entire solar system today.

As the nebular matter that formed the Sun contracted, small amounts of it were left behind in eddies, similar to whirlpools in a stream. The material in these eddies was the beginning of the **protoplanets** (*proto* = original, *planetes* = wanderer) and their orbiting satellites, which later consolidated into the present planets and their moons.

The Protoearth   The **Protoearth** looked very different from Earth today. Its size was larger than today's Earth and there were neither oceans nor any life on the planet. In addition, the structure of the deep Protoearth is thought to have been **homogenous** (*homo* = alike, *genous* = producing), which means that it had a uniform composition throughout. The structure of Protoearth changed, however, when its heavier constituents migrated toward the center to form a heavy core.

During this early formation of the protoplanets and their satellites, the Sun condensed into such a hot, concentrated mass that forces within its interior began releasing energy through a process known as a **fusion** (*fusus* = to melt) **reaction**. A fusion reaction occurs when temperatures reach tens of millions of degrees and hydrogen **atoms** (*a* = not, *tomos* = cut) are converted to helium atoms, releasing large amounts of energy. In addition to light energy, the Sun also emits

ionized (electrically charged) particles. In the early stages of our solar system, these ionized particles blew away the nebular gas that remained from the formation of the planets and their satellites.

Meanwhile, the protoplanets closest to the Sun (including Earth) were heated so intensely by solar radiation that their initial atmospheres (mostly hydrogen and helium) boiled away. Additionally, the combination of ionized solar particles and internal warming of these protoplanets caused them to drastically shrink in size. As the protoplanets continued to contract, heat was produced deep within their cores from the spontaneous disintegration of atoms, called *radioactivity* (*radio* = radiation, *acti* = a ray).

Density Stratification   The release of internal heat became so intense that Earth's surface became molten. Once Earth became a ball of hot liquid rock, the elements were able to segregate according to their **densities**[3] in a process called **density stratification** (*strati* = a layer, *fication* = making). The highest density materials (primarily iron and nickel) concentrated in the core,

---

[3]Density is defined as mass per unit volume. An easy way to remember this is that density is a measure of *how heavy something is for its size*. For instance, an object that has a low density is light for its size (like a dry sponge, foam packing, or a surfboard). An object that has a high density is heavy for its size (like cement, most metals, or a large container full of water). Note that density has nothing to do with the *thickness* of an object: some objects (like a stack of foam packing) can be thick but have low density.

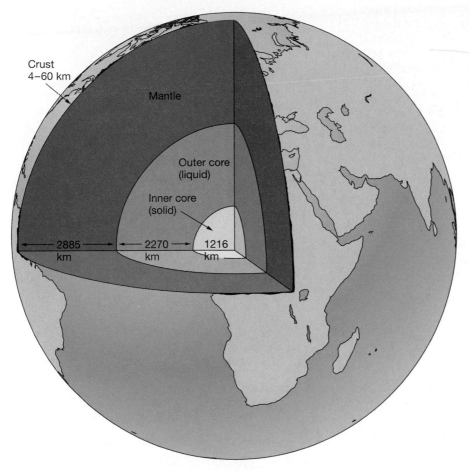

Crust
4–60 km

Mantle

Outer core
(liquid)

Inner core
(solid)

2885
km

2270
km

1216
km

Figure 1–13 Earth's internal structure. A cutaway view of Earth's interior, showing the major subdivisions of Earth structure. The inner core (highest density), outer core, and mantle are drawn to scale, but the thickness of the crust (lowest density) is exaggerated about five times.

whereas lower and lower density components (primarily rocky material) formed concentric spheres around the core. If you've ever noticed how oil-and-vinegar salad dressing settles out into a lower-density top layer (the oil) and a higher-density bottom layer (the vinegar), then you've seen how density stratification causes separate layers to form. Thus, Earth became a layered sphere based on density.

The cutaway view of Earth's interior in Figure 1–13 shows that the high-density **core** can be divided into a solid *inner core* and a liquid *outer core*. Surrounding the core is the lower-density **mantle** (a solid zone) and surrounding the mantle is the even lower-density **crust**. There are two kinds of crust—oceanic and continental. *Oceanic crust* underlies the ocean basins, has a higher density than continental crust, and is 4 to 10 kilometers (2.5 to 6.2 miles) thick. *Continental crust* underlies the continents, has a lower density than oceanic crust, and is 35 to 60 kilometers (22 to 37 miles) thick. The internal structure of Earth is discussed in more detail in Chapter 2, "Plate Tectonics and the Ocean Floor."

Origin of the Atmosphere and the Oceans

Where did the atmosphere come from? Most likely, an early atmosphere was expelled from *inside* Earth by a process called **outgassing**. During the period of density stratification, the lowest-density material contained

within Earth was various gases. These gases rose to the surface and were expelled to form Earth's early atmosphere. What was the composition of these gases? They are believed to have been similar to the gases emitted from volcanoes, geysers, and hot springs today; mostly water vapor (steam), with small amounts of carbon dioxide, hydrogen, and other gases. The composition of this early atmosphere was not, however, the same composition as today's atmosphere. The composition of the atmosphere changed over time because of the influence of life and possible changes in the mixing of material in the mantle.

Where did the oceans come from? Their origin is directly linked to the origin of the atmosphere. Figure 1–14 shows that as Earth cooled, the water vapor released to the atmosphere during outgassing condensed and fell to Earth. Evidence suggests that by at least 4 billion years ago, most of the water vapor from outgassing had accumulated to form the first permanent oceans on Earth.

Originally, Earth had no oceans. The oceans (and atmosphere) came from inside Earth as a result of outgassing and were present by at least 4 billion years ago.

Figure 1–14 Formation of Earth's oceans. Early in Earth's history, widespread volcanic activity released large amounts of water vapor (H_2O vapor) and smaller quantities of other gases. As Earth cooled, the water vapor **(a)** condensed into clouds and **(b)** fell to Earth's surface, where it accumulated to form the oceans **(c).**

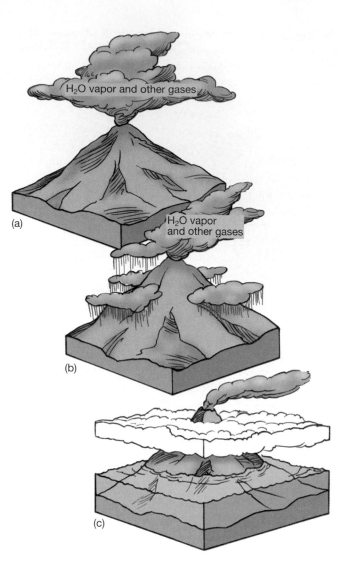

The Development of Ocean Salinity   The relentless rainfall that landed on Earth's rocky surface dissolved many elements and compounds and carried them into the newly forming oceans, producing the chemical composition our oceans have today.

Are the oceans growing more or less salty or has their salinity remained constant over time? We can answer this question by studying the proportion of water vapor to chloride ion, Cl^-, using ancient marine rocks. Chloride ion is important because it forms part of the most common salts in the ocean (e.g. sodium chloride, potassium chloride, and magnesium chloride). Also, chloride ion is produced by outgassing, like the water vapor that formed the oceans. Currently, there is no indication that the ratio of water vapor to chloride ion has fluctuated throughout geologic time, so it can be reasonably concluded that the oceans' salinity has been relatively constant through time. Further aspects of the oceans' salinity are explored in Chapter 5, "Water and Seawater."

Life Begins in the Oceans

How did life begin on Earth? One recent hypothesis is that the organic building blocks of life may have arrived embedded in meteors, comets, or cosmic dust. Alternatively, life may have originated around hydrothermal vents—hot springs—deep in the ocean. Yet another idea is that life originated in rock material deep below Earth's surface.

According to the fossil record on Earth, the earliest known life forms were primitive bacteria that lived about 3.5 billion years ago. This record indicates that the basic building blocks for the origin of life came from materials already present on Earth. The oceans were the most likely place for these materials to interact to produce life.

The Importance of Oxygen to Life

Oxygen, which comprises almost 21% of our present atmosphere, is essential to human life for two reasons. First, our bodies need oxygen to "burn" (*oxidize*) food, releasing energy to our cells. Second, oxygen in the form of ozone protects the surface of the Earth from most of the Sun's harmful ultraviolet radiation.

Evidence suggests that Earth's early atmosphere (the product of outgassing) was different from Earth's initial hydrogen-helium atmosphere and different from the mostly nitrogen-oxygen atmosphere of today. The early atmosphere probably contained large percentages of water vapor and carbon dioxide and smaller percentages of hydrogen, methane, and ammonia, but very little free oxygen (oxygen that is not chemically bound to other atoms). Why was there so little free oxygen in the early atmosphere? Oxygen may well have been outgassed, but oxygen and iron have a strong affinity for each other.[4] Iron in Earth's early crust would have reacted with the outgassed oxygen immediately, removing it from the atmosphere.

Without oxygen in Earth's early atmosphere, moreover, there would have been no ozone layer to block most of the Sun's ultraviolet radiation. The lack of a protective ozone layer may have been a key component in the development of life on Earth.

In 1952, **Stanley Miller** conducted a laboratory experiment by exposing a mixture of carbon dioxide, methane, ammonia, hydrogen, and water (the components of the early atmosphere and ocean) to ultraviolet light (from the Sun) and an electrical spark (to imitate lightning) (Figure 1–15). After a few weeks, the clear water turned pink and then brown, indicating the formation of a large assortment of organic molecules including amino acids, which are the basic components of life.

Miller's now-famous laboratory experiment of a simulated primitive "Earth in a bottle" demonstrated that vast amounts of organic molecules could have been produced in Earth's early oceans. What is unknown, however, is precisely how this organic material developed into more complex molecular structures—such as proteins and DNA—that are intrinsic to life. Through some process, however, the organic material must have become chemically self-reproductive and developed the ability to actively seek light, metabolize food, and grow toward a characteristic size and shape dictated by internal molecular codes.

Organic molecules were produced in a simulation of Earth's early atmosphere and ocean, suggesting that life most likely originated in the oceans.

Plants and Animals Evolve

The very earliest forms of life were probably **heterotrophs** (*hetero* = different, *tropho* = nourishment). Heterotrophs require an external food supply, which was abundantly available in the form of nonliving organic matter in the ocean around them. The **autotrophs** (*auto* = self, *tropho* = nourishment), which can manufacture their own food supply, evolved later. The first autotrophs were probably similar to our present-day **anaerobic** (*an* = without, *aero* = air) bacteria, which live without atmospheric oxygen. They may have been able to derive energy from inorganic compounds at deep-water hydrothermal vents using a process called **chemosynthesis** (*chemo* = chemistry, *syn* = with, *thesis* = an arranging).[5]

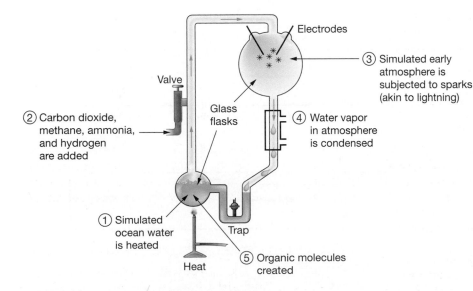

② Carbon dioxide, methane, ammonia, and hydrogen are added

Valve

Electrodes

Glass flasks

③ Simulated early atmosphere is subjected to sparks (akin to lightning)

④ Water vapor in atmosphere is condensed

① Simulated ocean water is heated

Trap

Heat

⑤ Organic molecules created

Figure 1–15 Creation of organic molecules. Laboratory apparatus used by Stanley Miller to simulate the conditions of the early atmosphere and the oceans. The experiment produced organic molecules and suggests that the basic components of life were created in the oceans.

---

[4]As an example of the strong affinity of iron and oxygen, consider how common rust—a compound of iron and oxygen—is on Earth's surface.

[5]In fact, the recent discovery (reported in 2000) of 3.2 billion year old microfossils of bacteria from deep-water marine rocks found in what is now Australia supports the idea of a high-temperature origin of life on the deep ocean floor in the absence of light.

Photosynthesis and Respiration   Eventually, more complex single-celled autotrophs evolved. They developed a green pigment called **chlorophyll** (*chloro* = green, *phyll* = leaf), which captures the Sun's energy through **photosynthesis** (*photo* = light, *syn* = with, *thesis* = an arranging). In photosynthesis (Figure 1–16), plants capture light energy and store it as sugars. A chemical reaction in which energy is captured or absorbed is said to be **endothermic** (*endo* = inside, *thermo* = heat). In **respiration** (*respir* = to breathe), the sugars are oxidized with oxygen so their stored energy can be used to carry on the life processes of the plant or the animal that eats the plant. A chemical reaction that releases energy is said to be **exothermic** (*exo* = outside, *thermo* = heat).

Not only are photosynthesis and respiration chemically opposite processes, but they are also complimentary because the products of photosynthesis (sugars and oxygen) are used during respiration and the products of respiration (water and carbon dioxide) are used in photosynthesis (see Figure 1–16, *bottom*). Thus, autotrophs (algae and plants) and heterotrophs (most bacteria and animals) have developed a mutual need for each other.

The oldest fossilized remains of organisms are primitive photosynthetic bacteria recovered from rocks formed on the sea floor about 3.5 billion years ago. However, the oldest rocks containing iron oxide (rust)—an indicator of an oxygen-rich atmosphere—do not appear until about 2 billion years ago. Thus, photosynthetic organisms took at least 1.5 billion years to develop and begin producing abundant free oxygen in the atmosphere. At the same time, when a large amount of oxygen-rich (ferric) iron sank to the base of the mantle, it may have been heated by the core, risen as a plume to the ocean floor and began releasing large amounts of oxygen about 2.5 billion years ago. By 1.8 billion years ago, the atmosphere's oxygen content had increased to such a high level that it began causing the extinction of many anaerobic organisms. The species that survived are the ancestors of the organisms on Earth today.

Evolution and Natural Selection   Every living organism that inhabits Earth today is the result of **evolution** by the process of **natural selection** that has been going on since these early times. Evolution is the theory that groups of organisms adapt and change with the passage of time. Descendants differ morphologically and physiologically from their ancestors. Certain advantageous traits are naturally selected and passed on from one generation to the next. Evolution is the process by which various **species** (*species* = a kind) have been able to inhabit increasingly numerous environments on Earth.

As species adapted to Earth's various environments, they also modified the environments in which they lived. For example, when plants emerged from the oceans and inhabited the land, they changed it from a harsh and bleak landscape (much like the Moon's surface today) to one that was green and lush. The ocean changed, too, as vast quantities of dead organisms accumulated on the sea floor. Some of these accumulations have been turned into rock and uplifted onto continents, sometimes at high elevations. Because these

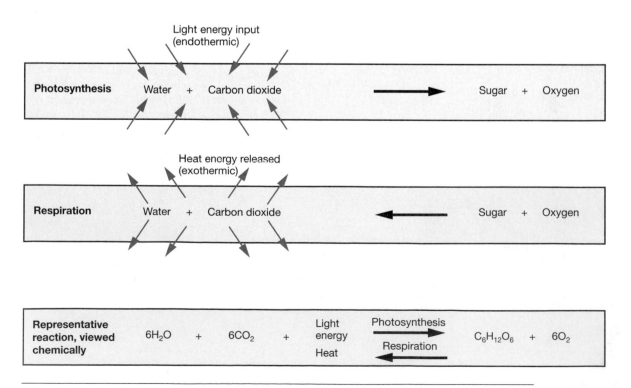

Figure 1–16 Photosynthesis (*top*), respiration (*middle*), and representative reactions viewed chemically (*bottom*).

rocks formed on the sea floor, there is much to be learned about the oceans of the past by studying these rocks.

Changes to Earth's Environment   The development and successful evolution of photosynthetic organisms is greatly responsible for the world as we know it today (Figure 1–17). These organisms reduced the amount of carbon dioxide in the early atmosphere from relatively high levels to 0.036% today. At the same time, they increased the amount of free oxygen in the early atmosphere from very low levels to 21% today. The increasing level of oxygen, in turn, made it possible for most present-day animals to develop.

The remains of ancient plants and animals buried in oxygen-free environments have become the oil, natural gas, and coal deposits of today. These so-called *fossil fuels* provide over 90% of the energy we consume to power modern society. We depend not only on the food

energy stored in today's plants, but also on the energy stored in plants during the geologic past.

Because of increased burning of fossil fuels for home heating, industry, power generation, and transportation during the industrial age, the atmospheric concentration of carbon dioxide and other gases that help warm the atmosphere has increased, too. Many people are concerned that increased global warming will cause serious problems in the future. This phenomenon, referred to as the *greenhouse effect*, is discussed in Chapter 6, "Air–Sea Interaction."

Radiometric Dating and the Geologic Time Scale

How can Earth scientists tell how old a rock is? It can be a difficult task to tell if a rock is thousands, millions, or even billions of years old—except if the rock contains telltale fossils. Fortunately, Earth scientists can

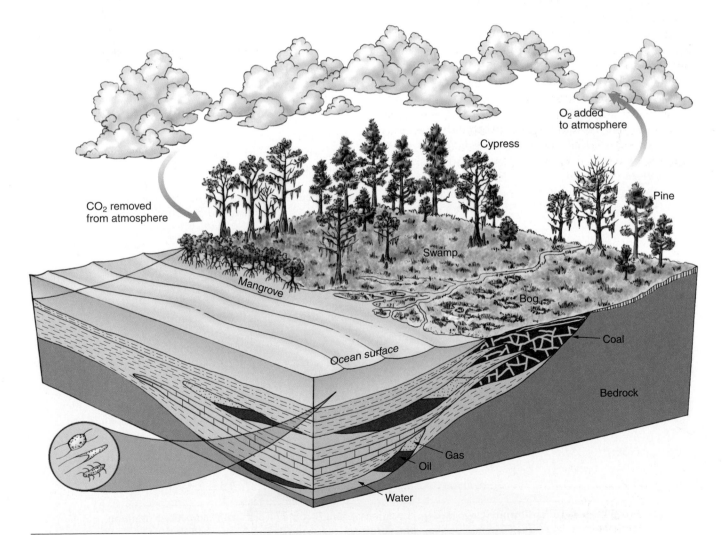

Figure 1–17 The effect of plants on Earth's environment. As microscopic photosynthetic cells (*inset*) became established in the ocean, Earth's atmosphere was enriched in oxygen and depleted in carbon dioxide. As organisms died and accumulated on the ocean floor, some of their remains were converted to oil and gas. The same process occurred on land, sometimes producing coal.

determine how old most rocks are by using the radioactive materials contained within rocks. In essence, this technique involves reading a rock's internal "rock clock."

Most rocks on Earth (as well as those from outer space) contain small amounts of radioactive materials such as uranium, thorium, and potassium. These radioactive materials spontaneously break apart or decay into atoms of other elements. Radioactive materials have a characteristic **half-life**, which is the time required for one-half of the atoms in a sample to decay to other atoms. The older the rock is, the more radioactive material will have been converted to decay product. Analytical instruments can accurately measure the amount of radioactive material and the amount of resulting decay product in rocks. By comparing these two quantities, the age of the rock can thus be determined. Such dating is referred to as **radiometric** (*radio* = radioactivity, *metri* = measure) **age dating** and is an extremely powerful tool for determining the age of rocks.

For example, Figure 1–18 shows two locations where rocks contain small amounts of a particular type of radioactive uranium, which has a half-life of 713 million years and decays to a stable form of lead. Measurements at Location A indicate that half of the radioactive uranium has been converted to lead, its decay product. Thus, one half-life has elapsed and the rock is 713 million years old. At Location B, three-quarters of the original radioactive material has converted to its decay product, so the rock here must be older than that at Location A. Two half-lives have elapsed and the rock is 1,426 million (1.4 billion) years old. Using this method, hundreds of thousands of rock samples have been age dated from around the world.

The ages of rocks on Earth are shown in the **geologic time scale** (Figure 1–19). It lists the names of the geologic time periods as well as important advances in the development of life forms on Earth. Initially, the divisions between geologic periods were based on major extinction episodes as recorded in the fossil record. As radiometric age dates became available, they were included on the geologic time scale as well.

> Scientists can accurately determine the age of most rocks by analyzing their radioactive components, some of which indicate that Earth is 4.6 billion years old.

? Students Sometimes Ask...

You mentioned that the oceans came from inside Earth. However, I recently heard that the oceans came from outer space as icy comets. Which one is true?

Recently, scientists have discovered what appear to be comets composed of ice (each about the size of a small house) that are entering Earth's atmosphere at a rate of up to 40,000 per day. Although these small comets disintegrate in the atmosphere upon entry, calculations show that they add about 2.5 centimeters (1 inch) of water every 10,000 years, which is enough to fill the oceans over the lifetime of Earth. It is still a matter of debate whether the objects are really icy comets, but this hypothesis suggests that the oceans may have originated as ice from outer space. Which hypothesis is correct? Because only small amounts of evidence are available to help support either one, it's difficult to say with certainty. Perhaps new

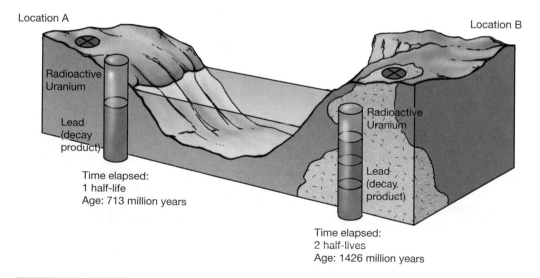

Figure 1–18 Radiometric age dating. Locations A and B contain rocks with radioactive uranium that has a half–life of 713 million years and decays to lead. At Location A, one–half of the radioactive uranium has converted to lead, indicating the one half–life has elapsed so the rocks are 713 million years old. At Location B, three–quarters of the radioactive uranium has converted to lead, indicating that two half–lives have elapsed and that the rocks are 1426 million (1.4 billion) years old.

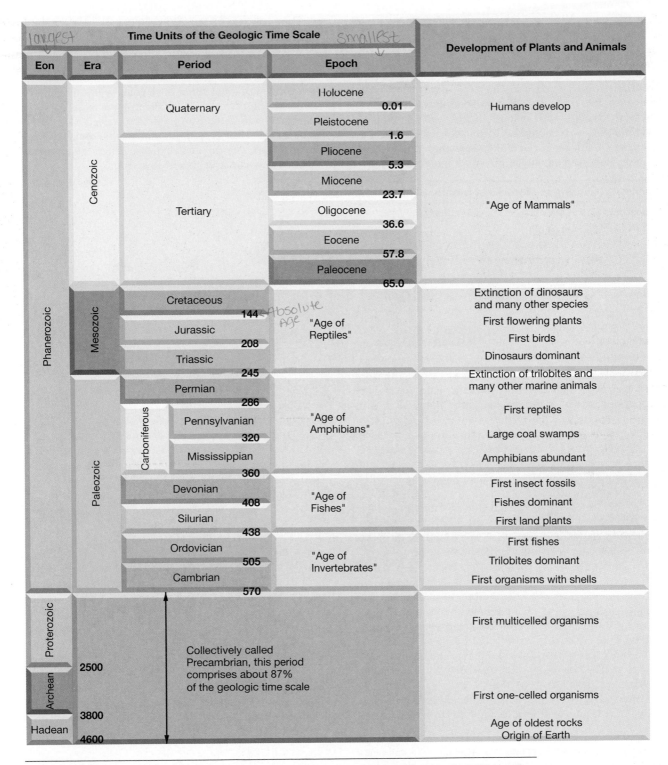

Time Units of the Geologic Time Scale				Development of Plants and Animals
Eon	Era	Period	Epoch	
Phanerozoic	Cenozoic	Quaternary	Holocene / 0.01 / Pleistocene / 1.6	Humans develop
		Tertiary	Pliocene / 5.3 / Miocene / 23.7 / Oligocene / 36.6 / Eocene / 57.8 / Paleocene / 65.0	"Age of Mammals"
	Mesozoic	Cretaceous / 144	"Age of Reptiles"	Extinction of dinosaurs and many other species
		Jurassic / 208		First flowering plants / First birds
		Triassic / 245		Dinosaurs dominant
	Paleozoic	Permian / 286	"Age of Amphibians"	Extinction of trilobites and many other marine animals
		Carboniferous — Pennsylvanian / 320		First reptiles / Large coal swamps
		Carboniferous — Mississippian / 360		Amphibians abundant
		Devonian / 408	"Age of Fishes"	First insect fossils / Fishes dominant
		Silurian / 438		First land plants
		Ordovician / 505	"Age of Invertebrates"	First fishes / Trilobites dominant
		Cambrian / 570		First organisms with shells
Proterozoic		2500	Collectively called Precambrian, this period comprises about 87% of the geologic time scale	First multicelled organisms
Archean		3800		First one-celled organisms
Hadean		4600		Age of oldest rocks / Origin of Earth

(handwritten annotations: "largest" above Eon, "smallest" above Epoch, "Absolute Age" next to 144)

Figure 1–19 The geologic time scale. Numbers on the time scale represent time in millions of years before the present; significant advances in the development of plants and animals on Earth are also shown.

developments in technology will fuel discoveries that will support one or the other of these hypotheses (or even an entirely new idea). To us, this is what makes science so interesting!

You've talked about the limits of compressed-air diving. What is the record depth for a dive from the surface with one breath?

The problem with any ocean dive is how the increased pressure with depth affects the human body. However, some people are pushing the limits of human endurance by holding onto a heavy weight at the surface and descending rapidly into the ocean. At the deepest depth they can tolerate, they let go of the weight and return to the surface. The world's record for these so-called "free dives" was set by Francisco Ferreras in

Box 1–2
"Deep" Time

According to the timeline in Figure 1F, Earth is 4.6 billion years old. It is difficult to comprehend how old Earth is, however, because 4.6 billion (that's 4600 million, or 4,600,000,000) is so enormous. To gain some idea of the immensity of a number in the billions, how high would a stack of one billion fresh one-dollar bills be? (Table 1A)

Table 1A. Thickness of stacks of fresh one-dollar bills.

Thickness of a stack of...	Would be...
100 fresh one-dollar bills	1 centimeter (0.4 inch) high
1000 fresh one-dollar bills	10 centimeters (4 inches) high
1,000,000 (one million) fresh one-dollar bills	100 meters (330 feet) high
1,000,000,000 (one billion) fresh one-dollar bills	100 kilometers (62 miles) high

The problem with an example such as this is that the mind becomes immune to comprehending such large amounts of money— or any such large numbers. For instance, how long would it take to count to 4.6 billion if you counted one number every second for 24 hours a day, seven days a week, 365 days a year? The answer can be easily calculated, remembering that there are 60 seconds in a minute and 60 minutes in an hour. The rather surprising answer is hidden below if you need help.

Another way to try to visualize the enormity of geologic time is by representing its entirety with a roll of toilet paper.[6] If you use a standard 500-sheet roll and round off the age of Earth to 5 billion years, then each sheet on the roll represents 10 million years. As you unroll the paper, keep in mind that you are progressing through all of geologic time and that the last sheet on the roll (the last 10 million years) is about four times longer than the time humans have existed on Earth. In fact, all of recorded human history is represented by the last $1/2000$ of a sheet, and a long human lifetime of 100 years is only $1/100,000$ of a sheet! It is indeed quite humbling to realize how insignificant the length of time human existence on Earth has been.

Can one really imagine the space taken up by $1 billion, or the lifetimes it would take to count to 4.6 billion, or how many sheets of toilet paper have gone by in 4.6 billion years of Earth history? The same scale problem exists with visualizing millions or billions of years, often called "deep" time or geologic time. But keep trying, and maybe someday you will get a glimpse into the huge expanse of time represented by the geologic time scale. It will amaze you.

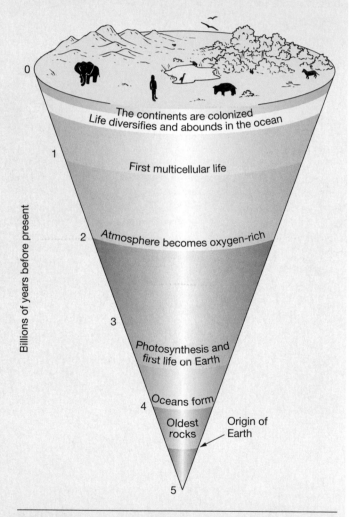

Figure 1F A timeline of major events in Earth's development.

[6]Toilet paper has many advantages: It is something with which most people are familiar, it is inexpensive and readily available, it is long and linear (like geologic time), and it is even perforated into individual sheets.

Answer to question posed above: 4.6 billion seconds corresponds to 145.9 years (which amounts to about two human lifetimes!).

1996 near Cabo San Lucas, Mexico. He dove to 130 meters (428 feet) and was under water for 2 minutes 11 seconds.

Is technology available to allow underwater divers to breathe a substance called "liquid air"?

Yes, the technology exists, but it won't be readily available for some time. The concept of breathing an inert liquid filled with a large amount of dissolved oxygen gas is called "liquid breathing" or "liquid ventilation." The technique uses a chemical called perfluorocarbon, which is a colorless, odorless, nontoxic fluid that readily absorbs oxygen and was initially used to cool electronic equipment. The fluid is infused with oxygen and breathed into the lungs, where it exchanges oxygen directly into the lung tissue. The first experiments were conducted successfully in 1966 to save ill animals. In 1989, it was used on a prematurely born infant (the first human) in an unsuccessful effort to save her life. Since then, it has been used as an emergency medical technique with good results to help children and adults who are in danger of respiratory failure. Although still in its experimental stage, the use of liquid breathing may enable people to dive to record depths in the future. This technology was featured in the 1989 movie, *The Abyss*.

I've heard stories about unusual incidences at sea in the Bermuda Triangle. Where is the Bermuda Triangle and are the stories for real?

The "Bermuda Triangle" (also known as the "Devil's Triangle" or the "Hoodoo Sea") is located off the southeastern Atlantic coast of the United States (Figure 1G). It is reputedly noted for having a high incidence of unexplained losses of ships, small boats, and aircraft—such as the disappearances of an entire squadron of TBM Avengers shortly after take off from Fort Lauderdale, Florida, in 1945 and the sinking of the vessel SS *Marine Sulphur Queen* without a trace in the Florida Straits in 1963. Reports such as these have fueled the popular belief that there is something mysterious or supernatural about the Bermuda Triangle.

Far-fetched explanations (including alien abductions) have been proposed to explain the many disappearances. However, experts have found that the most likely explanations for these incidences are related to the following factors:

- The magnetic north and true north directions overlap, which can confuse navigators if they do not properly compensate for it.
- The character of the Gulf Stream current, which is strong and turbulent.
- The unpredictable weather, including sudden and unexpected thunderstorms and water spouts.
- The navigational hazard of extensive shoals and reefs around many islands.
- The presence of large sea floor deposits of gas hydrates (discussed in Chapter 4, "Marine Sediments"), which, if mobilized, could cause ships to sink and aircraft to be immobilized.
- Human error by those who attempt to cross the waters between Florida and the Bahamas in too small a boat and with insufficient knowledge of the area's hazards, unsuitable maps, and a lack of good seamanship.

The U.S. Coast Guard is unimpressed with any of the supernatural explanations of incidences in the Bermuda Triangle and has exposed the mystery as a myth. Instead, it has been their experience that the unique environmental conditions combined with the unpredictability of humankind are the main factors responsible for the disasters that occur in the Bermuda Triangle.

I've heard of the U.S. Geological Survey and the National Biological Service. Is there a U.S. Oceanographical Survey that serves as the branch of the U.S. government dedicated to researching oceanographic phenomena?

Not quite. The branch of the U.S. government that oversees oceanographic research is the Commerce Department's National Oceanic and Atmospheric Administration (NOAA, pronounced "noah"). Scientists at NOAA work to ensure wise use of ocean resources through the National Ocean Service, the National Oceanographic Data Center, the National Marine Fisheries Service, and the National Sea Grant Office branches. Other U.S. government agencies that work with oceanographic data include the U.S. Naval Oceanographic Office, the Office of Naval Research, the U.S. Coast Guard, and the U.S. Geological Survey (coastal processes and marine geology).

I'm considering majoring in oceanography. Which universities have graduate-level programs in oceanography?

Numerous schools both in the United States and abroad offer undergraduate and graduate degrees in oceanography or marine science. In the United States, the three most prominent graduate-level oceanographic institutions are:

- Scripps Institution of Oceanography of the University of California San Diego in La Jolla, California
- Woods Hole Oceanographic Institution at Woods Hole, Massachusetts
- Lamont-Doherty Earth Observatory of Columbia University in Palisades, New York

If you are considering working as an oceanographer, see Appendix VI, "Careers in Oceanography" for some tips and practical advice.

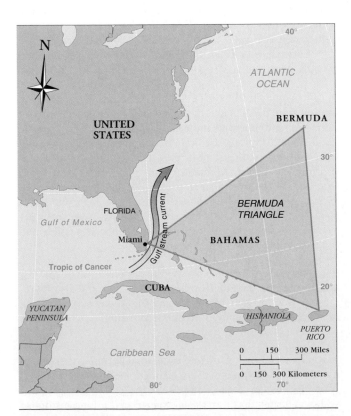

Figure IG The Bermuda Triangle.

Does Earth's continuing volcanic activity (and outgassing) mean that the oceans will gradually cover more and more of the surface?

Not necessarily, because the surface area of the oceans depends directly on the volume of the basin in which they form. During the initial solidification of Earth's crust, there may have been no distinct continents or ocean basins. Instead, both may have formed gradually as the oceans themselves formed. If the continents formed gradually, then the capacity of the oceans' basins to hold water may have gradually increased, too. It is likely the surface area of the oceans has been relatively constant for a long period of geologic time and the only major change in the ocean's character has been an increase in depth.

I've seen a movie where all land on Earth was flooded by the melting of the ice caps. Could this really happen?

You can't always believe what you see in the movies. All the ice in the world represents only a little more than 2% of all the world's water and the oceans already contain more than 97% of the total water (groundwater, surface water, and the atmosphere hold the rest). If all the ice melted, sea level would rise an estimated 60 meters (200 feet), which would flood coastal areas, but would still leave plenty of land above sea level on which to live. Was it the 1997 movie *Waterworld*?

Chapter in Review

• Water covers 70.8% of Earth's surface. The world ocean is a single interconnected body of water, which is large in size and volume. It can be divided into four principal oceans (the Pacific, Atlantic, Indian, and Arctic Oceans) plus an additional ocean (the Southern or Antarctic Ocean). Even though there is a technical distinction between a sea and an ocean, the two terms are used interchangeably. In comparing the oceans to the continents, it is apparent that the average land surface does not rise very far above sea level and that there is not a mountain on Earth that is as tall as the ocean is deep.

• In the Pacific, people who populated the Pacific Islands may have been the first great navigators. In the Western world, the Phoenicians were making remarkable voyages as well. Later the Greeks, Romans, and Arabs made significant contributions and advanced oceanographic knowledge. During the Middle Ages, the Vikings colonized Greenland and made voyages to North America. The Age of Discovery in Europe renewed the Western world's interest in exploring the unknown. It began with the voyage of Christopher Columbus in 1492 and ended in 1522 with the first circumnavigation of Earth by a voyage initiated by Ferdinand Magellan. Captain James Cook was one of the first to explore the ocean for scientific purposes.

• The scientific method is used to understand the occurrence of physical events or phenomena and can be stated as: science supports explanation of the natural world that best explains all available observations. Steps in the scientific method include making observations and establishing scientific facts; forming one or more hypotheses (a tentative, testable statement about the general nature of the phenomena observed); extensive testing and modification of hypotheses; and, finally, developing a theory (a well-substantiated explanation of some aspect of the natural world that can incorporate facts, laws, logical inferences, and tested hypotheses). Science never arrives at the absolute "truth;" rather, science arrives at what is probably true based on the available observations and can continually change because of new observations. Nevertheless, theories can be backed so solidly that the accuracy of a theory (such as the theory of plate tectonics) is no longer questioned.

• Our solar system, consisting of the Sun and nine planets, may have formed from a huge cloud of gas and space dust called a nebula. According to the nebular hypothesis, the nebular matter contracted to form the Sun, and the planets were formed from eddies of material that remained. The Sun, composed of hydrogen and helium, was massive enough and concentrated enough to emit large amounts of energy from a fusion reaction. The Sun also emitted ionized particles that swept away any nebular gas that remained from the formation of the planets and their satellites.

• The Protoearth, more massive and larger than Earth today, was molten and homogenous. The initial atmosphere, composed mostly of hydrogen and helium, was later driven off into space by intense solar radiation. The Protoearth began a period of rearrangement, forming a layered structure of core, mantle, and crust through density stratification. Also during this period, outgassing produced an early atmosphere rich in water vapor and carbon dioxide. Once Earth's surface cooled sufficiently, the water vapor condensed and accumulated to give Earth its oceans. Rainfall on the surface dissolved compounds which, when carried to the ocean, made it salty.

• Life is thought to have begun in the oceans. Ultraviolet radiation from the Sun and hydrogen, carbon dioxide, methane, ammonia, and inorganic molecules from the oceans may have combined to produce carbon-containing molecules. Certain combinations of these molecules eventually produced heterotrophic organisms (which cannot make their own food) that were probably similar to present-day anaerobic bacteria. Eventually, autotrophs evolved that had the ability to make their own food through chemosynthesis. Later, some cells developed chlorophyll, which made photosynthesis possible and led to the development of plants.

• Photosynthetic organisms altered the environment by extracting carbon dioxide from the atmosphere and also by releasing free oxygen, thereby creating today's oxygen-rich atmosphere. Eventually, both plants and animals evolved into forms that could survive on land.

• Radiometric age dating is used to determine the age of most rocks. Information from extinctions of organisms and from age dating rocks comprises the geologic time scale, which indicates that Earth has experienced a long history of changes since its origin 4.6 billion years ago.

Key Terms

Age of Discovery (p. 13)

Anaerobic (p. 23)

Antarctic Ocean (p. 9)

Arctic Ocean (p. 9)

Atlantic Ocean (p. 8)

Atom (p. 20)

Autotroph (p. 23)

Bathyscaphe (p. 10)

Chemosynthesis (p. 23)

Chlorophyll (p. 24)

Chronometer (p. 14)

Columbus, Christopher (p. 13)

Cook, James (p. 17)

Core (p. 21)

Crust (p. 21)

da Gama, Vasco (p. 13)

de Balboa, Vasco Núñez (p. 14)

Density (p. 20)

Density stratification (p. 20)

Diaz, Bartholomeu (p. 13)

Endothermic (p. 24)

Eratosthenes (p. 12)

Eriksson, Leif (p. 13)

Evolution (p. 24)

Exothermic (p. 24)

Fact (p. 18)

Fusion reaction (p. 20)

Geologic Time Scale (p. 25)

Global Positioning System (GPS) (p. 16)

Half-life (p. 26)

Harrison, John (p. 14)

Henry the Navigator (p. 13)

Herjolfsson, Bjarni (p. 13)

Heterotroph (p. 23)

Heyerdahl, Thor (p. 11)

Homogenous (p. 20)

Hypothesis (p. 18)

Indian Ocean (p. 8)

Kon Tiki (p. 11)

Latitude (p. 14)

Longitude (p. 14)

Magellan, Ferdinand (p. 14)

Mantle (p. 21)

Miller, Stanley (p. 23)

Natural selection (p. 24)

Nebula (p. 19)

Nebular hypothesis (p. 19)

Observation (p. 18)

Outgassing (p. 21)

Pacific Ocean (p. 7)

Phoenicians (p. 12)

Photosynthesis (p. 24)

Protoearth (p. 20)

Protoplanet (p. 20)

Ptolemy (p. 13)

Pytheas (p. 12)

Radiometric age dating (p. 26)

Respiration (p. 24)

Scientific method (p. 17)

Scuba (p. 7)

Solar system (p. 19)

Southern Ocean (p. 9)

Species (p. 24)

Stick chart (p. 14)

Theory (p. 18)

Thorvaldson, Erik "the Red" (p. 13)

Trieste (p. 10)

Vikings (p. 13)

Questions And Exercises

1. Discuss what problems the human body can experience as a result of diving underwater.

2. How did the view of the ocean by early Mediterranean cultures influence the naming of planet "Earth?"

3. What is the difference between an ocean and a sea? Which ones are the seven seas?

4. When humans first visited the deepest ocean floor, what was the voyage like and what did they see?

5. Describe the development of navigation techniques that have enabled sailors to navigate in the open ocean far from land.

6. Using a diagram, illustrate the method used by Pytheas to determine latitude in the Northern Hemisphere.

7. While the Arabs dominated the Mediterranean region during the Middle Ages, what were the most significant ocean-related events taking place in northern Europe?

8. Describe the important events in oceanography that occurred during the Age of Discovery in Europe.

9. List some of the major achievements of Captain James Cook.

10. What is the difference between a fact and a theory? Can either (or both) be revised?

11. Briefly comment on the phrase "scientific certainty." Is it an oxymoron (a combination of contradictory words) or are scientific theories considered to be the absolute truth?

12. Discuss the origin of the solar system using the nebular hypothesis.

13. How was the Protoearth different from today's Earth?

14. What is density stratification, and how did it change the Protoearth?

15. What is the origin of Earth's oceans and how is it related to the origin of Earth's atmosphere?

16. Have the oceans always been salty? Why or why not?

17. How does the presence of oxygen in our atmosphere help reduce the amount of ultraviolet radiation that reaches Earth's surface?

18. What was Stanley Miller's experiment, and what did it help demonstrate?

19. Earth has had three atmospheres (initial, early, and present). Describe the composition and origin of each one.

20. Construct a representation of the geologic time scale, using an appropriate quantity of any substance (other than dollar bills or toilet paper). Be sure to indicate some of the major changes that have occurred on Earth since its origin.

References

Borgese, E. M., and Ginsberg, N. 1993. *Ocean yearbook 10.* Chicago: University of Chicago Press.

Boror, D. J. 1960. *Dictionary of word roots and combining forms.* Mountain View, CA: Mayfield.

Brewer, P., ed. 1983. *Oceanography: The present and future.* New York: Springer-Verlag.

Deacon, M. 1971. *Scientists and the sea 1650–1900: A study of marine science.* London: Academic Press.

Duxbury, A. C. 1971. *The Earth and its oceans.* Reading, MA: Addison-Wesley.

Frank, L. A., and Sigwarth, J. B. 1993. Atmospheric holes and small comets. *Review of Geophysics* 31, 1–28.

———. 1997. Detection of atomic oxygen trails of small comets in the vicinity of Earth. *Geophysical Research Letters* 24:19, 2431–2434.

Glaessner, M. F. 1984. *The dawn of animal life: A biohistorical study.* Cambridge: Cambridge University Press.

Gregor, B. C., Garrels, R. M., Mackenzie, F. T., and Maynard, J. B., eds. 1988. *Chemical cycles in the evolution of the Earth.* New York: Wiley.

Hendrickson, R. 1984. *The ocean almanac.* New York: Doubleday.

Heyerdahl, T. 1979. *Early man and the ocean.* New York: Doubleday.

Holland, J. D. 1984. *The chemical evolution of the atmosphere and oceans.* Princeton, NJ: Princeton University Press.

Idyll, C. P., ed. 1969. *The science of the sea: A history of oceanography.* London: Thomas Y. Crowell.

Kennish, M. J., ed. 1994. *Practical handbook of marine science,* 2nd ed. Boca Raton, FL: CRC Press.

Kirch, P. V. 2000. *On the Road of the Winds.* Berkley: University of California Press.

Kirschvink, J. L., Ripperdan, R. L., and Evans, D. A. 1997. Evidence for a large-scale reorganization of early Cambrian continental masses by inertial interchange true polar wander. *Science* 277:5325, 541–545.

Kump, L. R., Kasting, J. F., and Barley, M. E. 2001. Rise of atmospheric oxygen and the "upside-down" Archean mantle, *Geochem. Geophys. Geosyst.,* vol. 2, Paper number 2000GC000114.

Marx, R. 1996. The early history of diving. In Pirie, R. G., ed., *Oceanography: Contemporary readings in ocean sciences,* 3rd ed. New York: Oxford University Press.

McGovern, T. H. and Perdikaris, S. 2000. The Viking's silent saga: What went wrong with the Scandinavian westward expansion? *Natural History* 109:8, 50-57.

McKay, D. S., et al. 1996. Search for past life on Mars: Possible relic biogenic activity in Martian meteorite ALH84001. *Science* 273:5277, 924–930.

Melamed, Y., Shupak, A., and Bitterman, H. 1996. Medical problems associated with underwater diving. In Pirie, R. G., ed., *Oceanography: Contemporary readings in ocean sciences*, 3rd ed. New York: Oxford University Press.

Mowat, F. 1965. *Westviking.* Boston: Atlantic–Little, Brown.

Ola, P. and D'Aulaire, E. 1999. Taking the measure of time. *Smithsonian* 30:9, 52–65.

Phillips, J. L. 1998. *The bends: Compressed air in the history of science, diving, and engineering.* Connecticut: Yale University Press.

Rasmussen, B. 2000. Filamentous microfossils in a 3,235-million-year-old volcanogenic massive sulphide deposit. *Nature* 405:6787, 676–679.

Rosing, M. T. 1999. ^{13}C-depleted carbon microparticles in >3700-Ma sea-floor sedimentary rocks from west Greenland. *Science* 283:5402, 674–676.

Rubey, W. W. 1951. Geologic history of seawater: An attempt to state the problem. *Geological Society of America Bulletin* 62:1110–1119.

Schopf, J. W. 1993. Microfossils of the early Archean Apex Chert: New evidence of the antiquity of life. *Science* 260:5108, 640–646.

Sears, M., and Merriman, D., eds. 1980. *Oceanography: The past.* New York: Springer-Verlag.

Sobel, D. 1995. *Longitude: The true story of a lone genius who solved the greatest scientific problem of his time.* New York: Walker.

Strom, K. M., and Strom, S. E. 1982. Galactic evolution: A survey of recent progress. *Science* 216:4546, 571–580.

Sverdrup, H. U., Johnson, M. W., and Fleming, R. H. 1942, reprinted 1970. *The oceans: Their physics, chemistry, and general biology.* Englewood Cliffs, NJ: Prentice-Hall.

Various authors. 1997. Special section on small comets (5 papers). *Geophysical Research Letters* 24:24, 3105–3124.

Waldrop, M. M. 1989. The (liquid) breath of life. *Science* 245:4922, 1043–1045.

Working group on teaching evolution. 1998. *Teaching about evolution and the nature of science.* Washington D.C.: National Academy of Sciences.

***Suggested Reading in* Scientific American**

Badash, L. 1989. The age-of-the-Earth debate. 261:2, 90–97. A history of the development of knowledge concerning Earth's age.

Bernstein, M. P., Sandford, S. A., and Allamandola, L. J. 1999. Life's far-flung raw materials. 281:1, 42–49. Examines evidence for the origin of life on Earth, including the idea that life may owe its start to complex organic molecules contained in extraterrestrial material.

Bothun, G. D. 1997. The ghostliest galaxies. 276:2, 56–61. Huge galaxies too diffuse to be seen until the 1980s were found to contain mass equal to that of previously known visible galaxies. They provide clues to a better understanding of how mass is distributed throughout the universe.

De Duve, C. 1996. The birth of complex cells. 274:4, 50–59. A discussion on the evolutionary process that may have led to the development of eukaryotic cells that are from 10 to 30 times larger than the prokaryotic cells from which they evolved.

Gibson, E. K., Jr., McKay, D. S., Thomas-Keprta, K., and Romanek, C. S. 1997. The case for relic life on Mars. 227:6, 58–65. A review of the compelling evidence from a Martian meteorite found in Antarctica suggesting that Mars has had (and may still have?) microbial life.

Hazen, R. M. 2001. Life's rocky start. 284:4, 77–85. New experiments suggest that minerals could have played a key role in the origin of life on Earth.

Herring, T. A. 1996. The global positioning system. 274:2, 44–50. A look at the technology involved in making the Global Positioning System (GPS) work.

Hoffman, P. G., and Schrag, D. P. 2000. Snowball Earth. 282:1, 68–75. Presents compelling evidence for ice covering all continents and oceans 600 million years ago. When Earth warmed and the ice melted, it may have led to the development of multicellular life.

Horgan, J. 1991. In the beginning. 264:2, 116–125. An overview of data relating to the validity of the Big Bang theory of the origin of the universe.

Kasting, J. F., Toon, O. B., and Pollack, J. B. 1988. How climate evolved on the terrestrial planets. 258:2, 90–97. A discussion of the possible sequence of events that culminated with the atmospheres that now exist on Mercury, Venus, Earth, and Mars.

Luu, J. X. and Jewitt, D. C. 1996. The Kuiper belt. 274:5, 46–53. Beyond Pluto lies a belt of objects a few tens to hundreds of kilometers across. This is where short-period comets may originate.

McMenamin, M. A. S. 1987. The emergence of animals. 256:4, 94–103. A discussion on how the explosive diversification of animal forms 570 million years ago may have been related to the breakup of a single large continental landmass.

Moon, R. E., Vann, R. D., and Bennett, P. B. 1995. The physiology of decompression illness. 273:2, 70–77. A summary of current knowledge about decompression illness and a look at research that is being conducted with pressure chambers.

Motani, R. 2000. Rulers of the Jurassic seas. 283:6, 52–61. Seagoing lizards called ichthyosaurs that were up to 15 meters long ruled the seas while dinosaurs ruled the land.

Richelson, J. T. 1998. Scientists in black. 278:2, 48–55. Scientists and intelligence officials are working together to use technology that enables them to analyze features below sea level.

Simpson, S. 2000. Looking for life below the bottom. 282:6, 94–101. An account of a voyage to sample life in the fractured rocks of the ocean crust.

Stebbins, G. L., and Ayala, F. J. 1985. The evolution of Darwinism. 253:1, 72–85. New advances in molecular biology and new interpretations of the fossil record add to the knowledge of evolution.

Various authors. 2001. Brave new cosmos. 284:1, 37–59. A series of articles updating our knowledge of the cosmos.

Whitehead, H. 1985. Why whales leap. 252:3, 84–93. An attempt to answer the enigmatic question, "Why do whales breach?"

Wilson, A. C. 1985. The molecular basis of evolution. 253:4, 164–175. Mutations within the genes of organisms play an important role in evolution at the organismal level.

Oceanography on the Web

Visit the Essentials of Oceanography home page for on-line resources for this chapter. There you will find an on-line study guide with review exercises, and links to oceanography sites to further your exploration of the topics in this chapter. Essentials of Oceanography is at: **http://www.prenhall.com/ thurman** (click on the Table of Contents menu and select this chapter).

HTTP://WWW.PRENHALL.COM/THURMAN •

CHAPTER

2

Plate Tectonics and the Ocean Floor

• What evidence did Alfred Wegener use to formulate his idea of continental drift?

• How did the early idea of continental drift differ from the more modern version of plate tectonics?

• What are the lines of evidence that support the theory of plate tectonics?

• How do structure, composition, and physical properties vary within the deep Earth?

• What types of features are found at the three main types of plate boundaries?

• How do mantle plumes and hotspots fit into the plate tectonic model?

• How have the features on Earth looked in the past and how will they look in the future?

"It is just as if we were to refit the torn pieces of a newspaper by matching their edges and then check whether the lines of print run smoothly across. If they do, there is nothing left but to conclude that the pieces were in fact joined in this way."

—*Alfred Wegener*, The Origin of Continents and Oceans *(1929)*

VOYAGES TO INNER SPACE: VISITING THE DEEP OCEAN FLOOR IN SUBMERSIBLES

For as long as people have been harvesting the bounty of the oceans and traveling across it in vessels, they have dreamed of plumbing the mysterious depths of the deep ocean, an area known as "inner space." However, the inaccessibility of the deep ocean and its harsh conditions have limited human exploration. One way to explore the deep ocean is in vessels called **submersibles** that can transport surface conditions to the intense pressure at depth.

History credits Alexander the Great with the first descent in a sealed waterproof container, which reportedly took place in 332 B.C. Unfortunately, there is no record of what Alexander's submersible looked like. Much later, the **submarine** was developed, which can submerge, propel itself underwater, and surface under its own power. Because submarines are difficult to detect, they can be used to sink enemy ships. The earliest report of a submarine used in warfare is from the early 16th century. Greenlanders used sealskins to waterproof a three-person, oar-powered submarine, which was used to drill holes in the sides of Norwegian ships.

In 1934, reaching even deeper depths for scientific exploration was the goal for naturalist William Beebe and his engineer-associate Otis Barton. They used a submersible called a **bathysphere** (*bathos* = depth, *sphere* = ball) to observe marine life in the clear waters off Bermuda. The bathysphere—a heavy steel ball with small windows—was suspended from a ship and lowered to a then-record depth of 923 meters (3028 feet), allowing the first descriptions of the deep.

In 1964, the research submersible *Alvin* (Figure 2A) from Woods Hole Oceanographic Institution began to explore the deep ocean. At 7.6 meters (25 feet) long, *Alvin* can carry a crew of one pilot and two scientists to a depth of 4000 meters (13,120 feet) and maneuver independently along the sea floor. Since it was commissioned, *Alvin* has made numerous dives to allow oceanographers to explore the deep ocean floor and retrieve samples. Under the direction of oceanographer **Robert Ballard**, some notable accomplishments of *Alvin* include discovering unique life forms along hot water springs at the mid-ocean ridge in 1977 and locating the sunken wreck of the RMS *Titanic* in 1985. Currently, the deepest-diving manned submersible is *Shinkai 6500*, a Japanese research vessel that can dive to 6500 meters (21,320 feet). Only the unmanned robotic vessel *Kaiko* can reach the deepest trenches.

Figure 2A The deep-diving submersible *Alvin*.

Several thousand earthquakes and dozens of volcanic eruptions occur on land and under the oceans each year, indicating that our planet is very dynamic. These events have occurred throughout history, constantly changing the surface of our planet, yet only a few decades ago most scientists believed the continents were stationary over geologic time. Since then, however, a bold new theory has been advanced that helps explain, for the first time, the following surface features and phenomena on Earth:

- The worldwide locations of volcanoes, faults, earthquakes, and mountain building
- Why mountains on Earth haven't been eroded away
- The origin of most landforms and ocean floor features
- How the continents and ocean floor formed, and why they are different
- The continuing development of Earth's surface
- The distribution of past and present life on Earth

This revolutionary new theory is called **plate tectonics** (*plate* = plates of the **lithosphere**; *tekton* = to build), or "the new global geology." According to the theory of plate tectonics, the outermost portion of Earth is composed of a patchwork of thin, rigid plates[1] that move horizontally with respect to one another, like icebergs floating on water. The interaction of these plates as they move builds features of Earth's crust (such as volcanoes, mountain belts, and ocean basins). As a result, the continents are mobile and move about on Earth's surface, controlled by forces deep within Earth.

In this chapter, we'll examine the early ideas about the movement of plates, the evidence for those ideas, and how they led to the theory of plate tectonics. Then, we'll look at Earth structure, plate boundaries, and some applications of plate tectonics, including what our planet may look like in the future.

Evidence for Continental Drift

Alfred Wegener (Figure 2–1), a German meteorologist and geophysicist, was the first to advance the idea of mobile continents in 1912. He envisioned that the continents were slowly drifting across the globe and called his idea **continental drift**. Let's examine the evidence that Wegener compiled that led him to formulate the idea of drifting continents.

Fit of the Continents

The idea that continents—particularly South America and Africa—fit together like pieces of a jigsaw puzzle originated with the development of reasonably accurate world maps. As far back as 1620, Sir Francis Bacon wrote about how the continents appeared to fit together. However, little significance was given to this idea until 1912, when Wegener used the shapes of matching shorelines on different continents as a supporting piece of evidence for continental drift.

Wegener suggested that during the geologic past, the continents collided to form a large landmass, which he named **Pangaea** (*pan* = all, *gaea* = Earth) (Figure 2–2). Further, a huge ocean surrounded Pangaea, called **Panthalassa** (*pan* = all, *thalassa* = sea). Panthalassa included several smaller seas, including the **Tethys** (*Tethys* = a Greek sea goddess) **Sea**. Wegener's evidence indicated that about 200 million years ago, Pangaea began to split apart, and the various continental masses started to drift toward their present geographic positions.

Wegener's attempt at matching shorelines revealed considerable areas of crustal overlap and large gaps. Some of the differences could be explained by material deposited by rivers or eroded from coastlines. What Wegener didn't know was the shallow parts of the ocean floor close to shore are closely related to the continents. In the early 1960s, Sir Edward Bullard and two associates used a computer to fit the continents together

---

[1]These thin, rigid plates are pieces of the lithosphere (*lithos* = rock, *sphere* = ball) that comprise Earth's outermost portion.

Figure 2–1 Alfred Wegener, circa 1912–1913.

(Figure 2–3). Instead of using the shorelines of the continents as Wegener had done, Bullard achieved the best fit (i.e. with minimal overlaps or gaps) by using a depth of 2000 meters (6560 feet) below sea level. This depth corresponds to halfway between the shoreline and the deep ocean basins and represents the true edge of the continents. By using this depth, the overall fit of the continents was even better than expected.

Matching Sequences of Rocks and Mountain Chains

If the continents were once together as Wegener had hypothesized, then evidence should appear in rock sequences that were originally continuous but now separated by large distances. To test the idea of drifting continents, geologists began comparing the rocks along the edges of continents with rocks found in adjacent positions on matching continents. They wanted to see if the rocks had similar types, ages, and structural styles (the type and degree of deformation). In some areas younger rocks had been deposited during the millions of years since the continents separated, covering the rocks that held the key to the past history of the continents. In other areas, the rocks had been eroded away. Nevertheless, in many other areas the key rocks were present.

These studies showed that many rock sequences from one continent matched up with identical rock sequences on a matching continent—although the two were separated by an ocean. In addition, mountain ranges that terminated abruptly at the edge of a continent continued on another continent across an ocean basin, with identical rock sequences, ages, and structural

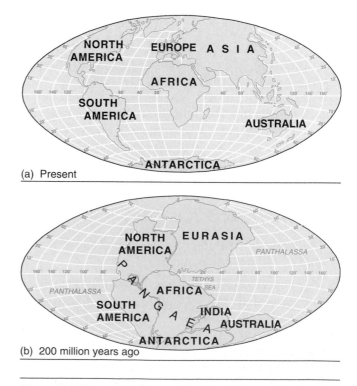

(a) Present

(b) 200 million years ago

Figure 2–2 Reconstruction of Pangaea. The positions of the continents about 200 million years ago, showing the supercontinent of Pangaea and the single large ocean, Panthalassa.

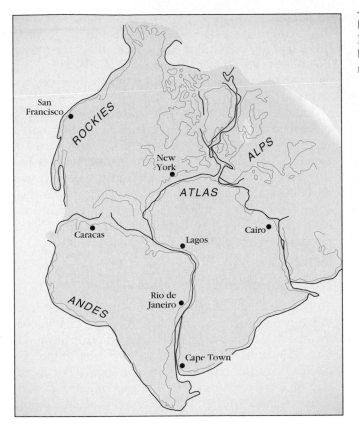

Figure 2–3 Computer fit of the continents. In 1965, Sir Edward Bullard used a depth of 2000 meters (6560 feet) for a best-fit match of the continents, which shows few gaps and minimal overlap.

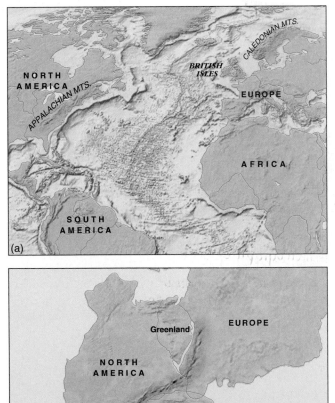

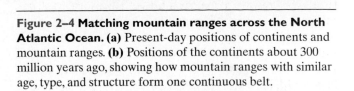

Figure 2–4 Matching mountain ranges across the North Atlantic Ocean. (a) Present-day positions of continents and mountain ranges. **(b)** Positions of the continents about 300 million years ago, showing how mountain ranges with similar age, type, and structure form one continuous belt.

styles. Figure 2–4 shows, for example, how similar rocks from the Appalachian Mountains in North America match up with identical rocks from the British Isles and the Caledonian Mountains in Europe.

Wegener noted the similarities in rock sequences on both sides of the Atlantic and used the information as a supporting piece of evidence for continental drift. He suggested that mountains such as those seen on opposite sides of the Atlantic formed during the collision when Pangaea was formed. Later, when the continents split apart, once-continuous mountain ranges were separated. Confirmation of this idea exists in a similar match with mountains extending from South America through Antarctica and across Australia.

Glacial Ages and Other Climate Evidence

Wegener also noticed the occurrence of past glacial activity in areas now tropical and suggested that it, too, provided supporting evidence for drifting continents. Currently, the only places in the world where thick continental **ice sheets** occur are in the polar regions of Greenland and Antarctica. However, evidence of ancient glaciation is found in the lower latitude regions of South America, Africa, India, and Australia.

These deposits, which have been dated at 300 million years old, indicate one of two possibilities: (1) There was a worldwide **ice age** and even tropical areas were covered by thick ice, or (2) some continents which are now in tropical areas were once located much closer to one of the poles. It is unlikely that the entire world was covered by ice 300 million years ago because coal deposits from the same geologic age now present in North America and Europe originated as vast semitropical swamps. Thus, a reasonable conclusion is that some of the continents must have been closer to the poles than they are today.

There is another type of glacial evidence that indicates certain continents have moved from more polar regions during the last 300 million years. When glaciers flow, they move and abrade the underlying rocks, leaving grooves that indicate the direction of flow. The blue arrows in Figure 2–5a show how the glaciers would have flowed away from the South Pole on Pangaea 300 million years ago. The direction of flow is consistent with the grooves found on many continents today (see Figure 2–5b), providing additional evidence for drifting continents.

Many examples of plant and animal fossils indicate very different climates than today. Two such examples are fossil palm trees in Arctic Spitsbergen and coal deposits in Antarctica. Earth's past environments can be interpreted from these rocks because plants and animals need specific environmental conditions in which to live. Corals, for example, generally need seawater above

18 degrees Centigrade (°C) or 64 degrees Fahrenheit (°F) in order to survive. When fossil corals are found in areas that are cold today, two explanations seem most plausible: (1) Worldwide climate has changed dramatically; or (2) the rocks have moved from their original location.

Latitude (distance north or south of the Equator), more than anything else, determines climate. Moreover, there is no evidence to suggest that Earth's axis of rotation has changed significantly throughout its history, so the climate at any particular latitude must not have changed significantly either. Thus, fossils that come from climates that seem out of place today must have moved from their original location through the movement of the continents as Wegener proposed.

Distribution of Organisms

To add credibility to his argument for the existence of the supercontinent of Pangaea, Wegener cited documented cases of several fossil organisms found on different landmasses that could not have crossed the vast oceans presently separating the continents. For example, the fossil remains of **Mesosaurus** (*meso* = middle, *saurus* = lizard), an extinct, presumably aquatic reptile

Figure 2–5 Ice age on Pangaea.
(a) Reconstruction of the supercontinent Pangaea, showing the area covered by glacial ice about 300 million years ago. Arrows indicate direction of ice flow. **(b)** The positions of the continents today.

① Geographic fit
② Geologic - similar rocks on sep. continents

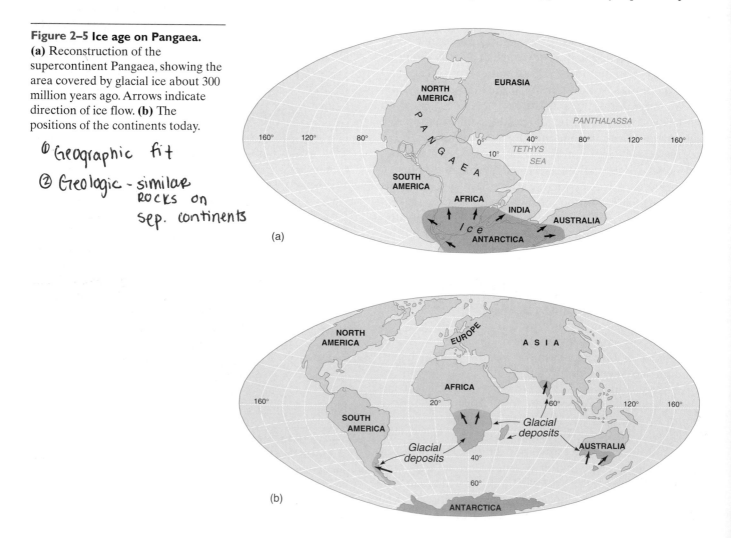

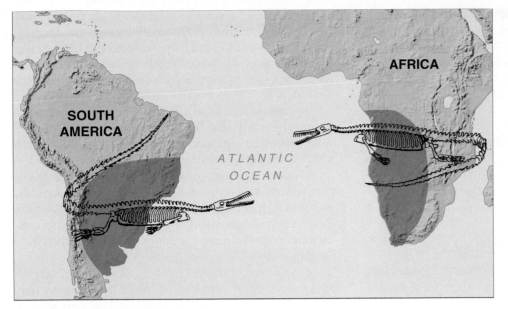

**Figure 2–6 Fossils of
Mesosaurus.** *Mesosaurus* fossils
are found only in South America
and Africa and appear to link
these two continents.

that lived about 250 million years ago, are located only in eastern South America and western Africa (Figure 2–6). If *Mesosaurus* had been strong enough to swim across an ocean, why aren't its remains more widely distributed?

Wegener's idea of continental drift provided an elegant solution to this problem. He suggested that the continents were closer together in the geologic past, so *Mesosaurus* didn't have to be a good swimmer to leave remains on two different continents. Later, after *Mesosaurus* became extinct, the continents moved to their present-day positions, and a large ocean now separates the once-connected landmasses. Other examples of similar fossils on different continents include those of plants, which would have had a difficult time traversing a large ocean.

Before continental drift, several ideas were proposed to help explain the curious pattern of these fossils, such as the existence of island stepping stones or a land bridge. It was even suggested that at least one pair of land-dwelling *Mesosaurus* survived the arduous journey across several thousand kilometers of open ocean by rafting on floating logs. However, there is no evidence to support the idea of island stepping stones or a land bridge and the idea of *Mesosaurus* rafting across an ocean seems implausible.

Wegener also cited the distribution of present-day organisms as evidence to support the concept of drifting continents. For example, modern organisms with similar ancestries clearly had to evolve in isolation during the last few million years. Most obvious of these are the Australian marsupials (such as the kangaroos, koalas, and wombats), which have a distinct similarity to the marsupial opossums found in the Americas.

Objections to the Continental Drift Model

In 1915, Wegener published his ideas in *The Origins of Continents and Oceans* but the book did not attract

much attention until it was translated into English, French, Spanish, and Russian in 1924. From that point until his death in 1930[2], his drift hypothesis received much hostile criticism—and sometimes open ridicule—from the scientific community because of the mechanism he proposed for the movement of the continents. Wegener suggested the continents plowed through the ocean basins to reach their present day positions and that the leading edges of the continents deformed into mountain ridges because of the drag imposed by ocean rocks. Further, the driving mechanism he proposed was a combination of the gravitational attraction of Earth's equatorial bulge and tidal forces from the Sun and Moon.

Scientists rejected the idea as too fantastic and contrary to the laws of physics. Material strength calculations showed that ocean rock was too strong for continental rock to plow through it and analysis of gravitational and tidal forces indicated that they were too small to move the great continental landmasses. Even without an acceptable mechanism, many geologists who studied rocks in South America and Africa accepted continental drift. North American geologists—most of whom were unfamiliar with these Southern Hemisphere rock sequences—remained highly skeptical.

As compelling as Wegener's evidence may seem today, he was unable to convince the scientific community as a whole of the validity of his ideas. Although his hypothesis was correct in principle, it contained several incorrect details, such as the driving mechanism for continental motion and how continents move across ocean basins. In order for any scientific viewpoint to gain wide acceptance, it must explain all available observations and have supporting evidence from a wide variety of scientific fields. This supporting evidence would not

---

[2]Wegener perished in 1930 during an expedition in Greenland while collecting data to help support his idea of continental drift.

come until more details of the nature of the ocean floor were revealed, which, along with new technology that enabled scientists to determine the original positions of rocks on Earth, provided additional observations in support of drifting continents.

> Alfred Wegener used a variety of interdisciplinary information from land to support continental drift. However, he did not have a suitable mechanism or any information about the sea floor.

Evidence for Plate Tectonics

Very little new information about Wegener's continental drift hypothesis was introduced between the time of Wegener's death in 1930 and the early 1950s. However, as the sea floor was explored and became better known, it provided critical evidence in support of drifting continents. In addition, technology unavailable in Wegener's time enabled scientists to analyze the way rocks retained the signature of Earth's **magnetic field**. These developments caused scientists to reexamine continental drift and advance it into the more encompassing theory of plate tectonics.

Earth's Magnetic Field and Paleomagnetism

Earth's magnetic field is shown in Figure 2–7. The invisible lines of magnetic force that originate within Earth and travel out into space resemble the magnetic field produced by a large bar magnet.[3] Similar to Earth's magnetic field, the ends of a bar magnet have opposite polarities (labeled either + and − or N for north and S for south) that cause magnetic objects to align parallel to its magnetic field. In addition, notice in Figure 2–7b that Earth's geographic north pole (the rotational axis) and Earth's magnetic north pole (magnetic north) do not coincide.

Rocks Affected by Earth's Magnetic Field   **Igneous** (*igne* = fire, *ous* = full of) **rocks** solidify from molten **magma** (*magma* = a mass) either underground or after volcanic eruptions at the surface, which produce **lava** (*lava* = to wash). Nearly all igneous rocks contain **magnetite**, a naturally magnetic iron mineral. Particles of magnetite in magma align themselves with Earth's magnetic field because magma and lava are fluid. Once

molten material cools and solidifies, the magnetite particles are frozen into position and record the angle of Earth's magnetic field at that place and time. In essence, grains of magnetite serve as tiny compass needles that record the strength and orientation of Earth's magnetic field. Unless the rock is heated to the temperature where magnetite grains are again mobile, these magnetite grains contain information about the magnetic field where the rock originated regardless of where the rock subsequently moves.

Magnetite is also deposited in sediments. As long as the sediment is surrounded by water, the magnetite particles can align themselves with Earth's magnetic field. After sediment is buried and solidifies into **sedimentary** (*sedimentum* = settling) rock, the particles are no longer able to realign themselves if they are subsequently moved. Thus, magnetite grains in sedimentary rocks also contain information about the magnetic field where the rock originated. Although other rock types have been successfully used to reveal information about Earth's ancient magnetic field, the most reliable ones are igneous rocks that have high concentrations of magnetite such as basalt.

Paleomagnetism   The study of Earth's ancient magnetic field is called **paleomagnetism** (*paleo* = ancient). The scientists who study paleomagnetism analyze magnetite particles in rocks to determine not only their north–south direction but also their angle relative to Earth's surface. The degree to which a magnetite particle points into Earth is called its **magnetic dip**, or **magnetic inclination**.

Magnetic dip is directly related to latitude. Figure 2–7b shows that a dip needle does not dip at all at Earth's magnetic equator. Instead, the needle lies horizontal to Earth's surface. At Earth's magnetic north pole, however, a dip needle points straight into the surface. A dip needle at Earth's south magnetic pole is also vertical to the surface, but it points out instead of in. Thus, magnetic dip increases with increasing latitude, from 0 degrees at the magnetic equator to 90 degrees at the magnetic poles. Because magnetic dip is retained in magnetically oriented rocks, measuring the dip angle reveals the latitude at which the rock initially formed. Done with care, paleomagnetism is an extremely powerful tool for interpreting where rocks first formed. Based on paleomagnetic studies, convincing arguments could finally be made that the continents had drifted relative to one another.

Apparent Polar Wandering   When magnetic dip data for rocks on the continents were used to determine the apparent position of the magnetic north pole over time, it appeared that the magnetic pole was wandering. For example, Figure 2–8a shows the **polar wandering curves** for North America and Eurasia. Both curves have a similar shape but, for all rocks older than about

---

[3]The properties of a magnetic field can be explored easily enough with a bar magnet and some iron particles. Place the iron particles on a table and place a bar magnet nearby. Depending on the strength of the magnet, you should get a pattern resembling Figure 2-7a.

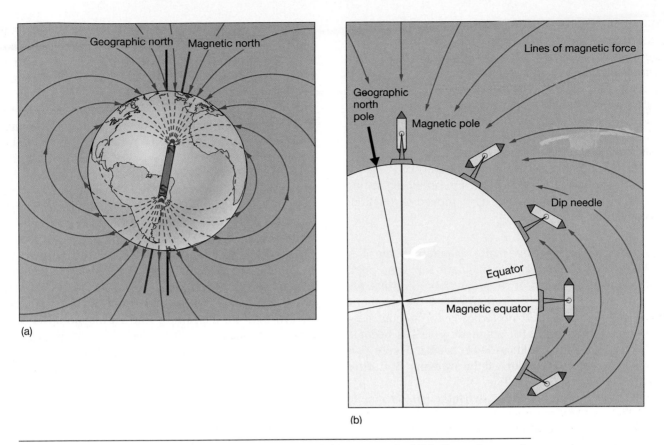

Figure 2–7 Earth's magnetic field. (a) Earth's magnetic field generates invisible lines of magnetic force similar to a large bar magnet. Note that magnetic north and true north are not in the same exact location. **(b)** Earth's magnetic field causes a dip needle to align parallel to the lines of magnetic force and change orientation with increasing latitude. Consequently, the latitude can be determined based on the dip angle.

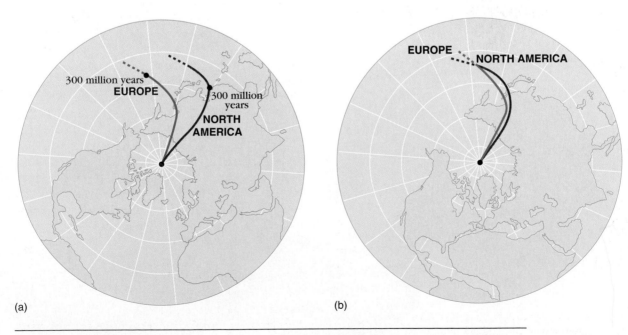

Figure 2–8 Apparent polar wandering paths. (a) Apparent polar wandering paths for North America and Eurasia resulted in a dilemma because they were not in alignment. **(b)** The positions of the polar wandering paths when the landmasses are assembled.

70 million years, the pole determined from North American rocks lies to the west of that determined from Eurasian rocks. There can be only one north and one south magnetic pole at any given time, however, and it is unlikely that their positions change with time. This discrepancy implies that magnetic poles remained stationary while North America and Eurasia moved relative to the pole and relative to each other. Figure 2–8b shows that when the continents are moved into the positions they occupied when they were part of Pangaea, the two wandering curves match up, providing strong evidence that the continents have moved throughout geologic time.

Magnetic Polarity Reversals Magnetic compasses on Earth today follow lines of magnetic force and point toward magnetic north. It turns out, however, that the **polarity** (the directional orientation of the magnetic field) has reversed itself periodically throughout geologic time. Thus, the north magnetic pole has become the south magnetic pole and vice versa. Figure 2–9 shows how rocks have recorded the switching of Earth's magnetic polarity through time.

The time during which a particular paleomagnetic condition existed ("normal" or "reversed") can be determined by radiometric dating. Over the last 76 million years, magnetic polarity has switched irregularly at the rate of about once or twice each million years. It takes a few thousand years for a change in polarity to occur and is identified in rock sequences by a gradual decrease in the intensity of the magnetic field of one polarity, followed by a gradual increase in the intensity of the magnetic field of opposite polarity. Earth's present magnetic field has been weakening during the last 150 years, which may be an indication that Earth's current "normal" polarity will reverse itself within the next 2000 years.

Paleomagnetism and the Ocean Floor Paleomagnetism had certainly proved its usefulness on land, but, up until the mid-1950s, it had only been conducted on continental rocks. Would the ocean floor also show variations in magnetic polarity? To test this idea, the United States Coast and Geodetic Survey in conjunction with scientists from Scripps Institution of Oceanography undertook an extensive deep-water mapping program off Oregon and Washington in 1955. Using a sensitive **magnetometer** (*magneto* = magnetism, *meter* = measure) towed behind a research ship, the scientists spent several weeks at sea moving back and forth in a regularly spaced pattern, measuring Earth's magnetic field and how it was affected by the magnetic properties of rocks on the ocean floor.

When the scientists analyzed their data, it revealed that the entire surveyed area had a pattern of north-south stripes in a surprisingly regular and alternating pattern of above-average and below-average magnetism. What was even more surprising was that the pattern appeared to be symmetrical with respect to a long mountain range that was fortuitously in the middle of their survey area.

Detailed paleomagnetism studies of this and other areas of the sea floor confirmed that a similar pattern of

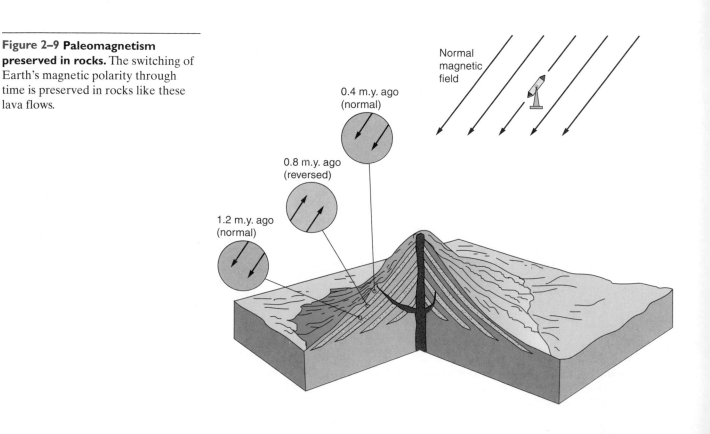

Figure 2–9 Paleomagnetism preserved in rocks. The switching of Earth's magnetic polarity through time is preserved in rocks like these lava flows.

alternating stripes of above-average and below-average magnetism. These stripes are called **magnetic anomalies** (*a* = without, *nomo* = law; an anomaly is a departure from normal conditions). The ocean floor had embedded in it a regular pattern of alternating magnetic stripes unlike anywhere on land.

Researchers had a difficult time explaining why the ocean floor had such a regular pattern of magnetic anomalies. Nor could they explain how the sequence on one side of the underwater mountain range matched the sequence on the opposite side—in essence, they were a mirror image of each other. To understand how this pattern could have formed, more information about ocean floor features and their origin was needed.

Sea Floor Spreading and Features of the Ocean Basins

Geologist **Harry Hess** (1906-1969), when he was a U.S. Navy captain in World War II, developed the habit of leaving his depth recorder on at all times while his ship was traveling at sea. After the war, compilation of these and many other depth records showed extensive mountain ridges near the centers of ocean basins and extremely deep, narrow trenches at the edges of ocean basins. In 1960, Hess published the idea of **sea floor spreading** with **convection** (*con* = with, *vect* = carried) **cells** as the driving mechanism (Figure 2–10). He

suggested that new ocean crust was created at the ridges, split apart, moved away from the ridges, and later disappeared back into the deep Earth at trenches. Mindful of the resistance of North American scientists to the idea of continental drift, Hess referred to his own work as "geopoetry."

As it turns out, Hess's initial ideas about sea floor spreading have been confirmed. The **mid-ocean ridge** (Figure 2–10) is a continuous underwater mountain range that winds through every ocean basin in the world and resembles the seam on a baseball. It is entirely volcanic in origin, wraps one-and-a-half times around the globe, and rises over 2.5 kilometers (1.5 miles) above the ocean floor. It even rises above sea level in places such as Iceland. New ocean floor forms at the crest, or *axis*, of the mid-ocean ridge. By the process of sea floor spreading, new ocean floor is split in two and carried away from the axis, replaced by the upwelling of volcanic material that fills the void with new strips of sea floor. Sea floor spreading occurs along the axis of the mid-ocean ridge, which is referred to as a **spreading center**. One way to think of the mid-ocean ridge is as a zipper that is being pulled apart. Thus, Earth's zipper (the mid-ocean ridge) is becoming unzipped!

At the same time, ocean floor is being destroyed at deep **ocean trenches**. Trenches are the deepest parts of the ocean floor and resemble a narrow crease or trough (Figure 2–10). Some of the largest earthquakes in the world occur near these trenches, caused by a

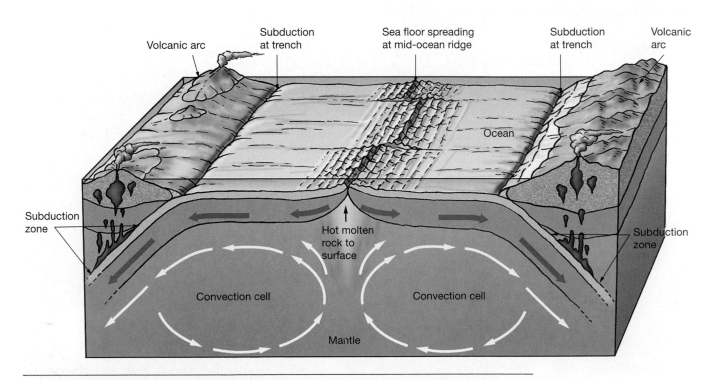

Figure 2–10 Processes of plate tectonics. Hot molten rock comes to the surface at the mid-ocean ridge and moves outward by the process of sea floor spreading. Eventually, sea floor is destroyed at the trenches, where the process of subduction occurs. Convection of material in the mantle produces convection cells.

Box 2–1 Do Sea Turtles (and Other Animals) Use Earth's Magnetic Field for Navigation?

Sea turtles travel great distances across the open ocean so they can lay their eggs on the island where they themselves were hatched. How do they know where the island is located and how do they navigate at sea during their long voyage? Studies have indicated that during their migration, green sea turtles (*Chelonia mydas*; Figure 2B) often travel in an essentially straight-line path to reach their destination. One hypothesis suggests that, like the Polynesian navigators, the sea turtles use wave direction to help them steer. However, sea turtles have been radio-tagged and tracked by satellites, which reveals that they continue along their straight-line path independent of wave direction.

Research in *magnetoreception*, the study of an animal's ability to sense magnetic fields, suggests that sea turtles may use Earth's magnetic field for navigation. For instance, turtles can distinguish between different magnetic inclination angles, which in effect would allow them to sense latitude. Sea turtles can also distinguish magnetic field intensity, a rough indication of longitude. By sensing these two magnetic field properties, a sea turtle could determine its position at sea and relocate a tiny island thousands of kilometers away. Like any good navigator, sea turtles may also use other tools, such as olfactory (scent) clues, Sun angles, local landmarks, and oceanographic phenomena.

Other animals may also use magnetic properties to navigate. For example, some whales and dolphins may detect and follow the magnetic stripes on the sea floor during their movements, which may help to explain why whales sometimes beach themselves. In addition, certain bacteria use the magnetic mineral magnetite to align themselves parallel to Earth's magnetic field. Subsequently, magnetite has been found in many other organisms that have a "homing" ability, including tuna, salmon, honeybees, pigeons, turtles, and even humans. What has been unclear is how these animals detect—and potentially use—Earth's magnetic field. Recent findings by a research team studying rainbow trout (close relatives of salmon) have traced magnetically receptive fibers of nerves back to the brain, more closely linking a magnetic sense with an organism's sensory system.

Do humans have an innate ability to use Earth's magnetic field for navigation? Studies conducted on humans indicate that the majority of people can identify north after being blindfolded and disoriented. Interestingly, many people point *south* instead of north, but this direction is along the lines of magnetic force. Similarly, migratory animals that rely on magnetism for navigation will not be confused by a reversal in Earth's magnetic field and will still be able to get to where they need to go. The detection of a directional sense in animals seems likely to remain an intriguing and elusive mystery in animal behavior.

Figure 2B Green sea turtle.

lithospheric plate that bends downward and slowly plunges back into Earth's interior. This process is called **subduction** (*sub* = under, *duct* = lead), and the sloping area from the trench along the downward plate is called a **subduction zone**.

In 1963, geologists **Fredrick Vine** and **Drummond Matthews** of Cambridge University combined the seemingly unrelated pattern of magnetic sea floor stripes with the process of sea floor spreading to explain the perplexing pattern of alternating and symmetric

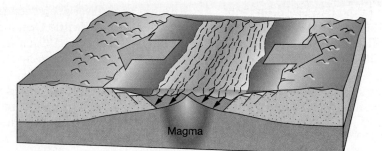

(a) Period of normal magnetism

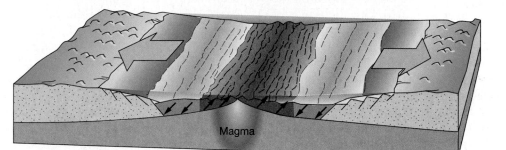

(b) Period of reverse magnetism

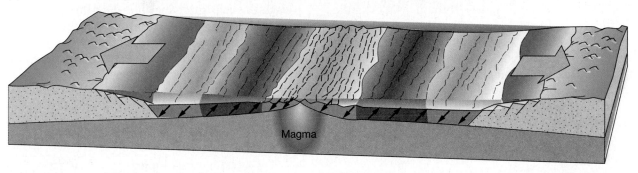

(c) Period of normal magnetism

Figure 2–11 Magnetic evidence of sea floor spreading. As new basalt is added to the ocean floor at mid-ocean ridges, it is magnetized according to Earth's existing magnetic field.

stripes on the sea floor (Figure 2–11). They proposed that the above-average magnetic stripes represented "normal" polarity sea floor rocks that enhanced Earth's current normal magnetic polarity and the below-average magnetic stripes represented the presence of "reverse" polarity sea floor rocks that subtracted from Earth's current polarity. The pattern could be created when newly formed rocks at the mid-ocean ridge are magnetized with whichever polarity exists on Earth during their formation. As those rocks are slowly moved away from the crest of the mid-ocean ridge, the periodic switching of Earth's magnetic polarity are recorded in subsequent rock. The result is an alternating pattern of magnetic polarity stripes that are symmetric with respect to the mid-ocean ridge.

The pattern of alternating reversals of Earth's magnetic field as recorded in the sea floor was the most convincing piece of evidence set forth to support the concept of sea floor spreading—and, as a result, continental drift. However, the continents weren't plowing through the ocean basins as Wegener had envisioned.

Instead, the ocean floor was a conveyer belt that was being continuously formed at the mid-ocean ridge and destroyed at the trenches, with the continents just passively riding along on the conveyer. By the late 1960s, most geologists had changed their stand on continental drift in light of this new evidence.

> The plate tectonic model states that new sea floor is created at the mid-ocean ridge where it moves outward by the process of sea floor spreading and is destroyed by subduction into an ocean trench.

Other Evidence from the Ocean Basins

Even though the tide of scientific opinion had indeed switched to favor a mobile Earth, additional evidence from the ocean floor would further support the ideas of continental drift and sea floor spreading.

Age of the Ocean Floor   In the late 1960s, an ambitious deep-sea drilling program was initiated to test the existence of sea floor spreading. One of the program's primary missions was to drill into and collect ocean floor rocks for radiometric age dating. If sea floor spreading does indeed occur, then the youngest sea floor rocks would be atop the mid-ocean ridge and the ages of rocks would increase on either side of the ridge in a symmetric pattern.

The map in Figure 2–12, showing the age of the ocean floor beneath deep-sea deposits, is based on the pattern of magnetic stripes verified with thousands of radiometrically age dated samples. It shows the ocean floor is youngest along the mid-ocean ridge, where new ocean floor is created, and the age of rocks increases with increasing distance in either direction away from the axis of the ridge. The symmetric pattern of ocean floor ages confirms that the process of sea floor spreading must indeed be occurring.

The Atlantic Ocean has the simplest and most symmetric pattern of age distribution in Figure 2–12. The pattern results from the newly formed Mid-Atlantic Ridge that rifted Pangaea apart. The Pacific Ocean has the least symmetric pattern because many subduction zones surround it. For example, ocean floor east of the East Pacific Rise that is older than 40 million years old has already been subducted. The ocean floor in the northwestern Pacific, about 180 million years old, has not yet been subducted. A portion of the East Pacific

Rise has even disappeared under North America. The age bands in the Pacific Ocean are wider than those in the Atlantic and Indian Oceans, which suggests the rate of sea floor spreading is greatest in the Pacific Ocean.

Recall from Chapter 1 that the ocean is at least 4 billion years old. However, the oldest ocean floor is only 180 million years old (or 0.18 billion years old), and the majority of the ocean floor is not even half that old (see Figure 2–12). How could the ocean floor be so incredibly young, while the oceans themselves are so phenomenally old? According to plate tectonic theory, new ocean floor is created at the mid-ocean ridge by sea floor spreading and moves off the ridge to eventually be subducted and remelted in the mantle. In this way, the ocean floor keeps regenerating itself. The floor beneath the oceans today is not the same one that existed beneath the oceans 4 billion years ago.

If the rocks that comprise the ocean floor are so young, why are continental rocks so old? Based on radiometric age dating, the oldest rocks on land are about 4 billion years old. Many other continental rocks approach this age, implying that the same processes that constantly renew the sea floor do not operate on land. Rather, evidence suggests that continental rocks do not get recycled by the process of sea floor spreading and thus remain at Earth's surface for long periods of time.

Heat Flow   The heat from Earth's interior is released to the surface as **heat flow**. Current models indicate this

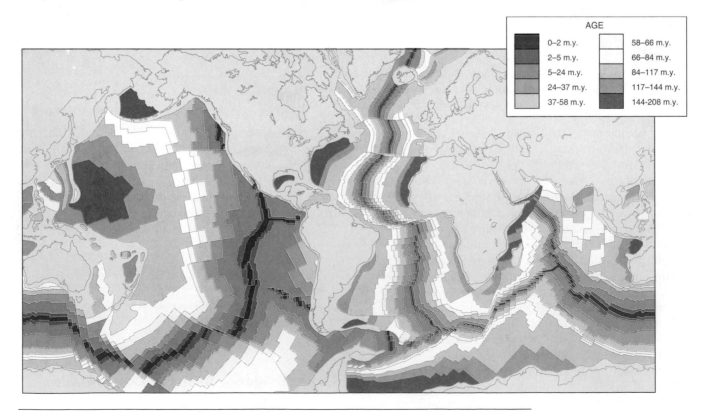

AGE

0–2 m.y.	58–66 m.y.
2–5 m.y.	66–84 m.y.
5–24 m.y.	84–117 m.y.
24–37 m.y.	117–144 m.y.
37-58 m.y.	144-208 m.y.

Figure 2–12 Age of the ocean crust beneath deep-sea deposits. The youngest rocks (*bright red areas*) are found along the mid-ocean ridge. Farther away from the mid-ocean ridge, the rocks increase linearly in age in either direction. Ages in millions of years.

heat moves to the surface with magma in convective motion. Most of the heat is carried to regions of the mid-ocean ridge spreading centers (see Figure 2–10). Cooler portions of the mantle descend in deep-sea trenches to complete each circular-moving convection cell.

Heat flow measurements show the amount of heat flowing to the surface along the mid-ocean ridge can be up to eight times greater than the average amount flowing to other parts of Earth's crust. Additionally, heat flow at deep-sea trenches, where ocean floor is subducted, can be as little as one-tenth the average. Increased heat flow at the mid-ocean ridge and decreased heat flow at subduction zones is what would be expected based on thin crust at the mid-ocean ridge and a double thickness of crust at the trenches (see Figure 2–10). Table 2–1 summarizes these relationships.

Worldwide Earthquakes Earthquakes are sudden releases of energy caused by fault movement or volcanic eruptions. The map in Figure 2–13a shows that most large earthquakes occur along ocean trenches, reflecting the energy released during subduction. Other earthquakes occur along the mid-ocean ridge, reflecting the energy released during sea floor spreading. Still others occur along major faults in the sea floor and on land, reflecting the energy released when moving plates contact other plates along their edges. The two maps in Figure 2–13 show that the distribution of worldwide earthquakes closely matches the locations of plate boundaries.

Evidence for plate tectonics includes many types of information from land and the sea floor, including the symmetric pattern of magnetic stripes relative to the mid-ocean ridge.

The Acceptance of a Theory

The accumulation of these and many other lines of evidence in support of moving continents have convinced scientists of the validity of continental drift. Since the late 1960s, the concepts of continental drift and sea floor spreading have been united into a much more encompassing theory known as plate tectonics, which describes the movement of the outermost portion of Earth and the resulting creating of continental and sea floor features.

Although several mechanisms have been proposed for the force or forces responsible for driving this motion, none of them are able to explain all aspects of plate motion. Nevertheless, it is clear that the unequal distribution of heat within Earth is the underlying driving force for this movement. The various properties of Earth's internal layers can help account for the movement of plates.

Earth Structure

Earth's internal structure consists of a series of nested spheres (similar to the layers of an onion) that differ in density. Let's examine these layers and discover their importance to plate tectonic processes.

Chemical Composition Versus Physical Properties

Earth is a layered sphere based on density, with the highest-density material found near the center of Earth and the lowest-density material located near the surface. The cross-sectional view of Earth in Figure 2–14 shows that Earth's inner structure can be subdivided according to its chemical composition (what the composition of the rocks are) or its physical properties (how the rocks respond to increased temperature and pressure at depth).

Table 2–1 Relationships relative to increasing distance in either direction from the axis of a mid-ocean ridge.

	Relationship	*Reason*
1.	Volcanic activity decreases	Magma chambers are concentrated only along the crest of the mid-ocean ridge
2.	The age of ocean crust increases	New ocean floor is created at the mid-ocean ridge
3.	The thickness of sea floor deposits increases	Older ocean floor has more time to accumulate a thicker deposit of debris due to settling
4.	The thickness of the lithospheric plate increases	As plates move away from the spreading center, they are cooler and gain thickness as material beneath the lithosphere attaches to the bottom of the plate
5.	Heat flow decreases	As the thickness of the lithospheric plate increases (see relationship 3), less heat can escape
6.	Water depth increases	The mid-ocean ridge is a topographically high feature, and as plate move away from there, they move into deeper water because plates undergo thermal contraction and isostatic adjustment downward

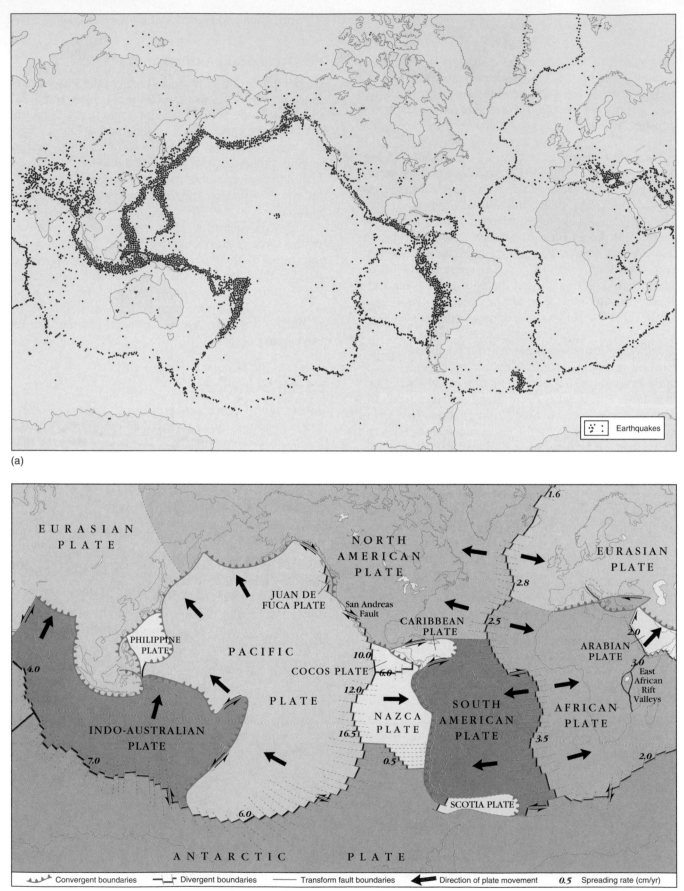

(a)

(b)

Convergent boundaries   Divergent boundaries   Transform fault boundaries   Direction of plate movement   *0.5* Spreading rate (cm/yr)

Figure 2–13 Earthquakes and lithospheric plates. (a) Distribution of earthquakes with magnitudes equal to or greater than $M_w = 5.0$ for the period 1980–1990. **(b)** Plate boundaries define the major lithospheric plates (*shaded*), with arrows indicating the direction of motion and numbers representing the rate of motion in centimeters per year. Notice how closely the pattern of major earthquakes follows plate boundaries.

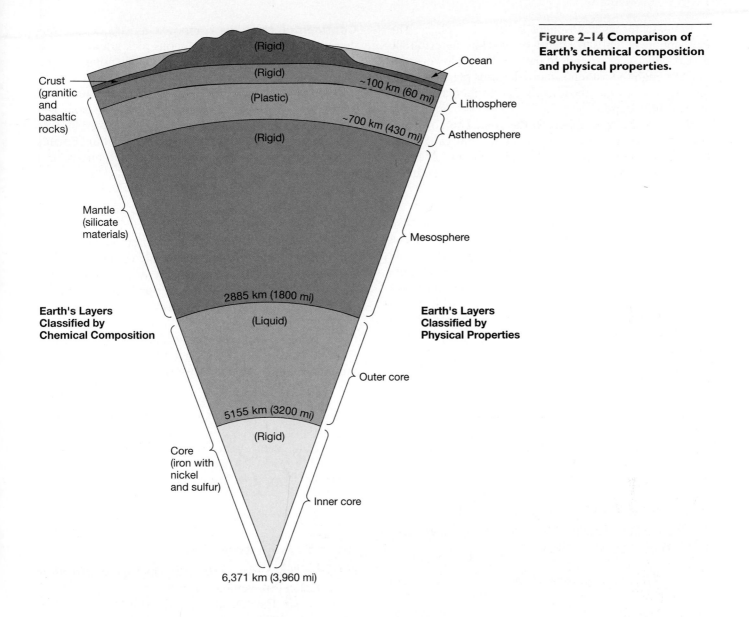

Chemical Composition Based on chemical composition, Earth consists of three layers: the **core**, the **mantle**, and the **crust** (Figure 2–14). If Earth were an apple, then the crust would be its thin skin. It extends from the surface to an average depth of 30 kilometers (20 miles). The crust is composed of relatively low-density rock, consisting mostly of various **silicate minerals** (common rock-forming minerals with silicon and oxygen). There are two types of crust, oceanic and continental, which will be discussed in the next section.

Immediately below the crust is the mantle. It occupies the largest volume of the three layers, and extends to a depth of 2900 kilometers (1800 miles). The mantle is composed of relatively high-density iron and magnesium silicate rock.

Beneath the mantle is the core. It forms a large mass from a depth of 2900 kilometers (1800 miles) to the center of Earth at 6370 kilometers (3960 miles). The core is composed of even higher-density metal (mostly iron and nickel).

Physical Properties Based on physical properties, Earth is composed of five layers: the **inner core**, the **outer core**, the **mesosphere** (*mesos* = middle, *sphere* = ball), the **asthenosphere** (*asthenos* = weak, *sphere* = ball), and, as previously mentioned, the lithosphere (Figure 2–14). The lithosphere is Earth's cool, rigid, outermost layer. It extends from the surface to an average depth of about 100 kilometers (62 miles) and includes the crust plus the topmost portion of the mantle. The lithosphere is **brittle** (*brytten* = to shatter), meaning that it will fracture when force is applied to it. The plates involved in plate tectonic motion are the plates of the lithosphere.

Beneath the lithosphere is the asthenosphere. The asthenosphere is **plastic** (*plasticus* = to mold), meaning it will flow when a gradual force is applied to it. It extends from about 100 kilometers (62 miles) to 700 kilometers (430 miles) below the surface, which is the base of the upper mantle. At these depths, it is hot enough to partially melt portions of most rocks.

Beneath the asthenosphere is the mesosphere. The mesosphere extends to a depth of 2900 kilometers (1800 miles), which corresponds to the middle and lower mantle. Although the asthenosphere deforms plastically, the mesosphere is rigid, most likely due to increased pressure at these depths.

Beneath the mesosphere is the core. The core consists of the outer core, which is liquid and capable of flowing; and the inner core, which is rigid and does not flow. Again, the increased pressure at the center of Earth keeps the inner core from flowing.

Knowledge of Earth's Inner Structure

How have scientists determined the inner structure of Earth? The deepest well in the world, which was drilled in the Kola Peninsula of Russia, reached a depth of 12,266 meters (40,478 feet, or nearly 8 miles) in 1992, but still never came close to penetrating the crust. Earth scientists have had to rely on *indirect* methods to understand what lies deep within Earth. For example, determining the gravitational attraction that Earth exerts on other bodies of the solar system gives scientists information about the density of deep Earth layers. Similarly, *seismologists* (*seismo* = earthquake, *ologist* = one who studies) analyze the pattern of seismic waves bouncing around within Earth (similar to listening through a doctor's stethoscope) to decipher the chemical composition and physical properties of Earth's internal layers.

Near the Surface

The top portion of Figure 2–15 shows an enlargement of Earth's layers closest to the surface.

Lithosphere

The lithosphere is a relatively cool, rigid shell that includes all of the crust and the topmost part of the mantle. In essence, the topmost part of the mantle is attached to the crust and the two act as a single unit, approximately 100 kilometers (62 miles) thick. The expanded view in Figure 2–15 shows that the crust portion of the lithosphere is further subdivided into oceanic crust and continental crust, which are compared in Table 2–2.

Oceanic crust

is composed of the igneous rock **basalt**, which is dark-colored and has a relatively high density of about 3.0 grams per cubic centimeter.[4] The average thickness of the oceanic crust is only about 8 kilometers (5 miles). Basalt originates as molten magma beneath Earth's crust. This magma comes to the surface mostly along the mid-ocean ridge, where it cools and hardens to form new oceanic crust.

Continental crust

is composed mostly of a lower-density and lighter-colored igneous rock **granite**.[5] It has a density of about 2.7 grams per cubic centimeter. The average thickness of the continental crust is about 35 kilometers (22 miles), but may reach a maximum of 60 kilometers (37 miles) beneath the highest mountain ranges. Most granite originates beneath the surface as molten magma that cools and hardens within Earth's crust. No matter which type of crust is at the surface, it is all part of the lithosphere.

Asthenosphere

The asthenosphere is a relatively hot, plastic region beneath the lithosphere. It extends from the base of the lithosphere to a depth of about 700 kilometers (430 miles) and is entirely contained within the upper mantle. The asthenosphere can deform without fracturing if a force is applied slowly. It has the ability to flow but has high **viscosity** (*viscos* = sticky). Viscosity is a measure of a substance's resistance to flow.[6] Studies indicate that the high-viscosity asthenosphere is flowing slowly through time, which has important implications in plate tectonic processes.

> Earth has differences in composition and physical properties that create layers such as the brittle lithosphere and the plastic asthenosphere, which is capable of flowing slowly over time.

Isostatic Adjustment

Isostatic (*iso* 5 equal, *stasis* 5 standing) **adjustment** —the vertical movement of crust—is the result of the buoyancy of Earth's lithosphere as it floats on the denser, plastic-like asthenosphere below it. The container ship in Figure 2–16 provides an example of isostatic adjustment. An empty ship floats high in the water. Once the ship is loaded with cargo, though, the ship undergoes isostatic adjustment and floats lower in the water (but hopefully won't sink!). When the cargo is unloaded, the ship isostatically adjusts itself and floats higher again.

Similarly, both continental and oceanic crust float on the denser mantle beneath. Oceanic crust is denser than continental crust, however, so oceanic crust floats lower in the mantle because of isostatic adjustment. Oceanic

---

[4]Water has a density of 1.0 grams per cubic centimeter. Thus, basalt with a density of 3.0 grams per cubic centimeter is three times the density of water.

[5]At the surface, continental crust is often covered by a relatively thin layer of surface sediments. Below these, granite can be found.
[6]Substances that have high viscosity (a high resistance to flow) include toothpaste, honey, tar, and silly putty; a common substance that has low viscosity is water. A substance's viscosity often changes with temperature. For instance, as honey is heated, it flows more easily.

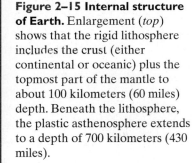

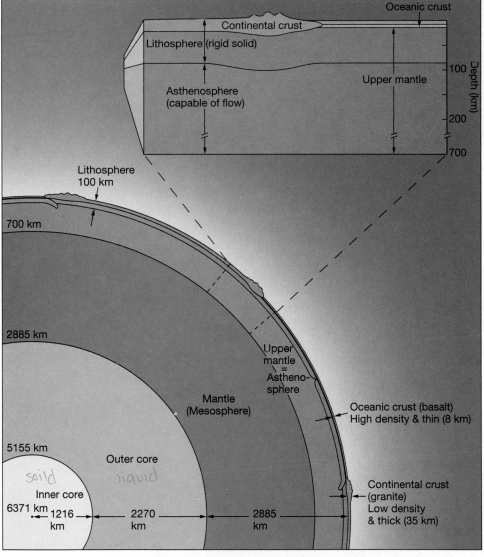

Figure 2–15 Internal structure of Earth. Enlargement (*top*) shows that the rigid lithosphere includes the crust (either continental or oceanic) plus the topmost part of the mantle to about 100 kilometers (60 miles) depth. Beneath the lithosphere, the plastic asthenosphere extends to a depth of 700 kilometers (430 miles).

crust is also thin, which creates low areas for the oceans to occupy. Areas where the continental crust is thickest (in large mountain ranges on the continents) float higher than continental crust of normal thickness, also because of isostatic adjustment. Thus, tall mountain ranges on Earth are composed of a great thickness of crustal material that in essence keeps them buoyed up.

Areas that are exposed to an increased or decreased load experience isostatic adjustment. For instance, during the most recent Ice Age (which occurred during the Pleistocene Epoch between 2 million and 10,000 years ago), massive ice sheets covered far northern regions such as Scandinavia and northern Canada. The additional weight of ice several kilometers thick caused

these areas to isostatically adjust themselves lower in the mantle. Since the end of the Ice Age, the reduced load on these areas caused by the melting of ice caused these areas to rise and experience **isostatic rebound**, which continues today. The rate at which isostatic rebound occurs gives scientists important information about the properties of the upper mantle.

Further, isostatic adjustment provides additional evidence for the movement of Earth's tectonic plates. Because continents isostatically adjust themselves by moving *vertically*, then they must not be firmly fixed in one position on Earth. If this is true, the plates that contain these continents should certainly be able to move *horizontally* across Earth's surface.

Table 2–2 Comparing oceanic and continental crust.

	Oceanic Crust	Continental Crust
Main rock type	Basalt (dark-colored igneous rock)	Granite (light-colored igneous rock)
Density (grams per cubic centimeter)	3.0	2.7
Average thickness	8 kilometers (5 miles)	35 kilometers (22 miles)

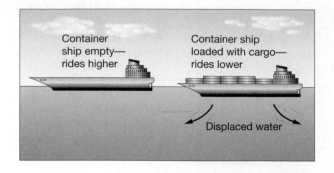

Figure 2–16 A container ship experiences isostatic adjustment. A ship will ride higher in water when it is empty and will ride lower in water when it is loaded with cargo.

Plate Boundaries

Plate boundaries—where plates interact with each other—are associated with a great deal of tectonic activity such as mountain building, volcanic activity, and earthquakes. In fact, the first clues to the locations of plate boundaries were the dramatic tectonic events that occur there. For example, Figure 2–13 shows the close correspondence between worldwide earthquakes and plate boundaries. Further, Figure 2–13b shows that Earth's surface is composed of seven major plates along with many smaller ones. Close examination of Figure 2–13b shows that the boundaries of plates do not always follow coastlines and, as a consequence, nearly all plates contain both oceanic and continental crust.

There are three types of plate boundaries, as shown in Figure 2–17. **Divergent** (*di* = apart, *vergere* = to incline) **boundaries** are found along oceanic ridges where new lithosphere is being added. **Convergent** (*con* = together, *vergere* = to incline) **boundaries** are found where plates are moving together and one plate subducts beneath the other. **Transform** (*trans* = across, *form* = shape) **boundaries** are found where lithospheric plates slowly grind past one another. Table 2–3 summarizes characteristics, tectonic processes, and examples of these plate boundaries.

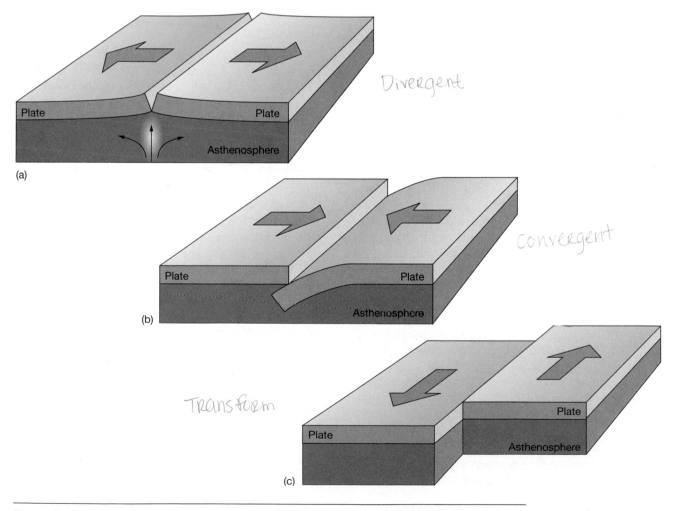

Figure 2–17 The three types of plate boundaries. (a) Divergent, where plates move away from each other. **(b)** Convergent, where plates approach each other. **(c)** Transform, where plates slide past each other.

Table 2–3   Characteristics, tectonic processes, and examples of plate boundaries.

Plate boundary	Plate movement	Crust type(s)	Sea floor created or destroyed?	Tectonic process	Geographic examples
Divergent plate boundaries "Constructive"	Apart ← →	Ocean-ocean	New sea floor created	Sea floor spreading	Mid-Atlantic Ridge, East Pacific Rise
		Continent-continent	As continent splits apart, new sea floor created	Continental rifting	East Africa Rift Valleys, Red Sea, Gulf of California
Convergent plate boundaries	Together → ←	Ocean-continent	Old sea floor destroyed	Subduction	Andes Mountains, Cascade Mountains
		Ocean-ocean	Old sea floor destroyed	Subduction	Aleutian Islands, Mariana Islands
		Continent-continent	N/A	Collision	Himalaya Mountains, Alps
Transform plate boundaries	Past each other → ←	Oceanic	Sea floor neither created nor destroyed	Transform faulting	Mendocino fault, Eltanin fault (between mid-ocean ridges)
		Continental	Sea floor neither created nor destroyed	Transform faulting	San Andreas fault, Alpine fault (New Zealand)

Divergent Boundaries

Divergent plate boundaries occur where two plates move apart, such as along the crest of the mid-ocean ridge where sea floor spreading creates new oceanic lithosphere (Figure 2–18). A common feature along the mid-ocean ridge is a central downdropped linear **rift valley** (Figure 2–19). Pull-apart faults located along the central rift valley show that the plates are being *continuously pulled apart* rather than being pushed apart by the upwelling of material beneath the mid-ocean ridge. Upwelling of magma beneath the mid-ocean ridge is simply filling in the void left by the separating plates of lithosphere. In the process, sea floor spreading produces about 20 cubic kilometers (4.8 cubic miles) of new ocean crust worldwide each year.

Figure 2–20 shows how the development of a mid-ocean ridge creates an ocean basin. Initially, molten material rises to the surface, causing upwarping and thinning of the crust. Volcanic activity produces vast quantities of high-density basaltic rock. As the plates begin to move apart, a linear rift valley is formed and volcanism continues. Further **rifting** of the land and more spreading cause the area to drop below sea level. When this occurs, the rift valley eventually floods with seawater and a young linear sea is formed. After millions of years of sea floor spreading, a full-fledged ocean basin is created with a mid-ocean ridge in the middle of the two landmasses.

Two different stages of ocean basin development are shown in the map of East Africa in Figure 2–21. First, the rift valleys are actively pulled apart and are at the rift valley stage of formation. Second, the Red Sea is at the linear sea stage. It has rifted apart so far that the land has dropped below sea level. The Gulf of California in Mexico is another linear sea. The Gulf of California and the Red Sea are two of the youngest seas in the world, having been created only a few million years ago. If plate motions continue rifting the plates apart in these areas, they will eventually become large oceans.

The rate at which the sea floor spreads apart varies along the mid-ocean ridge and affects its appearance. Figure 2–22a shows that faster spreading produces broader and less rugged segments of the global mid-ocean ridge system. The vast amount of rock being produced at a fast-spreading segment of the mid-ocean ridge does not experience the same thermal contraction and subsidence that occurs along slower spreading segments. In addition, central rift valleys on slow-spreading segments tend to be larger and better developed (Figure 2–22b).

The fast-spreading and gently sloping parts of the mid-ocean ridge are called **oceanic rises**. For example, the **East Pacific Rise** (Figure 2–22a) between the Pacific Plate and the Nazca Plate is a broad and low oceanic rise where the spreading rate is as high as 16.5 centimeters (6.5 inches) per year.[7] Conversely, slower-spreading and steeper-sloped areas of the mid-ocean ridge are called **oceanic ridges**. The **Mid-Atlantic Ridge** (Figure

---

[7]The spreading rate is the total widening rate of an ocean basin resulting from motion of *both* plates away from a spreading center.

Figure 2–18 Divergent boundary at the Mid-Atlantic Ridge. Most divergent plate boundaries occur along the crest of the mid-ocean ridge, where sea floor spreading creates new oceanic lithosphere.

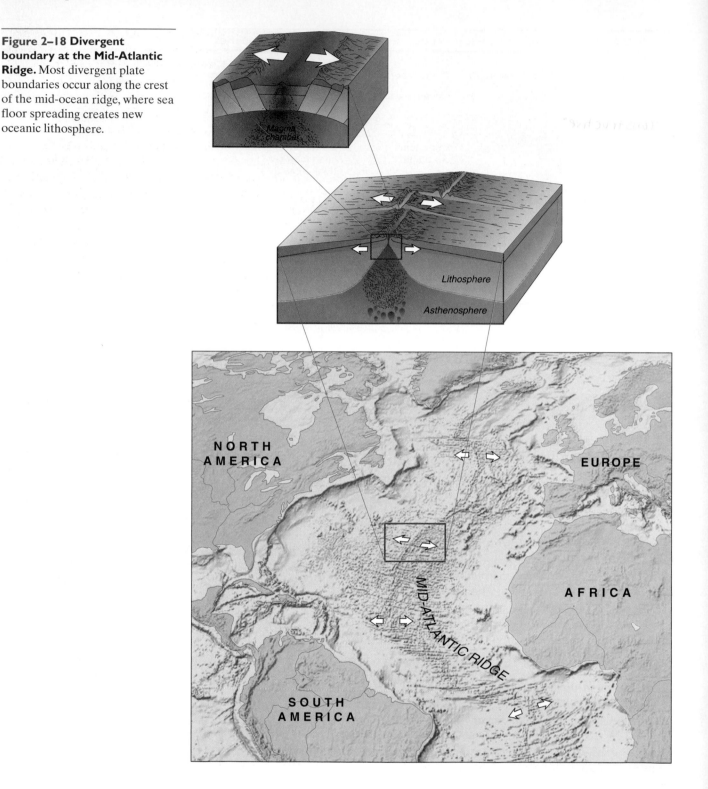

2–22b) between the South American Plate and the African Plate is an example of a steep, rugged oceanic ridge where the spreading rate is only 2 to 3 centimeters (0.8 to 1.2 inches) per year.

The amount of energy released by earthquakes along the divergent plate boundaries is closely related to the spreading rate. The faster the sea floor spreads, the less energy released in each earthquake. Earthquake inten-sity is usually measured on a scale called the **seismic moment magnitude**, which reflects the energy released to create very long-period seismic waves. Because it more adequately represents larger magnitude earth-quakes, the moment magnitude scale is increasingly used instead of the well-known Richter scale and is rep-resented by the symbol M_w. Earthquakes in the rift val-ley of the slow-spreading Mid-Atlantic Ridge reach a

Figure 2–19 Rift valley. Rift valley looking south from Laki volcano in Iceland, which sits atop the Mid-Atlantic Ridge.

maximum magnitude of about $M_w = 6.0$, whereas those occurring along the axis of the fast-spreading East Pacific Rise seldom exceed $M_w = 4.5$.[8]

Convergent Boundaries

Convergent boundaries—where two plates move together and collide—result in the destruction of ocean crust as one plate plunges below the other and is remelted in the mantle. The physiographic ocean floor feature associated with a convergent plate boundary is a deep-ocean trench. Trenches are deep linear scars where subduction occurs. Melting in the subduction zone causes an arc-shaped row of highly active and explosively erupting volcanoes that parallel the trench called a **volcanic arc**.

Figure 2–23 shows the three subtypes of convergent boundaries that result from interactions between two different types of crust (oceanic and continental).

Oceanic-Continental Convergence When an oceanic and a continental plate converge, the denser

oceanic plate is subducted (Figure 2–23a). The oceanic plate becomes heated as it is subducted into the asthenosphere and some of the basaltic rock is melted. This molten rock mixes with superheated gases (mostly water) and begins to rise to the surface through the overriding continental plate. The rising basalt-rich magma mixes with the granite of the continental crust, producing lava in volcanic eruptions at the surface that is intermediate in composition between basalt and granite. One type of volcanic rock with this composition is called **andesite**, named after the Andes Mountains of South America because it is so common there. Andesitic volcanic eruptions are usually quite explosive and have historically been very destructive because andesite contains such high gas content. The result of this volcanic activity on the continent above the subduction zone produces a type of volcanic arc called a **continental arc**. Continental arcs are created by andesitic volcanic eruptions and by the folding and uplifting associated with plate collision.

If the spreading center producing the subducting plate is far enough from the subduction zone, an oceanic trench becomes well developed along the margin of the continent. The Peru-Chile Trench is an example, and the Andes Mountains are the associated continental arc produced by melting of the subducting plate. If the spreading center producing the subducting plate is close to the subduction zone, however, the trench is not nearly as well developed. This is the case where the Juan de Fuca Plate subducts beneath the North American Plate off the coasts of Washington and Oregon to produce the Cascade Mountains continental arc (Figure 2–24). Here, the Juan de Fuca Ridge is so close to the North American Plate that the subducting lithosphere is less than 10 million years old and has not cooled enough to become very deep. In addition, the large amount of sediment carried to the ocean by the Columbia River has filled most of the trench with sediment. Many of the Cascade volcanoes of this continental arc have been active within the last 100 years. Most recently, Mount St. Helens erupted in May 1980, killing 62 people.

Oceanic-Oceanic Convergence When two oceanic plates converge, the denser oceanic plate is subducted (Figure 2–23b). Typically, the older oceanic plate is denser because it has had more time to cool and contract. This type of convergence produces the deepest trenches in the world, such as the Mariana Trench in the western Pacific Ocean. The subducting oceanic plate becomes heated and melts in the asthenosphere. Similar to oceanic-continental convergence, molten material mixes with superheated gases and rises to the surface to produce volcanoes. The molten material is mostly basaltic because there is no mixing with granitic rocks from the continents, and the eruptions are not quite so destructive. The result of this volcanic activity is a type of volcanic arc called an **island arc**. Examples of island

---

[8]Note that each one-unit increase of magnitude represents an increase of energy release of about 30 times.

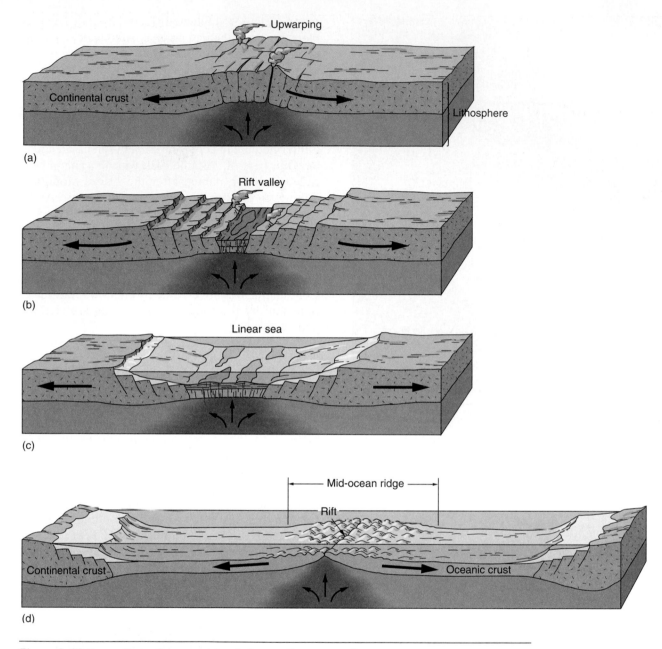

Figure 2–20 Formation of an ocean basin by sea floor spreading.

arc/trench systems are the West Indies' Leeward and Windward Islands/Puerto Rico Trench in the Caribbean Sea and the Aleutian Islands/Aleutian Trench in the North Pacific Ocean.

Continental-Continental Convergence When two continental plates converge, which one is subducted? You might expect the older of the two (which is probably the denser one) will be subducted. Continental lithosphere forms differently than oceanic lithosphere, however, and old continental lithosphere is no denser than young continental lithosphere. It turns out that *neither* subducts because both are too low in density to be pulled very far down into the mantle. Instead, a tall up-lifted mountain range is created by the collision of the two plates (Figure 2–23c). These mountains are com-

posed of folded and deformed sedimentary rocks origi-nally deposited on the sea floor that previously sepa-rated the two continental plates. The oceanic crust itself may subduct beneath such mountains. A prime example of continental-continental convergence is the collision of India with Asia (Figure 2–25). It began 45 million years ago and has created the Himalaya Mountains, presently the tallest mountains in the world.

Earthquakes Associated with Convergent Bound-aries Both spreading centers and trench systems are characterized by earthquakes, but in different ways. Spreading centers have shallow earthquakes, usually less than 10 kilometers (6 miles) deep. Earthquakes in the trenches, on the other hand, vary from near the sur-face down to 670 kilometers (415 miles) deep, which are

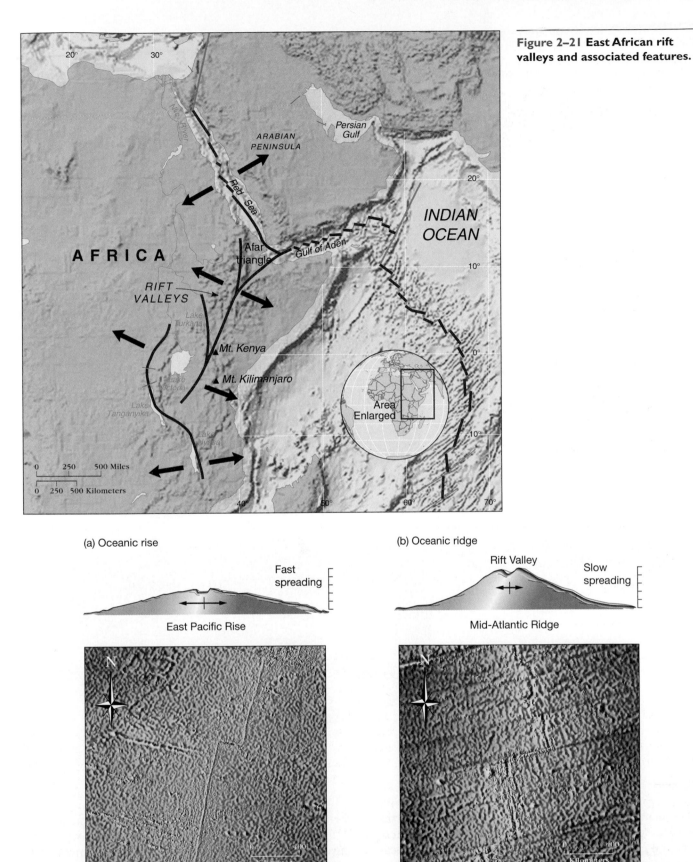

Figure 2–21 East African rift valleys and associated features.

(a) Oceanic rise

(b) Oceanic ridge

Fast spreading

East Pacific Rise

Rift Valley

Slow spreading

Mid-Atlantic Ridge

Figure 2–22 Comparing oceanic rises and ridges. Cross-sectional view (*above*) and map view (*below*) of: **(a)** The fast spreading East Pacific Rise. **(b)** The slow spreading Mid-Atlantic Ridge. On both cross-sections, vertical exaggeration is 50 times normal.

Figure 2–23 Three sub-types of convergent plate boundaries.
(a) Oceanic-continental convergence.
(b) Oceanic-oceanic convergence.
(c) Continental-continental convergence.

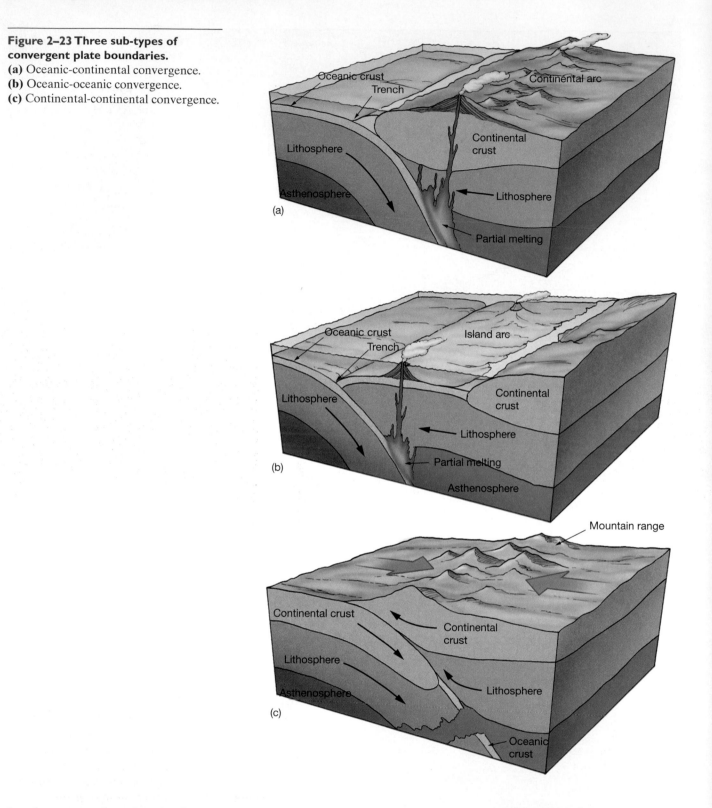

the deepest earthquakes in the world. These earthquakes are clustered in a band about 20 kilometers (12.5 miles) thick that closely corresponds to the location of the subduction zone. In fact, the subducting plate in a convergent plate boundary can be traced below the surface by examining the pattern of successively deeper earthquakes extending from the trench.

Many factors combine to produce large earthquakes at convergent boundaries. The forces involved in convergent-plate boundary collisions are enormous. Huge lithospheric slabs of rock are relentlessly pushing against each other, and the subducting plate must actually bend as it dives below the surface. In addition, thick crust associated with convergent boundaries tends to store more energy than the thinner crust at divergent boundaries. Also, mineral structure changes occur at the higher pressures encountered deep below the surface, which are thought to produce changes in volume that

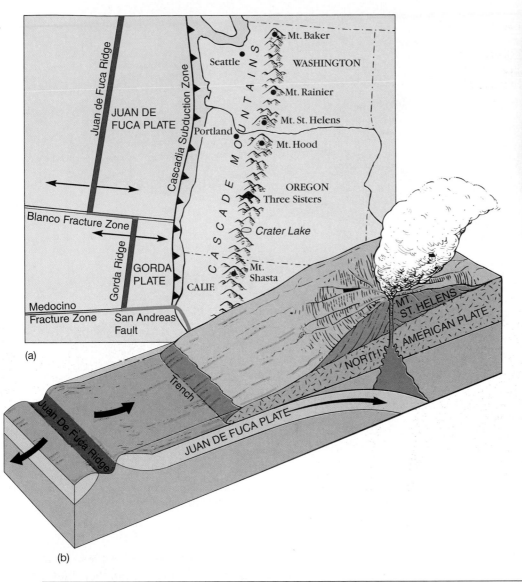

(a)

(b)

(c)

Figure 2–24 Convergent tectonic activity produces the Cascade Mountains. (a) Tectonic features of the Cascade Mountain Range and vicinity. **(b)** The volcanoes of the Cascade Mountains are created by the subduction of the Juan de Fuca Plate beneath the North American Plate. **(c)** The eruption of Mount St. Helens in 1980.

lead to some of the most powerful earthquakes in the world. In fact, the largest earthquake ever recorded was the 1960 Chilean earthquake near the Peru-Chile Trench, which had a magnitude of $M_w = 9.5$!

Transform Boundaries

A global sea floor map (such as the one inside the front cover of this book) shows that the mid-ocean ridge is offset by many large features oriented perpendicular (at right angles) to the crest of the ridge. What causes these offsets? They occur because spreading at a mid-ocean ridge only occurs perpendicular to the axis of a ridge and all parts of a plate must move together. As a result, offsets are oriented perpendicular to the ridge and par-

allel to each other to accommodate spreading of a linear ridge system on a spherical Earth. In addition, the offsets allow different segments of the mid-ocean ridge to spread apart at different rates. These offsets—called **transform faults**—give the mid-ocean ridge a zigzag appearance. There are thousands of these transform faults, some large and some small, which dissect the global mid-ocean ridge.

There are two types of transform faults. The first and most common type occurs wholly on the ocean floor and is called an **oceanic transform fault**. The second type cuts across a continent and is called a **continental transform fault**. Regardless of type, though, transform faults *always* occur between two segments of a mid-ocean ridge, as shown in Figure 2–26.

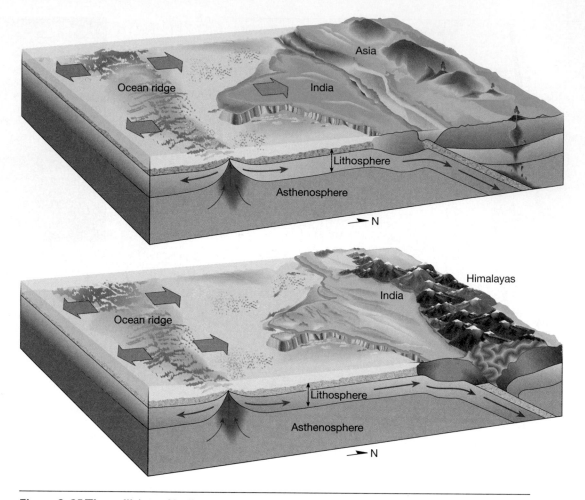

Figure 2–25 The collision of India with Asia. The ongoing collision of India with Asia began about 45 million years ago and has uplifted the Himalayan Mountains.

The movement of one plate past another—a process called **transform faulting**—produces shallow but often strong earthquakes in the lithosphere. Magnitudes of $M_w = 7.0$ have been recorded along some oceanic transform faults. One of the best studied faults in the world is California's **San Andreas Fault**, a continental transform fault that runs from the Gulf of California to near San Francisco and beyond into northern California. Because the San Andreas Fault cuts through continental crust, which is much thicker than oceanic crust, earthquakes are considerably larger than those produced by oceanic transform faults, sometimes up to $M_w = 8.5$.

Many people are mistakenly concerned that California will "fall off into the ocean" during a large earthquake along the San Andreas Fault because California experiences large periodic earthquakes. These earthquakes occur as the Pacific Plate continues to move to the northwest past the North American Plate at a rate of about 5 centimeters (2 inches) a year. At this rate, Los Angeles (on the Pacific Plate) will be adjacent to San Francisco (on the North American Plate) in about 18.5 million years—a length of time for about *one million generations of people* to live their lives. Although California will never fall into the ocean, people living

near this fault should be very aware they are likely to experience a large earthquake within their lifetime.

> The three main types of plate boundaries are divergent (plates moving apart such as at the mid-ocean ridge), convergent (plates moving together such as at an ocean trench), and transform (plates sliding past each other such as at a transform fault).

Testing the Model: Some Applications of Plate Tectonics

One of the strengths of plate tectonic theory is how it unifies so many seemingly separate events into a single consistent model. Let's look at a few examples that illustrate how plate tectonic processes can be used to explain the origin of features that, up until the acceptance of plate tectonics, were difficult to explain.

Mantle Plumes and Hotspots

Although the theory of plate tectonics helped explain the origin of many features near plate boundaries, it did

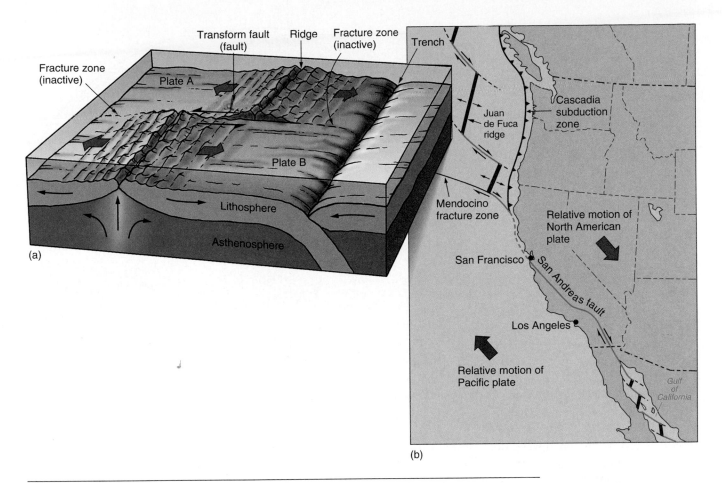

Figure 2–26 Transform faults. The Juan de Fuca ridge is offset by several oceanic transform faults. Also shown is the San Andreas fault, a continental transform fault connecting the Juan de Fuca ridge and the spreading center in the Gulf of California.

not seem to explain the origin of *intraplate* (*intra* = within, *plate* = plate of the lithosphere) *features* that are far from any plate boundary. For instance, how can plate tectonics explain volcanic islands near the middle of a plate?

According to the plate tectonic model, volcanism in the middle of a plate is caused by the presence of mantle plumes that are most likely related to the positions of convection cells in the mantle. **Mantle plumes** (*pluma* = a soft feather) are columnar areas of hot molten rock that arise from deep within the mantle. The areas where mantle plumes come to the surface are called **hotspots** and are marked by an abundance of volcanic activity.[9] For example, the continuing volcanism in Yellowstone National Park and Hawaii are caused by hotspots.

Worldwide, more than 100 hotspots have been active within the last 10 million years. Figure 2–27 shows the global distribution of prominent hotspots today. In general, hotspots do not coincide with plate boundaries.

Notable exceptions are those that are near divergent boundaries where the lithosphere is thin, such as at the Galápagos Islands and Iceland. In fact, Iceland straddles the Mid-Atlantic Ridge (a divergent plate boundary). It is also directly over a 150-kilometer (93-mile)-wide mantle plume, which accounts for its remarkable amount of volcanic activity—so much that it has caused Iceland to be one of the few areas of the global mid-ocean ridge above sea level.

Throughout the Pacific Plate, many island chains are oriented in a northwestward–southeastward direction. The most intensely studied of these is the **Hawaiian Islands–Emperor Seamount chain** in the northern Pacific Ocean (Figure 2–28). What created this chain of over 100 intraplate volcanoes that stretch over 5800 kilometers (3000 miles)? Further, what caused the prominent bend in the overall direction that occurs in the middle of the chain?

To help answer these questions, look at the ages of the volcanoes in the chain. Every volcano in the chain has long since become extinct, except the volcano Kilauea on the island of Hawaii, which is the southeasternmost island of the chain. The age of volcanoes progressively increases northwestward from Hawaii

---

[9]Note that hotspots are different than either a volcanic arc or a mid-ocean ridge, even though all are marked by a high degree of volcanic activity.

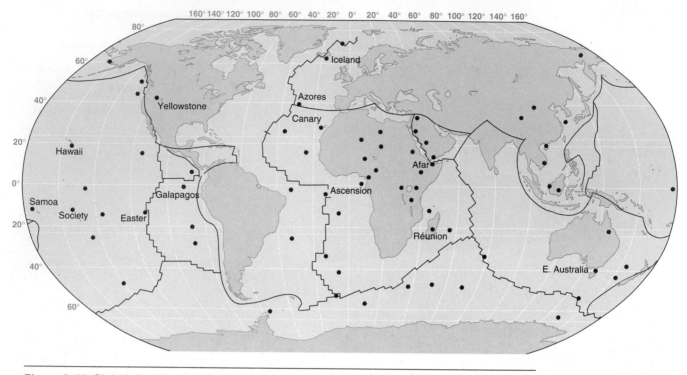

Figure 2–27 Global distribution of prominent hotspots. Lines represent the locations of plate boundaries.

(Figure 2–28). To the northwest, the volcanoes increase in age past Suiko Seamount (65 million years old) to Detroit Seamount (81 million years old) near the Aleutian Trench.

These age relationships suggest that the Pacific Plate has moved steadily northwestward while the underlying mantle plume remained relatively stationary. The resulting Hawaiian hotspot created each of the volcanoes in the chain. As the plate moved, it carried the active volcano off the hotspot and a new volcano began forming, younger in age than the previous one. A chain of extinct volcanoes that is progressively older as one travels away from a hotspot is called a **nematath** (*nema* = thread, *tath* = dung or manure[10]). Evidence suggests that about 40 million years ago, the Pacific Plate shifted from a northerly to a northwesterly direction with a rearrangement of plate boundaries along western North America. This change in plate motion accounts for the bend in Figure 2–28 about halfway through the chain, separating the Hawaiian Islands from the Emperor Seamounts.

If Hawaii is directly above the hotspot now, what will become of it in the future? It will be carried to the northwest off the hotspot, become inactive, and eventually be subducted into the Aleutian Trench like all the rest of the volcanoes in the chain to the north of it. In turn, other volcanoes will build up over the hotspot. In fact, a 3500-meter (11,500-foot) volcano already exists

32 kilometers (20 miles) southeast of Hawaii, named **Loihi**. Still 1 kilometer (0.6 mile) below sea level, Loihi is volcanically active and should reach the surface sometime between 30,000 and 100,000 years from now at its current rate of activity. As it builds above sea level, it will become the newest island in the long chain of volcanoes created by the Hawaiian hotspot.

> Mantle plumes create hotspots at the surface, which produce volcanic chains called nemataths that record the motion of plates.

Seamounts and Tablemounts

Many areas of the ocean floor (most notably on the Pacific Plate) contain tall volcanic peaks that are called **seamounts** if conical on top, and **tablemounts**—or **guyots**, after Princeton University's first geology professor Arnold Guyot[11]—if flat on top. Until the theory of plate tectonics, how seamounts and tablemounts formed was unclear. The theory explained why tablemounts were flat on top, and why the tops of some tablemounts had shallow-water deposits despite being located in very deep water.

The origin of many seamounts and tablemounts is related to the volcanic activity occurring at hotspots; others

[10]The highly descriptive scientific term nematath was obviously coined by someone with a sense of humor!

[11]Guyot is pronounced "GEE-oh" with a hard g as in "give."

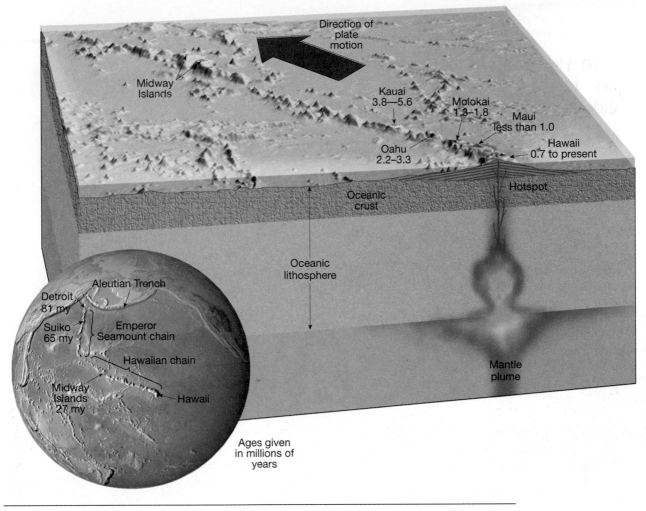

Figure 2–28 Hawaiian Islands—Emperor Seamount chain. The chain of volcanoes that extends from Hawaii to the Aleutian Trench results from the movement of the Pacific Plate over the relatively stationary Hawaiian hotspot. Numbers represent radiometric age dates in millions of years before present.

are related to processes occurring at the mid-ocean ridge (Figure 2–29). Because of sea floor spreading, active volcanoes (seamounts) occur along the crest of the mid-ocean ridge. Some may be built up so high they rise above sea level and become islands, at which point wave erosion becomes important. When sea floor spreading has moved the seamount off its source of magma (whether it is a mid-ocean ridge or a hotspot), the top of the seamount can be flattened by waves in just a few million years. This flattened seamount—now a table-mount—continues to be carried away from its source and, after millions of years, is submerged deeper into the ocean. Frequently, tops of tablemounts contain evidence of shallow-water conditions (such as ancient coral reef deposits) that were carried with it into deeper water.

Coral Reef Development

On his voyage aboard the HMS *Beagle*, the famous naturalist **Charles Darwin** noticed a progression of stages in **coral reef** development. He hypothesized that the origin of coral reefs depended on the subsidence (sinking) of volcanic islands (Figure 2–30) and published the concept in *The Structure and Distribution of Coral Reefs* in 1842. What Darwin's hypothesis lacked was a mechanism for how volcanic islands subside. Much later, advances in plate tectonic theory and samples of the deep structure of coral reefs provided evidence to help support Darwin's hypothesis.

Reef-building corals are colonial animals that live in shallow, warm, tropical seawater and produce a hard skeleton of limestone. Once corals are established in an area that has the conditions necessary for their growth, they continue to grow upward layer by layer with each new generation attached to the skeletons of its predecessors. Over millions of years, a thick sequence of coral reef deposits may develop if the conditions remain favorable.

The three stages of development in coral reefs are called fringing, barrier, and atoll. **Fringing reefs** (Figure

Figure 2–29 Formation of seamounts and tablemounts at a mid-ocean ridge.

sea floor deepins as it goes farther away from Ridge

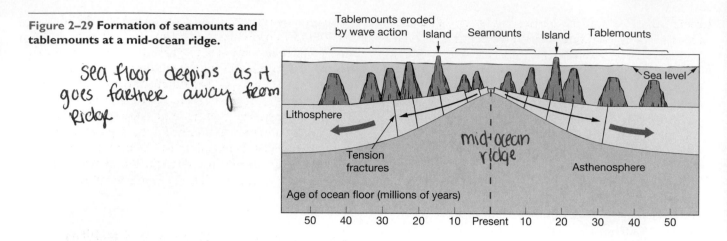

Tablemounts eroded by wave action · Island · Seamounts · Island · Tablemounts

Sea level

Lithosphere

Tension fractures

mid-ocean ridge

Asthenosphere

Age of ocean floor (millions of years)

50 40 30 20 10 Present 10 20 30 40 50

2–30a) initially develop along the margin of a landmass (an island or a continent) where the temperature, salinity, and turbidity (cloudiness) of the water are suitable for reef-building corals. Often, fringing reefs are associated with active volcanoes whose lava flows run down the flanks of the volcano and kill the coral. Thus, these fringing reefs are not very thick or well developed. Because of the close proximity of the landmass to the reef, runoff from the landmass can carry so much sediment that the reef is buried. The amount of living coral in a fringing reef at any given time is relatively small, with the greatest concentration in areas protected from sediment and salinity changes. If sea level does not rise or the land does not subside, the process stops at the fringing reef stage.

The **barrier reef** stage follows the fringing reef stage. Barrier reefs are linear or circular reefs separated from the landmass by a well-developed lagoon (Figure 2–30b). As the landmass subsides, the reef maintains its position close to sea level by growing upward. Studies of reef growth rates indicate most have grown 3 to 5 meters (10 to 16 feet) per 1000 years during the recent geologic past. Evidence suggests that some fast-growing reefs in the Caribbean have grown more than 10 meters (33 feet) per 1000 years. Note that if the landmass subsides at a rate faster than coral can grow upward, the coral reef will be submerged in water too deep for it to live.

The largest reef system in the world is Australia's **Great Barrier Reef**, a series of over 3000 individual

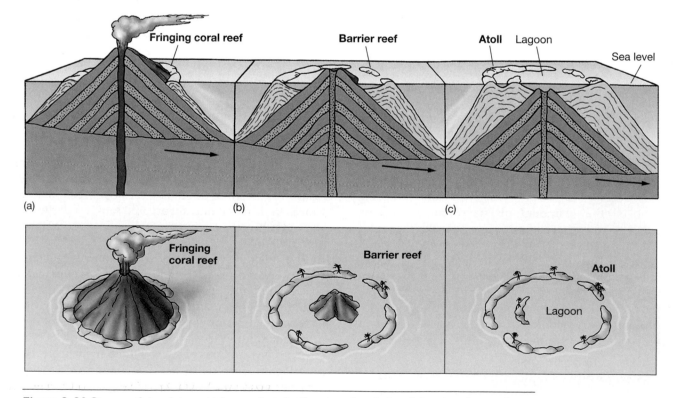

Fringing coral reef — **Barrier reef** — **Atoll** Lagoon — Sea level

(a) (b) (c)

Fringing coral reef — **Barrier reef** — **Atoll** — Lagoon

Figure 2–30 Stages of development in coral reefs. Cross-sectional view (*above*) and map view (*below*) of: **(a)** Fringing reef. **(b)** Barrier reef. **(c)** Atoll.

reefs collectively in the barrier reef stage of development, home to hundreds of coral species and thousands of other reef-dwelling organisms. The Great Barrier Reef lies 40 kilometers (25 miles) or more offshore, averages 150 kilometers (90 miles) in width, and extends for more than 2000 kilometers (1200 miles) along Australia's shallow northeastern coast. The effects of the Indian-Australian plate moving north toward the Equator from colder Antarctic waters are clearly visible in the age and structure of the Great Barrier Reef (Figure 2–31). It is oldest (around 25 million years old) and thickest at its northern end because the northern part of Australia reached water warm enough to grow coral before the southern parts did. In other areas of the Pacific, Indian, and Atlantic Oceans, smaller barrier reefs are found around the tall volcanic peaks that form tropical islands.

The **atoll** (*atar* = to be crowded together) stage (Figure 2–30c) comes after the barrier reef stage. As a barrier reef around a volcano continues to subside, coral builds up toward the surface. After millions of years, the volcano becomes completely submerged, but the coral reef continues to grow. If the rate of subsidence is slow enough for the coral to keep up, a circular reef called an atoll is formed. The atoll encloses a lagoon usually not more than 30 to 50 meters (100 to 165 feet) deep. The reef generally has many channels that allow circulation between the lagoon and the open ocean. Buildups of crushed-coral debris often form narrow islands that encircle the central lagoon (Figure 2–32) and are large enough to allow human habitation.

Detecting Plate Motion with Satellites

Since the late 1970s, orbiting satellites allow the accurate positioning of locations on Earth (this technique is also used for navigation by ships at sea; see Box 1–1). If the plates are moving, satellite positioning should show this movement over time. The map in Figure 2–33 shows locations that have been measured in this manner over a 20-year period. It demonstrates that locations on Earth are moving in good agreement with the *direction* and *rate of motion* predicted by plate tectonics. Successful prediction that locations on Earth are moving with respect to one another very strongly supports the plate tectonic theory.

The Past: Paleoceanography

The study of historical changes of *continental* shapes and positions is called **paleogeography** (*paleo* = ancient, *geo* = earth, *graphy* = the name of a descriptive science). **Paleoceanography** (*paleo* = ancient, *ocean* = the marine environment, *graphy* = the name of a descriptive science) is the study of changes in the physical shape, composition, and character of the *oceans* brought about by paleogeographical changes.

Figure 2–34 is a series of world maps showing the paleogeographical reconstruction of Earth at 60-million-year intervals. At 540 million years ago, many of the present-day continents are barely recognizable. North America was on the Equator and rotated 90 degrees clockwise. Antarctica was on the Equator, and was connected to many other continents.

Between 540 and 300 million years ago, the continents began to come together to form Pangaea. Notice that Alaska had not yet formed. Continents are thought to add material through the process of **continental accretion** (*ad* = toward, *crescere* = to grow). Like adding layers onto a snowball, bits and pieces of continents, islands, and volcanoes are added to the edges of continents and create larger landmasses.

From 180 million years ago to the present, Pangaea separated and the continents moved toward their present-day positions. North America and South America rifted away from Europe and Africa to produce the Atlantic Ocean. In the Southern Hemisphere, South

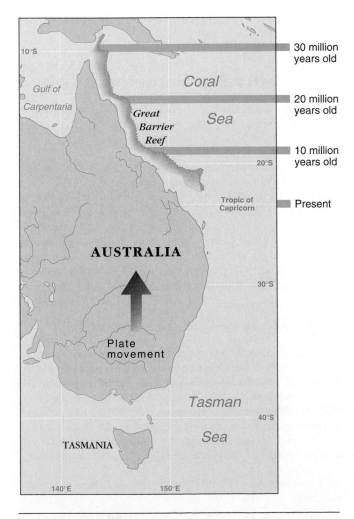

Figure 2–31 Australia's Great Barrier Reef records plate movement. About 30 million years ago, the Great Barrier Reef began to develop as northern Australia moved into tropical waters that allowed coral growth.

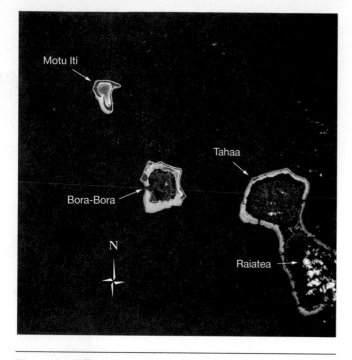

Figure 2–32 Barrier reefs and atoll. A portion of the Society Islands in the Pacific Ocean as viewed from the space shuttle. From lower right, the islands of Raiatea, Tahaa, and Bora-Bora are in the barrier reef stage of development, while Motu Iti is an atoll.

America and a continent composed of India, Australia, and Antarctica begin to separate from Africa.

By 120 million years ago, there was a clear separation between South America and Africa, and India had moved northward, away from the Australia-Antarctica mass, which began moving toward the South Pole. As the Atlantic Ocean continued to open, India moved rapidly northward and collided with Asia about 45 million years ago. Australia had also begun a rapid journey to the north since separating from Antarctica.

One major outcome of global plate tectonic events over the past 180 million years has been the creation of the Atlantic Ocean, which continues to grow as sea floor spreads along the Mid-Atlantic Ridge. At the same time, the Pacific Ocean continues to shrink due to subduction along the many trenches that surround it and continental plates that bear in from both the east and west.

The Future: Some Bold Predictions

Using plate tectonics, a prediction of the future positions of features on Earth can be made based on the assumption that the rate and direction of plate motion will remain the same. Although these assumptions may not be entirely valid, they do provide a framework for the prediction of the positions of continents and other Earth features in the future.

Figure 2–35 is a map of what the world may look like 50 million years from now, showing many notable differences as compared to today. For instance, the east African rift valleys may enlarge to form a new linear sea and the Red Sea may be greatly enlarged from rifting there. India may continue to plow into Asia, further uplifting the Himalaya Mountains. As Australia moves north toward Asia, it may use New Guinea like a snowplow to accrete various islands. North America and South America may continue to move west, enlarging the Atlantic Ocean but decreasing the size of the Pacific Ocean. The land bridge of Central America may no longer connect North and South America, which would dramatically alter ocean circulation. Lastly, the thin sliver of land that lies west of the San Andreas Fault may become an island in the North Pacific, soon to be accreted onto southern Alaska.

> The geographic positions of our continents and ocean basins are not fixed in time or place. Rather, they have changed in the past and will continue to change in the future.

 ## Students Sometimes Ask

Plate tectonics is interesting, but it is still just a theory. Are there other scientific ideas about the features on Earth for which there is better proof?

No, and don't discount the concept of plate tectonics because it is called a theory! Remember that scientific theories are supported by a wealth of evidence (some of which is presented in this chapter), are rigorously tested, serve as good working models for observable phenomena, and have been used to successfully predict occurrences. Plate tectonic theory is considered to be one of humankind's greatest scientific findings—along with physics' model of the atom, chemistry's periodic law, astronomy's Big Bang theory, and biology's theory of evolution.

How long has plate tectonics been operating on Earth?

It's difficult to say with much certainty because our planet has been so dynamic over the past several billion years. Evidence to support any conclusion about plate tectonics before 180 million years ago has been largely destroyed by ongoing subduction of the sea floor. However, very old marine rock sequences uplifted onto continents show certain characteristics of plate motion and suggest that plate tectonics has been operating for at least the last three billion years of Earth history.

Will plate tectonics eventually "turn off" and cease to operate on Earth?

Because plate tectonic processes are powered by heat released from within Earth (which is of a finite amount), it is likely the forces someday will decrease until plates will no longer move. However, the erosional work of water will

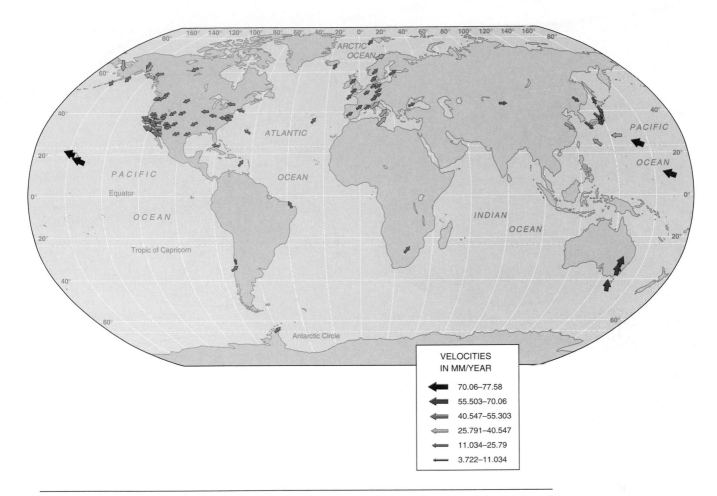

Figure 2–33 Satellite positioning of locations on Earth. Arrows show direction of motion based on satellite measurement of positions on Earth during the period 1979–1997. Rate of plate motion in millimeters per year is indicated with different colors.

continue to erode Earth's features. What a different world it will be then—an Earth with no earthquakes, no volcanoes, and no mountains. Flatness will prevail!

How fast do plates move, and have they always moved at the same rate?

Currently, plates move an average of 2 to 12 centimeters (1 to 5 inches) per year, which is about as fast as a person's fingernails grow. A person's fingernail growth is dependent on many factors, including heredity, gender, diet, and amount of exercise, but averages about 8 centimeters (3 inches) per year. This may not sound very fast, but the plates have been moving for millions of years. Even an object moving slowly will eventually travel a great distance over a very long time. For instance, fingernails growing at a rate of 8 centimeters (3 inches) per year for one million years would be 80 kilometers (50 miles) long!

Evidence shows the plates were moving faster millions of years ago. Geologists can determine the rate of plate motion in the past by analyzing the width of new oceanic crust produced by sea floor spreading since fast spreading produces more sea floor rock. (Using this relationship and by looking at Figure 2–12, you should be able to determine whether the Pacific Ocean or the Atlantic Ocean has a faster spreading rate.) Recent studies using this same technique

indicate that about 50 million years ago, India attained a speed of 19 centimeters (7.5 inches) per year. Other research indicates that about 530 million years ago, plate motions may have been as high as 30 centimeters (1 foot) per year! What caused these rapid bursts of plate motion? Geologists are not sure why plates moved more rapidly in the past, but increased heat release from Earth's interior is a likely mechanism.

If the continents move around on Earth's surface, do other features, such as the mid-ocean ridge, also move?

That's a good observation, and, yes, they do! In fact, nothing is permanently fixed in place on Earth's surface. When we talk about movement of features on Earth, we must consider the question, "Moving relative to what?" The mid-ocean ridge has been shown to move relative to the continents, which sometimes causes segments of the mid-ocean ridges to be subducted! In addition, the mid-ocean ridge is moving relative to a fixed location outside Earth. This means that an observer orbiting Earth would notice, after only a few million years, that most continental and sea floor features—even plate boundaries—are moving. Hotspots are the exception. They seem to be relatively stationary and can be used to determine the relative motions of other features.

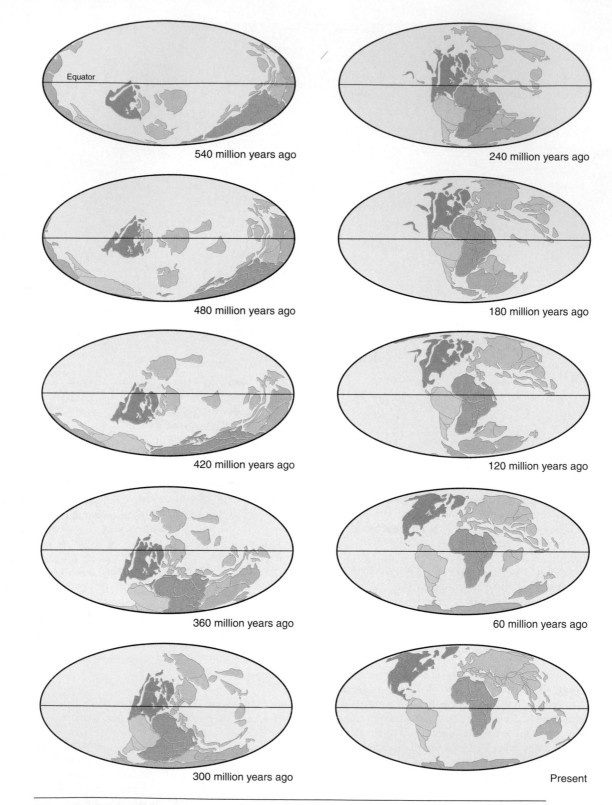

Figure 2–34 Paleogeographic reconstructions of Earth. The positions of the continents at 60-million-year intervals.

What causes Earth's magnetic field?

Studies of Earth's magnetic field and research in the field of *magnetodynamics* suggest that convective movement of fluids in Earth's liquid iron-nickel outer core is the cause of Earth's magnetic field. The most widely accepted view is that the core behaves like a self-sustaining *dynamo*, which converts the energy in convective motion into magnetic energy. Interestingly, most other planets (and even some planet's moons) have magnetic fields. The Sun has a strong magnetic field, which reverses its polarity about every 22 years and is closely tied to the Sun's 11-year sunspot cycle.

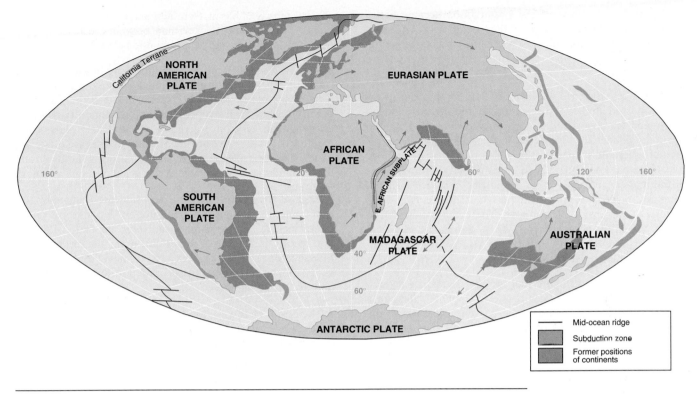

Figure 2–35 The world as it may look 50 million years from now. Purple shadows indicate the present-day positions of continents and tan shading indicates positions of the continents in 50 million years. Green arrows indicate direction of plate motion.

Will the continents come back together and form a single land-mass any time soon?

Yes, it is very likely that the continents will come back together, but not any time soon. Since all of the continents are on the same planetary body, a continent can travel only so far before it collides with other continents. Recent research suggests that the continents may form a supercontinent once every 500 million years or so. Since it has been 200 million years since Pangaea split up, we have only about 300 million years to establish world peace!

Can earthquakes be predicted?

Earthquake prediction—in particular short-term prediction—is a notoriously difficult task! Despite extensive efforts in several countries during the last several decades, scientists are no better at accurately predicting earthquakes. However, plate tectonics has given scientists a workable model to use in understanding why earthquakes occur in certain areas of the world but only rarely in others. By carefully analyzing the past seismic record, seismologists can then make a *forecast*, which is based on probability. A forecast is stated as the likelihood that a certain-magnitude earthquake will occur in a particular area over a certain period of time. For example, there is a 70% probability that a $M_w = 7.0$ earthquake will occur along a segment of the San Andreas Fault over the next 30 years.

Chapter in Review

• According to the theory of plate tectonics, the outermost portion of Earth is composed of a patchwork of thin, rigid lithospheric plates that move horizontally with respect to one another. The idea began as a hypothesis called continental drift proposed by Alfred Wegener at the start of the 20th century. He suggested that about 200 million years ago, all the continents were combined into one large continent (Pangaea) surrounded by a single large ocean (Panthalassa).

• Many lines of evidence were used to support the idea of continental drift, including the similar shape of nearby conti-nents, matching sequences of rocks and mountain chains, glacial ages and other climate evidence, and the distribution of fossil and present-day organisms. Although this evidence suggested that continents have drifted, other incorrect assumptions about the mechanism involved caused many geologists and geophysicists to discount this hypothesis throughout the first half of the twentieth century.

• More convincing evidence for drifting continents was introduced in the 1960s when paleomagnetism—the study of Earth's ancient magnetic field—was developed and the significance of features of the ocean floor became better known. The paleomagnetism of the ocean floor is permanently recorded in oceanic crust and reveals stripes of normal and reverse magnetic polarity in a symmetric pattern relative to the mid-ocean ridge.

• Harry Hess advanced the idea of sea floor spreading. New sea floor is created at the crest of the mid-ocean ridge and moves apart in opposite directions and is eventually destroyed by subduction into an ocean trench. This helps explain the pattern of magnetic stripes on the sea floor and why sea floor rocks increase linearly in age in either direction from the axis of the mid-ocean ridge. Other supporting evidence for plate tectonics includes oceanic heat flow measurements and the pattern of worldwide earthquakes. The combination of evidence convinced geologists of Earth's dynamic nature and helped advance the idea of continental drift into the more encompassing plate tectonic theory.

• Studies of Earth structure indicate that Earth is a layered sphere, with the brittle plates of the lithosphere riding on a plastic, high-viscosity asthenosphere. Near the surface, the lithosphere is composed of continental and oceanic crust. Continental crust consists mostly of granite, but oceanic crust consists mostly of basalt. Continental crust is lower in density, lighter in color, and thicker than oceanic crust. Both types float isostatically on the denser mantle below.

• As new crust is added to the lithosphere at the mid-ocean ridge (divergent boundaries where plates move apart), the opposite ends of the plates are subducted into the mantle at ocean trenches or beneath continental mountain ranges such as the Himalayas (convergent boundaries where plates come together). Additionally, oceanic ridges and rises are offset and plates slide past one another along transform faults (transform boundaries where plates slowly grind past one another).

• Tests of the plate tectonic model indicate that many features and phenomena provide support for shifting plates. These include mantle plumes and their associated hotspots that record the motion of plates past them, the origin of flat-topped tablemounts, stages of coral reef development, and the detection of plate motion by accurate positioning of locations on Earth using satellites.

• The positions of various sea floor and continental features have changed in the past, continue to change today, and will look very different in the future.

Key Terms

Alvin (p. 34)

Andesite (p. 55)

Asthenosphere (p. 49)

Atoll (p. 65)

Ballard, Robert (p. 34)

Barrier reef (p. 64)

Basalt (p. 50)

Bathysphere (p. 34)

Brittle (p. 49)

Continental accretion (p. 65)

Continental arc (p. 55)

Continental crust (p. 50)

Continental drift (p. 35)

Continental transform fault (p. 59)

Convection cell (p. 43)

Convergent boundary (p. 52)

Coral reef (p. 63)

Core (p. 49)

Crust (p. 49)

Darwin, Charles (p. 63)

Divergent boundary (p. 52)

East Pacific Rise (p. 53)

Fringing reef (p. 63)

Granite (p. 50)

Great Barrier Reef (p. 64)

Guyot (p. 62)

Hawaiian Islands-Emperor Seamount chain (p. 61)

Heat flow (p. 46)

Hess, Harry (p. 43)

Hotspot (p. 61)

Ice age (p. 37)

Ice sheet (p. 37)

Igneous rock (p. 40)

Inner core (p. 49)

Island arc (p. 55)

Isostatic adjustment (p. 50)

Isostatic rebound (p. 51)

Kaiko (p. 34)

Lava (p. 40)

Lithosphere (p. 35)

Loihi (p. 62)

Magma (p. 40)

Magnetic anomaly (p. 43)

Magnetic dip (p. 40)

Magnetic field (p. 40)

Magnetic inclination (p. 40)

Magnetite (p. 40)

Magnetometer (p. 42)

Mantle (p. 49)

Mantle plume (p. 61)

Matthews, Drummond (p. 44)

Mesosaurus (p. 38)

Mesosphere (p. 49)

Mid-Atlantic Ridge (p. 43)

Mid-ocean ridge (p. 43)

Nematath (p. 62)

Ocean trench (p. 43)

Oceanic crust (p. 50)

Oceanic ridge (p. 53)

Oceanic rise (p. 53)

Oceanic transform fault (p. 59)

Outer core (p. 49)

Paleoceanography (p. 65)

Paleogeography (p. 65)

Paleomagnetism (p. 65)

Pangaea (p. 35)

Panthalassa (p. 35)

Plastic (p. 49)

Plate tectonics (p. 35)

Polar wandering curve (p. 40)

Polarity (p. 42)

Rift valley (p. 53)

Rifting (p. 53)

San Andreas Fault (p. 60)

Sea floor spreading (p. 43)

Seamount (p. 62)

Sedimentary (p. 40)

Seismic moment magnitude (p. 54)

Shinkai 6500 (p. 34)

Silicate minerals (p. 49)

Spreading center (p. 43)

Subduction (p. 44)

Subduction zone (p. 44)

Submarine (p. 34)

Submersible (p. 34)

Tablemount (p. 62)

Tethys Sea (p. 35)

Transform boundary (p. 35)

Transform faulting (p. 60)

Transform fault (p. 59)

Vine, Frederick (p. 44)

Viscosity (p. 50)

Volcanic arc (p. 55)

Wegener, Alfred (p. 35)

Questions And Exercises

1. Describe three different types of submersibles and the differences between them.

2. When did the supercontinent of Pangaea exist? What was the ocean called that surrounded the supercontinent?

3. Cite the lines of evidence Alfred Wagener used to support his idea of continental drift. Why did scientists doubt that continents drifted?

4. Describe Earth's magnetic field, including how it has changed through time.

5. Describe how sea turtles use Earth's magnetic field for navigation.

6. Why was the pattern of alternating reversals of Earth's magnetic field as recorded in the sea floor rocks such an important piece of evidence for advancing plate tectonics?

7. Describe sea floor spreading and why it was an important piece of evidence in support of plate tectonics.

8. Describe the general relationships that exist among distance from the spreading centers, heat flow, age of the ocean crustal rock, and ocean depth.

9. Why does a map of worldwide earthquakes closely match the locations of worldwide plate boundaries?

10. If you could travel back in time with three figures (illustrations) from this chapter to help Alfred Wegener convince the scientists of his day that continental drift does exist, what would they be and why?

11. Discuss how the chemical composition of Earth's interior differs from its physical properties. Include specific examples.

12. What are differences between the lithosphere and the asthenosphere?

13. Describe differences between granite and basalt. Which property is responsible for making the granitic crust "float" higher in the mantle?

14. Describe a possible mechanism responsible for moving the continents over Earth's surface, including a discussion of the lithosphere and asthenosphere.

15. List and describe the three types of plate boundaries. Include in your discussion any sea floor features that are related to these plate boundaries, and include a real-world example of each. Construct a map view and cross section showing each of the three boundary types and direction of plate movement.

16. Most lithospheric plates contain both oceanic- and continental-type crust. Use plate boundaries to explain why this is true.

17. Describe the differences between oceanic ridges and oceanic rises. Include in your answer why these differences exist.

18. Convergent boundaries can be divided into three types based on the type of crust contained on the two colliding plates. Compare and contrast the different types of convergent boundaries that result from these collisions.

19. Describe the difference in earthquake magnitudes that occur between the three types of plate boundaries, and include why these differences occur.

20. How can plate tectonics be used to help explain the difference between a seamount and a tablemount?

21. How is the age distribution pattern of the Emperor Seamount and Hawaiian Island chains explained by the position of the Hawaiian hotspot? What could have caused the curious bend in between the two chains of seamounts/islands?

22. Describe the differences in origin between the Aleutian Islands and the Hawaiian Islands. Provide evidence to support your explanation.

23. What are differences between a mid-ocean ridge and a hotspot?

24. When does evidence for the following events first show up in the geologic record? (see Figure 2–34).

    a. North America lies on the Equator.

    b. The continents come together as Pangaea.

    c. The North Atlantic Ocean opens.

    d. India separates from Antarctica.

25. Assuming that a continent moves at a rate of 10 centimeters per year, how long would it take the continent to travel from your present location to a nearby large city?

References

Allmendinger, E. 1982. Submersibles: Past-present-future. *Oceanus* 25:1, 18–35.

Apperson, K. D. 1991. Stress fields of the overriding plate at convergent margins and beneath active volcanic arcs. *Science* 254:5032, 670–678.

Bercovici, D., Schubert, G., and Glatzmaier, G. A. 1989. Three-dimensional spherical models of convection in the earth's mantle. *Science* 244:4907, 950–954.

Bolt, B. 1993. *Earthquakes*. New York: W. H. Freeman and Co.

Bullard, Sir Edward. 1969. The origin of the oceans. *Scientific American* 221:66–75.

Burnett, M. S., Caress, D. W., and Orcutt, J. A. 1989. Tomographic image of the magma chamber at 12°50'N on the East Pacific Rise. *Nature* 339:206–208.

Carrigan, C. R., and Gubbins, D. 1979. The source of the earth's magnetic field. *Scientific American* 240:2, 118–133.

Davies, P. J., Symonds, P. A., Feary, D. A., and Pigram, C. J. 1987. Horizontal plate motion: A key allocyclic factor in the evolution of the Great Barrier Reef. *Science* 238:1697–1700.

Dietz, R. S., and Holden, J. C. 1970. Reconstruction of Pangaea: Breakup and dispersion of continents, Permian to present. *Journal of Geophysical Research* 75:26, 4939–4956.

———. 1970. The breakup of Pangaea. *Scientific American* 223:4 30–41.

Hamilton, W. B. 1988. Plate tectonics and island arcs. *Bulletin of the Geological Society of America* 100:1503–1526.

Hess, H. H. 1962. History of ocean basins. In Engel, A. E. J., Lames, H. L., and Leonard, B. F., eds., *Petrologic studies: A volume to honor A. F. Buddington*, 559–620. New York: Geological Society of America.

Keller, R.A., Fisk, M. R., and White, W. M. 2000. Isotopic evidence for Late Cretaceous plume–ridge interaction at the Hawaiian hotspot. *Nature* 405:6787, 673–676.

Kious, W. J., and Tilling, R. I. 1996. *This dynamic Earth: The story of plate tectonics*. Washington D.C.: U.S. Geological Survey.

Liu, M., Yuen, D. A., Zhao, W., and Honda, S. 1991. Development of diapiric structures in the upper mantle due to phase transitions. *Science* 252:1836–1839.

Lohmann, K. J., and Lohmann, C. M. F. 1996. Detection of magnetic field intensity by sea turtles. *Nature* 380:59–61.

Lowman, P. et. al. 1999. A digital tectonic activity map of the earth. *Journal of Geoscience Education* 47:5 428–437.

Marx, R. F. 1978. An illustrated history of submersibles: From Alexander the Great to Alvin. *Oceans* 11:6, 22–29.

Menard, H. W. 1986. *The ocean of truth: A personal history of global tectonics*. Princeton, NJ: Princeton University Press.

Olson, P., Silver, P. G., and Carlson, R. W. 1990. The large-scale structure of convection in the earth's mantle. *Nature* 344:209–214.

Schmidt, V. A. 1986. *Planet Earth and the new geoscience*. DuBuque, IA: Kendall/Hunt.

Sclater, J. G., Parsons, B., and Jaupart, C. 1981. Oceans and continents: Similarities and differences in the mechanisms of heat loss. *Journal of Geophysical Research* 86:11, 535–552.

Scotese, C. R. 1997. *Phanerozoic plate tectonic reconstructions*, PALEOMAP Progress Report #36 (Edition 7), Dept. of Geology. Arlington: University of Texas.

Seachrist, L. 1996. Sea turtles master migration with magnetic memories. In Pirie, R. G., ed., *Oceanography: Contemporary Readings in Ocean Sciences*, 3rd ed. New York: Oxford University Press.

Shepard, F. P. 1977. *Geological oceanography*. New York: Crane, Russak.

Smith, D. K., and Cann, J. R. 1993. Building the crust at the Mid-Atlantic Ridge. *Nature* 365:6448, 707–715.

Sullivan, W. 1991. *Continents in motion: The new earth debate*, 2nd ed. New York: American Institute of Physics

Tarbuck, E. J., and Lutgens, F. K. 1999. *The earth: An introduction to physical geology*, 6th ed. Upper Saddle River, NJ: Prentice-Hall.

———. 1997. *Earth science*, 8th ed. Upper Saddle River, NJ: Prentice-Hall.

The Open University Course Team. 1989. *The ocean basins: Their structure and evolution*. Oxford: Pergamon Press.

van der Hilst, R., Engdahl, R., Spakman, W., and Nolet, G. 1991. Tomographic imaging of subducted lithosphere below northwest Pacific island arcs. *Nature* 353:37–42.

Various authors. 1992. Special issue featuring several articles focusing on mid-ocean ridges. *Oceanus* 34:4, 1–111.

Various authors. 1993. Special issue featuring several articles on marine geology and geophysics. *Oceanus* 35:4, 1–104.

Walker, M. M., et al., 1997. Structure and function of the vertebrate magnetic sense. *Nature* 390:6658, 371–376.

Wessel, P., and Kroenke, L. 1997. A geometric technique for relocating hotspots and refining absolute plate motions. *Nature* 387:365–369.

Wiltschko, R., and Wiltschko, W. 1995. *Magnetic orientation in animals*. Berlin: Springer-Verlag.

Wolfe, C. J., Bjarnason, I. T., VanDecar, J. C., and Solomon, S. C. 1997. Seismic structure of the Iceland mantle plume. *Nature* 385:245–247.

Wynn, C. M., and Wiggins, A. W. 1997. *The five biggest ideas in science*. New York: Wiley.

Suggested Reading in **Scientific American**

Bloxham, I., and Bubbins, D. 1989. The evolution of the earth's magnetic field. 261:6, 68–75. The theory of how Earth's magnetic field is associated with motions in Earth's liquid outer core.

Bonatti, E. 1994. The earth's mantle below the oceans. 270:3, 44–51. An analysis of how sea floor spreading processes affect such diverse phenomena as the shape of the mid-ocean ridge, the composition of rocks along the mid-ocean ridge, the migratory behavior of sea turtles, and even the spin axis of Earth.

Bonatti, E., and Crane, K. 1984. Oceanic fracture zones. 250:5, 40–51. The role of oceanic fracture zones in the plate tectonics process is related to the spherical shape of Earth.

Burke, K. C., and Wilson, J. T. 1976. Hot spots on the earth's surface. 235:2 46–57. A review of how hotspots create island chains that record plate motion, their role in the breakup of continents, and their importance in failed rifting events.

Dietz, R. S., and Holden, J. C. 1970. The breakup of Pangaea. 223:4 30–41. A discussion of how Pangaea broke up into the separate continents that we know today; includes maps of the positions of the continents through time.

Gurnis, M. 2001. Sculpting the earth from inside out. 284:3, 40–47. The mantle records geologic features such as ancient subduction zones and has areas of hot rising rock called superplumes that cause continents to rise and fall over time.

Hawkes, G. S. 1997. Microsubs go to sea. 277:4, 132–135. Written by the co-inventor of *Deep Flight I*, the author explains the reasoning and engineering behind developing a single-passenger, maneuverable microsub.

Herbert, S. 1986. Darwin as a geologist. 254:5, 116–123. Before publication of *The Origin of Species*, Charles Darwin considered himself to be primarily a geologist. This article outlines his contributions in this field, with special attention given to his theory of coral reef formation.

Jeanloz, R., and Lay, T. 1993. The core-mantle boundary. 261:5, 48–55. The boundary between the liquid core that generates Earth's magnetic field and the overlying solid mantle is surprisingly active. Core fluids may rise into the mantle hundreds of meters above the average position of the core-mantle boundary.

Macdonald, K. C., and Fox, P. J. 1990. The mid-ocean ridge. 262:6, 72–95. An interesting and comprehensive overview of present knowledge of the oceanic ridges and rises is presented.

Pinter, N, and Brandon, M. T. 1997. How erosion builds mountains. 276:4, 74–79. A new look at how erosion can help construct mountains, which includes a good description of isostatic adjustment.

Schneider, D. 1997. Hot-spotting. 276:4 22–24. A new method is presented that may help accurately locate hotspots beneath the ocean floor.

Schneider, D., and Gibbs, W. W. 1997. Flight of fancy. 276:1, 22–24. A critical analysis of whether the new microsub *Deep Flight* I would be a real asset to the oceanographic research community.

White, R. S., and McKenzie, D. 1989. Volcanism at rifts. 261:1. 62–71. An analysis of the different types of volcanic activity at rift valleys along the mid-ocean ridge.

Oceanography on the Web

 Visit the Essentials of Oceanography home page for on-line resources for this chapter. There you will find an on-line study guide with review exercises, and links to oceanography sites to further your exploration of the topics in this chapter. Essentials of Oceanography is at: **http://www.prenhall.com/ thurman** (click on the Table of Contents menu and select this chapter).

CHAPTER
3
Marine Provinces

• How do scientists collect information about the depth and shape of the sea floor?

• What is the difference between a passive and an active continental margin?

• How are submarine canyons formed?

• Where are the deepest areas of the ocean?

• What kinds of features are found at the mid-ocean ridge?

• What are differences between transform faults and fracture zones?

EXPERIMENTS IN UNDERWATER LIVING

The shallow sea floor of the continental shelf, which adjoins the continents, covers about 26 million square kilometers (10 million square miles), an area almost as large as Africa. Because of its shallow depth and abundant food, minerals, and other resources, the continental shelf has often been considered an ideal place to establish permanent underwater living habitats.

Many attempts have been made to advance humankind's settlement of the continental shelf. One of the first trials in underwater living took place in 1962, when Albert Falco and Claude Wesly lived at a depth of 26 meters (85 feet) for one week in **Jacques-Yves Cousteau's** Continental Shelf Station (Conshelf) offshore Marseilles, France in the Mediterranean Sea. Cousteau's Conshelf project culminated with five "aquanauts" staying a month at 11 meters (36 feet) below the surface in the Red Sea. In 1964, the **Sealab** project was initiated by the U.S. Navy to study the feasibility of living underwater for long periods. A U.S. Navy team of four aquanauts remained 11 days at a depth of 59 meters (193 feet) off Bermuda in Sealab I. In 1965, Sealab II (Figure 3A) provided a habitat for three

Figure 3A *Sealab II*. The underwater habitat *Sealab II*, which measured 17.4 meters (57 feet) long by 3.6 meters (12 feet) in diameter.

"At the same time that we are earnest to explore and learn all things, we require that all things be mysterious and unexplorable, that land and sea be infinitely wild, unsurveyed and unfathomed by us because unfathomable."

—*Henry David Thoreau (1854)*

10-person teams to stay 15 days each at a depth of 62 meters (205 feet) off La Jolla, California. This project generated much public interest because it included M. Scott Carpenter, the second U.S. astronaut to orbit Earth, and a trained Navy dolphin named Tuffy who, by responding to signals, could swim errands and reach lost divers.

In all these habitats, the aquanauts made daily dives without incident because the habitats were at the same pressure as the surrounding water. How could these aquanauts stay at these high pressures for so long? Experiments using high-pressure chambers in the late 1950s revealed that certain mixtures of gases increased the length of time divers could stay at depth. It was estimated that divers could stay at high pressure almost indefinitely once they became saturated with these mixtures. In the Sealab project, for instance, the deleterious effects of nitrogen were avoided by replacing it with helium gas within the habitat. However, the helium-rich atmosphere distorted the aquanauts' voices, making them high-pitched and cartoon-like. This pro-

vided some interesting moments, such as when President Lyndon Johnson called to congratulate the aquanauts in Sealab II.

Unfortunately, the Sealab project was hampered by unrealistic goals, poor administration, and questionable safety practices. After many lengthy delays, Sealab III came to a halt in 1969 when aquanaut Berry Cannon died while establishing the habitat at a depth of 186 meters (610 feet) off San Clemente Island, California.

Today, only three small underwater habitats remain functional, all near Key Largo, Florida. They are the Jules' Undersea Lodge, which at a depth of 9 meters (30 feet) is the world's only underwater hotel; the MarineLab Undersea Laboratory, a research and education lab; and Aquarius, a research facility operated by the National Oceanic and Atmospheric Administration (NOAA) and the National Undersea Research Program at the University of North Carolina at Wilmington. The Aquarius accommodates a crew of six at a depth of 19 meters (63 feet) on 10-day missions to study coral reef communities.

Over a century ago, most scientists believed that the ocean floor was completely flat and carpeted with a thick layer of muddy sediment containing little of scientific interest. Further, it was believed that the ocean was deepest somewhere in the middle of the ocean basins. However, as more and more vessels crisscrossed the seas to map the ocean floor and to lay transoceanic cables, scientists found the terrain of the sea floor was highly varied and included deep troughs, ancient volcanoes, submarine canyons, and great mountain chains. It was unlike anything on land and, as it turned out, the deepest parts of the oceans were actually close to land!

As marine geologists and oceanographers began to analyze the features of the ocean floor, they realized that certain features had profound implications not only for the history of the ocean floor, but also for the history of Earth. How could all these remarkable features have formed, and how can their origin be explained? Over long periods of time, the shape of the ocean basins has changed as continents have ponderously migrated across Earth's surface in response to forces within Earth's interior. The ocean basins as they presently exist reflect the processes of plate tectonics (the topic of the previous chapter), which help explain the origin of sea floor features.

Bathymetry

Bathymetry (*bathos* = depth, *metry* = measurement) is the measurement of ocean depths and the charting of the shape or topography (*topos* = place, *graphy* = description of) of the ocean floor. Determining bathymetry involves measuring the vertical distance from the

ocean surface down to the mountains, valleys, and plains of the sea floor.

Bathymetric Techniques

The first recorded attempt to measure the ocean's depth was conducted in the Mediterranean Sea in about 85 B.C. by a Greek named Posidonius. His mission was to answer an age-old question: How deep is the ocean? Posidonius' crew made a **sounding** by letting out nearly 2 kilometers (1.2 miles) of line before the heavy weight on the end of the line touched bottom. Sounding lines were used for the next 2000 years by voyagers the world over to probe the ocean's depths. The standard unit of ocean depth is the **fathom** (*fathme* = outstretched arms[1]) and is equal to 1.8 meters (6 feet).

The first systematic bathymetric measurements of the oceans were made in 1872 aboard the HMS *Challenger* during its historic 3$\frac{1}{2}$ year voyage. Every so often, *Challenger*'s crew stopped and measured the depth, along with many other ocean properties. These measurements indicated that the deep ocean floor was not flat but had significant *relief* (variations in elevation), just as dry land does. However, determining bathymetry by making occasional soundings rarely gives a complete picture of the ocean floor. For instance, imagine trying to determine

---

[1]This term is derived from how depth sounding lines were brought back on board a vessel by hand. While hauling in the line, the workers counted the number of arm-lengths collected. By measuring the length of the person's outstretched arms, the amount of line taken in could be calculated. Much later, the distance of 1 fathom was standardized to equal exactly 6 feet.

Figure 3–1 An echosounder record.
An echosounder record from the east coast U.S. shows the provinces of the sea floor. Vertical exaggeration (the amount of expansion of the vertical scale) is 12 times. The scattering layer probably represents a concentration of marine organisms.

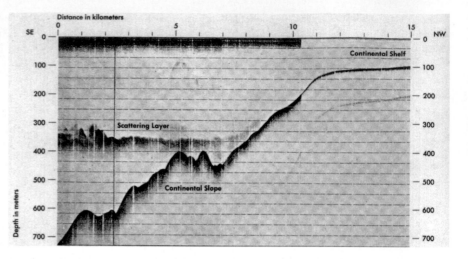

what the surface features on land look like while flying in a blimp at an elevation of several kilometers on a foggy night using only a long weighted rope to determine your height above the surface. This is analogous to how bathymetric measurements were collected from ships using sounding lines.

It wasn't until the early 1900s that ships began to use **echosounders** to more clearly delineate ocean floor features. An echosounder sends a sound signal (called a **ping**) into the ocean to produce echoes when the sound bounces off any density difference, such as marine organisms or the ocean floor (Figure 3–1). Water is a good transmitter of sound, so the time it takes for the echoes to return[2] is used to determine the depth and corresponding shape of the ocean floor. In 1925, for example, the German vessel *Meteor* used echosounding to identify a mountain range running through the center of the South Atlantic Ocean.

Echosounding, however, lacks detail and often gives an inaccurate view of the relief of the sea floor. For instance, the sound beam emitted from a ship 4000 meters (13,100 feet) above the ocean floor widens to a diameter of about 4600 meters (15,000 feet) at the bottom. Consequently, the first echoes to return from the bottom are usually from the closest (highest) peak within this broad area. Nonetheless, most of our knowledge of ocean bathymetry has been provided by the echosounder.

Because sound from echosounders bounces off any density difference, it was discovered that echosounders could detect and track submarines. During World War II, anti-submarine warfare inspired many improvements in the technology of "seeing" into the ocean with sound.

The **precision depth recorder (PDR)**, developed in the 1950s, uses a focused high-frequency sound beam to measure depths to a resolution of about 1 meter (3.3 feet). Throughout the 1960s, PDRs were used extensively and provided a reasonably good representation of the ocean floor. From thousands of research vessel tracks, the first reliable global maps of sea floor bathymetry were produced. These maps helped confirm the ideas of sea floor spreading and plate tectonics.

Today, multibeam echosounders (which use multiple frequencies of sound simultaneously) and side-scan **sonar** (an acronym for **so**und **na**vigation and **r**anging) give oceanographers a more precise picture of the ocean floor. The first multibeam echosounder, **Seabeam**, made it possible for a survey ship to map the features of the ocean floor along a strip up to 60 kilometers (37 miles) wide. The system uses sound emitters directed away from both sides of the ship, with receivers permanently mounted on the hull of the ship. Side-scan sonar systems such as **Sea MARC** (**Sea M**apping and **R**emote **C**haracterization) and **GLORIA** (**G**eological **Lo**ng-**R**ange **I**nclined **A**coustical instrument) can be towed behind a survey ship to produce a strip map of ocean floor bathymetry (Figure 3–2). To make a more detailed picture of the ocean floor, a side-scan instrument can be towed behind a ship on a cable so that it "flies" just above the ocean floor.

Although multibeam and side-scan sonar produce fairly detailed bathymetric maps, mapping the sea floor by ship is a time-consuming process. A research vessel must tediously travel back and forth throughout an area (called "mowing the lawn") to produce an accurate map of bathymetric features. A satellite, on the other hand, can observe large areas of the ocean at one time. Consequently, satellites are increasingly used to determine ocean properties. Remarkably, satellite measurements allow the ocean floor to be mapped from space (Box 3–1). Recent United States oceanographic satellite missions and their objectives are shown in Table 3–1.

Oceanographers who want to know about ocean structure beneath the sea floor use strong low-frequency sounds produced by explosions or air guns, as shown in Figure 3–3. These sounds penetrate beneath the sea floor and reflect off the boundaries between different rock or sediment layers, producing **seismic reflection profiles**, which have applications in mineral and petroleum exploration.

---

[2]This technique uses the speed of sound in seawater, which is 1507 meters (4945 feet) per second.

Sending pings of sound into the ocean (echosounding) is commonly used to determine ocean bathymetry. More recently, satellites have also been used to map the sea floor.

=EIO=

For more information and on-line exercises about the Environmental Issue in Oceanography (EIO) "Sonar and Whales," visit the EIO Web site at **http://www.prenhall. com/oceanissues** and select Issue #2.

Provinces of the Ocean Floor

Figure 3–4 illustrates the ocean's **hypsographic** (*hypsos* = height, *graphic* = drawn) **curve**, which shows the relationship between the height of the land and the depth of the oceans. The bar graph (Figure 3–4, *left*) gives the percentage of Earth's surface area at various ranges of elevation and depth. The cumulative curve (Figure 3–4, *right*) gives the percentage of surface area from the highest peaks to the deepest depths of the oceans. Together, they show that 70.8% of Earth's surface is covered by oceans and that the average depth of the ocean is 3729 meters (12,234 feet) while the average height of the land is only 840 meters (2756 feet).

The ocean floor can be divided into three major provinces (Figure 3–5): **continental margins** (shallow-water areas close to continents), **deep-ocean basins** (deep-water areas farther from land), and the **mid-ocean ridge** (shallower areas near the middle of an ocean). Plate tectonic processes are integral to the formation of these provinces. Through the process of sea floor spreading, mid-ocean ridges and deep-ocean basins are created. Elsewhere, as a continent is split apart, new continental margins are formed.

Features of Continental Margins

Passive and Active Continental Margins Continental margins can be classified as either passive or active depending on their proximity to plate boundaries. **Passive margins** (Figure 3–6, *left*) are imbedded within the interior of lithospheric plates and are therefore *not* in close proximity to any plate boundary. Thus, passive margins usually lack major tectonic activity (large earthquakes, eruptive volcanoes, and mountain building). The east coast of the United States, where there is no plate boundary, is an example of a passive continental margin. Passive margins are usually produced by rifting

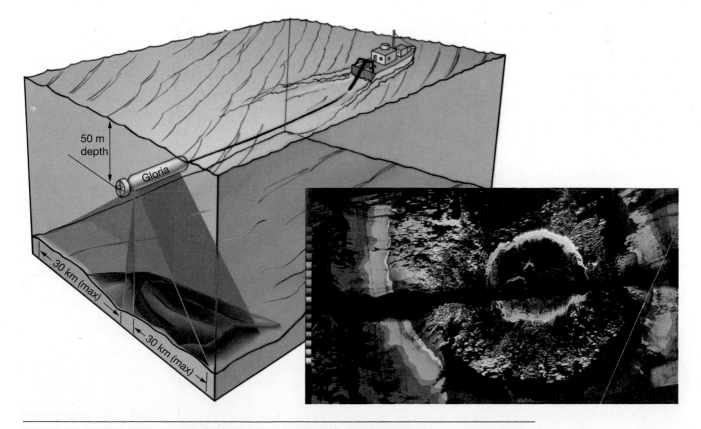

Figure 3–2 Side-scanning sonar. The side-scan sonar system GLORIA (*left*) is towed behind a survey ship and can map a strip of ocean floor (a swath) with a gap in data directly below the instrument. Side-scan sonar image of a volcano (*right*) with a summit crater about 2 kilometers (1.2 miles) in diameter in the Pacific Ocean. Black stripe through middle of image is the data gap.

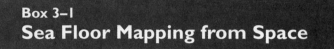

Box 3–1
Sea Floor Mapping from Space

Recently, satellite measurements of the ocean surface have been used to make maps of the sea floor. How does a satellite—which orbits at a great distance above the planet and can only view the ocean's *surface*—obtain a picture of the sea *floor*?

The answer lies in the fact that sea floor features directly influence Earth's gravitational field. Deep areas such as trenches correspond to a lower gravitational attraction and large undersea objects such as seamounts exert an extra gravitational pull. These differences affect sea level, causing the ocean surface to bulge outward and sink inward mimicking the relief of the ocean floor. A 2000-meter (6500-foot)-high seamount, for example, exerts a small but measurable gravitational pull on the water around it, creating a bulge 2 meters (7 feet) high on the ocean surface. These irregularities are easily detectable by satellites, which use microwave beams to measure sea level to within 4 centimeters (1.5 inches) accuracy. After corrections are made for waves, tides, currents, and atmospheric effects, the resulting pattern of lumps and bulges at the ocean surface can be used to indirectly reveal ocean floor bathymetry (Figure 3B). For example, Figure 3C compares two different maps of the same area: one based on bathymetric data from ships (*left*) and the other based on satellite measurements (*right*), which shows a much higher resolution.

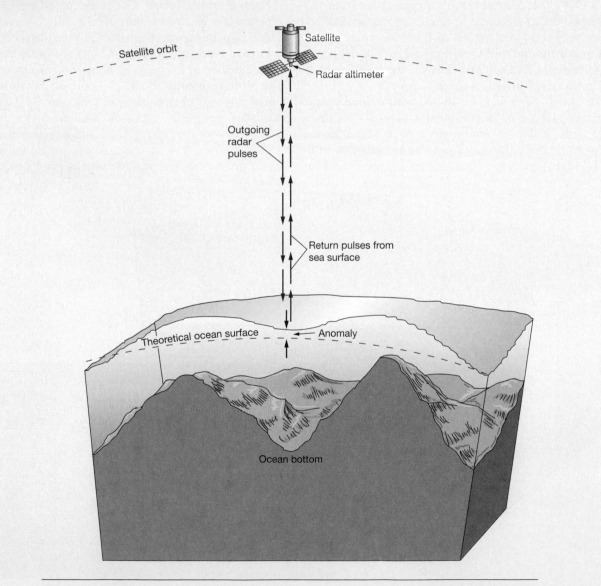

Figure 3B Satellite measurements of the ocean surface. A satellite measures the variation of ocean surface elevation, which is caused by gravitational attraction and mimics the shape of the sea floor. The sea surface anomaly is the difference between the measured and theoretical ocean surface.

Data from the European Space Agency's ERS-1 satellite and from Geosat, a U.S. Navy satellite, were collected during the 1980s. When this information was recently declassified, Walter Smith of the National Oceanic and Atmospheric Administration and David Sandwell of Scripps Institution of Oceanography began producing sea floor maps based on the shape of the sea surface. What is unique about these researchers' maps is that they provide a view of Earth similar to being able to drain the oceans and view the ocean floor

directly. Their map of ocean surface gravity (Figure 3D) uses depth soundings to calibrate the gravity measurements. Although gravity is not exactly bathymetry, this new map of the ocean floor clearly delineates many ocean floor features, such as the mid-ocean ridge, trenches, seamounts, and nemataths (island chains). In addition, this new mapping technique has revealed sea floor bathymetry in areas where research vessels have not conducted surveys.

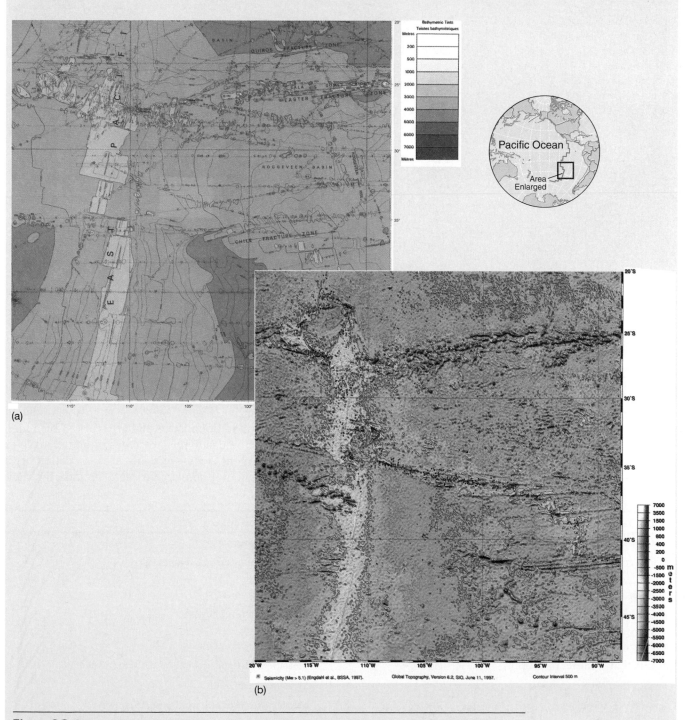

Figure 3C Comparing maps of the sea floor. Both maps show the same portion of the East Pacific Rise. **(a)** A map made using conventional echosounder records. **(b)** A map from satellite data made using measurements of the ocean surface. Red dots are earthquake locations.

(continued)

(continued)

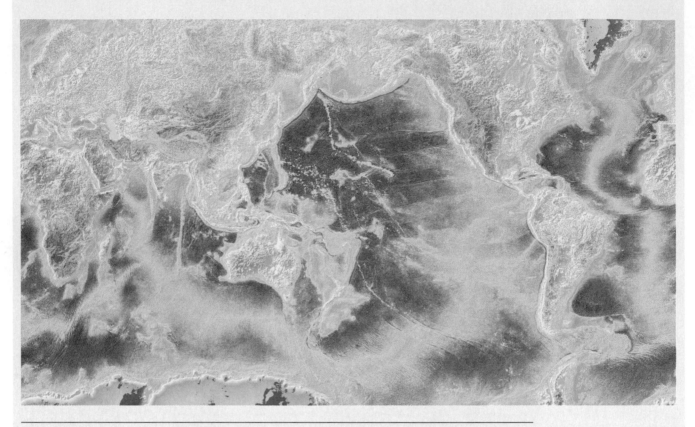

Figure 3D Global sea surface elevation map. Map showing the global gravity field, which, when adjusted using measured depths, closely corresponds to ocean depth. Light green color indicates shallowest water; purple indicates deep water. Map also shows land surface elevations, with yellow color indicating low elevations and light pink color indicating high elevations.

of continental landmasses and continued sea floor spreading over geologic time. Features of passive continental margins include the continental shelf, the continental slope, and the continental rise that extends toward the deep-ocean basins (Figures 3–6 and 3–7).

Active margins (Figure 3–6, *right*) are associated with lithospheric plate boundaries and are marked by a high degree of tectonic activity. Two types of active margins exist. **Convergent active margins** are associated with oceanic-continental convergent plate boundaries. Fea-

Table 3–I   Recent oceanographic satellite missions sponsored by the United States.

Satellite	*Dates*	*Mission objective*
Seasat A	1978 (3 months)	Measure ocean surface temperature
Nimbus-7/Coastal Zone Color Scanner (CZCS)	1978–1986	Measure ocean color
Geosat	1985–1989	Measure sea level heights
TOPEX/Poseidon (joint mission with France)	1992–	Measure sea level heights
NASA Scatterometer (NSCAT)	1996–1997	Measure atmospheric winds and ocean waves
SeaStar/SeaWiFS	1997–	Measure ocean color
QuikSCAT/SeaWinds	1999–	Measure atmospheric winds and ocean waves
Terra (flagship of NASA's Earth Observing System)	1999–	Monitor Earth's climate, including measuring ocean temperature, productivity, and ice cover
Jason-1 (joint mission with France)	2001–	Replace TOPEX/Poseidon; measure sea level heights

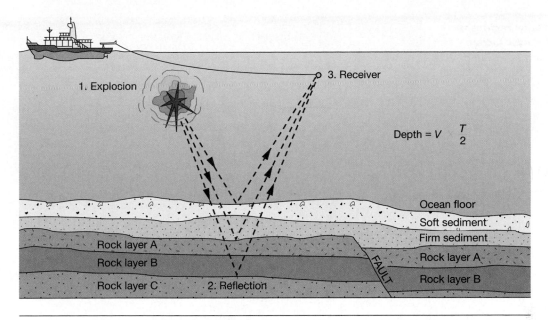

Figure 3–3 Seismic profiling. An air-gun explosion emits low frequency sounds (1) that can penetrate bottom sediments and rock layers. The sound reflects off the boundaries between these layers (2) and returns to the receiver (3).

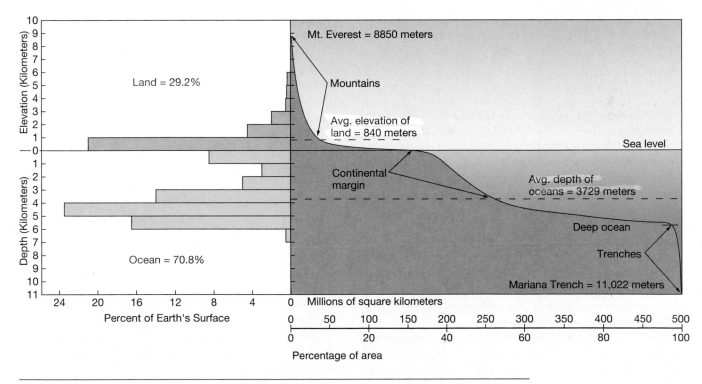

Figure 3–4 The hypsographic curve. The bar graph (*left*) gives the percentage of Earth's surface area at various ranges of elevation and depth. The cumulative hypsographic curve (*right*) gives the percentage of surface area from the highest peaks to the deepest depths of the oceans. Also shown is the average ocean depth and land elevation.

tures include a continental arc onshore, a narrow shelf, a steep slope, and an offshore trench that delineates the plate boundary. Western South America, where the Nazca Plate is being subducted beneath the South American Plate, is an example of a convergent active margin. **Transform active margins** are less common and are associated with transform plate boundaries. At these locations, faults that parallel the transform plate boundary create linear islands, banks (shallowly submerged areas), and deep basins close to shore. Coastal

Figure 3–5 Major regions of the North Atlantic Ocean floor. Map view above and profile view below, showing that the ocean floor can be divided into three major provinces: continental margins, deep ocean basins, and mid-ocean ridge.

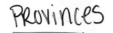

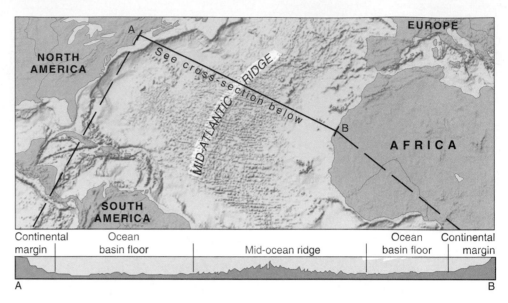

California along the San Andreas Fault is an example of a transform active margin.

> Passive continental margins lack a plate boundary and have different features than active continental margins, which include a plate boundary (either convergent or transform).

Continental Shelf The **continental shelf** is defined as a generally flat zone extending from the shore beneath the ocean surface to a point at which a marked increase in slope angle occurs, called the **shelf break** (Figure 3–7). It is usually flat and relatively featureless because of marine sediment deposits but can contain coastal islands, reefs, and raised banks. The underlying rock is granitic continental crust, so the continental shelf is geologically part of the continent. The general bathymetry of the continental shelf can usually be predicted by examining the topography of the adjacent coastal region. With few exceptions, the coastal topography extends beyond the shore and onto the continental shelf.

The average width of the continental shelf is about 70 kilometers (43 miles), but it varies from a few tens of meters to 1500 kilometers (930 miles). The broadest shelves occur off the northern coasts of Siberia and North America in the Arctic Ocean. The average depth at which the shelf break occurs is about 135 meters (443 feet). Around the continent of Antarctica, however, the shelf break occurs at 350 meters (2200 feet). The average slope of the continental shelf is only about a tenth of a degree, similar to the slope given to a large parking lot for drainage purposes.

Sea level has fluctuated over the history of Earth, causing the shoreline to migrate back and forth across the continental shelf. When colder climates prevailed during the Ice Age, for example, more of Earth's water

was frozen as glaciers on land, so sea level was lower than it is today. During this time, more of the continental shelf was exposed.

The type of continental margin will determine the shape and features associated with the continental shelf. For example, the east coast of South America has a broader continental shelf than its west coast. The east coast is a passive margin, which typically has a wider shelf. In contrast, the convergent active margin present along the west coast of South America is characterized by a narrow continental shelf and a shelf break close to shore. For transform active margins such as along California, the presence of offshore faults produces a continental shelf that is not flat. Rather, it is marked by a high degree of relief (islands, shallow banks, and deep basins) called a **continental borderland**.

Continental Slope The **continental slope**, which lies beyond the shelf break, is where the deep-ocean basins begin. Total relief in this region is similar to that found in mountain ranges on the continents. The break at the top of the slope may be from 1 to 5 kilometers (0.6 to 3 miles) above the deep-ocean basin at its base. Along convergent active margins where the slope descends into submarine trenches, even greater vertical relief is measured. Off the west coast of South America, for instance, the total relief from the top of the Andes Mountains to the bottom of the Peru-Chile Trench is about 15 kilometers (9.3 miles).

Worldwide, the slope of the continental slopes averages about 4 degrees, but varies from 1 to 25 degrees.[3] A study that compared the average of five different continental slopes in the United States revealed that it is just over 2 degrees. Around the margin of the Pacific Ocean, the continental slopes average more than 5 degrees because of the presence of convergent active

---

[3]For comparison, the windshield of an aerodynamically designed car has a slope of about 25 degrees.

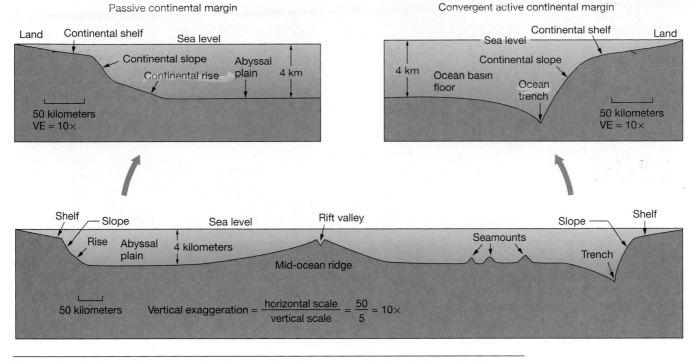

Figure 3–6 Passive and active continental margins. Cross-sectional view (*below*) of typical features across an ocean basin, including a passive continental margin (*left enlargement*) and a convergent active continental margin (*right enlargement*). Vertical exaggeration is 50 times.

margins that drop directly into deep offshore trenches. The Atlantic and Indian Oceans, on the other hand, contain many passive margins, which lack plate boundaries. Thus, the amount of relief is lower and slopes in these oceans average about 3 degrees.

Submarine Canyons and Turbidity Currents The continental slope—and, to a lesser extent, the continental shelf—exhibit **submarine canyons**. Submarine canyons are V-shaped in profile view and have branches or tributaries with steep to overhanging walls (Figure 3–8). They resemble canyons formed on land that are carved by rivers and can be quite large. In fact, the Monterey Canyon off California is comparable in depth, steepness, and length to Arizona's Grand Canyon.

How are submarine canyons formed? Initially it was thought submarine canyons were ancient river valleys created by the erosive power of rivers when sea level was lower and the continental shelf was exposed. Although some canyons are directly offshore from where rivers enter the sea, the majority of them are not. Many, in fact, are confined exclusively to the continental slope. Additionally, submarine canyons continue to the base of the continental slope, which averages some 3500 meters (11,500 feet) below sea level. There is no evidence, however, that sea level has ever been lowered by that much.

Side-scan sonar surveys along the Atlantic coast indicate that the continental slope is dominated by submarine canyons from Hudson Canyon near New York City to Baltimore Canyon in Maryland. Canyons confined to

the continental slope are straighter and have steeper canyon floor gradients than those that extend into the continental shelf. These characteristics suggest the canyons are created on the continental slope by some marine process and enlarge into the continental shelf through time.

Both indirect and direct observation of the erosive power of **turbidity** (*turbidus* = disordered) **currents** (Box 3–2) has suggested they are responsible for carving submarine canyons. Turbidity currents are underwater avalanches of muddy water mixed with rocks and other debris. When sediment moves across the continental shelf into the head of the canyon and accumulates there, turbidity currents may result from shaking by an earthquake, the oversteepening of sediment that accumulates on the shelf, hurricanes passing over the area, or the rapid input of sediment from flood waters. The mass moves down the slope under the force of gravity when set in motion, carving the canyon as it goes, analogous to a flash flood on land. Turbidity currents are strong enough to carry large rocks down submarine canyons and do a considerable amount of erosion over time.

Turbidity currents are underwater avalanches of muddy water mixed with sediment that move down the continental slope and are responsible for carving submarine canyons.

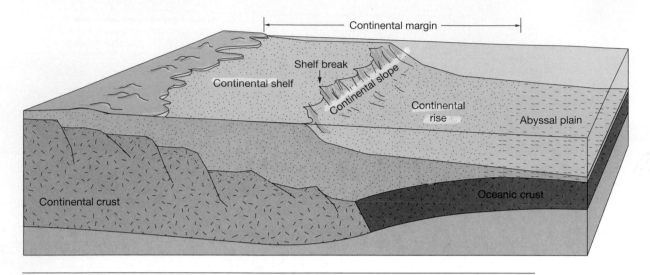

Figure 3–7 Features of a passive continental margin. Schematic view showing the main features of a passive continental margin.

Continental Rise The **continental rise** is a transition zone between the continental margin and the deep-ocean floor comprised of a huge submerged pile of debris. Where did all this debris come from, and how did it get there?

The existence of turbidity currents suggests that the material transported by these currents is responsible for the creation of continental rises. When a turbidity current moves through and erodes a submarine canyon, it exits through the mouth of the canyon. The slope angle decreases and the turbidity current slows, causing suspended material to settle out in a distinctive type of layering called **graded bedding** that *grades in size upwards* (Figure 3–8a, *inset*). As the energy of the turbidity current dissipates, larger pieces settle first, then progressively smaller pieces settle, and eventually even very fine pieces settle out, which may occur weeks or months later.

An individual turbidity current deposits one graded bedding sequence. The next turbidity current may partially erode the previous deposit and then deposit another graded bedding sequence on top of the previous one. After some time, a thick sequence of graded bedding deposits can develop one on top of another. These stacks of graded bedding are called **turbidite deposits**, of which the continental rise is composed.

As viewed from above, the deposits at the mouths of submarine canyons are fan-, lobate-, or apron-shaped, as shown in Figure 3–8a. Consequently, these deposits are called **deep-sea fans** or **submarine fans**. These deep-sea fans create the continental rise when they merge together along the base of the continental slope. Along convergent active margins, however, the steep continental slope leads directly into a deep-ocean trench. Sediment from turbidity currents accumulates in the trench and there is no continental rise.

Features of the Deep-Ocean Basin

The deep-ocean floor lies beyond the continental margin province (the shelf, slope, and the rise).

Abyssal Plains Extending from the base of the continental rise into the deep-ocean basins are flat depositional surfaces with slopes of less than a fraction of a degree that cover extensive portions of the deep-ocean basins. These **abyssal** (*a* = without, *byssus* = bottom) **plains** average between 4500 meters (15,000 feet) and 6000 meters (20,000 feet) deep. They are not literally bottomless, but they are some of the deepest (and flattest) regions on Earth.

Abyssal plains are formed by fine particles of sediment slowly drifting onto the deep-ocean floor. Over millions of years, a thick blanket of sediment is produced by **suspension settling** where fine particles (analogous to "marine dust") accumulate on the ocean floor. With enough time, these deposits cover most irregularities of the deep ocean, as shown in Figure 3–9. In addition, sediment traveling in turbidity currents from land adds to the sediment load.

The type of continental margin determines the distribution of abyssal plains. For instance, few abyssal plains are located in the Pacific Ocean; instead, most occur in the Atlantic and Indian Oceans. The deep-ocean trenches found on the convergent active margins of the Pacific Ocean prevent sediment from moving past the continental slope. In essence, the trenches act like a gutter that traps sediment transported off the land by turbidity currents. On the passive margins of the Atlantic and Indian Oceans, however, turbidity currents travel directly down the continental margin and deposit sediment on the abyssal plains.

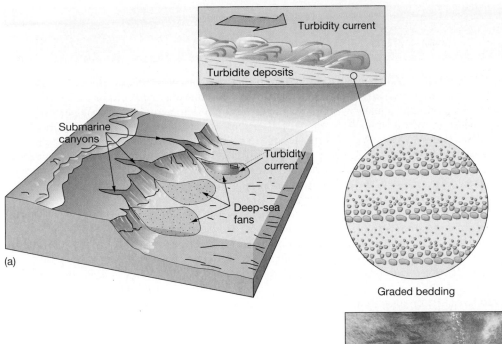

Figure 3–8 Submarine canyons and deep-sea fans. **(a)** Turbidity currents move downslope, eroding the continental margin to enlarge submarine canyons. Deep-sea fans are created by turbidite deposits, which consist of sequences of graded bedding (*inset*). **(b)** A diver descends into La Jolla submarine canyon, offshore California.

Volcanic Peaks of the Abyssal Plains Poking through the sediment cover of the abyssal plains are a variety of volcanic peaks, which extend to various elevations above the ocean floor (see Figure 3–2). Some even extend above sea level to form islands. Those that are below sea level but rise more than 1 kilometer (0.6 mile) above the deep-ocean floor are called seamounts. If they have flattened tops, they are called tablemounts, or guyots. The origin of seamounts and tablemounts was discussed as a piece of supporting evidence for plate tectonics in Chapter 2 (see Figure 2–29).

Volcanic features on the ocean floor that are less than 1000 m (0.6 mile) tall—the minimum height of a seamount—are called **abyssal hills** or **seaknolls**. Abyssal hills are one of the most abundant features on the planet (several hundred thousand have been identified)

and cover a large percentage of the entire ocean basin floor. Many are gently rounded in shape and they have an average height of about 200 meters (650 feet). Many abyssal hills are found buried beneath the sediments of the abyssal plains of the Atlantic and Indian Oceans. In the Pacific Ocean, the abundance of active margins means the rate of sediment deposition is lower. Consequently, extensive regions dominated by abyssal hills have resulted, which are called **abyssal hill provinces**. The evidence of volcanic activity on the bottom of the Pacific Ocean is particularly widespread—more than 20,000 volcanic peaks exist there.

Ocean Trenches Along passive margins, the continental rise commonly occurs at the base of the continental slope and merges smoothly into the abyssal

Box 3–2
A Grand "Break": Evidence for Turbidity Currents

How do earthquakes and telephone cables help explain how turbidity currents move across the ocean floor and carve submarine canyons? In 1929, the Grand Banks earthquake in the North Atlantic Ocean severed some of the trans-Atlantic telephone and telegraph cables that lay across the sea floor south of Newfoundland near the earthquake epicenter (Figure 3E). At first, it was assumed that sea floor movement caused all these breaks. However, analysis of the data revealed that the cables closest to the earthquake broke simultaneously with the earthquake, but cables that crossed the slope and deeper ocean floor at greater distances from the earthquake were broken progressively later in time. It seemed unusual that certain cables were affected by the failure of the slope due to ground shaking, but others were broken several minutes later.

Reanalysis of the pattern several years later suggested that a turbidity current moving down the slope could ac-

count for the pattern of cable breaks. Based on the sequence of breaks, the turbidity current must have reached speeds approaching 80 kilometers (50 miles) per hour on the steep portions of the continental slope, and about 24 kilometers (15 miles) per hour on the more gently sloping continental rise. Thus, turbidity currents reach high speeds and are strong enough to break underwater cables, suggesting that they must be powerful enough to erode submarine canyons.

Further evidence of turbidity currents comes from several studies that have documented turbidity currents using sonar. For instance, a study of Rupert Inlet in British Columbia, Canada, monitored turbidity currents moving through an underwater channel. These studies indicate that submarine canyons are carved by turbidity currents over long periods of time, just as canyons on land are carved by running water.

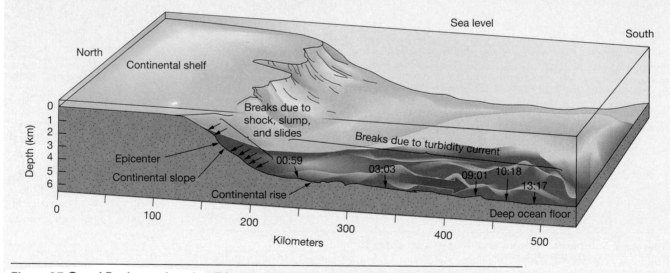

Figure 3E Grand Banks earthquake. Diagrammatic view of the sea floor showing the sequence of events for the 1929 Grand Banks Earthquake. The epicenter is the point on Earth's surface directly above the earthquake.

plain. In convergent active margins, however, the slope descends into a long, narrow, steep-sided **ocean trench**. Ocean trenches are deep linear scars in the ocean floor, caused by the collision of two plates along convergent plate margins (as discussed in Chapter 2). The landward side of the trench rises as a **volcanic arc** that may produce islands (such as the islands of Japan, an **island arc**) or a volcanic mountain range along the margin of a continent (such as the Andes Mountains, a **continental arc**).

The deepest portions of the world's oceans are found in these trenches. In fact, the deepest point on Earth's surface—11,022 meters (36,163 feet)—is found in the Challenger Deep area of the Mariana Trench. The majority of ocean trenches are found along the margins of

the Pacific Ocean (Figure 3–10) while only a few exist in the Atlantic and Indian Oceans. Table 3–2 compares the depth, average width, and length of selected trenches.

The **Pacific Ring of Fire** occurs along the margins of the Pacific Ocean. It has the majority of Earth's active volcanoes and large earthquakes because of the prevalence of convergent plate boundaries along the Pacific Rim. A part of the Pacific Ring of Fire is South America's western coast, including the Andes Mountains and the associated Peru-Chile Trench. Figure 3–11 shows a cross-sectional view of this area, illustrating the tremendous amount of relief at convergent plate boundaries where deep ocean trenches are associated with tall volcanic arcs.

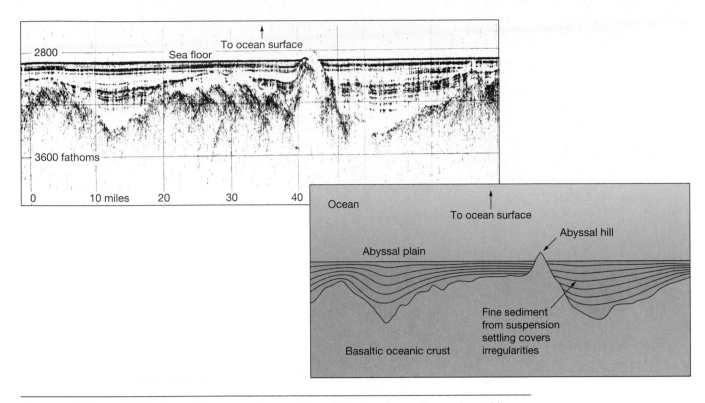

Figure 3–9 Abyssal plain formed by suspension settling. Seismic cross-section and matching drawing across part of the deep Madeira abyssal plain in the eastern Atlantic Ocean showing irregular volcanic terrain buried by sediments.

Deep ocean trenches and volcanic arcs are a result of the collision of two plates at convergent plate boundaries and mostly occur along the margins of the Pacific Ocean (Pacific Ring of Fire).

Features of the Mid-Ocean Ridge

The global mid-ocean ridge is a continuous, fractured-looking mountain ridge that extends through all the ocean basins. The portion of the mid-ocean ridge found in the North Atlantic Ocean is called the Mid-Atlantic Ridge and is shown in Figure 3–12. As discussed in Chapter 2, the mid-ocean ridge results from sea floor spreading along divergent plate boundaries. The mid-ocean ridge forms Earth's longest mountain chain, extending across some 75,000 kilometers (46,600 miles) of the deep-ocean basin. The width of the mid-ocean ridge varies along its length, but averages about 1000 kilometers (620 miles). The mid-ocean ridge is a topographically high feature, extending an average of 2.5 kilometers (1.5 miles) above the surrounding sea floor. In some areas, such as in Iceland, the mid-ocean ridge even extends above sea level. Remarkably, the mid-ocean ridge covers 23% of Earth's surface.

The mid-ocean ridge is entirely volcanic and is composed of basaltic lavas characteristic of the oceanic

crust. Along its crest is a central downdropped **rift valley** (Figure 3–13) created by seafloor spreading (rifting) where two plates diverge. Cracks called **fissures** (*fissus* = split) and faults are commonly observed in the central rift valley. Swarms of small earthquakes occur along the central rift valley caused by underground movement of magma or rifting along faults.

Volcanic features include volcanoes (seamounts) and recent underwater lava flows. When hot basaltic lava spills onto the sea floor, it is exposed to cold seawater that chills the margins of the lava. This creates **pillow lavas** or **pillow basalts**, which are smooth, rounded lobes of rock that resemble a stack of bed pillows (Figure 3–14). Frequent volcanic activity is common along the mid-ocean ridge. For example, bathymetric studies along the Juan de Fuca Ridge off Washington and Oregon revealed that 50 million cubic meters (1800 million cubic feet) of new lava was released sometime between 1981 and 1987 and in 1993, a research vessel was actually on-site during an underwater eruption. Subsequent surveys of the area indicated many changes along the mid-ocean ridge, including new volcanic features, recent lava flows, and depth changes of up to 37 meters (121 feet).

Other features in the central rift valley include hot springs called **hydrothermal** (*hydro* = water, *thermo* = heat) **vents**. Seawater seeps along fractures in the ocean crust and is heated when it comes in contact with underlying magma (Figure 3–15). It then rises back toward

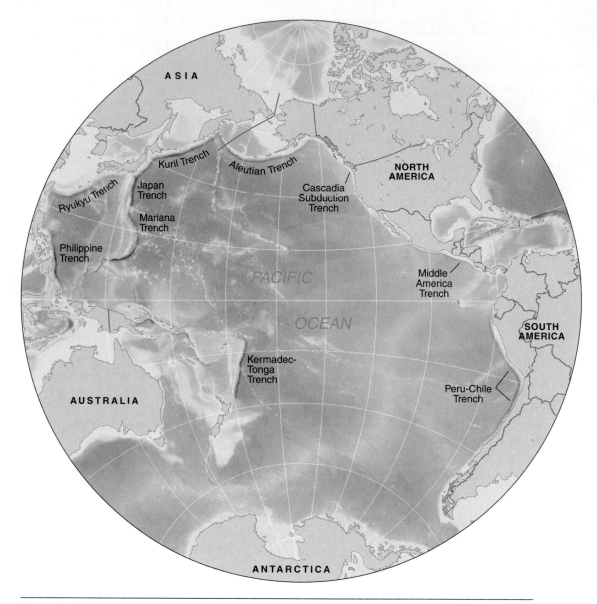

Figure 3–10 Location of ocean trenches. The majority of ocean trenches are along the margins of the Pacific Ocean where plates are being subducted. Most of the world's large earthquakes (due to subduction) and active volcanoes (as volcanic arcs) occur around the Pacific Rim, which is why the area is also called the Pacific Ring of Fire.

Figure 3–11 Profile across the Peru-Chile Trench and the Andes Mountains. Over a distance of 200 kilometers (125 miles), a change in elevation occurs of more than 14,900 meters (49,000 feet). Vertical scale is exaggerated 10 times.

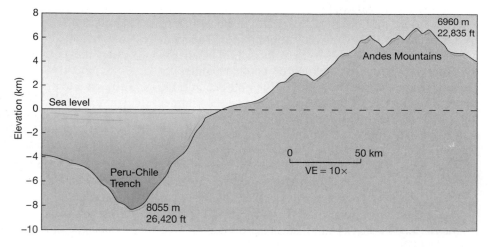

Table 3–2   Dimensions of selected trenches.

Trench	Ocean	Depth (kilometers)	Average width (kilometers)	Length (kilometers)
Middle America	Pacific	6.7	40	2800
Java	Indian	7.5	80	4500
Aleutian	Pacific	7.7	50	3700
Peru-Chile	Pacific	8.0	100	5900
South Sandwich	Atlantic	8.4	90	1450
Japan	Pacific	8.4	100	800
Puerto Rico	Atlantic	8.4	120	1550
Kermadec	Pacific	10.0	40	1500
Tonga	Pacific	10.0	55	1400
Philippine	Pacific	10.5	60	1400
Kuril	Pacific	10.5	120	2200
Mariana	Pacific	11.0	70	2550

Figure 3–12 Floor of the North Atlantic Ocean. The global mid-ocean ridge traverses near the center of the Atlantic Ocean.

the surface and exits through the sea floor. The temperature of the water that rushes out of a particular hydrothermal vent determines its appearance:

- **Warm-water vents** are below 30°C (86°F) and generally emit water clear in color.
- **White smokers** are between 30° and 350°C (86° to 662°F) and emit water that is white because of the presence of various light-colored compounds, including barium sulfide.
- **Black smokers** are above 350°C (662°F) and emit water that is black because of the presence of dark-

colored **metal sulfides** including iron, nickel, copper, and zinc.

Many black smokers spew out of chimney-like structures up to 30 meters (100 feet) high (Figure 3–15b). The dissolved metal particles often come out of solution, or **precipitate**,[4] when the hot water mixes with cold seawater, creating coatings of mineral deposits on nearby rocks. Chemical analyses of these deposits

---

[4]A chemical precipitate is formed whenever dissolved materials change from existing in the dissolved state to existing in the solid state.

Figure 3–13 **Rift valley fissures.** **(a)** A fissure in the rift valley of the Mid-Atlantic Ridge photographed by the submersible *Alvin*. **(b)** A larger fissure in the rift valley of Iceland.

reveal that they are composed of various metal sulfides and sometimes even silver and gold.

In addition, most hydrothermal vents support unusual biological communities, including large clams, mussels, tubeworms, and many other animals—many of which were new to science when they were first discovered. These organisms are able to survive in the absence of sunlight because the vents discharge hydrogen sulfide gas. Archaeons[5] and bacteria oxidize the hydrogen sulfide gas to provide a food source for other organisms in the community. The interesting associations of these organisms are discussed in Chapter 15, "Animals of the Benthic Environment."

Segments of the mid-ocean ridge called **oceanic ridges** have a prominent rift valley and steep, rugged slopes and **oceanic rises** have slopes that are gentler and less rugged. As explained in Chapter 2, the differences in overall shape are caused by the fact that oceanic ridges (such as the Mid-Atlantic Ridge) spread more slowly than oceanic rises (such as the East Pacific Rise).

The mid-ocean ridge is created by plate divergence and typically includes a central rift valley, faults and fissures, seamounts, pillow basalts, hydrothermal vents, and metal sulfide deposits.

[5]Archaeons are microscopic bacteria-like organisms—a newly discovered domain of life.

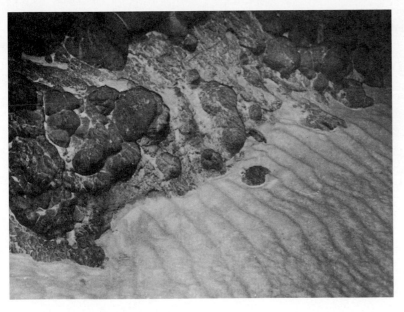

Figure 3–14 Pillow lava. Recently formed pillow lava on the sea floor along the East Pacific Rise. Photo shows an area of the sea floor about 3 meters (10 feet) across that also displays ripple marks from deep ocean currents.

Fracture Zones and Transform Faults The mid-ocean ridge is cut by a number of **transform faults**, which offset the spreading zones. Oriented perpendicular to the spreading zones, transform faults give the mid-ocean ridge the zigzag appearance seen in Figure 3–12. As described in Chapter 2, transform faults occur to accommodate spreading of a linear ridge system on a spherical Earth and because different segments of the mid-ocean ridge spread apart at different rates.

In the Pacific Ocean, where scars are less rapidly covered by sediment than in other ocean basins, transform faults can be seen to extend for thousands of kilometers away from the mid-ocean ridge and have widths of up to 200 kilometers (120 miles). These extensions, however, are not transform faults. Instead, they are **fracture zones**.

What is the difference between a transform fault and a fracture zone? Figure 3–16 shows that both run along the same long linear zone of weakness in Earth's crust. In fact, by following the same zone of weakness from one end to the other, it changes from a fracture zone to a transform fault and back again to a fracture zone. A transform fault is a seismically active area that offsets the axis of a mid-ocean ridge. A fracture zone, on the other hand, is a seismically inactive area that shows evidence of past transform fault activity. A helpful way to visualize the difference is that transform faults occur *between* offset segments of the mid-ocean ridge, while fracture zones occur *beyond* the offset segments of the mid-ocean ridge.

The relative direction of plate motion across transform faults and fracture zones further differentiates these two features. Across a transform fault, two lithospheric plates are moving in opposite directions. Across a fracture zone (which occurs entirely within a plate), there is no relative motion because the parts of the lithospheric plate cut by a fracture zone are moving in the same direction (Figure 3–16). Transform faults are actual plate boundaries, whereas fracture zones are not. Rather, fracture zones are ancient, inactive fault scars embedded in a plate.

In addition, earthquake activity is different in transform faults and fracture zones. Earthquakes shallower than 10 kilometers (6 miles) are common when plates move in opposite directions along transform faults. Along fracture zones, where plate motion is in the same direction, seismic activity is almost completely absent. Table 3–3 summarizes the differences between transform faults and fracture zones.

> Transform faults are plate boundaries that occur *between* offset segments of the mid-ocean ridge while fracture zones are intraplate features that occur *beyond* the offset segments of the mid-ocean ridge.

 Students Sometimes Ask

I didn't know that Jacques Cousteau was involved in developing underwater habitats. What ever happened to him?

Ocean explorer Jacques Cousteau did much to popularize oceanography. His widely acclaimed television show—*The Undersea World of Jacques Cousteau*—and documentaries on ocean life, many of them filmed while at sea aboard his ship *Calypso*, have inspired a generation of oceanographers and environmentalists. Always an advocate of human presence in the ocean, he experimented with many types of scuba equipment, underwater habitats, and submersibles (the most notable of which was his "diving saucer"). Cousteau died in 1997, at the age of 87.

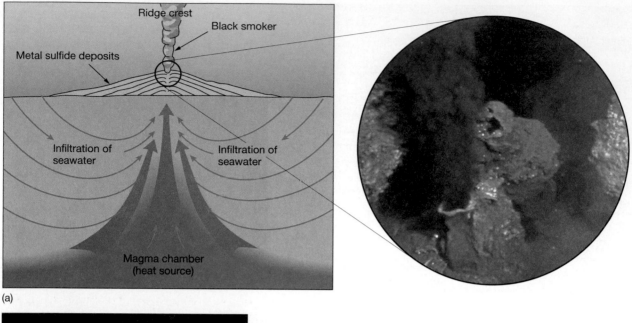

(a)

(b)

Figure 3–15 Hydrothermal vents. **(a)** Diagram showing hydrothermal circulation along the mid-ocean ridge and the creation of black smokers. Photo (*inset*) shows a close-up view of a black smoker along the East Pacific Rise. **(b)** Black smoker chimney and fissure at Susu north active site, Manus Basin, western Pacific Ocean. Chimney is about 3 meters (10 feet) tall.

Has anyone ever seen pillow lava forming?

Amazingly, yes! In the 1960s, an underwater film crew ventured to Hawaii during an eruption of the volcano Kilauea where lava spilled into the sea. They braved high water temperatures and risked being burned on the red-hot lava, but filmed some incredible footage. Underwater, the formation of pillow lava occurs where a tube emits molten lava directly into the ocean. When hot lava comes into contact with cold seawater, it forms the characteristic smooth and rounded margins of pillow basalt. The divers also experimented with a hammer on newly formed pillows and were able to initiate new lava outpourings.

What effect does all this volcanic activity along the mid-ocean ridge have at the ocean's surface?

Sometimes the underwater volcanic eruption is large enough to create what is called a "*megaplume*" of warm, mineral-rich water that is lower in density than the surrounding seawater and thus rises to the surface. Remarkably, a few research vessels have reported experiencing the effects of a megaplume at the surface while directly above an erupting sea floor volcano! Researchers on board describe bubbles of gas and steam at the surface, a marked increase in water temperature, and the presence of enough volcanic material to turn the water cloudy. In terms of warming the ocean, the heat released into the

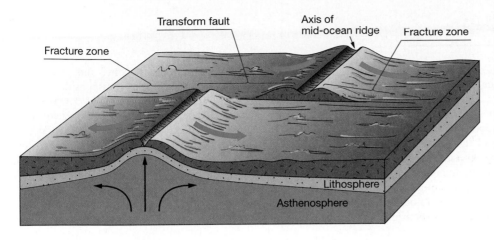

Transform fault

Axis of
mid-ocean ridge

Fracture zone

Fracture zone

Lithosphere

Asthenosphere

Figure 3–16 Transform faults and fracture zones. Transform faults are active transform plate boundaries that occur *between* the segments of the mid-ocean ridge. Fracture zones are inactive intraplate features that occur *outside of* the segments of the mid-ocean ridge.

ocean at mid-ocean ridges is probably not very significant, mostly because the ocean is so good at absorbing and redistributing heat.

Do the metal sulfides associated with hydrothermal vents create any economically interesting deposits?

Yes, these metal sulfides are prime exploration targets for mining the ocean floor in the future. However, not all sea floor deposits always remain on the sea floor: The deposition of metallic sulfides around sea floor hydrothermal vents is also a major source of *continental* sulfide ore deposits. Initially formed at the mid-ocean ridges and transported by sea floor spreading, they may be *obducted* (uplifted) onto continents during subduction of an oceanic plate. In fact, the copper deposits in Cyprus that have been so influential in Mediterranean cultures throughout history were created by hydrothermal vents on the sea floor. In addition, rifting that begins on land but becomes inactive can also create metal sulfide deposits underneath the ground surface. The large metal sulfide deposits in the states of Missouri and New Jersey were also formed by hydrothermal vents.

All this sea floor mapping is interesting, but haven't we found everything on the sea floor already?

Certainly not! The world's ocean floor—equal in area to almost two moons plus two Mars-sized planets—is one of the most poorly mapped surfaces in the solar system. Recent satellite missions to other planets and their moons have produced stunning high-resolution images of these worlds. In contrast, the ocean floor image produced from sea surface height data shown in (Figure 3D) is still an order of magnitude lower in resolution that the optic images obtained from a satellite flyby of Mercury in the early 1970s!

Despite recent advances in mapping technology, only 5% of the ocean floor has been mapped as precisely as the surface of the moon. In fact, there are areas of the ocean floor as large as the state of Oklahoma that have not been mapped in much detail at all and even well-mapped areas are based on widely separated ship tracks. The great depth of the oceans along with seawater's opaque character has hindered mapping efforts, leaving room for major discoveries in the future (such as new sea floor features, shipwrecks, and mineral deposits).

If black smokers are so hot, why isn't there steam coming out of them instead of hot water?

Indeed, black smokers emit water that can be up to three-and-a-half times the boiling point of water at the surface. However, the depth where black smokers are found results in much higher pressure than at the surface. At these higher pressures, water has a higher boiling point. Thus, water from hydrothermal vents remains in the liquid state instead of turning into water vapor (steam).

Table 3–3   Comparison between transform faults and fracture zones.

	Transform faults	*Fracture zones*
Plate boundary?	Yes—a transform plate boundary	No—an intraplate feature
Relative movement across feature	Movement in opposite directions	Movement in the same direction
	← ------------------------------- →	← ------------------------------- ←
Earthquakes?	Many	Few
Relationship to mid-ocean ridge	Occur *between* offset mid-ocean ridge segments	Occur *beyond* offset mid-ocean ridge segments
Geographic examples	San Andreas Fault, Alpine Fault, Dead Sea Fault	Mendocino Fracture Zone, Molokai Fracture Zone

Chapter in Review

• Bathymetry is the measurement of ocean depths and the charting of ocean floor topography. The varied bathymetry of the ocean floor was first determined using soundings to measure water depth. Later, the development of the echosounder gave ocean scientists a more detailed representation of the sea floor. Today, much of our knowledge of the ocean floor has been obtained using various multibeam echosounders or side-scan sonar instruments (to make detailed bathymetric maps of a small area of the ocean floor), satellite measurement of the ocean surface (to produce maps of the world ocean floor), and seismic reflection profiles (to examine ocean structure beneath the sea floor).

• Continental margins can be either passive (not associated with any plate boundaries) or active (associated with convergent or transform plate boundaries). Extending from the shoreline is the generally shallow, low relief, and gently sloping continental shelf that can contain various features such as coastal islands, reefs, and banks. The boundary between the continental slope and the continental shelf is marked by an increase in slope that occurs at the shelf break. Cutting deep into the slopes are submarine canyons, which resemble canyons on land but are created by erosive turbidity currents. Turbidity currents deposit their sediment load at the base of the continental slope, creating deep-sea fans that merge to produce a gently sloping continental rise. The deposits from turbidity currents (called turbidite deposits) have characteristic sequences of graded bedding. Active margins have similar features, although they are modified by their associated plate boundary.

• The continental rises gradually become flat, extensive, deep-ocean abyssal plains, which form by suspension settling of fine sediment. Poking through the sediment cover of the abyssal plains are numerous volcanic peaks, including volcanic islands, seamounts, tablemounts, and abyssal hills. In the Pacific Ocean, where sedimentation rates are low, abyssal plains are not extensively developed, and abyssal hill provinces cover broad expanses of ocean floor. Along the margins of many continents—especially those around the Pacific Ring of Fire—are deep linear scars called ocean trenches that are associated with convergent plate boundaries and volcanic arcs.

• The mid-ocean ridge is a continuous mountain range that winds through all ocean basins and is entirely volcanic in origin. Common features associated with the mid-ocean ridge include a central rift valley, faults and fissures, seamounts, pillow basalts, hydrothermal vents, deposits of metal sulfides, and unusual life forms. Segments of the mid-ocean ridge are either oceanic ridges if steep with rugged slopes (indicative of slow sea floor spreading) or oceanic rises if sloped gently and less rugged (indicative of fast spreading).

• Long linear zones of weakness—fracture zones and transform faults—cut across vast distances of ocean floor and offset the axes of the mid-ocean ridge. Fracture zones and transform faults are differentiated from one another based on the direction of movement across the feature. Fracture zones (an intraplate feature) have movement in the same direction, while transform faults (a transform plate boundary) have movement in opposite directions.

Key Terms

Abyssal hill (p. 85)

Abyssal hill province (p. 85)

Abyssal plain (p. 84)

Active margin (p. 78)

Bathymetry (p. 75)

Black smoker (p. 89)

Continental arc (p. 86)

Continental borderland (p. 82)

Continental margin (p. 77)

Continental rise (p. 84)

Continental shelf (p. 79)

Continental slope (p. 82)

Convergent active margin (p. 78)

Cousteau, Jacques-Yves (p. 74)

Deep-ocean basin (p. 77)

Deep-sea fan (p. 84)

Echosounder (p. 76)

Fathom (p. 75)

Fissure (p. 87)

Fracture zone (p. 91)

GLORIA (p. 76)

Graded bedding (p. 84)

Hydrothermal vent (p. 87)

Hypsographic curve (p. 77)

Island arc (p. 86)

Metal sulfides (p. 89)

Mid-ocean ridge (p. 77)

Ocean trench (p. 86)

Oceanic ridge (p. 90)

Oceanic rise (p. 90)

Pacific Ring of Fire (p. 86)

Passive margin (p. 77)

Pillow basalt (p. 87)

Pillow lava (p. 87)

Ping (p. 76)

Precipitate (p. 89)

Precision depth recorder (p. 76)

Rift valley (p. 87)

Seabeam (p. 76)

Sealab (p. 74)

Seaknoll (p. 85)

Sea MARC (p. 76)

Seismic reflection profile (p. 76)

Shelf break (p. 79)

Sonar (p. 76)

Sounding (p. 76)

Submarine canyon (p. 83)

Submarine fan (p. 84)

Suspension settling (p. 84)

Transform active margin (p. 78)

Transform fault (p. 91)

Turbidite deposit (p. 84)

Turbidity current (p. 83)

Volcanic arc (p. 86)

Warm-water vent (p. 89)

White smoker (p. 89)

Questions And Exercises

1. What are some human physiological problems with living in an underwater habitat?

2. What is bathymetry?

3. Discuss the development of bathymetric techniques, indicating significant advancements in technology.

4. Describe the differences between passive and active continental margins. Be sure to include how these features relate to plate tectonics, and include an example of each type of margin.

5. Describe the major features of a passive continental margin: continental shelf, continental slope, continental rise, submarine canyon, and deep-sea fans.

6. Explain how submarine canyons are created.

7. What are differences between a submarine canyon and an ocean trench?

8. Explain what graded bedding is and how it forms.

9. Describe the process by which abyssal plains are created.

10. Discuss the origin of the various volcanic peaks of the abyssal plains: seamounts, tablemounts, and abyssal hills.

11. In which ocean basin are most ocean trenches found? Use plate tectonic processes to help explain why.

12. Describe characteristics and features of the mid-ocean ridge, including the difference between oceanic ridges and oceanic rises.

13. List and describe the different types of hydrothermal vents.

14. What kinds of unusual life can be found associated with hydrothermal vents? How do these organisms survive?

15. Use pictures and words to describe the differences between a fracture zone and a transform fault.

References

Anderson, R. N. 1986. *Marine-geology: A planet earth perspective*. New York: Wiley.

Auzende, J., et. al. 2000. Extensive magmatic and hydrothermal activity documented in Manus Basin. *Eos Transactions* 81:39, 449–453.

Baker, E. T., Massoth, G. J., and Feely, R. A. 1987. Cataclysmic hydrothermal venting on the Juan de Fuca Ridge. *Nature* 329:6135, 149–151.

Carlowicz, M. 1996. New map of seafloor mirrors surface. *Eos Transactions* 76:44, 441–442.

Damuth, J. E., et. al. 1983. Distributary channel meandering and bifurcation patterns on the Amazon deep-sea fan as revealed by long-range side-scan sonar (GLORIA). *Geology* 11:2, 94–98.

Field, M. E., Gardner, J. V., Jennings, A. E., and Edwards, D. E. 1982. Earthquake-induced sediment failures on a 0.25° slope. Klamath River delta, California. *Geology* 10:10, 542–545.

Hay, A. E., Burling, R. W., and Murray, J. W. 1982. Remote acoustic detection of a turbidity current surge. *Science* 217:4562, 833–835.

Heezen, B. C., and Ewing, M. 1952. Turbidity currents and submarine slumps, and the 1929 Grand Banks earthquake. *American Journal of Science* 250, 849–873.

Heezen, B. C., and Hollister, C. D. 1971. *The face of the deep.* New York: Oxford University Press.

Lawrence, D. M. 1999. Mountains under the sea. *Mercator's World* 4:6, 36–43.

Lay, T. and Wallace, T. C. 1995. *Modern global seismology.* New York: Academic Press.

Ménard, Y. 2000. Cruising the ocean from space with Jason-1. *Eos Transactions* 81:34, 381–391.

Pratson, L. F., and Haxby, W. F. 1996. What is the slope of the U.S. continental slope? *Geology* 24:1, 3–6.

Small, C., and Sandwell, D. T. 1996. Sights unseen. *Natural History* 105:3, 28–33.

Smith, W. H. F., and Sandwell, D. T. 1997. Global sea floor topography from satellite altimetry and ship depth soundings. *Science* 277:5334, 1956–1962.

Tarbuck, E. J., and Lutgens, F. K. 1999. *The earth: An introduction to physical geology*, 6th ed. Upper Saddle River, NJ: Prentice-Hall.

———. 1997. *Earth science*, 8th ed. Upper Saddle River, NJ: Prentice-Hall.

The Open University Course Team. 1989. *The ocean basins: Their structure and evolution*. Oxford: Pergamon Press.

Twichell, D. C., and Roberts, D. G. 1982. Morphology, distribution and development of submarine canyons on the United States Atlantic continental slope between Hudson and Baltimore canyons. *Geology* 10:8, 408–412.

Various authors. 1986. Special issue on mapping the sea floor (13 articles). *Journal of Geophysical Research*, 91:B3.

Various authors. 1995. Special section: The 1993 volcanic eruption on the CoAxial segment, Juan de Fuca Ridge (14 articles). *Geophysical Research Letters*, 22:2.

Weirich, F. H. 1984. Turbidity currents: Monitoring their occurrence and movement with a three-dimensional sensor network. *Science* 224:4647, 384–387.

Suggested Reading in Scientific American

Brimhall, G. 1991. The genesis of ores. 264:5, 84–91. After being emplaced into Earth's crust at oceanic ridges, metal deposits undergo a number of processes before they become mineable ores embedded in continental rocks.

Edmonds, P. J., and Burton, D. 1996. Ten days under the sea. 275:4, 88–95. Living underwater in the world's only habitat devoted to science, six aquanauts studied juvenile corals and fought off "the funk."

Emery, K. O. 1969. The continental shelves. 221:3, 106–125. The nature of the continental shelves and the effect of the advance and retreat of the shoreline across them as a result of glaciation are discussed.

Heezen, B. C. 1956. The origin of submarine canyons. 195:2, 36–41. Theories explaining the origin of submarine canyons are presented along with data on the location and nature of such features.

King, M. D. and Herring, D. D. 2000. Monitoring Earth's vital signs. 282:4, 92–97. Information about Terra, the flagship of NASA's Earth Observing System fleet of satellites that monitor Earth's vital signs including ocean properties.

Menard, H. W. 1969. The deep-ocean floor. 221:3, 126–145. A summary of the dynamic effects of sea floor spreading is presented with a description of related sea floor features.

Pratson, L. F., and Haxby, W. F. 1997. Panoramas of the seafloor. 276:6, 82–87. A description of how modern sea floor images are produced and used to map features of the continental margins.

Wilson, A. C. 1985. The molecular basis of evolution. 253:4, 164–175. Mutations within the genes of organisms play an important role in evolution at the organismal level.

Oceanography on the Web

Visit the Essentials of Oceanography home page for on-line resources for this chapter. There you will find an on-line study guide with review exercises, and links to oceanography sites to further your exploration of the topics in this chapter. *Essentials of Oceanography* is at: **http://www.prenhall.com/ thurman** (click on the Table of Contents menu and select this chapter).

CHAPTER

4

Marine Sediments

COLLECTING THE HISTORICAL RECORD OF THE DEEP-OCEAN FLOOR

During early exploration of the oceans, a bucket-like device called a **dredge** was used to scoop up sediment from the deep-ocean floor for analysis. This technique, however, was limited to gathering samples from the *surface* of the ocean floor. Later, the **gravity corer**—a hollow steel tube with a heavy weight on top—was thrust into the sea floor to collect the first **cores** (cylinders of sediment and rock). Although the gravity corer could sample below the surface, its depth of penetration was limited.

In 1963, the National Science Foundation of the United States funded a program that borrowed drilling technology from the offshore oil industry to obtain long sections of core from deep below the surface of the ocean floor. The program united four leading oceanographic institutions (Scripps Institution of Oceanography in California; Rosenstiel School of Atmospheric and Oceanic Studies at the University of Miami, Florida; Lamont-Doherty Earth Observatory of Columbia University in New York; and the Woods Hole Oceanographic Institution in Massachusetts) to form the **Joint Oceanographic Institutions for Deep Earth Sampling (JOIDES)**. JOIDES was later joined by the oceanography departments of several other leading universities.

The first phase of the **Deep Sea Drilling Project (DSDP)** was initiated in 1968 when the specially designed drill ship *Glomar Challenger* was launched. It had a tall drilling rig resembling a steel tower. Cores could be collected by drilling into the ocean floor in water up to 6000 meters (3.7 miles) deep. From the initial cores collected, scientists confirmed the existence of sea floor spreading by documenting that: (1) the age of the ocean floor increased progressively with distance from the mid-ocean ridge; (2) sediment thickness increased progressively with distance from the mid-ocean ridge; and (3) Earth's magnetic field polarity reversals were recorded in ocean floor rocks.

Although the oceanographic research program was initially financed by the U.S. government, it became international in 1975 when West Germany, France, Japan, the United Kingdom, and the Soviet Union also provided financial and scientific support. In 1983, the Deep Sea Drilling Project became the **Ocean Drilling Program (ODP)** with 20 participating countries under the supervision of Texas A&M University and a broader objective of drilling the thick sediment layers near the continental margins.

In 1985, the *Glomar Challenger* was decommissioned and replaced by the drill ship *JOIDES Resolution* (Figure 4A). The new ship also has a tall metal drilling rig to conduct **rotary drilling**. The drill pipe is in individual sections of 9.5 meters (31 feet) and sections can be screwed together to make a single string of pipe up to 8200 meters (27,000 feet) long (Figure 4B). The drill bit, located at the end of the pipe string, rotates as it is

- How are marine sediments collected?
- What do marine sediments indicate about past environmental conditions on Earth?
- How are the four main types of marine sediment formed?
- Where can each of the four main types of marine sediment be found?
- Which types of marine sediment comprise coastal and deep-sea deposits?
- What resources do ocean sediments provide?

When I think of the floor of the deep sea, the single, overwhelming fact that possesses my imagination is the accumulation of sediments. I see always the steady, unremitting, downward drift of materials from above, flake upon flake, layer upon layer.... For the sediments are the materials of the most stupendous snowfall the Earth has ever seen.
—*Rachel Carson,*
The Sea Around Us *(1956)*

Figure 4A The drill ship *JOIDES Resolution*.

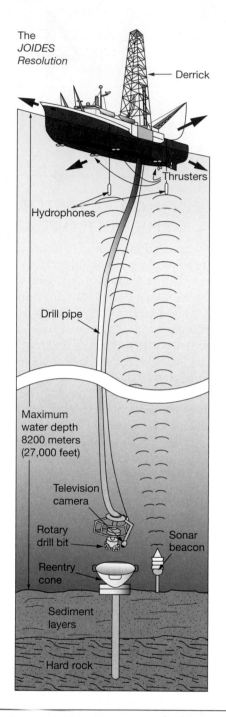

Figure 4B Rotary drilling from the *JOIDES Resolution*.

pressed against the ocean bottom and can drill up to 2100 meters (6900 feet) below the sea floor. Like twirling a soda straw into a layer cake, the drilling operation crushes the rock around the outside and retains a cylinder of rock (a core sample) on the inside of the hollow pipe, which can then be raised on board the ship. Cores are retrieved from inside the pipe and analyzed with state-of-the-art laboratory facilities on board the *Resolution*. Worldwide, more than 2000 holes have been drilled into the sea floor using this method, allowing the collection of cores that provide scientists with valuable information about Earth history as recorded in sea floor sediments.

The ODP is scheduled to be replaced in 2003 by the **Integrated Ocean Drilling Program (IODP)**, which will be led by the United States and Japan. Two new drill ships will be constructed: one for drilling shallow high-resolution cores, and one with advanced technology that can drill up to 7000 meters (23,000 feet) below the

sea floor. The primary objective of the new program is to collect cores that will allow scientists to better understand Earth history and Earth system processes, including the properties of the deep crust, climate change patterns, earthquake mechanisms, and the microbiology of the deep ocean floor.

Why are **sediments** (*sedimentum* = settling) interesting to oceanographers? Although ocean sediments are little more than eroded particles and fragments of dirt, dust, and other debris that have settled out of the water and accumulated on the ocean floor (Figure 4–1), they reveal much about Earth's history. For example, sediments provide clues to past climates, movements of the ocean floor, ocean circulation patterns, and nutrient supplies for marine organisms. By examining cores of sediment retrieved from ocean drilling (Figure 4–2) and interpreting them, oceanographers can ascertain the timing of major extinctions, global climate change, and the movement of plates. In fact, most of what is known of Earth's past geology, climate, and biology has been learned through studying ancient marine sediments.

Over time, sediments can become **lithified** (*lithos* = stone, *fic* = making)—turned to rock—and form **sedimentary rock**. More than half of the rocks exposed on the continents are sedimentary rocks deposited in ancient ocean environments and uplifted onto land by plate tectonic processes. Even in the tallest mountains on the continents, far from any ocean, telltale marine fossils indicate that these rocks originated on the ocean floor in the geologic past.

Particles of sediment come from worn pieces of rocks, as well as living organisms, minerals dissolved in

Marine sediments accumulate on the ocean floor and contain a record of Earth history including past environmental conditions. Cores of marine sediment are collected by rotary drilling.

Figure 4–2 Examination of deep-ocean sediment cores. Long cylinders of sediment and rock called cores are cut in half and examined, revealing interesting aspects of Earth history.

Figure 4–1 Oceanic sediment. View of the deep-ocean floor from a submersible. Most of the deep-ocean floor is covered with particles of material that have settled out through the water.

water, and outer space. Table 4–1 show a classification of marine sediments according to type, composition, sources, and main locations found. Clues to sediment origin are found in its mineral composition and its **texture** (the size and shape of its particles).

Lithogenous Sediment

Lithogenous (*lithos* = stone, *generare* = to produce) **sediment** is derived from preexisting rock material. Because most lithogenous sediment comes from the landmasses, it is also called **terrigenous** (*terra* = land, *generare* = to produce) **sediment**. Volcanic islands in

the open ocean are also important sources of lithogenous sediment.

Origin

Lithogenous sediment begins as rocks on continents or islands. Over time, **weathering** agents such as water, temperature extremes, and chemical effects break rocks into smaller pieces, as shown in Figure 4–3. When rocks are in smaller pieces, they can be more easily **eroded** (picked up) and transported. This eroded material is the basic component of which all lithogenous sediment is composed.

Table 4–1   Classification of marine sediments.

Type	Composition			Sources		Main locations found
Lithogenous	*Continental Margin*	Rock fragments		Rivers; Coastal erosion; Landslides		Continental shelf
		Quartz sand		Glaciers		Continental shelf in high latitudes
		Quartz silt		Turbidity currents		Continental slope and rise; Ocean basin margins
		Clay				
	Oceanic	Quartz silt		Wind-blown dust; Rivers		Deep-ocean basins
		Clay				
		Volcanic ash		Volcanic eruptions		
Biogenous	*Calcium carbonate ($CaCO_3$)*	Calcareous ooze (microscopic)		*Warm surface water*	Coccolithophores (algae); Foraminifers (protozoans)	Low-latitude regions; Sea floor above CCD; Along mid-ocean ridges & the tops of volcanic peaks
		Shell/coral fragments (macroscopic)			Macroscopic shell-producing organisms	Continental shelf; Beaches
					Coral reefs	Shallow low-latitude regions
	Silica ($SiO2.nH_2O$)	Siliceous ooze		*Cold surface water*	Diatoms (algae); Radiolarians (protozoans)	High-latitude regions; Sea floor below CCD; Surface current divergence near the Equator
Hydrogenous	Manganese nodules (manganese, iron, copper, nickel, cobalt)			Precipitation of dissolved materials directly from seawater due to chemical reactions		Abyssal plain
	Phosphorite (phosphorous)					Continental shelf
	Oolites ($CaCO_3$)					Shallow shelf in low-latitude regions
	Metal sulfides (iron, nickel, copper, zinc, silver)					Hydrothermal vents at mid-ocean ridges
	Evaporites (gypsum, halite, other salts)					Shallow restricted basins where evaporation is high in low-latitude regions
Cosmogenous	Iron-nickel spherules / Tektites (silica glass)			Space dust		In very small proportions mixed with all types of sediment and in all marine environments
	Iron-nickel meteorites / Silicate chondrites			Meteors		Localized near meteor impact structures

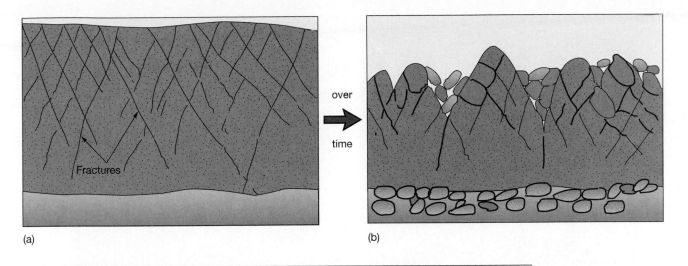

(a)

over

time

(b)

Figure 4–3 Weathering. Weathering often occurs along fractures in rock, breaking the rocks into smaller fragments over time.

Eroded material from the continents is carried to the oceans by streams, wind, glaciers, and gravity (Figure 4–4). The sediment can be deposited in many environments, including bays or lagoons near the ocean, along beaches at the shoreline, or further offshore across the continental margin. It can also be carried beyond the continental margin to the deep-ocean basin by turbidity currents, as discussed in Chapter 3.

The greatest quantity of lithogenous material by far is found around the margins of the continents, where it is constantly moved by high-energy currents along the shoreline and in deeper turbidity currents. Lower-energy currents distribute finer components that settle out onto the deep-ocean basins. Microscopic particles from wind-blown dust or volcanic eruptions can even be carried far out over the open ocean by prevailing winds. These particles either settle into fine layers as the velocity of the wind decreases or disperse into the ocean when they serve as nuclei around which raindrops and snowflakes form.

Composition

The composition of lithogenous sediment reflects the material from which it was derived. All rocks are composed of discrete crystals of naturally occurring compounds called **minerals**. One of the most abundant, chemically stable, and durable minerals in Earth's crust is **quartz**, composed of silicon and oxygen in the form of SiO_2—the same composition as ordinary glass. Quartz is the major component of nearly all rocks. Because quartz is resistant to abrasion, it can be transported long distances and deposited far from its source area. The majority of lithogenous deposits—such as beach sands—are composed primarily of quartz (Figure 4–5).

A large percentage of lithogenous particles that find their way into deep-ocean sediments far from conti-

nents are transported by prevailing winds that remove small particles from the continents' subtropical desert regions. The map in Figure 4–6 shows a close relationship between the location of microscopic fragments of lithogenous quartz in the surface sediments of the ocean floor and the strong prevailing winds in the desert regions of Africa, Asia, and Australia. Satellite observations of dust storms (Figure 4–6, *inset*) confirm this relationship.

Sediment Texture

One of the most important properties of lithogenous sediment is its texture, including its **grain[1] size**. The **Wentworth scale of grain size** (Table 4–2) indicates that particles can be classified as boulders (largest), cobbles, pebbles, granules, sand, silt, or clay (smallest). Sediment size is proportional to the energy needed to lay down a deposit. Deposits laid down where wave action is strong (areas of high energy) may be composed primarily of larger particles—cobbles and boulders. Fine-grained particles, on the other hand, are deposited where the energy level is low and the current speed is minimal. When clay-sized particles—which are flat—are deposited, they tend to stick together by cohesive forces. Consequently, higher-energy conditions than what would be expected based on grain size alone are required to erode and transport clays.

The texture of lithogenous sediment also depends on its **sorting**. Sorting is a measure of the uniformity of grain sizes and indicates the selectivity of the transportation process. For example, sediments composed of particles that are primarily the same size are well sorted—such as in coastal sand dunes, where winds can

[1]Sediment grains are also known as particles, fragments, or clasts.

Figure 4–4 Sediment-transporting media. Transporting media include: **(a)** Streams. **(b)** Wind. **(c)** Glaciers. **(d)** Gravity, which creates landslides.

Figure 4–5 Lithogenous beach sand. Lithogenous beach sand is composed mostly of particles of white quartz, plus small amounts of other minerals. This sand is from North Beach, Hampton, New Hampshire and is magnified approximately 107 times.

only pick up a certain size particle. Poorly sorted deposits, on the other hand, contain a variety of different sized particles and indicate a transportation process capable of picking up clay- to boulder-sized particles. An example of poorly sorted sediment is that which is carried by a glacier and left behind when the glacier melts.

The texture of lithogenous sediment also depends on its **maturity**. Sediment maturity increases as: (1) clay content decreases; (2) sorting increases; (3) non-quartz minerals decrease; and (4) grains within the deposit become more rounded. Particles increase in maturity as they are carried from the source to their point of deposition because more time is available during transportation to: (1) remove clays (which are carried in suspension and washed out to sea); (2) sort the sediment; (3) eliminate non-quartz minerals (which lack durability); and (4) round particles through abrasion.

A poorly sorted glacial deposit, which contains relatively large quantities of clay-sized particles and poorly rounded larger particles, is immature. Well-sorted beach sand, on the other hand, which contains well-rounded particles and very little clay, is a mature sedimentary deposit. Figure 4–7 illustrates the difference between mature and immature sediments.

Distribution

Marine sedimentary deposits can be categorized as neritic or pelagic. **Neritic** (*neritos* = of the coast) **deposits**

are found along continental margins and near islands, and **pelagic** (*pelagios* = of the sea) **deposits** are found in the deep-ocean basins. Lithogenous sediment in the ocean is ubiquitous: At least a small percentage of lithogenous sediment is found nearly everywhere on the ocean floor.

Neritic Deposits Lithogenous sediment dominates most neritic deposits. Lithogenous sediment is derived from rocks on nearby landmasses, consists of coarse-grained deposits, and accumulates rapidly on the continental shelf, slope, and rise. Examples of lithogenous neritic deposits include beach deposits, continental shelf deposits, turbidite deposits, and glacial deposits.

Beach Deposits Beaches are made of whatever materials are locally available. Beach materials are composed mostly of quartz-rich sand that is washed down to the coast by rivers but can also be composed of a wide variety of sizes and compositions. This material is transported by waves that crash against the shoreline, especially during storms.

Continental Shelf Deposits At the end of the last Ice Age (about 18,000 years ago), glaciers melted and sea level rose. As a result, many rivers of the world today deposit their sediment in drowned river mouths rather than carry it onto the continental shelf as they did during the geologic past. In many areas, the sediments that cover the continental shelf—called **relict** (*relict* = left behind) **sediments**—were deposited from 3000 to 7000 years ago and have not yet been covered by more recent deposits. These sediments presently cover about 70% of the world's continental shelves. In other areas, deposits of sand ridges on the continental shelves appear to have been formed more recently than the Ice Age and at present water depths.

Turbidite Deposits As discussed in Chapter 3, **turbidity currents** are underwater avalanches that periodically move down the continental slopes and carve submarine canyons. Turbidity currents also carry vast amounts of neritic material. This material spreads out as deep-sea fans, comprises the continental rise, and gradually thins toward the abyssal plains. These deposits are called **turbidite deposits** and are composed of characteristic layering called *graded bedding* (see Figure 3–8).

Glacial Deposits Poorly sorted deposits containing particles ranging from boulders to clays may be found in the high-latitude[2] portions of the continental shelf. These **glacial deposits** were laid down after the Ice Age when glaciers that covered the continental shelf eventually melted. Glacial deposits are currently forming around the continent of Antarctica and around Greenland by

---

[2]High-latitude regions are those far from the Equator (either north or south); low latitudes are areas close to the Equator.

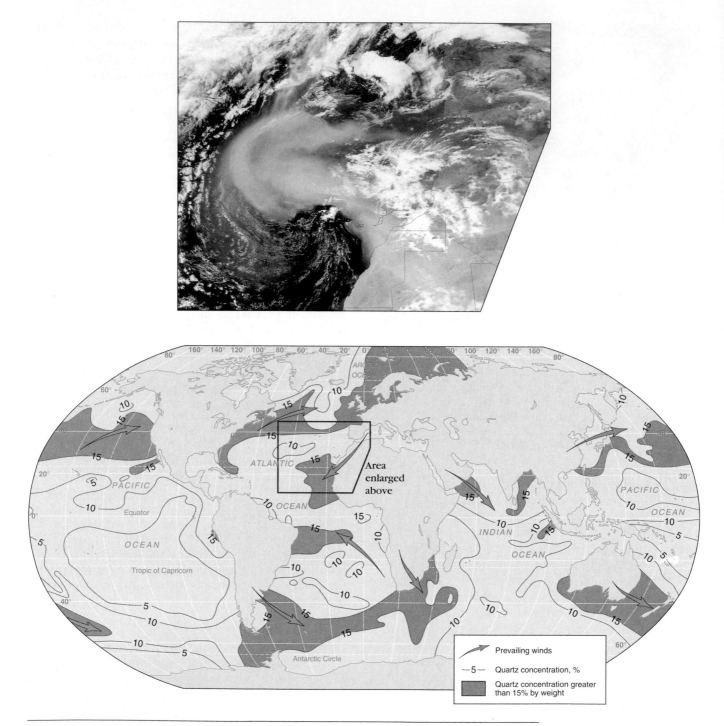

Figure 4–6 Lithogenous quartz in surface sediments of the world's oceans. High amounts of microscopic lithogenous quartz in deep-sea sediment match prevailing winds from land (*arrows*). SeaStar SeaWiFS satellite photo (*inset*) on February 26, 2000, shows a Sahara dust storm off the northwest coast of Africa that has spread out for 1000 miles (1600 kilometers) across the Atlantic Ocean.

ice rafting. In this process, rock particles trapped in glacial ice are carried out to sea by icebergs that break away from coastal glaciers. As the icebergs melt, lithogenous particles of many sizes are released and settle onto the ocean floor.

Pelagic Deposits Turbidite deposits of neritic sediment on the continental rise can spill over into the deep-ocean basin. However, most pelagic deposits are composed of fine-grained material that accumulates slowly on the deep-ocean floor. Pelagic lithogenous sediment includes particles that have come from volcanic eruptions, wind-blown dust, and fine material that is carried by deep ocean currents.

Abyssal Clay **Abyssal clay** is composed of at least 70% (by weight) fine clay-sized particles from the continents. Even though they are far from land, deep abyssal plains

Table 4–2 Wentworth scale of grain size for sediments.

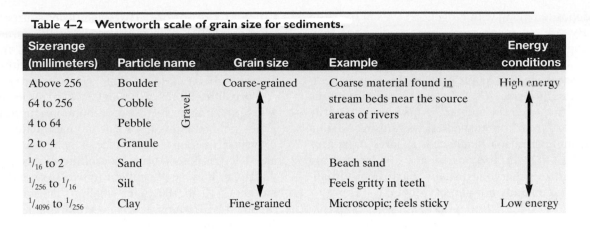

Size range (millimeters)	Particle name		Grain size	Example	Energy conditions
Above 256	Boulder		Coarse-grained ↑	Coarse material found in stream beds near the source areas of rivers	High energy ↑
64 to 256	Cobble	Gravel			
4 to 64	Pebble				
2 to 4	Granule				
$1/16$ to 2	Sand			Beach sand	
$1/256$ to $1/16$	Silt			Feels gritty in teeth	
$1/4096$ to $1/256$	Clay		Fine-grained	Microscopic; feels sticky	Low energy ↓

0 10 20 30 40 50 60
Scale in millimeters

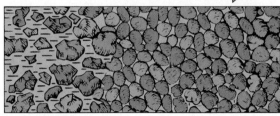

Maturity increases
Degree of sorting increases
Clay content decreases
Rounding of sand particles increases
Time increases

— Clay particle

Figure 4–7 Sediment maturity. As sediment maturity increases with time (*left to right*), the degree of sorting and rounding of particles increases, whereas clay content decreases.

contain thick sequences of abyssal clay deposits composed of particles transported great distances by winds or ocean currents and deposited on the deep ocean floor. Because abyssal clays contain oxidized iron, they are commonly red-brown or buff in color and are sometimes referred to as **red clays**. The predominance of abyssal clay on abyssal plains is caused not by an abundance of clay settling on the ocean floor, but by the absence of other material that would otherwise dilute it.

Lithogenous sediment is produced from preexisting rock material, is found on most parts of the ocean floor, and can occur as thick deposits close to land.

Biogenous Sediment

Biogenous (*bio* = life, *generare* = to produce) **sediment** is derived from the remains of hard parts of once-living organisms.

Origin

Biogenous sediment begins as the hard parts (shells, bones, and teeth) of living organisms ranging from minute algae and protozoans to fish and whales. When organisms that produce hard parts die, their remains settle onto the ocean floor and can accumulate as biogenous sediment.

Biogenous sediment can be classified as either macroscopic or microscopic. **Macroscopic biogenous sediment** is large enough to be seen without the aid of a microscope and includes shells, bones, and teeth of large organisms. Except in certain tropical beach localities where shells and coral fragments are numerous, this type of sediment is relatively rare in the marine environment, especially in deep water where fewer organisms live. Much more abundant is **microscopic biogenous sediment**, which contains particles so small they can only be seen through a microscope. Microscopic organisms produce tiny shells called **tests** (*testa* = shell) that begin to sink after the organisms die and continually rain down in great numbers onto the ocean floor. These microscopic tests can accumulate on the deep-ocean floor and form deposits called **ooze** (*wose* = juice). As its name implies, ooze resembles very fine-grained mushy material.[3] Technically, biogenous ooze must contain at least 30% biogenous test material by weight. What comprises the other part—up to 70% of an ooze? Commonly, it is fine-grained lithogenous clay

---

[3]Ooze has the consistency of toothpaste mixed about half and half with water. To help you remember this term, imagine walking barefoot across the deep-ocean floor and how the sediment would *ooze* between your toes.

that is deposited along with biogenous tests in the deep ocean. By volume, much more microscopic ooze than macroscopic biogenous sediment exists on the ocean floor.

The organisms that contribute to biogenous sediment are chiefly **algae** (*alga* = seaweed) and **protozoans** (*proto* = first, *zoa* = animal). Algae are primarily aquatic, eukaryotic,[4] photosynthetic organisms, ranging in size from microscopic single cells to large organisms like the giant kelp. Protozoans are any of a large group of single-celled, eukaryotic, usually microscopic organisms that are generally not photosynthetic.

Composition

The two most common chemical compounds in biogenous sediment are **calcium carbonate** ($CaCO_3$, which forms the mineral **calcite**) and **silica** (SiO_2). Often, the silica is chemically combined with water to produce $SiO_2 \cdot nH_2O$, the hydrated form of silica, which is called *opal*.

Silica Most of the silica in biogenous ooze comes from microscopic algae called **diatoms** (*diatoma* = cut in half) and protozoans called **radiolarians** (*radio* = a spoke or ray).

Because diatoms photosynthesize, they need strong sunlight and are found only within the upper sunlit surface waters of the ocean. Most diatoms are free-floating or **planktonic** (*planktos* = wandering). The living organism builds a glass greenhouse out of silica as a protective covering and lives inside. Most species have two parts to their test that fit together like a petri dish or pillbox (Figure 4–8a). The tiny tests are perforated with small holes in intricate patterns to allow nutrients to pass in and waste products to pass out. Where diatoms are abundant at the ocean surface, thick deposits of diatom-rich ooze can accumulate below on the ocean floor. When this ooze lithifies, it becomes **diatomaceous earth**,[5] a lightweight white rock composed of diatom tests and clay (Box 4–1).

Radiolarians are microscopic single-celled protozoans, most of which are also planktonic. As their name implies, they often have long spikes or rays of silica protruding from their siliceous shell (Figure 4–8b). They do not photosynthesize but rely on external food sources such as bacteria and other plankton. Radiolarians typically display well-developed symmetry, which is why they have been described as living snowflakes of the sea.

The accumulation of siliceous tests of diatoms, radiolarians, and other silica-secreting organisms produces **siliceous ooze** (Figure 4–8c).

Calcium Carbonate Two significant sources of calcium carbonate biogenous ooze are the **foraminifers** (*foramen* = an opening)—close relatives of radiolarians—and microscopic algae called **coccolithophores** (*coccus* = berry; *lithos* = stone; *phorid* = carrying).

Coccolithophores are single-celled algae, most of which are planktonic. Coccolithophores produce thin plates or shields made of calcium carbonate, 20 or 30 of which overlap to produce a spherical test (Figure 4–9a). Like diatoms, coccolithophores photosynthesize, so they need sunlight to live. Coccolithophores are about 10 to 100 times smaller than most diatoms (Figure 4–9b), which is why coccolithophores are often called **nannoplankton** (*nanno* = dwarf, *planktos* = wandering).

When the organism dies, the individual plates (called **coccoliths**) disaggregate and can accumulate on the ocean floor as coccolith-rich ooze. When this ooze lithifies over time, it forms a white deposit called **chalk**, which is used for a variety of purposes (including writing on chalkboards). The White Cliffs of southern England are composed of hardened coccolith-rich calcium carbonate ooze, which was deposited on the ocean floor and has been uplifted onto land (Figure 4–10). Deposits of chalk the same age as the White Cliffs are so common throughout Europe, North America, Australia, and the Middle East that the geologic period in which these deposits formed is named the Cretaceous (*creta* = chalk) Period.

Foraminifers are single-celled protozoans, many of which are planktonic, ranging in size from microscopic to macroscopic. They do not photosynthesize, so they must ingest other organisms for food. Foraminifers produce a hard calcium carbonate test in which the organism lives (Figure 4–9c). Most foraminifers produce a segmented or chambered test, and all tests have a prominent opening in one end. Although very small in size, the tests of foraminifers resemble the large shells that one might find at a beach.

Deposits comprised primarily of tests of foraminifers, coccoliths, and other calcareous-secreting organisms are called **calcareous ooze** (Figure 4–9d).

Distribution

Biogenous sediment is commonly found in pelagic deposits but only rarely found as neritic deposits. The distribution of biogenous sediment on the ocean floor depends on three fundamental processes: productivity, destruction, and dilution.

Productivity is the number of organisms present in the surface water above the ocean floor. Surface waters with high biologic productivity contain many living and reproducing organisms—conditions that are likely to produce biogenous sediments. Conversely, surface waters with low biologic productivity contain too few organisms to produce biogenous oozes on the ocean floor.

---

[4]Eukaryotic (*eu* = good, *karyo* = the nucleus) cells contain a distinct membrane-bound nucleus.
[5]Diatomaceous earth is also called diatomite, tripolite, or kieselguhr.

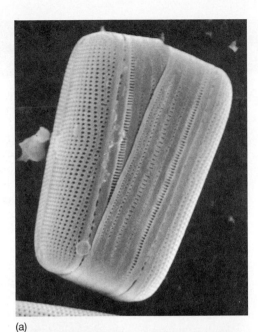

(a)

Figure 4–8 Microscopic siliceous tests.
Scanning electron micrographs: **(a)** Diatom
(length = 30 micrometers, equal to 30
millionths of a meter), showing how the
two parts of the diatom's test fit together.
(b) Radiolarian (length = 100 micrometers).
(c) Siliceous ooze, showing mostly fragments
of diatom tests (magnified 250 times).

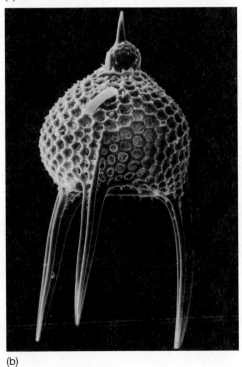

(b)

(c)

Destruction occurs when skeletal remains (tests) dissolve in seawater at depth.

Dilution occurs when other sediments are prevalent enough to keep the amount of biogenous test material below 30%. Dilution occurs most often because of the abundance of coarse-grained lithogenous material in neritic environments, so biogenous oozes are uncommon along continental margins.

Neritic Deposits Although neritic deposits are dominated by lithogenous sediment, both microscopic and macroscopic biogenous material may be incorporated into lithogenous sediment in neritic deposits. In addi-

tion, biogenous carbonate deposits are common in some areas.

Carbonate Deposits **Carbonate** minerals are those that contains CO_3 in its chemical formula—such as calcium carbonate, $CaCO_3$. Rocks from the marine environment composed primarily of calcium carbonate are called **limestones**. Most limestones contain fossil marine shells, suggesting a biogenous origin, while others appear to have formed directly from seawater without the help of any marine organism. Modern environments where calcium carbonate is currently forming (such as in the Bahama Banks, Australia's Great Barrier Reef, and the

Box 4–1 Diatoms: The Most Important Things You Have (Probably) Never Heard Of

Few objects are more beautiful than the minute siliceous cases of the diatomaceae: were these created that they might be examined and admired under the higher powers of the microscope? —Charles Darwin (1872)

Diatoms are microscopic single-celled photosynthetic organisms. Each one lives inside a protective silica test, most of which contain two halves that fit together like a shoebox and its lid. First described with the aid of a microscope in 1702, their tests are exquisitely ornamented with holes, ribs, and radiating spines unique to individual species. The fossil record indicates that diatoms have been on Earth since the Jurassic Period (180 million years ago) and at least 70,000 species of diatoms have been identified.

Diatoms live for a few days to as much as a week, can reproduce sexually or asexually, and occur individually or linked together into long communities. They are found in great abundance floating in the ocean and in certain freshwater lakes but can also be found in many diverse environments, such as on the undersides of polar ice, on the skins of whales, in soil, in thermal springs, and even on brick walls.

Figure 4C Products containing or produced using diatomaceous earth (diatom *Thalassiosira eccentrica*, inset).

When marine diatoms die, their tests rain down and accumulate on the sea floor as siliceous ooze. Hardened deposits of siliceous ooze, called diatomaceous earth, can be as much as 900 meters (3000 feet) thick. Diatomaceous earth consists of billions of minute silica tests and has many unusual properties: it is lightweight, has an inert chemical composition, is resistant to high temperatures, and has excellent filtering properties. Diatomaceous earth is used to produce a variety of common products (Figure 4C). The main uses of diatomaceous earth include:

- filters (for refining sugar, separating impurities from wine, straining yeast from beer, and filtering swimming pool water)
- mild abrasives (in toothpaste, facial scrubs, matches, and household cleaning and polishing compounds)
- absorbents (for chemical spills and pest control)

- chemical carriers (in pharmaceuticals, paint, and dynamite)

Other products from diatomaceous earth include optical-quality glass (because of the pure silica content of diatoms), space shuttle tiles (because they are lightweight and provide good insulation), an additive in concrete, a filler in tires, an anti-caking agent, and even building stone for constructing houses.

Further, the vast majority of oxygen that all animals breathe is a by-product of photosynthesis by diatoms. In addition, each living diatom contains a tiny droplet of oil. When diatoms die, their tests containing droplets of oil accumulate on the sea floor and are the beginnings of petroleum deposits, such as those found offshore California.

Given their many practical applications, it is difficult to imagine how different our lives would be without diatoms!

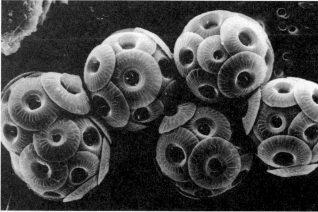
(a)

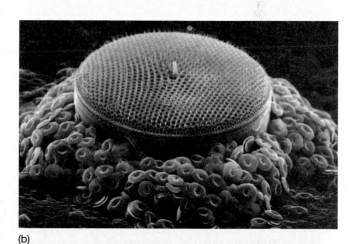
(b)

(c)

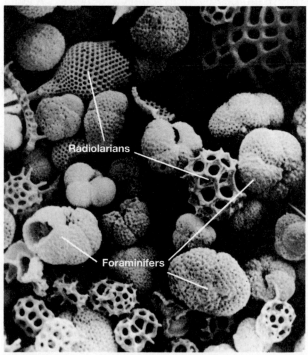
(d)

Figure 4–9 Microscopic calcareous tests. Scanning electron micrographs: **(a)** Coccolithophores (diameter of individual coccolithophores = 20 micrometers, equal to 20 millionths of a meter). **(b)** Diatom (siliceous) and coccoliths (diameter of diatom = 70 micrometers). **(c)** Foraminifers (most species 400 micrometers in diameter). **(d)** Calcareous ooze, which also includes some siliceous radiolarian tests (magnified 160 times).

Figure 4–10 The White Cliffs of southern England. The White Cliffs near Dover in southern England are composed of hardened coccolith-rich calcareous ooze (chalk).

Persian Gulf) suggest that these carbonate deposits occurred in shallow, warm water shelves and around tropical islands as coral reefs and beaches.

Ancient marine carbonate deposits constitute 2% of Earth's crust and 25% of all sedimentary rocks on Earth. These limestone deposits form the bedrock underlying Florida and many Midwestern states from Kentucky to Michigan and from Pennsylvania to Colorado. Percolation of groundwater through these deposits has dissolved the limestone to produce sinkholes and spectacular caverns.

Pelagic Deposits Microscopic biogenous sediment (ooze) is common on the deep-ocean floor because there is so little lithogenous sediment deposited at great distances from the continents that could dilute the biogenous material.

Siliceous Ooze Siliceous ooze contains at least 30% of the hard remains of silica-secreting organisms. When the siliceous ooze consists mostly of diatoms, it is called **diatomaceous ooze.** When it consists mostly of radiolarians, it is called **radiolarian ooze.** When it consists mostly of single-celled silicoflagellates—another type of alga—it is called **silicoflagellate ooze.**

The ocean is undersaturated with silica at all depths, so seawater slowly but continually dissolves silica. How can siliceous ooze accumulate on the ocean floor if it is being dissolved? One way is to accumulate the siliceous tests faster than seawater can dissolve them. For instance, many tests sinking at the same time will create a deposit of siliceous ooze on the sea floor below (Figure 4–11).[6] Once buried beneath other siliceous tests, they are no longer exposed to the dissolving effects of seawater. Thus, siliceous ooze is commonly found in areas below surface waters with high biologic productivity of silica-secreting organisms.

Calcareous ooze Calcareous ooze contains at least 30% of the hard remains of calcareous-secreting organisms. When it consists mostly of coccolithophores, it is called **coccolith ooze.** When it consists mostly of foraminifers, it is called **foraminifer ooze.** One of the most common types of foraminifer ooze is **Globigerina ooze**, named for a foraminifer that is especially widespread in the Atlantic and South Pacific Oceans. Other calcareous oozes include **pteropod oozes** and **ostracod oozes**.

The destruction (solubility) of calcium carbonate (calcite) varies with depth. At the warmer surface and in the shallow parts of the ocean, seawater is generally saturated with calcium carbonate, so calcite does not dissolve. In the deep ocean, however, the colder water contains greater amounts of carbon dioxide, which forms carbonic acid and causes calcareous material to dissolve. The higher pressure at depth also helps speed the dissolution of calcium carbonate.

The depth in the ocean at which the pressure is high enough and the amount of carbon dioxide in deep-ocean waters is great enough to begin dissolving calcium carbonate is called the **lysocline** (*lusis* = a loosening, *cline* = slope). Below the lysocline, calcium carbonate dissolves at an increasing rate with increasing depth until the **calcite compensation depth (CCD)**[7] is reached. At the CCD and greater depths, sediment does not usually contain much calcite because it readily

---

[6]An analogy to this is trying to get a layer of sugar to form on the bottom of a cup of hot coffee. If a few grains of sugar are slowly dropped into the cup, a layer of sugar won't accumulate. However, if a whole bowl full of sugar is dumped into the coffee, a thick layer of sugar will form on the bottom of the cup.

[7]Because the mineral calcite is composed of calcium carbonate, the calcite compensation depth is also known as the calcium carbonate compensation depth, or the carbonate compensation depth. All go by the handy abbreviation of CCD.

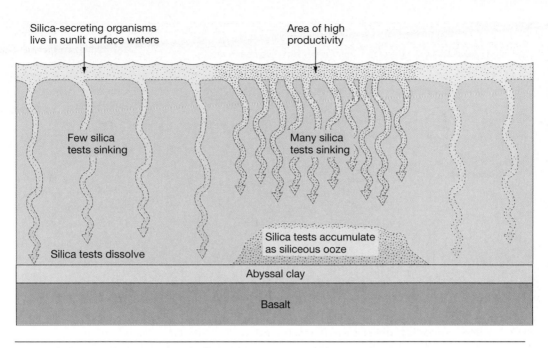

Figure 4–11 Accumulation of siliceous ooze. Siliceous ooze accumulates on the ocean floor beneath areas of high productivity, where the rate of accumulation of siliceous tests is greater than the rate at which silica is being dissolved.

dissolves—even the thick tests of foraminifers dissolve within a day or two. The CCD, on average, is 4500 meters (15,000 feet) below sea level, but, depending on the chemistry of the deep ocean, may be as deep as 6000 meters (20,000 feet) in portions of the Atlantic Ocean, or as shallow as 3500 meters (11,500 feet) in the Pacific Ocean. The depth of the lysocline also varies from ocean to ocean but averages about 4000 meters (13,100 feet).

Because of the CCD, modern carbonate oozes are generally rare below 5000 meters (16,400 feet). Still, buried deposits of ancient calcareous ooze are found beneath the CCD. How can calcareous ooze exist below the CCD? The necessary conditions are shown in Figure 4–12. The mid-ocean ridge is a topographically high feature that rises above the sea floor. It often pokes up above the CCD, even though the surrounding deep-ocean floor is below the CCD. Thus, calcareous ooze deposited on top of the mid-ocean ridge will not be dissolved. Sea floor spreading causes the newly created sea floor and the calcareous sediment on top of it to move into deeper water away from the ridge, eventually being transported below the CCD. This calcareous sediment will dissolve below the CCD, unless it is covered by a deposit that is unaffected by the CCD (such as siliceous ooze or abyssal clay).

The map in Figure 4–13 shows the percentage (by weight) of calcium carbonate in the modern surface sediments of the ocean basins. High concentrations of calcareous ooze (sometimes exceeding 80%) are found along segments of the mid-ocean ridge, but little is found in deep-ocean basins below the CCD. For example, in the northern Pacific Ocean—one of the deepest parts of the world ocean—there is very little calcium carbonate in the sediment. Calcium carbonate is also rare in sediments accumulating beneath cold, high-latitude waters where calcareous-secreting organisms are relatively uncommon. Table 4–3 compares the environmental differences that can be inferred from siliceous and calcareous oozes. It shows that siliceous ooze typically forms below cool surface water regions, including areas of **upwelling** where deep ocean water comes to the surface and supplies nutrients that stimulate high rates of biological productivity. Calcareous ooze, on the other hand, is found on the shallower areas of the ocean floor beneath warmer surface water.

Biogenous sediment is produced from the hard remains of once-living organisms. Microscopic biogenous sediment is especially widespread and forms deposits of ooze on the ocean floor.

Hydrogenous Sediment

Hydrogenous (*hydro* = water, *generare* = to produce) **sediment** is derived from the dissolved material in water.

Origin

Seawater contains many dissolved materials. Chemical reactions within seawater cause certain minerals to

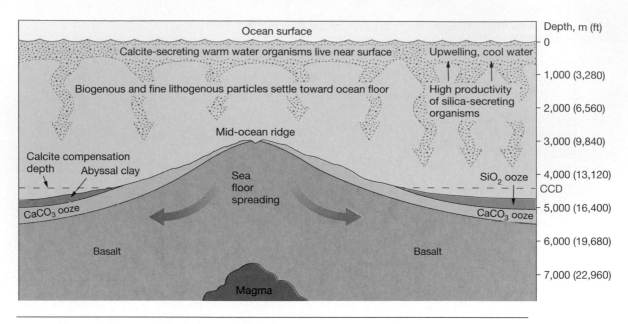

Figure 4–12 Sea floor spreading and sediment accumulation. Relationships among carbonate compensation depth, the mid-ocean ridge, sea floor spreading, productivity, and destruction that allow calcareous ooze to be preserved below the CCD.

come out of solution or **precipitate** (change from the dissolved to the solid state). Precipitation usually occurs when there is a *change in conditions*, such as a change in temperature or pressure or the addition of chemically active fluids. To make rock candy, for instance, a pan of water is heated and sugar is added. When the water is hot and the sugar dissolved, the pan is removed from the heat and the sugar water is allowed to cool. The *change in temperature* causes the sugar to become oversaturated, which causes it to precipitate. As the water cools, the sugar precipitates on anything that is put in the pan, such as pieces of string or kitchen utensils.

Composition and Distribution

Although hydrogenous sediments represent a relatively small portion of the overall sediment in the ocean, they have many different compositions and are distributed in diverse environments of deposition.

Manganese Nodules **Manganese nodules** are rounded, hard lumps of manganese, iron, and other metals typically 5 centimeters (2 inches) in diameter up to a maximum of about 20 centimeters (8 inches). When cut in half, they often reveal a layered structure formed by precipitation around a central nucleation object (Figure 4–14a). The nucleation object may be a piece of lithogenous sediment, coral, volcanic rock, a fish bone, or a shark's tooth. Manganese nodules are found on the deep-ocean floor at concentrations of about 100 nodules per square meter (square yard). In some areas, they occur in even greater abundance (Figure 4–14b),

resembling a scattered field of baseball-sized nodules. The formation of manganese nodules requires extremely low rates of lithogenous or biogenous input so that these sediments do not bury them.

The major components of these nodules are manganese dioxide (around 30% by weight) and iron oxide (around 20%). The element manganese is important for making high-strength steel alloys. Other accessory metals present in manganese nodules include copper (used in electrical wiring, pipe, and in making brass and bronze), nickel (used to make stainless steel), and cobalt (used as an alloy with iron to make strong magnets and steel tools). Although the concentration of these accessory metals are usually less than 1%, they can exceed 2% by weight, which may make them attractive exploration targets in the future.

Phosphates Phosphorus-bearing compounds (**phosphates**) occur abundantly as coatings on rocks and as nodules on the continental shelf and on banks at depths shallower than 1000 meters (3300 feet). Concentrations of phosphates in such deposits commonly reach 30% by weight and indicate abundant biological activity in surface water above where they accumulate. Phosphates are valuable as fertilizers, and ancient marine deposits on land are extensively mined to supply agricultural needs.

Carbonates The two most important carbonate minerals in marine sediment are **aragonite** and calcite. Both are composed of calcium carbonate ($CaCO_3$), but aragonite has a different crystalline structure that is less stable and changes into calcite over time.

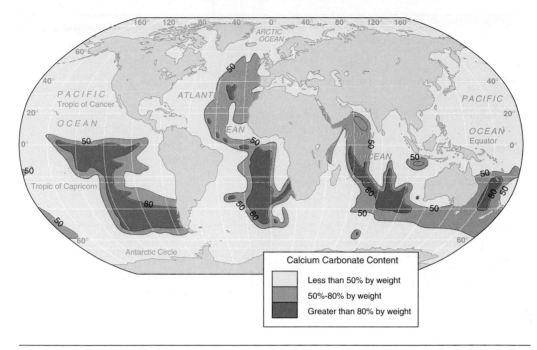

Calcium Carbonate Content

- Less than 50% by weight
- 50%-80% by weight
- Greater than 80% by weight

Figure 4–13 Distribution of calcium carbonate in modern surface sediments. High percentages of calcareous ooze closely follow the mid-ocean ridge, which is above the CCD.

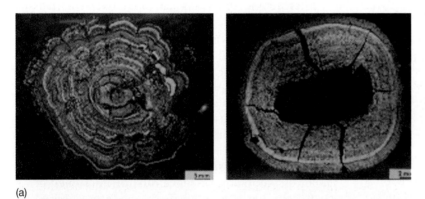

(a)

Figure 4–14 Manganese nodules.
(a) Manganese nodules cut in half, revealing their central nucleation object and layered internal structure. **(b)** A portion of the South Pacific Ocean floor about 4 meters (13 feet) across showing an abundance of manganese nodules.

(b)

Table 4–3   Comparison of environments interpreted from deposits of siliceous and calcareous ooze in surface sediments.

	Siliceous ooze	Calcareous ooze
Surface water temperature above sea floor deposits	Cool	Warm
Main location found	Sea floor beneath cool surface water in high latitudes	Sea floor beneath warm surface water in low latitudes
Other factors	Upwelling brings deep, cold, nutrient-rich water to the surface	Calcareous ooze dissolves below the CCD
Other locations found	Sea floor beneath areas of upwelling, including along the Equator	Sea floor beneath warm surface water in low latitudes along the mid-ocean ridge

As previously discussed, most carbonate deposits are of biogenous origin. However, hydrogenous carbonate deposits can precipitate directly from seawater in tropical climates to form aragonite crystals less than 2 millimeters (0.08 inch) long. Additionally, **oolites** (*oo* = egg, *ite* = stone) are small calcite spheres 2 millimeters (0.08 inch) in diameter or less that have layers like an onion and form in some shallow tropical waters where concentrations of $CaCO_3$ are high. Oolites precipitate around a nucleus and grow larger as they roll back and forth on beaches by wave action.

Metal Sulfides   Deposits of **metal sulfides** are associated with hydrothermal vents and black smokers along the mid-ocean ridge. These deposits contain iron, nickel, copper, zinc, silver and other metals in varying proportions. Transported away from the mid-ocean ridge by sea floor spreading, these deposits can be found throughout the ocean floor and can even be uplifted onto continents.

Evaporites   **Evaporite minerals** form where there is restricted open ocean circulation and where evaporation rates are high (Box 4–2). As water evaporates from these areas, the remaining seawater becomes saturated with dissolved minerals, which then begin to precipitate. Heavier than seawater, they sink to the bottom or form a white crust of evaporite minerals around the edges of these areas (Figure 4–15). Collectively termed "salts," some evaporite minerals taste salty, such as **halite** (common table salt, NaCl), and some do not, such as the calcium sulfate minerals **anhydrite** ($CaSO_4$) and **gypsum** ($CaSO_4 \cdot 2H_2O$).

Hydrogenous sediment is produced when dissolved materials precipitate out of solution and includes a variety of materials found in local concentrations on the ocean floor.

Cosmogenous Sediment

Cosmogenous (*cosmos* = universe, *generare* = to produce) **sediment** is derived from extraterrestrial sources.

Origin, Composition, and Distribution

Forming an insignificant portion of the overall sediment on the ocean floor, cosmogenous sediment consists of two main types: microscopic **spherules**, and macroscopic **meteor** debris.

Microscopic spherules are small globular masses. Some spherules are composed of silicate rock material and show evidence of being formed by extraterrestrial impact events on Earth or other planets that eject small molten pieces of crust into space. These **tektites** (*tektos* = molten) then rain down on Earth and can form *tektite fields*. Other spherules are composed mostly of iron and nickel (Figure 4–16) that form in the asteroid belt between the orbits of Mars and Jupiter and are produced when asteroids collide. This material constantly rains down on Earth as a general component of **space dust** or micrometeorites that float harmlessly through the atmosphere. Although about 90% of micrometeorites are destroyed by frictional heating as they enter the atmosphere, it has been estimated that as much as 300,000 metric tons (331,000 short tons) of space dust reach Earth's surface each year. The iron-rich space dust that lands in the oceans often dissolves in seawater. Glassy tektites, however, do not dissolve as easily and sometimes comprise minute proportions of various marine sediments.

Macroscopic meteor debris is rare on Earth but can be found associated with meteor impact sites. Evidence suggests that throughout time, meteors have collided with Earth at great speeds and that some larger ones have released energy equivalent to the explosion of multiple large nuclear bombs (Box 4–3). The debris from meteors—called **meteorite** material—settles out around the impact site and is either composed of silicate rock material (called **chondrites**) or iron and nickel (called **irons**).

Box 4–2
When a Sea Was Dry: Clues from the Mediterranean

The **Mediterranean** (*med* = middle, *terra* = land) **Sea** is surrounded by land (Figure 4D) except for its shallow connection to the Atlantic Ocean through the Strait of Gibraltar, which is only about 14 kilometers (9 miles) wide. Analysis of sea floor sediments from the Mediterranean Sea suggest that it must have nearly dried up at least once (and perhaps several times) in its history.

About 6 million years ago, a drop in sea level or tectonic activity at the Strait of Gibraltar cut off circulation from the Atlantic Ocean—in effect, creating a dam. With the inflow of the Atlantic Ocean eliminated, the arid climate and resulting high evaporation rates caused the Mediterranean Sea to nearly evaporate in just a few thousand years. As the seawater evaporated, the dissolved substances began to precipitate out of the water, leaving thick layers of evaporite minerals on the sea floor. Up to 4000 meters (13,100 feet) of salt can be found in parts of the Mediterranean, suggesting that the basin may have partially filled and dried up several times during this period. At the same time, unusual gravel deposits washed in from the continents and shallow-water carbonate algal mats called **stromatolites** (*stromat* = covering, *lithos* = stone) also formed. Other supporting evidence for the drying of the Mediterranean Sea includes changes in climate, fossil evidence, and even deep notches cut into the surrounding river valleys. Eventually, most of the water evaporated, leaving a hot, salty, desiccated basin floor far below sea level.

About a half million years later, erosion, further tectonic activity, or a rise in sea level caused the dam at Gibraltar to breach and the Mediterranean started to refill with seawater. The waterfall that spilled into the Mediterranean was probably the largest ever to exist and is estimated to have been 1000 times larger than the flow of all rivers in the world. At that rate, the Mediterranean would have again been full of seawater in only 100 years. Nonetheless, the clues to its history of drying out are preserved in its sea floor sediments.

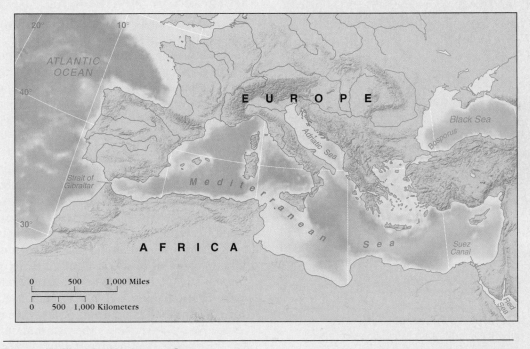

Figure 4D The Mediterranean Sea.

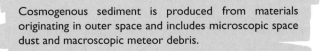

Cosmogenous sediment is produced from materials originating in outer space and includes microscopic space dust and macroscopic meteor debris.

Mixtures

Lithogenous and biogenous sediment rarely occur as an absolutely pure deposit that does not contain other types of sediment. For instance:

- Most calcareous oozes contain some siliceous material, and vice versa (see, for example, Figure 4–9d).

- The abundance of clay-sized lithogenous particles throughout the world and the ease with which they are transported by winds and currents means that these particles are incorporated into every sediment type.

- The composition of biogenous ooze includes up to 70% fine-grained lithogenous clays.

Figure 4–15 Evaporite salts. Due to a high evaporation rate, salts (white material) precipitate onto the floor of Death Valley, California.

Distribution of Neritic and Pelagic Deposits: A Summary

The map in Figure 4-17 shows the distribution of neritic and pelagic deposits in the world's oceans. Coarse-grained lithogenous neritic deposits dominate continental margin areas (*dark brown shading*). Although neritic deposits also contain biogenous, hydrogenous, and cosmogenous particles, these constitute only a minor percentage of the total sediment mass. Fine lithogenous pelagic deposits of abyssal clays (*yellow shading*) are common in deeper areas of the ocean basins.

However, pelagic deposits are dominated by biogenous calcareous oozes (*blue shading*), which are found on the relatively shallow deep-ocean areas along the mid-ocean ridge. Biogenous siliceous oozes are found beneath areas of unusually high biologic productivity such as the Antarctic (*light green shading*, where diatomaceous ooze occurs) and the equatorial Pacific Ocean (*dark green shading*, where radiolarian ooze occurs). Hydrogenous and cosmogenous sediment comprise only a small proportion of pelagic deposits in the ocean. Figure 4–18 shows the distribution of sediment across a passive continental margin.

While neritic deposits cover about one-quarter of the ocean floor, pelagic deposits cover the other three-quarters. The bar graph in Figure 4–19 shows the proportion of each ocean floor that is covered by pelagic calcareous ooze, siliceous ooze, and abyssal clay. Calcareous oozes predominate, covering almost 48% of the world's deep-ocean floor. Abyssal clay covers 38% and siliceous oozes 14% of the world ocean floor area. The graph also shows that the amount of ocean basin floor covered by calcareous ooze decreases in deeper basins because they generally lie beneath the CCD. The dominant oceanic sediment in the deepest basin—the Pacific—is abyssal clay (see also Figure 4–17). Calcareous

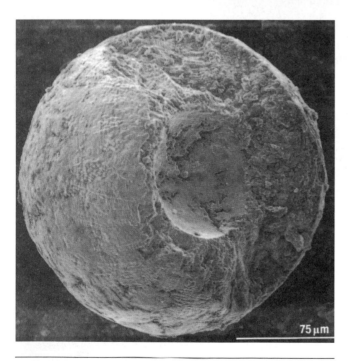

Figure 4–16 Microscopic cosmogenous spherule. Scanning electron micrograph of an iron-rich spherule of cosmic dust. Bar scale of 75 micrometers is equal to 75 millionths of a meter.

- Most lithogenous sediment contains small percentages of biogenous particles.
- Many types of sediment can be incorporated into hydrogenous sediment.
- Tiny amounts of cosmogenous sediment are mixed in with all other sediment types.

Each deposit is a mixture of different sediment types. Typically, however, one type of sediment dominates, which allows the deposit to be classified as primarily lithogenous, biogenous, hydrogenous, or cosmogenous.

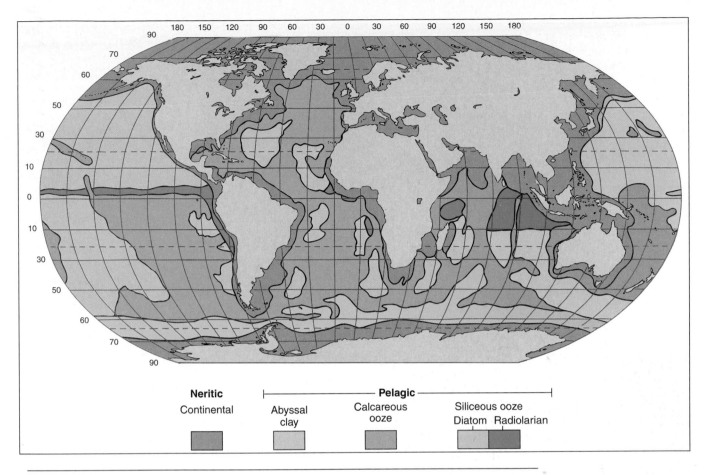

Neritic

Continental

Pelagic

Abyssal clay

Calcareous ooze

Siliceous ooze

Diatom   Radiolarian

Figure 4–17 Distribution of neritic and pelagic sediments.

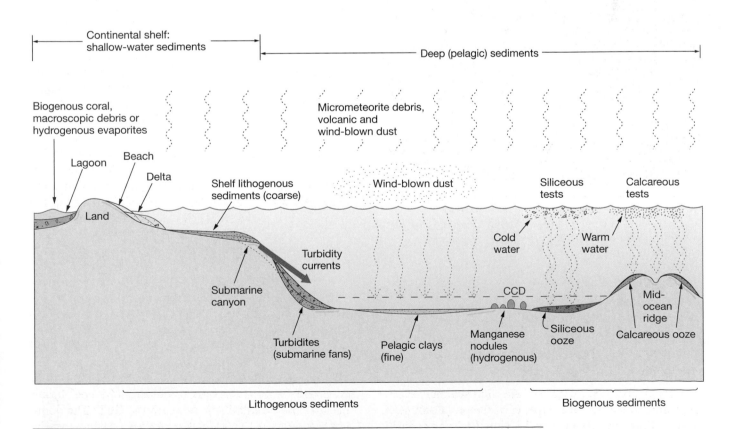

Continental shelf: shallow-water sediments

Deep (pelagic) sediments

Biogenous coral, macroscopic debris or hydrogenous evaporites

Micrometeorite debris, volcanic and wind-blown dust

Lagoon

Beach

Delta

Land

Shelf lithogenous sediments (coarse)

Wind-blown dust

Siliceous tests

Calcareous tests

Cold water

Warm water

Turbidity currents

Submarine canyon

CCD

Mid-ocean ridge

Turbidites (submarine fans)

Pelagic clays (fine)

Manganese nodules (hydrogenous)

Siliceous ooze

Calcareous ooze

Lithogenous sediments

Biogenous sediments

Figure 4–18 Distribution of sediment across a passive continental margin.

Box 4–3   When the Dinosaurs Died: The Cretaceous–Tertiary (K–T) Event

The extinction of the dinosaurs—and two-thirds of all plant and animal species on Earth (including many marine species)—occurred 65 million years ago. This extinction marks the boundary between the Cretaceous (K) and Tertiary (T) Periods of geologic time, and is known as the "**K–T event**." Did slow climate change lead to the extinction of these organisms, or was it a catastrophic event? Was their demise related to disease, diet, predation, or volcanic activity? Earth scientists have long sought clues to this mystery.

In 1980, geologist Walter Alvarez, his father, Nobel physics laureate Luis Alvarez, and two nuclear chemists, Frank Asaro and Helen Michel, reported that deposits collected in northern Italy from the K–T boundary contained a clay layer with high proportions of the metallic element iridium (Ir). Iridium is rare in rocks from Earth, but occurs in greater concentration in meteorites. Therefore, layers of sediment that contain unusually high concentrations of iridium suggest that the material may be of extraterrestrial origin. Additionally, the clay layer contained shocked quartz grains, which indicate that an event occurred with enough force to fracture and partially melt pieces of quartz. Other deposits from the K–T boundary revealed similar features, supporting the idea that Earth experienced an extraterrestrial impact at the same time that the dinosaurs died.

One problem with the impact hypothesis, however, was that volcanic eruptions on Earth could create similar clay deposits enriched in iridium and containing shocked quartz. In fact, large outpourings of basaltic volcanic rock in India (called the Deccan Traps) and other locations had occurred at about the same time as the dinosaur extinction. Also, if there was a catastrophic meteor impact, where was the crater?

In the early 1990s the **Chicxulub** (pronounced "SCHICK-sue-lube") **Crater** off the Yucatàn coast in the Gulf of Mexico was identified as a likely candidate because of its structure, age, and size. Its structure is comparable to other impact craters in the solar system, and its age matches the K–T event. At 190 kilometers (120 miles) in diameter, it is

the largest impact crater on Earth. To create a crater this large, a 10-kilometer (6-mile)-wide meteoroid composed of rock and/or ice traveling at speeds up to 72,000 kilometers (45,000 miles) per hour must have slammed into Earth (Figure 4E). The impact probably bared the sea floor in the area, and created huge waves—estimated to be up to 914 meters (3000 feet) high—that traveled throughout the oceans. This impact is thought to have kicked up so much dust that it blocked sunlight, chilled Earth's surface, and brought about the extinction of dinosaurs and other species. In addition, acid rains and global fires may have added to the environmental disaster. Such a large impact on Earth could easily cause major volcanic activity, so it seems likely that the basalt outpourings could be related to the impact.

Supporting evidence for the meteor impact hypothesis was provided in 1997 by the Ocean Drilling Program (ODP). Previous drilling close to the impact site did not reveal any K–T deposits. Evidently, the impact and resulting huge waves had stripped the ocean floor of its sediment. However, at 1600 kilometers (1000 miles) from the impact site, some of the telltale sediments were preserved on the sea floor. Drilling into the continental margin off Florida into an underwater peninsula called the Blake Nose, the ODP scientific party recovered cores from the K–T boundary that contain a complete record of the impact (Figure 4F).

The cores reveal that before the impact, the layers of Cretaceous age sediment are filled with abundant fossils of calcareous coccoliths and foraminifers and show signs of underwater landslide activity—perhaps the effect of an impact-triggered earthquake. Above this calcareous ooze is a 20-centimeter (8-inch) thick layer of rubble containing evidence of an impact: spherules, tektites, shocked quartz from hard-hit terrestrial rock—even a 2-centimeter (1-inch) piece of reef rock from the Yucatàn peninsula! This layer is also rich in iridium, just like other K–T boundary sequences. Atop this layer is a thick gray clay deposit containing meteor debris and severely reduced numbers of coccoliths and foraminifers. Life in the ocean apparently recovered slowly,

Figure 4E The K–T meteorite impact event.

taking at least 5,000 years before sediment teeming with new, Tertiary-age microorganisms began to be deposited.

Convincing evidence of the K–T impact from this and other cores along with the recent observation of Comet Shoemaker-Levy's 1994 collision with Jupiter suggests that Earth has experienced many such extraterrestrial impacts over geologic time. In fact, over 110 meteor impact sites—called *astroblemes*—have been identified on Earth. Statistics show that an impact the size of the K–T event should occur on Earth about once every 100 million years. Each impact would severely affect life on Earth as it did for the dinosaurs. Nevertheless, their extinction made it possible for mammals to eventually rise to the position of dominance they hold on Earth today.

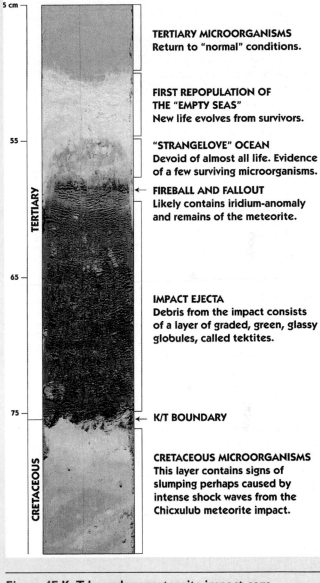

Cretaceous/Tertiary Boundary meteorite impact
ODP Leg 171B, Site 1049, Core 1049A, Section 17X-2

TERTIARY MICROORGANISMS
Return to "normal" conditions.

FIRST REPOPULATION OF THE "EMPTY SEAS"
New life evolves from survivors.

"STRANGELOVE" OCEAN
Devoid of almost all life. Evidence of a few surviving microorganisms.

← **FIREBALL AND FALLOUT**
Likely contains iridium-anomaly and remains of the meteorite.

IMPACT EJECTA
Debris from the impact consists of a layer of graded, green, glassy globules, called tektites.

← **K/T BOUNDARY**

CRETACEOUS MICROORGANISMS
This layer contains signs of slumping perhaps caused by intense shock waves from the Chicxulub meteorite impact.

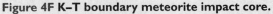

Figure 4F K–T boundary meteorite impact core.

ooze is the most widely deposited sediment in the shallower Atlantic and Indian Oceans. Siliceous oozes cover a smaller percentage of the ocean bottom in all the oceans because regions of high productivity of siliceous-secreting organisms are restricted to the equatorial region (for radiolarians) and high latitudes (for diatoms). Table 4–4 shows the average rates of deposition of selected marine sediments in neritic and pelagic deposits.

> Neritic deposits occur close to shore and are dominated by coarse lithogenous material. Pelagic deposits occur in the deep ocean and are dominated by biogenous oozes and fine lithogenous clay.

Microscopic biogenous tests should take from 10 to 50 years to sink from the ocean surface where the organisms lived to the abyssal depths where biogenous ooze accumulates. During this time, a horizontal ocean current of only 1 centimeter per second (about 0.02 mile per hour) could carry tests as much as 15,000 kilometers (9300 miles) before they settled onto the deep-ocean floor. Why, then, do biogenous tests on the deep-ocean floor closely reflect the population of organisms living in the surface water directly above? Remarkably, about 99% of the particles that fall to the ocean floor do so as part of **fecal pellets**, which are produced by tiny animals that eat algae and protozoans living in the water column, digest their tissues, and excrete their hard parts. These pellets are full of the remains of algae and

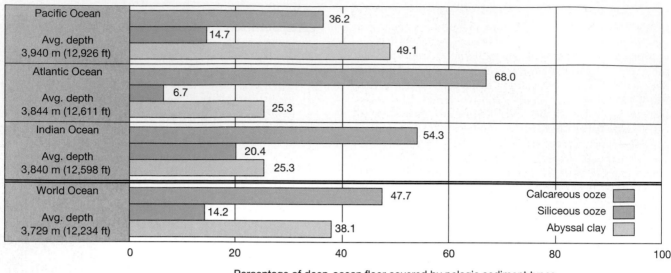

Figure 4–19 Percentage of pelagic sediment types within each ocean. Distribution of pelagic calcareous ooze, siliceous ooze, and abyssal clay in each ocean, and for the world ocean. The average depths shown exclude shallow adjacent seas, where little pelagic sediment accumulates.

Table 4-4   Average rates of deposition of selected marine sediments.

Type of sediment/deposit	Average rate of deposition (per 1000 years)	Thickness of deposit after 1000 years equivalent to ...
Coarse lithogenous sediment, neritic deposit	1 meter (3.3 feet)	A meter stick
Biogenous ooze, pelagic deposit	1 centimeter (0.4 inch)	The diameter of a dime
Abyssal clay, pelagic deposit	1 millimeter (0.04 inch)	The thickness of a dime
Manganese nodule, pelagic deposit	0.001 millimeter (0.00004 inch)	A microscopic dust particle

protozoans from the surface waters (Figure 4–20) and, though still small, are large enough to sink to the deep ocean floor in only 10 to 15 days.

Ocean Sediments as a Resource

The sea floor is rich in potential mineral and organic resources. Most of it, however, is not easily accessible and the recovery of these resources involves technological challenges and high cost. Nevertheless, let's examine some of the most appealing exploration targets.

Petroleum

The ancient remains of microscopic organisms, buried within marine sediments before they could decompose, are the source of today's **petroleum** (oil and natural gas) deposits. Of the nonliving resources extracted from the oceans, more than 95% of the economic value is in petroleum products.

The percentage of world oil produced from offshore regions has increased from trace amounts in the 1930s to over 30% in the 2000s. Most of this increase results from continuing technological advancements employed by offshore drilling platforms (Figure 4–21). Major offshore reserves exist in the Persian Gulf, in the Gulf of Mexico, off southern California, in the North Sea, and in the East Indies. Additional reserves are probably located off the north coast of Alaska and in the Canadian Arctic, Asian seas, Africa, and Brazil. With almost no likelihood of finding major new reserves on land, future offshore petroleum exploration will continue to be intense, especially in deeper waters of the continental margins. However, a major drawback to offshore petroleum exploration is the inevitable oil spills caused by inadvertent leaks or blowouts during the drilling process.

Gas Hydrates

Gas hydrates, which are also known as *clathrates* (*clathr* = a lattice), are unusually compact chemical structures made of water and natural gas. The most common type of natural gas is methane, which produces **methane**

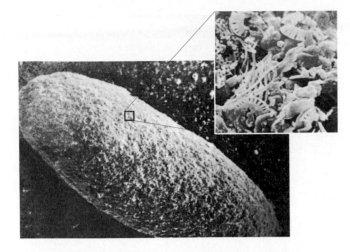

Figure 4–20 Fecal pellet. A 200-micrometer (0.008 inch)-long fecal pellet, which is large enough to sink rapidly from the surface to the ocean floor. Close-up of the surface of a fecal pellet (*inset*) shows the remains of coccoliths and other debris.

hydrate. Gas hydrates occur beneath permafrost areas on land and under the ocean floor, where they were discovered in 1976.

In deep-ocean sediments, where pressures are high and temperatures are low, water and natural gas combine in such a way that the gas is trapped inside a lattice-like cage of water molecules. Vessels that have drilled into gas hydrates have retrieved cores of mud mixed with chunks or layers of gas hydrates that fizzle and evaporate quickly when they are exposed to the relatively warm, low-pressure conditions at the ocean surface. Gas hydrates resemble chunks of ice, but ignite when lit by a flame because methane and other flammable gases are released as gas hydrates vaporize (Figure 4–22).

Most oceanic gas hydrates are created when bacteria break down organic matter trapped in sea floor sediments, producing methane gas with minor amounts of ethane and propane. These gases can be incorporated into gas hydrates under high pressure and low temperature conditions. Most ocean floor areas below 525 meters (1720 feet) provide these conditions, but gas hydrates seem to be confined to continental margin areas, where high productivity surface waters enrich ocean floor sediments below with organic matter.

Studies of the deep-ocean floor reveal that at least 50 sites worldwide may contain extensive gas hydrate deposits. Research suggests that at various times in the geologic past, changes in sea level or sea floor instability has allowed large quantities of methane to be released. The release of methane from the sea floor can affect global climate as methane—a potent greenhouse gas—increases in the atmosphere. Methane seeps also support a rich community of organisms, many of which are species new to science.

Figure 4–21 Offshore drilling rig. Constructed on tall stilts, rigs like this one in the Gulf of Mexico off Texas are important for extracting petroleum reserves from beneath the continental shelves.

Some estimates indicate that as much as 20 quadrillion cubic meters (700 quadrillion cubic feet) of methane are locked up in sediments containing gas hydrates. This is equivalent to about *twice* as much carbon as Earth's coal, oil, and conventional gas reserves combined (Figure 4–23), so gas hydrates may potentially be the world's largest source of usable energy. One of the major drawbacks in exploiting reserves of gas hydrate is that they rapidly decompose at surface temperatures and pressures. In the future, however, we may see the day when these vast sea floor reserves are used to power modern society.

▬EIO▬

For more information and on-line exercises about the Environmental Issue in Oceanography (EIO) "Methane Hydrates: Energy Boom or Climate Bust?", visit the EIO Web site at **http://www. prenhall.com/oceanissues** and select Issue #3.

Sand and Gravel

The offshore sand and gravel industry is second in economic value only to the petroleum industry. Sand and gravel, which includes rock fragments that are washed out to sea and shells of marine organisms, is mined by

(a)

(b)

Figure 4–22 Gas hydrates. **(a)** A sample retrieved from the ocean floor shows layers of white ice-like gas hydrate mixed with mud. **(b)** Gas hydrates evaporate when exposed to surface conditions and release natural gas, which can be ignited.

offshore barges using a suction dredge. This material is primarily used as an aggregate in concrete, as a fill material in grading projects, and on recreational beaches.

Offshore deposits are a major source of sand and gravel in New England, New York, and throughout the Gulf Coast. Many European countries, Iceland, Israel, and Lebanon also depend heavily on such deposits.

Some offshore sand and gravel deposits are rich in valuable minerals. Gem quality diamonds, for example, are recovered from gravel deposits on the continental shelf offshore South Africa and Australia, where waves reworked them during low stands of the sea. Sediments rich in tin have been mined offshore southeast Asia from Thailand to Indonesia. Platinum and gold have been found in deposits in gold mining areas throughout the world, and some Florida beach sands are rich in titanium.

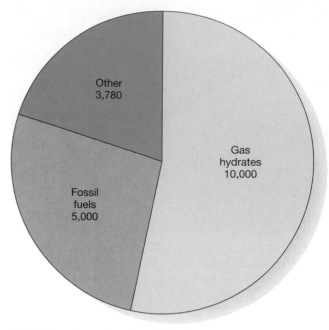

Distribution of organic carbon in Earth reservoirs (excluding dispersed carbon in rocks and sediments, which equals nearly 1,000 times this total amount). Numbers in gigatons (10^{25} tons) of carbon.

Figure 4–23 Organic carbon in Earth reservoirs. Gas hydrates contain twice as much organic carbon as all fossil fuels combined. Values in billions of tons of carbon; "other" includes sources such as soil, peat, and living organisms.

Evaporative Salts

When seawater evaporates, the salts increase in concentration until they can no longer remain dissolved, so they precipitate out of solution and form salt deposits (Figure 4–24). The most economically useful salts are gypsum and halite. Gypsum is used in plaster of Paris to make casts and molds, and is the main component in gypsum board (wallboard or sheet rock).

Halite—common table salt—is widely used for seasoning, curing, and preserving foods. It is also used to de-ice roads, in water conditioners, in agriculture, and in the clothing industry for dying fabric. Additionally, halite is used in the production of chemicals such as sodium hydroxide (to make soap products), sodium hypochlorite (for disinfectants, bleaching agents, and PVC piping), sodium chlorate (for herbicides, matches, and fireworks), and hydrochloric acid (for use in chemical applications and for cleaning scaled pipes). The manufacture and use of salt is one of the oldest chemical industries.[8]

---

[8]An interesting historical note about salt is that part of a Roman soldier's pay was in salt. That portion was called the *salarium*, from which the word *salary* is derived. If a soldier did not earn it, he was not worth his salt.

Phosphorite (Phosphate Minerals)

Phosphorite is a sedimentary rock consisting of various phosphate minerals containing the element phosphorus, an important plant nutrient. Consequently, phosphate deposits can be used to produce phosphate fertilizer. Although there is currently no commercial phosphorite mining occurring in the oceans, the marine reserve is estimated to exceed 45 billion metric tons (50 billion short tons). Phosphorite occurs in the ocean at depths of less than 300 meters (1000 feet) on the continental shelf and slope.

Some shallow sand and mud deposits contain up to 18% phosphate. Many phosphorite deposits occur as nodules, with a hard crust formed around a nucleus. The nodules may be as small as a sand grain or as large as 1 meter (3.3 feet) in diameter and may contain over 25% phosphate. For comparison, most land sources of phosphate have been enriched to more than 31% by groundwater leaching.

Manganese Nodules and Crusts

Manganese nodules contain significant concentrations of manganese, iron, and smaller concentrations of copper, nickel, and cobalt, all of which have a variety of economic uses. In the 1960s, mining companies began to assess the feasibility of mining manganese nodules from the deep-ocean floor (Figure 4–25). The map in Figure 4–26 shows that vast areas of the sea floor contain manganese nodules. Some of the most extensively covered areas occur in the Pacific Ocean.

Technologically, mining the deep-ocean floor for manganese nodules is possible. However, the political issue of determining international mining rights at great distances from land has hindered exploitation of this resource. Additionally, environmental concerns about mining the deep-ocean floor have not been fully addressed. Evidence suggests that it takes at least several million years for manganese nodules to form, and that their formation depends on a particular set of physical and chemical conditions that probably do not last long at any location. In essence, they are a nonrenewable resource that will not be replaced for a very long time once they are mined.

Of the five metals commonly found in manganese nodules, cobalt is the only metal deemed "strategic" (essential to national security) for the United States. It is required to produce dense, strong alloys with other metals for use in high-speed cutting tools, powerful permanent magnets, and jet engine parts. At present, the United States must import all of its cobalt from large deposits in southern Africa. However, the United States has considered deep-ocean nodules and **crusts** (hard coatings on other rocks) as a more reliable source of cobalt.

Figure 4–24 Mining sea salt. A salt mining operation at San Ignacio Lagoon, Baja California, Mexico. Low-lying areas near the lagoon are allowed to flood with seawater, which evaporates in the arid climate and leaves deposits of salt that are then collected.

Figure 4–25 Mining manganese nodules. Manganese nodules can be collected by dredging the ocean floor. The dredge is shown unloading nodules onto the deck of a ship.

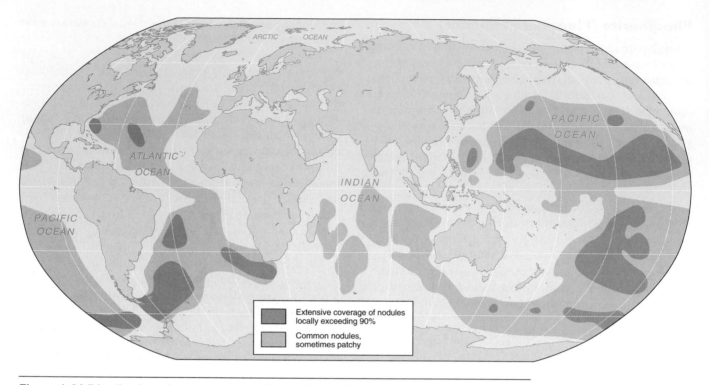

Figure 4–26 Distribution of manganese nodules on the sea floor.

Legend:
- Extensive coverage of nodules locally exceeding 90%
- Common nodules, sometimes patchy

In the 1980s cobalt-rich manganese crusts were discovered on the upper slopes of islands and seamounts that lie relatively close to shore and within the jurisdiction of the United States and its territories. The cobalt concentrations in these crusts are half again as rich as in the best African ores and at least twice as rich as in the deep-sea manganese nodules. However, interest in mining these deposits has faded because of lower metal prices from land-based sources.

> Ocean sediments contains many important resources, including petroleum, gas hydrates, sand and gravel, evaporative salts, phosphorite, and manganese nodules and crusts.

 Students Sometimes Ask...

Are there any areas of the ocean floor where no sediment is being deposited?

Various types of sediment accumulate on nearly all areas of the ocean floor in the same way dust accumulates in all parts of your home (which is why marine sediment is often referred to as "marine dust"). Even the deep-ocean floor far from land receives small amounts of wind-blown material, microscopic biogenous particles, and space dust.

There are a few places in the ocean, however, where very little sediment accumulates. One such place is along the continental slope, where there is active erosion by turbidity and other deep-ocean currents. Another place where very little sediment can be found is along the mid-ocean ridge. Here, the sea floor along the crest of the mid-ocean ridge is so young (because of sea floor spreading) and the rates of sediment accumulation far from land are so slow there hasn't been enough *time* for sediments to form.

How effective is wind as a transporting agent?

Any material that gets into the atmosphere—including dust from dust storms, soot from forest fires, specks of pollution, and ash from volcanic eruptions—is transported by wind and can be found as deposits on the ocean floor. Surprisingly, wind can transport huge volumes of silt- and clay-sized particles great distances. For example, a recent study indicated that each year an average of 11.5 million metric tons (12.6 million short tons) of dust from Africa's Sahara Desert is carried downwind up to 6500 kilometers (4000 miles) across the Atlantic Ocean (see Figure 4–6). Much of this dust falls in the Atlantic. That's why ships traveling downwind from the Sahara Desert often arrive at their destinations quite dusty. Some of it falls in the Caribbean (where it has been linked to stress and disease among coral reefs), the Amazon (where it fertilizes the nutrient-poor soil), and across the southern U.S. as far west as New Mexico.

I've been to Hawaii and seen a black sand beach. Because it forms by lava flowing into the ocean and being broken up by waves, is it hydrogenous sediment?

No. Many active volcanoes in the world have black sand beaches that are created when waves break apart dark-colored volcanic rock. The material that produces the black sand is derived from a continent or an island, so it is considered lithogenous sediment. Even though molten lava sometimes flows into the ocean, the resulting black sand could never be considered hydrogenous sediment because the lava was never *dissolved* in water.

How do manganese nodules form?

The answer to that question has puzzled oceanographers since manganese nodules were first discovered during the world's first great oceanographic voyage, begun in 1872 by the British research ship HMS *Challenger*. If manganese nodules are truly hydrogenous and precipitate from seawater, then how can they have such high concentrations of manganese (which occurs in seawater at concentrations often too small to measure accurately)? Furthermore, why are the nodules on *top* of ocean floor sediment and not buried by the constant rain of sedimentary particles?

Unfortunately, nobody has definitive answers to these questions. Perhaps the creation of manganese nodules is the result of one of the slowest chemical reactions known—on average, they grow at a rate of about 5 millimeters (0.2 inch) per *million years*. Recent research suggests the formation of manganese nodules may be aided by bacteria and an as-yet-unidentified marine organism that intermittently lifts and rotates them. Other studies reveal that the nodules don't form continuously over time but in spurts that are related to specific conditions such as a low sedimentation rate of lithogenous clay and strong deep-water currents. Interestingly, the larger the nodules are, the faster they grow. The mystery of manganese nodules is awaiting someone to investigate what is considered the most interesting unresolved problem in marine chemistry. If you do solve this mystery, you'll be famous throughout the field of oceanography!

When will we run out of oil?

Not any time soon. However, from an economic perspective, when the world runs completely out of oil—a finite resource—is not as relevant as when production begins to taper off. When this happens, we will run out of the *abundant* and *cheap* oil on which all industrialized nations depend. Several oil-producing countries are already past the peak of their production—including the United States and Canada, which topped out in 1972. Current estimates indicate that soon after 2010, more than half of all known and likely to-be-discovered oil will by gone. After that, it will be increasingly more costly to produce oil and prices will rise dramatically—unless demand declines proportionately or other sources such as extra-heavy oil, tar sands, gas hydrates, or other sources become readily available.

I've heard that some countries want to bury their nuclear waste in sediments at sea. Is this true?

Not only is this true, but the United States is one of the countries that has seriously considered high-level nuclear waste disposal at sea. Nuclear waste that remains from the production of nuclear weapons and from commercial power generation must be safely disposed of until it is no longer dangerous, which typically takes several hundred thousand years.

The disposal of nuclear waste has so far been focused on land sites because the materials remain accessible. From 1976 until 1986, however, research was conducted into disposal of high-level nuclear waste within the sea floor of the deep sea. Given the requirements that the waste must be kept away from human activities, kept safe from exposure by natural erosion, and kept away from seismically active regions, the centers of oceanic lithospheric plates are attractive disposal sites. In addition, the fine-grained abyssal clay that exists in the deep ocean serves as an ideal medium in which to place the high-level nuclear waste, presumably trapping any containment leaks.

Proposals specify that a drill-hole in the deep-ocean floor (similar to the holes drilled by ODP's drill ship *JOIDES Resolution* to recover deep-sea sediment cores) could be used for the disposal of canisters containing nuclear waste hundreds of meters beneath the sea floor. The plan calls for stacking the canisters within the drill hole, which would later be filled with mud.

In 1986, the United States halted research into seabed disposal of nuclear waste because of budget constraints. After a lengthy evaluation of potential sites nationwide, Yucca Mountain in Nevada was selected as the future repository site for U.S. nuclear waste. However, concerns about the safety of the Yucca Mountain site have recently been raised. In addition, it appears that the site is not large enough to hold all the nuclear waste that will be generated by the year 2020.

Given the political realities that may be faced as work proceeds toward land-based disposal, deep-sea disposal will most likely be considered again. The seabed disposal program did produce encouraging results, and, in fact, Japan is currently investigating disposing of its nuclear waste in the Mariana Trench. Studies of areas near where nuclear weapons and nuclear submarines have accidentally been lost at sea indicate that clays on the deep-ocean floor tend to hold fast to certain radioactive elements, effectively isolating them from the environment. It would seem that careful ocean disposal, if conducted safely, might be a disposal strategy with merit. However, environmental consequences must be carefully considered before ocean floor disposal is attempted.

Chapter in Review

• The existence of sea floor spreading was confirmed when the *Glomar Challenger* began the Deep Sea Drilling Project to sample ocean sediments and the underlying crust. Today, the Ocean Drilling Program's *JOIDES Resolution* continues the important work of retrieving sediments from the deep-ocean floor, which will be continued after 2003 by the Integrated Ocean Drilling Program. Analysis and interpretation of marine sediments reveals that Earth has had an interesting and complex history including mass extinctions and the drying of entire seas.

• Sediments that accumulate on the ocean floor are classified by origin as lithogenous (derived from rock), biogenous (derived from organisms), hydrogenous (derived from water), or cosmogenous (derived from outer space).

• Lithogenous sediments reflect the composition of the rock from which they were derived. Sediment texture—determined in part by the size, sorting, and rounding of particles—is affected greatly by how the particles were transported (by water, wind, ice, or gravity) and the energy conditions under which they were deposited. Coarse lithogenous material dominates neritic deposits that accumulate rapidly along the margins of continents while fine abyssal clays are found in pelagic deposits.

• Biogenous sediment consists of the hard remains (shells, bones, and teeth) of organisms. These are composed of either silica (SiO_2) from diatoms and radiolarians or calcium carbonate ($CaCO_3$) from foraminifers and coccolithophores. Accumulations of microscopic shells (tests) of organisms must comprise at least 30% of the deposit for it to be classified as biogenic ooze. Biogenous oozes are the most common type of pelagic deposits. The rate of biological productivity relative to the rates of destruction and dilution of biogenous sediment determines whether abyssal clay or oozes will form on the ocean floor. Siliceous ooze will only form below areas of high biologic productivity of silica-secreting organisms at the surface. Calcareous ooze will only form above the calcite compensation depth (CCD)—the depth where seawater dissolves calcium carbonate—although it can be covered and transported into deeper water through sea floor spreading.

• Hydrogenous sediment includes manganese nodules, phosphates, carbonates, metal sulfides, and evaporites that precipitate directly from water or are formed by the interaction of substances dissolved in water with materials on the ocean floor. Hydrogenous sediments represent a relatively small proportion of marine sediment and are distributed in many diverse environments.

• Cosmogenous sediment is composed of either macroscopic meteor debris (such as that produced during the K–T impact event) or microscopic iron-nickel and silicate spherules that result from asteroid collisions or extraterrestrial impacts. Minute amounts of cosmogenous sediment are mixed into most other types of ocean sediment.

• Although most ocean sediment is a mixture of various sediment types, it is usually dominated by lithogenous, biogenous, hydrogenous, or cosmogenous material.

• The distribution of neritic and pelagic sediment is influenced by many factors, including proximity to sources of lithogenous sediment, productivity of microscopic marine organisms, depth of the ocean floor, and the distribution of various sea floor features. Fecal pellets rapidly transport biogenous particles to the deep-ocean floor and cause the composition of sea floor deposits to match the organisms living in surface waters immediately above them.

• The most valuable nonliving resource from the ocean today is petroleum, which is recovered from below the continental shelves and used as a source of energy. Gas hydrates include vast deposits of ice-like material that may some day be used as a source of energy. Other important resources include sand and gravel, evaporative salts, phosphorite, and manganese nodules and crusts.

Key Terms

Abyssal clay (p. 104)

Algae (p. 106)

Anhydrite (p. 114)

Aragonite (p. 114)

Biogenous sediment (p. 105)

Calcareous ooze (p. 106)

Calcite (p. 106)

Calcite compensation depth (CCD) (p. 110)

Calcium carbonate (p. 106)

Carbonate (p. 107)

Chalk (p. 106)

Chicxulub Crater (p. 118)

Chondrite (p. 114)

Coccolith (p. 106)

Coccolith ooze (p. 109)

Coccolithophore (p. 106)

Core (p. 97)

Cosmogenous sediment (p. 114)

Crusts (p. 123)

Deep Sea Drilling Project (DSDP) (p. 97)

Diatom (p. 106)

Diatomaceous earth (p. 106)

Diatomaceous ooze (p. 108)

Dredge (p. 97)

Eroded (p. 100)

Evaporite mineral (p. 114)

Fecal pellet (p. 119)

Foraminifer (p. 106)

Foraminifer ooze (p. 109)

Gas hydrate (p. 120)

Glacial deposit (p. 103)

Globigerina ooze (p. 109)

Glomar Challenger (p. 97)

Gravity corer (p. 97)

Grain size (p. 101)

Gypsum (p. 114)

Halite (p. 114)

Hydrogenous sediment (p. 105)

Ice rafting (p. 104)

Integrated Ocean Drilling Program (IODP) (p. 98)

Irons (p. 114)

JOIDES Resolution (p. 97)

Joint Oceanographic Institutions for Deep Earth Sampling (JOIDES) (p. 97)

K–T event (p. 118)

Limestone (p. 107)

Lithified (p. 99)

Lithogenous sediment (p. 100)

Lysocline (p. 110)

Macroscopic biogenous sediment (p. 105)

Manganese nodule (p. 123)

Maturity (p. 103)

Mediterranean Sea (p. 115)

Metal sulfide (p. 114)

Meteor (p. 114)

Meteorite (p. 114)

Methane hydrate (p. 120)

Microscopic biogenous sediment (p. 105)

Mineral (p. 101)

Nannoplankton (p. 106)

Neritic deposit (p. 103)

Ocean Drilling Program (ODP) (p. 97)

Oolite (p. 114)

Ooze (p. 105)

Ostracod ooze (p. 109)

Pelagic deposit (p. 103)

Petroleum (p. 120)

Phosphate (p. 105)

Phosphorite (p. 123)

Planktonic (p. 106)

Precipitate (p. 110)

Protozoan (p. 106)

Pteropod ooze (p. 109)

Quartz (p. 101)

Radiolarian (p. 106)

Radiolarian ooze (p. 108)

Red clay (p. 105)

Relict sediment (p. 103)

Rotary drilling (p. 97)

Sediment (p. 99)

Sedimentary rock (p. 99)

Silica (p. 106)

Siliceous ooze (p. 106)

Silicoflagellate ooze (p. 108)

Sorting (p. 101)

Space dust (p. 114)

Spherules (p. 114)

Stromatolites (p. 115)

Tektites (p. 114)

Terrigenous sediment (p. 100)

Test (p. 105)

Texture (p. 100)

Turbidite deposits (p. 103)

Turbidity currents (p. 103)

Upwelling (p. 110)

Weathering (p. 100)

Wentworth scale of grain size (p. 101)

Questions And Exercises

1. Describe the process of how a drilling ship like the *JOIDES Resolution* obtains core samples from the deep-ocean floor.

2. What kind of information can be obtained by examining and analyzing core samples?

3. List and describe the characteristics of the four basic types of marine sediment.

4. How does lithogenous sediment originate?

5. Why is most lithogenous sediment composed of quartz grains? What is the chemical composition of quartz?

6. If a deposit has a coarse grain size, what does this indicate about the energy of the transporting medium? Give several examples of various transporting media that would produce such a deposit.

7. What characteristics of marine sediment indicate increasing maturity? Give an example of a mature and immature sediment.

8. List the two major chemical compounds of which most biogenous sediment is composed and the organisms that produce them. Sketch these organisms.

9. What are several reasons that diatoms are so remarkable? List products that contain or are produced using diatomaceous earth.

10. Describe manganese nodules, including what is currently known about how they form.

11. Describe the most common types of cosmogenous sediment and give the probable source of these particles.

12. Why is lithogenous sediment the most common neritic deposit? Why are biogenous oozes the most common pelagic deposits?

13. How do oozes differ from abyssal clay? Discuss how productivity, destruction, and dilution combine to determine whether an ooze or abyssal clay will form on the deep-ocean floor.

14. Describe the environmental conditions (e.g. surface water temperature, productivity, dissolution, etc.) that influence the distribution of siliceous and calcareous ooze.

15. If siliceous ooze is slowly but constantly dissolving in seawater, how can deposits of siliceous ooze accumulate on the ocean floor?

16. Explain the stages of progression that results in calcareous ooze existing below the CCD.

17. How do fecal pellets help explain why the particles found in the ocean surface waters are closely reflected in the particle composition of the sediment directly beneath? Why would one not expect this?

18. Discuss the present importance and the future prospects for the production of petroleum; sand and gravel; phosphorite; and manganese nodules and crusts.

19. What are gas hydrates, where are they found, and why are they important?

20. Describe the K–T event, including evidence for it and its effect on the environment.

References

Alvarez, L. W., Alvarez, W., Asaro, F., and Michel, H. V. 1980. Extraterrestrial cause for the Cretaceous-Tertiary extinction. *Science* 208:4484, 1095–1108.

———. 1982. Current status of the impact theory for the terminal Cretaceous extinction. In Silver, L. T., and Schultz, P. H., eds. *Geological implications of impacts of large asteroids and comets on the Earth*, Geological Society of America Special Paper 190. Boulder, CO: Geological Society of America.

Alvarez, W. 1997. *T. rex and the crater of doom.* Princeton, NJ: Princeton University Press.

Berger, W. H., Adelseck, C. G., Jr. and Mayer, L. A. 1976. Distribution of carbonate in surface sediments of the Pacific Ocean. *Journal of Geophysical Research* 81:15, 2617–2629.

Biscaye, P. E., Kolla, V., and Turedian, K. K. 1976. Distribution of calcium carbonate in surface sediments of the Atlantic Ocean. *Journal of Geophysical Research* 81:15, 2592–2602.

Brasier, M. D. 1980. *Microfossils.* London: George Allen & Unwin.

Brewer, P. G., et al. 1997. Deep-ocean field test of methane hydrate formation from a remotely operated vehicle. *Geology* 25:5, 407–410.

Broadus, J. M. 1987. Seabed minerals. *Science* 235:853–860.

Burnett, J. L. 1991. Diatoms—The forage of the sea. *California Geology* 44:4, 75–81.

Carson, R. L. 1956. *The sea around us.* London: Penguin.

Cronan, D. S., ed. 1999. *Handbook of marine mineral deposits.* Boca Raton: CRC Press.

Emery, K. O., and Uchupi, E. 1984. *The geology of the Atlantic Ocean.* New York: Springer-Verlag.

Frank, L. A., and Huyghe, P. 1990. *The big splash.* New York: Birch Lane Press.

Glassby. G. P., ed. 1977. *Marine manganese deposits.* New York: Elsevier.

Halbach, P., Friedrich, G., and von Stackelberg, U., eds. 1988. The manganese nodule belt of the Pacific Ocean: Geological environment, nodule formation, and mining aspects. Stuttgart, Germany: Ferdinand Enke Verlag.

Hallegraeff, G. M. 1988. *Plankton: A microscopic world.* New York: E. J. Brill.

Hildebrand, A. R., and Boynton, W. V. 1990. Proximal Cretaceous-Tertiary boundary impact deposits in the Caribbean. *Science* 248:4957, 843–846.

Honjo, S. 1992. From the Gobi to the bottom of the North Pacific. *Oceanus* 35:4, 45–53.

Howell, D. G., and Murray, R. W. 1986. A budget for continental growth and denudation. *Science* 233:4762, 446–449.

Hsü, K. 1983. *The Mediterranean was a desert: A voyage of the Glomar Challenger.* Princeton, NJ: Princeton University Press.

Inter-University Program of Research on Ferromanganese Deposits of the Ocean Floor. *Phase 1 Report.* Unpublished.

Kemper, S. 1999. Salt of the earth. *Smithsonian* 29:10, 70–78.

Kennett, J. P. 1982. *Marine geology.* Englewood Cliffs, NJ: Prentice-Hall.

Kolla, V., and Biscaye, P. E. 1976. Distribution of calcium carbonate in surface sediments of the Atlantic Ocean. *Journal of Geophysical Research* 81:15, 2602–2616.

Krajick, K. 1997. The crystal fuel. *Natural History* 106:4, 26–31.

Kvenvolden, K. A. 1996. Gas hydrates-geological perspective and climate change, *in* Pirie, R. G., ed., *Oceanography: Contemporary Readings in Ocean Sciences,* 3rd ed., New York: Oxford University Press.

Kyte, F. T., Zhou, L., and Wasson, J. T. 1988. New evidence on the size and possible effects of a late Pliocene oceanic asteroid impact. *Science* 241:4861, 63–65.

Leinen, M., Cwienk, D., Heath, G. R., Biscaye, P. E., Kolla, V., Thiede, J., and Dauphin, J. P. 1986. Distribution of biogenic silica and quartz in recent deep-sea sediments. *Geology* 14:3, 199–203.

Masters, C. D., Root, D. H., and Attanasi, E. D. 1991. Resource constraints in petroleum production potential. *Science* 253:146–152.

Olsson, R. K., Miller, K. G., Browning, J. V., Habib, D., and Sugarman, P. J. 1997. Ejecta layer at the Cretaceous-Tertiary boundary, Bass River, New Jersey (Ocean Drilling Program Leg 174AX). *Geology* 25:8, 759–762.

Power to the people: A survey of energy. 1994. *The Economist* 331:7868 (18-page insert following p. 60).

Rine, J. M., et. al. 1991. Generation of late Holocene sand ridges on the middle continental shelf of New Jersey, USA—evidence for formation in a mid-shelf setting based on comparisons with a nearshore ridge. *Special Publications of the International Association of Sedimentologists* 14, 395–423.

Ryder, G., Fastovsky, D., and Gartner, S., eds. 1996. *The Cretaceous-Tertiary event and other catastrophes in Earth history.* Geological Society of America Special Paper 307 (a compendium of 37 papers). Boulder, CO: Geological Society of America.

Sharpton, V. L., et al. 1992. New links between the Chicxulub impact structure and the Cretaceous/Tertiary boundary. *Nature* 359:6398, 819–821.

Shinn, E. A. et. al. 2000. African dust and the demise of Caribbean coral reefs. *Geophysical Research Letters* 113:4, 3029-3040.

Stanley, D. J. 1990. Med desert theory is drying up. *Oceanus* 33:1, 14–23.

Sverdrup, H. U., Johnson, M. W., and Fleming, R. H. 1942. Renewal 1970. *The oceans: Their physics, chemistry, and general biology.* Englewood Cliffs, NJ: Prentice-Hall.

The Open University Course Team. 1989. *Ocean chemistry and deep-sea sediments.* Oxford: Pergamon Press.

Udden, J. A. 1898. Mechanical composition of wind deposits. *Augustana Library Pub. 1.*

Various authors. 1994. Special issue featuring several articles focusing on 25 years of ocean drilling. *Oceanus.* 36:4, 1–136.

Watkins, D. K., Premoli Silva, I., and Erba, E. 1995. Cretaceous and Paleogene manganese-encrusted hardgrounds from central Pacific guyots. *Proceedings of the Ocean Drilling Program, Scientific Results* 144, 97–126.

Weaver, P. P. E., and Thomson, J., eds. 1987. *Geology and geochemistry of abyssal plains.* Palo Alto, CA: Blackwell Scientific.

Wentworth, C. K. 1922. A scale of grade and class terms for clastic sediments. *Journal of Geology* 30, 377–392.

Suggested Reading in **Scientific American**

Campbell, C. J., and Laherrère, J. H. 1998. The end of cheap oil. 278:3, 78–83. An updated look at the status of global petroleum reserves, including a forecast of when the world will run out of cheap oil.

Gehrelt, T. T. 1996. Collisions with comets and asteroids. 274:3, 54–61. The nature of asteroids and comets is reviewed. The age and size of some objects involved in major Earth-jarring hits by extraterrestrial bodies and the odds of hits by different size ranges of objects are discussed.

Hollister, C. D., and Nadis, S. 1998. Burial of radioactive waste under the seabed. 278:1, 60–65. Reexamines the merits of the disposal of nuclear waste by burying it within the sea floor.

Mack, W. N., and Leistikow, E. A. 1996. Sands of the world. 275:2, 62–67. A photo essay of sands of the world, showing a wide variety of beach sand material.

Nelson H., and Johnson, K. R. 1987. Whales and walruses as tillers of the sea floor. 256:2, 112–118. The 200,000 walruses and 16,000 gray whales that feed by scooping sediment from the Bering Sea continental shelf suspend large amounts of sediment that is transported by bottom currents.

Schneider, D. 1997. Not in my back yard: Could ocean mud trap nuclear waste from old Russian subs? 276:3, 20–22. A critical look at using the deep-ocean floor as a nuclear waste repository, with examples from accidental Russian submarine sinkings.

Suess, E., et. al. 1999. Flammable ice. 281:5, 76–83. Documents a research cruise by members of the Research Center for Marine Geosciences (GEOMAR) of Germany to study methane hydrates on the floor of the North Pacific Ocean.

Oceanography on the Web

Visit the *Essentials of Oceanography* home page for on-line resources for this chapter. There you will find an on-line study guide with review exercises, and links to oceanography sites to further your exploration of the topics in this chapter. *Essentials of Oceanography* is at: **http://www.prenhall.com/ thurman** (click on the Table of Contents menu and select this chapter).

CHAPTER
5

Water and Seawater

●

THE HMS *CHALLENGER* EXPEDITION: BIRTH OF OCEANOGRAPHY

Oceanography as a scientific discipline began in 1872 with the **HMS *Challenger*** Expedition, the first large-scale voyage with the express purpose of studying the ocean for scientific purposes. At that time, there were many unanswered questions about the oceans. For instance, did life forms live in the deep ocean? If so, what were the physical and chemical conditions there? What was the nature of sea floor deposits?

In 1871, the Royal Society of England recommended that funds be raised for an expedition to investigate the following:

1. The physical conditions of the deep sea in the great ocean basins
2. The chemical composition of seawater at all depths in the ocean
3. The physical and chemical characteristics of the deposits of the sea floor and the nature of their origin
4. The distribution of organic life at all depths in the sea, as well as on the sea floor

The British government agreed to sponsor an expedition to explore the world's deep oceans and conduct scientific investigations, in part because there was so much public enthusiasm for scientific journeys. In 1872, a reserve warship was refitted to support scientific studies and renamed the HMS *Challenger* (Figure 5A, *inset*). It contained a staff of six scientists under the direction of C. Wyville Thomson.

From time to time during the course of the voyage, the ship would stop to measure the water depth using a sounding line, the bottom temperature with newly developed thermometers that could withstand the high pressure at depth, and atmospheric and meteorologic conditions. In addition, a sample of the bottom water was collected and the bottom sediment was dredged. Other measurements included trawling the bottom for life using a net, collecting organisms at the surface, determining temperature at various depths, gathering samples of seawater from certain depths, and recording surface and deep-water currents.

Challenger returned in May 1876 after nearly $3^1/_2$ years of circumnavigating the world (Figure 5A). During the 127,500-kilometer (79,200-mile) voyage, the scientists performed 492 deep-sea soundings, dredged the bottom 133 times, trawled the open water 151 times, took 263 water temperature readings, and collected water samples from as deep as 1830 meters (6000 feet). As with most oceanographic expeditions, the real work of analyzing the data was just beginning. In fact, it took nearly 20 years to compile the expedition results into 50 volumes.

key questions

- Why does water have such unusual chemical properties?
- How have water's thermal properties been important for all life on Earth?
- How salty is the ocean?
- What is the pH of seawater and how does ocean buffering work?
- What processes affect seawater salinity?
- What factors affect seawater density?
- What are the halocline, pycnocline, and thermocline?

Chemistry...is one of the broadest branches of science, if for no other reason that, when we think about it, everything is chemistry.
　　　　　—*Luciano Caglioti,*
The Two Faces of Chemistry *(1985)*

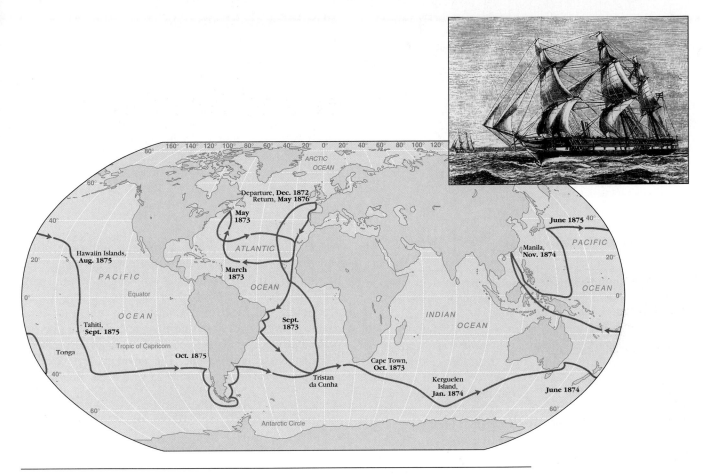

Figure 5A Route of the HMS *Challenger* and a block print of the ship (*inset*).

Major accomplishments of the voyage included verifying the existence of life at all ocean depths, classifying 4717 new marine species, measuring a record water depth of 8185 meters (26,850 feet) from the Mariana Trench in the western North Pacific Ocean, demonstrating that the ocean floor was not flat but had significant relief, and discovering manganese nodules.

Pioneering work on the chemistry of the oceans was completed, too. Analysis by chemist William Dittmar of 77 ocean water samples collected during the *Challenger* expedition revealed that the oceans had a remarkably consistent chemical composition, even down to minor dissolved substances. Not only were the ratios between various salts constant at the surface from ocean to ocean, but they were also constant at depth. This relationship established the *Principle of Constant Proportions*, which has contributed greatly to our understanding of ocean salinity.

Water is one of the most common substances on Earth, but it has remarkable properties. These properties, which stem from the arrangement of its atoms and how its molecules stick together, give water the ability to dissolve almost everything and to store vast quantities of heat.

The chemical properties of water are essential for sustaining all forms of life. In fact, the primary component of all living organisms is water. The water content of organisms, for instance, ranges from about 65% (humans) to 95% (most plants). Water is the ideal medium to have within our bodies because it facilitates chemical reactions. Our blood—so important for transporting nutrients and removing wastes within our bodies—is 83% water. Water, moreover, controls the distribution of heat on Earth, thus controlling Earth's climate. The very presence of water on our planet makes life possible, and the unusual properties of water make our planet livable.

Atomic Structure

Atoms (*a* = not, *tomos* = cut) are the basic building blocks of all matter. Every physical substance in our world—chairs, tables, books, people, the air we breathe—is composed of atoms. An atom resembles a

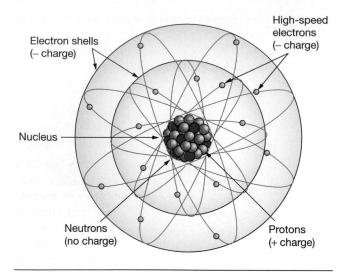

Figure 5–1 Simplified model of an atom. An atom consists of a central nucleus composed of protons and neutrons that is encircled by electrons.

microscopic sphere (Figure 5–1) and was originally thought to be the smallest form of matter. Additional study has revealed that atoms are composed of even smaller particles, called subatomic particles.[1] As shown in Figure 5–1, the **nucleus** (*nucleos* = a little nut) of an atom is composed of **protons** (*protos* = first) and **neutrons** (*neutr* = neutral), which are bound together by strong forces. Protons have a positive electrical charge, whereas neutrons have no electrical charge. The masses of protons and neutrons are about the same, but both are extremely small. Surrounding the nucleus are particles called **electrons** (*electro* = electricity), which have about $^1/_{2000}$ the mass of either protons or neutrons. Electrical attraction between positively-charged protons and negatively-charged electrons holds electrons in layers or shells around the nucleus.

The overall electrical charge of atoms is balanced because each atom contains an equal number of protons and electrons. An oxygen atom, for example, has eight protons and eight electrons. Most oxygen atoms also have eight neutrons, which do not affect the overall electrical charge because neutrons are electrically neutral. The number of protons is what distinguishes atoms of the 115 known chemical elements from one another. For example, an oxygen atom (and only an oxygen atom) has eight protons. Similarly, a hydrogen atom (and only a hydrogen atom) has one proton, a helium atom has two protons, and so on. In some cases, an atom will lose or gain one or more electrons and thus have an overall electrical charge. These atoms are called **ions** (*ienai* = to go).

---

[1] It has recently been discovered that subatomic particles themselves are composed of even smaller particles, called quarks.

The Water Molecule

A **molecule** (*molecula* = a mass) is a group of two or more atoms held together by mutually shared electrons. It is the smallest form of a substance that can exist yet still retain the original properties of that substance. When atoms combine with other atoms to form molecules, they share or trade electrons and establish chemical bonds. For instance, the chemical formula for water—H_2O—indicates that a water molecule is composed of two hydrogen atoms chemically bonded to one oxygen atom.

Geometry

Atoms can be represented as spheres of various sizes, with the more electrons the atom contains, the larger the sphere. It turns out that an oxygen atom (with eight electrons) is about twice the size of a hydrogen atom (with one electron). A water molecule consists of a central oxygen atom covalently bonded to the two hydrogen atoms, which are separated by an angle of about 105 degrees (Figure 5–2a). The **covalent** (*co* = with, *valere* = to be strong) **bonds** in water are due to the

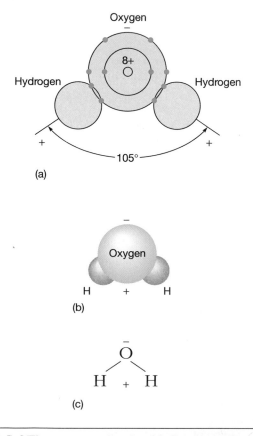

Figure 5–2 The water molecule. (a) Geometry of a water molecule. The oxygen end of the molecule is negatively charged, and the hydrogen regions exhibit a positive charge. Covalent bonds occur between the oxygen and the two hydrogen atoms **(b)** A three-dimensional representation of the water molecule. **(c)** The water molecule represented by letters (H = hydrogen, O = oxygen).

sharing of electrons between oxygen and each hydrogen atom. They are relatively strong chemical bonds, so a lot of energy is needed to break them. Figure 5–2b shows a water molecule in a more compact representation, and in Figure 5–2c, letter symbols are used to represent the atoms in water (O for oxygen, H for hydrogen). Instead of all atoms being in a straight line, *both hydrogen atoms are on the same side of the oxygen atom*. This curious bend in the geometry of the water molecule is the underlying cause of most of the unique and unusual properties of water.

Polarity

The bent geometry of the water molecule gives a slight overall negative charge to the side of the oxygen atom and a slight overall positive charge to the side of the hydrogen atoms (Figure 5–2a). This slight separation of charges gives the entire molecule an electrical **polarity** (*polus* = pole, *ity* = having the quality of), so water molecules are **dipolar** (*di* = two, *polus* = pole). Other common dipolar objects are flashlight batteries, car batteries, and bar magnets. Although the electrical charges are weak, water molecules behave as if they contain a tiny bar magnet.

Interconnections of Molecules

If you've ever experimented with bar magnets, you know they have polarity and orient themselves relative to one another such that the positive end of one bar magnet is attracted to the negative end of another. Water molecules have polarity, too, so they orient themselves relative to one another. In water, the positively charged hydrogen area of one water molecule interacts with the negatively charged oxygen end of an adjacent water molecule, forming a **hydrogen bond** (Figure 5–3).

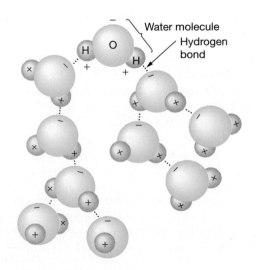

Figure 5–3 Hydrogen bonds. Dashed lines indicate locations of hydrogen bonds, which occur between water molecules.

The hydrogen bonds between water molecules are much weaker than the covalent bonds that hold individual water molecules together. In essence, weaker hydrogen bonds form *between* adjacent water molecules and stronger covalent bonds occur *within* water molecules.

Even though hydrogen bonds are weaker than covalent bonds, they are strong enough to cause water molecules to stick to one another and exhibit **cohesion** (*cohaesus* = to cling together). The cohesive properties of water cause it to "bead up" on a waxed surface, such as a freshly waxed car. They also give water its **surface tension**. Water's surface has a thin "skin" that allows a glass to be filled just above the brim without spilling any of the water. Surface tension results from the formation of hydrogen bonds between the outermost layer of water molecules and the underlying molecules. Water's ability to form hydrogen bonds causes it to have the highest surface tension of any liquid except the element mercury, the only metal that is a liquid at normal surface temperatures.

Water: The Universal Solvent

Water molecules stick not only to other water molecules, but also to other polar chemical compounds. In doing so, water molecules can reduce the attraction between ions of opposite charges by as much as 80 times. For instance, ordinary table salt—sodium[2] chloride, NaCl—consists of an alternating array of positively charged sodium ions and negatively charged chloride ions (Figure 5–4a). The **electrostatic** (*electro* = electricity, *statos* = standing) **attraction** between oppositely charged ions produces an **ionic** (*ienai* = to go) **bond**. When solid NaCl is placed in water, the electrostatic attraction (ionic bonding) between the sodium and chloride ions is reduced by 80 times. This, in turn, makes it much easier for the sodium ions and chloride ions to separate. When the ions separate, the positively charged sodium ions become attracted to the negative ends of the water molecules, the negatively charged chloride ions become attracted to the positive ends of the water molecules (Figure 5–4b), and the salt is dissolved in water. The process by which water molecules completely surround ions is called **hydration** (*hydra* = water, *ation* = action or process)

Because water molecules interact with other water molecules and other polar molecules, water is able to dissolve nearly everything.[3] Given enough time, water can dissolve more substances and in greater quantity than any other known substance. This is why water is

---

[2]Sodium is represented by the letters Na because the Latin term for sodium is *natrium*.

[3]If water is such a good solvent, why doesn't oil dissolve in water? As you might have guessed, the chemical structure of oil is remarkably nonpolar. With no positive or negative ends to attract the polar water molecule, oil will not dissolve in water.

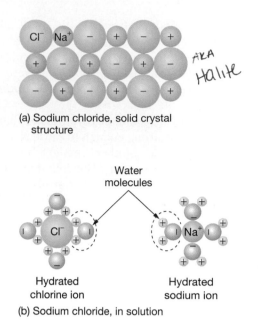

AKA
Halite

(a) Sodium chloride, solid crystal
structure

Water
molecules

Hydrated
chlorine ion

Hydrated
sodium ion

(b) Sodium chloride, in solution

Figure 5–4 Water as a solvent. (a) Table salt, composed of
sodium chloride (Na^+ = sodium ion, Cl^- = chlorine ion).
(b) As sodium chloride is dissolved, the positively charged
ends of water molecules are attracted to the negatively
charged Cl^- ion, while the negatively charged ends are
attracted to the positively charged Na^+ ion.

called "the universal solvent." It is also why the ocean
contains so much dissolved material—an estimated 50
quadrillion tons (50 million billion tons) of salt—which
make seawater taste "salty."

A water molecule has a bend in its geometry, with the two
hydrogen atoms on the same side of the oxygen atom,
which gives water its polarity and ability to form hydrogen
bonds.

Water's Thermal Properties

Water exists on Earth as a solid, liquid, and a gas, and
has the capacity to store and release great amounts of
heat. Water's thermal properties influence the world's
heat budget and are in part responsible for the development of tropical cyclones, worldwide wind belts, and
ocean surface currents.

Heat, Temperature, and Changes of State

Matter can exist in three states: solid, liquid, and gas.[4]
What must happen to change the state of a compound?
The attractive forces between molecules or ions in the

substance must be overcome if the state of the substance is to be changed from solid to liquid or from liquid to gas. These attractive forces include hydrogen
bonds and van der Waals forces. The **van der Waals
forces**—named for Dutch physicist Johannes Diderik
van der Waals (1837–1923)—are relatively weak interactions that become significant only when molecules
are very close together, as in the solid and liquid states
(but not the gaseous state). Energy must be added to
the molecules or ions so that they can move fast enough
to overcome these attractions.

What form of energy changes the state of matter?
Very simply, adding or removing heat is what causes a
substance to change its state of matter. For instance,
adding heat to ice cubes causes them to melt and removing heat from water causes ice to form. Before pro-
ceeding, you need to understand the difference between
heat and *temperature*:

- **Heat** is the *energy of moving molecules*. It is proportional to the energy level of molecules, and thus is the
total **kinetic** (*kinetos* = moving) **energy** of a substance. Water is a solid, liquid, or gas depending on
the amount of heat added. Heat may be generated by
combustion (a chemical reaction commonly called
"burning"), through other chemical reactions, by friction, or from radioactivity. A **calorie** (*calor* = heat) is
the amount of heat required to raise the temperature
of 1 gram of water[5] by 1 degree centigrade. The
calories used to measure heat are 1000 times smaller
than the calories used to measure the energy content
of food.

- **Temperature** is the *direct measure of the average
kinetic energy of the molecules that make up a
substance*. The greater the temperature, the greater
the kinetic energy of the substance. Temperature
changes when heat energy is added to or removed
from a substance. Temperature is usually measured in degrees Fahrenheit (°F) or degrees centigrade (°C).

Figure 5–5 shows water molecules in the solid, liquid,
and gaseous states. In the *solid state* (ice), water has a
rigid structure and does not flow. Intermolecular bonds
are constantly being broken and reformed, but the
molecules remain firmly attached. That is, the molecules
vibrate with energy but remain in relatively fixed positions. As a result, solids do not conform to the shape of
their container.

In the *liquid state* (water), water molecules still interact with each other, but they have enough kinetic
energy to flow past each other and take the shape of
their container. Intermolecular bonds are being formed
and broken at a much greater rate than in the solid
state.

---

[4]Plasma, an electrically neutral, highly ionized gas composed of ions,
electrons, and neutral particles, is a phase of matter distinct from
solids, liquids, and normal gases.

---

[5]One gram of water is equal to about 10 drops.

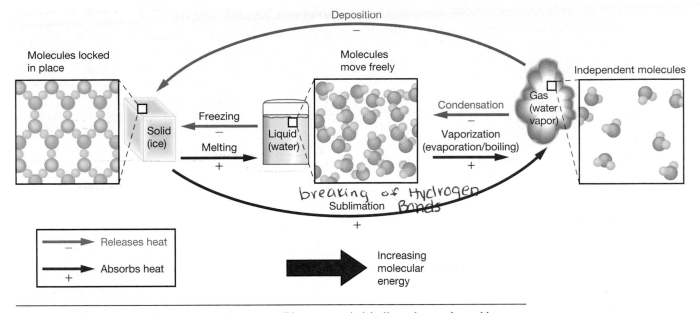

Figure 5–5 Water in the three states of matter. Blue arrows (−) indicate heat released by water (which warms the environment) as it changes state; red arrows (+) indicate heat absorbed by water (which cools the environment).

In the *gaseous state* (water **vapor**), water molecules no longer interact with one another except during random collisions. Water vapor molecules flow very freely, filling the volume of whatever container they are placed in.

Water's Freezing and Boiling Points

If enough heat energy is added to a solid, it melts to a liquid. The temperature at which melting occurs is the substance's **melting point**. If enough heat energy is removed from a liquid, it freezes to a solid. The temperature at which freezing occurs is the substance's **freezing point**, which is the same temperature as the melting point (Figure 5–5). For water, melting and freezing occur at 0°C (32°F).

If enough heat energy is added to a liquid, it converts to a gas. The temperature at which boiling occurs is the substance's **boiling point**. If enough heat energy is removed from a gas, it **condenses** to a liquid. The temperature at which condensation occurs is the substance's **condensation point**, which is at the same as the boiling point (Figure 5–5). For water, boiling and condensation occur at 100°C (212°F).

Based on the pattern of similar chemical compounds, a molecule as small and lightweight as water should melt at −90°C (−130°F) and boil at −68°C (−90°F). If that were the case, all water on Earth would be in the gaseous state. Instead, water melts and boils at the relatively high temperatures of 0°C (32°F) and 100°C (212°F)[6], respectively, because additional heat energy is required to overcome its hydrogen bonds and van der Waals forces. Thus, if not for the unusual geometry and resulting polarity of the water molecule, life as we know it would not exist on Earth.

Water's Heat Capacity

Heat capacity is *the amount of heat required to raise the temperature of 1 gram of any substance by 1 degree centigrade.*[7] A substance with a high heat capacity can absorb (or lose) large quantities of heat with only a small change in temperature. A substance that changes temperature rapidly when heat is applied has a low heat capacity (for instance, oil and all metals have low heat capacity). Water has a high heat capacity that is exactly 1 calorie per gram, whereas most other substances have a much lower heat capacity.

Water has the highest heat capacity of all common substances. It takes more energy to increase the kinetic energy of hydrogen-bonded water molecules than it does for substances in which the dominant intermolecular interaction is the much weaker van der Waals force. As a result, water gains or loses much more heat than other common substances while undergoing an equal temperature change. Additionally, water resists any change in temperature, as you may have observed when heating a large pot of water. When heat is applied to the pot, which is made of metal and has a low heat capacity, the pot heats up quickly. The water *inside* the pot, however, takes a long time to heat up (hence, the tale that a watched pot never boils but an unwatched pot boils

---

[6]Note that the temperature scale centigrade (*centi* = a hundred, *grad* = step) is based on 100 even divisions between the melting and boiling points of pure water.

[7]Note that the heat capacity of water is used as the unit of heat quantity, the calorie. Thus, water is the standard against which the heat capacities of other substances are compared.

over!). Making the water boil takes even more heat because all the hydrogen bonds must be broken. The exceptional capacity of water to absorb large quantities of heat helps explain why water is used in home heating, industrial and automobile cooling systems, and home cooking applications.

Figure 5–6 compares the average yearly temperature difference (summer high/winter low) for a coastal area (Monterey, California) with that of an inland area (Furnace Creek in Death Valley, California) at about the same latitude and elevation. The narrow range of temperature in Monterey is caused mostly by the high heat capacity of the ocean, which is capable of absorbing heat in the summer and releasing heat in the winter to moderate temperatures. Furnace Creek has a temperature range that is over three times that of Monterey and is related to the low heat capacity of rock material, which cools off more in the winter and dramatically heats up in the summer. In fact, Furnace Creek holds the record for the highest air temperature ever documented in North America: 56.6°C (134°F), which was recorded in July 1913.

Water's Latent Heats

When water undergoes a change of state—that is, when ice melts or water freezes, or when water boils or water vapor condenses—a large amount of heat is absorbed or released. The amount of heat absorbed or released is due to water's high latent (*latent* = hidden) heats and is closely related to water's unusually high heat capacity. As water evaporates from your skin, it cools your body by absorbing heat (this is why sweating cools your body). Conversely, if you ever have been scalded by water vapor—steam—you know that steam releases an enormous amount of latent heat.

Latent Heat of Melting The graph in Figure 5–7 shows how latent heat affects the amount of energy needed to increase water temperature and change its state. Beginning with 1 gram of ice (lower left), the addition of 20 calories of heat raises its temperature by 40 degrees, from −40°C to 0°C (point *a* on the graph). The temperature remains at 0°C (32°F) even though more heat is being added, as shown by the plateau on

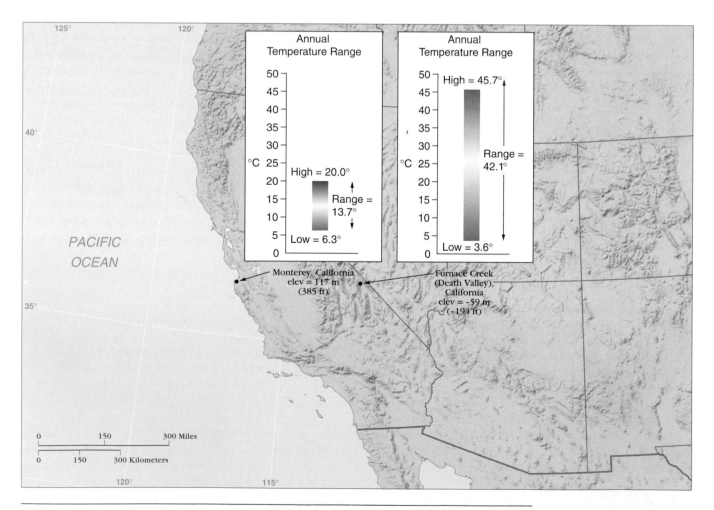

Figure 5–6 Comparison of yearly temperature difference between Monterey and Furnace Creek, California. The average yearly temperature range (summer high/winter low) at coastal Monterey (13.7°C) is about three times less than that for Furnace Creek (42.1°C) because of the higher heat capacity of the ocean as compared to continental regions. Both locations are at about the same latitude and elevation; temperature values are 30-year averages.

the graph between points *a* and *b*. The temperature of the water does not change until 80 more calories of heat energy have been added. The **latent heat of melting** is the energy needed to break the intermolecular bonds that hold water molecules rigidly in place in ice crystals. The temperature remains unchanged until most of the bonds are broken and the mixture of ice and water has changed completely to 1 gram of water.

After the change from ice to liquid water has occurred at 0°C, additional heat raises the water temperature between points *b* and *c* in Figure 5–7. As it does, it takes 1 calorie of heat to raise the temperature of water 1°C (or 1.8°F). Therefore, another 100 calories must be added before the gram of water reaches the boiling point of 100°C (212°F). So far, a total of 200 calories have been added to reach point *c*.

Latent Heat of Vaporization

The graph in Figure 5–7 flattens out again at 100°C, between points *c* and *d*. This plateau represents the **latent heat of vaporization**, which is 540 calories for water. This is the amount of heat that must be added to 1 gram of a substance at its boiling point to break the intermolecular bonds and complete the change of state from liquid to vapor (gas).

The drawings in Figure 5–8, which show the structure of water molecules in the solid, liquid, and gaseous states, help explain why the latent heat of vaporization is so much greater than the latent heat of melting. To go from a solid to a liquid, just enough hydrogen bonds must be broken to allow water molecules to slide past one another. To go from a liquid to a gas, however, all of the hydrogen bonds must be completely broken so that individual water molecules can move about freely.

Latent Heat of Evaporation

Sea-surface temperatures average 20°C (68°F) or less. How, then, does liquid water convert to vapor at the surface of the ocean? The conversion of a liquid to a gas below the boiling point is called **evaporation**. At ocean surface temperatures, individual molecules converted from the liquid to the gaseous state have less energy than do water molecules at 100°C. To gain the additional energy necessary to break free of the surrounding ocean water molecules, an individual molecule must capture heat energy from its neighbors. In other words, the molecules left behind have lost heat energy to those that evaporate, which explains the cooling effect of evaporation.

It takes more than 540 calories of heat to produce 1 gram of water vapor from the ocean surface at temperatures less than 100°C. At 20°C (68°F), for instance, the **latent heat of evaporation** is 585 calories per gram. More heat is required because more hydrogen bonds must be broken. At higher temperatures, liquid water has fewer hydrogen bonds because the molecules are vibrating and jostling about more.

Latent Heat of Condensation

When water vapor is cooled sufficiently, it condenses to a liquid and releases its **latent heat of condensation** into the surrounding air. On a small scale, the heat released is enough to cook food, which is how a "steamer" works. On a large scale, the heat released is sufficient to power large thunderstorms and hurricanes (see Chapter 6, "Air–Sea Interaction").

Latent Heat of Freezing

Heat is also released when water freezes. The amount of heat released when water freezes is the same amount that was absorbed when the water was melted in the first place. Thus, the **latent heat of freezing** is identical to the latent heat of melting. Similarly, the latent heats of vaporization and condensation are identical.

Importance of Latent Heat Exchange on Earth

The huge amount of heat energy exchanged in the evaporation–condensation cycle helps make life possible on Earth. The Sun radiates energy to Earth, where some is stored in the oceans. Evaporation removes this

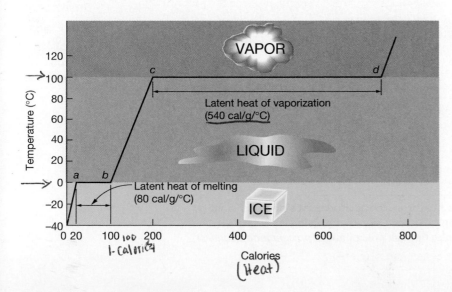

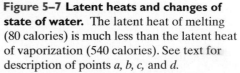

Figure 5–7 Latent heats and changes of state of water. The latent heat of melting (80 calories) is much less than the latent heat of vaporization (540 calories). See text for description of points *a*, *b*, *c*, and *d*.

Figure 5–8 Hydrogen bonds in H$_2$O. (a) In the solid state, water exists as ice, in which there are hydrogen bonds between all water molecules. **(b)** In the liquid state, there are some hydrogen bonds. **(c)** In the gaseous state, there are no hydrogen bonds and the water molecules are moving rapidly and independently.

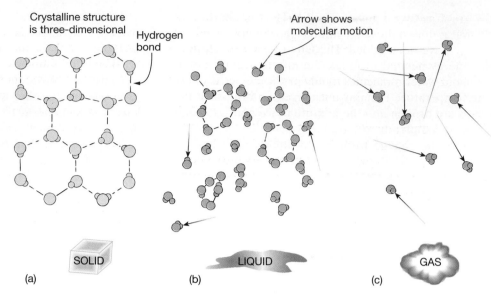

heat energy from the oceans and carries it high into the atmosphere. In the cooler upper atmosphere, water vapor condenses into clouds, which are the basis of **precipitation** (mostly rain and snow) that releases latent heat of condensation. The map in Figure 5–9 shows how this cycle of evaporation and condensation removes huge amounts of heat energy from the low latitude oceans and adds huge amounts of heat energy to the heat-deficient higher latitudes. In addition, the heat re-

leased when sea ice forms further moderates Earth's high-latitude regions.

The exchange of latent heat between ocean and atmosphere is very efficient. For every gram of water that condenses in cooler latitudes, the amount of heat released to warm these regions equals the amount of heat removed from the tropical ocean when that gram of water was evaporated initially. The end result is that the thermal properties of water have prevented wide varia-

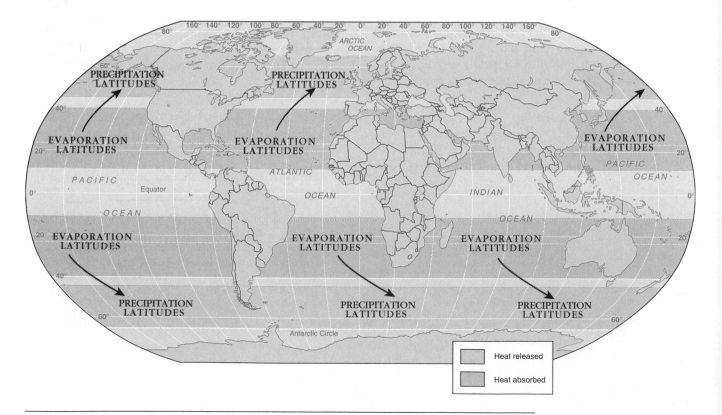

Figure 5–9 Atmospheric transport of surplus heat from low latitudes into heat-deficient high latitudes. The heat removed from the tropical ocean (*evaporation latitudes*) is carried toward the poles and is released at higher latitudes through precipitation (*precipitation latitudes*), thus moderating Earth's climate.

tions in Earth's temperature, moderating Earth's climate. Because rapid change is the enemy of all life, Earth's moderated climate is one of the main reasons life exists on Earth.

These same principles of heat exchange apply when you use ice for refrigeration. Put a block of ice in a cooler chest and it will lower the temperature of the food and drinks inside (Figure 5–10a). This occurs because heat energy travels from the food and drinks to the ice, where it changes the ice from a solid to a liquid. As the ice melts, it releases its latent heat of melting, thus keeping the food and drinks cold until all the ice melts.

An evaporative cooler (also called a swamp cooler), used in arid climates as an air conditioner, works on similar principles of heat exchange (Figure 5–10b). A fan blows hot dry air over a surface of liquid water. The hot air quickly loses heat energy to the water and becomes cooler, whereas the water releases its heat by evaporation.

> Water's unique thermal properties include water's latent heats and high heat capacity, which redistribute heat on Earth and have moderated Earth's climate.

Water Density

Recall from Chapter 1 that density is mass per unit volume, which can be thought of as *how heavy something is for its size*. Ultimately density is related to how tightly the molecules or ions of a substance are packed together. Typical units of density are grams per cubic centimeter (g/cm^3). Pure water, for example, has a density of 1.0 g/cm^3. Temperature, salinity, and pressure all affect water density.

The density of most substances increases as its temperature decreases. For example, cold air sinks and warm air rises because cold air is denser than warm air. Density increases as temperature decreases because the molecules lose energy and slow down, so the same number of molecules occupies less space. This shrinkage caused by cold temperatures, called **thermal contraction**, also occurs in water, but only to a certain point. As water cools to 4°C (39°F), its density increases. From 4°C down to 0°C (32°F), however, its density *decreases*. In other words, water stops contracting and actually expands, which is highly unusual among Earth's many substances. The result is that ice is less dense than liquid water, so ice floats on water. For most other substances, the solid state is denser than the liquid state, so the solid sinks.

Why is ice less dense than water? Figure 5–11 shows how molecular packing changes as water approaches its freezing point. From points *a* to *c* in the figure, the temperature decreases from 20°C (68°F) to 4°C (39°F) and the density increases from 0.9982 g/cm^3 to 1.000 g/cm^3. Density increases because the amount of thermal

motion decreases, so the water molecules occupy less volume. As a result, the window at point *c* contains more water molecules than the windows at points *a* or *b*. When the temperature is lowered below 4°C (39°F), the overall volume increases again because ice crystals become more abundant. Ice crystals are bulky, open, six-sided structures in which water molecules are widely spaced. Their characteristic hexagonal shape (Figure 5–12) mimics the hexagonal molecular structure resulting from hydrogen bonding between water molecules (see Figure 5–8a). By the time water fully freezes (point *e*), the density of the ice is much less than that of water at 4°C (39°F), the temperature at which water achieves its maximum density.

When water freezes, its volume increases by about 9%. Anyone who has put a beverage in a freezer for "just a few minutes" to cool it down and inadvertently forgotten about it has experienced the volume increase associated with water's expansion as it freezes—usually resulting in a burst beverage container. The force exerted when ice expands is powerful enough to break

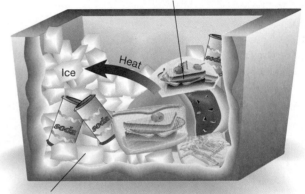

(a) Cooler chest

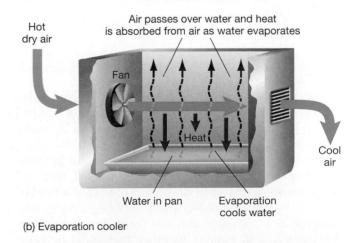

(b) Evaporation cooler

Figure 5–10 Principles of heat transfer. (a) The use of ice in a cooler chest cools beverages and food inside as the ice absorbs heat and melts. **(b)** An evaporative cooler uses the heat absorbing properties of water to cool air as the water evaporates.

Figure 5–11 The formation of ice. The formation of ice clusters in fresh water, showing how water reaches its maximum density at 4°C. Below 4°C, water becomes less dense as ice begins to form. At 0°C, ice forms, the crystal structure expands, and density decreases, causing ice to float.

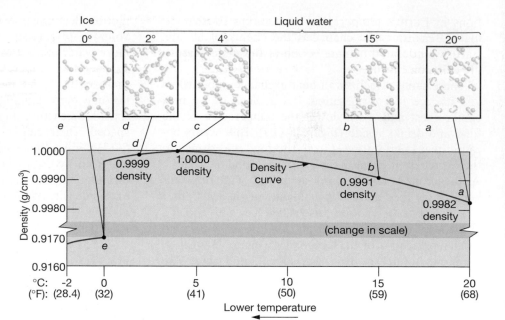

Figure 5–12 Snowflakes. Hexagonal snowflakes indicate the internal structure of water molecules held together by hydrogen bonds.

apart rocks, split pavement on roads and sidewalks, and crack water pipes.

Increasing the pressure or adding dissolved substances decreases the temperature of maximum density for fresh water because the formation of bulky ice crystals is inhibited. Increasing pressure increases the number of water molecules in a given volume and inhibits the number of ice crystals that can be created. Increasing amounts of dissolved substances inhibits the formation of hydrogen bonds, which further restricts the number of ice crystals that can form. To produce ice crystals equal in volume to those that could be produced at 4°C (39°F) in fresh water, more energy must be removed, causing a reduction in the temperature of maximum density.

Dissolved solids reduce the freezing point of water, too. It's the reason most seawater never freezes, except

near Earth's frigid poles (and even then, only at the surface). It's also why salt is spread on roads and sidewalks during the winter in cold climates. The salt lowers the freezing point of water, allowing ice-free roads and sidewalks at temperatures that are several degrees below freezing.

Table 5–1 summarizes the physical and biological significance of the unusual properties of water.

Seawater

What is the difference between pure water and seawater? One of the most obvious differences is that seawater contains dissolved substances that give it a distinctly salty taste. These dissolved substances are not simply sodium chloride (table salt)—they include various other salts, metals, and dissolved gases. The oceans contain enough salt to cover the entire planet with a layer more than 150 meters (500 feet) thick (about the height of a 50-story skyscraper). Unfortunately, the salt content of seawater makes it unsuitable for drinking or irrigating most crops and causes it to be highly corrosive to many materials.

Salinity

Salinity (*salinus* = salt) is the total amount of solid material dissolved in water, including dissolved gases, because even gases become solids at low enough temperatures. Salinity does *not* include fine particles being held in suspension (turbidity) or solid material in contact with water because these materials are not dissolved. Salinity is the ratio of the mass of dissolved substances to the mass of the water sample.

The salinity of seawater is typically about 3.5%, about 220 times saltier than fresh water. Seawater with a salinity of 3.5% indicates that it also contains 96.5% pure water, as shown in Figure 5–13. Because

Table 5–1   Summary of the properties of water and their significance.

Property	Comparison with other substances	Physical and/or biological significance
Physical states: gas (*water vapor*), liquid (*water*), solid (*ice*).	Water is the only substance that occurs as a gas, liquid, and solid within the range of surface temperatures on Earth.	*Water vapor* is an important component of the atmosphere, because water vapor transfers great quantities of heat from warm, low latitudes to cold, high latitudes. *Liquid water* runs across land, dissolving minerals and carrying them to the oceans. Most organisms are composed primarily of water. Serves as a transport medium and facilitates chemical reactions. *Ice* formation at the surface protects the water and life below it from freezing.
Solvent property: the ability to dissolve substances.	Water can dissolve more substances than any other liquid, hence it is often called "the universal solvent."	*Wide-ranging implications* in both physical and biological phenomena. Helps explain why seawater is "salty." In addition, seawater carries dissolved within it the nutrients required by marine algae and the oxygen needed by animals.
Surface tension: cohesive attraction of hydrogen bonds causes a molecule-thick "skin" to form on water surfaces.	Water has the greatest surface tension of all common liquids, which makes capillarity possible.	*Causes water to form drops*, to "bead up," and to overfill a glass without spilling. Water striders use this "skin" as a walking surface, while other organisms hang from its undersurface. Capillarity assists in raising water to the tops of tall plants such as trees.
Heat capacity: the quantity of heat required to change the temperature of 1 g of a substance by 1°C. The heat capacity of water is used as the unit of heat quantity, the *calorie* (*cal*).	Water has the highest heat capacity of all common liquids and gains or loses much more heat than other common substances while undergoing an equal temperature change.	*A major factor* in moderating Earth's climate, it helps explain the narrow range of temperature change occurring near the oceans as compared to interior regions.
Latent heat of melting: the quantity of heat gained or lost per gram by a substance changing from a solid to a liquid, or from a liquid to a solid, without a temperature change.	For water, it is 80 cal at 0°C (32°F), the highest of any common substance.	*Heat energy lost when ice forms* is mostly absorbed by the heat-deficient atmosphere at high latitudes. *Heat energy gained by water when ice melts* is manifested as molecular energy of the liquid water. This prevents the high-latitude ocean from becoming much warmer or colder than the freezing temperature of ocean water.
Latent heat of vaporization: the quantity of heat gained or lost per gram by a substance changing from a liquid to a gas, or from a gas to a liquid, without a temperature change.	It is greater for water, 540 cal at 100°C (212°F), than for any other common substance. At ocean surface temperatures of 20°C (68°F), it is 585 cal.	*Extremely important* in global heat and water transfer in the atmosphere. Evaporation from the low-latitude ocean removes a great amount of excess heat energy that is released through precipitation at heat-deficient higher latitudes. This greatly moderates temperatures at the poles and equator, which otherwise would be far more extreme.
Density: mass per unit volume (g/cm³).	Water's density increases as water cools, but it *decreases* below 4°C (39°F), which is highly unusual. Water density also increases as salinity and pressure increase.	*Plankton* that stay near the surface through buoyancy and frictional resistance to sinking are greatly influenced by the effect of temperature on density. In low-density warm water, plankton must be smaller or more ornate to obtain the increased ratio of surface area to body mass necessary to remain afloat. Also produces layering of ocean water.
Thermal expansion: the expansion of a substance when it is heated, and the contraction of a substance when it is cooled.	As water is cooled, it contracts. Below 4°C (39°F), it expands, which is highly unusual for a substance).	*Water expands* by 9% when frozen, causing ice to float. In the ocean at high latitudes where sea ice forms, ice provides a layer below which seawater does not freeze, protecting marine life.

seawater is mostly pure water, its physical properties are very similar to those of pure water, with only slight variations.

Figures 5–13 and 5–14 show that the elements chlorine, sodium, sulfur (as the sulfate ion), magnesium, calcium, and potassium account for over 99% of the dissolved solids in seawater. At least 89 other chemical elements have been identified in seawater, most in extremely small amounts. Probably all of Earth's naturally occurring elements exist in the sea. Remarkably, some of the trace amounts of dissolved components in seawater are vital for human survival (Box 5–1).

Salinity is often expressed in **parts per thousand (‰)**. Just as 1% is one part in 100, 1‰ is one part in 1000. When converting from percent to parts per thousand, the decimal is simply moved over one place to the right. For instance, typical seawater salinity of 3.5% is the same as 35‰. Advantages of expressing salinity in parts per thousand are that decimals are often avoided and values convert directly to grams of salt per kilogram of seawater.

Salinity Variations

Salinity varies within the oceans. In the open ocean far from land, it varies between about 33 and 38‰. In coastal areas, salinity variations can be extreme. In the

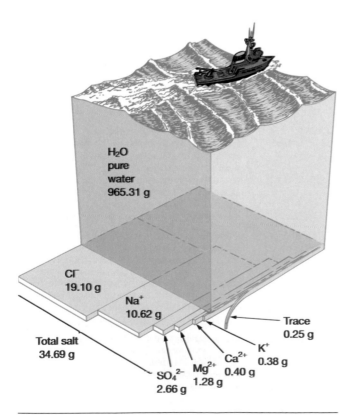

Figure 5–13 Constituents of ocean salinity. The composition of 1 kilogram of ocean water in grams (g), showing the concentrations of various dissolved substances. Average seawater salinity is approximately 35‰ or 3.5%.

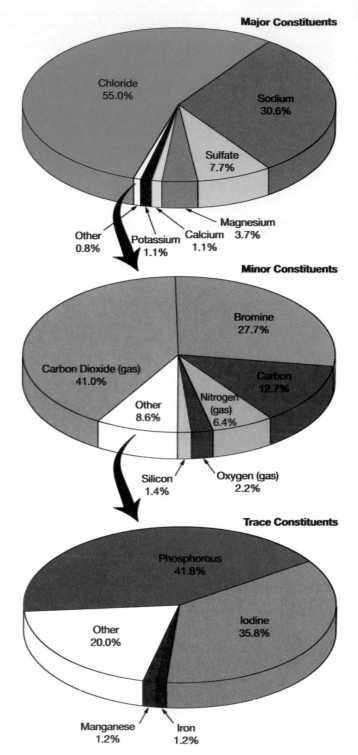

Figure 5–14 Major dissolved components in seawater. The percent of major constituents (*top*), minor constituents (*middle*), and trace constituents (*bottom*) dissolved in 35‰ seawater.

Baltic Sea, for example, salinity averages only 10‰ because physical conditions create **brackish** (*brak* = salt, *ish* = somewhat) water. Brackish water is produced in areas where fresh water (from rivers and high rainfall) and seawater mix. In the Red Sea, on the other hand,

Box 5–1
How to Avoid Goiters

The nutritional label on containers of salt usually proclaims "this product contains iodine, a necessary nutrient." Why is iodine necessary in our diets? It turns out that if a person's diet contains an insufficient amount of iodine, a potentially life-threatening disease called **goiters** (*guttur* = throat) may result (Figure 5B).

Iodine is used by the thyroid gland, which is a butterfly-shaped organ located in the neck in front of and on either side of the trachea (windpipe). The thyroid gland manufactures hormones that regulate cellular metabolism essential for mental development and physical growth. If people lack iodine in their diet, their thyroid glands cannot function properly. Often, this results in the enlargement or swelling of the thyroid gland. Severe symptoms include dry skin, loss of hair, puffy face, weakness of muscles, weight increase, diminished vigor, mental sluggishness, and a large nodular growth on the neck called a goiter. If proper steps are not taken to correct this disease, it can lead to cancer. Iodine ingested regularly often begins to reverse the effects. In advanced stages, surgery to remove the goiter or exposure to radioactivity is the only course of action.

How can you avoid goiters? Fortunately, goiters are preventable by a diet with just *trace amounts* of iodine. Where can you get iodine in your diet? All products from the sea contain trace amounts of iodine because iodine is one of the many elements dissolved in seawater. Sea salt, seafood, seaweed, and other sea products contain plenty of iodine to help prevent goiters. Although goiters are rarely a problem in developed nations like the United States, goiters pose a serious health hazard in many underdeveloped nations, especially those far from the sea. In the U.S., however, many people get too much iodine in their diet, leading to the overproduction of hormones by the thyroid gland. That's why most stores that sell iodized salt also carry noniodized salt for those people who have a *hyperthyroid* (*hyper* = excessive, *thyroid* = the thyroid gland) *condition* and must restrict their intake of iodine.

Figure 5B A woman with goiters.

salinity averages 42‰ because physical conditions produce **hypersaline** (*hyper* = excessive, *salinus* = salt) water. Hypersaline water is typical of seas and inland bodies of water that experience high evaporation rates and limited open-ocean circulation.

Some of the most hypersaline water in the world is found in inland lakes, which are often called seas because they are so salty. The Great Salt Lake in Utah, for example, has a salinity of 280‰, and the Dead Sea on the border of Israel and Jordan has a salinity of 330‰. The water in the Dead Sea, therefore, contains 33% dissolved solids and is almost *10 times saltier than seawater*. As a result, hypersaline waters are so dense that one can easily float—with arms and legs sticking up above water level! Hypersaline waters also taste much saltier than seawater.

Salinity of seawater in coastal areas also varies seasonally. For example, the salinity of seawater off Miami Beach, Florida, varies from about 34.8‰ in October to 36.4‰ in May and June when evaporation is high. Offshore of Astoria, Oregon, seawater salinity is always relatively low because of the vast fresh water input from the Columbia River. Here, seawater salinity varies from 0.3‰ in April and May (when the Columbia River is at its maximum flow rate) to 2.6‰ in October (the dry season).

Other types of water have much lower salinity. Tap water, for instance, has salinity somewhere below 0.8‰, and good-tasting tap water is usually below 0.6‰. Salinity of premium bottled water is on the order of 0.3‰, with the salinity often displayed prominently on its label, usually as total dissolved solids (TDS) in units of parts per million (ppm), where 1000 ppm equals 1‰.

> Average seawater salinity is 35‰ but varies widely from brackish (low salinity) to hypersaline (high salinity).

Determining Salinity

Early methods of determining seawater salinity involved evaporating a carefully weighed amount of seawater and weighing the salts that precipitated from it. However, the accuracy of this time-consuming method is limited because some water can remain bonded to salts that precipitate and some substances can evaporate along with the water.

Another way to measure salinity is to use the *Principle of Constant Proportions*, which was firmly established by chemist William Dittmar when he analyzed the water samples collected during the *Challenger* Expedition (see the beginning of this chapter). The **Principle of Constant Proportions** states that the major dissolved constituents responsible for the salinity of seawater occur nearly everywhere in the ocean in the exact same proportions, independent of salinity. The ocean, therefore, is well mixed. When salinity changes, moreover, the salts don't leave (or enter) the ocean but water molecules do. Seawater has *constancy of composition*, so the concentration of a single major constituent can be measured to determine the total salinity of a given water sample. The constituent that occurs in the greatest abundance and is the easiest to measure accurately is the chloride ion, Cl^-. The weight of this ion in a water sample is its **chlorinity.**

In any sample of ocean water worldwide, the chloride ion accounts for 55.04% of the total proportion of dissolved solids (Figure 5–14). Therefore, by measuring only the chloride ion concentration, the total salinity of a seawater sample can be determined using the following relationship:

$$\text{Salinity (‰)} = 1.80655 \times \text{chlorinity (‰)}^8$$

---

[8]The number 1.80655 comes from dividing 1 by 0.5504 (the chloride ion's proportion in seawater of 55.04%). However, if you actually divide this, you will get 1.81686, which is different from the original value by 0.57%. Empirically, oceanographers found that seawater's constancy of composition is an approximation and have agreed to use 1.080655 because it more accurately represents the total salinity of seawater.

For example, the average chlorinity of the ocean is 19.2‰, so the average salinity is 1.80655×19.2‰, which rounds to 34.7‰. In other words, on average there are 34.7 parts of dissolved material in every 1000 parts of seawater.

Standard seawater consists of ocean water analyzed for chloride ion content to the nearest ten-thousandth of a part per thousand by the Institute of Oceanographic Services in Wormly, England. It is then sealed in small glass vials called *ampules* and sent to laboratories throughout the world for use as a reference standard in calibrating analytical equipment.

The chloride content in seawater can be measured very accurately with advanced oceanographic instruments such as a **salinometer** (*salinus* = salt, *meter* = measure). A salinometer measures seawater's **electrical conductivity** (the ability of a substance to transmit electric current), which increases as more dissolved substances are dissolved in water (Figure 5–15). Salinometers can determine salinity to resolutions of better than 0.003‰.

Dissolved Components Added and Removed from Seawater

Why does seawater salinity vary from place to place? Interestingly, dissolved substances do not remain in the ocean forever. Instead, they are cycled into and out of seawater by the processes shown in Figure 5–16. These processes include stream **runoff** (river discharge) in which streams dissolve ions from continental rocks and carry them to the sea, and volcanic eruptions, both on the land and on the sea floor. Other sources include the atmosphere (which contributes gases) and biologic interactions.

Stream runoff is the primary method by which dissolved substances are added to the oceans. Table 5–2 compares the major components dissolved in stream water with those in seawater. It shows that streams have far lower salinity and a vastly different composition of dissolved substances than seawater. For example, carbonate ion (HCO_3^-) is the most abundant

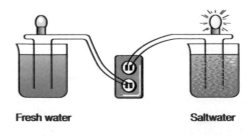

Fresh water   Saltwater

Figure 5–15 Salinity affects water conductivity. Increasing the amount of dissolved substances increases the conductivity of the water. A light bulb with bare electrodes shows that the higher the salinity, the more electricity is transmitted, and the brighter the bulb will be lit.

dissolved constituent in stream water yet is found in only trace amounts in seawater. Conversely, the most abundant dissolved component in seawater is the chloride ion (Cl^-), which exists in very small concentrations in streams.

If stream water is the main source of dissolved substances in seawater, why don't the components of the two match each other more closely? One of the reasons is that some dissolved substances stay in the ocean and accumulate over time. **Residence time** is the average length of time that an atom of an element resides in the ocean. Long residence times lead to higher concentrations of the dissolved substance. The ion sodium (Na^+), for instance, has a residence time of 260 million years. Others have residence times on the order of several thousand to several million years.

Even though various salts have long residence times in the ocean, the ocean is not becoming saltier over time. On average, the rate at which an element is added to the ocean equals the rate at which it is removed again. Several processes cycle dissolved substances out

of seawater. When waves break at sea, salt spray releases tiny salt particles into the atmosphere where they may be blown over land before being washed back to Earth. Along mid-ocean ridges, the infiltration of seawater near hydrothermal vents incorporates magnesium and sulfate ions into sea-floor mineral deposits. In fact, the *entire volume of ocean water* may be recycled through this hydrothermal circulation system at the mid-ocean ridge every 3 million years. Therefore, the chemical exchange between ocean water and the basaltic crust has a major influence on the composition of ocean water.

Dissolved substances are also removed from seawater in other ways. Calcium, carbonate, sulfate, sodium, and silicon are deposited in ocean sediments within the shells of dead microscopic organisms and animal feces. Vast amounts of dissolved substances can be removed when inland arms of seas dry up, leaving salt deposits called *evaporites* (such as those beneath the Mediterranean Sea; see Box 4–2. In addition, ions dissolved in ocean water are removed by adsorption (physical

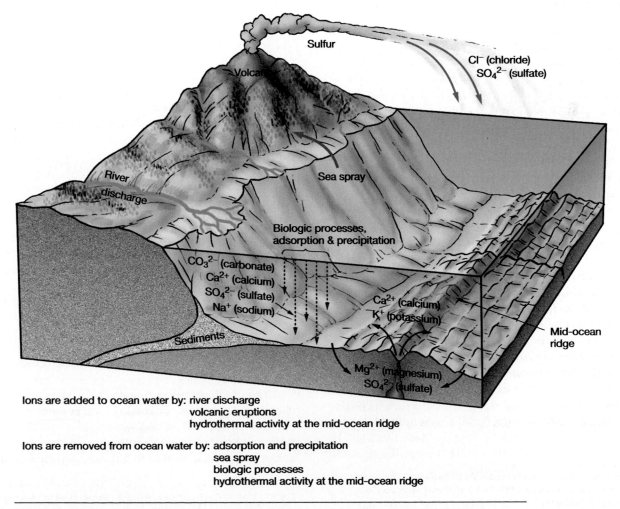

Ions are added to ocean water by: river discharge
volcanic eruptions
hydrothermal activity at the mid-ocean ridge

Ions are removed from ocean water by: adsorption and precipitation
sea spray
biologic processes
hydrothermal activity at the mid-ocean ridge

Figure 5–16 The cycling of dissolved components in seawater. Dissolved components are added to seawater primarily by river discharge and volcanic eruptions, while they are removed by adsorption, precipitation, ion entrapment in sea spray, and marine organisms that produce hard parts. Chemical reactions at the mid-ocean ridge add and remove various dissolved components.

Table 5–2   Comparison of major dissolved constituents in streams with those in seawater.

Constituent	Concentration in streams (parts per million)	Concentration in seawater (parts per million)
Carbonate ion (HCO_3^-)	58.4	trace
Calcium ion (Ca^{2+})	15.0	410
Silicate (SiO_2)	13.1	3.0
Sulfate ion (SO_4^{2-})	11.2	2700
Chloride ion (Cl^-)	7.8	19,300
Sodium ion (Na^+)	6.3	10,700
Magnesium ion (Mg^{2+})	4.1	1300
Potassium ion (K^+)	2.3	380
Total (parts per million)	119.2	34,793
Total (‰)	0.1192‰	34.8‰

attachment) to the surfaces of sinking clay and biologic particles.

Acidity and Alkalinity of Seawater

An **acid** is a compound that releases hydrogen ions (H^+) when dissolved in water. The resulting solution is said to be *acidic*. A strong acid readily and completely releases hydrogen ions when dissolved in water. An **alkaline** or **base** is a compound that releases hydroxide ion (OH^-) when dissolved in water. The resulting solution is said to be *alkaline* or *basic*. A strong base readily and completely releases hydroxide ions when dissolved in water.

Both hydrogen ions and hydroxide ions are present in extremely small amounts at all times in water because water molecules dissociate and reform. Chemically, this is represented as

$$H_2O \underset{\text{reform}}{\overset{\text{dissociate}}{\rightleftharpoons}} H^+ + OH^-$$

If the hydrogen ions and hydroxide ions in a solution are due only to the dissociation of water molecules, they are always found in equal concentrations, and the solution is consequently neutral.

If hydrochloric acid (HCl) is added to water, the resulting solution will be acidic because there will be a large excess of hydrogen ions from the dissociation of the HCl molecules. Conversely, if a base such as baking soda (sodium bicarbonate, $NaHCO_3$) is added to water, the resulting solution will be basic because there will be an excess of hydroxide ions (OH^-) from the dissociation of the $NaHCO_3$ molecules.

The pH Scale

Figure 5–17 shows the pH (potential of hydrogen) scale, which is a measure of the acidity or alkalinity of a solution. Values for pH range from 0 (strongly acidic) to 14

(strongly alkaline or basic). The pH of a **neutral** solution such as pure water[9] is 7.0, whereas the pH of an acidic solution is less than 7.0 and the pH of an alkaline solution is greater than 7.0.

Water in the ocean combines with carbon dioxide to form a weak acid, called carbonic acid (H_2CO_3), which dissociates and releases hydrogen ions (H^+).

$$H_2O + CO_2 \rightarrow H_2CO_3 \rightarrow H^+ + HCO_3^-$$

This reaction would seem to make the ocean slightly acidic. Carbonic acid, however, keeps the ocean slightly alkaline through the process of buffering.

The Carbonate Buffering System

The chemical reactions in Figure 5–18 show that carbon dioxide (CO_2) combines with water (H_2O) to form carbonic acid (H_2CO_3). Carbonic acid can then lose a hydrogen ion (H^+) to form the negatively charged bicarbonate ion (HCO_3^-). The bicarbonate ion can lose its hydrogen ion, too, though it does so less readily than carbonic acid. When the bicarbonate ion loses its hydrogen ion, it forms the double-charged negative carbonate ion (CO_3^{2-}), some of which combines with calcium ions to form calcium carbonate ($CaCO_3$). Some of the calcium carbonate is precipitated and deposited on the ocean floor, which can cycle back into the ocean by dissolving at depth.

The equations below Figure 5–18 show how these chemical reactions involving carbonate minimize changes in the pH of the ocean, in a process called buffering. Buffering protects the ocean from getting too acidic or too basic, similar to buffered aspirin protecting

---

[9]Water's neutral pH might seem surprising in light of water's tremendous ability to dissolve substances; it seem intuitive that water should be acidic and thus have a low pH. However, pH measures the activity of the hydrogen ion in solution, not the ability of a substance to dissolve by forming hydrogen bonds (as water does).

pH values of comon substances

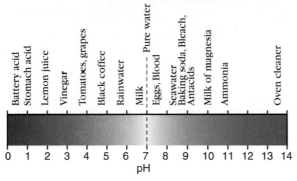

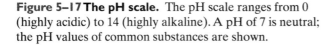

Figure 5–17 The pH scale. The pH scale ranges from 0 (highly acidic) to 14 (highly alkaline). A pH of 7 is neutral; the pH values of common substances are shown.

sensitive stomachs. If the pH of the ocean increases (becomes too basic), it causes H_2CO_3 to release H^+ and pH drops. If the pH of the ocean decreases (becomes too acidic), HCO_3^- combines with H^+ to remove it, causing pH to rise. Buffering maintains the pH of the ocean at an average of 8.1 with little variation through time.

Deep-ocean water contains more carbon dioxide than water at the surface because water can dissolve more of a gas at lower temperatures. Because carbon dioxide combines with water to form carbonic acid, why isn't the cold water of the deep ocean highly acidic? When microscopic marine organisms that make their shells out of calcium carbonate (calcite) die and sink

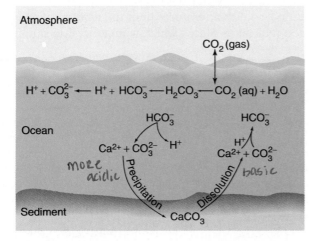

Figure 5–18 The carbonate buffering system. Atmospheric carbon dioxide (CO_2) enters the ocean and undergoes chemical reactions. If seawater is too basic, chemical reactions occur that release H^+ into seawater and lowers pH. If seawater is too acidic, chemical reactions occur that remove H^+ from seawater and cause pH to rise. Thus, buffering keeps the pH of seawater constant.

into the deep ocean, they neutralize the acid through buffering. In essence, these organisms act as an "antacid" for the deep ocean analogous to the way commercial antacids use calcium carbonate to neutralize excess stomach acid. As explained in Chapter 4, these shells are readily dissolved below the calcite (calcium carbonate) compensation depth (CCD).

Reactions involving carbonate chemicals serve to buffer the ocean and keep its pH at an average value of 8.1 (slightly alkaline).

EIO

For more information and on-line exercises about the Environmental Issue in Oceanography (EIO) "Toxic Chemicals in Seawater," visit the EIO Web site at **http:// www.prenhall. com/oceanissues** and select Issue #4.

Processes Affecting Seawater Salinity

Processes affecting seawater salinity change either the amount of dissolved substances within the water or the amount of water. Adding more water, for instance, dilutes the dissolved components, and lowers the salinity of the sample. Conversely, removing water increases salinity. Changing the salinity in these ways does not affect the *amount* or the *composition* of the dissolved components, which remain in constant proportions.

Processes That Decrease Seawater Salinity

Table 5–3 summarizes the processes affecting seawater salinity. Precipitation, runoff, melting icebergs, and melting sea ice *decrease* seawater salinity by adding more fresh water to the ocean. Precipitation is the way atmospheric water returns to Earth as rain, snow, sleet, and hail. Worldwide, about three-quarters of all precipitation falls directly back into the ocean and one-quarter falls onto land. Precipitation falling directly into the oceans adds fresh water, reducing seawater salinity.

Most of the precipitation that falls on land returns to the oceans indirectly as stream runoff. Even though this water dissolves minerals on land, the runoff is relatively pure water, as shown in Table 5–2. Runoff, therefore, adds mostly water to the ocean, causing seawater salinity to decrease.

Icebergs are chunks of ice that have broken free (*calved*) from a glacier when it flowed into an ocean or marginal sea and began to melt. Glacial ice originated as snowfall in high mountain areas, so icebergs are composed of fresh water. When icebergs melt in the ocean, they add fresh water, which reduces seawater salinity.

Sea ice, composed primarily of fresh water, forms when ocean water freezes in high-latitude regions. When warmer temperatures return to high-latitude regions in the summer, sea ice melts in the ocean, adding mostly fresh water with a small amount of salt to the ocean. Seawater salinity, therefore, is decreased.

Processes That Increase Seawater Salinity

The formation of sea ice and evaporation *increase* seawater salinity by removing water from the ocean (Table 5–3). Sea ice forms when seawater freezes. Depending on the salinity of seawater and the rate of ice formation, about 30% of the dissolved components in seawater are retained in sea ice. This means that 35‰ seawater creates sea ice with about 10‰ salinity (30% of 35‰ is 10‰). Consequently, the formation of sea ice removes mostly fresh water from seawater, increasing the salinity of the remaining unfrozen water. High-salinity water also has a high density, so it sinks below the surface.

Recall that evaporation is the conversion of water molecules from the liquid state to the vapor state at temperatures below the boiling point. Evaporation removes water from the ocean, leaving its dissolved substances behind. Evaporation, therefore, increases seawater salinity.

Various surface processes either decrease seawater salinity (precipitation, runoff, icebergs melting, or sea ice melting) or increase salinity (sea ice forming and evaporation).

The Hydrologic Cycle

Figure 5–19 shows how the **hydrologic** (*hydro* = water, *logus* = study of) **cycle** relates the processes that affect seawater salinity. These processes recycle water among the ocean, the atmosphere, and the continents, so water is in continual motion between the different components (*reservoirs*) of the hydrologic cycle. Earth's water supply exists in the following proportions:

- 97.2% in the world ocean
- 2.15% frozen in glaciers and ice caps
- 0.62% in groundwater and soil moisture
- 0.02% in streams and lakes
- 0.001% as water vapor in the atmosphere

Surface and Depth Salinity Variation

Average seawater salinity is 35‰, but it varies significantly from place to place at the surface and also with depth.

Surface Salinity Variation

Figure 5–20 shows how salinity varies at the surface with latitude. The red curve shows temperature, which decreases at high latitudes and increases near the Equator. The green curve shows salinity, which is lowest at high latitudes, highest at the Tropics of Cancer and Capricorn, and dips near the Equator.

Why does surface salinity vary this way? At high latitudes, abundant precipitation and runoff, and the melting of freshwater icebergs all decrease salinity. In addition, cool temperatures limit the amount of evaporation that takes place (which would increase salinity).

Table 5–3   Processes affecting seawater salinity.

Process	How accomplished	Adds or removes	Effect on salt in seawater	Effect on H_2O in seawater	Salinity increase or decrease?	Source of fresh water from the sea?
Precipitation	Rain, sleet, hail, or snow falls directly on the ocean	Adds very fresh water	None	More H_2O	Decrease	—
Runoff	Streams carry water to the ocean	Adds mostly fresh water	Negligible addition of salt	More H_2O	Decrease	—
Icebergs melting	Glacial ice calves into the ocean and melts	Adds very fresh water	None	More H_2O	Decrease	Yes, icebergs from Antarctic have been towed to South America
Sea ice melting	Sea ice melts in the ocean	Adds mostly fresh water and some salt	Adds a small amount of salt	More H_2O	Decrease	Yes, sea ice can be melted and is better than drinking seawater
Sea ice forming	Seawater freezes in cold ocean areas	Removes mostly fresh water	30% of salts in seawater are retained in ice	Less H_2O	Increase	Yes, through multiple freezings, called *freeze separation*
Evaporation	Seawater evaporates in hot climates	Removes very pure water	None (essentially all salts are left behind)	Less H_2O	Increase	Yes, through evaporation of seawater and condensation of water vapor, called *distillation*

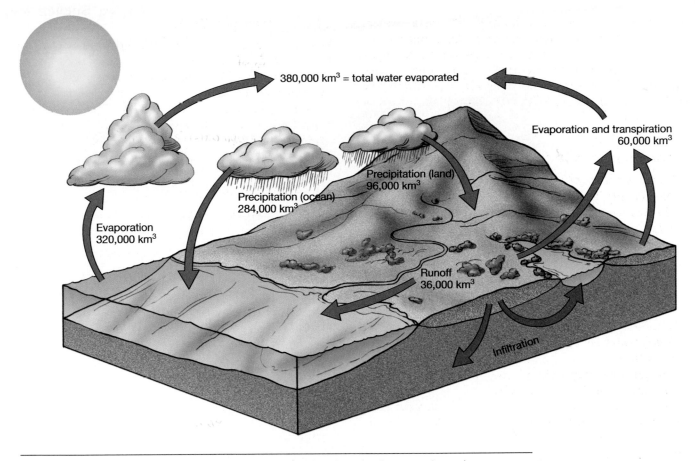

Figure 5–19 The hydrologic cycle. All water is in continual motion between the various components (reservoirs) of the hydrologic cycle. Volumes are Earth's average yearly amounts in cubic kilometers; ice not shown.

The formation and melting of sea ice balance each other out in the course of a year and are not a factor in changes in salinity.

The pattern of Earth's atmospheric circulation (see Chapter 6, "Air-Sea Interaction") causes warm dry air to descend at lower latitudes near the Tropics of Cancer and Capricorn, so evaporation rates are high and salinity increases. Additionally, little precipitation and runoff occur to decrease salinity. As a result, the regions near the Tropics of Cancer and Capricorn are the continental *and* maritime deserts of the world.

Temperatures are warm near the Equator, so evaporation rates are high enough to increase salinity. Increased precipitation and runoff partially offsets the high salinity, though. For example, daily rain showers are common along the Equator, adding water to the ocean and lowering its salinity.

The map in Figure 5–21 shows how ocean surface salinity varies worldwide. The overall pattern matches the graph in Figure 5–20.

Depth Salinity Variation

Figure 5–22 shows how seawater salinity varies with depth. The graph shows one curve for high-latitude regions and one curve for low-latitude regions.

For low-latitude regions, the curve begins at the surface with relatively high salinity (as was discussed in the preceding section). Even at the Equator, surface salinity is still relatively high. With increasing depth, the curve swings toward an intermediate salinity value.

For high-latitude regions, the curve begins at the surface with relatively low salinity (again, see the discussion in the preceding section). With increasing depth, the curve also swings toward an intermediate salinity value that approaches the value of the low-latitude salinity curve at the same depth.

These two curves, which together resemble the outline of a martini glass, show that salinity varies widely at the surface, but very little in the deep ocean. Why is this so? It occurs because all the processes that affect seawater salinity (precipitation, runoff, melting icebergs, melting sea ice, sea ice forming, and evaporation) occur at the *surface* and thus have no effect on the deep water below.

Halocline Both curves in Figure 5–22 show a rapid change in salinity between the depths of about 300 meters (980 feet) and 1000 meters (3300 feet). For the low-latitude curve, the change is a *decrease* in salinity. For the high-latitude curve, the change is an *increase* in salinity. In both cases, this layer of rapidly changing

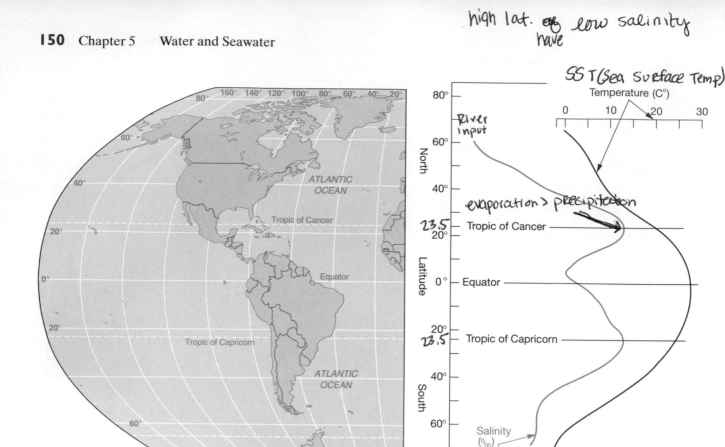

high lat. have low salinity

SST (Sea Surface Temp)

River input

evaporation > precipitation

Figure 5–20 Surface salinity variation. Surface seawater temperature (*red curve*) is lowest at the poles and highest at the Equator. Surface seawater salinity (*green curve*) is lowest at the poles, peaks at the Tropics of Cancer and Capricorn, and dips near the Equator.

salinity with depth is called a **halocline** (*halo* = salt, *cline* = slope). Haloclines separate layers of different salinity in the ocean.

Seawater Density

The density of pure water is 1.000 gram per cubic centimeter (g/cm^3) at 4°C (39°F). This value serves as a standard against which the density of all other substances can be measured. Seawater contains various dissolved substances that increase its density. In the open ocean, seawater density averages between 1.022 and 1.030 g/cm^3 (depending on its salinity). Thus, the density of seawater is 2 to 3% greater than pure water. Unlike fresh water, seawater continues to increase in density toward its freezing point, and reaches its maximum density at −1.3°C (29.6°F).

Density is an important property of ocean water because density differences make water masses sink or float, and determines the vertical position and movement of ocean water masses. For example, if seawater with a density of 1.030 g/cm^3 were added to fresh water with a density of 1.000 g/cm^3, the denser seawater would sink below the fresh water.

The ocean, like Earth's interior, is layered according to density. Low-density water exists near the surface and higher-density water occurs below. Except for some shallow inland seas with a high rate of evaporation, the highest-density water is found at the deepest ocean depths. Temperature, salinity, and pressure influence seawater density in the following ways:

- As temperature increases, density decreases[10] (due to thermal expansion)
- As salinity increases, density increases (due to the addition of more dissolved material)
- As pressure increases, density increases (due to the compressive effects of pressure)

Of these three factors, only temperature and salinity influence the density of surface water. Pressure influences seawater density only when very high pressures are encountered, such as in deep ocean trenches. Still, the density of seawater in the deep ocean is only about 5% greater than at the ocean surface, showing that despite tons of pressure per square centimeter, water is nearly incompressible. Unlike air, which can be compressed and put in a tank for scuba diving, the molecules in liquid water are already close together and cannot be compressed much more. Therefore, pressure

---

[10]A relationship where one variable *decreases* as a result of another variable's *increase* is known as an inverse relationship, in which the two variables are *inversely proportional*.

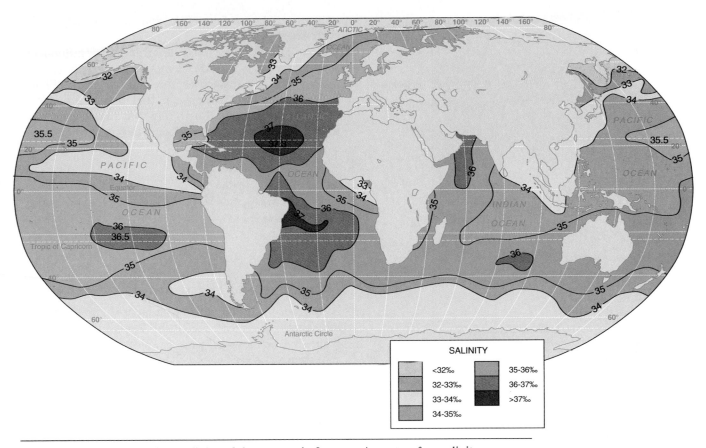

Figure 5–21 Average surface salinity of the oceans in August. August surface salinity map shows that the lowest salinities (*purple*) occur in high latitudes and the highest salinities (*red and pink*) occur near the tropics, while the Equator has a slightly reduced salinity. Values in parts per thousand (‰).

has the least effect on influencing the density of surface water and can be ignored as a factor that creates deep ocean currents.

Temperature, on the other hand, has the greatest influence on surface seawater density because the range of surface seawater temperature is greater than that of salinity. In fact, only in the extreme polar areas of the ocean, where temperatures are low and remain relatively constant, does salinity significantly affect density. Cold water that also has high salinity is some of the highest-density water in the world. The density of seawater—the result of its salinity and temperature—influences currents in the deep ocean because high-density water sinks below less-dense water.

The effects of temperature and salinity changes on density are shown in Figure 5–23. Temperature change affects density (*blue curving lines*) much more in warm, low-latitude regions than in high-latitude regions. Points a, b, c, and d mark different densities of water at a constant salinity of 35‰ Line a–b shows that the density change across a temperature range of 20 to 25°C (68 to 77°F) is 0.0012 g/cm³. Line c–d shows that the density change across a temperature range of 0 to 5°C (32 to 41°F) is only 0.0004 g/cm³. The difference shows that a change in temperature of warm, low-latitude

water has about three times the effect on density as an equal change in temperature of colder, high-latitude water.

The graph in Figure 5–24a shows how seawater density varies with depth, while the graph in Figure 5–24b shows how seawater temperature varies with depth. Each of these two graphs shows one curve for low-latitude regions (*colored line*) and one for high-latitude regions (*dashed line*).

The density curve for low-latitude regions in Figure 5–24a shows that density is relatively low at the surface. Density is low because temperature is high. (Remember that temperature has the greatest influence on density and temperature is inversely proportional to density.) Below the surface, density remains constant until a depth of about 300 meters (980 feet) because of good surface mixing mechanisms such as surface currents, waves, and tides. Below 300 meters (980 feet), the density increases rapidly until a depth of about 1000 meters (3300 feet). Below 1000 meters, the water's low density again remains constant down to the ocean floor.

The density curve for high-latitude regions shows very little variation with depth. Density is relatively high at the surface because water temperature is low.

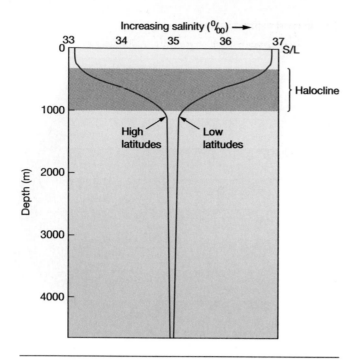

Figure 5–22 Salinity variation with depth. Vertical profile showing high- and low-latitude salinity variation (horizontal scale in ‰) with depth (vertical scale in meters with sea level at the top). The zone of rapidly changing salinity with depth is the halocline.

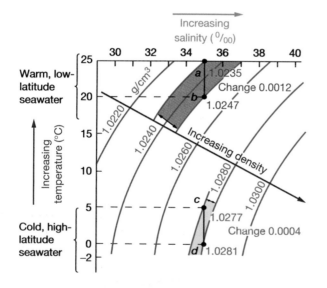

Figure 5–23 Seawater density varies with temperature and salinity. Blue curves show density, which increases with increasing salinity and decreases with increasing temperature. In high-latitude areas of cold water, temperature has less effect on density than in high-temperature, low-latitude areas.

Density is high below the surface, too, because water temperature is also low. The density curve for high-latitude regions, therefore, is a straight vertical line, which indicates uniform conditions at the surface and at depth. These conditions allow cold high-density water

to form at the surface, sink, and initiate deep ocean currents.

Temperature is the most important factor influencing seawater density, so the temperature graph (Figure 5–24b) strongly resembles the density graph (Figure 5–24a). The only difference is that they are a mirror image of each other, illustrating that temperature and density are inversely proportional to one another.

> Differences in ocean density cause the ocean to be layered. Seawater density increases with decreased temperature, increased salinity, and increased pressure.

Pycnocline and Thermocline

Analogous to the halocline (the layer of rapidly changing salinity in Figure 5–22), the low-latitude curve in Figure 5–24a shows a layer of rapidly changing density called a **pycnocline** (*pycno* = density, *cline* = slope), and the low-latitude graph in Figure 5–24b shows a layer of rapidly changing temperature called a **thermocline** (*thermo* = heat, *cline* = slope). Like the halocline, the pycnocline and thermocline typically occur between about 300 meters (980 feet) and 1000 meters (3300 feet) below the surface. The temperature difference between water above and below the thermocline can been used to generate electricity (Box 5–2).

When a pycnocline is established in an area, it presents an incredible barrier to mixing between low-density water above and high-density water below. A pycnocline has a high gravitational stability and thus physically isolates adjacent layers of water.[11] The pycnocline results from the combined effect of the thermocline and the halocline, because temperature and salinity influence density. The interrelation of these three layers determines the degree of separation between the upper-water and deep-water masses. Figure 5–24c shows that the **mixed surface layer** occurs above a strong permanent thermocline (and corresponding pycnocline). The water is uniform because it is well mixed by surface currents, waves, and tides. The thermocline and pycnocline occur in a relatively low-density layer called the **upper water**, which is well developed throughout the low and middle latitudes. Denser and colder **deep water** extends from below the thermocline/pycnocline to the deep-ocean floor.

At depths above the main thermocline, divers often experience lesser thermoclines (and corresponding pycnoclines) when descending into the ocean. Thermoclines can develop in swimming pools, ponds, and lakes, too. During the spring and fall, when nights are

---

[11]This is similar to a temperature inversion in the atmosphere, which traps cold (high-density) air underneath warm (low-density) air.

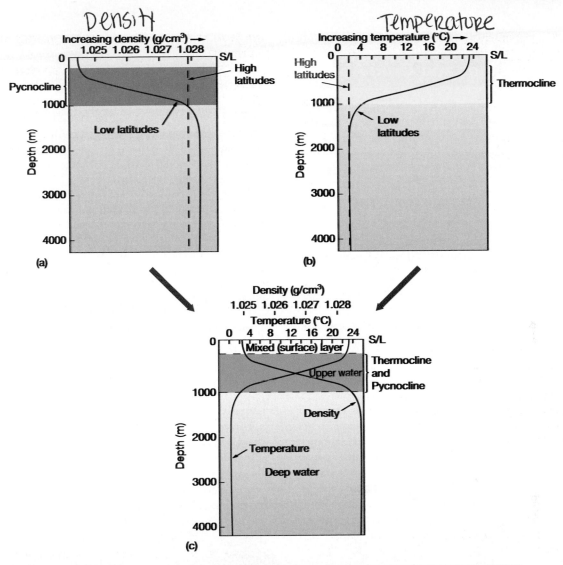

Figure 5–24 Density and temperature variations with depth. **(a)** Density variation with depth for high- and low-latitude regions. The zone of rapidly changing density with depth in low latitudes is the pycnocline. **(b)** Temperature variation with depth for high- and low-latitude regions. The zone of rapidly changing temperature with depth in low latitudes is the thermocline. **(c)** A typical ocean profile showing the variation of temperature and density with depth, showing the inverse relationship between temperature and density.

cool but days can be quite warm, the Sun heats the surface water of the pool yet the water below the surface can be quite cold. If the pool has not been mixed, a thermocline isolates the warm surface layer from the deeper cold water. The cold water below the thermocline can be quite a surprise for anyone who dives into the pool!

In high-latitude regions, the temperature of the surface water remains cold year round, so there is very little difference between the temperature at the surface and in deep water below. Thus, a thermocline and corresponding pycnocline rarely develop in high-latitude regions. Only during the short summer when the days are long does the Sun begin to heat the surface water. Even then, the water does not heat up very much. Nearly all year, then, the water column in high latitudes is **isothermal** (*iso* = same, *thermo* = heat) and **isopycnal** (*iso* =

same, *pycno* = density), allowing good vertical mixing between surface and deeper water.

> A halocline is a layer of rapidly changing salinity, a pycnocline is a layer of rapidly changing density, and a thermocline is a layer of rapidly changing temperature.

Comparing Pure Water and Seawater

Because seawater is 96.5% water, its physical properties are very similar to those of pure water (Table 5–4). For instance, the color of pure water and seawater is

Box 5–2
The Hot and Cold About OTEC Systems

The Sun imparts a large amount of radiant energy to Earth, where it is stored in the ocean and the atmosphere. Can this energy be harnessed for use by humans? The potential for extracting energy from the movement and heat distribution patterns of the atmosphere and ocean is attractive for the following reasons:

- It can be achieved without significant air or water pollution.

- The amount of energy available at any time is far greater than that in fossil fuels (coal, oil, natural gas) and nuclear fuel (uranium).

- The energy is renewable as long as the Sun continues to radiate energy.

- It can be collected any time because it depends on solar heat stored in the oceans and atmosphere over time, not on energy directly from the Sun.

The main sources of this type of renewable energy are:

1. Heat stored in the oceans
2. Kinetic energy of the winds
3. Potential and kinetic energy of waves

4. Potential and kinetic energy of tides and currents

Of these four sources, the one with the greatest amount of energy potential is the heat stored in the warm surface layer of the oceans because of water's unique thermal properties.

The ocean occupies 90% of Earth's surface in the *tropics* (the region between the Tropics of Cancer and Capricorn). What makes this warm tropical surface water such an abundant source of energy is the presence of much colder water beneath the thermocline. For instance, the temperature difference between tropical surface water and water below the thermocline can be 20°C (36°F) or more. A temperature difference of at least 17°C (31°F) is enough to provide a source of energy that can be converted to electricity using an **ocean thermal energy conversion (OTEC)** system (Figure 5C).

An OTEC system works like a typical home refrigeration system but in the opposite direction and on a much larger scale. It converts heat energy stored in the ocean by using warm surface water to heat a fluid (such as liquefied propane gas or ammonia), which evaporates (Figure 5C, *inset*). The vapor is directed through tubes and its increased volume allows it to turn a turbine, which powers an electrical

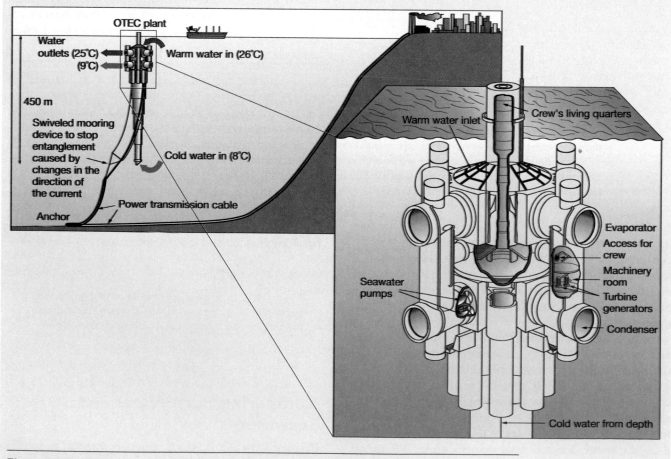

Figure 5C An OTEC system and detail of the main compartment (inset).

gcnerator that produces electricity. After passing through the turbine, cold water pumped up from the deep ocean condenses the fluid again, making it ready for heating by warm surface water to repeat the cycle.

By the late 1980s, a few small OTEC plants were tested using technology developed by the Japanese. In the U.S., only southern Florida and Hawaii have the necessary temperature difference to allow OTEC systems to successfully operate. In Florida, for example, an offshore warm surface current called the Gulf Stream provides an adequate temperature difference with colder deep water below.

The Pacific International Center for High Technology Research on the island of Hawaii currently operates the world's largest OTEC plant. In addition to producing up to 210 kilowatts of electricity, the cold water brought up from below the thermocline is used to condense atmospheric water, so the plant also produces a supply of fresh water. The plant is also experimenting with using the cold seawater to chill roots and stimulate growth in crops by channeling the cold seawater through pipelines below the soil. Another benefit is the enhancement of fishing grounds due to artificial upwelling, which results from cold, deep, nutrient-rich water being brought to the surface. Although OTEC requires a large initial investment, the Hawaii site has demonstrated that OTEC systems can be commercially successful.

identical, yet the dissolved substances in seawater give it a distinct odor and taste.

Recall that the density of water increases as its salinity increases because the amount of dissolved substances increases. The density of 35‰ seawater is 3% higher than the density of pure water.

Recall also that dissolved substances interfere with water changing state. The freezing points and boiling points in Table 5–4 show, for example, that dissolved substances decrease the freezing point and increase the boiling point of water. Thus, seawater freezes at a temperature 1.9°C (3.4°F) lower than pure water. Similarly, seawater boils at a temperature 0.6°C (1.1°F) higher than pure water. As a result, the salts in seawater extend the range of temperatures in which water is a liquid. The same principle applies to antifreeze used in automobile radiators. Antifreeze lowers the freezing point of the water in a radiator and increases the boiling point, thus extending the range over which the water remains in the liquid state. Antifreeze, therefore, protects your radiator from freezing in the winter and from boiling over in the summer.

Desalination

Earth's growing population uses fresh water in greater volumes each year. As fresh water becomes scarcer, several countries have begun to use the ocean as a source of water. **Desalination**, or salt removal from seawater, can provide fresh water for business, home, and agricultural use. Today, there are more than 12,500 desalination plants in existence worldwide that provide about 1% of the world's drinking water. The majority of these plants are located in the Middle East, the Caribbean, and the Mediterranean. The United States produces only about 10% of the world's desalted water, with the majority produced in Florida and California. More than half of the world's desalination plants use distillation to purify water, while the remaining plants use various membrane processes.

Distillation

Distillation is shown schematically in Figure 5–25. Saltwater is boiled and the resulting water vapor is passed through a cooling condenser, where it condenses and is collected as fresh water. This simple procedure is very efficient at purifying seawater. For instance, distillation of 35‰ seawater produces fresh water with a salinity of only 0.03‰, which is about 10 times fresher than bottled water and needs to be mixed with less pure water to make it taste better. Distillation is expensive, however, because it requires large amounts of heat energy to boil the saltwater. Because of water's high latent heat of vaporization, it takes 540 calories to convert only 1 gram of water at the boiling point to the vapor state.[12] Increased efficiency, such as using the waste heat from a power plant, is required to make distillation practical on a large scale.

Solar humidification or **solar distillation** does not require supplemental heating and has been used successfully in small-scale agricultural experiments in arid regions such as Israel, West Africa, and Peru. Solar humidification is similar to distillation in that saltwater is evaporated in a covered container, but the water is heated by direct sunlight, instead (Figure 5–25). Saltwater in the container evaporates, and the water vapor that condenses on the cover runs into collection trays. The major difficulty lies in effectively concentrating the energy of sunlight into a small area to speed evaporation.

Membrane Processes

Electrolysis can be used to desalinate seawater, too. In this method, two volumes of fresh water—one containing a positive electrode and the other a negative electrode—are placed on either side of a volume of seawater. The seawater is separated from each of the

---

[12] Assuming 100 percent efficiency, it takes a whopping *540,000 calories* of heat energy to make a half-liter (about one pint) bottle of distilled water.

Table 5–4 Comparison of selected properties of pure water and seawater.

Property	Pure water	35‰ seawater
Color (light transmission)		
–Small quantities of water	Clear (high transparency)	Same as for pure water
–Large quantities of water	Blue-green because water molecules scatter blue and green wavelengths best	Same as for pure water
Odor	Odorless	Distinctly marine
Taste	Tasteless	Distinctly salty
pH	7.0	8.1 (slightly alkaline)
Density at 4°C (39°F)	1.000 g/cm³	1.030 g/cm³
Freezing point	0°C (32°F)	−1.9°C (28.6°F)
Boiling point	100°C (212°F)	100.6°C (213°F)

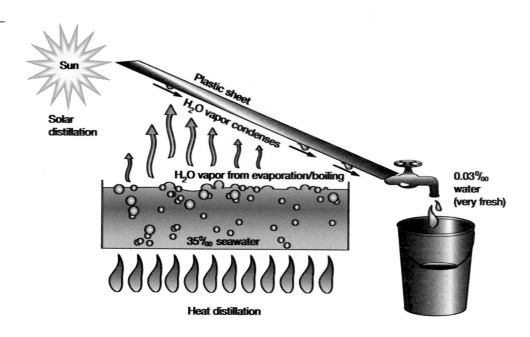

Figure 5–25 Distillation. The process of distillation requires boiling saltwater (heat distillation) or using the Sun's energy to evaporate seawater (solar distillation). In either case, the water vapor is captured and condensed, which produces very pure water.

fresh water reservoirs by semipermeable membranes. These membranes are permeable to salt ions but not to water molecules. When an electrical current is applied, positive ions such as sodium ions are attracted to the negative electrode, and negative ions such as chloride ions are attracted to the positive electrode. In time, enough ions are removed through the membranes to convert the seawater to fresh water. The major drawback to electrolysis is that it requires large amounts of energy.

Reverse osmosis (*osmos* = to push) may have potential for large-scale desalination. In osmosis, water molecules naturally pass through a thin, semipermeable membrane from a fresh water solution to a saltwater solution. In reverse osmosis, water on the salty side is highly pressurized to drive water molecules—but not salt and other impurities—through the membrane to the fresh water side (Figure 5–26). A significant problem with reverse osmosis is that the membranes are flimsy, become clogged, and must be replaced frequently. Advanced composite materials may help eliminate these problems because they are sturdier, provide better filtration, and last up to 10 years. Worldwide, at least 30 countries located in arid climates are operating reverse osmosis units. For example, Santa Barbara, California, operates a reverse osmosis plant that produces up to 34 million liters (9 million gallons) daily, which supplies up to 60% of its municipal water needs. This method is also used in many household water purification units and aquariums.

Other Methods of Desalination

Seawater selectively excludes dissolved substances as it freezes—a process called **freeze separation**. As a result, the salinity of sea ice (once it is melted) is typically 70% lower than seawater. To make this an effective desalination technique, though, the water must be

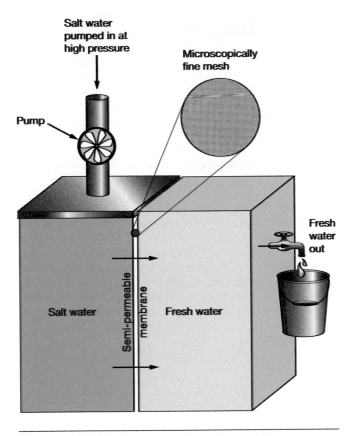

Figure 5–26 Reverse osmosis. The process of reverse osmosis involves applying pressure to salt water and forcing it through a semipermeable membrane, thus removing the salts and producing fresh water.

frozen and thawed multiple times, with the salts washed from the ice between each thawing. Like electrolysis, freeze separation requires large amounts of energy, so it may be impractical except on a small scale.

Yet another way to obtain fresh water is to melt naturally formed ice. Imaginative thinkers have proposed towing large icebergs to coastal waters off countries that need fresh water. Once there, the fresh water produced as the icebergs melt could be captured and pumped ashore. Studies have shown that towing large Antarctic icebergs to arid regions would be technologically feasible and, for certain Southern Hemisphere locations, economically feasible, too.

Other novel approaches to desalination include crystallization of dissolved components directly from seawater, solvent demineralization using chemical catalysts, and even making use of salt-eating bacteria!

 Students Sometimes Ask...

How can it be that water—a liquid at room temperature—can be created by combining hydrogen and oxygen—two gases at room temperature?

It is true that combining two parts hydrogen gas with one part oxygen gas produces liquid water. This can be accomplished as a chemistry experiment, although care should be taken because much energy is released during the reaction (don't try this at home!). Often times, when combining two elements together, the product has very different properties than the pure substances, which is what most people find amazing about chemistry. For instance, combining elemental sodium (Na), a highly reactive metal, with pure chlorine (Cl_2), a toxic nerve gas, produces cubes of harmless table salt (NaCl).

I've seen the labels on electric cords warning against using electrical appliances close to water. Are these warnings because water's polarity allows electricity to be transmitted through it?

Yes and no. Water molecules are polar, so you might assume that water is a good conductor of electricity. Pure water is a very poor conductor, however, because water molecules are neutral overall and will not move toward the negatively or positively charged pole in an electrical system. If an electrical appliance is dropped into a tub of absolutely pure water, the water molecules will transmit no electricity. Instead, the water molecules will simply orient their positively charged hydrogen ends toward the negative pole of the appliance and their negatively charged oxygen ends toward its positive pole, which tends to neutralize the electric field. Interestingly, it is the *dissolved substances* that transmit electrical current through water. Even slight amounts, such as those in tap water, allow electricity to be transmitted. That's why there are warning labels on the electric cords of household appliances that are commonly used in the bathroom, such as blow dryers, electric razors, and heaters. That's also why it is recommended to stay out of any water—including a bathtub or shower—during a lightning storm!

What would happen to a person if he or she drank seawater?

It depends on the quantity. The salinity of seawater is about four times greater than that of your body fluids. In your body, seawater causes your internal membranes to lose water through *osmosis*, which transports water molecules from higher concentrations (the normal body chemistry of your internal fluids) to areas of lower concentrations (your digestive tract containing seawater). Thus, your natural body fluids would move into your digestive tract and eventually be expelled, causing dehydration.

Don't worry too much if you've inadvertently swallowed some seawater. As a nutritional drink, seawater provides seven important nutrients and contains no fat, cholesterol, or calories. Some people claim that drinking a small amount of seawater daily gives them good health! However, beware of microscopic contaminants that are often present in seawater.

You mentioned that when seawater freezes, it produces ice with about 10‰ salinity. Once that ice melts, can a person drink it with no ill effects?

Early Arctic explorers found out the answer to your question by necessity. Some of these explorers who traveled by ship in high-latitude regions became inadvertently or purposely entrapped by sea ice. Lacking other water sources, they used melted sea ice. Although newly formed sea ice contains little salt, it does trap a significant amount of brine (drops of salty water). Depending on the rate of freezing, newly formed ice may have a total salinity from 4 to 15‰. The more rapidly it forms, the more brine it captures and the higher the salinity. Melted sea ice with salinity this high doesn't taste very good, and it still causes dehydration, but not as quickly as drinking 35‰ seawater does. Over time, however, the brine will trickle down through the coarse structure of the sea ice, so its salinity decreases. By the time it is a year old, sea ice normally becomes relatively pure. Drinking melted sea ice enabled these early explorers to survive.

Would it be dangerous to drink chemically purified water?

For short periods of time, no. However, drinking absolutely pure water for extended periods of time is not advisable. Because water is so good at dissolving other substances, pure water would begin to steal some of the important trace elements that your body needs to survive. Pure water is tasteless, too, so some companies add dissolved substances in carefully controlled amounts to enhance the taste of their bottled water. It's best to stick with tap water or good bottled water.

If seawater is so good at dissolving things, why aren't marine organisms dissolving?

They are! Marine organisms live in a medium that is actively working to dissolve them, so they have developed special defenses, such as *mucus*, to prevent it from happening. Composed of an oily organic substance, mucus provides a protective coating because oil is insoluble in water. That's why fish feel slimy when you hold them.

What is the strategy behind adding salt to a pot of water when making pasta? Does it make the water boil faster?

Adding salt to water will not make the water boil faster. It will, however, make the water boil at a higher temperature because dissolved substances *raise* its boiling point (and *lower* its freezing point). Thus, the pasta will cook in less time. Additionally, the salt adds flavoring, so the pasta may taste better, too. Be sure to add the salt *after* the water has come to a boil, though, or it will take longer to reach a boil. This is a wonderful use of chemical principles—helping you to cook better!

Chapter in Review

• Water's remarkable properties help make life as we know it possible on Earth. These properties include the arrangement of its atoms, how its molecules stick together, its ability to dissolve almost everything, and its heat storage capacity.

• The water molecule is composed of one atom of oxygen and two atoms of hydrogen (H_2O). The two hydrogen atoms, which are covalently bonded to the oxygen atom, are attached to the same side of the oxygen atom and produce a bend in the geometry of a water molecule. This geometry makes water molecules polar, which allows them to form hydrogen bonds with other water molecules or other substances and gives water its remarkable properties. Water, for example, is "the universal solvent" because it can hydrate charged particles (ions), thereby dissolving them.

• Water is one of the few substances that exists naturally on Earth in all three states of matter (solid, liquid, gas). Hydrogen bonding gives water unusual thermal properties, such a high freezing point (0°C or 32°F) and boiling point (100°C or 212°F), a high heat capacity (1 calorie), a high latent heat of melting (80 calories per gram), and a high latent heat of vaporization (540 calories per gram).

• The density of water increases as temperature decreases, similar to most substances, and reaches a maximum density at 4°C (39°F). Below 4°C, however, water density decreases with temperature, due to the formation of bulky ice crystals. As water freezes, it expands by about 9% in volume, so ice floats on water.

• Salinity is the amount of dissolved solids in ocean water. It averages about 35 grams of dissolved solids per kilogram of ocean water [35 parts per thousand (‰)], but ranges from brackish to hypersaline. Six ions—chloride, sodium, sulfate,

magnesium, calcium, and potassium—account for over 99% of the dissolved solids in ocean water. These ions always occur in a constant proportion in any seawater sample, so salinity can be determined by measuring the concentration of only one—typically, the chloride ion. The residence time of various elements indicates how long they stay in the ocean.

• A natural buffering system exists in the ocean, based on the chemical reaction of carbon dioxide in water. This buffering system regulates any changes in pH, creating a stable ocean environment.

• Dissolved components in seawater are added and removed by a variety of processes. Precipitation, runoff, and the melting of icebergs and sea ice add fresh water to seawater and decrease its salinity. The formation of sea ice and evaporation remove fresh water from seawater and increase its salinity. The hydrologic cycle includes all the reservoirs of water on Earth, including the oceans, which contain 97% of Earth's water.

• The salinity of surface water varies considerably due to surface processes, with the maximum salinity found near the Tropics of Cancer and Capricorn and the minimum salinity found in high-latitude regions. Salinity also varies with depth down to about 1000 meters but below that, the salinity of deep water is very consistent. A halocline is a layer of rapidly changing salinity.

• Seawater density increases as temperature decreases and salinity increases, though temperature influences surface seawater density more strongly than salinity (the influence of pressure is negligible). Density and temperature vary considerably with depth in low-latitude regions, creating a pycnocline and corresponding thermocline, which are generally absent in high latitudes. Ocean thermal energy conversion (OTEC) is a process that uses the temperature difference between warm tropical surface water and the cold water below the thermocline to produce electricity.

• The physical properties of pure water and seawater are remarkably similar, with a few notable exceptions. Compared to pure water, seawater has a higher pH, density, and boiling point (but a lower freezing point).

• Desalination of ocean water to provide fresh water for business, home, and agricultural use is of growing interest. Solar distillation (solar humidification), electrolysis, freeze separation, and reverse osmosis are methods currently used to desalinate seawater.

Key Terms

Acid (p. 146)

Alkaline (p. 146)

Atom (p. 131)

Base (p. 146)

Boiling point (p. 135)

Brackish (p. 142)

Buffering (p. 146)

Calorie (p. 134)

Challenger, HMS (p. 130)

Chlorinity (p. 144)

Cohesion (p. 133)

Condensation point (p. 135)

Condense (p. 135)

Covalent bond (p. 132)

Deep water (p. 153)

Desalination (p. 155)

Dipolar (p. 133)

Distillation (p. 155)

Electrolysis (p. 156)

Electron (p. 132)

Electrical conductivity (p. 144)

Electrostatic attraction (p. 133)

Evaporation (p. 137)

Freeze separation (p. 157)

Freezing point (p. 135)

Goiters (p. 143)

Halocline (p. 150)

Heat (p. 134)

Heat capacity (p. 135)

Hydration (p. 133)

Hydrogen bond (p. 133)

Hydrologic cycle (p. 148)

Hypersaline (p. 143)

Iceberg (p. 147)

Ionic bond (p. 133)

Ions (p. 132)

Isopycnal (p. 155)

Isothermal (p. 155)

Kinetic energy (p. 134)

Latent heat of condensation (p. 137)

Latent heat of evaporation (p. 137)

Latent heat of freezing (p. 137)

Latent heat of melting (p. 137)

Latent heat of vaporization (p. 137)

Melting point (p. 135)

Mixed surface layer (p. 153)

Molecule (p. 132)

Neutral (p. 146)

Neutron (p. 132)

Nucleus (p. 132)

Ocean thermal energy conversion (OTEC) (p. 154)

Parts per thousand (‰) (p. 142)

pH scale (p. 147)

Polarity (p. 133)

Precipitation (p. 138)

Principle of Constant Proportions (p. 131)

Proton (p. 144)

Pycnocline (p. 152)

Residence time (p. 145)

Reverse osmosis (p. 156)

Runoff (p. 150)

Salinity (p. 144)

Salinometer (p. 144)

Sea ice (p. 148)

Solar distillation (p. 155)

Solar humidification (p. 155)

Surface tension (p. 133)

Temperature (p. 134)

Thermal contraction (p. 139)

Thermocline (p. 152)

Upper water (p. 153)

van der Waals force (p. 134)

Vapor (p. 135)

Questions and Exercises

1. List some major achievements of the voyage of HMS *Challenger*.

2. Sketch a model of an atom, showing the positions of the subatomic particles protons, neutrons, and electrons.

3. Describe what condition exists in water molecules to make them dipolar.

4. Sketch several water molecules, showing all covalent and hydrogen bonds. Be sure to indicate the polarity of each water molecule.

5. How does hydrogen bonding produce the surface tension phenomenon of water?

6. Discuss how the dipolar nature of the water molecule makes it such an effective solvent of ionic compounds.

7. Describe the differences between the three states of matter, using the arrangement of molecules in your explanation.

8. Why are the freezing and boiling points of water higher than would be expected for a compound of its molecular makeup?

9. How does the heat capacity of water compare with that of other substances? Describe the effect this has on climate.

10. The heat energy added as latent heat of melting and latent heat of vaporization does not increase water temperature. Explain why this occurs, and where the energy is used.

11. Why is the latent heat of vaporization so much greater than the latent heat of melting?

12. Describe how excess heat energy absorbed by Earth's low-latitude regions is transferred to heat-deficient higher latitudes through a process that uses water's latent heat of evaporation.

13. As water cools, two distinct changes take place in the behavior of molecules: Their slower movement tends to increase density, whereas the formation of bulky ice crystals decreases density. Describe how the relative rates of their occurrence cause pure water to have a temperature of maximum density at 4°C (39.2°F) and make ice less dense than liquid water.

14. What is your state sales tax, in parts per thousand?

15. What are goiters? How can they be avoided?

16. What physical conditions create brackish water in the Baltic Sea and hypersaline water in the Red Sea?

17. What condition of salinity makes it possible to determine the total salinity of ocean water by measuring the concentration of only one constituent, the chloride ion?

18. Explain the difference between an acid and an alkali (base) substance. How does the ocean's buffering system work?

19. Describe the ways that dissolved components are added and removed from seawater.

20. List the components (reservoirs) of the hydrologic cycle that hold water on Earth and the percentage of Earth's water in each one. Describe the processes by which water moves among these reservoirs.

21. Explain why there is such a wide variation of surface salinity, but such a narrow range of salinity at depth.

22. Why is there such a close association between (a) the curve showing seawater density variation with ocean depth and (b) the curve showing seawater temperature variation with ocean depth?

23. Compare and contrast the following seawater desalination methods: distillation, solar humidification, and reverse osmosis.

24. Explain how an ocean thermal energy conversion (OTEC) unit generates electricity (a labeled diagram might help).

References

Bailey, H. S., Jr. 1953. The voyage of the *Challenger. Scientific American* 188:5, 88–94.

Behrman, A. S. 1968. *Water is everybody's business: The chemistry of water purification.* Garden City, NY: Doubleday.

Charnock, H. 1996. H.M.S. *Challenger* and the development of marine science. In Pirie, R. G., ed., *Oceanography: Contemporary readings in ocean sciences*, 3rd ed. New York: Oxford University Press.

Davis, K. S., and Day, J. S. 1961. *Water: The mirror of science.* Garden City, NY: Doubleday.

Friederici, P. 1996. Across a crowded sea: How (and what) whales hear. *Pacific Discovery* 49:1, 28–31.

Hammond, A. L. 1977. Oceanography: Geochemical tracers offer new insight. *Science* 195:164–166.

Harvey, H. W. 1960. *The chemistry and fertility of sea waters.* New York: Cambridge University Press.

Kuenen, P. H. 1963. *Realms of water.* New York: Science Editions.

MacIntyre, F. 1970. Why the sea is salt. *Scientific American* 223:5, 104–115.

Munk, W. 1993. The sound of oceans warming. *The Sciences* 33:5, 20–26.

Packard, G. L. 1975. *Descriptive physical oceanography*, 2nd ed. New York: Pergamon Press.

Revelle, R. 1963. Water. *Scientific American* 209:3, 92–108.

Swenson, H. 1993. *Why is the ocean salty?* Washington D.C.: U.S. Geological Survey pamphlet.

The Open University Course Team. 1989. *Seawater: Its composition, properties and behaviour.* Oxford: Pergamon Press.

Wadhams, P. 1996. The resource potential of Antarctic icebergs. In Pirie, R. G., ed., *Oceanography: Contemporary readings in ocean sciences*, 3rd ed. New York: Oxford University Press.

Yokohama, Y., Guichard, F., Reyss, J-L., and Van, N. H. 1978. Oceanic residence times of dissolved beryllium and aluminum deduced from cosmogenic tracers ^{10}Be and ^{26}Al. *Science* 201:1016–1017.

Suggested Reading in Scientific American

Bailey, H. S., Jr. 1953. The voyage of the *Challenger.* 188:5, 88–94. A summary of the accomplishments of the English oceanographic expedition of 1872.

Baker, J. A., and Henderson, D. 1981. The fluid phase of matter. 245:5, 130–139. The structure of gases and liquids is modeled using hard spheres.

Buckingham, M. J., Potter, J. R., and Epiphany, C. L. 1996. Seeing underwater with background noise. 274:2, 86–91. A technique called acoustic-daylight imaging uses ambient sound in the ocean to form images just as daylight is used for that purpose in the atmosphere. The resolution may be a bit less than that used for seeing in daylight, but it has tremendous potential for use in the ocean.

Gerstein, M., and Leavitt, M. 1998. Simulating water and the molecules of life. 279:5, 100–105. Computer modeling reveals how the unusual properties of water affect biological molecules such as proteins and DNA.

Liss, T. M., and Tipton, P. L. 1997. The discovery of the top quark. 277:3, 54–59.

A review of the quest to find the top (6th) quark, which involved the world's most energetic collisions and a cast of thousands.

MacIntyre, F. 1970. Why the sea is salt. 223:5, 104–115. A summary of what is known of the processes that add and remove the elements dissolved in the ocean.

Martindale, D. 2001. Sweating the small stuff. 284:2, 52–55. Reviews existing desalination techniques as part of an article on the state of Earth's fresh water supplies.

Wettlaufer, J. S. and Dash, G. 2000. Melting below zero. 282:2, 50–55. The significance of a sub-zero layer of water on ice has recently been discovered.

Oceanography on the Web

Visit the *Essentials of Oceanography* home page for on-line resources for this chapter. There you will find an on-line study guide with review exercises, and links to oceanography sites to further your exploration of the topics in this chapter. *Essentials of Oceanography* is at: **http://www.prenhall.com/ thurman** (click on the Table of Contents menu and select this chapter).

C H A P T E R

6

Air–Sea Interaction

•

RMS *TITANIC:* LOST (1912) AND FOUND (1985)

The early 20th century was a time of rapidly advancing technology and a widely held belief that science could cure diseases, contribute to a higher standard of living, and generally improve everyone's lives. It was during this time, and with great acclaim, that the luxury liner **RMS** *Titanic* (Figure 6A) was constructed. At 269 meters (882 feet) long (not quite the length of three football fields), the *Titanic* was the world's largest passenger ship and the world's largest movable object. She was designed to be unsinkable, with multiple compartments that would keep her afloat even if some of the compartments filled with water. As a result, the *Titanic* had only enough lifeboats for about half of its passengers. Operated by the White Star Line, she departed in April 1912 on her maiden voyage from England to the United States across the North Atlantic Ocean.

Glaciers on Greenland and Ellesmere Island produce icebergs, which become navigational hazards when ocean currents carry them into North Atlantic shipping lanes. The large size of some of these icebergs takes them several years to melt, and, in that time, they may be carried as far south as 40 degrees north latitude, the same latitude as Philadelphia.

On the evening of April 14, the *Titanic* was traveling under a clear sky and calm seas near the Grand Banks south of Newfoundland. She was running ahead of schedule and the officials of the White Star Line were hoping to make a good impression by arriving in New York early. Despite repeated warnings of icebergs in the area, the crew maintained an excessive speed of 41 kilometers (25.5 miles) per hour. About half an hour before midnight, she sideswiped a huge iceberg and sustained a 100-meter (330-foot) gash in her hull. As the *Titanic's* front compartments began to fill with water, the ship tipped and allowed more compartments to fill with water. Within $2^{1}/_{2}$ hours after striking the iceberg, the ship tipped so far that it broke in half and sank. Because of cold water temperatures, night-time conditions, lack of aid from other ships in the area, and a shortage of lifeboats, 1513 of the 2224 passengers perished.

The tragedy brought about more stringent safety rules for ships and the formation of an iceberg patrol. The ice patrol regularly surveys an area larger than the state of Pennsylvania to prevent further disasters of this kind. Originated by the U.S. Navy, the ice patrol became international in 1914. Today, the International Ice Patrol is maintained by the U.S. Coast Guard.

There had been much interest in locating the wreck of the *Titanic*, but the technology to find it in such deep water only recently became available. Further, the exact location of the collision was somewhat of a mystery. Whether she sank straight down onto the sea floor or if she drifted on the way to the bottom was also unclear. After several unsuccessful attempts, an expedition led by oceanographer Robert Ballard of the Woods

---

key questions

- What causes Earth's seasons?
- How does the Coriolis effect influence moving objects?
- Why does the atmosphere move?
- What are the characteristics of Earth's major wind belts and boundaries?
- How does a hurricane work?
- What causes the greenhouse effect?

When the still sea conspires an armor
And her sullen and aborted
Currents breed tiny monsters,
True sailing is dead.
Awkward instant
And the first animal is jettisoned...
　　　　　　—The Doors
　　Horse Latitudes *(1972)*

Figure 6A RMS *Titanic*, circa 1912, and at the bottom of the North Atlantic in 1985 (*inset*).

Hole Oceanographic Institution used the unmanned, tethered submersible *Argo* to locate the wreck in September 1985 (Figure 6A, *inset*).

The *Titanic* lay in two pieces at a depth of 3844 meters (12,612 feet), with debris scattered all around the wreckage. In 1986, the manned research submersible *Alvin* touched down on the deck of the *Titanic*, carry-ing the first humans to observe the *Titanic* in almost 75 years. Dr. Ballard believes the wreck should be consecrated as a burial ground, and that retrieval of any items constitutes grave robbing. Since Ballard's discovery of the *Titanic*, however, several other submersibles have visited the site and retrieved items, some of which have been available for auction on the Internet.

The atmosphere and the ocean act as one interdependent system. What happens in one causes changes in the other, and the two are linked by complex feedback loops. Surface currents in the oceans, for instance, are a direct result of Earth's atmospheric wind belts. Conversely, certain atmospheric weather phenomena are manifested in the oceans. If either the oceans or the atmosphere is to be understood, then their interactions and relationships must be studied.

Atmospheric winds create most of the surface currents and waves in the ocean. Solar energy, in turn, which heats the surface of the Earth, creates the winds. Radiant energy from the Sun, therefore, is responsible for motion in the atmosphere and the ocean. Recall from Chapter 5 that the atmosphere and ocean use the high heat capacity of water to constantly exchange this energy, shaping Earth's global weather patterns in the process.

Periodic extremes of atmospheric weather, such as droughts and profuse precipitation, are related to periodic changes in oceanic conditions. For instance, it was recognized as far back as the 1920s that El Niño—an ocean event—was tied to catastrophic weather events worldwide. What is as yet unclear, however, is whether changes in the ocean produce changes in the atmosphere that lead to the El Niño phenomenon—or vice versa. El Niño–Southern Oscillation events are discussed in Chapter 7, "Ocean Circulation."

Air–sea interactions influence the greenhouse effect, too. In the 1980s, studies confirmed that Earth's average temperature had risen over the past century. Global warming may be due to human-caused increases in carbon dioxide and other gases that absorb heat. However, atmospheric carbon dioxide has increased only half as much as predicted from human activities, so where did the rest of the carbon dioxide go? If the oceans are removing carbon dioxide from the atmosphere, how does carbon dioxide enter the oceans and where does it go?

These questions are answered slowly as new research is completed and more sophisticated computer models are developed. In this chapter, we examine the redistribution of solar heat by the atmosphere and its influence on oceanic conditions. First, large-scale phenomena that influence air-sea interactions are studied, and then smaller-scale phenomena are examined.

Uneven Solar Heating on Earth

The side of Earth facing the Sun (the daytime side) receives a tremendous dose of intense solar energy. This energy drives the global ocean–atmosphere engine, creating pressure and density differences that stir currents and waves in both the atmosphere and the ocean.

Distribution of Solar Energy

If Earth were a flat plate in space, with its flat side directly facing the Sun, then sunlight would fall equally on all parts of Earth. Earth is spherical, however, so the amount and intensity of solar radiation received at higher latitudes is much less than at lower latitudes. The following factors influence the amount of radiation received at low and high latitudes:

- Sunlight strikes low latitudes at a high angle, so the radiation is concentrated in a relatively small area (*area A* in Figure 6–1). Sunlight strikes high latitudes at a low angle, so the same amount of radiation is spread over a larger area (*area B* in Figure 6–1).

- Earth's atmosphere absorbs some radiation, so less radiation strikes Earth at high latitudes than at low latitudes, because sunlight passes through more of the atmosphere at high latitudes.

- The **albedo** (*albus* = white) of various Earth materials is the percentage of incident radiation that is reflected back to space. The average albedo of Earth's surface is about 30%. More radiation is reflected back to space at high latitudes because ice has a much higher albedo than soil or vegetation.

- The angle at which sunlight strikes the ocean surface determines how much is absorbed and how much is reflected. If the Sun shines down on a smooth sea from directly overhead, only 2% of the radiation is reflected, but if the Sun is only 5 degrees above the horizon, 40% is reflected back into the atmosphere (Table 6–1). Thus, the ocean reflects more radiation at high latitudes than at low latitudes.

Because of all these reasons, the intensity of radiation at high latitudes is greatly decreased compared with that falling in the equatorial region.

In addition, the amount of radiation received at Earth's surface varies *daily* because Earth rotates on its axis so the surface experiences daylight and darkness each day. The amount of radiation also varies *annually* due to Earth's seasons.

Earth's Seasons

Earth revolves around the Sun along an elliptical path (Figure 6–2). The plane traced by Earth's orbit is called the **ecliptic**. Earth's axis of rotation is not perpendicular ("upright") on the ecliptic; rather, it tilts at an angle of 23.5 degrees. Figure 6–2 shows that throughout the yearly cycle, Earth's axis *always points in the same direction*, which is toward Polaris, the North Star.

The tilt of Earth's rotational axis (and not its elliptical path) causes Earth to have seasons. Spring, summer, fall, and winter occur as follows:

- At the **vernal equinox** (*vernus* = spring; *equi* = equal, *noct* = night), which occurs on or about March 21, the Sun is directly overhead along the Equator. During this time, all places in the world experience equal lengths of night and day (hence the name *equinox*). In the Northern Hemisphere, the vernal equinox is also known as the spring equinox.

- At the **summer solstice** (*sol* = the Sun, *stitium* = a stoppage), which occurs on or about June 21, the Sun reaches its most northerly point in the sky, directly overhead along the **Tropic of Cancer**, at 23.5 degrees north latitude (Figure 6–2, *left inset*). To an observer on Earth, the noonday Sun reaches its northernmost or southernmost position in the sky at this time and appears to pause—hence the term *solstice*—before beginning it next 6-month cycle.

- At the **autumnal** (*autumnus* = fall) **equinox**, which occurs on or about September 23, the Sun is directly overhead along the Equator again. In the Northern Hemisphere, the autumnal equinox is also known as the fall equinox.

- At the **winter solstice**, which occurs on or about December 22, the Sun is directly overhead along the **Tropic of Capricorn,**[1] at 23.5 degrees south latitude (Figure 6–2, *right inset*). In the Southern Hemisphere, the seasons are reversed. Thus, the winter solstice is the time when the Southern Hemisphere is most directly facing the Sun, which is the beginning of the Southern Hemisphere summer.

Because Earth's rotational axis is tilted 23.5 degrees, the Sun's **declination** (angular distance from the equatorial plane) varies between 23.5 degrees north and 23.5 degrees south of the Equator on a yearly cycle. As a result, the region between these two latitudes (called the **tropics**) receives much greater annual radiation than polar areas.

---

[1]The tropics are named for their constellations: directly over the Tropic of Cancer is the constellation Cancer (the crab) and over the Tropic of Capricorn is Capricorn (the goat).

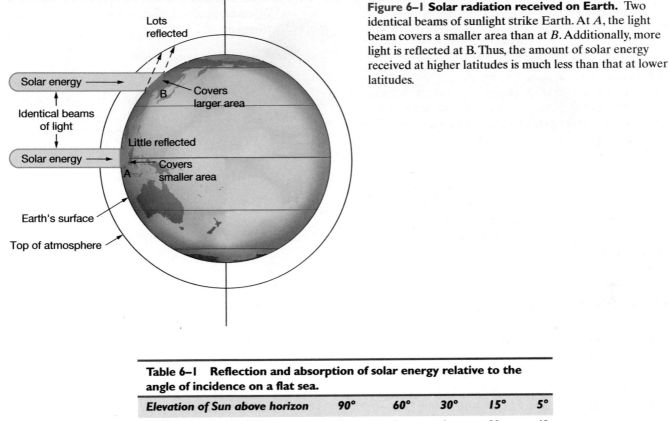

Figure 6–1 Solar radiation received on Earth. Two identical beams of sunlight strike Earth. At *A*, the light beam covers a smaller area than at *B*. Additionally, more light is reflected at B. Thus, the amount of solar energy received at higher latitudes is much less than that at lower latitudes.

Table 6–1   Reflection and absorption of solar energy relative to the angle of incidence on a flat sea.

Elevation of Sun above horizon	*90°*	*60°*	*30°*	*15°*	*5°*
Reflected radiation (%)	2	3	6	20	40
Absorbed radiation (%)	98	97	94	80	60

Seasonal changes in the angle of the Sun and the length of day profoundly influence Earth's climate. In the Northern Hemisphere, for example, the longest day occurs on the summer solstice and the shortest day on the winter solstice.

Daily heating of Earth also influences climate in most locations. Exceptions to this pattern occur north of the **Arctic Circle** (66.5 degrees north latitude) and south of the **Antarctic Circle** (66.5 degrees south latitude), which at certain times of the year do not experience daily cycles of daylight and darkness. For instance, during the Northern Hemisphere winter, the area north of the Arctic Circle receives no direct solar radiation at all and experiences up to six months of darkness. At the same time, the area south of the Antarctic Circle receives continuous radiation ("midnight Sun"), so it experiences up to six months of light. Half a year later, during the Northern Hemisphere summer (the Southern Hemisphere winter), the situation is reversed.

Earth's axis is tilted at an angle of 23.5 degrees, causing the Northern and Southern Hemispheres to take turns "leaning toward" the Sun every six months, resulting in the change of seasons.

Oceanic Heat Flow

Close to the poles, much incoming solar radiation strikes Earth's surface at low angles. Combined with the high albedo of ice, more energy is reflected back into space than absorbed. In contrast, between 35 degrees north latitude and 40 degrees south latitude[2], sunlight strikes Earth at much higher angles and more energy is absorbed than reflected back into space. The graph in Figure 6–3 shows how incoming sunlight and outgoing heat combine on a daily basis for a net heat gain in low-latitude oceans and a net heat loss in high-latitude oceans.

Based on Figure 6–3, you might expect the equatorial zone to grow progressively warmer and the polar regions to grow progressively cooler. The polar regions are always considerably colder than the equatorial zone, but the temperature *difference* remains the same because excess heat is transferred from the equatorial zone to the poles. How is this accomplished? Circulation in both the oceans and the atmosphere transfers the heat.

---

[2]The reason this latitudinal range extends farther into the Southern Hemisphere is because the Southern Hemisphere has more ocean surface area in the mid-latitudes than the Northern Hemisphere does.

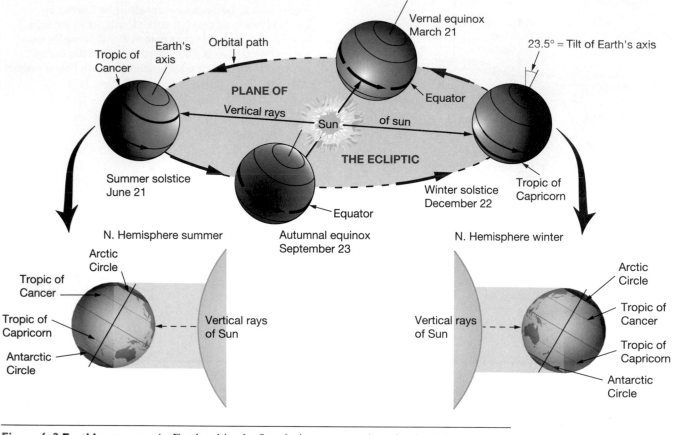

Figure 6–2 Earth's seasons. As Earth orbits the Sun during one year, its axis of rotation constantly tilts 23.5 degrees from perpendicular (relative to the plane of the ecliptic) and always points in the same direction, causing different areas to experience vertical rays of the Sun (*dark lines*). This tilt causes Earth to have seasons, such as when the Northern Hemisphere is tilted toward the Sun during the summer (*left inset*), and away from the Sun during the winter (*right inset*).

The Atmosphere: Physical Properties

The atmosphere transfers heat and water vapor from place to place on Earth. Within the atmosphere, complex relationships exist among air composition, temperature, density, water vapor content, and pressure. Before we apply these relationships, let's examine the atmosphere's composition and some of its physical properties.

Composition Table 6–2 shows that the atmosphere consists almost entirely of nitrogen and oxygen gases. Other gases include argon (an inert gas), carbon diox-

Table 6–2 Composition of dry air.	
Gas	**Concentration (%)**
Nitrogen (N₂)	78.1
Oxygen (O₂)	20.9
Argon (Ar)	0.9
Carbon dioxide (CO₂)	0.036
All others	Trace

ide, and others in trace amounts. Although these gases are present in very small amounts, they can trap significant amounts of heat within the atmosphere.

Temperature Figure 6–4 shows the temperature profile of the atmosphere. The lowermost portion of the atmosphere, which extends from the surface to about 12 kilometers (7 miles), is called the **troposphere** (*tropo* = turn, *sphere* = a ball) and is where all weather is produced. The abundance of atmospheric mixing in the troposphere gives this layer its name. Within the troposphere, temperature gets cooler with altitude. At high altitudes, the air temperature is well below freezing. If you have ever flown in a jet airplane, for instance, you may have noticed that any water on the wings or inside your window freezes during the flight.

Density It may seem surprising that air has density, but since air is composed of molecules, it certainly does. Temperature affects density dramatically. At higher temperatures air molecules move more quickly, take up more space, and density is decreased. Thus, the general relationship between density and temperature is as follows:

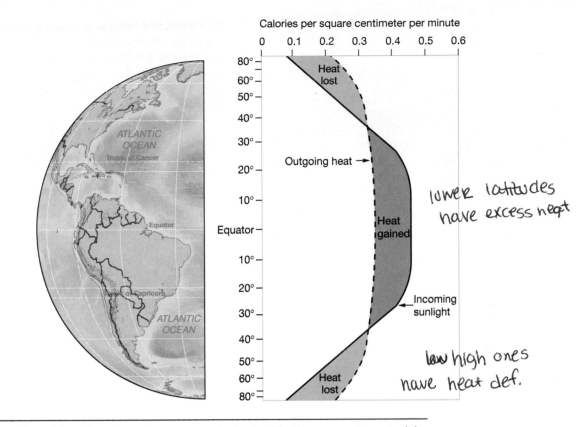

Figure 6–3 Heat gain and heat loss from the oceans. Heat gained by the oceans at equatorial latitudes (*red*) equals heat lost at polar latitudes (*blue*). On average, the two balance each other and the excess heat from equatorial latitudes is transferred to heat-deficient polar latitudes by both oceanic and atmospheric circulation.

- Warm air is less dense, so it rises, which is commonly expressed as "heat rises."
- Cool air is more dense, so it sinks.

Figure 6–5 shows how a radiator (heater) uses convection to heat a room. The heater warms the nearby air and causes it to expand. This expansion makes the air less dense, causing it to rise. Conversely, a cold window cools the nearby air and causes it to contract, thereby becoming more dense, which causes it to sink. A **convection cell** forms, composed of the rising and sinking air moving in a circular fashion, similar to the convection in Earth's mantle discussed in Chapter 2.

Water Vapor Content   The amount of water vapor in air depends in part on the air's temperature. Warm air, for instance, can hold more water vapor than cold air because the air molecules are moving more quickly and come into contact with more water vapor. Thus, warm air is typically moist, and, conversely, cool air is typically dry. As a result, a warm, breezy day speeds evaporation when you hang your laundry outside to dry.

Water vapor influences the density of air. The addition of water vapor decreases the density of air because water vapor has a lower density than air. Thus, humid air is less dense than dry air.

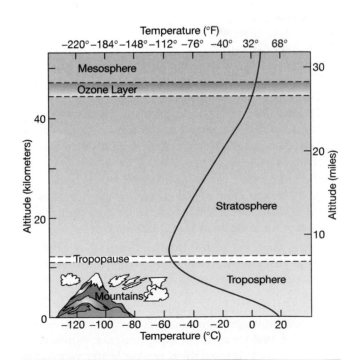

Figure 6–4 Temperature profile of the atmosphere. Within the troposphere, the atmosphere gets cooler with increasing altitude and above the troposphere, the atmosphere generally warms.

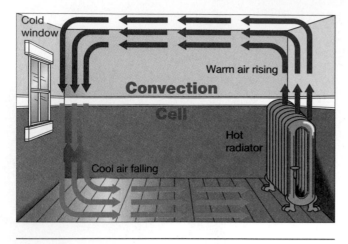

Figure 6–5 Convection in a room. A circular-moving loop of air (a *convection cell*) is caused by warm air rising and cool air sinking.

Pressure Atmospheric pressure is 1.0 atmosphere (14.7 pounds per square inch) at sea level and decreases with increasing altitude. Atmospheric pressure depends on the weight of the column of air above. For instance, a thick column of air produces higher atmospheric pressure than a thin column of air. An analogy to this is water pressure in a swimming pool: the thicker the column of water above, the higher the water pressure. Thus, the highest pressure in a pool is at the bottom of the deep end.

Similarly, the thicker column of air at sea level means air pressure is high at sea level and decreases with increasing elevation. When sealed bags of potato chips or pretzels are taken to a high elevation, the pressure is much lower than where they were sealed, sometimes causing the bags to burst! You may have also experienced this change in pressure when your ears "popped" during the takeoff or landing of an airplane or while driving on steep mountain roads.

Changes in atmospheric pressure cause air movement as a result of changes in the molecular density of the air. The general relationship is as follows (Figure 6–6):

- A column of cool, dense air causes high pressure at the surface, which will lead to sinking air (movement *toward* the surface and compression).
- A column of warm, less dense air causes low pressure at the surface, which will lead to rising air (movement *away from* the surface and expansion).

In addition, sinking air tends to warm because of its compression, while rising air tends to cool due to expansion. Note that there are complex relationships among air composition, temperature, density, water vapor content, and pressure.

Movement Air *always* moves from high-pressure regions toward low-pressure regions. This moving air is

called **wind.** If a balloon is inflated and let go, what happens to the air inside the balloon? It rapidly escapes, moving from a high-pressure region inside the balloon (caused by the balloon pushing on the air inside) to the lower-pressure region outside the balloon.

> The atmosphere's changing temperature, density, water vapor content, and pressure cause atmospheric movement, initiating wind.

An Example: A Nonspinning Earth

Imagine for a moment that Earth is not spinning on its axis but that the Sun rotates around Earth, with the Sun directly above Earth's Equator at all times (Figure 6–7). Because more solar radiation is received along the Equator than at the poles, the air at the Equator in contact with Earth's surface is warmed. This warm, moist air rises, creating low pressure at the surface. This rising air cools (see Figure 6–4) and releases its moisture as rain. Thus, a zone of low pressure and much precipitation occurs along the Equator.

As the air along the Equator rises, it reaches the top of the troposphere and begins to move toward the poles. Because the temperature is much lower at high altitudes, the air cools and its density increases. This cool, dense air sinks at the poles, creating high pressure at the surface. The sinking air is quite dry because cool air cannot hold much water vapor. Thus, there is high pressure and clear, dry weather at the poles.

Which way will surface winds blow? Air always moves from high pressure to low pressure, so air travels from the high pressure at the poles toward the low pressure at the Equator. Thus, there are strong northerly winds in the Northern Hemisphere and strong southerly winds in the Southern Hemisphere.[3] The air warms as it makes its way back to the Equator, completing the loop (called a *convection* or *circulation cell*; see Figure 6–5).

Is this fictional case of a nonspinning Earth a good analogy to what is really happening on Earth? Actually, it is not, even though the *principles* that drive the physical movement of air remain the same whether Earth is spinning or not. Let's now examine how Earth's spin influences atmospheric circulation.

The Coriolis Effect

The **Coriolis effect** changes the intended path of a moving body. Named after Gaspard Gustave de Coriolis, the French engineer who first calculated its influence in 1835, it is often incorrectly called the Coriolis *force*. It does not accelerate the moving body, so it does not

---

[3]Notice that winds are named based on the direction *from which they are moving.* As we'll see, this is also true for ocean currents.

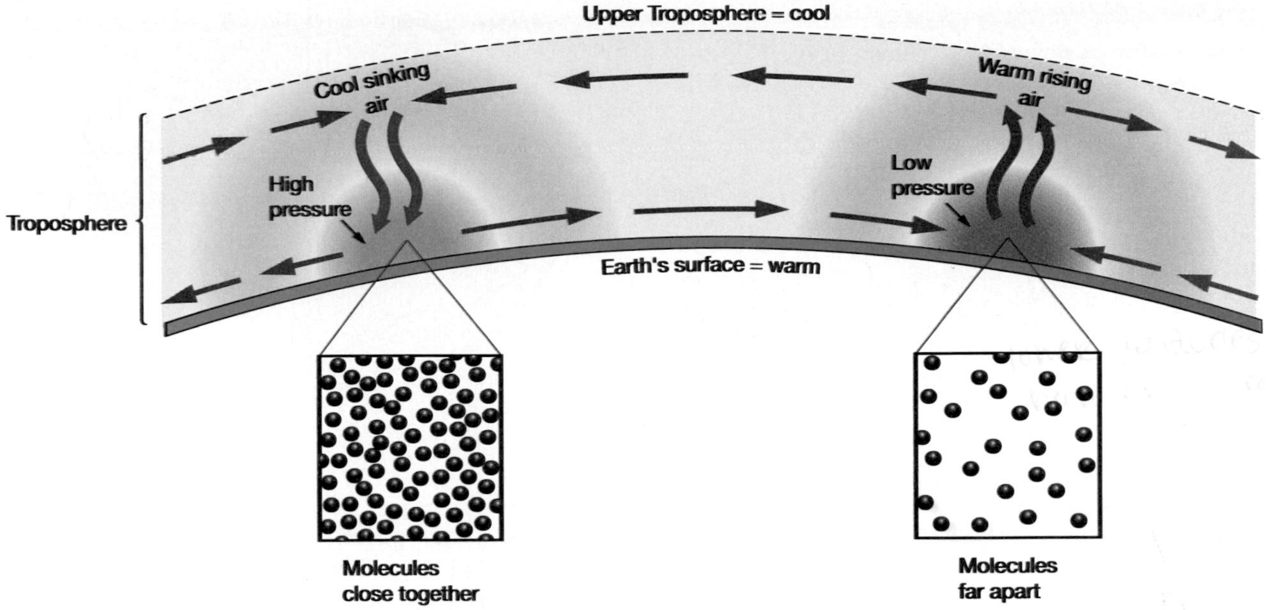

Upper Troposphere = cool

Cool sinking air

Warm rising air

Troposphere

High pressure

Low pressure

Earth's surface = warm

Molecules close together

Molecules far apart

Figure 6–6 **High and low atmospheric pressure zones.** A column of cool, dense air causes high pressure at the surface (*left*), which will lead to sinking air. A column of warm, less dense air causes low pressure at the surface (*right*), which will lead to rising air.

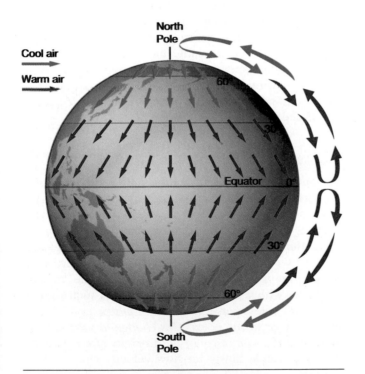

Cool air

Warm air

North Pole

60°

30°

Equator 0°

30°

60°

South Pole

Figure 6–7 **A non-spinning Earth.** A fictional non-spinning Earth with the Sun rotating around Earth directly above Earth's Equator at all times. Arrows show the pattern of winds that would develop due to uneven solar heating on Earth.

influence the body's speed. As a result, it is an effect, not a true force.

The Coriolis effect causes moving objects on Earth to follow curved paths. In the Northern Hemisphere, an object will follow a path to the *right* of its intended di-

rection; in the Southern Hemisphere, an object will follow a path to the *left* of its intended direction. The directions right and left are the *viewer's perspective looking in the direction in which the object is traveling*. For example, the Coriolis effect very slightly influences the movement of a ball thrown between two people. In the Northern Hemisphere, the ball will veer slightly to its right *from the thrower's perspective*.

The Coriolis effect acts on all moving objects. However, it is much more pronounced on objects traveling long distances, especially north or south. This is why the Coriolis effect has a dramatic effect on atmospheric circulation and the movement of ocean currents.

The Coriolis effect arises because the Earth rotates toward the east. More specifically, it is the difference in the speed of Earth's rotation at different latitudes that causes the Coriolis effect. In reality, objects travel along straight-line paths,[4] but the Earth rotates underneath them, making the objects appear to curve. Let's look at two examples to help clarify this.

Example 1: Perspectives and Frames of Reference on a Merry-Go-Round

A merry-go-round is a useful experimental apparatus with which to test some of the concepts of the Coriolis effect. A merry-go-round is a large circular wheel that rotates around its center. It has bars that people hang onto while the merry-go-round spins, as shown in Figure 6–8.

---

[4]Newton's first law of motion (the law of inertia) states that every body persists in its state of rest or of uniform motion in a straight line unless it is compelled to change that state by forces imposed on it.

Figure 6–8 A merry-go-round spinning counterclockwise as viewed from above. See text for description of paths A, B, and C.

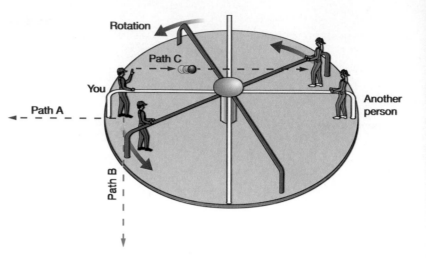

Imagine that you are on a merry-go-round that is spinning counterclockwise as viewed from above (Figure 6–8). As you are spinning, what will happen to you if you let go of the bar? If you guessed that you would fly off along a straight line path perpendicular to the merry-go-round (Figure 6–8, *path A*), that's not quite right. Your angular momentum would propel you in a straight line *tangent* to your circular path on the merry-go-round at the point where you let go (Figure 6–8, *path B*). The law of inertia states that a moving object will follow a straight-line path until it is compelled to change that path by other forces. Thus, you would follow a straight-line path (*path B*) until you collide with some object such as other playground equipment or the ground. From the perspective of another person on the merry-go-round, your departure along path B would *appear* to curve to the right due to the merry-go-round's rotation.

Imagine you are again on the merry-go-round, spinning counterclockwise, but you are now joined by another person who is facing you directly but on the opposite side of the merry-go-round. If you were to toss a ball to the other person, what path would it appear to follow? Even though you threw the ball straight at the other person, from *your perspective* the ball's path would appear to curve to the right. That's because the frame of reference (in this example, the merry-go-round) has rotated during the time that it took the ball to reach where the other person had been (Figure 6–8). A person viewing the merry-go-round from directly overhead would observe that the ball did indeed travel along a straight-line path (Figure 6–8, *path C*), just as your path was straight when you let go of the merry-go-round bar. Similarly, the perspective of being on the rotating Earth causes objects to appear to travel along curved paths. This is the Coriolis effect. The merry-go-round spinning in a counterclockwise direction is analogous to the Northern Hemisphere because, as viewed from above the North Pole, Earth is spinning counterclockwise. Thus, moving objects appear to follow curved paths to the *right* of their intended direction in the Northern Hemisphere.

If the other person on the merry-go-round had thrown a ball toward you, it would also appear to have curved. From the perspective of the other person, the ball would appear to curve to its right, just as the ball you threw curved. From your perspective, however, the ball thrown toward you would appear to curve to its *left*. The perspective to keep in mind when considering the Coriolis effect is the one *looking in the same direction that the object is moving*.

To simulate the Southern Hemisphere, the merry-go-round would need to rotate in a *clockwise* direction, which is analogous to the Earth when viewed from above the South Pole. Thus, moving objects appear to follow curved paths to the *left* of their intended direction in the Southern Hemisphere.

Example 2: A Tale of Two Missiles

The distance that a point on Earth has to travel in a day is shorter with increasing latitude. A location near the pole, for example, travels in a circle not nearly as far in a day as will an area near the Equator. Since both areas travel their respective distances in one day, the velocity of the two areas must not be the same. Figure 6–9a shows that as Earth rotates on its axis, the velocity decreases with latitude, ranging from over 1600 kilometers (1000 miles) per hour at the Equator to 0 kilometers per hour at the poles. *This change in velocity with latitude is the true cause of the Coriolis effect.* The following example illustrates how velocity changes with latitude.

Imagine we have two missiles that fly in straight lines toward their destinations. For simplicity, assume that the flight of each missile takes one hour regardless of the distance flown. The first missile is launched from the North Pole toward New Orleans, Louisiana, which is at 30 degrees north latitude (Figure 6–9b). Does the missile land in New Orleans? Actually, no. Earth rotates eastward at 1400 kilometers (870 miles) per hour along the 30 degrees latitude line (Figure 6–9a), so the missile lands somewhere near El Paso, Texas, 1400 kilometers west of

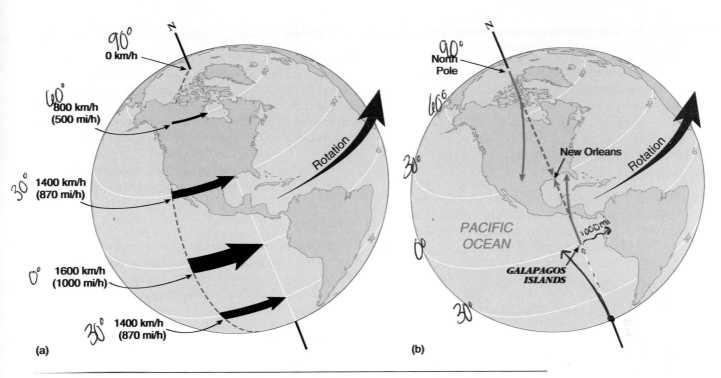

Figure 6–9 The Coriolis effect and missile paths. **(a)** The velocity of any point on Earth varies with latitude from about 1600 kilometers (1000 miles) per hour at the Equator to 0 kilometers per hour at either pole. **(b)** The paths of missiles shot toward New Orleans from the North Pole and from the Galápagos Islands on the Equator. Dashed lines indicate intended paths; solid lines indicate paths that the missiles would travel as viewed from Earth's surface.

its target. From your perspective at the North Pole, the path of the missile appears to curve *to its right* in accordance with the Coriolis effect. In reality, New Orleans has moved out of the line of fire due to Earth's rotation.

The second missile is launched toward New Orleans from the Galápagos Islands, which are directly south of New Orleans along the Equator (Figure 6–9b). From their position on the Equator, the Galápagos Islands are moving east at 1600 kilometers (1000 miles) per hour, 200 kilometers (124 miles) per hour faster than New Orleans (Figure 6–9a). At take-off, therefore, the missile is also moving toward the east 200 kilometers per hour faster than New Orleans. Thus, when the missile returns to Earth one hour later at the latitude of New Orleans, it will land offshore of Alabama, 200 kilometers east of New Orleans. Again, from your perspective on the Galápagos Islands, the missile appears to curve *to its right*. Keep in mind that both of these missile examples ignore friction, which would greatly reduce the amount the missiles deflect to the right of their intended courses.

Changes in the Coriolis Effect with Latitude

The first missile (shot from the North Pole) missed the target by 1600 kilometers (1000 miles), while the second missile (shot from the Galápagos Islands) missed its target by only 200 kilometers (124 miles). What was re-

sponsible for the difference? Not only does the rotational velocity of points on Earth range from 0 kilometers per hour at the poles to more than 1600 kilometers (1000 miles) per hour at the Equator, but the *rate of change* of the rotational velocity (per degree of latitude) increases as the pole is approached from the Equator.

For example, the rotational velocity differs by 200 kilometers (124 miles) per hour between the Equator (0 degrees) and 30 degrees north latitude. From 30 degrees north latitude to 60 degrees north latitude, however, the rotational velocity differs by 600 kilometers (372 miles) per hour. Finally, from 60 degrees north latitude to the North Pole (where the rotational velocity is zero), the rotational velocity differs by over 800 kilometers (500 miles) per hour.

Thus, the maximum Coriolis effect is at the poles, and there is no Coriolis effect at the Equator. The magnitude of the Coriolis effect depends much more, however, on the length of time the object (such as an air mass or ocean current) is in motion. Even at low latitudes, where the Coriolis effect is small, a large Coriolis deflection is possible if an object is in motion for a long time. In addition, because the Coriolis effect is caused by the *difference* in velocity of different latitudes on Earth, there is no Coriolis effect for those objects moving due east or due west along the Equator.

A summary of the Coriolis effect is presented in Table 6–3.

> The Coriolis effect causes moving objects to curve to the right in the Northern Hemisphere and to the left in the Southern Hemisphere. It is maximized at the poles and is zero at the Equator.

Atmospheric Circulation Cells on a Spinning Earth

Figure 6–10 shows atmospheric circulation and the corresponding wind belts on a spinning Earth, which presents a more complex pattern than that of the fictional non-spinning Earth (Figure 6–7).

Circulation Cells

The greater heating of the atmosphere over the Equator causes the air to expand, to decrease in density, and to rise. As the air rises, it cools by expansion because the pressure is lower, and the water vapor it contains condenses and falls as rain in the equatorial zone. The resulting dry air mass travels north or south of the Equator. Around 30 degrees north and south latitude, the air cools off enough to become denser than the surrounding air, so it begins to descend, completing the loop (Figure 6–10). These circulation cells are called **Hadley cells** after noted English meteorologist George Hadley (1685–1768).

In addition to Hadley cells, each hemisphere has a **Ferrel cell** between 30 and 60 degrees latitude, and a **polar cell** between 60 and 90 degrees latitude. The Ferrel cell—named after American meteorologist William Ferrel (1817–1891) who invented the three-cell per hemisphere model for atmospheric circulation—is not driven solely by differences in solar heating; if it were, air within it would circulate in the opposite direction. Similar to the movement of interlocking gears, the Ferrel cell moves in the direction that coincides with the movement of the two adjoining circulation cells.

Pressure

A column of cool, dense air moves *toward* the surface and creates high pressure. The descending air at about 30 degrees north and south latitude creates high-pressure zones called the **subtropical highs**. Similarly, descending air at the poles creates high-pressure regions called the **polar highs**.

What kind of weather is experienced in these high-pressure areas? Descending air is quite dry, and air tends to warm under its own weight, so these areas typically experience dry, clear, fair conditions. The conditions are not necessarily warm (such as at the poles)—just dry and associated with clear skies.

A column of warm, low density air rises *away* from the surface and creates low pressure. Thus, rising air creates a band of low pressure at the Equator—the **equatorial low**—and at about 60 degrees north and

Figure 6–10 Atmospheric circulation and wind belts of the world. The three-cell model of atmospheric circulation creates the major wind belts of the world. Boundaries between wind belts and surface atmospheric pressures are also shown. The general pattern of wind belts is modified by seasonal changes and the distribution of continents.

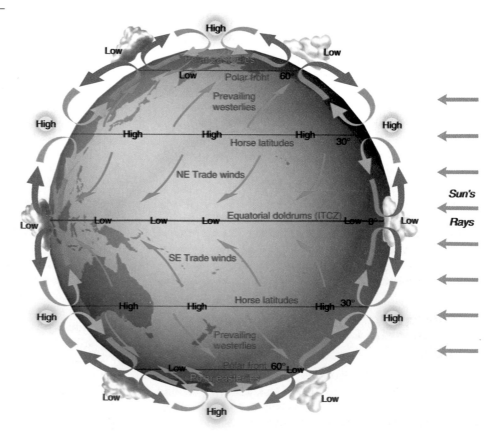

Table 6–3   The Coriolis effect: A summary.

- The Coriolis effect is caused by Earth's rotation and the resulting decrease in velocity with increasing latitude.
- The Coriolis effect influences all moving objects, especially those that move over large distances.
- The Coriolis effect changes only the direction of a moving object, never its speed.
- Coriolis deflection is to the right in the Northern Hemisphere and to the left in the Southern Hemisphere.
- The Coriolis effect is zero at the Equator and increases with increasing latitude, reaching a maximum level at the poles.

south latitude—the **subpolar low**. The weather in areas of low pressure is dominated by cloudy conditions with lots of precipitation, because rising air cools and cannot hold its water vapor.

Wind Belts

The lowermost portion of the circulation cells—that is, the part that is closest to the surface—generates the major wind belts of the world. The masses of air that move across Earth's surface from the subtropical high-pressure belts toward the equatorial low-pressure belt constitute the **trade winds**. These steady winds are named from the term *to blow trade*, which means to blow in a regular course. If Earth did not rotate, these winds would blow in a north-south direction. In the Northern Hemisphere, however, the **northeast trade winds** curve to the right due to the Coriolis effect and blow from the *northeast* to the *southwest*. In the Southern Hemisphere, on the other hand, the **southeast trade winds** curve to the left due to the Coriolis effect and blow from the *southeast* to the *northwest*.

Some of the air that descends in the subtropical regions moves along Earth's surface to higher latitudes as the **prevailing westerly wind belts**. Because of the Coriolis effect, the prevailing westerlies blow from southwest to northeast in the Northern Hemisphere and from northwest to southeast in the Southern Hemisphere.

Air moves away from the high pressure at the poles, too, producing the **polar easterly wind belts**. The Coriolis effect is maximized at high latitudes, so these winds are deflected strongly. The polar easterlies blow from the northeast in the Northern Hemisphere, and from the southeast in the Southern Hemisphere. When the cold polar easterlies meet the relatively warm prevailing westerlies near the subpolar low pressure belts (60 degrees north and south latitude), the warmer, less dense prevailing westerlies rise above the colder, more dense polar easterlies.

Boundaries

The boundary between the two trade wind belts along the Equator is known as the **doldrums** (*doldrum* = dull) because, long ago, sailing ships were becalmed there by the lack of winds. Sometimes stranded for days or weeks, the situation was unfortunate but not life-threatening: Daily rainshowers supplied sailors with

plenty of fresh water. Today, meteorologists refer to this region as the **Intertropical Convergence Zone (ITCZ)**, because it is the region between the tropics where the trade winds converge (Figure 6–10).

The boundary between the trade winds and the prevailing westerlies (centered at 30 degrees north or south latitude) is known as the **horse latitudes**. High pressure and clear, dry, fair conditions dominate these regions, where the air is sinking and surface winds are light and variable. For sailors on a ship powered only by sails, becoming stuck in the horse latitudes was a potentially serious problem because they could run out of fresh water. Legend has it that some ships tried to conserve water by throwing some of their livestock overboard. Hence, the name "horse latitudes."

The boundary between the prevailing westerlies and the polar easterlies at 60 degrees north or south latitude is known as the **polar front**. This is a battleground for different air masses, so cloudy conditions and lots of precipitation are common here.

Clear, dry, fair conditions are associated with the high pressure at the poles, so precipitation is minimal. The poles are often classified as cold deserts because the annual precipitation is so low.

Circulation Cells: Idealized or Real?

The three-cell model of atmospheric circulation first proposed by Ferrel provides a simplified model of the general circulation pattern on Earth. This circulation model is idealized and does not always match the complexities observed in nature, particularly for the location and direction of motion of the Ferrel and polar cells. Nonetheless, it generally matches the pattern of major wind belts of the world and provides a general framework for understanding why they exist.

Further, the following factors significantly alter the idealized wind, pressure, and atmospheric circulation patterns illustrated in Figure 6–10:

1. The tilt of Earth's rotation axis, which produces seasons
2. The lower heat capacity of continental rock compared to seawater,[5] which makes the air over

_____

[5]An object that has low heat capacity heats up quickly when heat energy is applied. Recall from Chapter 5 that water has high heat capacity.

Box 6-1
Why Christopher Columbus Never Set Foot on North America

The Italian navigator and explorer **Christopher Columbus** is widely credited with discovering North America in the year 1492. However, America was already populated with many natives, and the Vikings predated his voyage to North America by about 500 years. Moreover, the pattern of the major wind belts of the world prevented his sailing ships from reaching continental North America during his four voyages.

Rather than sailing east, Columbus was determined to reach the East Indies (today the country of Indonesia) by sailing *west* across the Atlantic Ocean. An astronomer in Florence, Italy, named Toscanelli was the first to suggest such a route in a letter to the king of Portugal. Columbus later contacted Toscanelli and was told how far he would have to sail west to reach India. Today, we know that this distance would have carried him just west of North America.

After years of difficulties in initiating the voyage, Columbus received the financial backing of the Spanish monarchs Ferdinand V and Isabella I. He set sail with 88 men and three ships (the *Niña*, the *Pinta*, and the *Santa María*) on August 3, 1492, from the Canary Islands off Africa (Figure 6B). The Canary Islands are located at 28 degrees north latitude and are within the northeast trade winds, which blow steadily from the northeast to the southwest. Instead of sailing directly east, which would have allowed Columbus to reach central Florida, the map in Figure 6B shows that Columbus sailed a more southerly route.

During the morning of October 12, 1492, the first land was sighted, which is generally believed to have been Watling Island in the Bahama Islands southeast of Florida. Based on the inaccurate information he had been given,

Figure 6B Route of Christopher Columbus's first voyage and a modern-day replica of the *Niña* (inset).

Columbus was convinced that he had arrived in the East Indies and was somewhere near India. Consequently, he called the inhabitants "Indians," and the area is known today as the West Indies. Later during this voyage, he landed on Cuba and Hispaniola.

On his return journey, he sailed to the northeast and picked up the prevailing westerlies, which transported him away from North America and toward Spain. Upon his return to Spain and the announcement of his discovery, additional voyages were planned. Columbus made three more trips across the Atlantic Ocean, following similar paths through the Atlantic. Thus, his ship's were controlled by the trade winds on the outbound voyage and the prevailing westerlies on the return trip. During his next voyage, in 1493, Columbus explored Puerto Rico and the Leeward Islands and established a colony on Hispaniola. In 1498, he explored Venezuela and landed on South America, unaware that it was a new continent to Europeans. On his last voyage in 1502, he reached Central America.

Although he is today considered a master mariner, he died in neglect in 1506, still convinced that he had explored islands near India. Even though he never set foot on North America, his journeys inspired other Spanish and Portuguese navigators to explore the "New World," including the coasts of North and South America.

continents colder in winter and warmer in summer than the air over adjacent oceans

3. The uneven distribution of land and ocean over Earth's surface, which particularly affects patterns in the Northern Hemisphere

During winter, therefore, the continents usually develop atmospheric high-pressure cells from the weight of cold air centered over them and, during the summer, they usually develop low-pressure cells (Figure 6–11). Otherwise, the pattern of atmospheric high- and low-pressure zones shown in Figure 6–11 corresponds closely to those shown in Figure 6–10.

Table 6–4 summarizes the characteristics of global wind belts and boundaries. The world's wind belts closely match the pattern of ocean surface currents, which are discussed in Chapter 7, "Ocean Circulation."

> The major wind belts in each hemisphere are the trade winds, the prevailing westerlies, and the polar easterlies. The boundaries between these wind belts include the doldrums, the horse latitudes, the polar front, and the polar high.

The Oceans, Weather, and Climate

Weather describes the conditions of the atmosphere at a given time and place. **Climate** is the long-term average of weather. If we observe the weather conditions in an area over a long period, we can begin to draw some conclusions about its climate. For instance, if the weather in an area is dry over many years, we can say that the area has an arid climate.

Winds

The movement of air is called *wind.* Air always moves from high pressure to low pressure. However, as air moves away from high-pressure regions and toward low-pressure regions, the Coriolis effect modifies its direction. In the Northern Hemisphere, for example, air moving from high to low pressure curves to the right and results in a counterclockwise[6] flow of air around low-pressure cells [called **cyclonic** (*cyclo* = a circle) **flow**]. Similarly, as the air leaves the high-pressure region and curves to the right, it establishes a clockwise flow of air around high-pressure cells (called **anticyclonic flow**). Figure 6–12 shows how a screwdriver can help you remember how air moves around high and low pressure regions. High pressures are similar to a high screw that needs tightening, so a screwdriver would be turned clockwise; low pressures are similar to a tightened screw that needs loosening, so a screwdriver would be turned counterclockwise. Because winter high-pressure cells are replaced by summer low-pressure cells over the continents, wind patterns associated with continents often reverse themselves seasonally.

Other factors that influence regional winds, especially in coastal areas, are **sea breezes** and **land breezes** (Figure 6–13). When an equal amount of solar energy is applied to both land and ocean, the land heats up about five times more due to its lower heat capacity. The land heats the air around it and, during the afternoon, the warm, low-density air over the land rises. Rising air creates a low-pressure region over the land, pulling the cooler air over the ocean toward land, creating what is known as a sea breeze. At night, the land surface cools about five times more rapidly than the ocean and cools the air around it. This cool, high-density air sinks, creating a high-pressure region that causes the wind to blow from the land. This is known as a land breeze, and is most prominent in the late evening and early morning hours.

Storms

At high and low latitudes, there is little daily and minor seasonal change in weather.[7] Equatorial regions are usually warm, damp, and typically calm, because the dominant direction of air movement in the doldrums is upward. Midday rains are common, even during the

---

[6]These directions are reversed in the Southern Hemisphere.
[7]In fact, in equatorial Indonesia, the vocabulary of Indonesians doesn't include the word *seasons.*

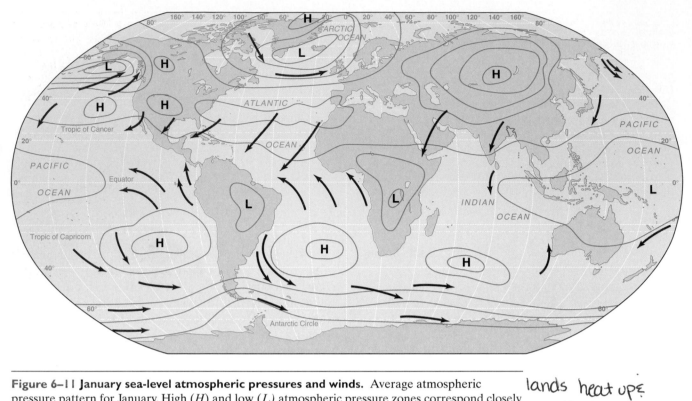

Figure 6–11 January sea-level atmospheric pressures and winds. Average atmospheric pressure pattern for January. High (*H*) and low (*L*) atmospheric pressure zones correspond closely to those shown in Figure 6–10, but are modified by the change of seasons and the distribution of continents. Arrows show direction of winds, which move from high to low pressure regions.

lands heat up & cool down faster

Table 6–4   Characteristics of wind belts and boundaries

Region (north or south latitude)	Wind belt or boundary name	Atmospheric pressure	Characteristics
Equatorial (0–5 degrees)	Doldrums	Low	Light, variable winds. Abundant cloudiness and much precipitation. Breeding ground for hurricanes.
5–30 degrees	Trade winds	—	Strong, steady winds generally from the east.
30 degrees	Horse latitudes	High	Light, variable winds. Dry, clear, fair weather with little precipitation. Major deserts of the world.
30–60 degrees	Prevailing westerlies	—	Winds generally from the west. Brings storms that influence weather across U.S.
60 degrees	Polar front	Low	Variable winds. Stormy, cloudy weather year-round.
60–90 degrees	Polar easterlies	—	Cold, dry winds generally from the east.
Poles (90 degrees)	Polar high pressure	High	Variable winds. Clear, dry, fair conditions, cold temperatures, and minimal precipitation. Cold deserts.

supposedly "dry" season. It is within the mid-latitudes where storms are common.

Storms are atmospheric disturbances characterized by strong winds, precipitation, and often thunder and lightning. Due to the seasonal change of pressure systems over continents, air masses from the high and low latitudes may move into the mid-latitudes, meet, and produce severe storms. **Air masses** are large volumes of air that have a definite area of origin and distinctive characteristics. Several air masses influence the United States, including polar air masses and tropical air

masses (Figure 6–14). Some air masses originate over land (c = continental) and are therefore dryer, but most originate over the sea (m = maritime) and are moist. Some are colder (P = polar; A = Arctic) and some are warm (T = tropical). Typically, the United States is influenced more by polar air masses during the winter and more by tropical air masses during the summer.

As polar and tropical air masses move into the mid-latitudes, they also move gradually in an easterly direction. A **warm front** is the contact between a warm air mass moving into an area occupied by cold air. A **cold**

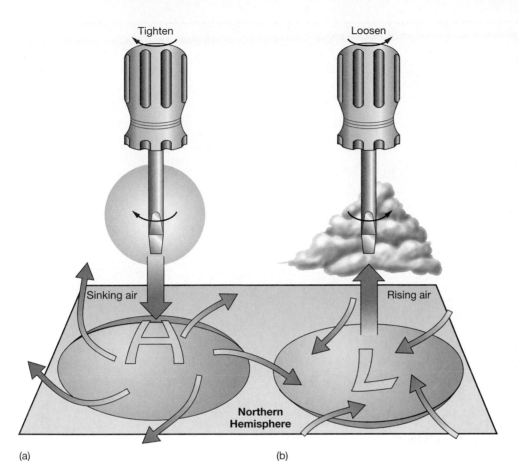

(a)   (b)

Figure 6–12 High- and low-pressure regions and air flow. As air moves away from a high-pressure region **(a)** toward a low-pressure region **(b),** the Coriolis effect causes the air to curve to the right in the Northern Hemisphere. This results in clockwise winds around high-pressure regions (anticyclonic flow) and counterclockwise winds around low-pressure regions (cyclonic flow). It can be thought of as high pressures need tightening by a screwdriver (clockwise) and low pressures need loosening (counterclockwise).

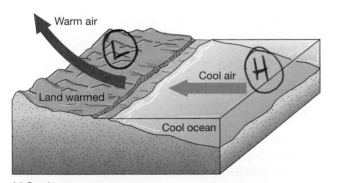

(a) Sea breeze

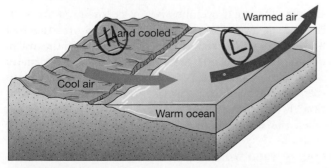

(b) Land breeze

Figure 6–13 Sea and land breezes. (a) Sea breezes occur when air warmed by the land rises and is replaced by cool air from the ocean. **(b)** Land breezes occur when the land has cooled, causing dense air to sink and flow toward the warmer ocean.

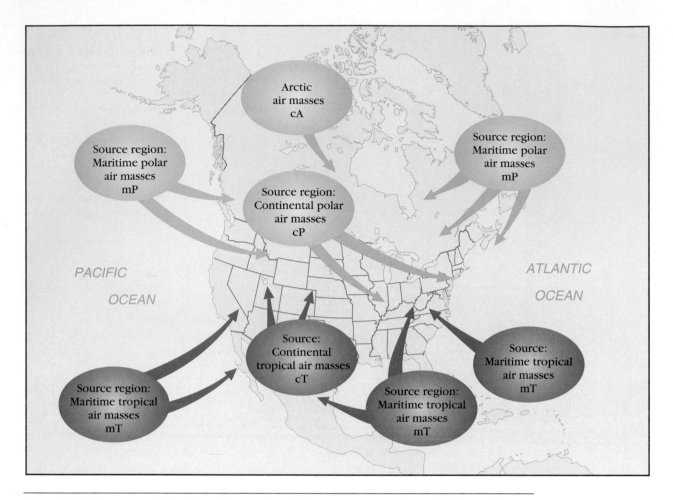

Figure 6–14 Air masses that affect U.S. weather. Polar air masses are shown in blue, and tropical air masses are shown in red. Air masses are classified based on their source region: The designation continental (c) or maritime (m) indicates moisture content, whereas polar (P), Arctic (A), and tropical (T) indicate temperature conditions.

front is the contact between a cold air mass moving into an area occupied by warm air (Figure 6–15).

These confrontations are brought about by the movement of the **jet stream**, a narrow, fast-moving, easterly-flowing air mass. It exists above the mid-latitudes just below the top of the troposphere, centered at an altitude of about 10 kilometers (6 miles). It usually follows a wavy path and may cause unusual weather by steering a polar air mass far to the south or a tropical air mass far to the north.

Regardless of whether a warm front or cold front is produced, the warmer, less-dense air always rises above the denser cold air. The warm air cools as it rises, so its water vapor condenses as precipitation. A cold front is usually steeper, and the temperature difference across it is greater than a warm front. Therefore, rainfall along a cold front is usually heavier and briefer than rainfall along a warm front.

Tropical Cyclones (Hurricanes)

Tropical cyclones (*kyklon* = moving in a circle) are huge rotating masses of low pressure characterized by

strong winds and torrential rain. They are the largest storm systems on Earth, though they are not associated with any fronts. In North and South America, tropical cyclones are called **hurricanes** (*Huracan* = Taino god of wind); in the western North Pacific Ocean, they are called **typhoons** (*tai-fung* = great wind); and in the Indian Ocean, they are called **cyclones**. No matter what they are called, tropical cyclones can be highly destructive. The energy contained in *a single* hurricane is greater than that generated by all energy sources over the last 20 years in the United States.

Origin Tropical cyclones begin as low-pressure cells that break away from the equatorial low-pressure belt and grow as they pick up heat energy from the warm ocean. Surface winds feed moisture (in the form of water vapor) into the storm. When water evaporates, it stores tremendous amounts of latent heat of evaporation[8], which is released as latent heat of condensation when the water vapor condenses to form liquid water

---

[8]For a discussion of water's latent heat of evaporation, see Chapter 5.

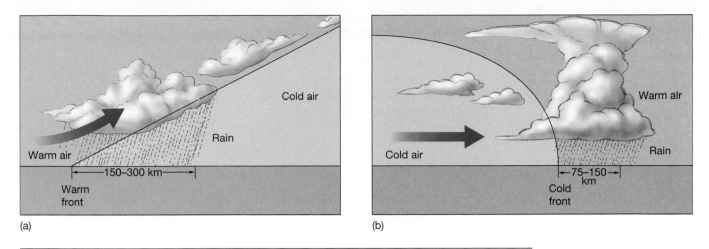

(a)

(b)

Figure 6–15 Warm and cold fronts. Cross-sections through a gradually rising warm front **(a)** and a steeper cold front **(b)**. With both fronts, warm air rises and precipitation is produced.

(rain). The release of vast amounts of water's latent heat of condensation powers tropical cyclones.

Tropical storms are classified according to their maximum sustained wind speed:

- If winds are less than 61 kilometers (38 miles) per hour, the storm is classified as a *tropical depression*.
- If winds are between 61 and 120 kilometers (38 and 74 miles) per hour, the storm is called a *tropical storm*.
- If winds exceed 120 kilometers (74 miles) per hour, the storm is a *tropical cyclone*.

The **Saffir–Simpson Scale** of hurricane intensity further divides tropical cyclones into categories based on wind speed and damage (Table 6–5). In some cases, the wind in tropical cyclones attains speeds as high as 400 kilometers (250 miles) per hour!

Worldwide, about 100 storms grow to hurricane status each year. The conditions needed to create a hurricane are as follows:

- Ocean water with a temperature greater than 25°C (77°F), which provides an abundance of water vapor to the atmosphere through evaporation.

- Warm moist air, which supplies vast amounts of latent heat as the water vapor in the air condenses and fuels the storm.
- The Coriolis effect, which causes the hurricane to spin counterclockwise in the Northern Hemisphere and clockwise in the Southern Hemisphere. No hurricanes occur directly on the Equator because the Coriolis effect is negligible there.

These conditions are found during the late summer and early fall, when the tropical and subtropical oceans are at their maximum temperature. Hurricane season in the Atlantic Basin is officially from June 1 to November 30, but hurricanes can form outside of this time frame, although only rarely.

Movement When hurricanes are initiated, they typically remain in the tropics. On rare occasions, they pass through the tropics and affect the mid-latitudes (Figure 6–16). If a hurricane moves over land, its energy source is cut off and it eventually dissipates. The trade winds drive hurricanes, so they move from east to west across the oceans, and typically last five to ten days.

The diameter of a hurricane can exceed 800 kilometers (500 miles); more typically, though, the diameter is

Table 6–5 The Saffir–Simpson scale of hurricane intensity.			
Category	Wind speed		Damage
	kilometers per hour	miles per hour	
1	120–153	74–95	Minimal
2	154–177	96–110	Moderate
3	178–209	111–130	Extensive
4	210–249	131–155	Extreme
5	>250	>155	Catastrophic

Figure 6–16 Paths of major tropical cyclones. Cyclones originate in tropical areas where ocean surface temperatures are warm (*shaded areas*) and migrate across the tropics (*red arrows*), sometimes reaching the mid-latitudes. Depending on the area, cyclones may also be called hurricanes or typhoons.

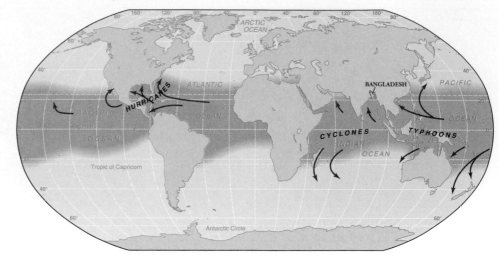

less than 200 kilometers (124 miles). As air moves across the ocean surface toward the low-pressure center, it is drawn up around the **eye of the hurricane** (Figure 6–17). The air in the vicinity of the eye spirals upward, so horizontal wind speeds may be less than 15 kilometers (25 miles) per hour. The eye of the hurricane, therefore, is usually calm. Hurricanes are composed of spiral rain bands where intense rainfall caused by severe thunderstorms can drop tens of centimeters (several inches) of rainfall in an hour.

Types of Destruction Destruction from hurricanes is caused by high winds and flooding from intense rainfall. **Storm surge**, however, causes the majority of a hurricane's coastal destruction. In fact, storm surge is responsible for 90% of the deaths associated with hurricanes.

When a hurricane develops over the ocean, its low-pressure center produces a low "hill" of water (Figure 6–18). As the hurricane migrates across the open ocean, the hill moves with it. As the hurricane approaches shallow water near shore, the portion of the hill over which the wind is blowing shoreward produces a mass of elevated, wind-driven water. This mass of water—the storm surge—can be as high as 12 meters (40 feet), resulting in a dramatic increase in sea level at the shore, large storm waves, and tremendous destruction to low-lying coastal areas.

Historic Destruction Periodic destruction from hurricanes occurs along the East Coast and the Gulf Coast regions of the United States. The most deadly natural disaster in U.S. history was caused by a hurricane that hit Galveston, Texas, in September 1900 (Box 6–2). Flooding caused by the hurricane's storm surge combined with intense rainfall and 135 kilometer (84 mile) per hour winds killed at least 6000 people in Galveston, 1000 people in other parts of low-lying Galveston Island, and another 1000 people on the mainland.

Category 4 hurricanes like the one that hit Galveston have been surpassed by category 5 hurricanes only three times in the United States. In 1935, on Labor Day, one flattened the Florida Keys and in 1969, Hurricane Camille roared through Mississippi. In August 1992, Hurricane Andrew came ashore in southern Florida and did more than $20 billion in damage before crossing the Gulf of Mexico and inflicting another $2 billion in damage along the Gulf Coast (Figure 6–19). Winds reached 258 kilometers (160 miles) per hour as Andrew crossed the Everglades, ripping down every tree in its path. More property was damaged by this hurricane than in any other natural disaster in U.S. history and over 250,000 people were left homeless. Fortunately, many people heeded the warnings to evacuate, so only 54 were killed.

In October 1998, Hurricane Mitch proved to be one of the most devastating tropical cyclones to affect the western Hemisphere. At its peak, it was estimated to have winds of 290 kilometers (180 miles) per hour—a strong category 5 hurricane. It hit Central America with 160 kilometer (100 miles) per hour winds and as much as 130 centimeters (51 inches) of total rainfall, causing widespread flooding and mudslides in Honduras and Nicaragua that destroyed entire towns. The hurricane resulted in more than 11,000 deaths, left more than 2 million homeless, and caused more than $10 billion in damage across the region.

The majority of the world's tropical cyclones are formed in the waters north of the Equator in the western Pacific Ocean. These storms, called typhoons, do enormous damage to coastal areas and islands in Southeast Asia (see Figure 6–16).

Other areas of the world such as Bangladesh experience tropical cyclones on a regular basis. Bangladesh borders the Indian Ocean and is particularly vulnerable because it is a highly populated and low-lying country, much of it only 3 meters (10 feet) above sea level. In 1970, a 12-meter (40-foot)-high storm surge from a tropical cyclone killed an estimated 1 million people.

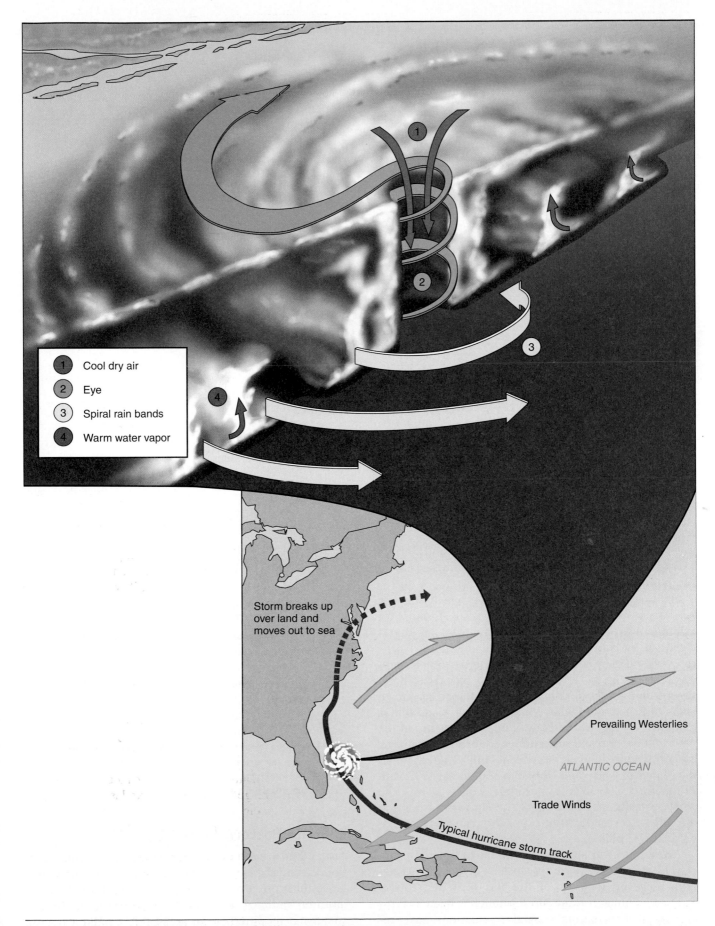

Figure 6–17 Typical hurricane storm track and internal structure (*inset*).

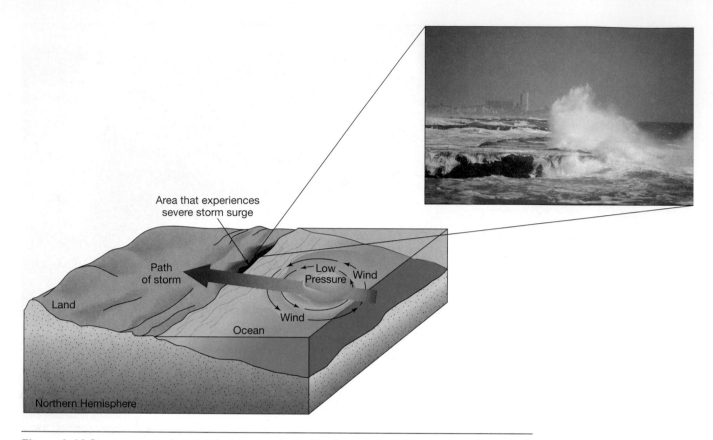

Figure 6–18 Storm surge. As a tropical cyclone in the Northern Hemisphere moves ashore, the low-pressure center around which the storm winds blow combined with strong onshore winds produce a high-water storm surge that floods and batters the coast. A storm surge caused by Hurricane Felix in August 1995 along New Jersey (*inset*).

Another tropical cyclone hit the area in 1972 that caused up to 500,000 deaths. In 1991, Hurricane Gorky's 145-mile-per-hour winds and storm surge caused extensive damage and killed 200,000 people.

Even islands near the centers of ocean basins can be struck by hurricanes. The Hawaiian Islands, for example, were hit hard by Hurricane Dot in August 1959 and by Hurricane Iwa in November 1982. Hurricane Iwa hit very late in the hurricane season and produced winds up to 130 kilometers (81 miles) per hour. Over $100 million of damage was sustained on the islands of Kauai and Oahu. Niihau, a small island that is inhabited by 230 native Hawaiians, was directly in the path of the storm and suffered severe property damage, but no serious injuries. Hurricane Iniki roared across the islands of Kauai and Niihau in September 1992, with 210-kilometer (130-mile)-per-hour winds. It was the most powerful hurricane to hit the Hawaiian Islands in the last 100 years, with property damage that approached $1 billion.

Hurricanes will always be a threat to life and property. Because of accurate forecasts and prompt evacuation, however, the loss of life has been decreasing. Property damage, on the other hand, has been *increasing* because increasing coastal populations have resulted in more and more construction along the coasts.

Inhabitants of areas subject to a hurricane's destructive force must be made aware of the danger so that they can be prepared for its eventuality.

> Hurricanes are intense and sometimes destructive tropical storms that form where water temperatures are high, where there is an abundance of warm moist air, and where they can spin.

EIO

For more information and on-line exercises about the Environmental Issue in Oceanography (EIO) "The Lasting Influences of Hurricanes," visit the EIO Web site at **http://www.prenhall.com/oceanissues** and select Issue #5.

Climate Patterns in the Oceans

Just as land areas have climate patterns, so do regions of the oceans. The open ocean is divided into climatic regions that run generally east–west (parallel to lines of latitude) and have relatively stable boundaries that are somewhat modified by ocean surface currents (Figure 6–20).

(a)

(b)

Figure 6–19 Hurricane Andrew, the most destructive hurricane in U.S. history. (a) A satellite view of Hurricane Andrew, which had a diameter of about 200 kilometers (124 miles); its counterclockwise direction of flow and prominent central eye are also visible. **(b)** The hurricane struck Florida in 1992, causing extensive damage.

The **equatorial** region spans the Equator. The major air movement there is upward because heated air rises. Surface winds, therefore, are weak and variable, which is why this region is called the doldrums. Surface waters are warm, and the air is saturated with water vapor.

Daily rainshowers are common, which keeps surface salinity relatively low.

Tropical regions extend north or south of the equatorial region up to the Tropic of Cancer and the Tropic of Capricorn, respectively. They are

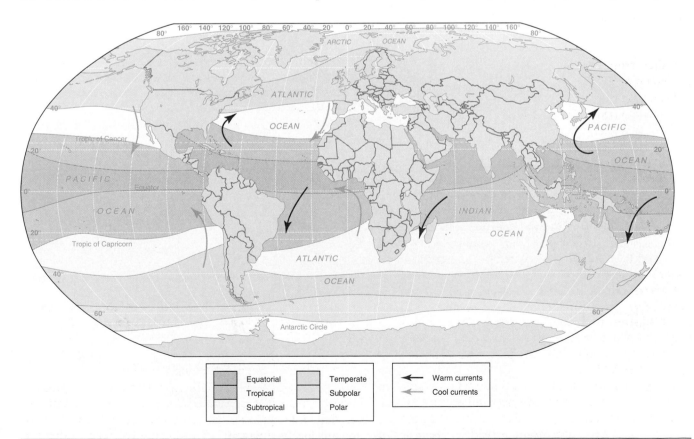

	Equatorial		Temperate		Warm currents
	Tropical		Subpolar		Cool currents
	Subtropical		Polar		

Figure 6–20 Climatic regions of the ocean. The ocean's climatic regions are defined primarily by latitude but modified by ocean currents and wind belts. Red arrows indicate warm surface currents; blue arrows indicate cool surface currents.

Box 6–2
The Storm of the Century: Galveston, Texas (1900)

Among the natural disasters in U.S. history, many have resulted in large numbers of deaths. The Chicago fire of 1871, for instance, killed 250 people; the San Francisco earthquake of 1906 killed 503; a 1928 hurricane in Florida killed 1836; and the 1889 Johnstown flood killed 2209. The single deadliest U.S. natural disaster, however, was a hurricane that struck Galveston Island, Texas, on September 8, 1900, resulting in over 6000 deaths.

Galveston Island is a thin strip of sand called a barrier island located in the Gulf of Mexico off Texas (Figure 6C, *inset*). In 1900, it averaged only 1.5 meters (5 feet) above sea level and its highest point was 2.7 meters (9 feet) above sea level. During construction of the city of Galveston, protective sand dunes up to 4.6 meters (15 feet) high had been removed to improve beach access. Galveston became a popular beach resort that boasted a milder climate than mainland areas, so it drew many summer visitors. The beach sloped off into deep water so gradually that most believed any destructive waves from the Gulf would break and dissipate before reaching shore.

The loss of life resulted not from lack of warning but from a failure to take the threat seriously. Two days earlier, a strong storm was reported moving westward into the Gulf of Mexico off Cuba, and ships returning from sea reported encountering the storm offshore the day before it made landfall. Furthermore, the Local Forecaster for the U.S. Weather Bureau in Galveston, Isaac Cline, was tracking the storm and predicted that it would move onshore. But warnings of the impending landfall seem to have been largely unheeded, in part because most U.S. meteorologists erroneously believed it was virtually impossible for a storm in the Caribbean to move across the Gulf.

Despite Cline's warnings (he even rode through Galveston shortly before the storm, urging residents to evacuate), few evacuated. In fact, Cline's warnings may have had the opposite effect, prompting curious residents to go the beach and observe the powerful waves crashing against the shore! In spite of the rain, a holiday atmosphere prevailed in Galveston, with most people believing that the flooding already apparent was the result of another minor storm.

By the time the hurricane neared Galveston and people realized they were in eminent danger, the escape routes to the mainland had been severed and the people were trapped on the island. Within hours the hurricane's 6-meter

Figure 6C Destruction from the 1900 hurricane at Galveston and location map of Galveston, Texas (inset).

(20-foot)-high storm surge came ashore and completely submerged the island. The high water combined with intense wind—measured at 160 kilometers (100 miles) per hour until the weather station's anemometer blew away—and large waves destroyed many structures on the island (Figure 6C). Even Cline's house, which was built to withstand a hurricane, collapsed with 50 people inside. The hurricane claimed the lives of at least 6000 people, including Cline's wife. Most of the people who survived—including Cline, his three daughters, and his brother—rode out the storm by clinging to floating debris.

The horror did not end when the hurricane passed. Thousands of dead bodies were found in the wreckage of the town and more washed ashore. Unable to bury them in the saturated ground, some were taken out to sea on barges and unceremoniously dumped, only to wash back to shore. Ultimately, most of the bodies were cremated where they were found.

Today, Galveston has been shored up and a 5.2-meter (17-foot)-high protective seawall has been built to prevent similar hurricane disasters. In addition, satellite tracking techniques have led to much more accurate forecasts of approaching hurricanes and timely warnings. The dreadful catastrophe of Galveston, however, is a reminder to never underestimate a hurricane's destructive power.

characterized by strong trade winds, which blow from the northeast in the Northern Hemisphere and from the southeast in the Southern Hemisphere. These winds push the equatorial currents and create moderately rough seas. Relatively little precipitation falls at higher latitudes within tropical regions, but precipitation increases toward the Equator. The tropical regions are the breeding grounds for tropical cyclones, which transfer large quantities of heat to higher latitudes.

Beyond the tropics are the **subtropical** regions. Belts of high pressure are centered there, so the dry, descending air produces little precipitation and a high rate of evaporation, resulting in the highest surface salinities in the open ocean (see Figure 5–21). Winds are weak and currents are sluggish, typical of the horse latitudes. However, strong boundary currents (along the boundaries of continents) flow north and south, particularly along the western margins of the subtropical oceans.

The **temperate** regions (also called the midlatitudes) are characterized by strong westerly winds (the prevailing westerlies) blowing from the southwest in the Northern Hemisphere and from the northwest in the Southern Hemisphere (see Figure 6–10). Severe storms are common, especially during winter, and precipitation is heavy. In fact, the North Atlantic is noted for fierce storms in this zone. These storms have claimed many ships and numerous lives over the centuries.

The **subpolar** region experiences extensive precipitation due to the subpolar low. Sea ice covers the subpolar ocean in winter, but it melts away, for the most part, in summer. Icebergs are common, and the surface temperature seldom exceeds 5°C (41°F) in the summer months.

Surface temperatures remain at or near freezing in the **polar** regions, which are covered with ice throughout most of the year. The polar high pressure dominates the area, which includes the Arctic Ocean and the ocean adjacent to Antarctica. There is no sunlight during the winter and constant daylight during the summer.

The Atmosphere's Greenhouse Effect

The worldwide average temperature of Earth and the troposphere is about 15°C (59°F). If the atmosphere contained no water vapor, carbon dioxide, methane, or other trace gases, however, the average worldwide temperature would be –18°C (–4°F). At this temperature, all of the water on Earth would be frozen and it would be too cold to support the present distribution of life. Instead, the greenhouse effect of gases in the atmosphere helps create and sustain the more pleasant temperatures to which life has grown accustomed.

The **greenhouse effect** gets its name because it keeps Earth's surface and lower atmosphere warm the same way a greenhouse keeps plants warm enough to grow regardless of the outside conditions (Figure 6–21). Energy radiated by the Sun covers the full electromagnetic spectrum, but most of the energy that reaches Earth's surface is short wavelengths, in and near the visible portion of the spectrum. In a greenhouse, short-wave sunlight passes through the glass or plastic covering, where it strikes the plants, the floor, and other objects inside, and is converted into longer-wavelength infrared radiation (heat). The heat energy tries to escape the greenhouse, but the glass or plastic covering is opaque (impenetrable) to infrared radiation, so the greenhouse traps the heat and the temperature inside increases. The greenhouse effect is the same process that causes the temperature inside a car parked in the Sun to increase so dramatically.

Figure 6–22 shows that most of the energy coming to Earth from the Sun is within the visible spectrum and peaks at a wavelength of 0.48 micrometers[9] (0.0002 inch). The atmosphere is transparent to much of this radiation, but it is absorbed by materials such as water and rocks at Earth's surface. These materials reradiate this energy back toward space as longer-wavelength infrared (heat) radiation, with a peak at a wavelength of 10 micrometers (0.004 inch). Earth has maintained a constant average temperature over long periods of

---

[9]A micrometer (μm) is one-millionth of a meter.

Figure 6–21 How a greenhouse works. The glass of a greenhouse allows incoming sunlight to pass through but traps heat. Similarly, gases like water, carbon dioxide, and methane in Earth's atmosphere act just like the glass of a greenhouse by allowing sunlight to pass through but trapping heat.

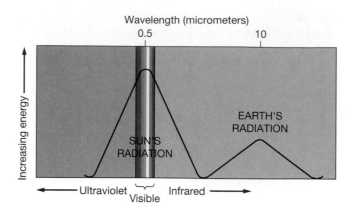

Figure 6–22 Energy radiated by the Sun and Earth. The intensity of energy radiated by the Sun peaks at a wavelength of 0.48 micrometer (0.0002 inch), which is in the visible part of the spectrum. Some of this energy is absorbed or reflected while some re-radiates from Earth in the infrared (heat) range at a wavelength of 10 micrometers (0.004 inch).

time, so the rates of energy absorption and reradiation back into space must be equal.

Figure 6–23 shows that most of the solar radiation that is not reflected back to space passes through the atmosphere and is absorbed at Earth's surface. Earth's surface, in turn, emits longer wavelength infrared radiation (heat). A portion of this energy is absorbed by certain heat-trapping gases in the atmosphere such as water vapor, carbon dioxide, and other gases. These gases then reradiate some of the infrared energy, producing the greenhouse effect. Thus, *the change of wavelengths from visible to infrared is the key to how the greenhouse effect works.*

Some of the infrared energy absorbed in the atmosphere becomes reabsorbed by Earth to continue the process (the rest is lost to space). The solar radiation received at the surface, therefore, is retained for a time within our atmosphere, where it moderates temperature fluctuations between night and day and also between seasons.

Which Gases Contribute to the Greenhouse Effect?

Water vapor contributes more to the greenhouse effect than any other gas. It occurs naturally in the atmosphere, however, and human activities are not thought to

affect the amount of water vapor in the atmosphere. Table 6–6 shows the greenhouse gases that have been increasing as a result of human activities. Some of these gases also occur naturally and have been in the atmosphere prior to human activities (such as carbon dioxide) whereas others are clearly human induced (such as chlorofluorocarbons).

Of the gases increased as a result of human activities, carbon dioxide makes the greatest relative contribution to the greenhouse effect (Table 6–6). As a result of human activities, the atmospheric concentration of carbon dioxide has increased 30% over the past 200 years (Figure 6–24). Currently 360 parts per million, the concentration increases by 1.2 parts per million each year.

The other trace gases—methane, nitrous oxide, tropospheric ozone, and chlorofluorocarbons—are present in far lower concentrations. They are important, however, because they absorb many times more infrared radiation per molecule than carbon dioxide (Table 6–6, *last column*). Still, these gases have a smaller overall contribution to increasing the greenhouse effect because their concentrations are so low. Nevertheless, all of these gases must be taken into account when considering the total amount of greenhouse warming.

The greenhouse effect is caused by gases such as water vapor and carbon dioxide that allow sunlight to pass through but trap heat energy before it is reradiated back to space.

What Changes Will Occur as a Result of Increased Global Warming?

Earth's average surface temperature has risen at least 0.6°C (1.1°F) in the last 130 years (Figure 6–25).

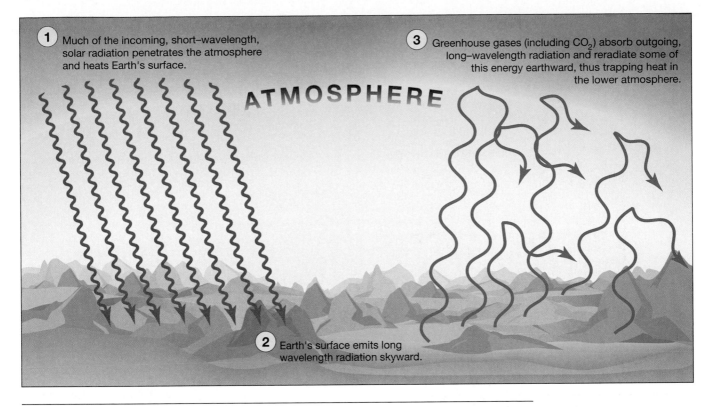

① Much of the incoming, short–wavelength, solar radiation penetrates the atmosphere and heats Earth's surface.

③ Greenhouse gases (including CO₂) absorb outgoing, long–wavelength radiation and reradiate some of this energy earthward, thus trapping heat in the lower atmosphere.

ATMOSPHERE

② Earth's surface emits long wavelength radiation skyward.

Figure 6–23 The heating of Earth's atmosphere. Most of the solar radiation that is not reflected back to space passes through the atmosphere and is absorbed at Earth's surface (*1*). Earth's surface, in turn, emits longer wavelength infrared (heat) radiation (*2*). A portion of this energy is absorbed by certain gases in the atmosphere and reradiated back to Earth, thus trapping heat and warming Earth (*3*).

Is this warming attributable to the increase in atmospheric carbon dioxide and other greenhouse gases, or is it related to a long-term natural climate cycle? Unfortunately, there is no clear proof to answer this question and many scientists debate the relative merits of information in support of human-induced greenhouse warming. In fact, some scientists who develop sophisticated computer models of climate have evidence that additional greenhouse warming may evaporate more seawater, using up much of the excess heat and generating more cloud cover, which will block the Sun's rays and significantly reduce the warming effect.

If global temperatures continue to increase as a result of global warming, researchers believe there will be many changes, but not all agree as to what those changes will be or how severe they might become. Some of the predicted changes include:

Table 6–6   Main greenhouse gases (excluding water) and their contribution to increasing the greenhouse effect.

Gas	Concentration (ppbv[a])	Rate of increase (% per year)	Relative contribution to increasing the greenhouse effect (%)	Infrared radiation absorption per molecule (number of times greater than CO₂)
Carbon dioxide (CO₂)	353,000	0.5	60	1
Methane (CH₄)	1,700	1	15	25
Nitrous oxide (N₂O)	310	0.2	5	200
Tropospheric ozone (O₃)	10–50	0.5	8	2,000
Chlorofluorocarbon (CFC-11)	0.28	4	4	12,000
Chlorofluorocarbon (CFC-12)	0.48	4	8	15,000
Total	—	—	100	

[a]ppbv = parts per billion by volume (not by weight)

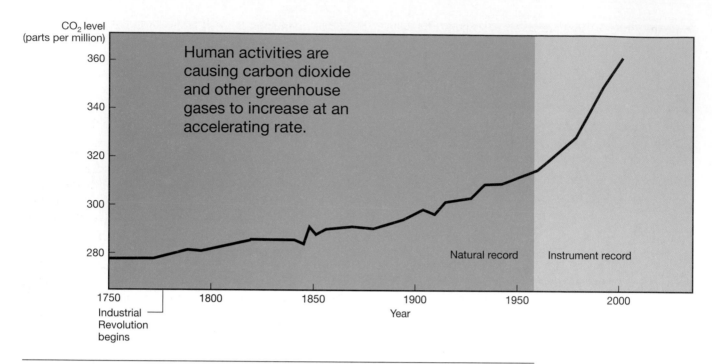

Figure 6–24 Amount of carbon dioxide in the atmosphere since 1750. There has been a dramatic increase of average worldwide atmospheric carbon dioxide since the Industrial Revolution began in the late 1700s. Values for 1958–present are from instrumental measurement of CO_2 at Mauna Loa Observatory in Hawaii; natural record values prior to 1958 are estimated from air bubbles in polar ice cores.

$CO_2 \uparrow$ Temp $\uparrow$

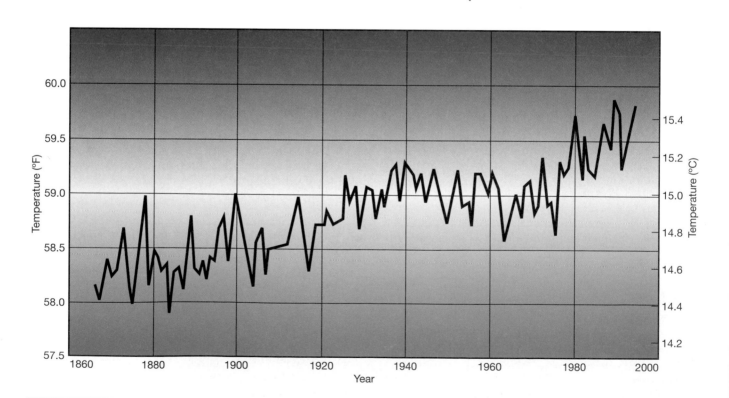

Figure 6–25 Instrumental temperature record since 1865. The record of global average surface air temperature from thermometer readings indicates a global warming of at least 0.6°C (1.1°F) over the past 130 years. The peaks and troughs indicate natural year-to-year variability of climate.

- Higher than normal sea surface temperatures, which cause more frequent and more intense tropical storms. In addition, high water temperatures affect temperature-sensitive marine organisms (such as corals) and may alter the ocean's deep-water circulation pattern, profoundly affecting Earth's climate.

- More severe droughts in certain areas and more precipitation in other areas, leading to increased chances of flooding.

- Water contamination issues that lead to larger outbreaks of water-borne diseases such as malaria, yellow fever, and dengue fever.

- Longer and more intense heat waves.

- Shifts in the distribution of plant and animal communities.

- The melting of polar ice caps, resulting in a rise in sea level that could flood low-lying coastal areas.

If global temperatures increase, it is also possible that the ice caps may actually *enlarge*. How can global warming cause ice caps to enlarge? If the atmosphere is warmer, then evaporation will occur more quickly (which, incidentally, will probably make tropical storms more intense). Increased evaporation will cause more water vapor to be present in the atmosphere. Much of this water vapor will fall as precipitation on land, thus potentially increasing the amount of snowfall for the formation of ice caps. An increase in the amount of water stored in ice caps (instead of in the ocean) will lower sea level.

In addition, not all predicted changes have negative consequences. For instance, increased warming will provide a longer growing season for some crops and increased atmospheric carbon dioxide helps promote productivity in plants. However, a number of uncertainties remain in understanding regional effects of climate change, and various systems may respond to these changes in unanticipated ways.

The global climate system is complex and contains many poorly understood feedback loops, including the role of clouds and atmospheric water. The successful modeling of Earth's climate, therefore, is one of the biggest scientific challenges today, even with some of the world's most powerful computers.

The IPCC and the Kyoto Protocol

In 1988, the United Nations Environment Programme and the World Meteorological Organization sponsored the **Intergovernmental Panel on Climate Change (IPCC)**, a group of over 200 scientists worldwide, to study human effects on global warming. The group's report, published in 1995, states that "the balance of evidence suggests a discernable human influence on global climate" and that global warming "is unlikely to be entirely due to natural causes." Since the Industrial Revolution in the 1700s, humans have been burning large quantities of fossil fuels—coal, oil, and gas—that had been buried in sedimentary rocks. Mechanization in agriculture has led to the removal of thousands of acres of forests. Both of these are natural storage places for Earth's organic carbon. Deforestation and fossil fuel burning releases previously stored carbon, increasing the amount of CO_2 in the atmosphere (see Figure 6–24).

The mounting scientific evidence that global warming and some of its side effects are occurring has led to international efforts to address the human contribution to the greenhouse effect. A number of international conferences have resulted in an agreement amongst 60 nations to voluntarily limit greenhouse gas emissions. This agreement, called the **Kyoto Protocol** because in was created at an international conference held in 1997 in Kyoto, Japan, sets target reductions for each country. For example, the U.S. is committed to reducing all greenhouse gas emissions to 7% below 1992 emissions by 2007. The protocol also establishes processes by which technology may be transferred to developing nations to enable them to industrialize without becoming producers of large quantities of greenhouse gases.

In 2001, a second IPCC group published its report, which was prepared by 426 authors and unanimously accepted by more than 160 delegates from 100 countries. The report states that recent regional climate changes already have affected many physical and biological systems on Earth and that projected climate change—as well as changes in climate extremes—could have major consequences. The report also increased the estimate of the world's temperature rise between 1990 and 2100 from 1.0 to 3.5°C (1.8 to 6.3°F) to 1.4 to 5.8°C (2.5 to 10.4°F) according to new climate models.

The Ocean's Role in Reducing the Greenhouse Effect

The ocean is extremely important for reducing the amount of carbon dioxide in the atmosphere, thus reducing the greenhouse effect. In fact, the vast majority of carbon dioxide in the ocean–atmosphere system is found in the ocean because carbon dioxide is approximately 30 times more soluble in water than are other common gases.

What happens to the carbon dioxide that enters the ocean? Most of it is incorporated into organisms through photosynthesis and through their secretion of carbonate shells. Over geologic time, more than 99% of the carbon dioxide added to the atmosphere by volcanic activity has been removed by the ocean and deposited in marine sediments as biogenous calcium carbonate and fossil fuels (oil and natural gas). Thus, the ocean acts as a *repository* (or *sink*) for carbon dioxide, soaking it up and removing it from the environment as sea floor deposits.

In light of the historic Kyoto Protocol, the ocean's ability to remove excess carbon dioxide has been investigated by many countries. Experiments have been conducted that capture emissions before they are released into the atmosphere and pumping the gas into the deep ocean or underground reservoirs (thereby removing carbon dioxide and reducing global warming). However, there may be unwanted side effects when vast amounts of carbon dioxide are added to the deep ocean.

In addition, the thermal properties of the ocean make it ideal for minimizing changes in temperature that may be brought about by global warming. The oceans act as a "thermal sponge," absorbing heat but not increasing much in temperature. To determine if the oceans are warming, the global monitoring of ocean temperature has been initiated (Box 6–3).

What Should We Do About the Increasing Greenhouse Gases?

Stimulating productivity in the ocean has been proven to remove carbon dioxide from the atmosphere. Through photosynthesis, microscopic marine algae convert carbon dioxide dissolved in the ocean to carbohydrate and oxygen gas. By removing more carbon dioxide from the ocean, the ocean can, in turn, absorb more carbon dioxide from the atmosphere.

Areas of the ocean that have relatively low productivity, such as the tropical oceans, are a good place to stimulate productivity and thus increase the amount of carbon dioxide removed from the atmosphere. In 1987, oceanographer John Martin determined that the absence of iron limited productivity in tropical oceans, so he proposed fertilizing the ocean with iron to increase its productivity (the **"iron hypothesis"**). In 1993, Martin's associates added finely ground iron to a test area of the ocean near the Galápagos Islands in the Pacific Ocean. Their results and the results of other "Iron Ex" open-ocean experiments in 1995 and 1999 showed that adding iron to the ocean increased productivity up to 30 times.

Although these results are promising, there were problems grinding the iron fine enough, dispersing the iron, and keeping it in suspension for long periods of time. Moreover, the long-term global environmental effects of adding additional iron and large amounts of carbon dioxide to the ocean are as yet unknown.

What is very clear, however, is that our agricultural and industrial activities are changing the environment. Although we cannot completely eliminate the changes, we can at least reduce our impact on Earth. For instance, we can reduce the rate of increasing combustion of fossil fuels and the widespread deforestation that accompanies agricultural development. In order to halt further damage, we must also preserve Earth's plant communities and replace much of what has been removed. Additionally, we must modify activities that affect the environment most severely, while continuing to pursue research that improves our understanding of how Earth's systems work.

Students Sometimes Ask...

How effective has the ice patrol been in preventing accidents like the Titanic disaster?

Since the International Ice Patrol was established in 1914, not one ship has been lost to ice within the patrolled area (except during wartime). However, accidents still do occur outside the patrolled area. For example, in 1989 the Soviet cruise ship *Maxim Gorky* accidentally rammed an iceberg in the Arctic Ocean well north of the Arctic Circle between Greenland and Spitzbergen. Fortunately, quick action by the Norwegian Coast Guard prevented any loss of life during evacuation of the more than 1300 passengers and crew on board.

I've heard that gigantic icebergs the sizes of small U.S. states are found near Antarctica. How are they formed?

In Antarctica, the edges of glaciers form large floating sheets of ice called shelf ice that break off and produce vast plate-like icebergs. In March 2000, for example, a Connecticut-sized iceberg (11,000 square kilometers or 4250 square miles) known as B–15 and nicknamed "Godzilla" broke loose from the Ross Ice Shelf into the Ross Sea. The largest iceberg ever recorded in Antarctic waters measured an incredible 335 by 97 kilometers (208 by 60 miles)—about the same size as Connecticut and Massachusetts combined.

The icebergs have flat tops that may stand up to 200 meters (650 feet) above the ocean surface, although most rise less than 100 meters (330 feet) above sea level, and as much as 90% of their mass is below waterline. Once created, ocean currents driven by strong winds carry the icebergs north, where they eventually melt. Because this region is not a major shipping route, the icebergs pose little serious navigation hazard except for supply ships traveling to Antarctica. Ships sighting these gigantic bergs have often mistaken them for land!

Is the Coriolis effect what makes a curve ball curve in baseball?

No, and you can't blame the Coriolis effect when your jump shot doesn't go through the basketball hoop either. Basketballs and baseballs travel too quickly over distances that are too short to be influenced significantly by the Coriolis effect. What makes a curve ball curve in baseball is the spin put on the ball, which creates high- and low-pressure regions around the ball that influence its path. The Coriolis effect has a much greater influence over objects that travel long distances, such as ocean currents and air masses. That's why oceanographers and meteorologists (as well as missile launchers and airline pilots) must correct for it.

Is it true that the Coriolis effect causes water to drain one way in the Northern Hemisphere and the other way in the Southern?

In most cases, no. If all other effects are nullified, however, the Coriolis effect comes into play and makes draining water spiral counterclockwise north of the Equator and the other way in the Southern Hemisphere (the same direction that hurricanes spin). But the Coriolis effect is extremely weak on small systems like a basin of water. The shape and irregularities of the basin, local slopes, or any external movement can easily outweigh the Coriolis effect in determining the direction water drains.

You mentioned that there is no Coriolis effect for those objects moving due east or west along the Equator. Do objects moving due east or west at other latitudes experience the Coriolis effect?

Yes. Any moving object in the Northern Hemisphere will still curve slightly to the right regardless of its direction of travel. This is because the route traveled by an object across Earth's curved surface is a straight line, not along a path that parallels a line of latitude (which would be a curved path). Considering a more mathematical approach, the Coriolis acceleration acting perpendicular to the direction of motion causes any mass that is in motion in any direction to be deflected to the right in the Northern Hemisphere. The magnitude of this acceleration is:

$$\text{Coriolis acceleration} = 2 \times \begin{pmatrix} \text{velocity of} \\ \text{mass relative to} \\ \text{Earth's surface} \end{pmatrix} \times \begin{pmatrix} \text{rotational} \\ \text{velocity} \\ \text{of Earth} \end{pmatrix} \times \begin{pmatrix} \text{sine of} \\ \text{the} \\ \text{latitude} \end{pmatrix}$$

Although this deflection is small, any object traveling due east or west will still be influenced by the Coriolis effect (except at the Equator, where the sine of 0 degrees latitude = 0).

If Earth is spinning so fast, why don't we feel it?

In spite of Earth's constant rotation, we have the illusion that Earth is still. The reason that we don't feel the motion is because Earth rotates smoothly and quietly, with no bumps, no accelerations or decelerations and the atmosphere moves along with us. Thus, all sensations we receive tell us there is no motion and the ground is comfortably at rest—even though most of the United States is continually moving at speeds greater than 800 kilometers (500 miles) per hour!

Is a water spout the same thing as a hurricane?

No, water spouts more closely resemble tornadoes, but are usually smaller in scale and can form anywhere there are strong enough winds over water. **Water spouts** are funnel-shaped vortices of water associated with extremely low atmospheric pressure. The low pressure causes air and water to spiral rapidly inward and upward, producing a spectacular funnel that extends from the ocean surface to the clouds (Figure 6D).

Since scientists don't know for sure that the greenhouse effect is actually increasing global temperature, why should we do anything about it?

Figure 6D A water spout near the Florida Keys.

The kinds of measures that prevent the greenhouse effect—being more fuel-efficient, protecting plant communities, eliminating chlorofluorocarbons, reducing carbon dioxide emissions—are all sound practices for preserving the environment regardless of reducing the greenhouse effect. Besides, if we find out in the next few decades that there really has been increased global warming due to the addition of human-caused gases, we'll wish that we had started making changes much earlier.

Chapter in Review

• The atmosphere and the ocean act as one interdependent system, linked by complex feedback loops. There is a close association between many atmospheric and oceanic phenomena.

• The Sun heats Earth's surface unevenly due to the change of seasons (caused by the tilt of Earth's rotational axis, which is 23.5 degrees from vertical) and the daily cycle of sunlight and darkness (Earth's rotation on its axis). The uneven distribution of solar energy on Earth is responsible for creating the temperature, density, water vapor content, and pressure differences that produce atmospheric and oceanic movement.

• The Coriolis effect influences the paths of objects moving in a north or south direction on Earth and is caused by Earth's rotation. Because Earth's surface rotates at different velocities at different latitudes, objects in motion tend to veer to the right in the Northern Hemisphere and to the left in the Southern Hemisphere. The Coriolis effect is nonexistent at the Equator, but increases with latitude, reaching a maximum at the poles.

• More solar energy is received than is radiated back into space at low latitudes than at high latitudes. On the spinning Earth, this creates three circulation cells in each hemisphere: a Hadley cell between 0 and 30 degrees latitude, a Ferrel cell between 30

Box 6–3
The ATOC Experiment: SOFAR So Good?

Worldwide, there is a layer in the ocean at a depth of about 1000 meters (3300 feet) caused by temperature and pressure conditions that causes sound originating above and below it to become refracted, or bent, into the layer (Figure 6E, *lower inset*). Once in this layer, called the **SOFAR channel** (an acronym for SOund Fixing And Ranging), sound is efficiently trapped and transmitted long distances. For instance, certain whales may use the SOFAR channel to send sounds across entire ocean basins.

Walter Munk of the Scripps Institution of Oceanography (Figure 6E, *upper inset*) came up with the idea of sending sounds through the SOFAR channel in an effort to detect the amount of ocean warming as a result of the greenhouse effect. His **Acoustic Thermometry of Ocean Climate (ATOC)** experiment is designed to accurately measure the travel time of similar low-frequency sound signals through the SOFAR channel now and in the future. The speed of sound in seawater increases as temperature increases, so

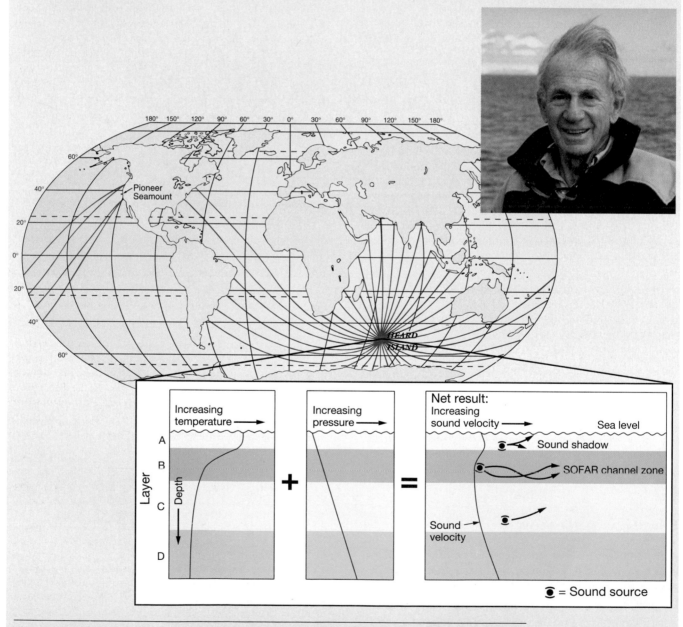

Figure 6E The ATOC experiment. Map showing Heard Island sound travel paths, oceanographer Walter Munk (*upper inset*), and the SOFAR channel (*lower inset*). Temperature and pressure conditions in the ocean combine to produce the SOFAR channel (*layer B*) that traps and transmits sound energy. Sound waves created in layers A and C are bent into the SOFAR channel and remain within it, allowing the sound to be transmitted across entire ocean basins.

sound should take less time to travel the same distance in the future if, in fact, the oceans are warming.

In 1991, Munk's group successfully tested ATOC at Heard Island in the southern Indian Ocean, from which sound can reach many different receiving sites along straight-line paths (Figure 6E). The researchers used an underwater array similar to a series of loudspeakers deployed from a ship to send *acoustical* (*akouein* = to hear) signals for six days that were refracted into the SOFAR channel and transmitted throughout the oceans. These signals were received at shipboard recording stations up to three-and-a-half hours later after traveling as far as 19,000 kilometers (11,800 miles), where the precise time of their arrival was noted.

The success of ship-based testing at Heard Island led to the establishment of a fixed sound source near Pioneer Seamount off California and an array of fixed receivers throughout the Pacific Ocean. In 1995, ATOC sound signals were sent again. Even though many precautions were taken to avoid any unwanted effects on marine mammals, three humpback whales were found dead in the area a few days after the sound transmissions had begun. The sounds were halted while the U.S. National Marine Fisheries Service (NMFS) conducted research to determine whether the sounds affected the hearing of nearby whales and contributed to their deaths. After extensive study, their results indicated that marine mammals are unaffected by the transmissions and that the dead whales were an unfortunate coincidence. The long-term effect of ATOC signals on marine mammals—including marine mammal communication through the SOFAR channel—is poorly understood, but is also likely to be minimal. The NMFS approved the project and transmissions from offshore California and Hawaii have been conducted without incident, establishing an important baseline for comparison with future measurements.

A similar but smaller-scale operation, the **Transarctic Acoustic Propagation Experiment (TAP)**, was successfully undertaken in 1994. Scientists sent sound signals through Arctic waters from near Spitzbergen in the North Atlantic Ocean to receivers off the north coast of Baffin Island and in the Beaufort Sea. Because heat exchange mechanisms are so important in high latitudes, it is thought that any global change in temperature would be experienced first in high-latitude waters. When compared with 10-year-old temperature measurements, it was discovered that the sound traveled faster than predicted, indicating that Arctic waters have warmed. Further research suggests that the warming has resulted from an influx of water from the North Atlantic. What remains unclear, however, is if this influx occurred because of global warming or if it is part of a natural cycle in the Atlantic Ocean.

━EIO━

For more information and on-line exercises about the Environmental Issue in Oceanography (EIO) "Sonar and Whales," visit the EIO Web site at **http://www.prenhall. com/oceanissues** and select Issue #2.

and 60 degrees latitude, and a polar cell between 60 and 90 degrees latitude. High-pressure regions, where dense air descends, are located at about 30 degrees north or south latitude and at the poles. Belts of low pressure, where air rises, are generally found at the Equator and at about 60 degrees latitude.

• The movement of air within the circulation cells produces the major wind belts of the world. The air at Earth's surface that is moving away from the subtropical highs produces trade winds moving toward the Equator and prevailing westerlies moving toward higher latitudes. The air moving along Earth's surface from the polar high to the subpolar low creates the polar easterlies.

• Calm winds characterize the boundaries between the major wind belts of the world. The boundary between the two trade wind belts is called the doldrums, which coincides with the Intertropical Convergence Zone (ITCZ). The boundary between the trade winds and the prevailing westerlies is called the horse latitudes. The boundary between the prevailing westerlies and the polar easterlies is called the polar front.

• The tilt of Earth's axis of rotation, the lower heat capacity of rock material compared to seawater, and the distribution of continents modify the wind and pressure belts of the idealized three-cell model. However, the three-cell model closely matches the pattern of the major wind belts of the world.

• Weather describes the conditions of the atmosphere at a given place and time, while climate is the long-term average of weather. Atmospheric motion (wind) always moves from high- to low-pressure regions. In the Northern Hemisphere, therefore, there is a counterclockwise cyclonic movement of air around low-pressure cells and a clockwise anticyclonic movement around high-pressure cells. Coastal regions commonly experience sea and land breezes, due to the daily cycle of heating and cooling.

• Many storms are due to the movement of air masses. In the mid-latitudes, cold air masses from higher latitudes meet warm air masses from lower latitudes and create cold and warm fronts that move from west to east across Earth's surface. Tropical cyclones (hurricanes) are large powerful storms that mostly affect tropical regions of the world. Destruction caused by hurricanes is caused by storm surge, high winds, and intense rainfall.

• Climate patterns in the ocean are closely related to the distribution of solar energy and the wind belts of the world. Ocean surface currents somewhat modify oceanic climate patterns.

• Energy reaching Earth from the Sun is mostly in the ultraviolet and visible regions of the electromagnetic spectrum, whereas energy radiated back to space from Earth is primarily in the infrared (heat) region. Water vapor, carbon dioxide, and other trace gases absorb infrared radiation and heat the atmosphere, creating the greenhouse effect. Earth's average surface temperature has warmed over the last century and there is concern that the human-caused increases of certain heat-trapping gases have enhanced the greenhouse effect.

• A low-velocity SOFAR sound channel can transmit sound over great distances in the oceans. This sound channel is being used in an experiment to document global warming in the ocean.

Key Terms

Acoustic Thermometry of Ocean Climate (ATOC) (p. 192)

Air mass (p. 176)

Albedo (p. 164)

Antarctic Circle (p. 165)

Anticyclonic flow (p. 175)

Arctic Circle (p. 165)

Autumnal equinox (p. 164)

Climate (p. 175)

Cold front (p. 176)

Columbus, Christopher (p. 174)

Convection cell (p. 167)

Coriolis effect (p. 168)

Cyclone (p. 178)

Cyclonic flow (p. 175)

Declination (p. 164)

Doldrums (p. 173)

Ecliptic (p. 164)

Equatorial (p. 183)

Equatorial low (p. 172)

Eye of the hurricane (p. 179)

Ferrel cell (p. 172)

Greenhouse effect (p. 185)

Hadley cell (p. 172)

Horse latitudes (p. 173)

Hurricane (p. 178)

Intergovernmental Panel on Climate Change (IPCC) (p. 189)

Intertropical Convergence Zone (ITCZ) (p. 173)

Iron hypothesis, the (p. 190)

Jet stream (p. 178)

Kyoto Protocol (p. 189)

Land breeze (p. 175)

Northeast trade winds (p. 173)

Polar (p. 185)

Polar cell (p. 172)

Polar easterly wind belt (p. 173)

Polar front (p. 173)

Polar high (p. 172)

Prevailing westerly wind belt (p. 173)

Saffir–Simpson Scale (p. 179)

Sea breeze (p. 175)

SOFAR channel (p. 192)

Southeast trade winds (p. 173)

Storm surge (p. 180)

Storm (p. 176)

Subpolar (p. 185)

Subpolar low (p. 172)

Subtropical (p. 185)

Subtropical high (p. 172)

Summer solstice (p. 164)

Temperate (p. 185)

Titanic, RMS (p. 162)

Trade winds (p. 173)

Transarctic Acoustic Propagation Experiment (TAP) (p. 193)

Tropic of Cancer (p. 164)

Tropic of Capricorn (p. 164)

Tropical (p. 185)

Tropical cyclone (p. 178)

Tropics (p. 164)

Troposphere (p. 166)

Typhoon (p. 178)

Vernal equinox (p. 164)

Warm front (p. 176)

Water spout (p. 191)

Weather (p. 175)

Wind (p. 168)

Winter solstice (p. 164)

Questions And Exercises

1. What caused the sinking of RMS *Titanic*? What international organization was formed after the sinking of the *Titanic* to prevent other such disasters?

2. Describe the effect on Earth as a result of Earth's axis of rotation being angled 23.5 degrees from perpendicular relative to the ecliptic. What would happen if Earth were not tilted on its axis?

3. Along the Arctic Circle, how would the Sun appear during the summer solstice? During the winter solstice?

4. Since there is a net annual heat loss at high latitudes and a net annual heat gain at low latitudes, why does the temperature difference between these regions not increase?

5. Describe the physical properties of the atmosphere, including its composition, temperature, density, water vapor content, pressure, and movement.

6. Describe the Coriolis effect in the Northern and Southern Hemispheres and include a discussion of why the effect increases with increased latitude.

7. Sketch the pattern of surface wind belts on Earth, showing atmospheric circulation cells, zones of high and low pressure, the names of the wind belts, and the names of the boundaries between wind belts.

8. Why are there high-pressure caps at each pole and a low-pressure belt in the equatorial region?

9. Discuss why the idealized belts of high and low atmospheric pressure shown in Figure 6–10 are modified (see Figure 6–11).

10. What is the difference between weather and climate? If it rains in a particular area during a day, does that mean that the area has a wet climate? Explain.

11. Describe the difference between cyclonic and anticyclonic flow, and show how the Coriolis effect is important in producing both a clockwise and a counterclockwise flow pattern.

12. How do sea breezes and land breezes form? During a hot summer day, which one would be most common and why?

13. Name the polar and tropical air masses that affect U.S. weather. Describe the pattern of movement across the continent and patterns of precipitation associated with warm and cold fronts.

14. What are the conditions needed for the formation of a tropical cyclone? Why do most mid-latitude areas only rarely experience a hurricane? Why are there no hurricanes at the Equator?

15. Describe the types of destruction caused by hurricanes. Of those, which one causes the majority of fatalities and destruction?

16. How are the ocean's climate belts (Figure 6–20) related to the broad patterns of air circulation described in Figure 6–10? What are some areas where the two are not closely related?

17. Describe the fundamental difference between solar radiation absorbed at Earth's surface and the radiation that is primarily responsible for heating Earth's atmosphere.

18. Discuss the greenhouse gases in terms of their relative concentrations and relative contributions to any increased greenhouse effect.

19. Describe the iron hypothesis, and discuss the relative merits and dangers of undertaking a project that could cause dramatic changes in the global environment.

20. What physical conditions produce a SOFAR, or sound channel, below the ocean's surface? How is the SOFAR channel being used to determine if global warming has occurred in the oceans?

References

Aguado, E., and Burt, J. E. 2001. *Understanding weather and climate*, 2nd ed. Upper Saddle River, NJ: Prentice-Hall.

Boyd, P. W. et. al. 2000. A mesoscale phytoplankton bloom in the polar Southern Ocean stimulated by iron fertilization. *Nature* 407:6805, 695–702.

Changing climate and the oceans. 1987. *Oceanus* 29:4, 1–93.

Charlson, R. J., et al. 1992. Climate forcing by anthropogenic aerosols. *Science* 255:5043, 423–430.

Coale, K. H., et al. 1996. A massive phytoplankton bloom induced by an ecosystem-scale iron fertilization experiment in the equatorial Pacific Ocean. *Nature* 383:6493, 495–501.

Coale, K. H., Worsfold. P. , and de Baar, H. 1999. Iron age in oceanography. *Eos Transactions AGU* 80:34, 377–382.

de la Mare, W. 1997. Abrupt mid-twentieth-century decline in Antarctic sea-ice extent from whaling records. *Nature* 389:6646, 57–60.

Dopyera, C. 1996. The iron hypothesis. *Earth* 5:5, 26–33.

Fox, S. 1999 "For a while...it was fun" *Smithsonian* 30:6, 128–142.

Golden, J. H. 1969. The Dinner Key "tornadic waterspout" of June 7, 1968. *Mariner's Weather Log* 13:4, 139–147.

Hartman, D. 1997. Our changing climate. *University Corporation for Atmospheric Research, Reports to the Nation on Our Changing Planet* 4: 24.

Intergovernmental Panel on Climate Change (IPCC) 1995. *Second assessment report: Climate change*. IPCC: Geneva, Switzerland.

Intergovernmental Panel on Climate Change (IPCC) 2001. *Climate change 2001: Impacts, adaptation, and vulnerability*. IPCC: Geneva, Switzerland.

Keeling, C. D., Whorf, T. P., Wahlen, M., and van der Plicht, J. 1995. Interannual extremes in the rate of rise of atmospheric carbon dioxide since 1980. *Nature* 375:6533, 666–670.

Ledley, T. S., et. al. 1999. Climate change and greenhouse gases. *Eos Transactions AGU* 80:39, 453–458.

Martin, J. H., Gordon, R. M., and Fitzwater, S. E. 1990. Iron in Antarctic waters. *Nature* 345:3271, 156–158.

Martin, J. H., et al. 1994. Testing the iron hypothesis in ecosystems of the equatorial Pacific Ocean. *Nature* 371:6493, 123–129.

Ocean energy. 1979. *Oceanus* 22:4, 1–68.

Oceans and climate. 1978. *Oceanus* 21:4, 1–70.

Philander, S. G. 1998. *Is the temperature rising? The uncertain science of global warming*. Princeton NJ: Princeton University Press.

Pickard, G. L. 1975. *Descriptive physical oceanography,* 2nd ed. New York: Pergamon Press.

Rodhe, H. 1990. A comparison of the contribution of various gases to the greenhouse effect. *Science* 248:4960, 1217–1219.

Ryan, P. R. 1985. The Titanic: Lost and found, 1985. *Oceanus* 28:4, 1–112.

———. 1986. The Titanic revisited. *Oceanus* 29:3, 2–17.

The Open University Course Team. 1989. *Ocean circulation*. Oxford: Pergamon Press.

Thieler, E. R., and Bush, D. M., 1991. Hurricanes Gilbert and Hugo send powerful messages for coastal development. *Journal of Geological Education* 39:4, 291–298.

Various authors. 1989. The oceans and global warming. *Oceanus*. 32:2, 1–75.

Various authors. 1994. The enigma of weather—A collection of works exploring the dynamics of meteorological phenomena. *Scientific American* (Special Publication).

Suggested Reading in Scientific American

Alley, R. B., and Bender, M. L. 1998. Greenland ice cores: Frozen in time. 278:2, 80–85. An analysis of the techniques that scientists use to study ice cores from Greenland and interpret clues to Earth's past—and future—climate.

Beardsley, T. 2000. Dissecting a hurricane. 282:3, 80–85. Illustrates how scientists study hurricanes, including an eyewitness description of flying through Hurricane Dennis to collect data in 1999.

Epstein, P. R. 2000. Is global warming harmful to health? 283:2, 50–57. A look at the hidden health risks of water borne diseases such as malaria and the West Nile virus are associated with increased global temperatures.

Gregg, M. 1973. Microstructure of the ocean. 228:2, 64–77. A discussion of the methods of studying the detailed movements of ocean water by observing temperature and salinity changes over distances of 1 centimeter (0.4 inch) and the motions they reveal.

Herzog, H., Eliasson, B., and Kaarstad, O. 2000. 282:2, 72–79. A look at techniques used to sequester (remove) carbon dioxide from the atmosphere by storing it underground and in the deep ocean.

Houghton, R. A., and Woodwell, G. M. 1989. Global climate change. 260:4, 36–44. Presents evidence that global warming has begun and looks at what the future might hold in a warmer climate.

Jones, P. D., and Wigley, T. M. 1990. Global warming trends. 263:2, 84–91. Data relating to evidence of global warming over the past 100 years are presented.

Karl, T. R., Nicholls, N., and Gregory, J., 1997. The coming climate. 276:5, 78–83. An analysis of meteorological records and computer models indicate how a warmer climate will affect plant and animal life in the future.

Karl, T. R., and Trenberth, K. E. 1999. The human impact on climate. 281:6, 100–105. Outlines the steps needed to be taken to undertake long-term monitoring of climate so that we will truly know the human impact on climate.

King, M. D. and Herring, D. D. 2000. Monitoring Earth's vital signs. 282:4, 92–97. The NASA satellite Terra has sensors to monitor with increased precision Earth's radiation budget, concentration of greenhouse gases, types of cloud and aerosols, and photosynthetic productivity.

MacIntyre, F. 1974. The top millimeter of the ocean. 230:5, 62–77. Processes that are confined to a thin film at the surface of ocean water and their role in the overall nature of the oceans are discussed.

Nadis, S. 1998. Fertilizing the sea. 278: 4, 33. A news item about a U. S. company that is planning to stimulate fishery production in the Marshall Islands by using John Martin's iron hypothesis.

Penney, T. R., and Bharathan, D. 1987. Power from the sea. 256:1, 86–93. A prediction that generating electricity by ocean thermal energy conversion will be competitive with fossil fuel plants as the price of oil rises.

Pollack, H. N., and Chapman, D. S. 1993. Underground records of changing climate. 268:6, 44–53. Direct measurement of temperature shows that Earth has warmed over the past 150 years. Data from boreholes reveal ancient temperatures that may give us a more complete picture of the history of global climate.

Revelle, R. 1982. Carbon dioxide and world climate. 247:2, 35–43. Some of the possible effects of increasing atmospheric temperature due to carbon dioxide accumulation are considered.

Stanley, S. M. 1984. Mass extinctions in the oceans. 250:6, 64–83. Geological evidence suggests that most major periods of species extinction over the last 700 million years occurred during the brief intervals of ocean cooling.

Stolarski, R. S. 1988. The Antarctic ozone holes. 258:1, 30–37. The discovery of the Antarctic ozone hole and its possible significance are discussed.

White, R. M. 1990. The great climate debate. 263:1, 36–45. The controversy over the degree of global warming that can be expected in the future is discussed.

Oceanography on the Web

Visit the *Essentials of Oceanography* home page for on-line resources for this chapter. There you will find an on-line study guide with review exercises, and links to oceanography sites to further your exploration of the topics in this chapter. *Essentials of Oceanography* is at: **http://www.prenhall.com/ thurman** (click on the Table of Contents menu and select this chapter).

CHAPTER
7
Ocean Circulation

BENJAMIN FRANKLIN: THE WORLD'S MOST FAMOUS PHYSICAL OCEANOGRAPHER

Benjamin Franklin was an inventor, a statesman, and one of the founding fathers of our country. He was also deputy postmaster general of the colonies from 1753 to 1774 and a physical oceanographer—one who studies physical phenomena in the ocean, including waves, tides, and currents. Franklin contributed greatly to the understanding of the Gulf Stream, a North Atlantic Ocean surface current that influenced mail routes between the colonies and England.

Franklin became interested in the North Atlantic Ocean circulation patterns because it took mail ships coming from Europe two weeks *longer* to reach New England by a northerly route than it did by a more southerly route. In about 1769 or 1770, Franklin mentioned this dilemma to his cousin, a Nantucket sea captain named Timothy Folger. Folger told Franklin that a strong current with which the mail ships were unfamiliar was impeding their journey because it flowed *against* them. The whaling ships were familiar with the current because they often hunted whales along its boundaries. The whalers often met the mail ships within the current and told their crews they would make swifter progress if they avoided the current. The British captains of the mail ships, however, would not accept advice from simple American fishers, so they continued to make slow progress within the current. If the winds were light, their ships were actually carried *backwards*!

Folger sketched the current on a map for Franklin, including directions for avoiding it when sailing from Europe to North America. Franklin then asked other ship captains for information concerning the movement of surface waters in the North Atlantic Ocean. Franklin inferred that there was a significant current moving northward along the eastern coast of the United States, which then headed east across the North Atlantic. He concluded that this current was responsible for aiding the progress of ships traveling through the North Atlantic to Europe and slowing ships traveling in the reverse direction. He subsequently published a map of the current in 1777 based on these observations (Figure 7A) and distributed it to the captains of the mail ships (who initially ignored it). This strong current was named the Gulf Stream because it carried warm water from the Gulf of Mexico, and because it was narrow and well defined—similar to a stream, but in the ocean.

In 1969, six scientists studied the Gulf Stream for a month aboard a vessel that was allowed to float wherever the current took it. During the vessel's 2640-kilometer (1650-mile) journey, the scientists observed and measured the properties of and cataloged life forms within the Gulf Stream. Appropriately enough, the vessel was named the *Ben Franklin*.

- How are ocean currents measured?
- How are surface currents organized in each ocean basin?
- What is western intensification?
- Why is upwelling associated with abundant marine life?
- What environmental effects do El Niños and La Niñas produce?
- What is thermohaline circulation?

The coldest winter I ever spent was a summer in San Francisco.
—*Anonymous, but often attributed to Mark Twain*

Ocean currents are masses of ocean water that flow from one place to another. The amount of water can be large or small, currents can be at the surface or deep below, and the phenomena that create them can be simple or quite complex. Simply put, currents are *water masses in motion.*

Huge current systems dominate the surfaces of the major oceans. These currents transfer heat from warmer to cooler areas on Earth, just as the major wind belts of the world do. Wind belts transfer about two-thirds of the total amount of heat from the tropics to the poles; ocean surface currents transfer the other third. Ultimately, energy from the Sun drives surface currents and they closely follow the pattern of the world's major wind belts.

More locally, surface currents affect the climates of coastal continental regions. Cold currents flowing toward the Equator on the western sides of continents produce arid conditions. Conversely, warm currents flowing poleward on the eastern sides of continents produce warm, humid conditions. Additionally, ocean currents contribute to the mild climate of northern Europe and Iceland, whereas conditions at similar latitudes along the Atlantic coast of North America (such as Labrador) are much colder.

Currents profoundly affect ocean life, especially those organisms in the deep sea where currents provide a continuing supply of oxygen. This oxygen is carried there by cold, dense water that sinks in polar regions and spreads across the deep-ocean floor. Ocean currents influence the abundance of life in surface waters by affecting the growth of microscopic algae, which is the basis of most oceanic food chains. Currents have also aided the travel of prehistoric peoples from Europe and Africa to the New World and throughout the Pacific Ocean islands.

Measuring Ocean Currents

Ocean currents are either *wind-driven* or *density-driven*. Moving air masses—particularly the major wind belts of the world—set wind-driven currents in motion. This motion is parallel to the surface (horizontal) and occurs primarily in the ocean's surface waters, so these currents are called **surface currents**. Density-driven circulation, on the other hand, moves vertically and accounts for the thorough mixing of the deep masses of ocean water. Temperature and salinity conditions at the surface that produce high-density water initiate density-driven circulation. The dense water sinks and spreads slowly beneath the surface, so these currents are called **deep currents**.

Surface currents rarely flow in the same direction and at the same rate for very long, so measuring average flow rates can be difficult. Some consistency, however, exists in the *overall* surface current pattern worldwide. Surface currents can be measured directly or indirectly.

Two main methods are used to *directly* measure currents. In one, a floating device is released into the current and tracked through time. Typically, radio-transmitting float bottles or other devices are used (Figure 7–1a), but other accidentally released items also

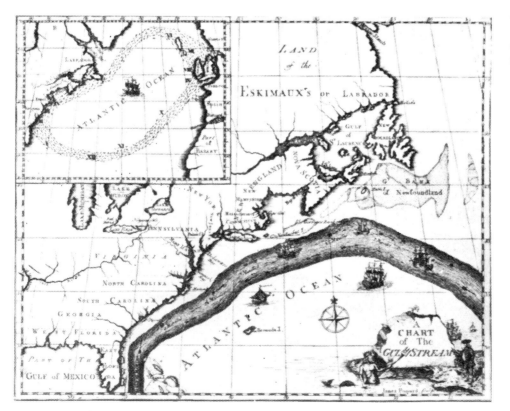

Figure 7A Benjamin Franklin's chart of the Gulf Stream (1777).

make good drift meters (Box 7–1). The other method is done from a fixed position (such as a pier) where a current-measuring device, such as the propeller flow meter shown in Figure 7–1b, is lowered into the water. Propeller devices can also be towed behind ships, and the ship's speed is then subtracted to determine a current's true flow rate.

Three different methods can be used to *indirectly* measure surface currents. Water flows parallel to a pressure gradient, so one method is to determine the internal distribution of density and the corresponding pressure gradient across an area of the ocean. A second method uses radar altimeters such as the one launched aboard the TOPEX/Poseidon satellite in 1992 to determine the lumps and bulges at the ocean surface, which are a result of the shape of the underlying sea floor (see Box 3–1) and current flow. From these data, *dynamic topography* maps can be produced that show the speed and direction of surface currents (Figure 7–2). A third method uses a *Doppler flow meter* to transmit low-frequency sound signals through the water. The flow meter measures the shift in frequency between the sound waves emitted and those backscattered by particles in the water to determine current movement.

The location of deep currents deep below the surface makes them even more difficult to measure. Often, they are mapped using devices that are carried with the current or by tracking telltale chemical tracers. Some tracers are naturally absorbed into seawater, while others are intentionally added. Some useful tracers that have inadvertently been added to seawater include tritium (a radioactive isotope of hydrogen produced by nuclear bomb tests in the 1950s and early 1960s) and chlorofluorocarbons (freons and other gases now thought to be depleting the ozone layer). Other techniques used to identify deep currents include measuring the distinctive temperature and salinity characteristics of a deep-water mass.

> Wind-induced surface currents are measured with floating objects, by satellites, or by other techniques. Density-induced deep currents are measured using tracers or other devices.

Surface Currents

Surface currents develop from friction between the ocean and the wind that blows across its surface. Only about 2% of the wind's energy is transferred to the ocean surface, so a 50-knot[1] wind will create a 1-knot current. You can simulate this on a tiny scale simply by blowing gently and steadily across a cup of coffee.

[1]A knot is one nautical mile per hour. A nautical mile is defined as the distance of one minute of latitude, and is equivalent to 1.15 statute (land) miles or 1.85 kilometers.

(a)

(b)

Figure 7–1 **Current-measuring devices.** **(a)** Drift current meter. Depth of metal vanes is 1 meter (3.3 feet).
(b) Propeller-type flow meter. Length of instrument is 0.6 meter (2 feet).

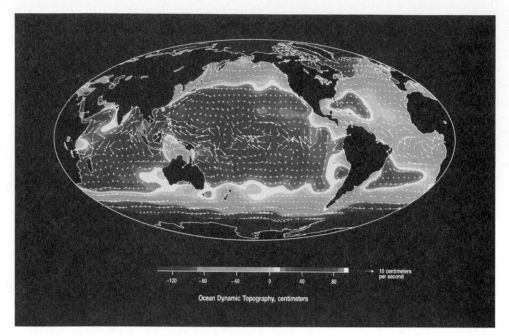

Figure 7–2 Satellite view of ocean dynamic topography. Map showing TOPEX/Poseidon radar altimeter data in centimeters from September 1992 to September 1993. Red colors are areas that have higher than normal sea level; blue colors are areas that are lower than normal. White arrows indicate the flow direction of currents, with longer arrows indicating faster flow rates.

If there were no continents on Earth, the surface currents would simply follow the major wind belts of the world. In each hemisphere, therefore, a current would flow between 0 and 30 degrees latitude due to the trade winds, a second would flow between 30 and 60 degrees latitude due to the prevailing westerlies, and a third would flow between 60 and 90 degrees latitude due to the polar easterlies.

However, the distribution of continents on Earth influences the nature and the direction of flow of surface currents. For example, Figure 7–3 shows how the trade winds and prevailing westerlies create large circular-moving loops of water in the Atlantic Ocean. Other ocean basins show a similar pattern, and surface currents are also influenced by gravity, friction, and the Coriolis effect.

Surface currents occur within and above the *pycnocline* (the layer of rapidly changing density) to a depth of about 1 kilometer (0.6 mile) and affect only about 10% of the world's ocean water.

Equatorial Currents, Boundary Currents, and Gyres

The trade winds, which blow from the southeast in the Southern Hemisphere and from the northeast in the Northern Hemisphere, set in motion the water masses between the tropics. The resulting currents are called **equatorial currents**, which travel westward along the Equator (Figure 7–4). They are called north or south equatorial currents, depending on their position relative to the Equator.

When equatorial currents reach the western portion of an ocean basin, they must turn because they cannot cross land. The Coriolis effect deflects these currents away from the Equator as **western boundary currents**.

The name means they are currents traveling along the western boundary of their ocean basin.[2] The Gulf Stream and the Brazil Current, which are shown in Figure 7–4, are western boundary currents. They come from equatorial regions, where water temperatures are warm, so they carry warm water to higher latitudes. Figure 7–4 shows warm currents as red arrows.

Between 30 and 60 degrees latitude, the prevailing westerlies blow from the northwest in the Southern Hemisphere and from the southwest in the Northern Hemisphere. These winds direct ocean surface water in an easterly direction across the ocean basin, as shown in Figure 7–4 by the North Atlantic Current and the West Wind Drift.[3]

When currents flow back across the ocean basin, the Coriolis effect and continental barriers turn them toward the Equator, creating **eastern boundary currents** along the eastern boundary of the ocean basins. The Canary Current and the Benguela Current, which are shown in Figure 7–4, are eastern boundary currents.[4] They come from high-latitude regions where water temperatures are cool, so they carry cool water to lower latitudes. Figure 7–4 shows cold currents as blue arrows.

The equatorial, western boundary, prevailing westerly, and eastern boundary currents combine to create a

---

[2]Notice that the western boundary currents are off the *eastern* coasts of continents. This sounds confusing, but is a result of the fact that we have a land-based perspective. From an oceanic perspective, the western side of the ocean basin is where the western boundary current resides.

[3]Currents, like winds, are often named based on the direction from which they flow. Thus, the West Wind Drift is initiated by the westerlies and flows from the west.

[4]Currents are sometimes named for a prominent geographic location near where they pass. For instance, the Canary Current passes the Canary Islands; the Benguela Current is named for the Benguela Province in Angola, Africa.

Figure 7–3 Atlantic Ocean surface circulation pattern. The trade winds *(blue arrows)* in conjunction with the prevailing westerlies *(green arrows)* create circular-moving loops of water *(underlying purple arrows)* at the surface in both parts of the Atlantic Ocean basin. If there were no continents, the ocean's surface circulation pattern would closely match the major wind belts of the world.

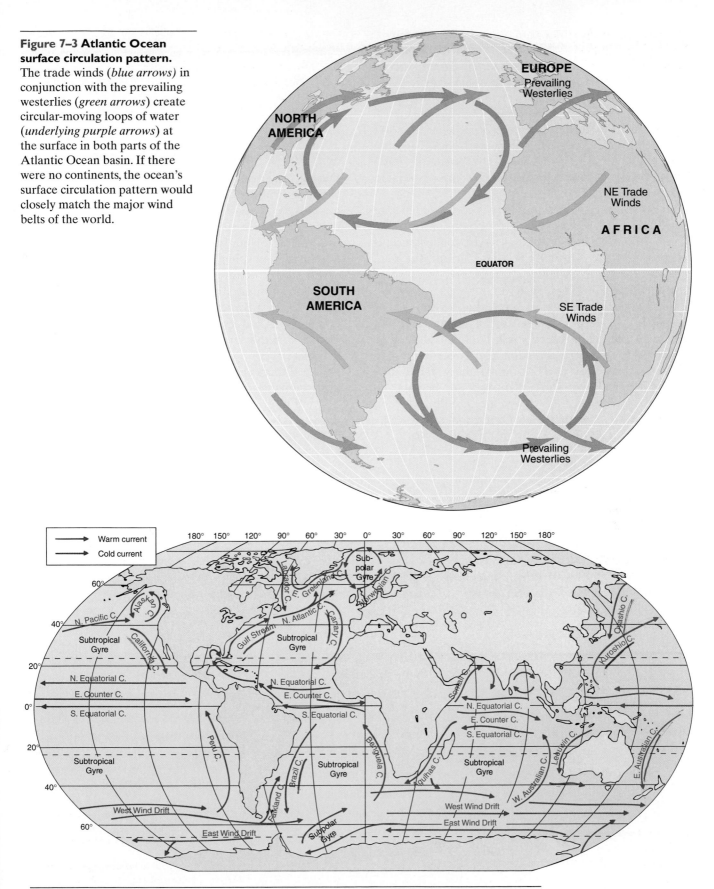

Figure 7–4 Wind-driven surface currents. Major wind-driven surface currents of the world's oceans during February-March. The five major subtropical gyres are the North and South Pacific Ocean Gyres, the North and South Atlantic Ocean Gyres, and the Indian Ocean Gyre. The smaller subpolar gyres rotate in the reverse direction of the adjacent subtropical gyres.

circular flow within an ocean basin called a **gyre** (*gyros* = a circle). Figure 7–4 shows the world's five **subtropical gyres**: the *North Atlantic Gyre*, the *South Atlantic Gyre*, the *North Pacific Gyre*, the *South Pacific Gyre*, and the *Indian Ocean Gyre* (which is mostly within the Southern Hemisphere). The center of each subtropical gyre coincides with the subtropics at 30 degrees north or south latitude. As shown in Figures 7–3 and 7–4, subtropical gyres rotate clockwise in the Northern Hemisphere and counterclockwise in the Southern Hemisphere.

Generally, each subtropical gyre is composed of four main currents that flow progressively into one another (Table 7–1). The North Atlantic Gyre, for instance, is composed of the North Equatorial Current, the Gulf Stream, the North Atlantic Current, and the Canary Current (Figure 7–4).

Surface currents moving eastward as a result of the prevailing westerlies approach subpolar latitudes (about 60 degrees north or south latitude). Here, they are driven in a westerly direction by the polar easterlies, producing **subpolar gyres** that rotate opposite the adjacent subtropical gyres. Subpolar gyres are smaller and fewer than subtropical gyres, and are best developed in the Atlantic Ocean between Greenland and Europe and the Weddell Sea off Antarctica (Figure 7–4).

> The principal ocean surface current pattern consists of subtropical and subpolar gyres that are large circular-moving loops of water powered by the major wind belts of the world.

Ekman Spiral and Ekman Transport

During the voyage of the *Fram* (Box 7–2), Norwegian explorer Fridtjof Nansen observed that Arctic Ocean ice moved 20 to 40 degrees to the *right* of the wind blowing across its surface (Figure 7–5). Surface water in the Northern Hemisphere behaves similarly and, in the Southern Hemisphere, surface currents move to the left of the wind direction. Why does surface water move in a direction different than the wind? **V. Walfrid Ekman** (1874–1954), a Swedish physicist, developed a circulation model called the **Ekman spiral** (Figure 7–6) that explains Nansen's observations in accordance with the Coriolis effect.

The Ekman spiral describes the speed and direction of flow of surface waters at various depths. It is caused by wind blowing across the surface and is modified by the Coriolis effect. Ekman's model assumes that a uniform column of water is set in motion by wind blowing

Table 7–1 Subtropical gyres and surface currents.

Pacific Ocean	*Atlantic Ocean*	*Indian Ocean*
North Pacific Gyre	**North Atlantic Gyre**	**Indian Ocean Gyre**
North Pacific Current	North Atlantic Current	South Equatorial Current
California Current[a]	Canary Current[a]	Agulhas Current[b]
North Equatorial Current	North Equatorial Current	West Wind Drift
Kuroshio (Japan) Current[b]	Gulf Stream[b]	West Australian Current[a]
South Pacific Gyre	**South Atlantic Gyre**	**Other Major Currents**
South Equatorial Current	South Equatorial Current	Equatorial Countercurrent
East Australian Current[b]	Brazil Current[b]	North Equatorial Current
West Wind Drift	West Wind Drift	Leeuwin Current
Peru (Humboldt) Current[a]	Benguela Current[a]	Somali Current
Other Major Currents	**Other Major Currents**	
Equatorial Countercurrent	Equatorial Countercurrent	
Alaskan Current	Florida Current	
Oyashio Current	East Greenland Current	
	Labrador Current	
	Falkland Current	

[a]Denotes an eastern boundary current of a gyre, which is relatively *slow*, *wide*, and *shallow* (and is also a *cold water* current).
[b]Denotes a western boundary current of a gyre, which is relatively *fast*, *narrow*, and *deep* (and is also a *warm water* current).

Box 7–1
Running Shoes as Drift Meters: Just Do It

Any floating object can serve as a makeshift drift meter, as long as it is known where the object entered the ocean and where it was retrieved. The path of the object can then be inferred, providing information about the movement of surface currents. If the time of release and retrieval are known, the speed of currents can also be determined. Oceanographers have long used **drift bottles** (a floating "message in a bottle" or a radio-transmitting device set adrift in the ocean) to track the movement of currents.

Many objects have inadvertently become drift meters when ships have lost some (or all) of their cargo at sea. In this way, Nike athletic shoes and colorful floating bathtub toys (Figure 7B, *inset*) have advanced the understanding of current movement in the North Pacific Ocean.

In May 1990, the container vessel *Hansa Carrier* was en route from Korea to Seattle, Washington, when it encountered a severe North Pacific storm. The ship was transporting 12.2-meter (40-foot)-long rectangular metal shipping containers, many of which were lashed to the ship's deck for the voyage. During the storm, the ship lost 21 deck contain-

ers overboard, including five that held Nike athletic shoes. The shoes floated, so those that were released from their containers were carried east by the North Pacific Current. Within six months, thousands of the shoes began to wash up along the beaches of Alaska, Canada, Washington, and Oregon (Figure 7B), over 2400 kilometers (1500 miles) from the site of the spill. A few shoes were found on beaches in northern California, and over two years later shoes from the spill were even recovered from the north end of the Big Island of Hawaii!

Even though the shoes had spent considerable time drifting in the ocean, they were in good shape and wearable (after barnacles and oil were removed). Because the shoes were not tied together, many beachcombers found individual shoes or pairs that did not match. Many of the shoes retailed for around $100, so people interested in finding matching pairs placed ads in newspapers or attended local swapmeets.

With help from the beachcombing public (as well as lighthouse operators), information on the location and number

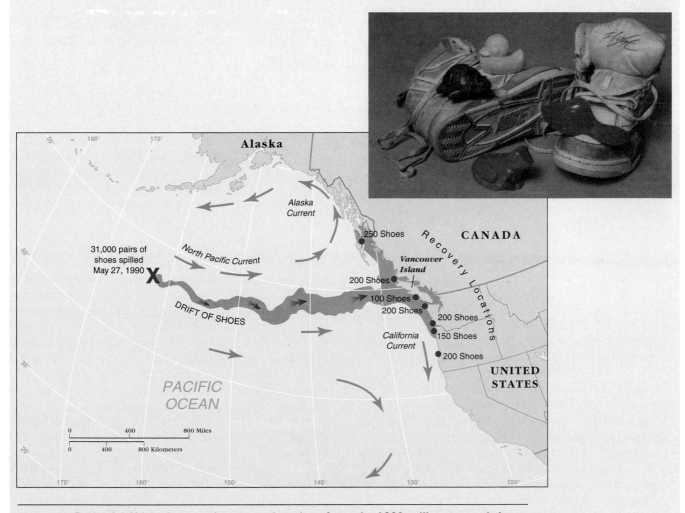

Figure 7B **Path of drifting shoes and recovery locations from the 1990 spill; recovered shoes and plastic bathtub toys (*inset*).**

(*continued*)

(continued)

of shoes collected was compiled during the months following the spill. Serial numbers inside the shoes were traced to individual containers and they indicated that only four of the five containers had released their shoes; evidently, one entire container sank without opening. Thus, a maximum of 30,910 pairs of shoes (61,820 individual shoes) were released. The almost instantaneous release of such a large number of drift items helped oceanographers refine computer models of North Pacific circulation. Before the shoe spill, the largest number of drift bottles purposefully released at one time by oceanographers was about 30,000. Although only 2.6% of the shoes were recovered, this compares favorably with the 2.4% recovery rate of drift bottles released by oceanographers conducting research.

In January 1992, another cargo ship lost 12 containers during a storm to the north of where the shoes had previously spilled. One of these containers held 29,000 packages of small, floatable, colorful plastic bathtub toys in the shapes of blue turtles, yellow ducks, red beavers, and green frogs (Figure 7B, *inset*). Even though the toys were housed in plastic packaging glued to a cardboard backing, studies showed that after 24 hours in seawater, the glue deteriorated and over 100,00 of the toys were released.

The floating bathtub toys began to come ashore in southeast Alaska 10 months later, verifying the computer models. The models indicate that many of the bathtub toys will continue to be carried by the Alaska Current, eventually dispersing throughout the North Pacific Ocean. Some may find their way into the Arctic Ocean, where they could spend time within the Arctic Ocean ice pack. From there, the toys may drift into the North Atlantic, eventually washing up on beaches in northern Europe, thousands of kilometers from where they were accidentally released into the ocean.

Since 1992, oceanographers have continued to study ocean currents by tracking other floating items spilled from cargo ships, including 34,000 hockey gloves, 5 million plastic Lego pieces, and an unidentified number of small plastic doll parts.

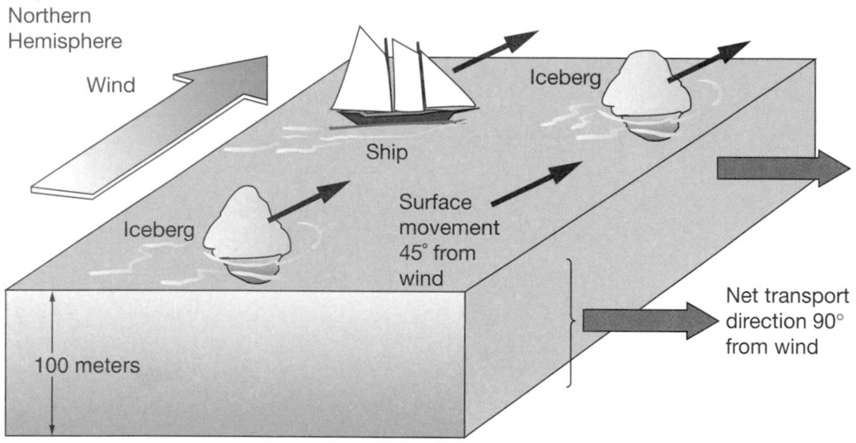

Figure 7–5 Transport of floating objects. Fridtjof Nansen first noticed that floating objects, such as icebergs and ships, were carried to the right of the wind direction in the Northern Hemisphere.

across its surface. Because of the Coriolis effect, the immediate surface water moves in a direction 45 degrees to the right of the wind (in the Northern Hemisphere). The surface water moves as a thin "layer" on top of deeper layers of water. As the surface layer moves, it also sets in motion other layers beneath it, thus passing the energy of the wind down through the water column.

Current speed decreases with increasing depth, however, and the Coriolis effect increases curvature to the right (like a spiral). Thus, each successive layer of water is set in motion at a progressively slower velocity, and in a direction progressively to the right of the one above it. At some depth, a layer of water may move in a direction *exactly opposite to the wind direction that initiated it!* If the water is deep enough, friction will consume the energy imparted by the wind and no motion will occur below that depth. Although it depends on wind speed and latitude, this stillness normally occurs at a depth of about 100 meters (330 feet).

Figure 7–6 shows the spiral nature of this movement with increasing depth from the ocean's surface. The length of each arrow in Figure 7–6 is proportional to the velocity of the individual layer, and the direction of each arrow indicates the direction it moves.[5] Under ideal conditions, therefore, the surface layer should move at an angle of 45 degrees from the direction of the wind. All the layers combine, however, to create a net water movement that is 90 degrees from the direction of the wind. This average movement, called **Ekman transport**, is 90 degrees to the right in the Northern Hemisphere and 90 degrees to the left in the Southern Hemisphere.

"Ideal" conditions rarely exist in the ocean, so the actual movement of surface currents deviates slightly from the angles shown in Figure 7–6. Generally, surface

---

[5]The name Ekman *spiral* refers to the spiral observed by connecting the tips of the arrows shown in Figure 7–6.

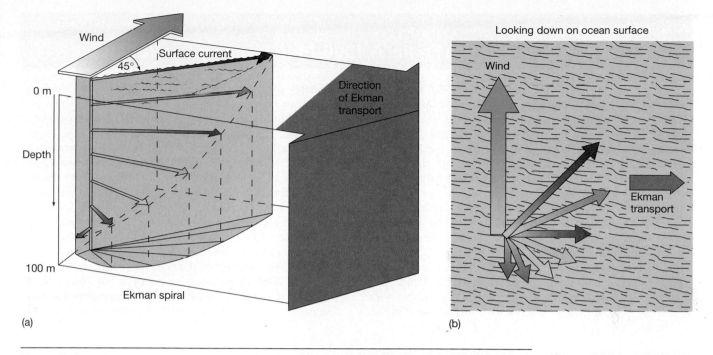

Figure 7–6 Ekman spiral. Perspective view **(a)** and top view **(b)** of Ekman spiral and Ekman transport. Wind drives surface water in a direction 45 degrees to the right of the wind in the Northern Hemisphere. Deeper water continues to deflect to the right and moves at a slower speed with increased depth, causing the Ekman spiral. Ekman transport, which is the net water movement, is at a right angle (90 degrees) to the wind direction.

currents move at an angle somewhat less than 45 degrees from the direction of the wind and Ekman transport in the open ocean is typically about 70 degrees from the wind direction. In shallow coastal waters, Ekman transport may be very nearly the same direction as the wind.

Geostrophic Currents

Ekman transport deflects surface water to the right in the Northern Hemisphere, so a clockwise rotation develops within an ocean basin and produces a **subtropical convergence** of water in the middle of the gyre, causing water literally to pile up in the center of the subtropical gyre. Thus, there is a hill of water within all subtropical gyres that is as much as 2 meters (6.6 feet) high.

Surface water in a subtropical convergence tends to flow downhill in response to gravity. The Coriolis effect opposes gravity, however, deflecting the water to the right in a curved path (Figure 7–7a) into the hill again. When these two factors balance, the net effect is a **geostrophic** (*geo* = earth, *strophio* = turn) **current** that moves in a circular path around the hill.[6] In Figure 7–7a it is labeled as the *path of ideal geostrophic flow*. Friction between water molecules, however, causes the water to gradually move down the slope of the hill as it

flows around it. This is the *path of actual geostrophic flow* labeled in Figure 7–7a.

Western Intensification

Figure 7–7a shows that the apex (top) of the hill formed within a rotating gyre is closer to the western boundary than the center of the gyre. As a result, the western boundary currents of the subtropical gyres are faster, narrower, and deeper than their eastern boundary current counterparts. For example, the Kuroshio Current (a western boundary current) of the North Pacific Subtropical Gyre is up to 15 times faster, 20 times narrower, and five times as deep as the California Current (an eastern boundary current). This phenomenon is called **western intensification**, and currents affected by this phenomenon are said to be western intensified. *The western boundary currents of all subtropical gyres are western intensified, even in the Southern Hemisphere.*

A number of factors cause western intensification, including the Coriolis effect. The Coriolis effect increases toward the poles, so eastward-flowing high-latitude water turns toward the Equator more strongly than westward-flowing equatorial water turns toward higher latitudes. This causes a wide, slow, and shallow flow of water toward the Equator across most of each subtropical gyre, leaving only a narrow band through which the poleward flow can occur along the western margin of the ocean basin. If a constant volume of water

---

[6]The term *geostrophic* for these currents is appropriate, since the currents behave as they do because of Earth's rotation.

Box 7–2
The Voyage of the *Fram*: A 1000-Mile Journey Locked in Ice

Norwegian explorer **Fridtjof Nansen** (1861-1930) was interested in a voyage to determine the existence of land in the unexplored Arctic. A previous expedition that piqued Nansen's curiosity was the ill-fated voyage of American George Washington DeLong and his crew on the *Jeanette* in 1879. DeLong had attempted to sail through the Bering Strait between North America (Alaska) and Asia (Russia) to Wrangell Island. From there, he planned to trek overland to the North Pole. At the time, it was believed that Wrangell Island might be the tip of a peninsula extending southward from an as-yet-undiscovered Arctic continent that lay beneath Arctic Ocean ice.

The *Jeanette* became stuck in the polar ice pack on September 6, 1879, and drifted north of Wrangell Island. This proved that the island was not a peninsula of an Arctic continent, because the ship could not have drifted past it in the ice. After two years of drifting, the *Jeanette* was crushed while still embedded within the ice off the New Siberian Islands. Unfortunately, DeLong and many of his crew perished in the Lena Delta region of eastern Siberia.

Five years later, a number of items from the *Jeanette* were found frozen in the pack ice off the southwestern coast of Greenland, over 4800 kilometers (3000 miles) from where she sank. Based on this unusual evidence, Nansen reasoned that the articles must have drifted with Arctic ice from the wreck of the *Jeanette* to Greenland by following a path that would have taken them near the North Pole. If Arctic ice could be verified to drift that far, it would help prove that an Arctic continent did not exist. Arctic ice could also provide a means of transporting an expedition to the North Pole, which had never been visited before. Nansen envisioned that a ship, once locked in the ice, might be able to follow a path

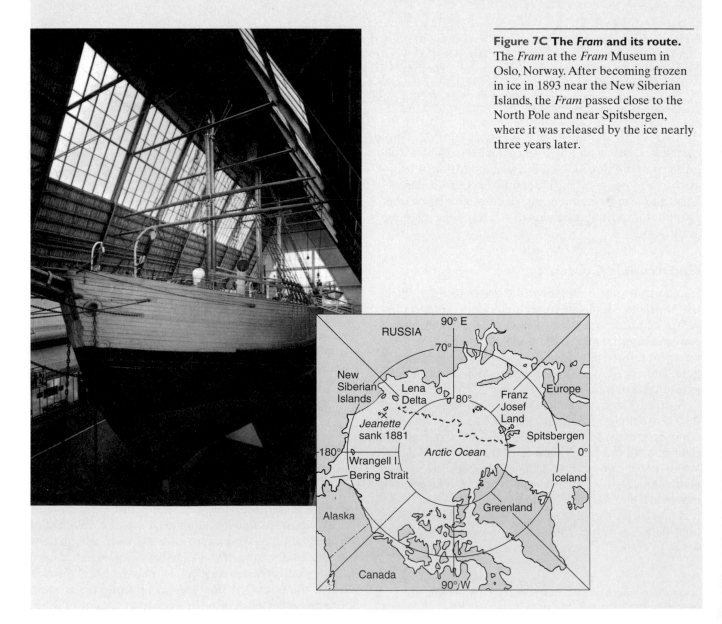

Figure 7C The *Fram* and its route.
The *Fram* at the *Fram* Museum in Oslo, Norway. After becoming frozen in ice in 1893 near the New Siberian Islands, the *Fram* passed close to the North Pole and near Spitsbergen, where it was released by the ice nearly three years later.

similar to the transported articles, carrying the vessel near the North Pole.

Nansen eventually raised enough money to build the *Fram*, a 39-meter (128-foot)-long wooden ship (Figure 7C). The *Fram* was designed with extra reinforcing in its hull so that the expanding ice would not crush it but rather would force it up to the surface, free of the grip of the growing ice.

Provisions for 13 men for five years were stored on the small ship, and the crew set sail from Oslo, Norway, on June 24, 1893. On September 21, after an arduous voyage along the northern coast of Siberia, the ship became entrapped in ice at 78.5 degrees north latitude, 1100 kilometers (683 miles) from the North Pole. The crew had begun the long and lonely endeavor of drifting slowly with the pack ice as it was moved by winds and currents (Figure 7C, *inset*). During their time locked in the Arctic ice, the explorers were able to collect depth soundings, seawater samples, and ice measurements. On November 15, 1895, the *Fram* reached its northernmost point, only 394 kilometers (244 miles) from the pole, but the explorers were unable to get to the North Pole.

On August 13, 1896, after being locked in ice for almost three years, the *Fram* broke free and floated again in the open sea. She had drifted 1658 kilometers (1028 miles) during an amazingly successful journey that proved no continent existed under the Arctic Ocean. It also showed that the polar ice was not of glacial origin but formed by freezing seawater. During the voyage, the crew had found that the depth of the Arctic Ocean exceeded 3000 meters (9840 feet). They also discovered a body of relatively warm water with temperatures as high as 1.5°C (35°F) between the depths of 150 and 900 meters (500 and 3000 feet), which Nansen correctly described this as being a mass of Atlantic Ocean water that had sunk below the less saline Arctic Ocean water.

Nansen's accomplishments include laying the groundwork for future Arctic exploration and developing a sampling apparatus called the Nansen bottle for collecting water samples at depth to determine water temperature and salinity. Further, his observations of the direction of ice drift relative to the wind direction helped V. Walfrid Ekman, a Swedish physicist, develop the mathematical explanation of this phenomenon, known today as the Ekman spiral. Nansen (who was awarded the Nobel Peace Prize in 1922), Ekman, and other Scandinavian scientists led the way in developing the discipline of physical oceanography in the early twentieth century.

rotates around the apex of the hill in Figure 7–7b, then the velocity of the water along the western margin will be much greater than the velocity around the eastern side.[7] In Figure 7–7b, the lines are close together along the western margin, indicating the faster flow. The end result is a high-speed western boundary current that flows along the hill's steeper westward slope and a slow drift of water toward the Equator along the more gradual eastern slope. Table 7–2 summarizes the differences between western and eastern boundary currents of subtropical gyres.

> Western intensification is a result of Earth's rotation and causes the western boundary current of all subtropical gyres to be fast, narrow, and deep.

Equatorial Countercurrents

A large volume of water is driven westward due to the north and south equatorial currents. The Coriolis effect is minimal near the Equator, so much of the water is not turned toward higher latitudes. Instead, it piles up at the western margins of the ocean basins, which causes average sea level on the western side of the basin to be as much as 2 meters (6.6 feet) higher than on the eastern side. The water on the western margins then flows downhill under the influence of gravity, creating narrow **equatorial countercurrents** that flow to the east *counter to* and *between* the adjoining equatorial currents.

Figure 7–4 shows that an equatorial countercurrent is particularly apparent in the western Pacific Ocean, where a dome of equatorial water is trapped in the island-filled embayment between Australia and Asia. This dome of water with very weak current flow has the highest year-round ocean surface temperatures found anywhere in the world ocean, as shown in Figure 7–8. Continual influx of water from equatorial currents builds the dome and creates an eastward countercurrent that stretches across the Pacific toward South America.

The hills of water within the subtropical gyres of the Atlantic Ocean are clearly visible in the satellite image of sea surface elevation shown in Figure 7–2. The hill in the North Pacific is visible, as well, but the low equatorial elevations expected between the northern and southern subtropical gyres is missing. These data were recorded during a moderate El Niño event (which will be explained shortly), so the high stand of equatorial water is likely due to a well-developed equatorial countercurrent.

Very little definition between the two Pacific Ocean gyres exists in part because the South Pacific subtropical gyre is less intense than other gyres. It is less intense because it covers such a large area, it lacks confinement by continental barriers along its western margin, and it has numerous islands (really the tops of tall sea floor mountains). The South Indian Ocean hill is rather well developed, although its northeastern boundary stands high because of the influx of warm Pacific Ocean water through the East Indies islands.

[7]A good analogy for this phenomenon is a funnel: In the narrow end of a funnel, the flow rates are speeded up (such as in western intensified currents); in the wide end, the flow rates are sluggish (such as in eastern boundary currents).

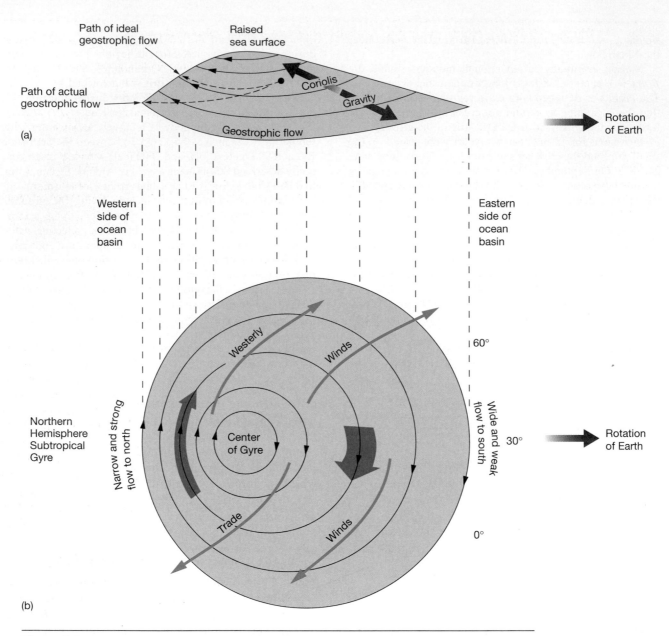

Figure 7–7 Geostrophic current and western intensification. **(a)** A cross-sectional view of a subtropical gyre showing how water literally piles up in the center, forming a hill up to 2 meters (6.6 feet) high. Gravity and the Coriolis effect balance to create an ideal geostrophic current that flows in equilibrium around the hill. However, friction makes the current gradually run downslope (*actual geostrophic flow*). **(b)** A map view of the same subtropical gyre, showing that the flow pattern is restricted (lines are closer together) on the western side of the gyre, resulting in western intensification.

Ocean Currents and Climate

Ocean surface currents directly influence the climate of adjoining landmasses. For instance, warm ocean currents warm the nearby air. This warm air can hold more water vapor, which puts more moisture (high humidity) in the atmosphere. When this warm, moist air travels over a continent, it releases its water vapor in the form of precipitation. Continental margins that have warm ocean currents offshore (Figure 7–8, *red arrows*) typically have a humid climate. The presence of a warm cur-

rent off the east coast of the United States helps explain why the area experiences such high humidity, especially in the summer.

Conversely, cold ocean currents cool the nearby air, which cannot hold as much water vapor. When the cool, dry air travels over a continent, it results in very little precipitation. Continental margins that have cool ocean currents offshore (Figure 7–8, *blue arrows*) typically have a dry climate. The presence of a cold current off California is part of the reason why it has such an arid climate.

Table 7–2   Characteristics of western and eastern boundary currents of subtropical gyres.

Current type (examples)	Width	Depth	Speed	Transport volume (millions of cubic meters per second[a])	Comments
Western boundary currents (Gulf Stream, Brazil Current, Kuroshio Current)	Narrow, usually less than 100 kilometers (60 miles)	Deep, to depths of 2 kilometers (1.2 miles)	Fast, hundreds of kilometers per day	Large, as much as 100 Sv[a]	Waters derived from low latitudes and are warm; little or no upwelling
Eastern boundary currents (Canary Current, Benguela Current, California Current)	Wide, up to 1000 kilometers (600 miles)	Shallow, to depths of 0.5 kilometer (0.3 mile)	Slow, tens of kilometers per day	Small, typically 10 to 15 Sv[a]	Waters derived from mid-latitudes and are cool; coastal upwelling common

[a]One million cubic meters per second is a flow rate equal to one Sverdrup (Sv).

Upwelling and Downwelling

Upwelling is the vertical movement of cold, deep, nutrient-rich water to the surface; **downwelling** is the vertical movement of surface water to deeper parts of the ocean. Upwelling hoists chilled water to the surface. This cold water, rich in nutrients, creates high **productivity** (an abundance of microscopic algae) that establishes the base of the food web and in turn, supports incredible numbers of larger marine life like fish and whales. Downwelling, on the other hand, is associated with much lower amounts of surface productivity but carries necessary dissolved oxygen to those organisms living on the deep sea floor.

Upwelling and downwelling provide important mixing mechanisms between surface and deep waters and are accomplished by a variety of methods.

Diverging Surface Water

Current divergence occurs when surface waters move *away from* an area on the ocean's surface, such as along the Equator. As shown in Figure 7–9, the South Equatorial Current occupies the area along the *geographical Equator* (most notably in the Pacific Ocean; see Figure 7–4), while the *meteorological Equator* (where the doldrums exist) typically occurs a few degrees of latitude to the north. As the southeast trade winds blow across this region, Ekman transport causes surface water north of the Equator to veer to the right (northward) and water south of the Equator to veer to the left (southward). The net result is a divergence of surface currents along the geographical Equator, which causes upwelling of cold, nutrient-rich water. Since this type of upwelling is common along the Equator—especially in the Pacific—it is called **equatorial upwelling** and creates areas of high productivity that are some of the most prolific fishing grounds in the world.

Converging Surface Water

Current convergence occurs when surface waters move *toward* each other. In the North Atlantic Ocean, for instance, the Gulf Stream, the Labrador Current, and the East Greenland Current all come together in the same vicinity. When currents converge, water stacks up and has no place to go but downward. The surface water slowly sinks in a process called downwelling (Figure 7–10). Unlike upwelling, areas of downwelling are not associated with prolific marine life because the necessary nutrients are not continuously replenished from the cold, deep, nutrient-rich water below the surface. Consequently, downwelling areas have low productivity.

Coastal Upwelling and Downwelling

Coastal winds can cause upwelling or downwelling due to Ekman transport. Figure 7–11 shows a coastal region along the west coast of a continent in the Southern Hemisphere with winds moving parallel to the coast. If the winds are from the south (Figure 7–11a), Ekman transport moves the coastal water to the left of the wind direction, causing the water to flow *away from* the shoreline. Water rises from below to replace the water moving away from shore in a process called **coastal upwelling**. Areas where coastal upwelling occurs, such as the West Coast of the United States, are characterized by high concentrations of nutrients, resulting in high biological productivity and rich marine life. This coastal upwelling also creates low water temperatures in areas such as San Francisco that provide a natural form of air conditioning (and much cool weather and fog) in the summer.

If the winds are from the north, Figure 7–11b shows that Ekman transport still moves the coastal water to the left of the wind direction but, in this case, the water flows *toward* the shoreline. This causes the water to stack up along the shoreline, where it has nowhere to go

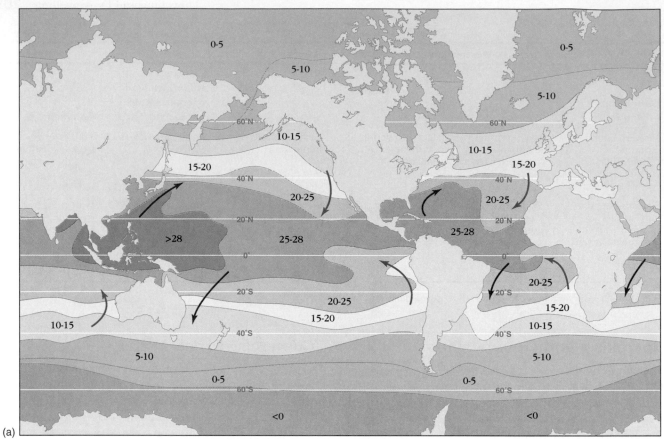

(a)

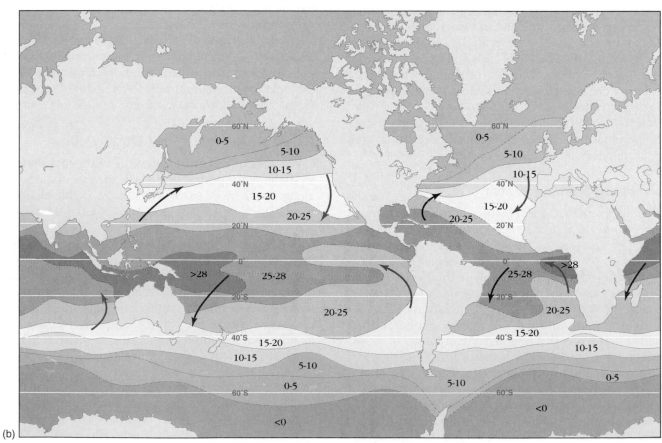

(b)

Figure 7–8 Surface temperature of the world ocean. Average sea surface temperature distribution in degrees centigrade for August **(a)** and for February **(b).** Note that temperatures migrate north–south with the seasons. Red arrows indicate warm surface currents; blue arrows indicate cool surface currents.

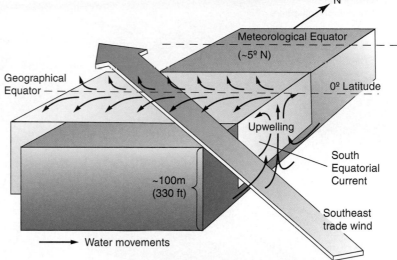

Figure 7–9 Equatorial upwelling. As the southeast trade winds pass over the geographical Equator to the meteorological Equator, they cause water within the South Equatorial Current north of the Equator to veer to the right (northward) and water south of the Equator to veer to the left (southward). Thus, surface water diverges, which causes equatorial upwelling.

but down, in a process called **coastal downwelling**. Areas where coastal downwelling occurs have low productivity. Coastal downwelling can occur in areas that typically experience coastal upwelling when the winds reverse.

Other Upwelling

Figure 7–12 shows how upwelling can be created by offshore winds, sea floor obstructions, or a sharp bend in a coastline. Upwelling also occurs in high-latitude regions, where there is no pycnocline (a layer of rapidly changing density). The absence of a pycnocline allows significant vertical mixing between high-density cold surface water and high-density cold deep water below. Thus, both upwelling and downwelling are common in high latitudes.

> Upwelling and downwelling cause vertical mixing between surface and deep waters. Upwelling brings cold, deep, nutrient-rich water to the surface, which results in high productivity.

Surface Currents of the Oceans

The pattern of surface currents varies from ocean to ocean depending upon the geometry of the ocean basin, the pattern of major wind belts, seasonal factors, and other periodic changes.

Antarctic Circulation

Antarctic circulation is dominated by the movement of water masses in the southern Atlantic, Indian, and Pacific Oceans, south of 50 degrees south latitude. The **Antarctic Convergence** of water occurs near this latitude, marking the northern boundary of the Southern or Antarctic Ocean (Figure 7–13). The **East Wind Drift**, a surface current propelled by the polar easterlies, moves in an easterly direction around the margin of the Antarctic continent. The East Wind Drift is most extensively developed to the east of the Antarctic Peninsula in the Weddell Sea region and in the area of the Ross Sea (Figure 7–13).

The main current in Antarctic waters is the **West Wind Drift**, which encircles Antarctica and flows from west to east at approximately 50 degrees south latitude but varies between 40 and 65 degrees south latitude. It is driven by the powerful prevailing westerly wind belt, which creates winds so strong that these latitudes have been called the "Roaring Forties," "Furious Fifties," and "Screaming Sixties."

The West Wind Drift (also called the *Antarctic Circumpolar Current*) is the only current that completely circumscribes Earth and is allowed to do so because of

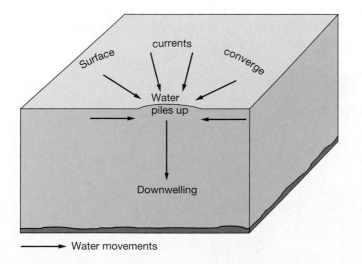

Figure 7–10 Downwelling caused by convergence of surface currents. Where surface currents converge, water piles up and slowly sinks downward, creating downwelling.

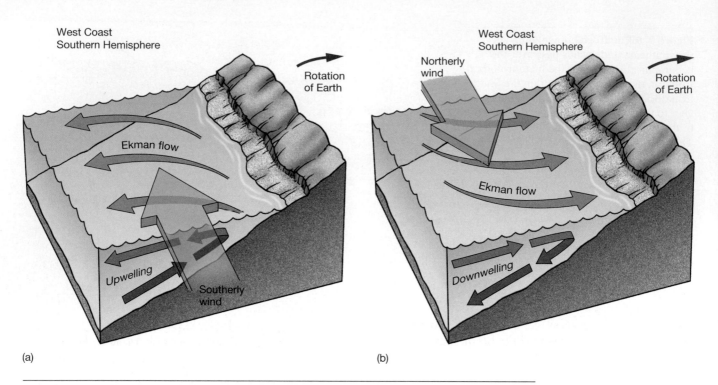

(a)

(b)

Figure 7–11 Coastal upwelling and downwelling. (a) Where southerly coastal winds blow parallel to a west coast in the Southern Hemisphere, Ekman transport carries surface water away from the continent. Upwelling of deeper water occurs to replace the surface water that has moved away from the coast. **(b)** A reversal of the direction of the winds that cause upwelling causes water to pile up against the shore and causes downwelling.

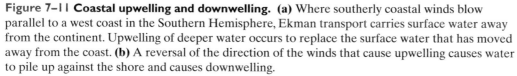

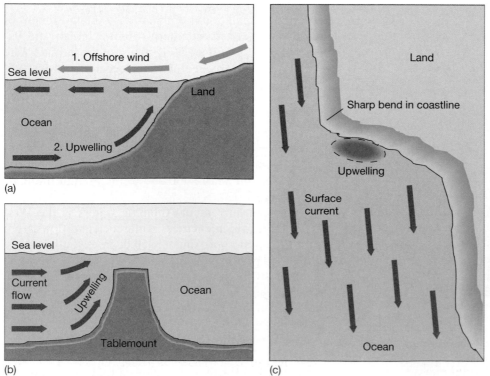

(a)

(b)

(c)

Figure 7–12 Other types of upwelling. Upwelling can be caused by: **(a)** Offshore winds; **(b)** A sea floor obstruction, in this case, a tablemount; **(c)** A sharp bend in coastal geometry.

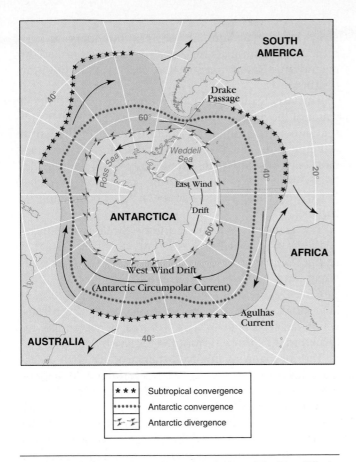

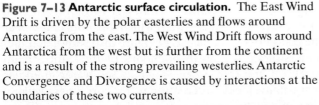

```
* * *   Subtropical convergence
•••••  Antarctic convergence
⊥—⊥  Antarctic divergence
```

Figure 7–13 Antarctic surface circulation. The East Wind Drift is driven by the polar easterlies and flows around Antarctica from the east. The West Wind Drift flows around Antarctica from the west but is further from the continent and is a result of the strong prevailing westerlies. Antarctic Convergence and Divergence is caused by interactions at the boundaries of these two currents.

the lack of land at high southern latitudes. It meets its greatest restriction as it passes through the Drake Passage (named for explorer Sir Francis Drake) between the Antarctic Peninsula and the southern islands of South America, which is about 1000 kilometers (600 miles) wide. Although the current is not speedy [its maximum surface velocity is about 2.75 kilometers (1.65 miles) per hour], it transports more water (an average of about 130 million cubic meters per second[8]) than any other surface current.

Two main currents flow around Antarctica in opposite directions. Recall that the Coriolis effect deflects moving masses to the left in the Southern Hemisphere, so the East Wind Drift is deflected toward the continent and the West Wind Drift is deflected away from it. This creates the **Antarctic Divergence**, which occurs between the two currents. The Antarctic Divergence has abundant marine life in the Southern Hemisphere summer because of the upwelling and mixing of these two currents, which supplies nutrients.

---

[8]One million cubic meters per second is a useful flow rate for describing ocean currents, so it has become a standard unit, named the **Sverdrup (Sv)** after Norwegian explorer Otto Sverdrup.

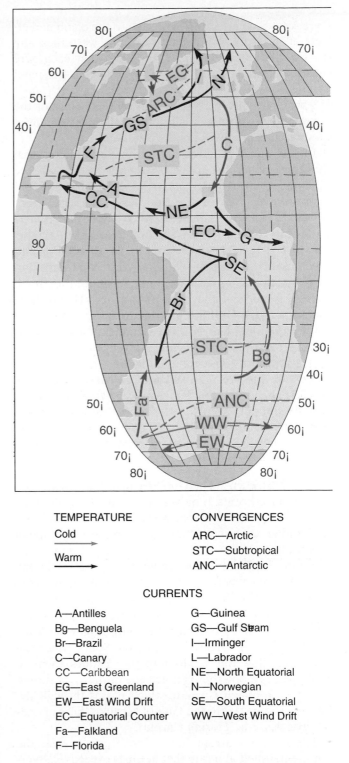

TEMPERATURE	CONVERGENCES
Cold ⟶	ARC—Arctic
	STC—Subtropical
Warm ⟶	ANC—Antarctic

CURRENTS

A—Antilles	G—Guinea
Bg—Benguela	GS—Gulf Stream
Br—Brazil	I—Irminger
C—Canary	L—Labrador
CC—Caribbean	NE—North Equatorial
EG—East Greenland	N—Norwegian
EW—East Wind Drift	SE—South Equatorial
EC—Equatorial Counter	WW—West Wind Drift
Fa—Falkland	
F—Florida	

Figure 7–14 Atlantic Ocean surface currents. Atlantic Ocean surface circulation is composed primarily of two subtropical gyres.

Atlantic Ocean Circulation

Figure 7–14 shows Atlantic Ocean surface circulation, which consists of two large subtropical gyres.

The North and South Atlantic Gyres The **North Atlantic Gyre** rotates clockwise and the **South Atlantic**

Gyre rotates counterclockwise, due to the combined effects of the trade winds, the prevailing westerlies, and the Coriolis effect. Figure 7–14 shows that each gyre consists of a poleward-moving warm current (*red*) and an equatorward-moving cold "return" current (*blue*). The two gyres are partially offset by the shapes of the surrounding continents and the **Atlantic Equatorial Countercurrent** moves in between them.

In the South Atlantic Gyre, the **South Equatorial Current** reaches its greatest strength just below the Equator, where it encounters the coast of Brazil and splits in two. Part of the South Equatorial Current moves off along the northeastern coast of South America toward the Caribbean Sea and the North Atlantic. The rest is turned southward as the **Brazil Current**, which ultimately merges with the West Wind Drift and moves eastward across the South Atlantic. The Brazil Current is much smaller than its Northern Hemisphere counterpart, the Gulf Stream, due to the splitting of the South Equatorial Current. The **Benguela Current**, slow-moving and cold, flows toward the Equator along Africa's western coast, completing the gyre.

Outside the gyre, the **Falkland Current** (Figure 7–14), which is also called the *Malvinas Current*, moves a significant amount of cold water along the coast of Argentina as far north as 25 to 30 degrees south latitude, wedging its way between the continent and the southbound Brazil Current.

The Gulf Stream

The **Gulf Stream** is the best studied of all ocean currents. It moves northward along the East Coast of the United States, warming coastal states and moderating winters in these and northern European regions.

Figure 7–15 shows the network of currents in the North Atlantic Ocean that contribute to the flow of the Gulf Stream. The **North Equatorial Current** moves parallel to the Equator in the Northern Hemisphere, where it is joined by the portion of the South Equatorial Current that turned northward along the South American coast. This flow then splits into the **Antilles Current**, which passes along the Atlantic side of the West Indies, and the **Caribbean Current**, which passes through the Yucatán Channel into the Gulf of Mexico. These masses reconverge as the **Florida Current**.

The Florida Current flows close to shore over the continental shelf at a rate that at times exceeds 35 Sverdrups. As it moves off North Carolina's Cape Hatteras and flows across the deep ocean in a northeasterly direction, it is called the Gulf Stream. The Gulf Stream is a western boundary current, so it is subject to western intensification. Thus, it is only 50 to 75 kilometers (31 to 47 miles) wide, but it reaches depths of 1.5 kilometers (1 mile), and speeds from 3 to 10 kilometers (2 to 6 miles) per hour, making it the fastest current in the world ocean.

The western boundary of the Gulf Stream is usually abrupt, but it periodically migrates closer to and farther away from the shore. Its eastern boundary is very difficult to identify because it is usually masked by meandering water masses that change their position continuously.

The Gulf Stream gradually merges eastward with the water of the **Sargasso Sea**. The Sargasso Sea is the water that circulates around the rotation center of the North Atlantic gyre. The Sargasso Sea, therefore, is the stagnant eddy of the North Atlantic Gyre. Its name is derived from a type of floating marine alga called *Sargassum* (*sargassum* = grapes) that abounds on its surface.

The transport rate of the Gulf Stream off Chesapeake Bay is about 100 Sverdrups[9], which suggests that a large volume of water from the Sargasso Sea has combined with the Florida Current to produce the Gulf Stream. By the time the Gulf Stream nears Newfoundland, however, the transport rate is only 40 Sverdrups, which suggests that a large volume of water has returned to the diffuse flow of the Sargasso Sea.

The mechanisms involved that produce such a dramatic loss of water are yet to be determined. Meanders, however, may cause much of it. *Meanders* (*Menderes* = a river in Turkey that has a very sinuous course) are snake-like bends in the current that often disconnect from the Gulf Stream and form large rotating masses of water called *vortexes* (*vertere* = to turn), which are more commonly known as *eddies* or *rings*. Figure 7–16 shows several of these rings, which are noticeable near the center of each image. The figure also shows that meanders along the north boundary of the Gulf Stream pinch off and trap warm Sargasso Sea water in eddies that rotate clockwise, creating **warm core rings** (*yellow*) surrounded by cooler (*blue and green*) water. These warm rings contain shallow, bowl-shaped masses of warm water about 1 kilometer (0.6 mile) deep, with diameters of about 100 kilometers (60 miles). Warm core rings remove large volumes of water as they disconnect from the Gulf Stream.

Cold near-shore water spins off to the south of the Gulf Stream as counterclockwise-rotating **cold core rings** (*green*) surrounded by warmer (*yellow and red-orange*) water (Figure 7–16). The cold rings consist of spinning cone-shaped masses of cold water that extend over 3.5 kilometers (2.2 miles) deep. These rings may exceed 500 kilometers (310 miles) in diameter at the surface. The diameter of the cone increases with depth, and sometimes reaches all the way to the sea floor, where they have a tremendous impact on seafloor sediment. Cold rings move southwest at speeds of 3 to 7 kilometers (2 to 4 miles) per day toward Cape Hatteras, where they often rejoin the Gulf Stream.

---

[9]The Gulf Stream's flow of 100 Sverdrups equates to a volume of about 100 major league football stadiums passing by the southeast U.S. coast *each second* and is more than 100 times greater than the combined flow of *all* the world's rivers!

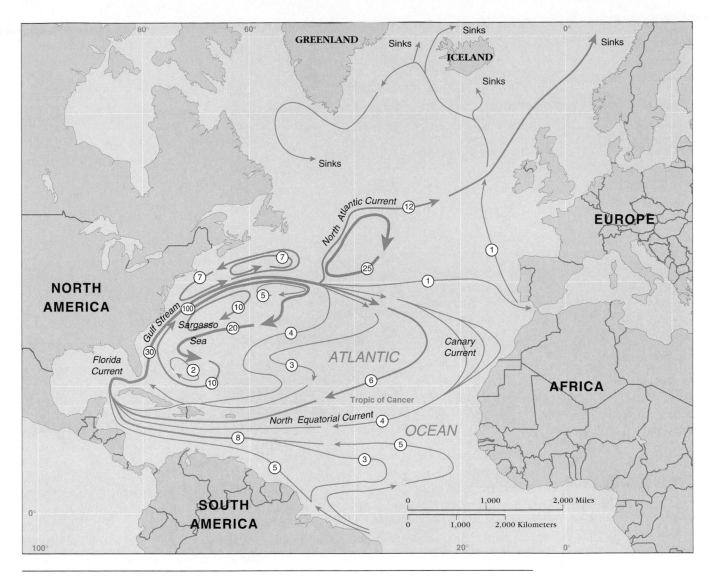

Figure 7–15 North Atlantic Ocean circulation. Average flow rate in Sverdrups (1 Sverdrup = 1 million cubic meters per second) for the North Atlantic Gyre. The four major currents include the western intensified Gulf Stream, the North Atlantic Current, the Canary Current, and the North Equatorial Current. Some water splits off in the North Atlantic, where it becomes cold and dense, so it sinks. The Sargasso Sea occupies the stagnant eddy in the middle of the subtropical gyre.

Southeast of Newfoundland, the Gulf Stream continues in an easterly direction across the North Atlantic (Figure 7–15). Here the Gulf Stream breaks into numerous branches, many of which become cold and dense enough to sink beneath the surface. As shown in Figure 7–14, one major branch combines the cold water of the **Labrador Current** with the warm Gulf Stream, producing abundant fog in the North Atlantic. This branch eventually breaks into the **Irminger Current**, which flows along Iceland's west coast, and the **Norwegian Current**, which moves northward along Norway's coast. The other major branch crosses the North Atlantic as the **North Atlantic Current** (also called the *North Atlantic Drift*, emphasizing its sluggish nature) and turns southward to become the cool **Canary Current**. The Canary Cur-

rent is a broad, diffuse southward flow that eventually joins the North Equatorial Current, thus completing the gyre.

Climatic Effects of North Atlantic Currents The warming effects of the Gulf Stream are far ranging. It not only moderates temperatures along the East Coast of the United States, but also in Northern Europe (in conjunction with heat transferred by the atmosphere). Thus, the temperatures across the Atlantic at different latitudes are much higher in Europe than in North America because of the effects of heat transfer from the Gulf Stream to Europe. For example, Spain and Portugal have warm climates yet they are at the same latitude as the New England states, which are known for severe winters. The warming that

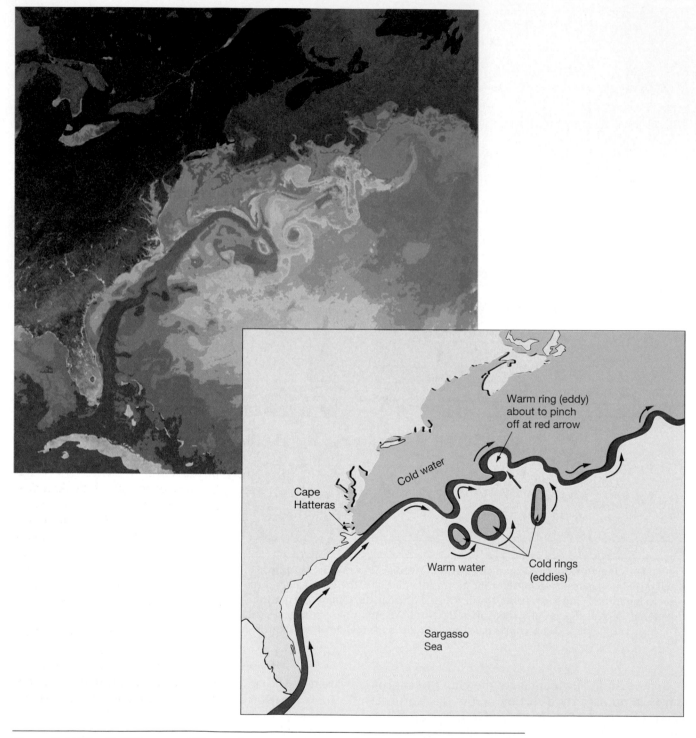

Figure 7–16 The Gulf Stream and sea surface temperatures. A NOAA satellite false-color image of sea surface temperature (*top*) and a schematic diagram of the same area (*bottom*). The warm waters of the Gulf Stream are shown in red and orange; colder waters are shown in green, blue, and purple. As the Gulf Stream meanders northward, some of its meanders pinch off and form either warm core or cold core rings.

Northern Europe experiences because of the Gulf Stream is as much as 9°C (20°F), which is enough to keep high-latitude Baltic ports ice free throughout the year.

The warming effects of western boundary currents in the North Atlantic Ocean can be seen on the aver- age sea-surface temperature map for February shown in Figure 7–8b. Off the east coast of North America from latitudes 20 degrees north (the latitude of Cuba) to 40 degrees north (the latitude of Philadel- phia), for example, there is a 20°C (36°F) difference in sea surface temperatures. On the eastern side of

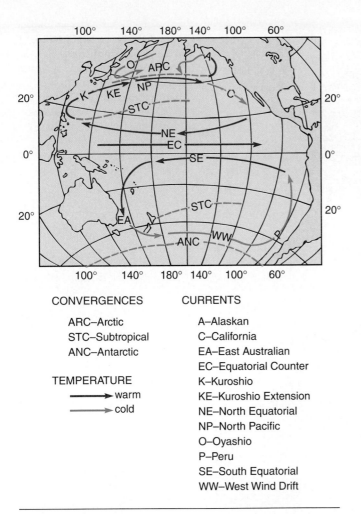

CONVERGENCES

ARC–Arctic
STC–Subtropical
ANC–Antarctic

TEMPERATURE

→ warm
→ cold

CURRENTS

A–Alaskan
C–California
EA–East Australian
EC–Equatorial Counter
K–Kuroshio
KE–Kuroshio Extension
NE–North Equatorial
NP–North Pacific
O–Oyashio
P–Peru
SE–South Equatorial
WW–West Wind Drift

Figure 7–17 Pacific Ocean surface currents. Similar to the Atlantic Ocean, the Pacific contains two large subtropical gyres. However, the equatorial countercurrent is more strongly developed than in smaller ocean basins.

the North Atlantic, on the other hand, there is only a 5°C (9°F) difference in temperature between the same latitudes, indicating the moderating effect of the Gulf Stream.

The average sea-surface temperature map for August (see Figure 7–8a) also shows how the North Atlantic and Norwegian Currents (branches of the Gulf Stream) warm northwestern Europe compared with the same latitudes along the North American coast. On the western side of the North Atlantic, the southward-flowing Labrador Current—which is cold and often contains icebergs from western Greenland—keeps Canadian coastal waters much cooler. During the Northern Hemisphere winter (Figure 7–8b), North Africa's coastal waters are cooled by the southward-flowing Canary Current and are much cooler than waters near Florida and the Gulf of Mexico.

Pacific Ocean Circulation

Two large subtropical gyres dominate the circulation pattern in the Pacific Ocean, resulting in surface water

movement and climatic effects similar to those found in the Atlantic. However, the Equatorial Countercurrent is much better developed in the Pacific Ocean than in the Atlantic (Figure 7–17), because the Pacific Ocean basin is larger and more unobstructed than the Atlantic Ocean basin.

Normal Conditions Figure 7–17 shows the **North Pacific Gyre**, which consists in part of the North Equatorial Current, which flows westward into the western intensified **Kuroshio Current**[10] near Asia. Because of its proximity to Japan, the Kuroshio Current is also called the *Japan Current*. Its warm waters make Japan's climate warmer than would be expected for its latitude. This current flows into the **North Pacific Current**, which connects to the cool-water **California Current**. The California Current flows south along the coast of California to complete the loop. Some North Pacific Current water also flows to the north and merges into the **Alaskan Current** in the Gulf of Alaska.

Figure 7–17 also shows the **South Pacific Gyre**, which consists in part of the South Equatorial Current, which flows westward into the western intensified **East Australian Current**.[11] From there, it joins the West Wind Drift, and completes the gyre as the **Peru Current** (also called the *Humboldt Current*, after German naturalist Friedrich Heinrich Alexander von Humboldt).

The cool water of the Peru Current has historically been one of Earth's richest fishing grounds. What conditions produce such an abundance of fish? Figure 7–18a shows that along the west coast of South America, coastal winds create Ekman transport that moves water away from shore, causing upwelling of cool, nutrient-rich water. This upwelling increases productivity and results in an abundance of marine life, including small silver-colored fish called *anchovetas* (anchovies) that become particularly plentiful near Peru and Ecuador. Anchovies provide a food source for many larger marine organisms and also supply Peru's commercial fishing industry, which was established in the 1950s. Anchovies are so abundant in the waters off South America that by 1970, Peru was the largest producer of fish from the sea in the world, with a peak production of 12.3 million metric tons (13.5 million short tons), accounting for about one-quarter of *all* fish from the sea worldwide.

Figure 7–18a shows that high pressure and sinking air dominates the coastal region of South America, resulting in clear, fair, and dry weather. On the other side of the Pacific, a low-pressure region and rising air creates cloudy conditions with plentiful precipitation in Indonesia, New Guinea, and northern Australia. This

---

[10]Kuroshio is pronounced "kuhr–ROH–shee–oh."

[11]Note that the western intensified East Australian Current was named because it lies off the *East Coast* of Australia, even though it occupies a position along the *western* margin of the Pacific Ocean basin.

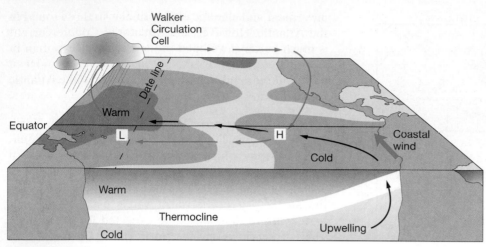

(a) Normal conditions

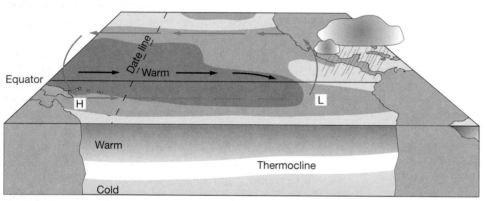

(b) El Niño conditions

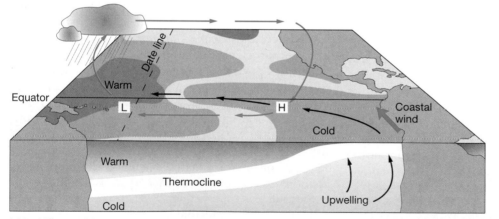

(c) La Niña conditions

Figure 7–18 Normal, El Niño, and La Niña conditions.
(a) Normal oceanic and atmospheric conditions in the equatorial Pacific. **(b)** El Niño (ENSO warm phase) conditions. **(c)** La Niña (ENSO cool phase) conditions.

pressure difference causes the strong southeast trade winds to blow across the equatorial South Pacific. The resulting atmospheric circulation cell in the equatorial South Pacific Ocean is named the **Walker Circulation Cell** (*green arrows*) after Sir Gilbert T. Walker, the British meteorologist who first described the effect in the 1920s.

The southeast trade winds set ocean water in motion, which also moves across the Pacific towards the west.

The water warms as it flows in the equatorial region and creates a wedge of warm water on the western side of the Pacific Ocean, called the **Pacific Warm Pool** (see Figure 7–8). Due the movement of equatorial currents to the west, the Pacific Warm Pool is thicker along the western side of the Pacific than along the eastern side. The *thermocline* beneath the warm pool in the western equatorial Pacific occurs below 100 meters (330 feet) depth. In the eastern Pacific, however, the thermocline is within 30 meters (100 feet) of the surface. The difference in depth of the thermocline can be seen by the sloping boundary between the warm surface water and the cold deep water in Figure 7–18a.

El Niño–Southern Oscillation (ENSO) Conditions

Historically, Peru's residents knew that every few years, a current of warm water reduced the population of anchovies in coastal waters. The decrease in anchovies caused a dramatic decline not only in the fishing industry, but also in marine life such as sea birds and seals that depended on anchovies for food. The warm current also brought about changes in the weather—usually intense rainfall—and even brought such interesting items as floating coconuts from tropical islands near the Equator. At first, these events were called *años de abundancia* (years of abundance) because the additional rainfall dramatically increased plant growth on the normally arid land. What was once thought of as a joyous event, however, soon became associated with the ecological and economic disaster that is now a well-known consequence of the phenomenon.

This warm-water current usually occurred around Christmas and thus was given the name **El Niño**, Spanish for "the child," in reference to baby Jesus. In the 1920s, Walker was the first to recognize that an east-west atmospheric pressure seesaw accompanied the warm current and called the phenomenon the **Southern Oscillation**. Today, the combined oceanic and atmospheric effects are called **El Niño–Southern Oscillation (ENSO)**, which periodically alternate between warm and cold phases and cause dramatic environmental changes.

ENSO Warm Phase (El Niño)

Figure 7–18b shows the atmospheric and oceanic conditions during an ENSO warm phase, which is known as El Niño. The high pressure along the coast of South America weakens, reducing the difference between the high- and low-pressure regions of the Walker Circulation Cell. This, in turn, causes the southeast trade winds to diminish. In very strong El Niño events, the trade winds actually blow in the *reverse* direction.

Without the trade winds, the Pacific warm pool that has built up on the western side of the Pacific begins to flow back across the ocean toward South America. Aided by an increase in the flow of the Equatorial Countercurrent, the Pacific warm pool creates a band of warm water that stretches across the equatorial Pacific Ocean (Figure 7–19a). The warm water usually begins to move in September of an El Niño year, and reaches South America by December or January. During strong to very strong El Niños, the water temperature off Peru can increase up to 10°C (18°F) compared to normal conditions. In addition, the average sea level can increase as much as 20 centimeters (8 inches), simply due to thermal expansion of the warm water along the coast.

As the warm water increases sea surface temperatures across the equatorial Pacific, temperature-

sensitive corals are decimated in Tahiti, the Galápagos, and other tropical Pacific islands. In addition, many other organisms are affected by the warm water (Box 7–3). Once the warm water reaches South America, it moves north and south along the west coast of the Americas, increasing average sea level and the number of tropical hurricanes formed in the eastern Pacific.

The flow of warm water across the Pacific also causes the sloped thermocline boundary between warm surface waters and the cooler waters below to flatten out and become more horizontal (Figure 7–18b). Near Peru, upwelling brings warmer, nutrient-depleted water to the surface instead of cold, nutrient-rich water. In fact, *downwelling* can sometimes occur as the warm water stacks up along coastal South America. Productivity diminishes and most types of marine life in the area are dramatically reduced.

As the warm water moves to the east across the Pacific, the low-pressure zone also migrates. In a strong to very strong El Niño event, the low pressure can move entirely across the Pacific and remain over South America. The low pressure substantially increases precipitation along coastal South America. Conversely, high pressure replaces the Indonesian low, bringing dry conditions or, in strong to very strong El Niño events, drought conditions to Indonesia and northern Australia.

ENSO Cool Phase (La Niña)

In some instances, conditions opposite of El Niño prevail in the equatorial South Pacific, which is known as ENSO cool phase or La Niña (Spanish for "the female child"). Figure 7–18c shows La Niña conditions, which are similar to normal conditions but more intensified because there is a larger pressure difference across the Pacific Ocean. This larger pressure difference creates stronger Walker Circulation and stronger trade winds, which in turn causes more upwelling, a shallower thermocline in the eastern Pacific, and a band of cooler than normal water that stretches across the equatorial South Pacific (Figure 7–19b).

La Niña conditions commonly occur following an El Niño. For instance, the 1997–1998 El Niño was followed by several years of persistent La Niña conditions. The alternating pattern of El Niño–La Niña conditions since 1950 is shown by the multivariate **ENSO index** (Figure 7–20), which is calculated using a weighted average of atmospheric and oceanic factors including atmospheric pressure, winds, and sea surface temperatures. Positive ENSO index numbers indicate El Niño conditions whereas negative numbers reflect La Niña conditions. Normal conditions are indicated by a value near zero, and the greater the index value differs from zero (either negative or positive), the stronger the respective condition.

How Often do El Niño Events Occur?

Records of sea surface temperatures over the past 100 years reveal that throughout the twentieth century, El Niño

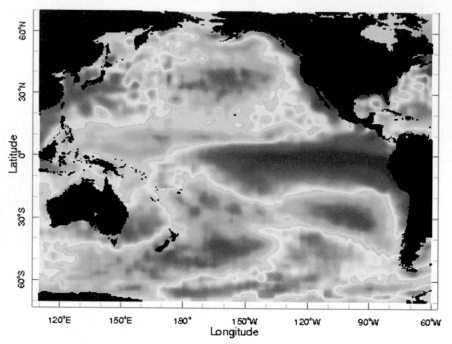

(a) Jan 1998

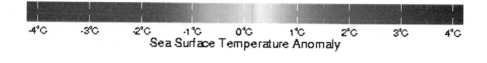

Sea Surface Temperature Anomaly

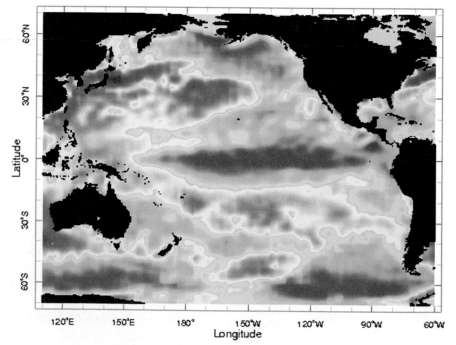

(b) Jan 2000

Figure 7–19 Sea-surface temperature anomaly maps. Maps showing sea-surface temperature anomalies, which represent departures from normal conditions. Red colors indicate water warmer than normal and blue colors represent water cooler than normal. **(a)** Map of the Pacific Ocean in January 1998, showing the anomalous warming during the 1997–1998 El Niño. **(b)** Map of the same area in January 2000, showing cooling in the equatorial Pacific related to La Niña.

conditions occur on average about every 2 to 10 years, but in a highly irregular pattern. In some decades, for instance, there has been an El Niño event every few years, while in others there may have been only one. Figure 7–20 shows the pattern since 1950, revealing that the

equatorial Pacific fluctuates between El Niño and La Niña conditions, with only a few years that could be considered "normal" conditions (represented by an ENSO index value close to zero). Typically, El Niño events last for 12–18 months and are followed by La

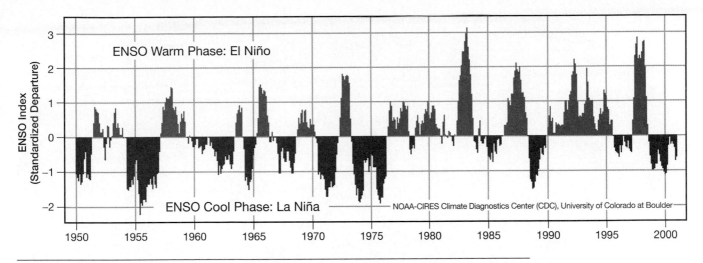

Figure 7–20 Multivariate ENSO index 1950–present. The multivariate ENSO index is calculated using a variety of atmospheric and oceanic factors. ENSO index values greater that zero (*red areas*) indicate El Niño conditions while ENSO index values less than zero (*blue areas*) indicate La Niña conditions. The greater the value is from zero, the stronger the corresponding El Niño or La Niña.

Niña conditions that exist for a similar length of time. However, some El Niño or La Niña conditions can last for several years.

El Niño events—especially severe ones—may occur more frequently as a result of increased global warming. For instance, the two most severe El Niño events in the 20th century occurred in 1982–1983 and 1997–1998. Presumably, increased ocean temperatures could more easily trigger more frequent and more severe El Niños. However, this pattern could also be a part of a long-term natural climate cycle. Recently, oceanographers have recognized a phenomenon called the **Pacific Decadal Oscillation (PDO)** that lasts 20 to 30 years and appears to influence Pacific sea surface temperatures. Analysis of TOPEX/Poseidon satellite date suggests that the Pacific Ocean has been in the warm phase of the PDO from 1977 to 1999 and that it has just entered the cool phase, which may suppress the initiation of El Niño events.

Effects of El Niños and La Niñas

Mild El Niño events influence only the equatorial South Pacific Ocean while strong to very strong El Niño events can influence worldwide weather. Typically, stronger El Niños alter the atmospheric jet steam and produce unusual weather in most parts of the globe. Sometimes the weather is drier than normal; at other times, it is wetter. The weather may also be warmer or cooler than normal. It is still difficult to predict exactly how a particular El Niño will affect any region's weather.

Figure 7–21 shows how very strong El Niño events can result in flooding, erosion, droughts, fires, tropical storms, and effects on marine life worldwide. These weather perturbations also affect the production of corn, cotton, and coffee. More locally, the satellite images in Figure 7–22 show that sea surface temperatures off southern California are significantly higher during an El Niño year.

Even though severe El Niños are typically associated with vast amounts of destruction, they can be beneficial in some areas. Tropical hurricane formation, for instance, is generally suppressed in the Atlantic Ocean, some desert regions receive much-needed rain, and organisms adapted to warm-water conditions thrive in the Pacific.

La Niña events are associated with sea surface temperatures and weather phenomena opposite to those of El Niño. Indian Ocean monsoons, for instance, are typically drier than usual in El Niño years but wetter than usual in La Niña years.

Examples from Recent El Niños

Recent El Niños provide an indication of the variability of the effects of El Niño events. For instance, in the winter of 1976, a moderate El Niño event coincided with northern California's worst drought of the last century, showing that El Niño events don't always bring torrential rains to the western United States. During that same winter, the eastern United States experienced record cold temperatures.

The 1982–1983 El Niño is the strongest ever recorded, causing far-ranging effects across the globe. Not only was there anomalous warming in the tropical Pacific, but the warm water spread along the coast of North America, influencing sea surface temperatures from California (Figure 7–22) to Alaska. Sea level was higher than normal (due to thermal expansion of the water), which, when high surf was experienced, caused

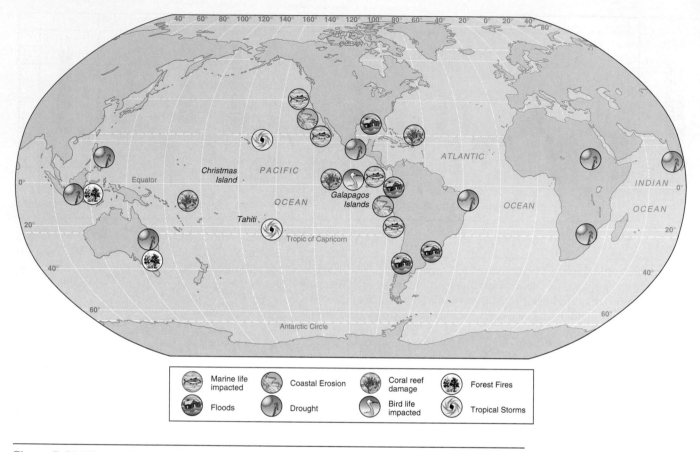

Figure 7–21 Effects of severe El Niños. Flooding, erosion, droughts, fires, tropical storms, and effects on marine life are all associated with severe El Niño events.

damage to coastal structures and increased coastal erosion. In addition, the jet stream swung much farther south than normal across the United States, bringing a series of powerful storms that resulted in three times normal rainfall across the southwestern United States. The increased rainfall caused severe flooding and landslides as well as higher than normal snowfall in the Rocky Mountains. Alaska and western Canada had a relatively warm winter, and the eastern United States had its mildest winter in 25 years.

The full strength of El Niño was experienced in western South America. Normally arid Peru was drenched with more than 3 meters (10 feet) of rain, causing extreme flooding and landslides. Sea surface temperatures were so high and for so long that temperature-sensitive coral reefs across the equatorial Pacific were decimated. Marine mammals and sea birds, which depend on the food normally available in the highly productive waters along the west coast of South America, went elsewhere or died. In the Galápagos Islands, for example, over half of the island's seals and sea lions died of starvation during the 1982–1983 El Niño.

French Polynesia had not experienced a hurricane in 75 years; in 1983, it endured six. The Hawaiian Island of Kauai also experienced a rare hurricane. Meanwhile, in Europe, severe cold weather prevailed. Worldwide,

droughts occurred in Australia, Indonesia, China, India, Africa, and Central America. In all, over 2000 deaths and at least $10 billion in property damage ($2.5 billion in the United States) were attributed to the 1982–1983 El Niño event.

The 1997–1998 El Niño event began several months earlier than normal and peaked in January 1998. The amount of Southern Oscillation and sea surface warming in the equatorial Pacific was initially as strong as the 1982–1983 El Niño, which caused a great deal of concern. However, the 1997–1998 El Niño weakened in the last few months of 1997 before reintensifying in early 1998. The impact of the 1997–1998 El Niño was felt mostly in the tropical Pacific, where surface water temperatures in the eastern Pacific averaged more than 4°C (7°F) warmer than normal, and, in some locations, reached up to 9°C (16°F) above normal (see Figure 7–19a). High pressure in the western Pacific brought drought conditions that caused wildfires to burn out of control in Indonesia. Also, the warmer than normal water along the west coast of Central and North America increased the number of hurricanes off Mexico.

In the United States, the 1997–1998 El Niño caused killer tornadoes in the southeast, massive blizzards in the upper Midwest, and flooding in the Ohio River Valley. Most of California received twice the normal

Figure 7–22 Sea-surface temperatures off southern California. Sea-surface temperature maps (in °C) for southern California from data collected by the satellite-mounted Advanced Very High Resolution Radiometer. Blue is cold water; red is warm water. **(a)** Water temperatures in January 1982, a non-El Niño year. **(b)** Water temperatures a year later, during an El Niño event.

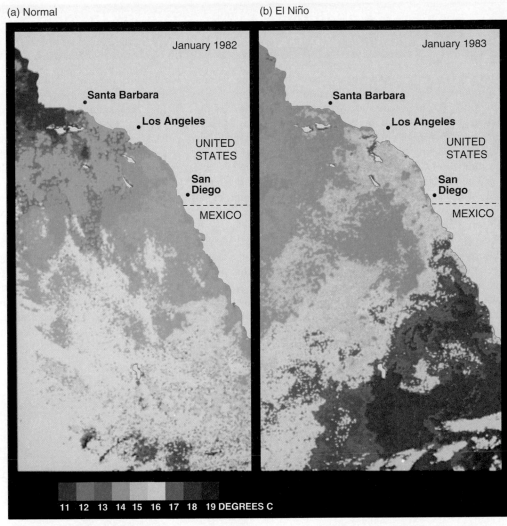

(a) Normal

(b) El Niño

rainfall, which caused flooding and landslides in many parts of the state. The lower Midwest, the Pacific Northwest, and the eastern seaboard, on the other hand, had relatively mild weather. In all, the 1997–1998 El Niño caused 2100 deaths and $33 billion in property damage worldwide.

Predicting El Niño Events   The 1982–1983 El Niño event was not predicted, nor was it recognized until it was near its peak. Because it affected weather worldwide and caused such extensive damage, the **Tropical Ocean—Global Atmosphere (TOGA)** program was initiated in 1985 to study how El Niño events developed. The goal of the TOGA program was to monitor the equatorial South Pacific Ocean during El Niño events to enable scientists to model and predict future El Niño events. The 10-year program studied the ocean from research vessels, analyzed surface and subsurface data from radio-transmitting sensor buoys, monitored oceanic phenomena by satellite, and developed computer models.

These models made it possible to predict the El Niño events of 1987–1988, 1991–1995, and 1997–1998 as much as one year in advance. Since the completion of TOGA, the **Tropical Atmosphere and Ocean (TAO)** project

(sponsored by the United States, Canada, Australia, and Japan) has continued to monitor the equatorial Pacific Ocean with a series of 70 moored buoys, providing real-time information about the conditions of the tropical Pacific that is available on the Internet. Although monitoring has improved, the causes of El Niño events are still not fully understood.

> El Niño is a combined oceanic–atmospheric phenomenon that occurs periodically in the tropical Pacific Ocean, bringing warm water to the east. La Niña describes conditions opposite of El Niño.

Indian Ocean Circulation

From November to March, equatorial circulation in the Indian Ocean is similar to that in the other oceans, with two westward-flowing equatorial currents (North and South Equatorial Currents) separated by an eastward-flowing Equatorial Countercurrent. Unlike the Atlantic and Pacific systems, however, the Equatorial Countercurrent flows between 2 and 8 degrees south of the Equator instead of north. This flow occurs because the

Box 7–3 El Niño and the Incredible Shrinking Marine Iguanas of the Galápagos Islands

The world's only marine lizards are the marine iguanas (*Amblyrhynchus cristatus*) of Ecuador's Galápagos Islands (Figure 7D). The iguanas are vegetarians that eat only marine algae and have adaptations that enable them to spend long periods foraging for food in the ocean. The islands they inhabit are severely affected by El Niño events, which cause periodic food shortages for the iguanas.

When El Niño conditions occur, warm water from the Pacific Warm Pool moves to the east along the Equator. This change, accompanied by a decrease in upwelling along the eastern Pacific, causes ocean surface temperatures to rise by as much as 10°C (18°F). In the Galápagos Islands, these conditions cause severe hardship for species that like cooler water temperatures, such as green and red algae that are the iguana's preferred food. Instead, these types of algae are replaced by brown algae, which are harder for the iguanas to digest. Unlike many marine animals that can migrate to other areas where food supplies are plentiful, the iguanas are confined to the islands where food supplies become limited.

During a severe El Niño, up to 90% of the iguana population can die of starvation.

Two studies covering 8 and 18 years reveal that the iguana's main adaptation for this lack of food is to shrink in size, allowing them to utilize meager food supplies more efficiently. During the 1997–1998 El Niño, for example, Galápagos marine iguanas shrank by as much as 20%, with larger animals shrinking the most. Also, females shrank more than similar-sized males, probably as a result of the additional energy females expended producing eggs the previous year.

About half of the shrinkage can be attributed to decreases in the mass of cartilage and connective tissue, while bone absorption may account for the remainder. In addition, the iguanas don't forage for food and thus get little exercise during El Niño events, which may result in additional shrinkage (similar to the decrease in weight associated with inactivity experienced by astronauts that spend long periods in weightlessness).

Figure 7D Marine iguana of the Galápagos Islands.

Following El Niño conditions, upwelling is reestablished, which creates cooler surface water temperatures. As a result, the iguana's preferred food supply again becomes abundant and the iguanas grow back to normal size. Remarkably, the marine iguanas of the Galápagos Islands repeatedly shrink and regrow in response to El Niño-induced changes in the environment.

Indian Ocean lies mostly in the Southern Hemisphere (extending only to about 20 degrees north latitude).

The winds of the northern Indian Ocean have a seasonal pattern called the **monsoon** (*mausim* = season) winds. During winter, air over the Asian mainland rapidly cools, creating high pressure, which forces atmospheric masses off the continent and out over the ocean (*green arrows* in Figure 7–23a). These northeast trade winds are called the *northeast monsoon.*

During summer, the winds reverse. Because of the lower heat capacity of rocks and soil compared with water, the Asian mainland warms faster than the adjacent ocean creating low pressure over the continent. This forces air over the Indian Ocean onto the Asian landmass, giving rise to the *southwest monsoon* (*green arrows* in Figure 7–23b), which may be thought of as a continuation of the southeast trade winds across the Equator.

The North Equatorial Current disappears during the summer and is replaced by the *Southwest Monsoon Current.* It flows from west to east across the North Indian Ocean. The **Somali Current**, which flows northward from the Equator along the coast of Africa with velocities approaching 4 kilometers (2.5 miles) per hour, feeds the Southwest Monsoon Current. In September or October, the northeast trade winds are reestablished, and the North Equatorial Current reappears (Figure 7–23a).

Surface circulation in the southern Indian Ocean (the **Indian Ocean Gyre**) is similar to subtropical gyres observed in other southern oceans. When the northeast trade winds blow, the South Equatorial Current provides water for the Equatorial Countercurrent and the **Agulhas Current**, which flows southward along Africa's eastern coast and joins the West Wind Drift. Turning northward out of the West Wind Drift is the **West Australian Current**, an eastern boundary current that merges with the South Equatorial Current, completing the gyre.

Eastern boundary currents in other subtropical gyres are cold drifts toward the Equator that produce arid coastal climates [that is, they receive less than 25 centimeters (10 inches) of rain per year]. In the southern Indian Ocean, however, the **Leeuwin Current** displaces the West Australian Current offshore. The Leeuwin Current is driven southward along the Australian coast from the warm-water dome piled up in the East Indies by the Pacific equatorial currents.

The Leeuwin Current produces a mild climate in southwestern Australia, which receives about 125 centimeters (50 inches) of rain per year. During El Niño events, however, the Leeuwin Current weakens, so the cold Western Australian Current brings drought instead.

Deep Currents

Deep currents occur in the deep zone below the pycnocline, so they influence about 90% of all ocean water. Density differences create deep currents. Although these density differences are usually small, they are large enough to cause denser waters to sink. Because the density variations that cause deep ocean circulation are caused by differences in temperature and salinity, deep ocean circulation is also referred to as **thermohaline** (*thermo* = heat, *haline* = salt) **circulation**.

Origin of Thermohaline Circulation

Recall from Chapter 5 that an increase in seawater density can be caused by a *decrease* in temperature or an *increase* in salinity. Temperature, though, has the greater influence on density. Density changes due to salinity are important only in very high latitudes, where water temperature remains low and relatively constant.

Most water involved in deep-ocean currents (thermohaline circulation) begins in high latitudes *at the surface.* In these regions, surface water becomes cold and its salinity increases as sea ice forms. When this surface water becomes dense enough, it sinks, initiating deep-ocean currents. Once this water sinks, it is removed from the physical processes that increased its density in the first place, and so its temperature and salinity remain largely unchanged for the duration it spends in the deep ocean. Thus, a **temperature–salinity (T–S) diagram** can be used to identify deep water masses based on their characteristic temperature, salinity, and resulting density. Figure 7–24 shows a T–S diagram for the North Atlantic Ocean.

As these surface-water masses become dense and are sinking (downwelling) in high-latitude areas, deep-water masses are also rising to the surface (upwelling). Because the water temperature in high latitude regions is the same at the surface as it is down below, the water column is isothermal, there is no thermocline and associated pycnocline (see Chapter 5), and upwelling and downwelling can easily be accomplished.

Deep-water currents move larger volumes of water and are much slower than surface currents. Typical speeds of deep currents range from 10 to 20 kilometers (6 to 12 miles) per *year.* Thus, it takes a deep current *an*

WINTER: November–March,
Northeast monsoon wind season

SUMMER: May–September,
Southwest monsoon wind season

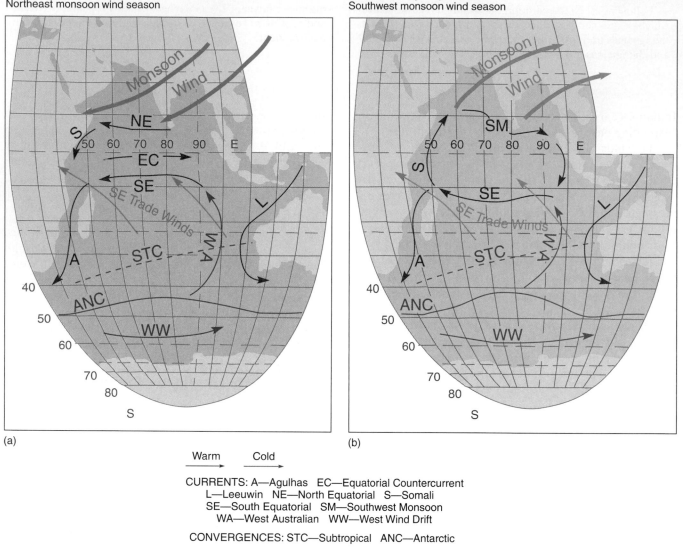

Warm ⟶   Cold ⟶

CURRENTS: A—Agulhas   EC—Equatorial Countercurrent
L—Leeuwin   NE—North Equatorial   S—Somali
SE—South Equatorial   SM—Southwest Monsoon
WA—West Australian   WW—West Wind Drift

CONVERGENCES: STC—Subtropical   ANC—Antarctic

Figure 7–23 Indian Ocean surface currents. Surface currents in the Indian Ocean are
influenced by the seasonal monsoons.

entire year to travel the same distance that a western intensified surface current can move in *one hour*.

Sources of Deep Water

In southern subpolar latitudes, huge masses of deep water
form beneath sea ice along the margins of the Antarctic
continent. Here, rapid winter freezing produces very cold,
high-density water that sinks down the continental slope
of Antarctica and becomes **Antarctic Bottom Water**, the
densest water in the open ocean (Figure 7–25). Antarctic
Bottom Water slowly sinks beneath the surface and
spreads into all the world's ocean basins, eventually returning to the surface perhaps 1000 years later.

In the northern subpolar latitudes, large masses of
deep water form in the Norwegian Sea. From there, it
flows as a subsurface current into the North Atlantic,
where it becomes part of the **North Atlantic Deep
Water**. North Atlantic Deep Water also comes from the
margins of the Irminger Sea off southeastern Green-

land, the Labrador Sea, and the dense, salty Mediterranean Sea. Like Antarctic Bottom Water, North Atlantic Deep Water spreads throughout the ocean basins.
It is less dense, however, so it layers on top of the
Antarctic Bottom Water (Figure 7–25).

Surface-water masses converge within the subtropical gyres and in the Arctic and Antarctic. Subtropical
convergences do not produce deep water, however, because the density of warm surface waters is too low for
them to sink. Major sinking does occur, however, along
the **Arctic Convergence** and Antarctic Convergence
(Figure 7–25, *inset*). The deep water mass formed from
sinking at the Antarctic convergence is called the
Antarctic Intermediate Water mass (Figure 7–25). Scientists have not yet thoroughly studied Antarctic Intermediate Water, so it remains a true frontier of
knowledge, awaiting further exploration.

Figure 7–25 also shows that the highest-density water
is found along the ocean bottom, with less-dense water
above. In low-latitude regions, the boundary between

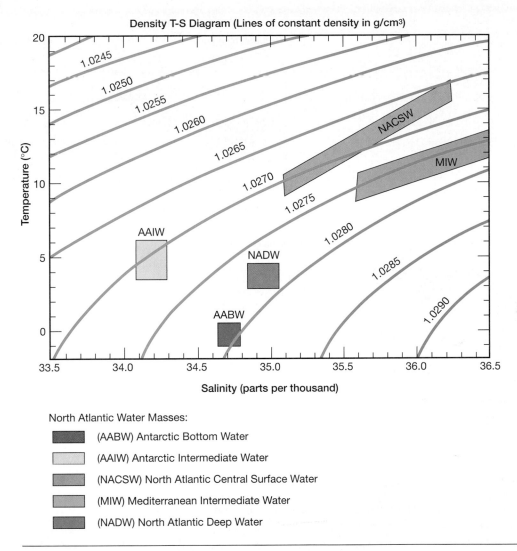

Density T-S Diagram (Lines of constant density in g/cm³)

North Atlantic Water Masses:

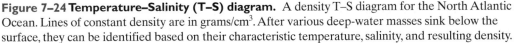

- (AABW) Antarctic Bottom Water
- (AAIW) Antarctic Intermediate Water
- (NACSW) North Atlantic Central Surface Water
- (MIW) Mediterranean Intermediate Water
- (NADW) North Atlantic Deep Water

Figure 7–24 Temperature–Salinity (T–S) diagram. A density T–S diagram for the North Atlantic Ocean. Lines of constant density are in grams/cm³. After various deep-water masses sink below the surface, they can be identified based on their characteristic temperature, salinity, and resulting density.

the warm surface water and the deeper cold water is marked by a prominent thermocline and corresponding pycnocline that prevent vertical mixing. There is no pycnocline in high-latitude regions, so substantial vertical mixing (upwelling and downwelling) occurs.

This same general pattern of layering based on density occurs in the Pacific and Indian Oceans as well. They have no source of Northern Hemisphere deep water, however, so they lack a deep-water mass. In the northern Pacific Ocean, the low salinity of surface waters prevents it from sinking into the deep ocean. In the northern Indian Ocean, surface waters are too warm to sink. **Oceanic Common Water**, which is created when Antarctic Bottom Water and North Atlantic Deep Water mix, lines the bottoms of these basins.

Worldwide Deep-Water Circulation

For every liter of water that sinks from the surface into the deep ocean, a liter of deep water must return to the surface somewhere else. It is difficult to identify specifically *where* this vertical flow to the surface is occurring.

It is generally believed that it occurs as a gradual, uniform upwelling throughout the ocean basins. It may be somewhat greater in low-latitude regions, where surface temperatures are higher.

Figure 7–26 shows a general deep-water circulation model of the world ocean. The most intense deep-water flow in each ocean basin is along the western side, due to the Coriolis effect and bathymetric features along the sea floor (such as the mid-ocean ridge).

Conveyer-Belt Circulation   An integrated model combining deep thermohaline circulation and surface currents is shown in Figure 7–27. Because the overall circulation pattern resembles a large conveyer belt, the model is called **conveyer-belt circulation**. Beginning in the North Atlantic, surface water carries heat to high latitudes via the Gulf Stream. During the cold winter months, this heat is transferred to the overlying atmosphere, warming northern Europe.

Cooling in the North Atlantic increases the density of this surface water to the point where it sinks to the bottom and flows southward, initiating the lower limb

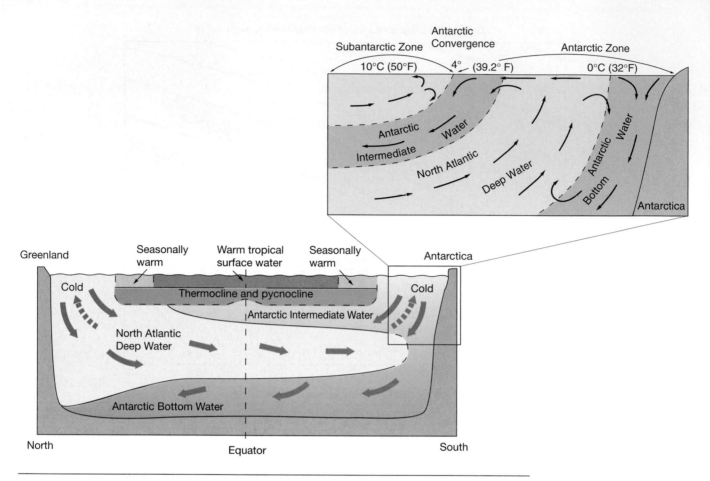

Figure 7–25 Atlantic Ocean subsurface water masses. Schematic diagram of the various water masses in the Atlantic Ocean. Similar but less distinct layering based on density occurs in the Pacific and Indian Oceans as well. Upwelling and downwelling occurs in the North Atlantic and near Antarctica (*inset*), creating deep-water masses.

of the "conveyor." Here, seawater flows downward at a rate equal to 100 Amazon Rivers and begins its long journey into the deep basins of all of the world's oceans. This limb extends all the way to the southern tip of Africa, where it joins the deep water that encircles Antarctica. The deep water that encircles Antarctica includes deep water that descends along the margins of the Antarctic continent. This mixture of deep waters flows northward into the deep Pacific and Indian Ocean basins, where it eventually surfaces and completes the conveyer belt by flowing west and then north again into the North Atlantic Ocean.

Dissolved Oxygen in Deep Water Cold water can dissolve more oxygen than warm water. Thus, deep-water circulation brings dense, cold, oxygen-enriched water from the surface to the deep ocean. During its time in the deep ocean, deep water becomes enriched in nutrients, as well, due to decomposition of dead organisms and the lack of organisms using nutrients there.

At various times in the geologic past, warmer water probably constituted a larger proportion of deep oceanic waters. As a result, the oceans had a lower oxygen concentration than today because warm water

cannot hold as much oxygen. In essence, the oxygen content of the oceans has probably fluctuated widely throughout time.

If high-latitude surface waters did not sink and eventually return from the deep sea to the surface, the distribution of life in the sea would be considerably different. There would be no life in the deep ocean, for instance, because there would be no oxygen for organisms to breathe. Life in surface waters would be significantly reduced and confined to the extreme margins of the oceans, where the only source of oxygen and nutrients would be runoff from streams.

Thermohaline Circulation and Climate Change
Evidence from deep-sea sediments and recently developed computer models indicate that changes in the global deep-water circulation pattern can dramatically and abruptly change climate. If surface waters stopped sinking, for instance, the oceans would absorb and redistribute heat from solar radiation less efficiently. This might cause much warmer surface water temperatures and much higher land temperatures than we have now.

Alternatively, the buildup of greenhouse gases in the atmosphere may change ocean circulation. Warmer

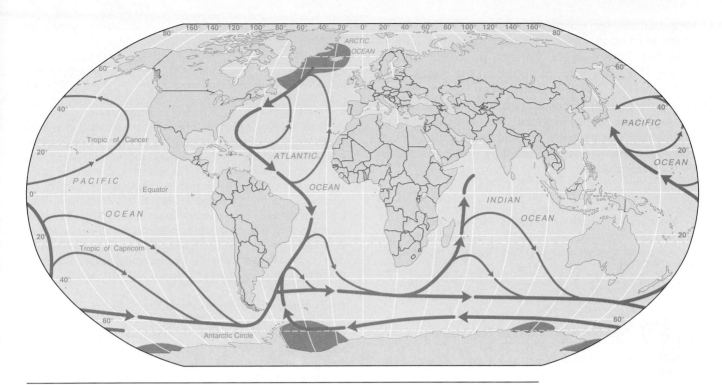

Figure 7–26 Deep-water circulation model. Schematic model of deep-water circulation first developed by oceanographer Henry Stommel in 1958. Heavy lines mark the major western boundary currents, which result from the same forces that produce western intensified surface currents. The purple shaded areas indicate source areas for deep water.

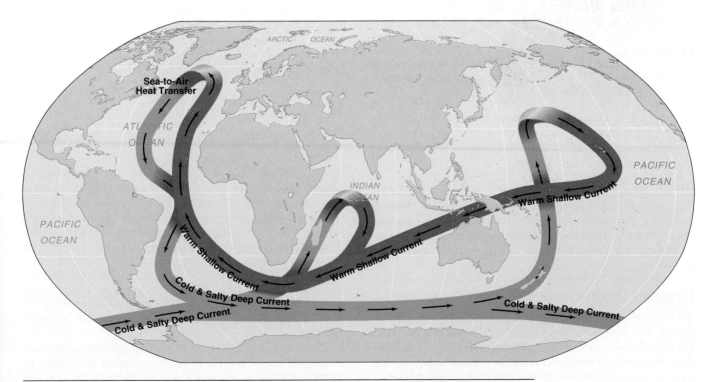

Figure 7–27 Conveyer-belt circulation. Conveyer-belt circulation is initiated in the North Atlantic Ocean, where warm water cools and sinks below the surface. This water moves southward as a subsurface flow and joins water near Antarctica. This deep water spreads into the Indian and Pacific Oceans, where it slowly rises and completes the conveyer as it travels along the surface into the North Atlantic Ocean.

temperatures, for example, may increase the rate at which glaciers in Greenland melt, forming a pool of fresh, low-density surface water in the North Atlantic Ocean. This fresh water could inhibit the downwelling that generates North Atlantic Deep Water, altering global deep-water circulation patterns and causing a corresponding change in climate. Because changes such as these can occur rapidly, it would be difficult for plant and animal life to adapt successfully to the new conditions on the planet.

> Thermohaline circulation describes the movement of deep currents, which are created at the surface in high latitudes where they become cold and dense, so they sink.

 ## Students Sometimes Ask...

What does an Ekman spiral look like at the surface? Is it strong enough to disturb ships?

The Ekman spiral creates different layers of surface water that move in slightly different directions at slightly different speeds. It is too weak to create eddies or whirlpools (vortexes) at the surface, and so is no danger to ships. In fact, the Ekman spiral is unnoticeable at the surface. It can be observed, however, by lowering oceanographic equipment over the side of a vessel. At various depths, the equipment can be observed to drift at various angles from the wind direction according to the Ekman spiral.

When the westerlies blow offshore from the east coast of the United States, does Ekman transport move water in the direction opposite the Gulf Stream?

The prevailing westerlies blow from the west, and Ekman transport in the Northern Hemisphere is to the right of the wind direction, so you might think that Ekman transport would move water in a direction opposite that of the Gulf Stream (to the south). However, Ekman transport is relatively weak compared to powerful ocean currents such as the western intensified Gulf Stream. The flow of the Gulf Stream overwhelms any Ekman transport. The presence of strong, western intensified currents also prevents coastal winds from producing coastal upwelling or downwelling along the east coasts of most continents.

I know that objects expand when they are heated. Does thermal expansion cause currents to flow in the ocean?

That's alert thinking, but it is not observed to happen in the oceans. The intensity of solar radiation is much greater in the equatorial region than at high latitudes regions. You might expect, therefore, that equatorial surface water would expand and move away from the Equator toward the poles, spreading over the colder, denser, high-latitude waters. You might further expect some return flow from the high latitudes toward the Equator. This exchange does not occur, however, because the energy imparted to surface waters by winds greatly exceeds the effect of density changes due to heating. Thus, wind over-

comes any tendency for this kind of circulation pattern to develop.

I've heard that unusual weather—such as severe storms, droughts, flooding, and warm weather—are a result of El Niño conditions. Is this true?

Because El Niño is blamed for many phenomena that are unrelated to actual El Niño conditions, this may or may not be true. Strong to very strong El Niño events are the largest single weather disturbance on the planet and cause unusual weather to occur nearly worldwide. Many people incorrectly attribute *any* kind of unusual weather to an El Niño, even during years when El Niño conditions are entirely absent. In reality, many different phenomena influence weather conditions at a particular location. To determine if an El Niño is occurring, it's best to check the sea-surface temperature patterns in the tropical Pacific Ocean for anomalous warming.

Do El Niño events occur in other ocean basins?

Yes, the Atlantic and Indian Oceans both experience events similar to the Pacific's El Niño. These events are not nearly as strong, however, nor do they influence worldwide weather phenomena to the same extent as those that occur in the equatorial Pacific Ocean. The great width of the Pacific Ocean in equatorial latitudes is the main reason that El Niño events occur more strongly in the Pacific.

In the Atlantic Ocean, this phenomenon is related to the North Atlantic Oscillation (NAO), which is a periodic change in atmospheric pressure between Iceland and Portugal. This pressure difference determines the strength of the prevailing westerlies in the North Atlantic, which in turn affects ocean surface currents there. The Atlantic Ocean periodically experiences NAO events, which sometimes cause unusual weather in Europe and heavy rainfall along the normally arid coast of southwest Africa.

The amount of anchovies produced by Peru is impressive! Besides a topping for pizza, what are some other uses of anchovies?

Anchovies are an ingredient in certain dishes, hors d'oeuvres, sauces, and salad dressing, and they are also used as bait by fishers. Historically, however, most of the *anchoveta* caught in Peruvian waters were exported and used as fishmeal (consisting of ground anchovies). The fishmeal, in turn, was used largely in pet food and as a high-protein chicken feed. As unbelievable as it may seem, El Niños affected the price of eggs! Prior to the collapse of the Peruvian *anchoveta* fishing industry in 1972–1973, El Niño events significantly reduced the availability of *anchoveta*. This drastically cut the export of anchovies from Peru, causing U.S. farmers to pursue more expensive options for chicken feed. Thus, egg prices typically increased.

The collapse of the *anchoveta* fishing industry in Peru was triggered by the 1972–1973 El Niño event but was caused by chronic overfishing in prior years. Interestingly, the shortage of fishmeal after 1972–1973 led to an increased demand for soyameal, an alternative source of high-quality protein. Increased demand for soyameal increased the price of soy commodities, thereby encouraging U.S. farmers to plant soybeans instead of wheat. Reduced production of wheat, in turn, caused a major global food crisis—all triggered by an El Niño event.

Is the Gulf Stream rich in life?

The Gulf Stream *itself* isn't, but its *boundaries* often are. The oceanic areas that have abundant marine life are typically associated with cool water—either in high-latitude regions, or in any region where upwelling occurs. These areas are constantly resupplied with oxygen- and nutrient-rich water, which results in high productivity. Warm-water areas develop a prominent thermocline that isolates the surface water from colder, nutrient-rich water below. Nutrients used up in warm waters tend not to be resupplied. The Gulf Stream, therefore, which is a western intensified, warm-water current, is associated with low productivity and an absence of marine life. The reason New England fishers knew about the Gulf Stream (see chapter opener) was because they sought their catch along the *sides* of the current, where mixing and upwelling occur.

Actually, all western intensified currents are warm and are associated with low productivities. The Kuroshio Current in the North Pacific Ocean, for example, is named for its conspicuous absence of marine life. In Japanese, *Kuroshio* means "black current," in reference to its clear, lifeless waters.

If high-salinity water has high density, why doesn't surface water sink in the subtropics, where salinities are the highest?

The highest-salinity water in the open ocean is found in the subtropical regions, but both salinity *and* temperature affect density, with temperature being the dominant factor. Water temperatures are high enough in the subtropics to reduce the density of the surface water, thereby preventing it from sinking. In such areas, a strong *halocline* (a layer of rapidly changing salinity) develops, separating a relatively thin surface layer with salinity exceeding 37‰ from deeper water with a salinity of about 35‰

One exception to the lack of sinking water in the subtropics is in the enclosed Mediterranean Sea where evaporation rates are high. These conditions result in the formation of Mediterranean Intermediate Water, which has high enough salinity to increase its density despite being warm. This high-density water flows out beneath the surface through the Strait of Gibraltar and sinks into the Atlantic Ocean. It reaches equilibrium at a depth of about 1000 meters (3300 feet), where it spreads out and remains as a distinct layer throughout much of the Atlantic (see the section on Mediterranean Circulation in Chapter 11 "The Coastal Ocean" for more information).

Chapter in Review

- Ocean currents are masses of water that flow from one place to another, and can be divided into surface currents that are wind driven or deep currents that are density driven. Currents can be measured directly or indirectly.

- Surface currents occur within and above the pycnocline. They consist of circular-moving loops of water called gyres, set in motion by the major wind belts of the world. They are modified by the positions of the continents, the Coriolis effect, and other factors. There are five major subtropical gyres in the world, which rotate clockwise in the Northern Hemisphere and counterclockwise in the Southern Hemisphere. Water is pushed toward the center of the gyres, forming low "hills" of water.

- The Ekman spiral influences shallow surface water and is caused by winds and the Coriolis effect. The average net flow of water affected by the Ekman spiral causes the water to move at 90-degree angles to the wind direction. At the center of a gyre, the Coriolis effect deflects the water so that it tends to move into the hill, whereas gravity moves the water down the hill. When gravity and the Coriolis effect balance, a geostrophic current flowing parallel to the contours of the hill is established.

- The apex (top) of the hill is located to the west of the geographical center of the gyre due to Earth's rotation. A phenomenon called western intensification occurs in which western boundary currents of subtropical gyres are faster, narrower, and deeper than their eastern boundary counterparts.

- Upwelling and downwelling help vertically mix deep and surface waters. Upwelling—the movement of cold, deep, nutrient-rich water to the surface—stimulates biologic productivity and creates large amount of marine life. Upwelling and downwelling can occur in a variety of ways.

- Antarctic circulation is dominated by a single large current, the West Wind Drift, which flows in a clockwise direction around Antarctica and is driven by the Southern Hemisphere's prevailing westerly winds. Between the West Wind Drift and the Antarctic continent is a current called the East Wind Drift, which is powered by the polar easterly winds. The two currents flow in opposite directions, so the Coriolis effect deflects them away from each other, creating the Antarctic Divergence, an area of abundant marine life due to upwelling and current mixing.

- The North Atlantic Gyre and the South Atlantic Gyre dominate circulation in the Atlantic Ocean. A poorly developed equatorial countercurrent separates these two subtropical gyres. The highest-velocity and best-studied ocean current is the Gulf Stream, which carries warm water along the southeastern U.S. Atlantic coast. Meanders of the Gulf Stream produce warm- and cold-core rings. The warming effects of the Gulf Stream extend along its route and reach as far away as Northern Europe.

- Circulation in the Pacific Ocean consists of two subtropical gyres, too, the North Pacific Gyre and the South Pacific Gyre, which are separated by a well-developed equatorial countercurrent.

- A periodic disruption of normal sea surface and atmospheric circulation patterns in the Pacific Ocean is called El Niño–Southern Oscillation (ENSO). The warm phase of ENSO (El Niño) is associated with the eastward movement of the Pacific Warm Pool, halting or reversal of the trade winds, a rise in sea level along the Equator, a decrease in productivity along the west coast of South America, and, in very strong El Niños, worldwide changes in weather. El Niños fluctuate with the cool phase of ENSO (La Niña conditions), which are associated with cooler than normal water in the eastern tropical Pacific.

• The Indian Ocean consists of one gyre, the Indian Ocean Gyre, which exists mostly in the Southern Hemisphere. The monsoon wind system, which changes direction with the seasons, dominates circulation in the Indian Ocean. The monsoons blow from the northeast in the winter and from the southwest in the summer.

• Deep currents occur below the pycnocline. They affect much larger amounts of ocean water and move much more slowly than surface currents. Changes in temperature and/or salinity at the surface create slight increases in density, which set deep currents in motion. Deep currents, therefore, are called thermohaline circulation.

• The deep ocean is layered based on density. Antarctic Bottom Water, the densest deep-water mass in the oceans, forms near Antarctica and sinks along the continental shelf into the South Atlantic Ocean. Farther north, at the Antarctic Convergence, the low-salinity Antarctic Intermediate Water sinks to an intermediate depth dictated by its density. Sandwiched between these two masses is the North Atlantic Deep Water, rich in algal nutrients after hundreds of years in the deep ocean. Layering in the Pacific and Indian oceans is similar, except there is no source of Northern Hemisphere deep water.

• Worldwide circulation models that include both surface and deep currents resemble a conveyer belt. Deep currents carry oxygen into the deep ocean, which is extremely important for life on the planet. Recent investigations indicate that worldwide deep-water circulation is closely linked to global climate change.

Key Terms

Agulhas Current (p. 225)

Alaskan Current (p. 217)

Antarctic Bottom Water (p. 226)

Antarctic Convergence (p. 211)

Antarctic Divergence (p. 213)

Antarctic Intermediate Water (p. 226)

Antilles Current (p. 214)

Arctic Convergence (p. 226)

Atlantic Equatorial Countercurrent (p. 214)

Benguela Current (p. 214)

Brazil Current (p. 214)

California Current (p. 217)

Canary Current (p. 215)

Caribbean Current (p. 214)

Coastal downwelling (p. 211)

Coastal upwelling (p. 209)

Cold core ring (p. 214)

Conveyer-belt circulation (p. 227)

Deep currents (p. 198)

Downwelling (p. 209)

Drift bottle (p. 203)

East Australian Current (p. 217)

East Wind Drift (p. 211)

Eastern boundary current (p. 200)

Ekman spiral (p. 202)

Ekman transport (p. 204)

Ekman, V. Walfrid (p. 202)

El Niño (p. 219)

El Niño–Southern Oscillation (ENSO) (p. 219)

ENSO index (p. 219)

Equatorial countercurrent (p. 207)

Equatorial current (p. 200)

Equatorial upwelling (p. 209)

Falkland Current (p. 214)

Florida Current (p. 214)

Fram (p. 207)

Franklin, Benjamin (p. 197)

Geostrophic current (p. 205)

Gulf Stream (p. 214)

Gyre (p. 202)

Indian Ocean Gyre (p. 225)

Irminger Current (p. 215)

Kuroshio Current (p. 217)

La Niña (p. 219)

Labrador Current (p. 215)

Leeuwin Current (p. 225)

Monsoon (p. 225)

Nansen, Fridtjof (p. 206)

North Atlantic Current (p. 215)

North Atlantic Deep Water (p. 226)

North Atlantic Gyre (p. 213)

North Equatorial Current (p. 214)

North Pacific Current (p. 217)

North Pacific Gyre (p. 217)

Norwegian Current (p. 215)

Ocean currents (p. 198)

Oceanic Common Water (p. 227)

Pacific Decadal Oscillation (PDO) (p. 221)

Pacific Warm Pool (p. 218)

Peru Current (p. 217)

Productivity (p. 209)

Sargasso Sea (p. 214)

Somali Current (p. 225)

South Atlantic Gyre (p. 213)

South Equatorial Current (p. 214)

South Pacific Gyre (p. 217)

Subpolar gyre (p. 202)

Subtropical convergence (p. 205)

Subtropical gyre (p. 202)

Surface currents (p. 198)

Sverdrup (Sv) (p. 213)

Temperature–salinity (T–S) diagram (p. 225)

Thermohaline circulation (p. 225)

Tropical Atmosphere and Ocean (TAO) (p. 223)

Tropical Ocean—Global Atmosphere (TOGA) (p. 223)

Upwelling (p. 209)

Warm core ring (p. 214)

Walker Circulation Cell (p. 218)

West Australian Current (p. 225)

West Wind Drift (p. 211)

Western boundary current (p. 200)

Western intensification (p. 205)

Questions And Exercises

1. Why did Benjamin Franklin want to know about the surface current pattern in the North Atlantic Ocean?

2. Compare the forces that are directly responsible for creating horizontal and deep vertical circulation in the oceans. What is the ultimate source of energy that drives both circulation systems?

3. Describe the different ways in which currents are measured.

4. What would the pattern of ocean surface currents look like if there were no continents on Earth?

5. On a base map of the world, plot and label the major currents involved in the surface circulation gyres of the oceans. Use colors to represent warm versus cool currents and indicate which currents are western intensified. On an overlay, superimpose the major wind belts of the world on the gyres.

6. What atmospheric pressure is associated with the centers of subtropical gyres? With subpolar gyres? Explain why the subtropical gyres in the Northern Hemisphere move in a clockwise fashion while the subpolar gyres rotate in a counterclockwise pattern.

7. Diagram and discuss how Ekman transport produces the "hill" of water within subtropical gyres that causes geostrophic current flow. As a starting place on the diagram, use the wind belts (the trade winds and the prevailing westerlies).

8. Describe the voyage of the *Fram* and how it helped prove that there was no continent beneath the Arctic ice pack.

9. What causes the apex of the geostrophic "hills" to be offset to the west of the center of the ocean gyre systems?

10. Draw or describe several different oceanographic conditions that produce upwelling.

11. During flood stage, the largest river in the world—the mighty Amazon River—dumps 200,000 cubic meters of water into the Atlantic Ocean each second. Compare its flow rate with the volume of water transported by the West Wind Drift

and the Gulf Stream. How many times larger than the Amazon is each of these two ocean currents?

12. Observing the flow of Atlantic Ocean currents in Figure 7–14, offer an explanation as to why the Brazil Current has a much lower velocity and volume transport than the Gulf Stream.

13. Explain why Gulf Stream eddies that develop northeast of the Gulf Stream rotate clockwise and have warm-water cores, whereas those that develop to the southwest rotate counterclockwise and have cold-water cores.

14. Describe changes in oceanographic phenomena including Walker Circulation, the Pacific Warm Pool, trade winds, equatorial countercurrent flow, upwelling/downwelling, and the abundance of marine life that occur during an El Niño event. What are some global effects of El Niño?

15. How often do El Niño events occur? Using Figure 7–20, determine how many years since 1950 have been El Niño years. Has the pattern of El Niño events occurred at regular intervals?

16. How is La Niña different from El Niño? Describe the pattern of La Niña events in relation to El Niños since 1950 (see Figure 7–20).

17. Describe the relationship between atmospheric pressure, winds, and surface currents during the monsoons of the Indian Ocean.

18. Discuss the origin of thermohaline vertical circulation. Why do deep currents form only in high-latitude regions?

19. Name the two major deep-water masses and give the locations of their formation at the ocean's surface.

20. The Antarctic Intermediate Water can be identified throughout much of the South Atlantic based on its temperature, salinity, and dissolved oxygen content. Why is it colder and less salty—and contain more oxygen—than the surface-water mass above it and the North Atlantic Deep Water below it?

References

Ackerman, J. 2000. New eyes on the oceans. *National Geographic* 198:4, 86–115.

Behringer, D., et al., 1982. A demonstration of ocean acoustic tomography. *Nature* 299: 121–125.

Boyle, E., and Weaver, A. 1994. Conveying past climates. *Nature* 372:41–42.

Broecker, W. S. 1997. Thermohaline circulation, the Achilles heel of our climate system: Will man-made CO_2 upset the current balance? *Science* 278:5343, 1582–1588.

Ebbesmeyer, C. C., and Ingraham, W. J., 1992. Shoe spill in the North Pacific. *Eos Trans. AGU* 73:27, 361.

————. 1994. Pacific toy spill fuels ocean current pathways research. *Eos Trans. AGU* 75:37, 425–427.

Genthe, H. 1998. The Sargasso Sea. *Smithsonian* 29:8, 82–93.

Gill, A. 1982. *Atmosphere-ocean dynamics*. Orlando, FL: Academic Press.

Glantz, M. H. 1996. *Currents of change: El Niño's impact on climate and society*. Cambridge: Cambridge University Press.

Goldstein, R., Barnctt, T., and Zebker, H. 1989. Remote sensing of ocean currents. *Science* 246:4935, 1282–1286.

Gordon, A. L. 1986. Interocean exchange of thermocline water. *Journal of Geophysical Research* 91:C4, 5037–5046.

Ledwell, J. R., Watson, A. J., and Law, C. S. 1993. Evidence for slow mixing across the pycnocline from an open-ocean tracer-release experiment. *Nature* 364:6439, 701–703.

McCartney, M. 1997. Is the ocean at the helm? *Nature* 388:6642, 521–522.

McCartney, M. S. 1994. Towards a model of Atlantic Ocean circulation: The plumbing of the climate's radiator. *Oceanus* 37:1, 5–8.

McDonald, A. M., and Wunsch, C. 1996. An estimate of global ocean circulation and heat fluxes. *Nature* 382:6590, 436–439.

Montgomery, R. 1940. The present evidence of the importance of lateral mixing processes in the ocean. *American Meteorological Society Bulletin* 21:87–94.

National Research Council. 1996. *Learning to predict climate variations associated with El Niño and the Southern Oscillation: Accomplishments and legacies of the TOGA program*. Washington, DC: National Academy Press.

Pedlosky, J. 1990. The dynamics of the oceanic subtropical gyres. *Science* 248:4935, 316–322.

Philander, G. 1996. El Niño and La Niña. In Pirie, R. G., ed., *Oceanography: Contemporary readings in ocean sciences*, 3rd ed., New York: Oxford University Press.

Philander, S. G. 1992. El Niño. *Oceanus* 35:2, 56–61.

————. 1998. Who is El Niño? *Eos Trans. AGU* 79:13, 170.

Philander, S. G. H. 1983. El Niño Southern Oscillation phenomena. *Nature* 302:5906, 295–301.

Pickard, G. L. 1975. *Descriptive physical oceanography*, 2nd ed. New York: Pergamon Press.

Quinn, W. H., Neal, V. T., and Antunez de Mayolo, S. E. 1987. El Niño occurrences over the past four and a half centuries. *Journal of Geophysical Research* 92:C13, 14449–14461.

Schmitz, W. J., and McCartney, M. S. 1993. On the North Atlantic circulation. *Reviews of Geophysics* 31:1, 29–49.

Stommel, H. 1958. The abyssal circulation. Letters to the editor. *Deep sea research* 5, New York: Pergamon Press.

————. 1987. *A view of the sea*. Princeton, NJ: Princeton University Press.

Street-Perrott, A. F., and Perrott, R. 1990. Abrupt climate fluctuations in the tropics: The influence of Atlantic Ocean circulation. *Nature* 343:6259, 607–612.

Stuiver, M., Quay, P. D., and Ostlund, H. G. 1983. Abyssal water carbon-14 distribution and the age of the world oceans. *Science* 219:4586, 849–851.

Sverdrup, H. U., Johnson, W., and Fleming, R. H. 1942. Renewal 1970. *The oceans*. Englewood Cliffs, NJ: Prentice-Hall.

The Open University Course Team. 1989. *Ocean circulation*. Oxford: Pergamon Press.

University Corporation for Atmospheric Research. 1994. *El Niño and climate prediction*. Reports to the Nation on our Changing Planet, Spring 1994, no. 3.

Wikelski, M., and Thom, C. 2000. Marine iguanas shrink to survive El Niño. *Nature* 403:6765, 37–38.

Wood, J. D. 1985. The world ocean circulation experiment. *Nature* 314:6011, 501–511.

Suggested Reading in Scientific American

Baker, D. J., Jr. 1970. Models of ocean circulation. 222:1, 114–121. This article discusses observations of a model depicting a segment of the surface of Earth over which fluids move and helps explain geostrophic flow within ocean gyres. Some knowledge of basic physics is necessary for full comprehension of the material presented.

Hollister, C. D., and Nowell, A. 1984. The dynamic abyss. 250:3, 42–53. Submarine "storms" are associated with deep, cold currents flowing away from polar regions toward the Equator.

Munk, W. 1955. The circulation of the oceans. 191:1, 96–108. A physical oceanographer who has been highly honored by the scientific community presents his views of ocean circulation.

Rhyther, J. H. 1956. The Sargasso Sea. 192:1, 98–104. A look at the ecology and physical oceanography of this biological desert.

Spindel, R. C., and Worcester, P. F. 1990. Ocean acoustic tomography. 263:4, 94–99. The theoretical basis and practical application of ocean acoustic tomography are clearly presented.

Stewart, R. W. 1969. The atmosphere and the ocean. 221:3, 76–105. The exchange of energy between the atmosphere and ocean and the resulting phenomena, currents, and waves are covered in this readable, comprehensive article.

Stommel, H. 1955. The anatomy of the Atlantic. 190:30-35. Another vintage article by one of the world's preeminent physical oceanographers.

Webster, P. J. 1981. Monsoons. 245:5, 108–119. The mechanism of the monsoons and their role in bringing water to half Earth's population are discussed.

Welbe, P. H. 1982. Rings of the Gulf Stream. 246:3, 60–79. The biological implications of the large cold-water rings are considered.

Whitehead, J. A. 1989. Giant ocean cataracts. 260:2, 50–57. An examination of the effect on ocean circulation of giant ocean cataracts, which are large and fast flows of deep water that play a crucial role in maintaining the chemistry and climate of the deep ocean.

Oceanography on the Web

Visit the *Essentials of Oceanography* home page for on-line resources for this chapter. There you will find an on-line study guide with review exercises, and links to oceanography sites to further your exploration of the topics in this chapter. *Essentials of Oceanography* is at: **http://www.prenhall.com/ thurman** (click on the Table of Contents menu and select this chapter).

CHAPTER

8

Waves and Water Dynamics

- What causes waves?
- How do waves move?
- What are differences between deep- and shallow-water waves?
- How do destructive and constructive interference affect waves?
- What physical changes occur in waves as they approach shore and break?
- How is wave refraction different from wave reflection?
- How are tsunamis formed and what kinds of destruction do they cause?

key

There's no time to put on survival suits or grab a life vest; the boat's moving through the most extreme motion of her life and there isn't even time to shout. The refrigerator comes out of the wall and crashes across the galley. Dirty dishes cascade out of the sink. The TV, the washing machine, the VCR tapes, the men, all go flying. And, seconds later, the water moves in.
—*Sebastian Junger,*
The Perfect Storm *(1997)*

THE BIGGEST WAVE IN RECORDED HISTORY: LITUYA BAY, ALASKA (1958)

Lituya Bay is located in southeast Alaska about 200 kilometers (125 miles) west of Juneau, Alaska's capital. It is a deep, T-shaped, 11-kilometer (7-mile)-long bay with a sand bar named La Chaussee Spit that separates it from the Pacific Ocean (Figure 8A). The largest wave ever authentically recorded occurred in Lituya Bay. Remarkably, the wave was witnessed by six people on board three small fishing boats that were near the bay's entrance (Figure 8A, *top*).

At about 10:00 P.M. on July 9, 1958, an earthquake of magnitude $M_w = 7.9$[1] occurred along the Fairweather Fault, which runs along the top of the "T" portion of the bay. The earthquake didn't produce the wave directly, but it triggered an enormous rockslide that dumped at least 90 million tons of rock—some of it from as high as 914 meters (3000 feet) above sea level—into the upper part of the bay. The rockslide created a huge **splash wave** (a long-wavelength wave produced when an object splashes into water) that swept over the ridge facing the rockslide area and uprooted, de-barked, or snapped off trees up to 530 meters (1740 feet) above the water level of the bay—a full 87 meters (285 feet) *higher* than the world's tallest building, the Sears Tower in Chicago. As the giant wave raced down the bay toward the boats at a speed of over 160 kilometers (100 miles) per hour, it continued to snap off trees and completely overtopped the island in the middle of the bay.

During the summer in Alaska, it was still light enough at 10:00 P.M. for the people on board the boats to see the rockslide occur—and the giant wave bearing down on them. The *Badger*, a 13.4-meter (44-foot) fishing vessel, had its anchor chain snapped and was lifted up bow-first into the oncoming wave. Amazingly, the vessel surfed the wave over the sand bar! The two people on board reported looking down from a height of 24 meters (80 feet) above the tops of the trees on the sand bar, in an area where trees reach heights of 30 meters (100 feet). The *Badger* plunged into the Pacific Ocean on the other side of the sand bar stern-first, where it foundered and eventually sank. The people on board were able to launch a small skiff before the *Badger* sank and were rescued a few hours later, shaken but alive.

The *Edrie* was at anchor in the bay when the wave arrived. Its anchor chain snapped, and the vessel (including two people on board) was washed onto land. After the wave passed, the withdrawal of water washed it back into the bay, leaving the vessel largely undamaged. The two people on board the *Sunmore* were not nearly so lucky. The wave hit their vessel

---

[1]The symbol M_w indicates the *moment magnitude* of an earthquake, as discussed in Chapter 2.

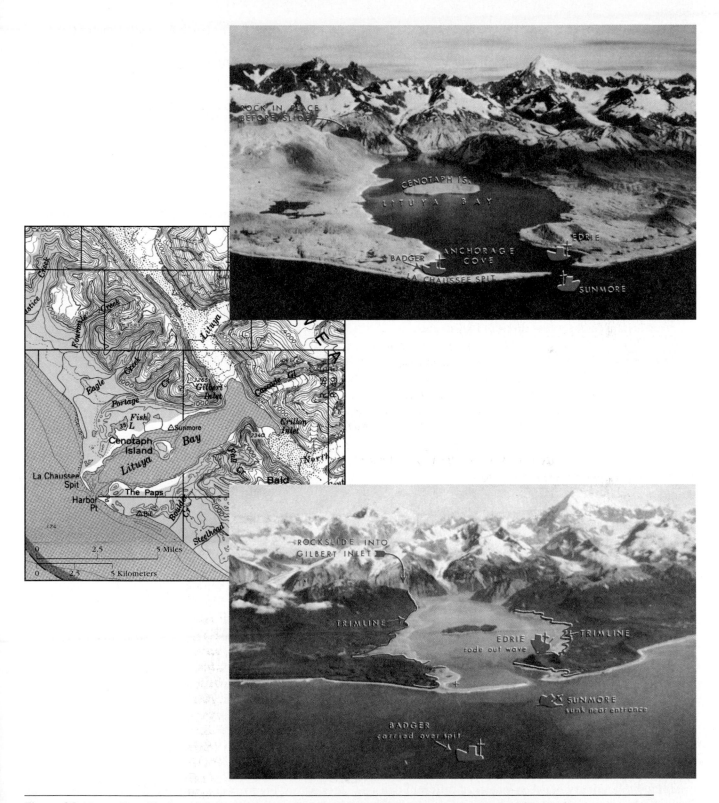

Figure 8A Lituya Bay, Alaska, with aerial view before the 1958 splash wave (*top*) and after (*bottom*).

broadside, which capsized and sank the *Sunmore*, killing both people on board. The wave spread out into the Pacific, and was even detected over six hours later at a tide-recording station in Hawaii, where the wave was only 10 centimeters (4 inches) tall.

The most noticeable damage to the shoreline of the bay included a trimline of trees extending around the bay and across the island (Figure 8A, *bottom*). The wave also knocked down all the trees on the sand bar and killed most of the shellfish living near the water's edge. Additionally, floating logs from the destruction filled Lituya Bay for many years.

Older knocked-down trees suggest that Lituya Bay periodically experiences rockslides that generate giant splash waves. For instance, there is evidence of a 120-meter (395-foot) wave in 1853, a 61-meter (200-foot) wave in 1899, and a 150-meter (490-foot) wave on October 27, 1936. Even though other events may have

produced larger waves (such as the 914-meter (3000-foot) wave created by a meteorite impact in the Gulf of Mexico about 65 million years ago[2]), the 1958 splash wave in Lituya Bay stands as the largest wave in recorded history.

[2]See Box 4–3 in Chapter 4 for a description of the Cretaceous-Tertiary (K–T) impact event.

Even on calm days, the ocean is in continual motion as waves travel across its surface. Most waves are driven by the wind and release their energy gently, although ocean storms can push waves ashore hard and fast, with devastating effects. Waves are *moving energy* traveling along the interface between ocean and atmosphere, often transferring energy from a storm far out at sea over distances of several thousand kilometers.

What Causes Waves?

All waves begin as *disturbances*. A rock thrown into a still pond creates waves that radiate out in all directions. Releases of energy, similar to the rock hitting the water, are the cause of all waves.

Wind blowing across the surface of the ocean generates most ocean waves. The waves radiate out in all directions, just as when the rock is thrown into the pond, but on a much larger scale.

The movement of fluids with different densities can also create waves. These waves travel along the interface (boundary) between the two different fluids. Both the air and the ocean are fluids, so waves can be created along interfaces *between* and *within* these fluids as follows:

- Along an *air–water interface*, the movement of air across the ocean surface creates **ocean waves** (simply called *waves*).
- Along an *air–air interface*, the movement of different air masses creates **atmospheric waves**, which are often represented by ripple-like clouds in the sky. Atmospheric waves are especially common when cold fronts (high-density air) invade an area.
- Along a *water–water interface*, the movement of water of different densities creates **internal waves**, as shown in Figure 8–1a. Because these waves travel along the boundary between waters of different density, they are associated with a *pycnocline*.[3] Internal waves can be much larger than surface waves, with heights exceeding 100 meters (330 feet). Tidal movement, turbidity currents, wind stress, or even passing ships at the surface create internal waves, which can sometimes be observed from space (Figure 8–1b).

[3]As discussed in Chapter 5, a *pycnocline* is a layer of rapidly changing density.

Internal waves can even be a hazard for submarines because they can carry submarines to depths exceeding their designed pressure limits. At the surface, parallel slicks caused by a film of surface debris may indicate the presence of internal waves below. On a smaller scale, internal waves are prominently featured sloshing back and forth in "desktop oceans," which contain two fluids that do not mix.

Mass movement into the ocean, such as coastal landslides and calving icebergs, also create waves. These waves are commonly called *splash waves* (see the chapter-opening feature for a description of a large splash wave).

Sea floor movement, which changes the shape of the ocean floor and can release large amounts of energy to the entire water column (compared to wind-driven waves, which affect only surface water), can create very large waves. Examples include underwater avalanches (turbidity currents), volcanic eruptions, and fault slippage. The resulting waves are called *seismic sea waves* or *tsunami*. Fortunately, tsunami occur infrequently. When they do, however, they can flood coastal areas and cause large amounts of destruction.

The gravitational pull of the Moon and the Sun tug on every part of Earth's oceans and create vast, low, highly predictable waves called *tides*. Tides are discussed in Chapter 9.

Human activities also cause ocean waves. When ships travel across the ocean, they leave behind a wake, which is a wave. In fact, smaller boats are often carried along in the wake of larger ships, and marine mammals sometimes play there. Also, the detonation of nuclear devices at or near sea level releases huge amounts of energy that creates waves.

In all cases, though, some type of energy release creates waves. Figure 8–2 shows the distribution of energy in waves, indicating that most ocean waves are wind-generated.

Most ocean waves are caused by wind, but many other waves are created by releases of energy in the ocean, including internal waves, splash waves, tsunami, tides, and human-induced waves.

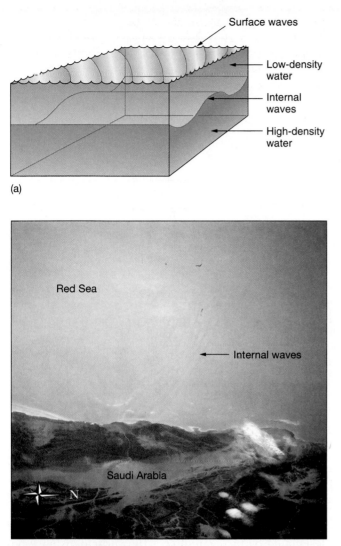

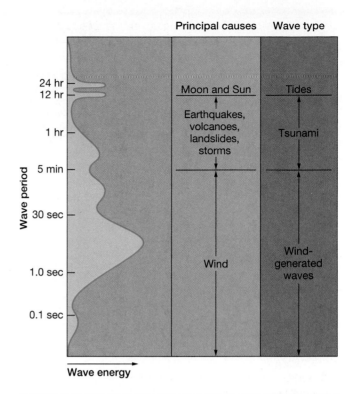

Figure 8–1 **Internal wave.** (a) An internal wave moving along the density interface (*pycnocline*) below the ocean surface. (b) A view from the space shuttle of internal waves off Saudi Arabia in the Red Sea. The internal waves are the large waves extending across the middle of the image and are moving to the right (north).

How Waves Move

Waves are energy in motion. Waves transmit energy by means of cyclic movement through matter. The medium itself (solid, liquid, or gas) does not actually travel in the direction of the energy that is passing through it. The particles in the medium simply oscillate, or cycle, back-and-forth, up-and-down, or around-and-around, transmitting energy from one particle to another. If you thump your fist on a table, for example, the energy travels through the table as waves that someone sitting at the other end can feel, but the table itself does not move.

Waves move in different ways. Simple *progressive waves* (Figure 8–3a) are waves that oscillate uniformly and *progress* or travel without breaking. Progressive

Figure 8–2 **Distribution of energy in ocean waves.** Most of the energy possessed by ocean waves exists as wind-generated waves while other peaks of wave energy represent tsunami and ocean tides.

waves may be *longitudinal*, *transverse*, or a combination of the two motions, called *orbital*.

In **longitudinal waves** (also known as push-pull waves), the particles that vibrate "push and pull" in the same direction that the energy is traveling, like a spring whose coils are alternately compressed and expanded. The shape of the wave (called a *waveform*) moves through the medium by compressing and decompressing as it goes. Sound, for instance, travels as longitudinal waves. Clapping your hands initiates a percussion that compresses and decompresses the air as the sound moves through a room. Energy can be transmitted through all states of matter—gaseous, liquid, or solid—by this longitudinal movement of particles.

In **transverse waves** (also known as side-to-side waves), energy travels at right angles to the direction of the vibrating particles. If one end of a rope is tied to a doorknob while the other end is moved up and down by hand, for example, a waveform progresses along the rope and energy is transmitted from the motion of the hand to the doorknob. The waveform moves up and down with the hand, but at right angles to the direction in which energy is transmitted (from the hand to the doorknob). Generally, transverse waves transmit energy only through solids, because the particles in solids are bound to one another strongly enough to transmit this kind of motion.

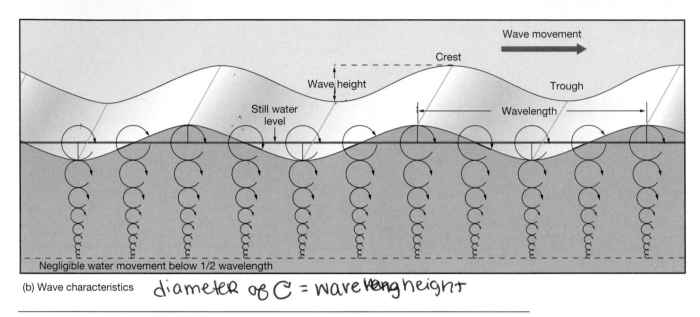

LONGITUDINAL WAVE
Particles (color) move back and forth in direction of energy transmission. These waves transmit energy through all states of matter.

TRANSVERSE WAVE
Particles (color) move back and forth at right angles to direction of energy transmission. These waves transmit energy only through solids.

ORBITAL WAVE
Particles (color) move in orbital path. These waves transmit energy along interface between two fluids of different density (liquids and/or gases).

(a) Types of progressive waves

(b) Wave characteristics   *diameter of C = wave length height*

Figure 8–3 Types and characteristics of progressive waves. (a) Types of progressive waves. (b) A diagrammatic view of an idealized ocean wave showing its characteristics.

Longitudinal and transverse waves are called *body waves* because they transfer energy through a body of matter. Ocean waves, on the other hand, transmit energy along an interface between the atmosphere and the ocean. The movement of particles along the interface involves components of *both* longitudinal and transverse waves, so particles move in circular orbits. Thus, waves at the ocean surface are **orbital waves** (also called *interface waves*).

Wave Characteristics

Figure 8–3b shows the characteristics of an idealized ocean wave. The simple, uniform, moving waveform transmits energy from a single source and travels along the ocean–atmosphere interface. These waves are also called *sine waves* because their uniform shape resembles the oscillating pattern expressed by a sine curve. Even though idealized waveforms do not exist in nature, they help us understand wave characteristics.

As the idealized wave passes a permanent marker, such as a pier piling, a succession of high parts of the waves, called **crests**, alternate with low parts, called **troughs**. Halfway between the crests and the troughs is the **still water level**, or *zero energy level*. This is the level of the water if there were no waves. The **wave height**, designated by the symbol H, is the vertical distance between a crest and a trough.

The horizontal distance between any two corresponding points on successive waveforms, such as from crest to crest or from trough to trough, is the **wavelength**, L. **Wave steepness** is the ratio of wave height to wavelength:

$$\text{Wave steepness} = \frac{\text{wave height } (H)}{\text{wavelength } (L)}$$

If the wave steepness exceeds $1/7$, the wave *breaks* (spills forward) because the wave is too steep to support itself. A wave can break anytime the 1:7 ratio is exceeded, either along the shoreline or out at sea. This ratio also dictates the maximum height of a wave. For example, a wave 7 meters long can only be 1 meter high or it will break.

The time it takes one full wave—one wavelength—to pass a fixed position (like a pier piling) is the **wave period**, T. Typical wave periods range between 6 and 16 seconds. The **frequency** (f) is defined as the number of wave crests passing a fixed location per unit of time and is the inverse of the period:

$$\text{Frequency } (f) = \frac{1}{T}$$

For instance, consider waves with a period of 12 seconds. These waves have a frequency of $^{1}/_{12}$ or 0.083 waves per second, which converts to 5 waves per minute.

Circular Orbital Motion

Waves can travel great distances across ocean basins. In one study, waves generated near Antarctica were tracked as they traveled through the Pacific Ocean basin. After more than 10,000 kilometers (over 6000 miles), the waves finally expended their energy a week later along the shoreline of the Aleutian Islands of Alaska. The water itself doesn't travel the entire distance, but the waveform does. As the wave travels, the water passes the energy along by moving in a circle. This movement is called **circular orbital motion**.

Observation of an object floating in the waves reveals that it moves not only up-and-down, but also slightly forward and backward with each successive wave. Figure 8–4 shows that a floating object moves up and backward as the crest approaches, up and forward as the crest passes, down and forward after the crest, down and backward as the trough approaches, and rises and moves backward again as the next crest advances. When the movement of the rubber duck shown in Figure 8–4 is traced as a wave passes, it can be seen that the duck moves in a circle and returns close to its original position.[4] This motion allows a waveform (the wave's shape) to move forward through the water while the individual water particles that transmit the wave move around in a circle and return to essentially the same place. Wind moving across a field of wheat causes a similar phenomenon: The wheat itself doesn't travel across the field, but the waves do.

The circular orbits of an object floating at the surface have a diameter equal to the wave height (Figure 8–3b). Figure 8–5 shows that circular orbital motion dies out quickly below the surface. At some depth below the surface, the circular orbits become so small that movement is negligible. This depth is called the **wave base**, and it is equal to one-half the wavelength ($L/2$) *measured from still water level*. Only wavelength controls the depth of the wave base, so the longer the wave, the deeper the wave base.

The decrease of orbital motion with depth has many practical applications. For instance, submarines can avoid large ocean waves simply by submerging below the wave base. Even the largest storm waves will go unnoticed if a submarine submerges to only 150 meters (500 feet). Floating bridges and floating oil rigs are constructed so that most of their mass is below wave base, so they will be unaffected by wave motion. In fact, off-

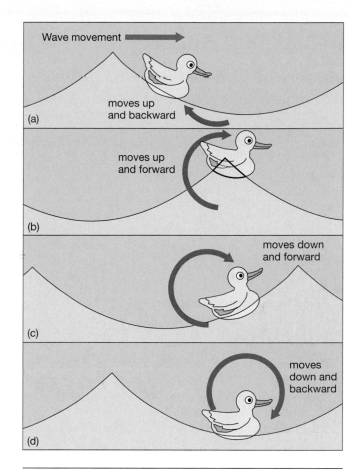

Figure 8–4 A rubber duck in water. As waves pass, the motion of a floating rubber duck resembles that of a circular orbit.

shore floating airport runways have been designed using similar principles. Additionally, seasick scuba divers find relief when they submerge into the calm, motionless water below wave base. Finally, as you walk from the beach into the ocean, you reach a point where it is easier to dive under an incoming wave than to jump over it. It is easier to swim through the smaller orbital motion below the surface than to fight the large waves at the surface.

> The ocean transmits wave energy by circular orbital motion, where the water particles move in circular orbits and return to approximately the same location.

Deep-Water Waves

If the water depth (d) is greater than the wave base ($L/2$), the waves are called **deep-water waves** (Figure 8–6a). Deep-water waves have no interference with the ocean bottom, so they include all wind-generated waves in the open ocean, where water depths far exceed wave base.

---

[4]Actually, the circular orbit does not quite return the floating object to its original position because the half of the orbit accomplished in the trough is slower than the crest half of the orbit. This slight forward movement is called *wave drift*.

Figure 8–5 Orbital motion in waves.
The orbital motion of water particles in waves extends to a depth of one-half the wavelength, measured from still water level, which is the wave base.

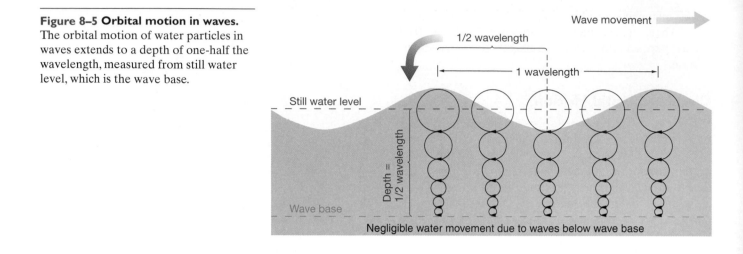

Still water level

Depth = 1/2 wavelength

Wave base

Negligible water movement due to waves below wave base

Wave movement

1/2 wavelength

1 wavelength

Wave speed (*S*) is defined as:

$$\text{Wave speed } (S) = \frac{\text{wavelength } (L)}{\text{period } (T)}$$

Wave speed is more correctly known as *celerity* (*C*), which is different from the traditional concept of speed. Celerity is used only in relation to waves where no mass is in motion, just the wave form.

According to progressive wave theory, the general formula for the speed of a deep-water wave is:

$$\text{Wave speed } (S) = \sqrt{\frac{gL}{2\pi}}$$

where *g* is the acceleration due to gravity [9.8 meters (32.2 feet) per second per second] and *L* is the wavelength. Filling in the numbers gives:

Wave speed (*S*) in meters per second = $1.25\sqrt{L}$, in meters

or:

Wave speed (*S*) in feet per second = $2.26\sqrt{L}$, in feet

Since wave speed (*S*) is defined as *L/T*, we can replace *S* with *L/T* and square both sides of the equation as follows:

$$\frac{L^2}{T^2} = \frac{gL}{2\pi}$$

Then, using algebra to reduce terms in the equation, wave speed becomes:

$$S = \frac{L}{T} = \frac{gT}{2\pi}$$

Filling in numbers gives:

$$S = 1.56T, \text{ in meters per second}$$

or:

$$S = 5.12T, \text{ in feet per second}$$

The graph in Figure 8–7 uses these equations to relate the wavelength, period, and speed of deep-water waves. Of the three variables, the wave period is usually easiest to measure. Since all three variables are related, the other two can be determined using Figure 8–7. For example, the vertical red line in Figure 8–7 shows that a wave with a period of 8 seconds has a wavelength of 100 meters. Thus, the speed of the wave is shown by horizontal red line in Figure 8–7, which is:

$$\text{Speed } (S) = \frac{L}{T} = \frac{100 \text{ meters}}{8 \text{ seconds}} = 12.5 \text{ meters per second}$$

The general relationship shown by Figure 8–7 is the *longer the wavelength, the faster the wave travels*. A fast wave does not necessarily have a large wave height, however, because wave speed depends *only* on wavelength.

Shallow-Water Waves

Waves in which depth (*d*) is less than $^1/_{20}$ of the wavelength (*L*/20) are called **shallow-water waves**, or *long waves* (Figure 8–6b). Shallow-water waves are said to *touch bottom* or *feel bottom* because the ocean floor interferes with their orbital motion.

The speed of shallow-water waves is influenced only by gravitational attraction (*g*) and the water depth (*d*):

$$\text{Wave speed } (S) = \sqrt{gd}$$

Since gravitational attraction remains constant, the equation gives:

Wave speed (*S*) in meters per second = $3.13\sqrt{d}$, in meters

or:

Wave speed (*S*) in feet per second = $5.67\sqrt{d}$, in feet

Thus, wave speed in shallow-water waves is determined *only* by water depth.

Shallow-water waves include wind-generated waves that have moved into shallow near-shore areas; *tsunami* (seismic sea waves), generated by earthquakes in the ocean floor; and the *tides*, which are a type of wave generated by the gravitational attraction of the Moon and the Sun. Tsunami and tides are very long-wavelength waves, which far exceeds even the deepest ocean water depths.

Particle motion in shallow-water waves is in a very flat elliptical orbit that approaches horizontal (back-and-forth) oscillation. The vertical component of

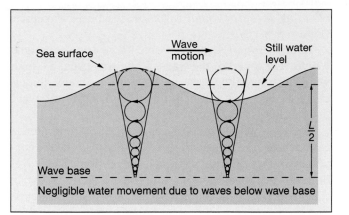

(a) Deep-water wave

(b) Shallow-water wave

(c) Relationship of wavelength to water depth

Figure 8–6 Deep-water and shallow-water waves. (a) Wave profile and water-particle motions of a deep-water wave, showing the diminishing size of the orbits with increasing depth. **(b)** Motions of water particles in shallow-water waves, where water motion extends to ocean floor. **(c)** Relationship of wavelength to water depth.

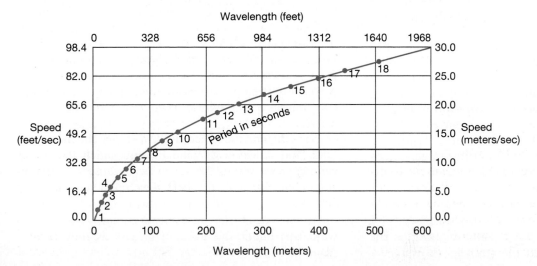

Figure 8–7 Speed of deep-water waves. Ideal relations among wavelength, period (*blue line*), and wave speed for deep-water waves. Red lines show an example wave with a wavelength of 100 meters, a period of 8 seconds, and a speed of 12.5 meters per second.

particle motion decreases with increasing depth below sea level, causing the orbits to become even more flattened.

Transitional Waves

Waves that have some characteristics of shallow-water waves and some of deep-water waves are called **transitional waves**. The wavelengths of transitional waves are between two times and 20 times the water depth (Figure 8–6c). The wave speed of shallow-water waves is a function of water depth; for deep-water waves, wave speed is a function of wavelength. Thus, the speed of transitional waves depends partially on water depth and partially on wavelength.

> Deep-water waves exist in water that is deeper than wave base and move at speeds controlled by wavelength; shallow-water waves occur in water shallower than wave base and move at speeds controlled by water depth.

Wind-Generated Waves

The life history of a wind-generated wave includes its origin in a windy region of the ocean, its movement across great expanses of open water without subsequent aid of wind, and its termination when it breaks and releases its energy, either in the open ocean or against the shore.

"Sea"

As the wind blows over the ocean surface, it creates pressure and stress. These factors deform the ocean surface into small, rounded waves with V-shaped troughs and wavelengths less than 1.74 centimeters (0.7 inch). Commonly called *ripples*, oceanographers call them **capillary waves** (Figure 8–8, *left*). The name comes from *capillarity*, a property that results from the surface tension of water. Capillarity is the dominant **restoring force** that works to destroy these tiny waves, restoring the smooth ocean surface once again.

As capillary wave development increases, the sea surface takes on a rougher appearance. The water "catches" more of the wind, allowing the wind and ocean surface to interact more efficiently. As more energy is transferred to the ocean, **gravity waves** develop, which are symmetric waves that have wavelengths exceeding 1.74 centimeters (0.7 inch) (Figure 8–8, *middle*). Because they reach greater height at this stage, gravity replaces capillarity as the dominant restoring force, giving these waves their name.

The length of gravity waves is generally 15 to 35 times their height. As additional energy is gained, wave height increases more rapidly than wavelength. The crests become pointed and the troughs are rounded,

resulting in a *trochoidal* (*trokhos* = wheel) waveform (Figure 8–8, *right*).

Energy imparted by the wind increases the height, length, and speed of the wave. When wave speed equals wind speed, neither wave height nor length can change because there is no net energy exchange and the wave has reached its maximum size.

The area where wind-driven waves are generated is called **"sea"** or the *sea area*. It is characterized by choppiness and waves moving in many directions. The waves have a variety of periods and wavelengths (most of them short) due to frequently changing wind speed and direction.

Factors that determine the amount of energy in waves are (1) wind speed, (2) the length of time during which the wind blows in one direction, and (3) the *fetch*—the distance over which the wind blows in one direction, as shown in Figure 8–9.

Wave height is directly related to the energy in a wave. Wave heights in a sea area are usually less than 2 meters (6.6 feet), but waves with heights of 10 meters (33 feet) and periods of 12 seconds are not uncommon. As "sea" waves gain energy, their steepness increases. When steepness reaches a critical value of $1/7$, open ocean breakers—called *whitecaps*—form.

Figure 8–10 is a map based on satellite data of average wave heights during October 3–12, 1992. The waves in the Southern Hemisphere are particularly large because the prevailing westerlies between 40 and 60 degrees south latitude reach the highest average wind speeds on Earth, creating the latitudes called the "Roaring Forties," "Furious Fifties," and "Screaming Sixties."

The largest wind-generated waves authentically measured occurred during a typhoon in the western Pacific Ocean in 1935. The 152-meter (500-foot)-long U.S. Navy tanker USS *Ramapo* encountered 108-kilometer (67-mile)-per-hour winds en route from the Philippines to San Diego, California. The resulting waves were symmetrical, uniform, and had a period of 14.8 seconds. The vessel's officers carefully measured the waves, using the dimensions of the ship including the *eye height* of an observer on the ship's bridge (Figure 8–11). The waves were 34 meters (112 feet) high, taller than an 11-story building. Fortunately, the *Ramapo* was traveling in the same direction as the waves, so the ship was largely undamaged. Other ships traveling in heavy seas aren't always so lucky (Figure 8–12).

For a given wind speed, Table 8–1 lists the maximum fetch and duration of wind beyond which the waves cannot grow. Waves cannot grow because an equilibrium condition, called a **fully developed sea**, has been achieved. Waves can grow no further in a fully developed sea because they lose as much energy breaking as whitecaps under the force of gravity as they receive from the wind.

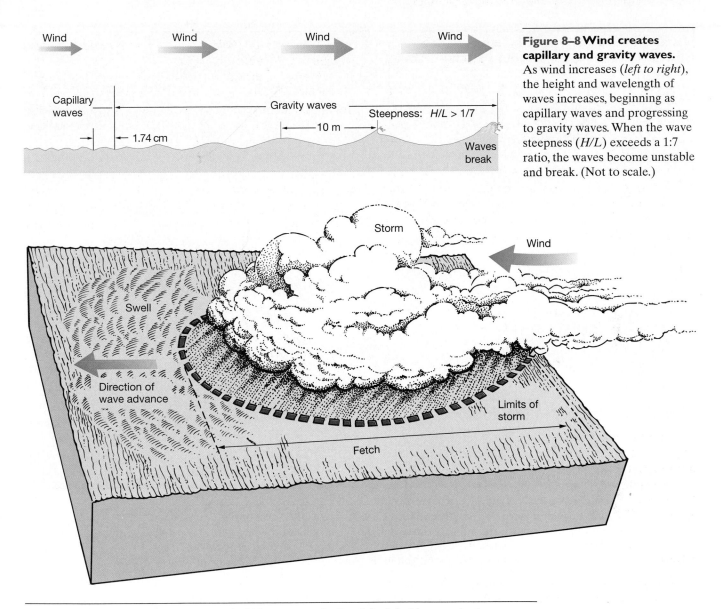

Figure 8–8 Wind creates capillary and gravity waves. As wind increases (*left to right*), the height and wavelength of waves increases, beginning as capillary waves and progressing to gravity waves. When the wave steepness (*H/L*) exceeds a 1:7 ratio, the waves become unstable and break. (Not to scale.)

Figure 8–9 The "sea" and swell. As wind blows across the "sea" (*red dash*), wave size increases with increasing wind speed, duration, and fetch. As waves advance beyond their area of origination, they advance across the ocean surface and become sorted into uniform, symmetric swell.

Swell

As waves generated in a sea area move toward its margins, wind speeds diminish and the waves eventually move faster than the wind. When this occurs, wave steepness decreases, and they become long-crested waves called **swells** (*swellan* = swollen). Swells are uniform, symmetrical waves that have traveled out of the area where they originated. Swells moves with little loss of energy over large stretches of the ocean surface, transporting energy away from one sea area and depositing it in another. Thus, there can be waves at distant shorelines where there is no wind.

Waves with longer wavelengths travel faster, and thus leave the sea area first. They are followed by slower, shorter **wave trains**, or groups of waves. The progression from long, fast waves to short, slow waves illustrates the principle of **wave dispersion** (*dis* = apart, *spargere* = to scatter)—the sorting of waves by their wavelength. Waves of many wavelengths are present in the generating area. Wave speed depends on wavelength in deep water (see Figure 8–7), however, so the longer waves "outrun" the shorter ones. The distance over which waves change from a choppy "sea" to uniform swell is called the **decay distance**, which can be up to several hundred kilometers.

Interference Patterns When swells from different storms run together, the waves clash, or *interfere* with one another, giving rise to **interference patterns**. An interference pattern produced when two or more wave systems collide is the sum of the disturbance that each wave would have produced individually. Figure 8–13 shows that the result may be a larger or smaller trough or crest, depending on conditions.

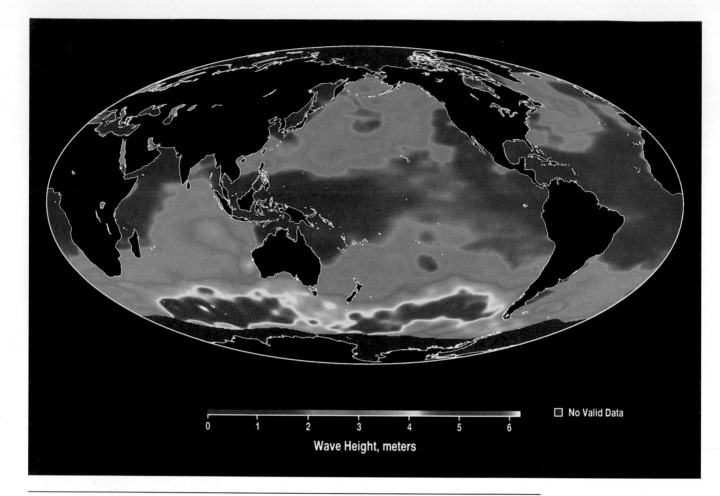

Figure 8–10 TOPEX/Poseidon wave height, October 3–12, 1992. The TOPEX/Poseidon satellite receives a return of stronger signals from calm seas and weaker signals from seas with large waves. Based on these data, a map of wave height can be produced. The largest average wave heights (*red areas*) are in the prevailing westerly wind belt in the Southern Hemisphere. (Scale in meters.)

Figure 8–11 USS *Ramapo* in heavy seas. Bridge officers aboard the USS *Ramapo* measured the largest authentically recorded wave by sighting from the bridge across the crow's nest to the horizon while the ship's stern was directly in the trough of a large wave. A wave height of 34 meters (112 feet) was calculated based on geometric relationships of the vessel and the waves.

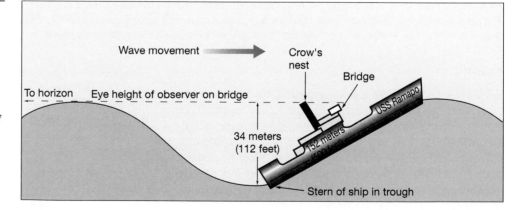

When swells from two storm areas collide, the interference pattern may be constructive or destructive, but it is more likely to be mixed. **Constructive interference** occurs when wave trains having the same wavelength come together *in phase*, meaning crest to crest and trough to trough. If the displacements from each wave are added together, the interference pattern results in a wave with the same wavelength as the two overlapping wave systems, but with a wave height equal to the sum of the individual wave heights (Figure 8–13, *left*).

Destructive interference occurs when wave trains having the same wavelength come together *out of phase*, meaning the crest from one wave coincides with the trough from a second wave. If the waves have

Figure 8–12 Wave damage on the aircraft carrier *Bennington*. The *Bennington* returns from heavy seas encountered in a typhoon off Okinawa in 1945 with part of its reinforced steel flight deck bent down over the bow. Damage to the flight deck, which is 16.5 meters (54 feet) above still water level, was caused by large waves.

identical heights, the sum of the crest of one and the trough of another is zero, so the energy of these waves cancel each other (Figure 8–13, *center*).

It is more likely, however, that the two swells consist of waves of various heights and lengths that come together with a mixture of constructive and destructive interference. A more complex **mixed interference** pattern develops (Figure 8–13, *right*), which explains the varied sequence of high and lower waves (called **surf beat**) and other irregular wave patterns that occur when swell approaches the seashore. In the open ocean, several swell systems often interact, creating complex wave patterns (Figure 8–14) and, sometimes, large waves that can be hazardous to ships (Box 8–1).

Constructive interference results from in phase overlapping of waves and creates larger waves, while destructive interference results from waves overlapping out of phase, reducing wave height.

Surf

Most waves generated in the sea area by storm winds move across the ocean as swell. These waves then release their energy along the margins of continents in the **surf zone**, which is the zone of breaking waves. Breaking waves exemplify power and persistence, sometimes

Table 8–1   Description of a fully developed sea for a given wind speed.

Wind speed in km/h (mi/h)	Average height in m (ft)	Average length in m (ft)	Average period in sec	Highest 10% of waves in m (ft)
20 (12)	0.33 (1.0)	10.6 (34.8)	3.2	0.75 (2.5)
30 (19)	0.88 (2.9)	22.2 (72.8)	4.6	2.1 (6.9)
40 (25)	1.8 (5.9)	39.7 (130.2)	6.2	3.9 (12.8)
50 (31)	3.2 (10.5)	61.8 (202.7)	7.7	6.8 (22.3)
60 (37)	5.1 (16.7)	89.2 (292.6)	9.1	10.5 (34.4)
70 (43)	7.4 (24.3)	121.4 (398.2)	10.8	15.3 (50.2)
80 (50)	10.3 (33.8)	158.6 (520.2)	12.4	21.4 (70.2)
90 (56)	13.9 (45.6)	201.6 (661.2)	13.9	28.4 (93.2)

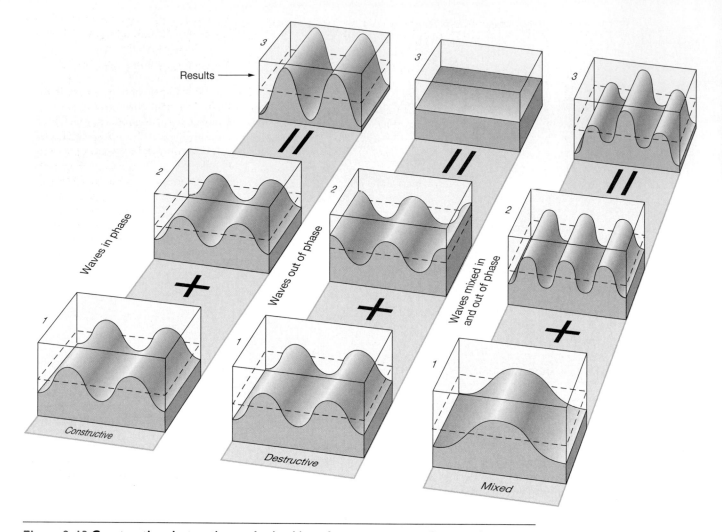

Results →

Waves in phase

Constructive

Waves out of phase

Destructive

Waves mixed in and out of phase

Mixed

Figure 8–13 Constructive, destructive, and mixed interference patterns. Constructive interference (*left*) occurs when waves of the same wavelength come together in phase (crest to crest and trough to trough), producing waves of greater height. Destructive interference (*center*) occurs when overlapping waves have identical characteristics but come together out of phase, resulting in a canceling effect. More commonly, waves of different lengths and heights encounter one another and produce a complex, mixed interference pattern (*right*).

moving objects weighing several tons. In doing so, energy from a distant storm can travel thousands of kilometers until it is finally expended along a distant shoreline in a few wild moments.

As deep-water waves of swell move toward continental margins over gradually **shoaling** (*shold* = shallow) water, they eventually encounter water depths that are less than one-half of their wavelength (Figure 8–15) and become transitional waves. Actually, any shallowly submerged obstacle (such as a coral reef, sunken wreck, or sand bar) will cause waves to release some energy. Navigators have long known that breaking waves indicate dangerously shallow water.

Many physical changes occur to a wave as it encounters shallow water, becomes a shallow-water wave, and breaks. The shoaling depths interfere with water particle movement at the base of the wave, so the *wave speed decreases*. As one wave slows, the following waveform, which is still moving at its original speed, moves closer

to the wave that is being slowed, causing a *decrease in wavelength*. The energy in the wave, which remains the same, must go somewhere, so *wave height increases*. This increase in wave height combined with the decrease in wavelength causes an *increase in wave steepness* (H/L). When the wave steepness reaches the 1:7 ratio, the waves break as surf (Figure 8–15).

If the surf is swell that has traveled from distant storms, breakers will develop relatively near shore in shallow water. The horizontal motion characteristic of shallow-water waves moves water alternately toward and away from the shore as an oscillation. The surf will be characterized by parallel lines of relatively uniform breakers.

If the surf consists of waves generated by local winds, the waves may not have been sorted into swell. The surf may be mostly unstable, deep-water, high-energy waves with steepness already near the 1:7 ratio. In this case, the waves will break shortly after feeling bottom some

Box 8–1
Rogue Waves: Ships Beware!

Rogue waves are massive, solitary waves that can reach enormous height and often occur at times when normal ocean waves are not unusually large. In a sea of 2-meter (6.5-foot) waves, for example, a 20-meter (65-foot) rogue wave may suddenly appear. *Rogue* means "unusual" and, in this case, the waves are unusually large. Rogue waves—sometimes called **superwaves**—can be quite destructive and have been popularized in literature and movies such as "*The Perfect Storm*."

In the open ocean, one wave in 23 will be over twice the height of the wave average, one in 1175 will be three times as high, and one in 300,000 will be four times as high. The chances of a truly monstrous wave, therefore, are only one in several billion. Nevertheless, rogue waves do occur, though no one knows specifically when or where they will arise. For instance, the 17-meter (56-foot) NOAA research vessel R/V *Ballena* was flipped and sunk in November 2000 by a 4.6-meter (15-foot) rogue wave off the California coast while conducting a survey in shallow, calm water. Fortunately, the three people on board survived the incident.

Even with satellites that can measure average wave size and forecast storms, about 10 large ships each year are reported missing without a trace. Worldwide, the total number of vessels lost of all sizes may reach 1000 per year, some of which are the victims of rogue waves. Recent satellites designed to observe the ocean (see Table 3–1) have provided a wealth of data about ocean waves, but still don't allow the prediction of rogue waves.

The main cause of rogue waves is theorized to be an extraordinary case of constructive wave interference, where multiple waves overlap in phase to produce an extremely large wave. Rogue waves also tend to occur more frequently downwind from islands or shoals. In addition, rogue waves can occur when storm-driven waves move against strong ocean currents, causing the waves to steepen, shorten, and become larger. These conditions exist along the "Wild Coast" off the southeast coast of Africa, where the Agulhas Current flows directly against large Antarctic storm waves, creating rogue waves that can crash onto the bow of a ship, overcome its structural capacity, and cause the ship to sink (Figure 8b).

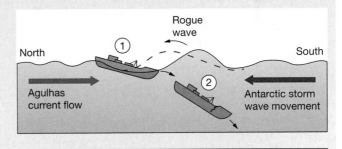

Figure 8B Rogue waves along Africa's "Wild Coast."

Figure 8–14 Mixed interference pattern. The observed wave pattern in the ocean (*above*) is often the result of mixed interference of many different overlapping wave sets (*below*).

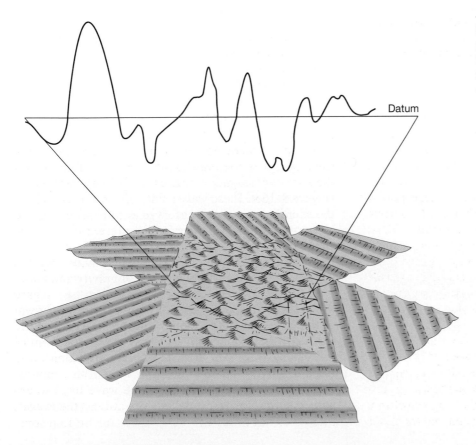

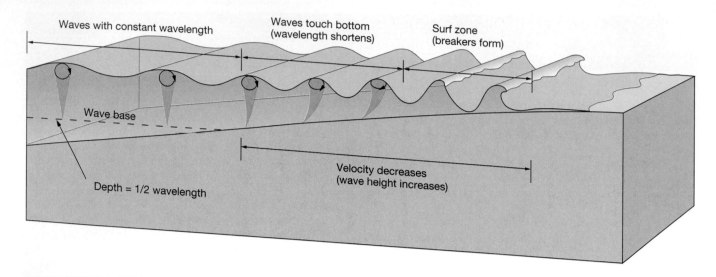

Figure 8–15 Physical changes of a wave in the surf zone. As waves approach the shore and encounter water depths of less than one-half wavelength, the waves "feel bottom." The *wave speed decreases* and waves stack up against the shore, causing the *wavelength to decrease*. This results in an *increase in wave height* to the point where the *wave steepness is increased* beyond the 1:7 ratio, causing the wave to pitch forward and break in the surf zone.

distance from shore, and the surf will be rough, choppy, and irregular.

When the water depth is about one and one-third times the wave height, the crest of the wave breaks, producing surf.[5] When the water depth becomes less than $1/20$ the wavelength, waves in the surf zone begin to behave as shallow-water waves (see Figure 8–6). Particle motion is greatly impeded by the bottom, and a significant transport of water toward the shoreline occurs (Figure 8–15).

Waves break in the surf zone because particle motion near the bottom of the wave is severely restricted, slowing the waveform. At the surface, however, individual orbiting water particles have not yet been slowed because they have no contact with the bottom. In addition, the wave height increases in shallow water. The difference in speed between the top and bottom parts of the wave cause the top part of the wave to overrun the lower part, which results in the wave toppling over and breaking. Breaking waves are analogous to a person who leans too far forward. If you don't catch yourself, you may also "break" something when you fall.

As waves come into shallow water and feel bottom, their speed and wavelength decrease while their wave height and wave steepness increase, causing the wave to break.

Breakers and Surfing Figure 8–16a shows a **spilling breaker**. Spilling breakers result from a gently sloped ocean bottom, which extracts energy from the wave more gradually, producing a turbulent mass of air and water that runs down the front slope of the wave instead of producing a spectacular cresting curl. Spilling breakers have a longer life span and give surfers a long—but somewhat less exciting—ride than other breakers.

Figure 8–16b shows a **plunging breaker**, which has a curling crest that moves over an air pocket. The curling crest occurs because the particles in the crest literally outrun the wave, and there is nothing beneath them to support their motion. Plunging breakers form on moderately steep beach slopes, and are the best waves for surfing.

When the ocean bottom has an abrupt slope, the wave energy is compressed into a shorter distance and the wave will surge forward, creating a **surging breaker** (Figure 8–16c). These waves build up and break right at the shoreline, so board surfers tend to avoid them. For body surfers, however, these waves present the greatest challenge.

Surfing is analogous to riding a gravity-operated water sled by balancing the forces of gravity and buoyancy. The particle motion of ocean waves (see Figure 8–3b) shows that water particles move up into the front of the crest. This force, along with the buoyancy of the surfboard, helps maintain a surfer's position in front of a breaking wave. The trick is to perfectly balance the force of gravity (directed downward) with the buoyant force (directed perpendicular to the wave face) to enable a surfer to be propelled forward by the wave's

---

[5]This is a handy way of estimating water depth in the surf zone: The depth of the water where waves are breaking is one and one-third times the breaker height.

(a)

(b)

(c)

Figure 8–16 Types of breakers. (a) Spilling breaker, resulting from a gradual beach slope.
(b) Plunging breaker at Oahu, Hawaii, resulting from a steep beach slope. **(c)** Surging breaker,
resulting from an abrupt beach slope.

energy. A skillful surfer, by positioning the board properly on the wave front, can regulate the degree to which the propelling gravitational forces exceed the buoyancy forces, and speeds up to 40 kilometers (25 miles) per hour can be obtained while moving along the face of a breaking wave. When the wave passes over water that is too shallow to allow the upward movement of water particles to continue, the ride is over.

Wave Refraction

Waves seldom approach a shore at a perfect right angle (90 degrees). Instead, some segment of the wave will "feel bottom" first and will slow before the rest of the

wave. This results in the **refraction** (*refringere* = to break up) or *bending* of each wave crest (also called a wave front) as the waves approach the shore.

Figure 8–17 shows how waves coming toward a straight shoreline are refracted and tend to align themselves *nearly* parallel to the shore. This explains why all waves come almost straight in toward a beach, no matter what their original orientation was.

Figure 8–18 shows how waves coming toward an irregular shoreline refract so that they, too, nearly align with the shore. However, the refraction of waves along an irregular shoreline distributes wave energy unevenly along the shore.

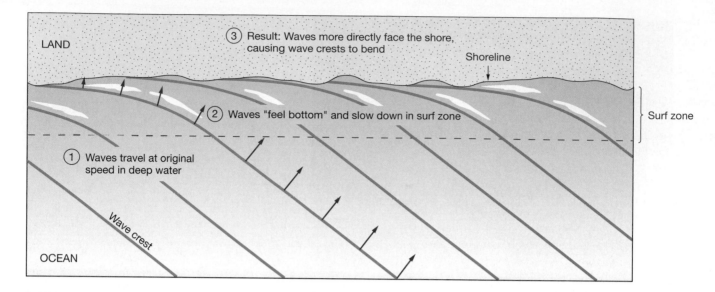

Figure 8–17 Refraction along a straight shoreline. Waves approaching the shore at an angle first "feel bottom" close to shore. This causes the segment of the wave in shallow water to slow, causing the crest of the wave to refract or bend so that the waves arrive at the shore nearly parallel to the shoreline. Red arrows represent direction and speed of the wave.

Figure 8–18 Refraction along an irregular shoreline. As waves first "feel bottom" in the shallows off the headlands, they are slowed, causing the waves to refract and align nearly parallel to the shoreline. Evenly spaced orthogonal lines (*black arrows*) show that wave energy is concentrated on headlands (causing erosion) and dispersed in bays (resulting in deposition).

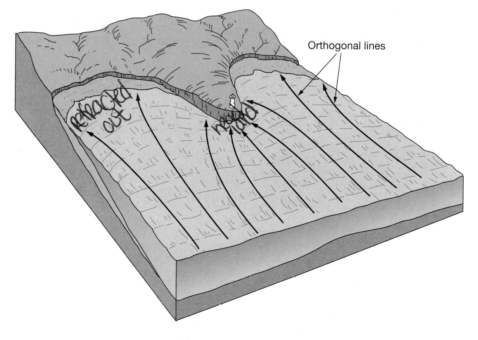

The long black arrows in Figure 8–18 are called **or-thogonal** (*ortho* = straight, *gonia* = angle) **lines**. Orthogonal lines are drawn perpendicular to the wave fronts (so they indicate the direction that waves travel) and are spaced so that the energy between lines is equal at all times. They help determine how energy is distributed along the shoreline by breaking waves.

The orthogonals in Figure 8–18 are equally spaced far from shore. As they approach the shore, however, the orthogonals *converge* on headlands that jut into the ocean, and *diverge* in bays. This means that wave energy is concentrated against the headlands, but dispersed in bays. The result is heavy erosion of headlands and depo-sition of sediment in bays. The greater energy of waves breaking on headlands is reflected in an increased wave height.[6] Conversely, the smaller waves in bays provide areas for good boat anchorages.

Wave Reflection

Not all of the energy of waves is expended as they rush onto the shore. A vertical barrier, such as a seawall or a rock ledge, can reflect waves back into the ocean with

---

[6]Sailors have long known that "the points draw the waves." Surfers also know how wave refraction causes good "point breaks."

little loss of energy—a process called **wave reflection** (*reflecten* = to bend back), which is similar to how a mirror reflects (bounces) back light. If the incoming wave strikes the barrier at a right (90-degree) angle, the wave energy is reflected back parallel to the incoming wave, often interfering with the next incoming wave and creating unusual waveforms. More commonly, waves approach the shore at an angle, causing wave energy to be reflected at an angle equal to the angle at which the wave approached the barrier.

An outstanding example of wave reflection occurs in an area called "The Wedge," which develops west of the jetty that protects the harbor entrance at Newport Harbor, California (Figure 8–19). The jetty is a solid man-made object that extends into the ocean 400 meters (1300 feet) and has a near-vertical side facing the waves. As incoming waves strike the vertical side of the jetty at an angle, they are reflected at an equivalent angle. Because the original waves and the reflected waves have the same wavelength, a constructive interference pattern develops, creating plunging breakers that may exceed 8 meters (26 feet) in height (Figure 8–19, *inset*). Too dangerous for board surfers, these waves present a fierce challenge to the most experienced body surfers. The Wedge has crippled or even killed many who have come to try it.

Standing waves (or *stationary waves*) can be produced when waves are reflected at right angles to a barrier. Standing waves are the sum of two waves with the same wavelength moving in opposite directions, resulting in no net movement. Although the water particles continue to move vertically and horizontally, there is none of the circular motion that is characteristic of a progressive wave.

Figure 8–20 shows the movement of water during the wave cycle of a standing wave. Lines along which there is no vertical movement are called *nodes* (*node* = knot), or nodal lines. *Antinodes*, crests that alternately become troughs, are the points of greatest vertical movement within a standing wave.

No particle motion exits when an antinode is at its greatest vertical displacement, and the maximum particle movement occurs when the water surface is level. At this time, the maximum movement of the water is in a horizontal direction directly beneath the nodal lines. The movement of water particles beneath the antinodes is entirely vertical.

We consider standing waves further when we discuss tidal phenomena in Chapter 9, "Tides." Under certain conditions, the development of standing waves significantly affects the tidal character in coastal regions.

> Wave refraction is the bending of waves caused when waves slow in shallow water; wave reflection is the bouncing back of wave energy caused when waves strike a hard barrier.

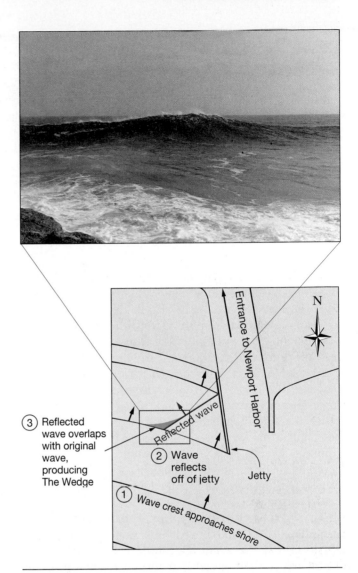

Figure 8–19 Wave reflection at The Wedge, Newport Harbor, California. As waves approach the shore (1), some of the wave energy is reflected off the long jetty at the entrance to the harbor (2). The reflected wave overlaps and constructively interferes with the original wave (3), resulting in a wedge-shaped wave (*dark blue triangle*) that may reach heights exceeding 8 meters (26 feet). Photo of The Wedge (*inset*) shows three dots in front of the wave that are the heads of body surfers.

Tsunami

The Japanese term for the large, sometimes destructive waves that occasionally roll into their harbors is **tsunami** (*tsu* = harbor, *nami* = wave). Tsunami originate from sudden changes in the topography of the sea floor caused by slippage along underwater faults, underwater avalanches, or underwater volcanic eruptions. Many people mistakenly call them "tidal waves," but tsunami are unrelated to the tides. The mechanisms that trigger tsunami are typically seismic events, so tsunami are *seismic sea waves*.

The majority of tsunami are caused by *fault movement*. Underwater fault movement displaces Earth's

Figure 8–20 Sequence of motion in a standing wave. In a standing wave, water is motionless when antinodes reach maximum displacement (*a, c,* and *e; a* and *e* are identical). Water movement is at a maximum (*blue arrows*) when the water is horizontal (*b* and *d*). Movement is vertical beneath the antinodes, and maximum horizontal movement occurs beneath the node. After *e*, cycle begins again at *b*.

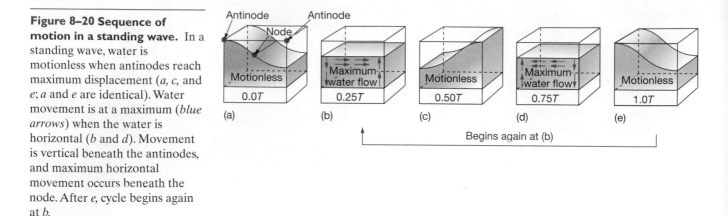

crust, generates earthquakes, and, if it ruptures the sea floor, produces a sudden change in water level at the ocean surface (Figure 8–21a). Faults that produce *vertical* displacement (the uplift or downdropping of ocean floor) change the volume of the ocean basin, which affect the entire water column and generate tsunami. Conversely, faults that produce *horizontal* displacement (such as the lateral movement associated with transform faulting) generally do not generate tsunami because the side-to-side movement of these faults does not change the volume of the ocean basin. Much less common events, such as underwater avalanches triggered by shaking, meteorite impacts, or underwater volcanic eruptions—which create the largest waves—also produce tsunami.

The wavelength of a typical tsunami exceeds 200 kilometers (125 miles), so it is a shallow-water wave everywhere in the ocean.[7] Because tsunami are shallow-water waves, their speed is determined only by water depth. In the open ocean, tsunami move at well over 700 kilometers (435 miles) per hour—they could easily keep pace with a jet airplane—and have heights of only about 0.5 meter (1.6 feet). Even though they are fast, tsunami are small in the open ocean and pass unnoticed in deep water until they reach shore, where they slow in the shallow water, and the water begins to pile up.

Coastal Effects

A tsunami does not form a huge breaking wave at the shoreline. Instead, it is a strong flood or surge of water that causes the ocean to advance (or, in certain cases, retreat) dramatically. In fact, a tsunami resembles a sudden, *extremely* high tide, which is why they are misnamed "tidal waves." It takes several minutes for the tsunami to express itself fully, during which time sea level can rise up to 40 meters (131 feet) above normal, with normal waves superimposed on top of the higher

---
[7]Recall that the depth of the wave base is equal to one-half a wave's wavelength. Thus, tsunami can typically be felt to depths of 100 kilometers (62 miles), which is deeper than even the deepest ocean trenches.

sea level. The strong surge of water can rush into low-lying areas with destructive results (Figure 8–21b).

As the trough of the tsunami arrives at the shore, the water will rapidly drain off the land. In coastal areas, it will look like a sudden and *extremely* low tide, where sea level is many meters lower than even the lowest low tide. Because tsunami are typically a series of waves, there are often an alternating series of dramatic surges and withdrawals of water, separated by only a few minutes. The first surge may not always be the largest; the third, fourth, or even seventh surge may be the largest, instead.

In some cases, the trough of a tsunami arrives at the coast first, exposing parts of the lowermost shoreline that are rarely seen. For people at the shoreline, the temptation is to explore these newly exposed areas and catch stranded fish. Within a few minutes, however, a strong surge of water (the crest of the tsunami) is due to arrive.

The alternating surges and retreats of water by tsunami can severely damage coastal structures. Tsunami can be deadly as well. The speed of the advance—up to 4 meters (13 feet) per second—is faster than a person can run. Those who are trapped by tsunami are often drowned or crushed by floating debris (Figure 8–22).

Historic Tsunami

Many small tsunami are created each year, and go largely unnoticed. On average, 57 tsunami occur every decade, with a large tsunami occurring somewhere in the world every two to three years and an extremely large and damaging one occurring every 15 to 20 years. About 86% of all great waves are generated in the Pacific Ocean because large-magnitude earthquakes occur along the series of trenches that ring its ocean basin where oceanic plates are subducted along convergent plate boundaries. Volcanic activity is also common along the Pacific "Ring of Fire," and the large earthquakes that occur along its margin are capable of producing extremely large tsunami (Box 8–2).

One of the most destructive tsunami ever generated came from the eruption of the volcanic island of

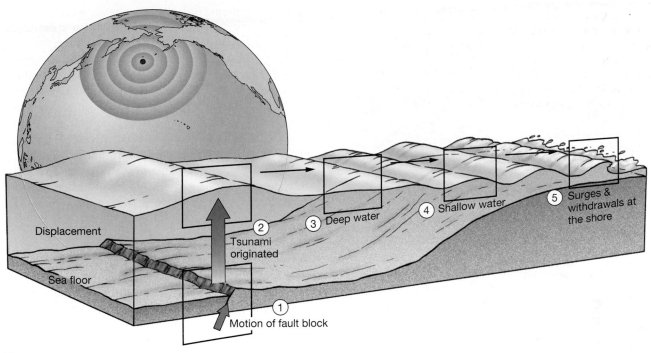

(a)

(b)

Figure 8–21 Origin of a tsunami. (a) Abrupt vertical movement along a fault on the sea floor raises or drops the ocean water column above a fault, creating a tsunami that travels from deep to shallow water where it is experienced as alternating surges and withdrawals of water at the shore. **(b)** Sequence of photos of a 1983 tsunami in northern Japan that surges toward fleeing spectators in a harbor. Red arrows show stationary motorcycle.

Figure 8–22 Tsunami damage in Hilo, Hawaii. Flattened parking meters in Hilo, Hawaii, caused by the 1946 tsunami that resulted in more than $25 million in damage and 159 deaths.

Krakatau[8] on August 27, 1883. Approximately the size of a small Hawaiian Island in what is now Indonesia, Krakatau exploded with the greatest release of energy from Earth's interior observed in historic times. The island was nearly obliterated and the sound of the explosion was heard up to 4800 kilometers (2980 miles) away in Australia. Dust from the explosion ascended into the atmosphere and circled Earth on high-altitude winds, producing unusual and beautiful sunsets for nearly a year.

Not many were killed by the outright explosion of the volcano because the island was uninhabited. However, the displacement of water from the energy released during the explosion was enormous, creating a tsunami that exceeded 35 meters (116 feet)—as high as a 12-story building. It devastated the coastal region of the Sunda Strait between the nearby islands of Sumatra and Java, drowning over 1000 villages and taking more than 36,000 lives. The energy carried by this wave reached every ocean basin, and was detected by tide recording stations as far away as London and San Francisco.

Several ships were along the coast of Java during the eruption of Krakatau, carrying passengers that were eyewitnesses to the tsunami and its destruction. N. van Sandick, an engineer aboard the Dutch vessel *Loudon*, gave the following account:

> Suddenly we saw a gigantic wave of prodigious height advancing from the sea-shore with considerable speed. Immediately the crew set to under considerable pressure and managed after a fashion to set sail in face of the imminent danger; the ship had just enough time to meet with the wave from the front.

After a moment, full of anguish, we were lifted up with a dizzy rapidity. The ship made a formidable leap, and immediately afterwards we felt as though we had plunged into the abyss. But the ship's blade went higher and we were safe. Like a high mountain, the monstrous wave precipitated its journey towards the land. Immediately afterwards another three waves of colossal size appeared. And before our eyes this terrifying upheaval of the sea, in a sweeping transit, consumed in one instant the ruin of the town; the lighthouse fell in one piece, and all the houses of the town were swept away in one blow like a castle of cards. All was finished. There, where a few moments ago lived the town of Telok Betong, was nothing but the open sea.

Another strong tsunami was experienced in the port of Hilo, Hawaii, on April 1, 1946. The tsunami was from a magnitude $M_w = 7.3$ earthquake in the Aleutian Trench off the island of Unimak, Alaska, over 3000 kilometers (1850 miles) away. The bathymetry in horseshoe-shaped Hilo Bay tends to focus a tsunami's energy directly toward town, building up waves to tremendous heights. In this case, the tsunami expressed itself as a strong recession followed by a surge of water nearly 17 meters (55 feet) above normal high tide, causing more than $25 million in damage and killing 159 people. Remarkably, it stands as Hawaii's worst natural disaster (Figure 8–22).

Closer to the source of the earthquake, the tsunami was considerably larger. The tsunami struck Scotch Cap, Alaska, on Unimak Island, where a two-story reinforced concrete lighthouse stood 14 meters (46 feet) above sea level at its base. The lighthouse was destroyed by a wave that is estimated to have reached 36 meters (118 feet), killing all five people inside the lighthouse at the time. Vehicles on a nearby mesa 31 meters (103 feet) above water level were also moved by the onrush of water.

Figure 8–23 shows that since 1990, ten destructive tsunamis along the Pacific Ring of Fire have claimed more than 4000 lives. Of these tsunami, the one that caused the greatest number of casualties occurred in Papua New Guinea in July 1998. An offshore magnitude $M_w = 7.1$ earthquake was followed shortly thereafter by a 15-meter (49-foot) tsunami, which was up to five times larger than expected for a quake that size. The tsunami completely overtopped a heavily populated low-lying sand bar, destroying three entire villages and resulting in at least 2200 deaths. Researchers who mapped the sea floor after the tsunami discovered the remains of a huge underwater landslide, which was apparently triggered by the shaking and generated the deadliest tsunami in 65 years.

Tsunami Warning System

In response to the tsunami that struck Hawaii in 1946, a tsunami warning system was established throughout the Pacific Ocean. It led to what is now the **Pacific**

[8]The volcanic island Krakatau (which is *west* of Java) is also called Krakatoa.

Box 8–2 The Big Shake: A Tsunami from the Cascadia Subduction Zone Hits Japan

The Juan de Fuca Plate slides beneath the North America Plate offshore of the U.S. Pacific Northwest, creating the Cascadia subduction zone. The February 28, 2001, magnitude M_w = 6.8 Seattle Earthquake is a reminder that the Cascadia subduction zone is capable of producing large, damaging earthquakes.

Paleo-earthquake and paleo-tsunami evidence suggest that much larger earthquakes have occurred along the Cascadia subduction zone in the past. The paleo-earthquake evidence consists of offshore turbidity current deposits, accumulations of organic matter including entire forests that were buried during sudden coastal downdropping, and sediments that have liquefied on land due to ground shaking. The paleo-tsunami evidence consists of sand layers deposited in low-lying coastal areas. The Cascadia subduction zone is the presumed source of the earthquakes, although other major faults on land (such as the Seattle Fault) might also produce them. The evidence indicates that many large earthquakes have occurred in the area during the last 1000 years, but for the most part, the exact size of the earthquakes is unknown and the dates of occurrence can be resolved only to within a decade.

Evidence for large Cascadia earthquakes also comes from the historic records of tsunami damage in Japan. The records indicate that a tsunami of unknown origin hit the coast of Japan on January 27–28 in 1700. Based on the time of arrival and the height of the tsunami in different locations in Japan, other Pacific Rim regions (South America, Alaska, and the Kamchatka Peninsula in Russia) can be eliminated as potential sources. Even a locally generated tsunami does not fit the reported effects, leaving the Cascadia subduction zone as the most likely source (Figure 8C). Because Japan received tsunami heights of 2 to 3 meters (6.5 to 10 feet), the earthquake must have been magnitude M_w = 9.0. Only two earthquakes of this size have ever been recorded: one along the coast of Chile in 1960, and the other in southern Alaska in 1964. An earthquake this large must have ruptured the entire length of the Cascadia subduction zone and is estimated to have a recurrence interval of 300 to 700 years.

Based on the historic records in Japan (which indicate when the tsunami hit different parts of the coast) and based on how fast tsunami travel, the time of the earthquake can be determined: it occurred at about 9:00 P.M. Pacific Northwest local time on January 26, 1700. This estimate is consistent with American Indian legends, which indicate that a large earthquake occurred at about this date during a winter night.

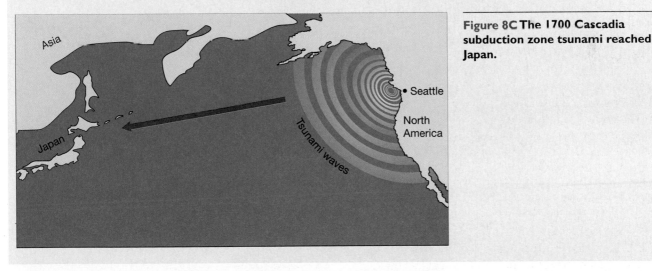

Figure 8C The 1700 Cascadia subduction zone tsunami reached Japan.

Tsunami Warning Center (PTWC), which coordinates information from 25 Pacific Rim countries and is headquartered in Ewa Beach (near Honolulu), Hawaii. In the open ocean, tsunami have small wave heights and are difficult to detect, so the tsunami warning system uses seismic waves—some of which travel through Earth at speeds 15 times faster than tsunami—to forecast destructive tsunami.[9] When a seismic disturbance occurs beneath the ocean surface that is large enough to be tsunamigenic (capable of producing a tsunami), a *tsunami watch* is is-

sued. At this point, a tsunami may or may not have been generated, but the potential for one exists.

The PTWC is linked to over 50 tide-measuring stations throughout the Pacific, so the recording station nearest the earthquake is closely monitored for any indication of unusual wave activity. If unusual wave activity is verified, the tsunami watch is upgraded to a *tsunami warning*. Generally, earthquakes smaller than magnitude M_w = 6.5 are not tsunamigenic because they lack the duration of ground shaking necessary to initiate a tsunami. Additionally, transform faults do not usually produce tsunami because lateral movement does not offset the ocean floor and impart energy to the

[9]A new method of tracking tsunami that is currently being tested uses a series of sensitive pressure sensors on the ocean floor that can detect the passage of tsunami in the open ocean.

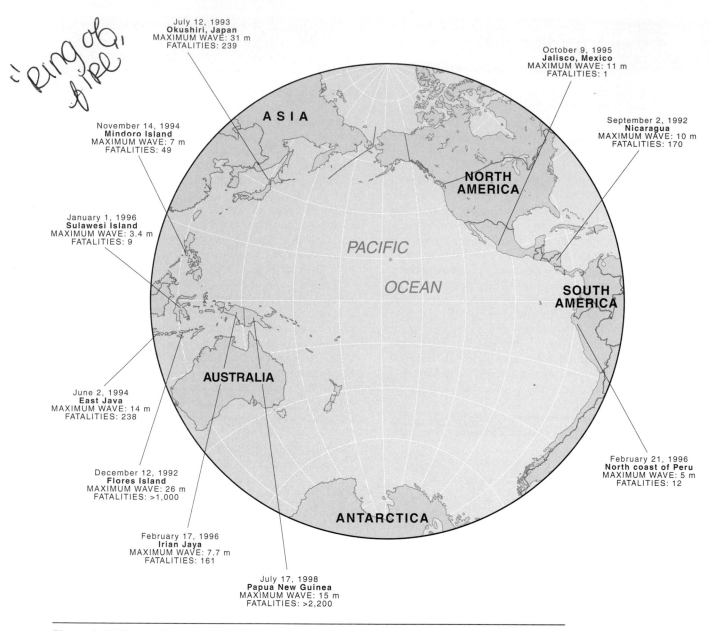

"Ring of fire"

July 12, 1993
Okushiri, Japan
MAXIMUM WAVE: 31 m
FATALITIES: 239

October 9, 1995
Jalisco, Mexico
MAXIMUM WAVE: 11 m
FATALITIES: 1

November 14, 1994
Mindoro Island
MAXIMUM WAVE: 7 m
FATALITIES: 49

September 2, 1992
Nicaragua
MAXIMUM WAVE: 10 m
FATALITIES: 170

January 1, 1996
Sulawesi Island
MAXIMUM WAVE: 3.4 m
FATALITIES: 9

ASIA

NORTH
AMERICA

PACIFIC

OCEAN

SOUTH
AMERICA

June 2, 1994
East Java
MAXIMUM WAVE: 14 m
FATALITIES: 238

AUSTRALIA

February 21, 1996
North coast of Peru
MAXIMUM WAVE: 5 m
FATALITIES: 12

December 12, 1992
Flores Island
MAXIMUM WAVE: 26 m
FATALITIES: >1,000

February 17, 1996
Irian Jaya
MAXIMUM WAVE: 7.7 m
FATALITIES: 161

ANTARCTICA

July 17, 1998
Papua New Guinea
MAXIMUM WAVE: 15 m
FATALITIES: >2,200

Figure 8–23 Tsunamis since 1990. Ten destructive tsunamis have claimed more than 4000 lives since 1990. These killer waves are most often generated by earthquakes along colliding tectonic plates of the Pacific Rim, although the deadly 1998 Papua New Guinea tsunami that killed more than 2200 was generated by an underwater landslide.

water column in the same way that vertical fault movements do.

When a tsunami is detected, warnings are sent to all the coastal regions that might encounter the destructive wave, along with its estimated time of arrival. This warning, usually just a few hours in advance of the tsunami, makes it possible to evacuate people from low-lying areas and remove ships from harbors before the waves arrive. If the disturbance is nearby, however, there is not enough time to issue a warning because a tsunami travels so rapidly. Unlike hurricanes, whose high winds and waves threaten ships at sea and send them to the protection of a coastal harbor, a tsunami washes ships from their coastal moorings into the open ocean or onto

shore. The best strategy during a tsunami warning is to get ships out of coastal harbors and into deep water, where tsunami are not easily felt.

Since the PTWC was established in 1948, it has effectively prevented loss of life due to tsunami when people have heeded the evacuation warnings. Property damage, however, has increased as more buildings have been constructed close to shore. To combat the damage caused by tsunami, countries that are especially prone to tsunami like Japan have invested in shoreline barriers, seawalls, and other coastal fortifications.

Perhaps one of the best strategies to limit tsunami damage and loss of life is to restrict construction projects in low-lying coastal regions where tsunami have frequently struck in the past. However, the long time interval between large tsunami can lead people to forget past disasters.

> Most tsunami are generated by underwater fault movement, which transfers energy to the entire water column. When these fast and long waves surge ashore, they can do considerable damage.

Power from Waves

Moving water has a huge amount of energy, which is why there are so many hydroelectric power plants on rivers. Even greater energy exists in ocean waves, but significant problems must be overcome for the power to be harnessed efficiently. The most likely locations for power generation are where waves refract and converge, focusing wave energy, such as around headlands (see Figure 8–18). Using this advantage, a wave power plant might extract up to 10 megawatts of power[10] per kilometer (0.6 mile) of shoreline.

One of the disadvantages of wave energy is that the system produces significant power only when large storm waves break against it, so the system could serve only as a power supplement. In addition, a series of one hundred or more of these structures along the shore would be required. Structures of this type could have a significant impact on the environment, with negative effects on marine organisms that rely on wave energy for dispersal, transporting food supplies, or removing wastes. Also, harnessing wave energy might alter the transport of sand along the coast, causing erosion in areas deprived of sediment.

Internal waves are a potential source of energy, too. Along shores that have favorable ocean floor topography for focusing wave energy, internal waves may be effectively concentrated by refraction. This energy could power an energy-conversion device that would produce electricity.

In November 2000, the world's first commercial wave power plant began generating electricity. The small plant, called **LIMPET 500** (Land Installed Marine Powered Energy Transformer), is located on Islay, a small island off the west coast of Scotland. The plant was constructed at a cost of about $1.6 million and allows waves to compress air in a chamber that, in turn, rotates a turbine for the generation of power. Under peak operating capacity, the facility is capable of producing 500 kilowatts of power, which is capable of lighting about 400 homes. Economic conditions in the future may lead to the construction of larger wave plants that are capable of using this renewable source of energy.

---

[10]Ten megawatts of power is comparable to the electricity consumed by 20,000 average U.S. households.

Figure 8–24 shows the average wave height experienced along coastal regions and indicates the sites most favorable for wave energy generation (*red areas*). The map shows that west-to-east movement of storm systems in the mid-latitudes between 30 and 60 degrees north or south latitude causes the western coasts of continents to be struck by larger waves than eastern coasts. Thus, more wave energy is generally available along western than eastern shores. Furthermore, some of the largest waves (and greatest potential for wave power) are associated with the prevailing westerly wind belt in the mid-latitude Southern Hemisphere.

 ## Students Sometimes Ask...

Do waves always travel in the same directions as currents?

Not always. Surface currents and most waves are created by winds blowing across the ocean surface, so it seems logical that they should move in the same direction, too. Most surface waves do travel in the same direction as the wind blows, but waves radiate outward in all directions from the disturbance (release of energy) that creates them. In addition, as waves move away from the sea area where they were generated, they enter areas where other currents exist. Consequently, the direction of wave movement is often unrelated to that of currents. Fundamentally, movement of waves is independent of currents, so wave trains can travel in various directions relative to currents. In fact, waves can even travel in a direction *completely opposite* to that of a current. A rip current, for example, moves away from shoreline, opposite to the direction of incoming waves.

Can internal waves break?

Internal waves do not break in the way that surface waves break in the surf zone. When internal waves approach the edges of continents, however, they do undergo similar physical changes as waves in the surf zone. This causes the waves to build up and expend their energy with much turbulent motion, in essence "breaking" against the continent.

Can a wave break twice?

Certainly. This is commonly seen anywhere offshore obstacles such as coral reefs, rock reefs, or sand bars are shallowly submerged. Waves approaching these obstacles undergo physical changes just as breaking waves do in the surf zone, expending some of their energy by breaking. Some of the energy moves over the top of the obstacle, however, and the wave—now diminished in size—continues to move toward shore. The wave will expend the last of its energy against the shore, effectively breaking twice.

I know that swell is what surfers hope for. Is swell always big?

Not necessarily. Swell is defined as waves that have moved out of their area of origination, so these waves do not have to

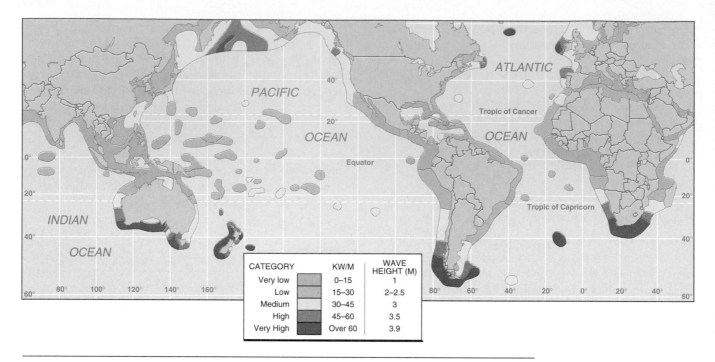

Figure 8–24 Global coastal wave energy resources. Distribution of coastal wave energy shows that more wave energy is available along western shores of continents, especially in the Southern Hemisphere. KW/M is kilowatts per meter (for example, every meter of "red" shoreline is a potential site for generating over 60 kilowatts of electricity); average wave height is in meters.

be a certain wave height to be classified as swell. It is true, however, that the uniform and symmetrical shape of most swell delights surfers.

In your opinion, where is the best surfing?

The answer to that question is highly subjective, with each surfer claiming his or her own favorite spot—often revealed to very few others. Some of the factors that determine an ideal surfing location must include size and regularity of waves, climate, cost, accessibility, and seclusion (not necessarily in that order). Figure 8–10, which compares worldwide wave height, shows that some of the largest waves are found in the south Indian and South Pacific Oceans. Because the Pacific Ocean has the greatest fetch (which allows for the possibility of bigger waves to develop), continents and islands in the South Pacific seem like ideal locations if wave size is the most important factor.

Surfing can often be enhanced, however, when waves have been sorted into swell, so locations a bit farther from the Southern Hemisphere might have slightly smaller waves but ones that are more regular. In this case, some of the tropical Pacific islands (such as Hawaii, Fiji, and some of the more remote Indonesian islands) might offer better surfing. Besides their location in warm climates, they also have the advantage of being exposed to North Pacific Ocean swell.

If seclusion is the highest priority—and you don't mind donning a full wetsuit—some coastal locations with large waves in Alaska have recently been surfed, and surfing in Antarctica can't be far behind!

Why is surfing so much better along the west coast of the United States than along the east coast?

There are three main reasons why the west coast has better surfing conditions:

- The waves are generally bigger in the Pacific. The Pacific is larger than the Atlantic, so the fetch is larger, allowing bigger waves to develop in the Pacific (see Figure 8–24).

- The beach slopes are generally steeper along the west coast. Along the east coast, the gentle slopes often create spilling breakers, which are not as favorable for surfing. The steeper beach slopes along the west coast cause plunging breakers, which are better for surfing.

- The wind is more favorable. Most of the United States is influenced by the prevailing westerlies, which blow toward shore and enhance waves along the west coast. Along the east coast, the wind blows away from shore.

What is the record height of a tsunami?

Japan holds the record because Japan's proximity to subduction zones causes it to endure more tsunami than any other place on Earth (followed by Chile and Hawaii). The largest documented tsunami occurred in the Ryukyu Islands of southern Japan in 1971, when one raised normal sea level by 85 meters (278 feet). In low-lying coastal areas, such an enormous vertical rise can send water many kilometers inland, causing flooding and widespread damage. The most deadly tsunami was probably the one that hit Aura, Japan, in 1703 and was responsible for an estimated 100,000 deaths.

Since 5-foot surf is common at the beach where I live, why should I be worried about a 5-foot tsunami?

A tsunami is quite different from a normal ocean wave, even though both may have the same wave height. A tsunami carries much more energy than a normal ocean wave because a tsunami influences the *entire* column of water, while an ocean wave is confined to surface waters. A tsunami has a much greater wavelength than a wave, so it will reach much farther ashore, too. The biggest storm breakers expend their energy within the surf zone while a tsunami can reach several kilometers inland in low-lying areas. Five-foot surf may be fun to play in, but a 5-foot tsunami can knock down the building you are in, drag you out to sea, and batter you to death.

If there is a tsunami warning issued, what is the best thing to do?

The *smartest* thing to do is to stay out of coastal areas, but people often want to see a tsunami for themselves. For instance, when an earthquake of magnitude $M_w = 7.7$ occurred offshore of Alaska in May 1986, a tsunami warning was issued for the west coast of the United States. In southern California, people flocked to the beach to observe this natural phenomenon. Fortunately, the tsunami was only a few centimeters by the time it reached southern California, so it went unnoticed.

If you must go to the beach to observe the tsunami, expect crowds, road closings, and general mayhem. It would be a good idea to stay at least 30 meters (100 feet) above sea level. If you happen to be at a remote beach where the water suddenly withdraws, evacuate immediately to higher ground (Figure 8D). And, if you happen to be at a beach where an earthquake occurs and shakes the ground so hard that you can't stand up, *RUN*—don't walk—for high ground as soon as you *can* stand up!

Figure 8D Warning sign advises residents to evacuate low-lying areas during a tsunami.

After the first surge of the tsunami, stay out of low-lying coastal areas for several hours because several more surges (and withdrawals) can be expected. There have been many documented cases where curious people have been killed when they are trapped by the third or fourth surge of a tsunami.

Chapter in Review

• All ocean waves begin as disturbances caused by releases of energy. The releases of energy include wind, the movement of fluids of different densities (which create internal waves), mass movement into the ocean, underwater sea floor movements, the gravitational pull of the Moon and the Sun on Earth, and human activities in the ocean.

• Once initiated, waves transmit energy through matter by setting up patterns of oscillatory motion in the particles that make up the matter. Progressive waves are longitudinal, transverse, or orbital, depending on the pattern of particle oscillation. Particles in ocean waves move primarily in orbital paths.

• Waves are described according to their wavelength (L), wave height (H), wave steepness (H/L), wave period (T), frequency (f), and wave speed (S). As a wave travels, the water passes the energy along by moving in a circle, called circular orbital motion. This motion advances the waveform, not the water particles themselves. Circular orbital motion decreases with depth, ceasing entirely at wave base, which is equal to one-half the wavelength measured from still water level.

• If water depth is greater than one-half the wavelength, a progressive wave travels as a deep-water wave with a speed that is directly proportional to wavelength. If water depth is less than $1/20$ wavelength ($L/20$), the wave moves as a shallow-water wave with a speed that is directly proportional to water depth. Transitional waves have wavelengths between deep- and shallow-water waves, with speeds that depend on both wavelength and water depth.

• As wind-generated waves form in a sea area, capillary waves with rounded crests and wavelengths less than 1.74 centimeters (0.7 inch) form first. As the energy of the waves increases, gravity waves form, with increased wave speed, wavelength, and wave height. Factors that influence the size of wind-generated waves include wind speed, duration (time), and fetch (distance). An equilibrium condition called a fully developed sea is reached when the maximum wave height is achieved for a particular wind speed, duration, and fetch.

• Energy is transmitted from the sea area across the ocean by uniform, symmetrical waves called swell. Different wave trains of swell can create either constructive, destructive, or mixed interference patterns. Constructive interference produces unusually large waves called rogue waves or superwaves.

• As waves approach shoaling water near shore, they undergo many physical changes. Waves release their energy in the surf zone when their steepness exceeds a 1:7 ratio and break. If waves break on a relatively flat surface, they produce spilling breakers. The curling crests of plunging breakers, which are the best for surfing, form on steep slopes and abrupt beach slopes create surging breakers.

• When swell approaches the shore, segments of the waves that first encounter shallow water are slowed. The parts of the waves in deeper water move at their original speed, causing each wave to refract, or bend. Refraction concentrates wave energy on headlands, while low-energy breakers are characteristic of bays.

• Reflection of waves off seawalls or other barriers can cause an interference pattern called a standing wave. The crests of standing waves do not move laterally as in progressive waves but alternate with troughs at antinodes. Between the antinodes are nodes, where there is no vertical movement of the water.

• Sudden changes in the elevation of the sea floor, such as from fault movement or volcanic eruptions, generate tsunami, or seismic sea waves. These waves often have lengths exceeding 200 kilometers (125 miles) and travel across the open ocean with undetectable heights of about 0.5 meter (1.6 feet) at speeds in excess of 700 kilometers (435 miles) per hour. Upon approaching shore, a tsunami produces a series of rapid withdrawals and surges, some of which may increase the height of sea level by over 30 meters (100 feet). Most tsunami occur in the Pacific Ocean, where they have caused millions of dollars of coastal damage and taken tens of thousands of lives. The Pacific Tsunami Warning Center (PTWC) has dramatically reduced fatalities by successfully predicting tsunami using real-time seismic information.

• Ocean waves can be harnessed to produce hydroelectric power, but significant problems must be overcome to make this a practical source of energy.

Key Terms

Atmospheric wave (p. 238)

Capillary wave (p. 244)

Circular orbital motion (p. 241)

Constructive interference (p. 246)

Crest (p. 240)

Decay distance (p. 245)

Deep-water wave (p. 241)

Destructive interference (p. 246)

Frequency (p. 240)

Fully developed sea (p. 244)

Gravity wave (p. 244)

Interference pattern (p. 245)

Internal wave (p. 238)

LIMPET 500 (p. 259)

Longitudinal wave (p. 239)

Mixed interference (p. 247)

Ocean wave (p. 238)

Orbital wave (p. 240)

Orthogonal line (p. 252)

Pacific Tsunami Warning Center (PTWC) (p. 256)

Plunging breaker (p. 250)

Refraction (p. 251)

Restoring force (p. 244)

Rogue wave (p. 249)

Sea (p. 244)

Shallow-water wave (p. 242)

Shoaling (p. 248)

Spilling breaker (p. 250)

Splash wave (p. 236)

Standing wave (p. 253)

Still water level (p. 240)

Superwave (p. 249)

Surf beat (p. 247)

Surf zone (p. 247)

Surging breaker (p. 250)

Swell (p. 245)

Transitional wave (p. 244)

Transverse wave (p. 239)

Trough (p. 240)

Tsunami (p. 253)

Wave base (p. 241)

Wave dispersion (p. 245)

Wave height (p. 240)

Wave period (p. 240)

Wave reflection (p. 253)

Wave speed (p. 242)

Wave steepness (p. 240)

Wave train (p. 245)

Wavelength (p. 240)

Questions and Exercises

1. How large was the largest wave ever authentically recorded? Where did it occur, and how did it form?

2. Discuss several different ways in which waves form. How are most ocean waves generated?

3. Why is the development of internal waves likely within the pycnocline?

4. Discuss longitudinal, transverse, and orbital wave phenomena, including the states of matter in which each can transmit energy.

5. Draw a diagram of a simple progressive wave. From memory, label the crest, trough, wavelength, wave height, and still water level.

6. Can a wave with a wavelength of 14 meters ever be more than 2 meters high? Why or why not?

7. What physical feature of a wave is related to the depth of the wave base? On the diagram that you drew for Question 5, add the wave base. What is the difference between the wave base and still water level?

8. Explain why the following statements for deep-water waves are either true or false:

    a. The longer the wave, the deeper the wave base.

    b. The greater the wave height, the deeper the wave base.

    c. The longer the wave, the faster the wave travels.

    d. The greater the wave height, the faster the wave travels.

    e. The faster the wave, the greater the wave height.

9. Calculate the speed (S) in meters per second for deep-water waves with the following characteristics:

    a. $L = 351$ meters, $T = 15$ seconds

    b. $T = 12$ seconds

    c. $f = 0.125$ wave per second

10. Using the information about the giant waves experienced by the USS *Ramapo*, what were the waves' wavelength and speed?

11. Define swell. Does swell necessarily imply a particular wave size? Why or why not?

12. Waves from separate sea areas move away as swell and produce an interference pattern when they come together. If Sea A has wave heights of 1.5 meters (5 feet) and Sea B has wave heights of 3.5 meters (11.5 feet), what would be the height of waves resulting from constructive interference and destructive interference? Illustrate your answer (see Figure 8–13).

13. Describe the physical changes that occur to a wave's wave speed (S), wavelength (L), height (H), and wave steepness (H/L) as a wave moves across shoaling water to break on the shore.

14. Describe the three different types of breakers and indicate the slope of the beach that produces the three types. How is the energy of the wave distributed differently within the surf zone by the three types of breakers?

15. Using examples, explain how wave refraction is different from wave reflection.

16. Using orthogonal lines, illustrate how wave energy is distributed along a shoreline with headlands and bays. Identify areas of high and low energy release.

17. Define the terms *node* and *antinode* as they relate to standing waves.

18. Why is it more likely that a tsunami will be generated by faults beneath the ocean along which vertical rather than horizontal movement has occurred?

19. While shopping in a surf shop, you overhear some surfing enthusiasts mention that they would really like to ride the curling wave of a tidal wave at least once in their life, because it is a single breaking wave of enormous height. What would you say to these surfers?

20. Explain what it would look like at the shoreline if the trough of a tsunami arrives there first. What is the impending danger?

21. What ocean depth would be required for a tsunami with a wavelength of 220 kilometers (136 miles) to travel as a deep-water wave? Is it possible that such a wave could become a deep-water wave any place in the world ocean? Explain.

22. Explain how the tsunami warning system in the Pacific Ocean works. Why must the tsunami be verified at the closest tide recording station?

23. Describe the different types of evidence used to support the idea that a large earthquake along the Cascadia subduction zone produced a tsunami in 1700 that was felt in Japan.

24. Discuss some environmental problems that might result from developing facilities for conversion of wave energy to electrical energy.

References

Bascom, W. 1980. *Waves and beaches: The dynamics of the ocean surface* (revised edition). New York: Anchor Books (Doubleday).

Bowditch, N. 1958. *American practical navigator*, rev. ed. H.O. Pub. 9. Washington, DC: U.S. Naval Oceanographic Office.

Brown, J. 1989. Rogue waves. *Discover* 10:4, 47–52.

Folger, T. 1994. Waves of destruction. *Discover* 15:5, 66–73.

Furbish, D. J., and Parker, W. C. 2000. Corks, buoyancy, and wave-particle orbits. *Journal of Geoscience Education* 48:4, 500–507.

Gill, A. E. 1982. *Atmosphere-ocean dynamics*. International Geophysics Series, Vol. 30. Orlando, FL; Academic Press.

Kinsman, B. 1965. *Wind waves: Their generation and propagation on the ocean surface*. Englewood Cliffs, NJ: Prentice-Hall.

Knamori, H., and Kikuchi, M. 1993. The 1992 Nicaragua earthquake: A slow tsunami earthquake associated with subducted sediments. *Nature* 361:6414, 714–716.

McCredie, S. 1994. When nightmare waves appear out of nowhere to smash the land. *Smithsonian* 24:12, 28–39.

Melville, W., and Rapp, R. 1985. Momentum flux in breaking waves. *Nature* 317:6037, 514–516.

Miller, D. J. 1960. *Giant waves in Lituya Bay, Alaska*. U. S. Geological Survey Professional Paper 354-C.

Pickard, G. L. 1975. *Descriptive physical oceanography: An introduction*, 2nd ed. New York: Pergamon Press.

Satake, K., Shimazaki, K., Tsuji, Y., and Ueda, K. 1996. Time and size of a giant earthquake in Cascadia inferred from Japanese tsunami records of January 1700. *Nature* 379: 6562, 246–249.

Schneider, D. 1995. Tempest on the high sea: The Ocean Drilling Program narrowly averts catastrophe. *Scientific American* 273:6, 14–16.

Stewart, R. H. 1985. *Methods of satellite oceanography*. Berkeley, CA: University of California Press.

Sverdrup, H. U., Johnson, M. W., and Fleming, R. H. 1942. Renewal 1970. *The oceans: Their physics, chemistry, and general biology*. Englewood Cliffs, NJ: Prentice-Hall.

Tappin, D. R., et. al. 1999. Sediment slump likely caused 1998 Papua New Guinea tsunami. *Eos Trans. AGU* 80:30, 329–334.

van Arx, W. S. 1962. *An introduction to physical oceanography*. Reading, MA: Addison-Wesley.

Suggested Reading in **Scientific American**

Bascom, W. 1959. Ocean waves. 201:2, 89–97. An informative discussion by a renowned expert of the nature of wind-generated waves, tsunami, and tides.

Greenberg, D. A. 1987. Modeling tidal power. 257:5, 44–55. Examines what effect a dam to extract tidal power would have in upper New England.

González, F. I. 1999. Tsunami! 280:5, 56–65. One of the world's leading tsunami researchers examines the physics and results of these killer waves, with special attention to descriptions of recent tsunami.

Hyndman, R. D. 1995. Giant earthquakes of the Pacific Northwest. 273:6, 68–75. A review of the evidence that supports the idea that the Cascadia subduction zone can produce extremely large earthquakes, with special attention to the January 1700 tsunami felt in Japan.

Koehl, M. A. R. 1982. The interaction of moving water and sessile organisms. 274:6, 124–135. A discussion of the adaptations of benthic shore-dwelling animals to the stresses of strong currents and breaking waves.

Oceanography on the Web

Visit the *Essentials of Oceanography* home page for on-line resources for this chapter. There you will find an on-line study guide with review exercises, and links to oceanography sites to further your exploration of the topics in this chapter. *Essentials of Oceanography* is at: **http://www.prenhall.com/ thurman** (click on the Table of Contents menu and select this chapter).

A BRIEF HISTORY OF SOME SUCCESSFUL TIDAL POWER PLANTS

Throughout history, tides have been used as a source of power by trapping water during high tide and using its fluidity and weight to do work as it flows back to the sea. In the 12th century, for example, water wheels driven by the tides were used to power gristmills and sawmills. During the 17th and 18th centuries, much of Boston's flour was produced at a tidal mill. Today, tidal water trapped in bays in estuaries (*aestus* = tide; an *estuary* is an inland arm of the sea along a river) can be used to turn turbines and generate electrical energy.

One successful tidal power plant is operating in the estuary of La Rance River near the English Channel in northern France (Figure 9A). The estuary has a surface area of approximately 23 square kilometers (9 square miles), and the **tidal range** (the vertical difference between high and low tides) is 13.4 meters (44 feet). Usable tidal energy increases as the area of the basin increases and as the tidal range increases.

The power-generating barrier was built across the estuary a little over 3 kilometers (2 miles) upstream to protect it from storm waves. The barrier is 760 meters (2500 feet) wide and supports a two-lane road (Figure 9A). Water passing through the barrier powers 24 electricity-generating units that operate beneath the power plant. Each unit can generate 10 million watts (10 megawatts) of electricity, enough to serve the energy needs of over 1500 homes for a year.

The plant generates electricity only when sufficient water height exists between the estuary and the ocean—about half of the time. Annual power production of about 540 million kilowatt-hours can be increased to 670 million kilowatt-hours by using the turbine generators as pumps to move water into the estuary at proper times.

Within the Bay of Fundy, which has the largest tidal range in the world, the Canadian province of Nova Scotia has constructed a small tidal power plant that has generated up to 40 million kilowatt-hours per year since its completion in 1984. It is the only North American tidal power plant in operation today, and is built on the Annapolis River estuary, an arm of the Bay of Fundy, where maximum tidal range is 8.7 meters (26 feet).

Some engineers think a tidal power plant could be made to generate electricity continually if it were located on the Passamaquoddy Bay near the U.S.–Canadian border at the south end of the Bay of Fundy. Although a tidal power plant across the Bay of Fundy has often been proposed, it has never been built. Potentially, the usable tidal energy seems great compared to the La Rance plant, because the flow volume is about 117 times greater.

- What causes the tides?
- How is a lunar day different from a solar day?
- Which body creates a larger tidal influence on Earth: The Moon or the Sun?
- How do the relative positions of the Earth-Moon-Sun affect the tidal range on Earth?
- What are differences between diurnal, semidiurnal, and mixed tidal patterns?
- Where is the world's largest tidal range?
- What types of tidal currents exist?

I derive from the celestial phenomena the forces of gravity with which bodies tend to the sun and several planets. Then from these forces, by other propositions which are also mathematical, I deduce the motions of the planets, the comets, the moon, and the sea.

—*Sir Isaac Newton,* Philosophiae Naturalis Principia Mathematica (Philosophy of Natural Mathematical Principles) *(1686)*

key questions

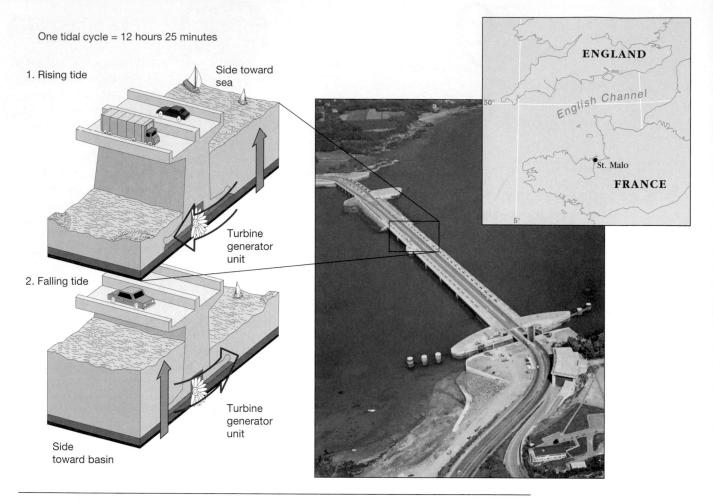

One tidal cycle = 12 hours 25 minutes

1. Rising tide

Side toward sea

Turbine generator unit

2. Falling tide

Turbine generator unit

Side toward basin

ENGLAND

English Channel

50°

St. Malo

FRANCE

5°

Figure 9A La Rance tidal power plant at St. Malo, France.

Tides are the periodic raising and lowering of average sea level that occurs throughout the oceans of the world. As sea level rises and falls, the edge of the sea slowly shifts landward and seaward daily, often destroying sand castles built during low tide. Tides are so important that accurate records have been kept at nearly every port for several centuries and there are many examples of the term *tide* in everyday vocabulary (for instance, "to tide someone over," "to go against the tide," or to wish someone "good tidings").

People have undoubtedly observed the tides for as long as they have inhabited the coastal regions of continents. No written record of tides exists before Herodotus' observations of the Mediterranean Sea in about 450 B.C. Even the earliest sailors knew the Moon had some connection with the tides because both followed a similar cyclic pattern. However, it wasn't until Sir **Isaac Newton** (1642–1727) developed the universal law of gravitation that the tides could be adequately explained.

Although the study of the tides can be complex, tides are fundamentally very long and regular shallow-water waves. Their wavelengths are measured in thousands of kilometers and their heights range to more than 15 meters (50 feet). The gravitational attraction of the Sun

and Moon generate ocean tides, thereby affecting every particle of water from the surface to the deepest ocean basin.

Generating Tides

Fundamentally, tides are generated by forces imposed on Earth that are generated by a combination of *gravity* and *motion* among Earth, the Moon, and the Sun.

Tide-Generating Forces

Newton's work on quantifying the forces involved in the Earth–Moon–Sun system led to the first understanding of why tides behave as they do. It is well known that gravity tethers the Sun, its planets, and their moons together. Most of us are taught that "the Moon orbits Earth," but it is not quite that simple. The two bodies actually rotate around a common center of mass called the **barycenter** (*barus* = heavy, *center* – center), which is located 1600 kilometers (1000 miles) beneath Earth's surface (Figure 9–1a). This can be visualized by imagining Earth and its Moon as ends of a sledgehammer, flung into space, tumbling slowly end over end about its balance point, which is closest to the hammer. The Earth–Moon system is involved in a mutual orbit held together by gravity and motion, which prevents the Moon and Earth from colliding. In this way, orbits are established which keep objects at more-or-less fixed distances. Gravity also tugs every particle of water on Earth toward the Moon and the Sun, thus creating tides on Earth.

Gravitational and Centripetal Forces in the Earth–Moon System

To understand how *tide-generating forces* influence the oceans, let's examine how *gravitational forces* and *centripetal forces* affect objects on Earth within the Earth–Moon system. (We'll ignore the influence of the Sun for the moment.)

The **gravitational force** is derived from Newton's law of universal gravitation, which states that *every particle of mass in the universe attracts every other particle of mass*. The attraction occurs with a force that is directly proportional to the product of their masses and inversely proportional to the square of the distance between the masses. This can be stated as:

> If mass increases (↑), then gravitational force increases (↑)

because objects with a large mass (such as the Sun) produce a large gravitational force. In addition:

If distance increases (↑), then gravitational force greatly decreases (↓↓)

Gravitational attraction varies with the square of distance, so even a small *increase* in the distance between two objects significantly *decreases* the gravitational force between them. Thus, the *greater* the mass of the objects and the *closer* they are together, the greater their gravitational attraction.

Figure 9–2 shows how gravitational forces for points on Earth (caused by the Moon) vary depending on their distances from the Moon. The greatest gravitational attraction (the longest arrow) is at *Z*, the **zenith** (*zenith* = a path over the head), which is the point closest to the Moon. The gravitational attraction is weakest at *N*, the **nadir** (*nadir* = opposite the zenith), which is the point farthest from the Moon. The direction of the gravitational attraction between most particles and the center of the Moon is at an angle relative to a line connecting the center of Earth and the Moon (Figure 9–2). This angle causes the force of gravitational attraction between each particle and the Moon to be slightly different.

Gravitational attraction is inversely proportional to the *square* of the distance between the masses. However, the tide-generating force is inversely proportional to the *cube* of the distance between each point on

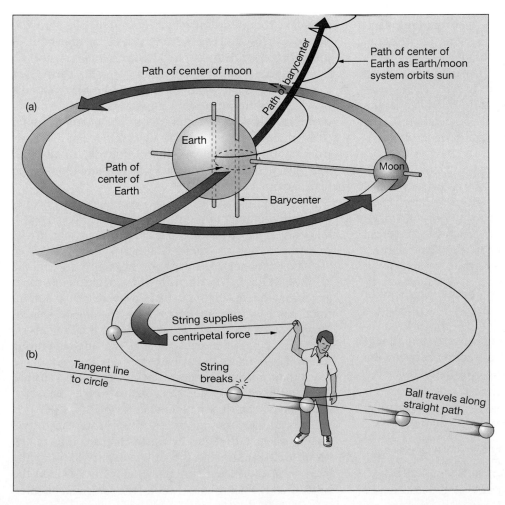

(a)

Path of center of moon

Path of center of Earth

Earth

Barycenter

Path of barycenter

Path of center of Earth as Earth/moon system orbits sun

Moon

(b)

String supplies centripetal force →

Tangent line to circle

String breaks

Ball travels along straight path

Figure 9–1 Earth–Moon system rotation. (a) The center of mass (barycenter) of the Earth–Moon system moves in a nearly circular orbit around the Sun. **(b)** If a ball with a string attached is swung overhead, it stays in a circular orbit because the string exerts a centripetal (center-seeking) force on the ball. If the string breaks, the ball will fly off along a straight path along a tangent to the circle.

Figure 9–2 Gravitational forces on Earth due to the Moon. The gravitational forces on objects located at different places on Earth due to the Moon are shown by arrows. The length and orientation of the arrows indicate the strength and direction of the gravitational force. Notice the length and angular differences of the arrows for different points on Earth. The letter Z represents the zenith; N represents the nadir. Distance between Earth and Moon not shown to scale.

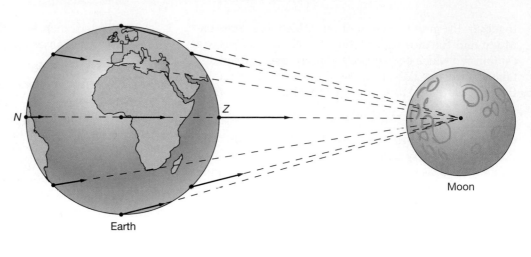

Earth and the *center* of the tide-generating body (Moon or Sun). The tide-generating force is similar to the gravitational force, but they are not linearly proportional.

The **centripetal** (*centri* = the center, *pet* = seeking) **force**[1] required to keep planets in their orbits is provided by the gravitational attraction between each of them and the Sun. Centripetal force "tethers" an orbiting body to its parent, pulling the object *inward* toward the parent, "seeking the center" of its orbit. For example, if you tie a string to a ball and swing the tethered ball around your head (Figure 9–1b), the string pulls the ball toward your hand. The string provides a *centripetal force* on the ball, forcing the ball to *seek the center* of its orbit. If the string should break, the force is gone and the ball can no longer maintain its circular orbit. The ball flies off in a *straight* line,[2] tangent (*tangent* = touching) to the circle (Figure 9–1b).

The Earth and Moon are tethered, too, not by strings but by gravity. Gravity provides the centripetal force that holds the Moon in its orbit around Earth. If all gravity in the solar system could be shut off, centripetal force would vanish, and the momentum of the celestial bodies would send them flying off into space along straight-line paths, tangent to their orbits.

Resultant and Tide-Generating Forces Particles of identical mass rotate in identical sized paths due to the Earth–Moon rotation system (Figure 9–3). Each particle requires an identical centripetal force to maintain it in its circular path. Gravitational attraction between the particle and the Moon supplies the centripetal force, but

the *supplied* force is different than the *required* force (because gravitational attraction varies with distance from the Moon) except at the center of Earth. This difference creates tiny **resultant forces**, which are the mathematical difference between the two sets of arrows shown in Figures 9–2 and 9–3.

Figure 9–4 combines Figures 9–2 and 9–3 to show that resultant forces are produced by the difference between the required centripetal (C) and supplied gravitational (G) forces. However, do not think that both of these forces are being applied to the points, because (C) is a force that would be required to keep the particles in a perfectly circular path, while (G) is the force actually provided for this purpose by gravitational attraction between the particles and the Moon. The resultant forces (*blue arrows*) are established by constructing an arrow from the tip of the centripetal (*red*) arrow to the tip of the gravity (*black*) arrow and located where the red and black arrows begin.

Resultant forces are small, averaging about one-millionth the magnitude of Earth's gravity. If the resultant force is vertical to Earth's surface, as it is at the zenith and nadir (oriented upward) and along an "equator" connecting all points halfway between the zenith and nadir (oriented downward), it has no tide-generating effect (Figure 9–5). However, if the resultant force has a significant *horizontal component*—that is, tangent to Earth's surface—it produces tidal bulges on Earth, creating what are known as the **tide-generating forces**. These tide-generating forces are quite small but reach their maximum value at points on Earth's surface at a "latitude" of 45 degrees relative to the "equator" between the zenith and nadir (Figure 9–5).

The tide-generating forces push water into two bulges: one on the side of Earth directed *toward* the Moon (the zenith) and the other on the side directed *away from* the Moon (the nadir) (Figure 9–6). On the side directly facing the Moon, the bulge is created

---

[1]This is not to be confused with the so-called *centrifugal* (*centri* = the center, *fug* = flee) *force*, an apparent force which is oriented outward.

[2]At the moment that the string breaks, the ball will continue along a straight-line path, obeying Newton's first law of motion (the law of inertia), which states that moving objects follow straight-line paths until they are compelled to change that path by other forces.

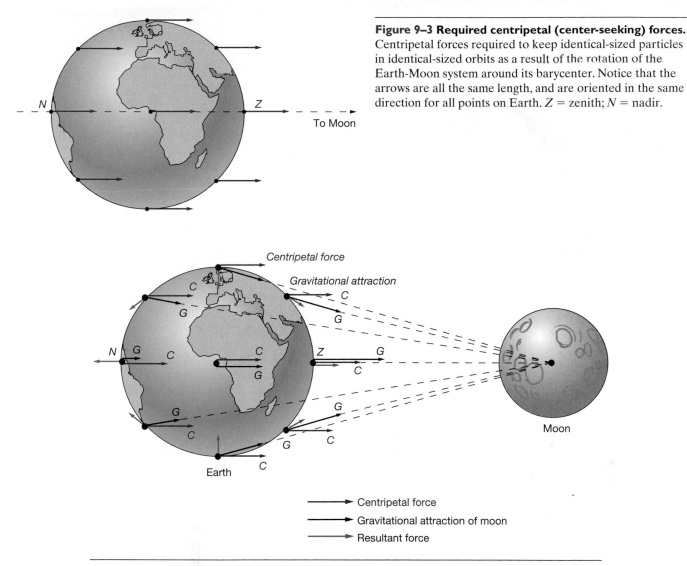

Figure 9–3 Required centripetal (center-seeking) forces. Centripetal forces required to keep identical-sized particles in identical-sized orbits as a result of the rotation of the Earth-Moon system around its barycenter. Notice that the arrows are all the same length, and are oriented in the same direction for all points on Earth. Z = zenith; N = nadir.

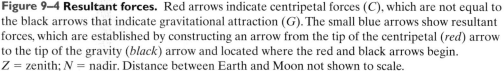

Figure 9–4 Resultant forces. Red arrows indicate centripetal forces (C), which are not equal to the black arrows that indicate gravitational attraction (G). The small blue arrows show resultant forces, which are established by constructing an arrow from the tip of the centripetal (*red*) arrow to the tip of the gravity (*black*) arrow and located where the red and black arrows begin. Z = zenith; N = nadir. Distance between Earth and Moon not shown to scale.

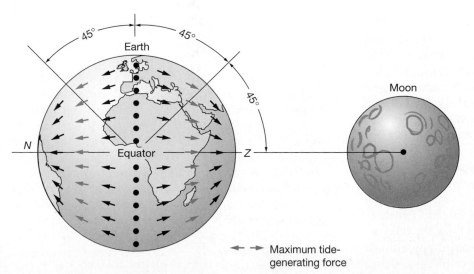

Figure 9–5 Tide-generating forces. Where the resultant force acts vertically relative to Earth's surface, the tide-generating force is zero. This occurs at the zenith (Z) and nadir (N), and along an "equator" connecting all points halfway between the zenith and nadir (*black dots*). However, where the resultant force has a significant *horizontal component*, it produces a tide-generating force on Earth. These tide-generating forces reach their maximum value at points on Earth's surface at a "latitude" of 45 degrees relative to the "equator" mentioned here (*blue arrows*). Distance between Earth and Moon not shown to scale.

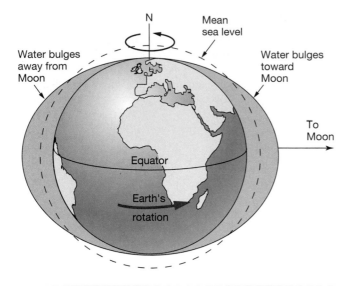

Figure 9–6 Idealized tidal bulges. In an idealized case, the Moon creates two bulges in the ocean surface: One that extends *toward* the Moon and the other *away from* the Moon. As Earth rotates, it carries various locations into and out of the two tidal bulges so that all points on its surface (except the poles) experience two high tides daily.

because the gravitational force is greater than the required centripetal force. Conversely, on the side facing away from the Moon, the bulge is created because the required centripetal force is greater than the gravitational force. Although the forces are oriented in opposite directions on the two sides of Earth, the resultant forces are equal in magnitude, so the bulges are equal, too.

> The tides are caused by an imbalance between the required centripetal and the provided gravitational forces acting on Earth. This difference produces residual forces, the horizontal component of which push ocean water into two equal tidal bulges on opposite sides of Earth.

Tidal Bulges: The Moon's Effect

It is easier to understand how tides on Earth are created if we consider an ideal Earth and an ideal ocean. The ideal Earth has two tidal bulges, one toward the Moon and one away from the Moon (called the **lunar bulges**) as shown in Figure 9–6. The ideal ocean has a uniform depth, with no friction between the seawater and the sea floor. Newton made these same simplifications when he first explained Earth's tides.

If the Moon is stationary and aligned with the ideal Earth's Equator, the maximum bulge will occur on the Equator on opposite sides of Earth. If you were standing on the Equator, you would experience two high tides each day. The time between high tides, the **tidal period**, would be 12 hours. If you moved to any latitude north or south of the Equator, you would experience

the same tidal period, but the high tides would be less high, because you would be at a lower point on the bulge.

In most places on Earth, however, high tides occur every 12 hours 25 minutes because tides depend on the lunar day, not the solar day. The **lunar day** is measured from the time the Moon is on the meridian of an observer—that is, directly overhead—to the next time the Moon is on that meridian, and is 24 hours 50 minutes.[3] The **solar day** is measured from the time the Sun is on the meridian of an observer to the next time the Sun is on that meridian, and is 24 hours. Why is the lunar day 50 minutes longer than the solar day? During the 24 hours it takes Earth to make a full rotation, the Moon has continued moving another 12.2 degrees to the east in its orbit around Earth (Figure 9–7). Thus, Earth must rotate an additional 50 minutes to "catch up" to the Moon so that the Moon is again on the meridian (directly overhead) of our observer.

The difference between a solar day and a lunar day can be seen in some of the natural phenomena related to the tides. For example, alternating high tides are normally 50 minutes *later* each successive day and the Moon rises 50 minutes *later* each successive night.

> A solar day (24 hours) is shorter than a lunar day (24 hours and 50 minutes). The extra 50 minutes is caused by the Moon's movement in its orbit around Earth.

Tidal Bulges: The Sun's Effect

The Sun affects the tides, too. Like the Moon, the Sun produces tidal bulges on opposite sides of Earth, one oriented *toward* the Sun and one oriented *away from* the Sun. Even though the Sun is much more massive than the Moon, it is much further away from Earth, so the **solar bulges** are only 46% the size of the lunar bulges (Figure 9–8). Thus, the Moon controls tides far more than the Sun.

Earth's Rotation

The tides appear to move water in toward shore (the **flood tide**) and to move water away from shore (the **ebb tide**). According to the nature of the idealized tides we have presented, *Earth's rotation carries various parts of Earth into and out of the tidal bulges,* which are in fixed positions relative to the Moon and the Sun. In essence, alternating high and low tides are created as Earth constantly rotates inside a fluid envelope of ocean whose watery bulges are supported by the Moon and the Sun.

---

[3] A lunar day is exactly 24 hours, 50 minutes, 28 seconds long.

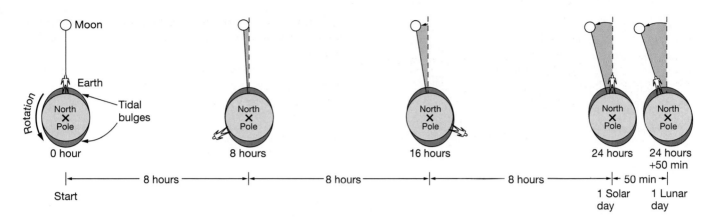

Figure 9–7 The lunar day. A lunar day is the time that elapses between when the Moon is directly overhead and the next time the Moon is directly overhead. During one complete rotation of Earth (the 24-hour solar day), the Moon moves eastward 12.2 degrees, and Earth must rotate an additional 50 minutes to place the Moon in the exact same position overhead. Thus, a lunar day is 24 hours 50 minutes long.

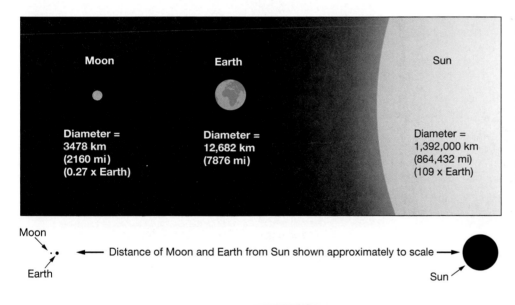

Figure 9–8 Relative sizes and distances of the Moon, Earth, and Sun. *Top*: The relative sizes of the Moon, Earth, and Sun, showing the diameter of the Moon is roughly one-fourth that of Earth, while the diameter of the Sun is 109 times the diameter of Earth. *Bottom*: The relative distances of the Moon, Earth, and Sun are shown to scale.

The lunar bulges are about twice the size of the solar bulges. In an idealized case, the rise and fall of the tides are caused by Earth's rotation carrying various locations into and out of the tidal bulges.

The Monthly Tidal Cycle

The monthly tidal cycle (called a *lunar cycle*) is $29^1/_2$ days because that's how long it takes the Moon to complete an orbit around Earth. During this time, the phase of the Moon changes dramatically. When the Moon is between Earth and the Sun, it cannot be seen at night, and it is called the **new moon**. When the Moon is on the side of Earth opposite the Sun, its entire disk is brightly visible, and it is called a **full moon**. A **quarter moon**—a moon that is half lit and half dark as viewed from Earth—occurs when the Moon is at right angles to the Sun relative to Earth.

Figure 9–9 shows the positions of the Earth, Moon, and Sun at various points during the $29^1/_2$-day lunar cycle. When the Sun and Moon are aligned, either with the Moon between Earth and the Sun (new moon; Moon in *conjunction*) or with the Moon on the side opposite the Sun (full moon; Moon in *opposition*), the tide-generating forces of the Sun and Moon combine (Figure 9–9, *top*). At this time, the tidal range is large (very *high* high tides and quite *low* low tides) because there is *constructive interference*[4] between the lunar and solar tidal bulges. The maximum tidal range is called a **spring** (*springen* = to rise up) **tide**,[5] because the tide is extremely large or "springs forth." When the

---

[4]As mentioned in Chapter 8, constructive interference occurs when two waves (or, in this case, two tidal bulges) overlap crest to crest and trough to trough.

[5]Spring tides have no connection with the spring season; they occur twice a month during the time when the Earth–Moon–Sun system is aligned.

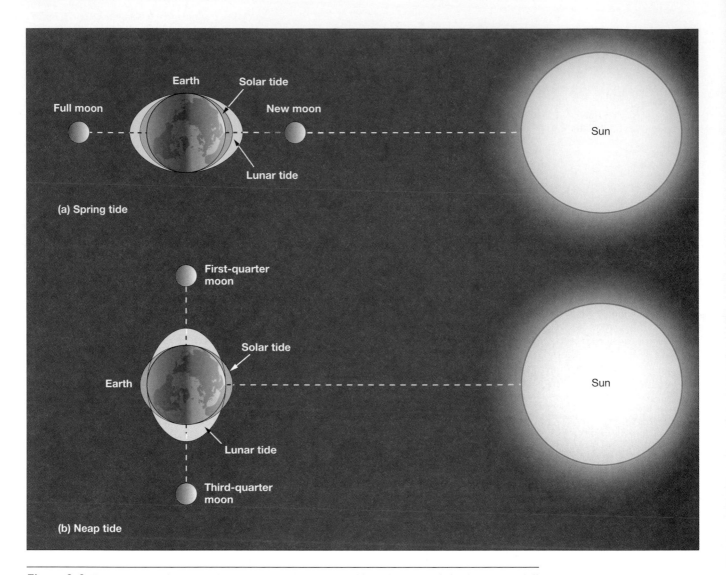

(a) Spring tide

(b) Neap tide

Figure 9–9 Earth–Moon–Sun positions and the tides. *Top*: When the Moon is in the new or full position, the tidal bulges created by the Sun and Moon are aligned, there is a large tidal range on Earth, and spring tides are experienced. *Bottom*: When the Moon is in the first- or third-quarter position, the tidal bulges produced by the Moon are at right angles to the bulges created by the Sun. Tidal ranges are smaller and neap tides are experienced.

Earth–Moon–Sun system is aligned, the Moon is said to be in **syzygy** (*syzygia* = union).

When the Moon is in either the first or third quarter[6] phase (Figure 9–9, *bottom*), the tide-generating force of the Sun is working at right angles to the tide-generating force of the Moon. The tidal range is small (*lower* high tides and *higher* low tides) because there is *destructive interference*[7] between the lunar and solar tidal bulges. This is called a **neap** (*nep* = scarcely or barely touching) **tide**,[8] and the Moon is said to be in **quadrature** (*quadra* = four).

The time between successive spring tides (full moon and new moon) or neap tides (first quarter and third quarter) is one-half the monthly lunar cycle, which is about two weeks. The time between a spring tide and a successive neap tide is one-quarter the monthly lunar cycle, which is about one week.

Figure 9–10 shows the pattern that the Moon experiences as it moves through its monthly cycle. As the Moon progresses from new moon to first quarter phase, the Moon is a **waxing crescent** (*waxen* = to increase,

> Spring tides occur during the full and new moon when the lunar and solar tidal bulges constructively interfere, producing a large tidal range. Neap tides occur during the quarter moon phases when the lunar and solar tidal bulges destructively interfere, producing a small tidal range.

---

[6]The third-quarter moon is often called the last-quarter moon, which is not to be confused with certain sports that have a fourth quarter.
[7]Destructive interference occurs when two waves (or, in this case, two tidal bulges) match up crest to trough and trough to crest.
[8]To help you remember a *neap* tide, think of it as one that has been "*nipped*" in the bud," indicating a small tidal range.

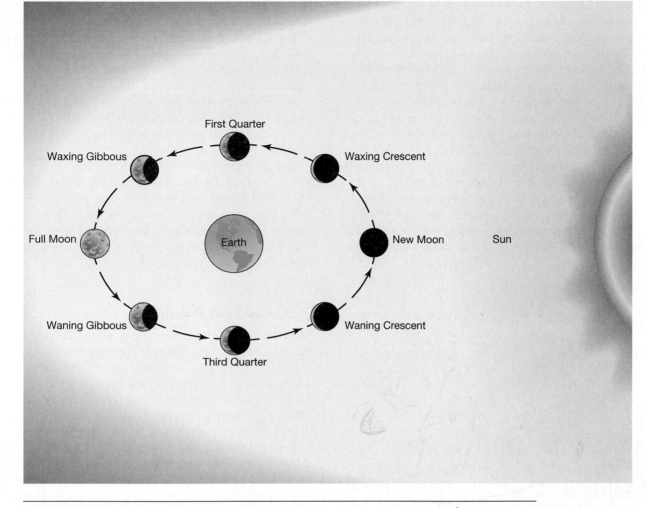

Figure 9–10 Phases of the Moon. As the Moon moves around Earth during its $29\frac{1}{2}$-day lunar cycle, its phase changes depending on its position relative to the Sun and Earth. During a new moon, the dark side of the Moon faces Earth while during a full moon, the lit side of the Moon faces Earth.

crescere = to grow). In between the first quarter and full moon phase, the Moon is a **waxing gibbous** (*gibbus* = hump). Between the Moon's full and third quarter phase, it is a **waning gibbous** (*wanen* = to decrease). And, in between the third quarter and new moon phase, the Moon is a **waning crescent**.

Other Factors

Besides Earth's rotation and the relative positions of the Moon and the Sun, many other factors influence tides on Earth. Two of the most prominent are the declination of the Moon and Sun and the elliptical shapes of Earth's and the Moon's orbits.

Declination of the Moon and Sun The Moon and Sun are not always aligned over the Equator. Most of the year, they are either north or south of the Equator. The angular distance of the Sun or Moon above or below Earth's equatorial plane is called **declination**.

Earth revolves around the Sun along an invisible ellipse in space. The imaginary plane that contains this ellipse is called the **ecliptic**. Recall from Chapter 6 that Earth's axis of rotation is tilted 23.5 degrees with respect to the ecliptic and that this tilt causes Earth's seasons. It also means the maximum declination of the Sun relative to Earth's Equator is 23.5 degrees.

To complicate matters further, the plane of the Moon's orbit is tilted 5 degrees with respect to the ecliptic. Thus, the maximum declination of the Moon's orbit relative to Earth's Equator is 28.5 degrees (5 degrees plus the 23.5 degrees of Earth's tilt). The declination changes from 28.5 degrees south to 28.5 degrees north and back to 28.5 degrees south of the Equator in one lunar cycle (one month). As a result, tidal bulges are rarely aligned with the Equator. Instead, they occur mostly north and south of the Equator. The Moon affects Earth's tides more than the Sun, so tidal bulges follow the Moon, ranging from a maximum of 28.5 degrees north to a maximum of 28.5 degrees south of the Equator (Figure 9–11).

Effects of Elliptical Orbits Earth revolves around the Sun in an elliptical orbit (Figure 9–12) such that Earth is 148.5 million kilometers (92.2 million miles) from the Sun during the Northern Hemisphere winter and 152.2 million kilometers (94.5 million miles) from the Sun during summer. Thus, the distance between Earth and the Sun varies by 2.5% over the course of a year. Tidal ranges are largest when Earth is near its closest point, called **perihelion** (*peri* = near, *helios* = Sun) and smallest near its most distant point, called **aphelion** (*apo* = away from, *helios* = Sun). Thus, the greatest tidal ranges typically occur in January each year.

The Moon revolves around Earth in an elliptical orbit, too. The Earth–Moon distance varies by 8% [between 375,000 kilometers (233,000 miles) and 405,800 kilometers (252,000 miles)]. Tidal ranges are largest when the Moon is closest to Earth, called **perigee** (*peri* = near, *geo* = Earth), and smallest when most distant, called **apogee** (*apo* = away from, *geo* = Earth) (Figure 9–12, *top*). The Moon cycles between perigee, apogee, and back to perigee every 27 1/2 days. When spring tides coincide with perigee, the tides—called **proxigean** (*proximus* = nearest, *geo* = Earth) or "closest of the close moon" tides—are especially large, which often result in the flooding of low-lying coastal areas during high tide.

The elliptical orbits of Earth around the Sun and the Moon around Earth change the distances between Earth, the Moon, and the Sun, thus affecting Earth's tides. The net result is that spring tides have greater ranges during the Northern Hemisphere winter than in the summer, and spring tides have greater ranges when they coincide with perigee.

Idealized Tide Prediction

The declination of the Moon determines the position of the tidal bulges. The example illustrated in Figure 9–13 shows that the Moon is directly overhead at 28 degrees north latitude when its declination is 28 degrees north of the Equator. If you stand at this latitude when the Moon is directly overhead, it will be high tide (Figure 9–13a). Low tide occurs six lunar hours later (6 hours 12 1/2 minutes solar time) (Figure 9–13b). Another high tide, but one much lower than the first, occurs six lunar hours later (Figure 9–13c). Another low tide occurs six lunar hours later (Figure 9–13d). Six lunar hours later, at the end of a 24-lunar-hour period (24 hours 50 minutes solar time), you will have passed through a complete lunar-day cycle of two high tides and two low tides.

The graphs in Figure 9–13e show the heights of the tides observed during the same lunar day at 28 degrees north latitude, the Equator, and 28 degrees south latitude when the declination of the Moon is 28 degrees north of the Equator. Tide curves for 28 degrees north and 28 degrees south latitude have identically timed highs and lows, but the *higher* high tides and *lower* low tides occur 12 hours later. The reason that they occur out of phase by 12 hours is because the bulges in the two hemispheres are on opposite sides of Earth in relation to the Moon. Table 9–1 summarizes the characteristics of the tides on the idealized Earth.

Tides in the Ocean

If tidal bulges are wave crests separated by a distance of one-half Earth's circumference—about 20,000 kilometers (12,420 miles)—one would expect the bulges to move across Earth at about 1600 kilometers (1000 miles) per hour. Tides, however, are an extreme example of shallow-water waves, so their speed is proportional to the water depth. For a tide wave to travel at

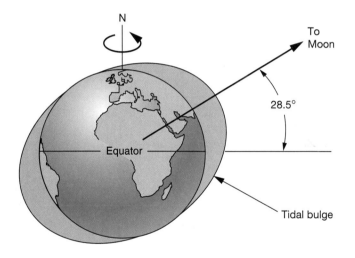

Figure 9–11 Maximum declination of tidal bulges from the Equator. The center of the tidal bulges may lie at any latitude from the Equator to a maximum of 28.5 degrees on either side of the Equator, depending on the season of the year (solar angle) and the Moon's position.

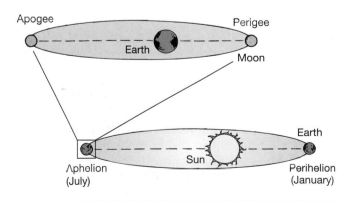

Figure 9–12 Effects of elliptical orbits. *Top:* The Moon moves from its most distant point (*apogee*) to its closest point to Earth (*perigee*), which causes greater tidal ranges every 27 1/2 days. *Bottom:* The Earth also moves from its most distant point (*aphelion*) to its closest point (*perihelion*), which causes greater tidal ranges every year in January.

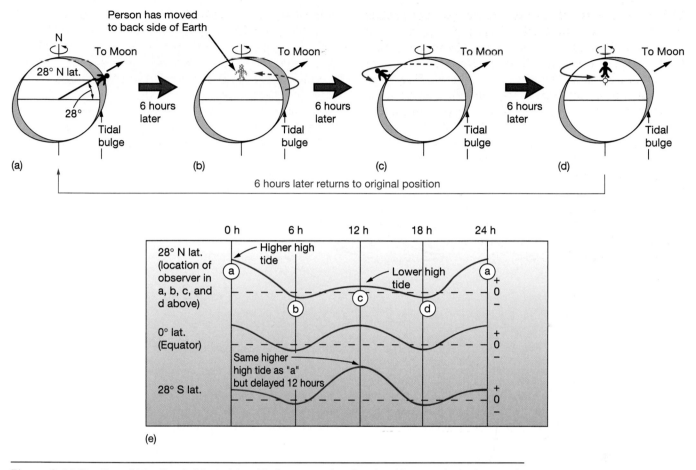

Figure 9–13 Predicted idealized tides. (a)—(d) Sequence showing the tide experienced every 6 lunar hours at 28 degrees north latitude when the declination of the Moon is 28 degrees north. **(e)** Tide curves for 28 degrees north, 0 degrees, and 28 degrees south latitudes during the lunar day shown in the sequence above. The tide curves for 28 degrees north and 28 degrees south latitude show that the higher high tides occur 12 hours later.

Table 9–1   Summary of characteristics of the tides on the idealized Earth.

- Any location (except the poles) will have two high tides and two low tides per lunar day.

- Neither the two high tides nor the two low tides are of the same height because of the declination of the Moon and the Sun (except for the rare occasions when the Moon and Sun are simultaneously above the Equator).

- Monthly and yearly cycles of tidal range are related to the changing distances of the Moon and Sun from Earth.

- Each week, there would be alternating spring and neap tides. Thus, in a lunar month, there are two spring tides and two neap tides.

1600 kilometers (1000 miles) per hour, the ocean would have to be 22 kilometers (13.7 miles) deep! Instead, the average depth of the ocean is only 3.7 kilometers (2.3 miles), so tidal bulges move as *forced waves*, with their speed determined by ocean depth.

Based on the average ocean depth, the average speed at which tide waves can travel across the open ocean is only about 700 kilometers (435 miles) per hour. Thus, the idealized bulges that are oriented toward and away from a tide-generating body cannot exist because they cannot keep up with the rotational speed of Earth. Instead, they break up into cells.

In the open ocean, the crests and troughs of the tide wave rotate around an **amphidromic** (*amphi* = around,

dromus = running) **point** near the center of each cell. There is essentially no tidal range here, but radiating from this point are **cotidal** (*co* = with, *tidal* = tide) **lines**, which connect points where high tide occurs simultaneously. The labels on the cotidal lines in Figure 9–14 indicate the time of high tide in hours after the Moon crosses the Greenwich Meridian.

The times in Figure 9–14 indicate that the tide wave rotates counterclockwise in the Northern Hemisphere and clockwise in the Southern Hemisphere. The wave must complete one rotation during the tidal period (usually 12 lunar hours), so this limits the size of the cells.

Low tide occurs six hours after high tide in an amphidromic cell. If high tide is occurring along the cotidal

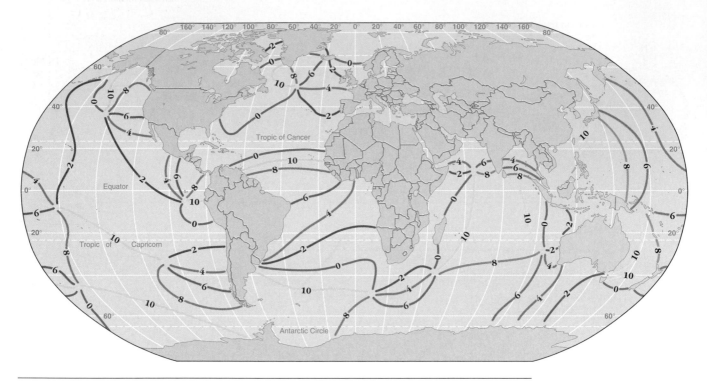

Figure 9–14 Cotidal map of the world. Cotidal lines indicate times of the main lunar daily high tide in lunar hours after the Moon has crossed the Greenwich Meridian (0 degrees longitude). Tidal ranges generally increase with increasing distance along cotidal lines away from the amphidromic points. Where cotidal lines terminate at both ends in amphidromic points, maximum tidal range will be near the midpoints of the lines.

line labeled "10," for example, then low tide is occurring along the cotidal line labeled "4."

The continents affect tides, too, because they interrupt the free movement of the tidal bulges across the ocean surface. The ocean basins between continents have free-standing waves set up within them. The positions and shapes of the continents modify the forced astronomical tide waves that develop within an ocean basin.

Over 150 different factors affect the tides at a particular coast, which are far more than can be adequately addressed here. One of the results of these factors, however, is that high tide rarely occurs when the Moon is at its highest point in the sky. Instead, the time between the Moon crossing the meridian and a corresponding high tide varies from place to place.

Tidal Patterns

In theory, most areas on Earth should experience two high tides and two low tides of unequal heights during a lunar day. In practice, however, the various depths, sizes, and shapes of ocean basins modify tides so they exhibit three different patterns in different parts of the world. The three tidal patterns, which are illustrated in Figure 9–15, are *diurnal* (*diurnal* = daily), *semidiurnal* (*semi* = twice, *diurnal* = daily), and *mixed*.[9]

A **diurnal tidal pattern** has a single high and low tide each lunar day. These tides are common in shallow inland seas such as the Gulf of Mexico and along the coast of Southeast Asia. Diurnal tides have a tidal period of 24 hours 50 minutes.

A **semidiurnal tidal pattern** has two high and two low tides each lunar day. The heights of successive high tides and successive low tides are approximately the same.[10] Semidiurnal tides are common along the Atlantic Coast of the United States. The tidal period is 12 hours 25 minutes.

A **mixed tidal pattern** may have characteristics of both diurnal and semidiurnal tides. Successive high tides and/or low tides will have significantly different heights, called *diurnal inequalities*. Mixed tides commonly have a tidal period of 12 hours 25 minutes, but they may also have diurnal periods. Mixed tides are the most common type in the world, including along the Pacific Coast of North America.

Figure 9–16 shows examples of monthly tidal curves for various coastal locations. Even though a tide at any particular location follows a single tidal pattern, it still may pass through stages of one or both of the other tidal patterns. Typically, however, the tidal pattern for a location remains the same throughout the year. Also,

[9]Sometimes a *mixed* tidal pattern is referred to as *mixed semidiurnal*.

[10]Since tides are always growing higher or lower at any location due to the spring-neap tide sequence, successive high tides and successive low tides can never be *exactly* the same at any location.

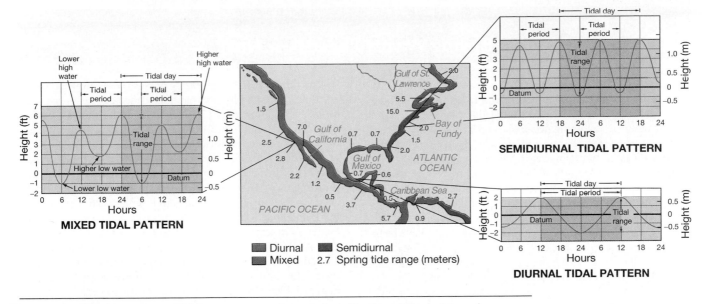

Figure 9–15 Tidal patterns. Tidal patterns experienced along North and Central American coasts. A diurnal tidal pattern (*lower right*) shows one high and low tide each lunar day. A semidiurnal pattern (*upper right*) shows two highs and lows of approximately equal heights during each lunar day. A mixed tidal pattern (*left*) shows two highs and lows of unequal heights during each lunar day.

the tidal curves in Figure 9–16 clearly show the weekly switching of the spring tide–neap tide cycle.

A diurnal tidal pattern exhibits one high and low tide each lunar day; a semidiurnal tidal pattern exhibits two high and low tides daily of about the same height; a mixed tidal pattern usually has two high and low tides daily of different heights, but may also exhibit diurnal qualities.

An Example of Tidal Extremes: The Bay of Fundy

When tide waves enter coastal waters, they are subject to reflection and amplification similar to what wind-generated waves experience. In certain locations reflected wave energy causes water to slosh around in a bay, producing standing waves (see Chapter 8 for a discussion of standing waves). If a standing wave has a period near that of the forced tide wave, it results in constructive interference that produces significant increases in tidal range.

Nova Scotia's **Bay of Fundy** is one such place, and it is here that the largest tidal range in the world is found. With a length of 258 kilometers (160 miles), the Bay of Fundy has a wide opening into the Atlantic Ocean. At its northern end, however, it splits into two narrow basins, Chignecto Bay and Minas Basin (Figure 9–17). The period of free oscillation in the bay—the oscillation that occurs when a body is displaced and then released—is very nearly that of the tidal period. The

resulting constructive interference—along with the narrowing and shoaling of the bay to the north—causes a buildup of tidal energy in the northern end of the bay. In addition, the bay curves to the right, so the Coriolis effect in the Northern Hemisphere adds to the extreme tidal range.

During maximum spring tide conditions, the tidal range at the mouth of the bay (where it opens to the ocean) is only about 2 meters (6.6 feet). However, the tidal range increases progressively from the mouth of the bay northward. In the northern end of Minas Basin, the maximum spring tidal range is 17 meters (56 feet), which leaves boats high and dry during low tide (Figures 9–17, *insets*).

The world's largest tides occur in the upper end of the Bay of Fundy, where reflection and amplification produce a maximum spring tidal range of 17 meters (56 feet).

Coastal Tidal Currents

The current that accompanies the slowly turning tide crest in a Northern Hemisphere basin rotates counterclockwise, producing a **rotary current** in the open portion of the basin. Friction increases in nearshore shoaling waters, so the rotary current changes to an alternating or **reversing current** that moves into and out of restricted passages along a coast.

Box 9–1
Tidal Bores: Boring Waves These Are Not!

A **tidal bore** (*bore* = crest or wave) is a wall of water that moves up certain low-lying rivers due to an incoming tide. Because it is a wave created by the tides, it is a *true* tidal wave. When an incoming tide rushes up a river, it develops a steep forward slope because the flow of the river resists the advance of the tide (Figure 9B). This creates a tidal bore, which may reach heights of 5 meters (16.4 feet) or more and move at speeds up to 22 kilometers (14 miles) per hour.

Tidal bores develop where there is a large tidal range and a low-lying coastal river. Although tidal bores do not attain the size of some waves in the surf zone, tidal bores have been successfully surfed. They can give a surfer a very long ride because the bore travels many kilometers upriver. If you miss the bore, though, you have to wait about half a day before the next one comes along because the incoming high tide occurs only twice a day. In some locations, tidal bore rafting is promoted as a draw for tourists.

The Amazon River probably possesses the longest estuary that is affected by oceanic tides. Tides can be measured as far as 800 kilometers (500 miles) from the river's mouth, although the effects are quite small at this distance. Tidal bores near the mouth of the Amazon River can be up to 5 meters (16.4 feet) high and are locally called *pororocas* (waterfalls). Other rivers that have notable tidal bores include the Chientang River in China [which has the largest tidal bores in the world, often reaching 8 meters (26 feet) high]; the Petitcodiac River in New Brunswick, Canada; the River Seine in France; the Trent River in England; and Cook Inlet near Anchorage, Alaska (where the largest tidal bore in the United States can be found). Although the Bay of Fundy has the world's largest tidal range, its tidal bore rarely exceeds 1 meter (3.3 feet), mostly because the bay is so wide.

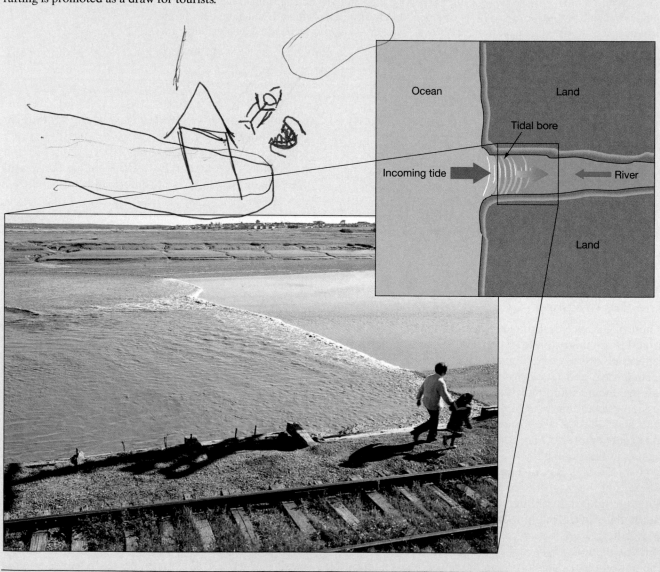

Figure 9B A tidal bore near Chignecto Bay, New Brunswick, Canada.

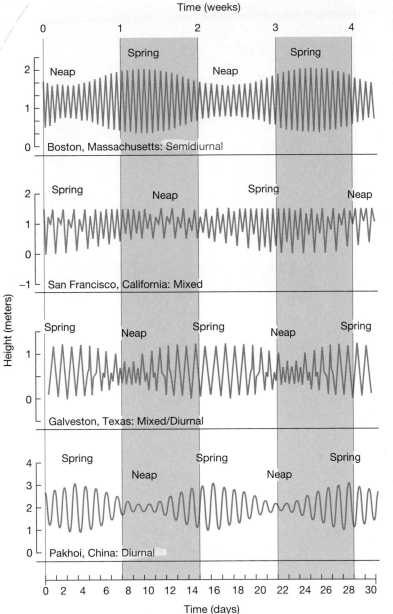

Time (weeks)

Height (meters)

Boston, Massachusetts: Semidiurnal

San Francisco, California: Mixed

Galveston, Texas: Mixed/Diurnal

Pakhoi, China: Diurnal

Time (days)

Figure 9–16 Monthly tidal curves. *Top*: Boston, Massachusetts, showing semidiurnal tidal pattern. *Upper middle*: San Francisco, California, showing mixed tidal pattern. *Lower middle*: Galveston, Texas, showing mixed tidal pattern with strong diurnal tendencies. *Bottom*: Pakhoi, China, showing diurnal tidal pattern.

The velocity of rotary currents in the open ocean is usually well below 1 kilometer (0.6 mile) per hour. Reversing currents, however, can reach velocities up to 44 kilometers (28 miles) per hour in restricted channels such as between islands of coastal waters.

Reversing currents also exist in the mouths of bays (and some rivers) due to the daily flow of tides. Figure 9–18 shows that a **flood current** is produced when water rushes into a bay (or river) with an incoming high tide. Conversely, an **ebb current** is produced when water drains out of a bay (or river) because a low tide is approaching. No currents occur for several minutes during either **high slack water** (which occurs at the peak of each high tide) or during **low slack water** (at the peak of each low tide).

Reversing currents in bays can sometimes reach speeds of 40 kilometers (25 miles) per hour, creating a navigation hazard for ships. On the other hand, the daily flow of these currents often keeps sediment from closing off the bay and resupplies the bay with new seawater and ocean nutrients.

Tidal currents can be significant even in deep ocean waters. For example, tidal currents were encountered shortly after the discovery of the remains of the *Titanic* at a depth of 3795 meters (12,448 feet) on the continental slope south of Newfoundland's Grand Banks in 1985. These tidal currents were so strong that they forced researchers to abandon the use of the camera-equipped tethered remotely operated vehicle, *Jason Jr.*

Rotary tidal currents occur in the deep ocean while reversing tidal currents occur close to shore, most notably in bays and rivers due to the change in the tides.

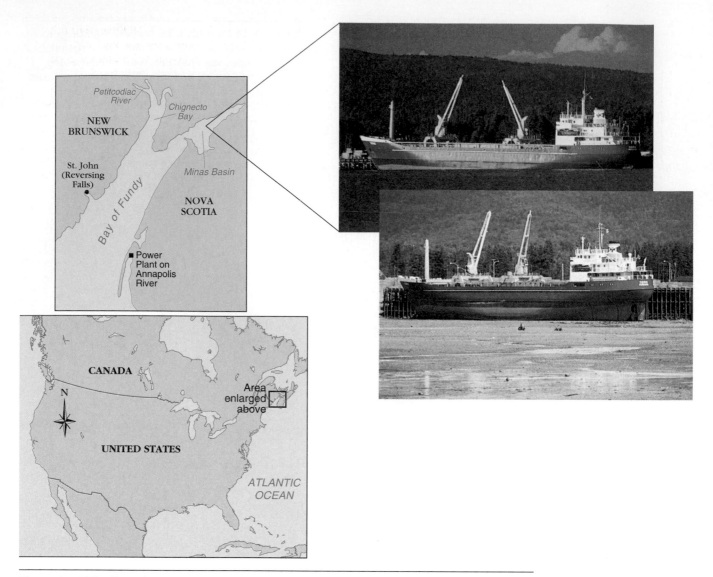

Figure 9–17 The Bay of Fundy, site of the world's largest tidal range. Even though the maximum spring tidal range at the mouth of the Bay of Fundy is only 2 meters (6.6 feet), amplification of tidal energy causes a maximum tidal range at the northern end of Minas Basin of 17 meters (56 feet), often stranding ships (*insets*).

Some Considerations of Tidal Power

Tidal power has long been considered a renewable resource with vast potential. The initial cost of building a tidal power-generating plant may be higher than a conventional thermal power plant, but the operating costs would be less because it does not use fossil fuels or radioactive isotopes to generate electricity.

One disadvantage of tidal power, however, is the periodicity of the tides, allowing power to be generated only during a portion of a 24-hour day. People operate on a solar period, but tides operate on a lunar period, so the energy available from the tides would coincide with need only part of the time. Power would have to be distributed to the point of need at the moment it was generated, which could be a great distance away, resulting in an expensive transmission problem. The power could be stored, but this alternative presents a large and expensive technical problem.

To generate electricity effectively, electrical turbines (generators) need to run at a constant speed, which is difficult to maintain when generated by the variable flow of tidal currents in two directions (flood tide and ebb tide). Specially designed turbines that allow both advancing and receding water to spin their blades are necessary to solve the problem of generating electricity from the tides.

Another disadvantage of tidal power is unwanted environmental effects resulting from the modification of tidal current flow. In addition, a tidal power plant would likely interfere with many traditional uses of coastal waters, such as transportation and fishing.

Currently, there are only a few small tidal power plants in the world (see chapter-opening feature). Whether or not tidal power stations are ever constructed on a large scale, this potential source of energy will likely receive increased attention as fossil fuels are

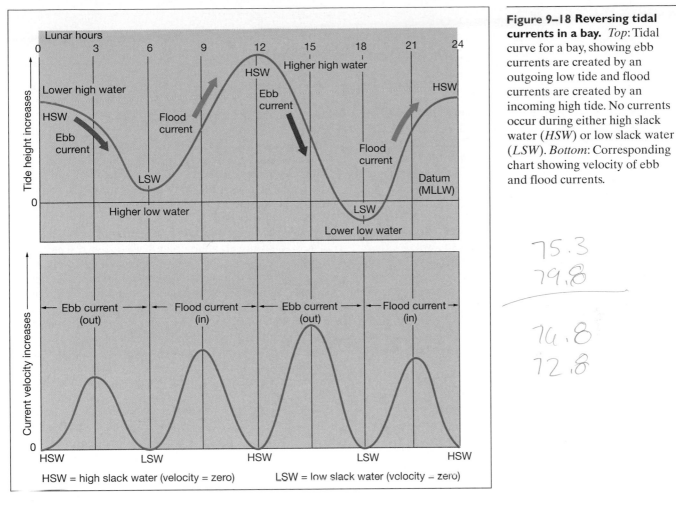

Figure 9–18 Reversing tidal currents in a bay. *Top:* Tidal curve for a bay, showing ebb currents are created by an outgoing low tide and flood currents are created by an incoming high tide. No currents occur during either high slack water (*HSW*) or low slack water (*LSW*). *Bottom:* Corresponding chart showing velocity of ebb and flood currents.

75.3
79.8

74.8
72.8

depleted and the cost of generating electricity by conventional means increases. Worldwide, many sites have the potential for tidal power generation (Figure 9–19).

? Students Sometimes Ask...

Are there also tides in other objects, such as lakes and swimming pools?

The Moon and the Sun act on all objects that have the ability to flow, so there are tides in lakes, wells, and swimming pools. In fact, there are even extremely tiny tidal bulges in a glass of water! However, the tides in the atmosphere and the "solid" Earth have greater significance. Tides in the atmosphere—called *atmospheric tides*—can be miles high. The tides inside Earth's interior—called *Earth tides*—cause a slight but measurable stretching of Earth's crust, typically only a few centimeters high.

I know that the Sun is much more massive than the Moon. Why is the Sun's effect on the tide-generating force only about half that of the Moon's?

The Sun is 27 million times more massive than the Moon, but its tide-generating force is not 27 million times greater than the Moon's because the Sun is 390 times farther from Earth than the Moon (see the bottom part of Figure 9–8). Tide-generating forces vary inversely as the *cube* of the distance between ob-

jects. Thus, the tide-generating force is reduced by the cube of 390, or about *59 million times* compared with that of the Moon. These conditions result in the Sun's tide-generating force being $^{27}/_{59}$ that of the Moon, or 46% (about one-half).

I've heard of a blue moon. Is the Moon really blue then?

No. It's just a phrase that has gained popularity and is synonymous with a rather unlikely occurrence. A blue moon is the second full moon of any calendar month, which occurs when the $29^1/_2$-day lunar cycle falls entirely within a 30- or 31-day month. Because the divisions between our calendar months were determined arbitrarily, a blue moon has no special significance, other than it occurs only once every 2.72 years (about 33 months). At that rate, it's certainly less common than a month of Sundays!

The origin of the term "blue moon" is not exactly known, but one likely explanation involves the Old English word *belewe*, meaning "to betray." Thus, the Moon is *belewe* because it betrays the usual perception of one full moon per month. Another explanation links the term to the Farmer's Almanac, which was first published in color in 1938 and included a calendar designating the first full moon of each month in red color and the second full moon in blue.

What are tropical tides?

Differences between successive high tides and successive low tides occur each lunar day. Because these differences occur

Box 9–2
Grunion: Doing What Comes Naturally on the Beach

From March through September, shortly after the maximum spring tide has occurred, the **grunion** (*Leuresthes tenuis*) come ashore along the beaches of southern California and Baja California to bury their fertilized eggs in the sand. Grunion—slender, silvery, and 12 to 15 centimeters (4.7 to 6 inches) long—are the only marine fish in the world that come completely out of water to spawn. The name "grunion" comes from the Spanish *gruñón*, which means "grunter," and refers to the faint noise they make during spawning.

A mixed tidal pattern occurs along southern California and Baja California beaches. On most tidal days (24 hours and 50 minutes), there are two high and two low tides. There is usually a significant difference in the heights of the two high tides that occur each day. During the summer months, the higher high tide occurs at night. The night high tide becomes higher each night as the maximum spring-tide range is approached, causing sand to be eroded from the beach (Figure 9C, *graph*). After the maximum spring tide has occurred, the night high tide diminishes each night. As neap tide is approached, sand is deposited on the beach.

Grunion spawn only after each night's higher high tide has peaked on the three or four nights following the night of the highest spring high tide. This assures that their eggs will be covered deeply in sand deposited by the receding higher

high tides each succeeding night. The fertilized eggs buried in the sand are ready to hatch nine days after spawning. By this time, another spring tide is approaching, so the night high tide is getting progressively higher each night again. The beach sand is eroding again, too, which exposes the eggs to the waves that break ever higher on the beach. The eggs hatch about three minutes after being freed in the water. Tests done in laboratories have shown that the grunion eggs will not hatch until agitated in a manner that simulates that of the eroding waves.

The spawning begins as the grunion come ashore immediately following an appropriate high tide, and it may last from one to three hours. Spawning usually peaks about an hour after it starts and may last an additional 30 minutes to an hour. Thousands of fish may be on the beach at this time. During a run, the females, which are larger than the males, move high on the beach. If no males are near, a female may return to the water without depositing her eggs. In the presence of males, she drills her tail into the semifluid sand until only her head is visible. The female continues to twist, depositing her eggs 5 to 7 centimeters (2 to 3 inches) below the surface.

The male curls around the female's body and deposits his milt against it (Figure 9C, *picture*). The milt runs down the

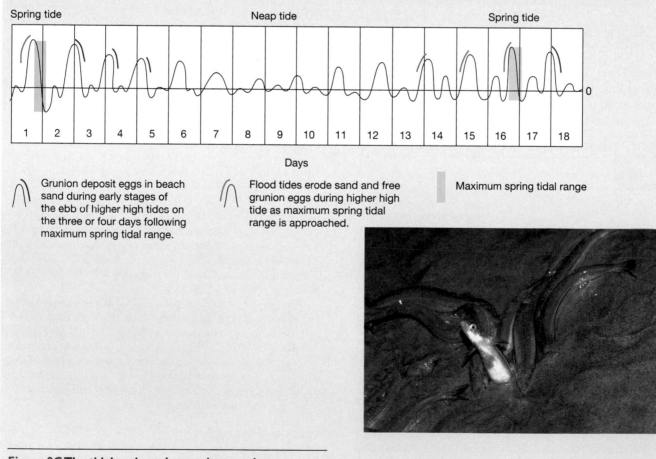

Figure 9C The tidal cycle and spawning grunion.

body of the female to fertilize the eggs. When the spawning is completed, both fish return to the water with the next wave.

Larger females are capable of producing up to 3000 eggs for each series of spawning runs, which are separated by the two-week period between spring tides. As soon as the eggs are deposited, another group of eggs begins to form within the female. They will be deposited during the next spring tide run. Early in the season, only older fish spawn. By May, however, even the one-year-old females are in spawning condition.

Young grunion grow rapidly and are about 12 centimeters (5 inches) long when they are a year old and ready for their first spawning. They usually live two or three years, but four-year-olds have been recovered. The age of a grunion

can be determined by the scales. After growing rapidly during the first year, they grow very slowly thereafter. There is no growth at all during the 6-month spawning season, which causes marks to form on each scale that can be used to identify the grunion's age.

It is not known how the grunion are able to time their spawning behavior so precisely with the tides. Some investigators believe the grunion are able to sense very small changes in the hydrostatic pressure caused by the changing level of the water associated with rising and falling sea level due to the tides. Certainly, a very dependable detection mechanism keeps the grunion accurately informed of the tidal conditions, because their survival depends on a spawning behavior precisely tuned to tidal motions.

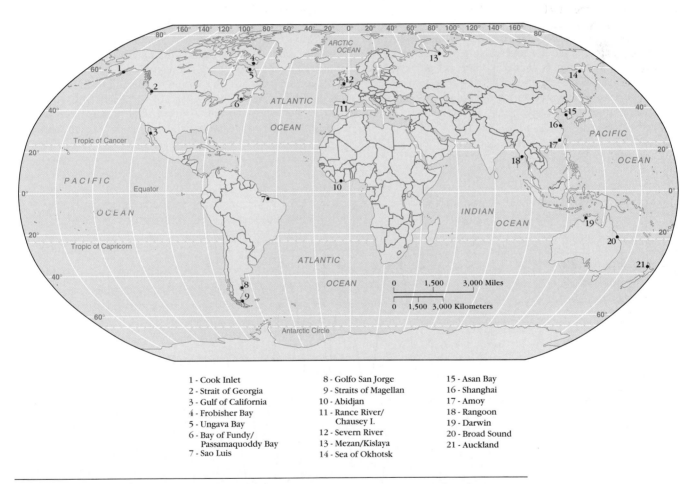

1 - Cook Inlet	8 - Golfo San Jorge	15 - Asan Bay
2 - Strait of Georgia	9 - Straits of Magellan	16 - Shanghai
3 - Gulf of California	10 - Abidjan	17 - Amoy
4 - Frobisher Bay	11 - Rance River/ Chausey I.	18 - Rangoon
5 - Ungava Bay	12 - Severn River	19 - Darwin
6 - Bay of Fundy/ Passamaquoddy Bay	13 - Mezan/Kislaya	20 - Broad Sound
7 - Sao Luis	14 - Sea of Okhotsk	21 - Auckland

Figure 9–19 Sites with high potential for tidal power generation. Twenty-one locations occur worldwide where tidal ranges are great enough to create potential for generating electricity. Where a large area exists for storing water behind a dam, a tidal range of 3 meters (10 feet) suffices; where smaller storage areas exist, greater tidal ranges are required.

within a period of one day, they are called diurnal (daily) inequalities. These inequalities are at their greatest when the Moon is at its maximum declination, and such tides are called *tropical tides* because the Moon is over one of Earth's tropics. When the Moon is over the Equator (*equatorial tides*), the difference between successive high tides and low tides is minimal.

I know that tide tables are published years in advance. How can tide tables be so accurate?

Often available free of charge at boating, fishing, and surf shops, tide tables list not only the exact time of all the high and low tides for a year, but also the height of the tides within 0.25 centimeter (a tenth of an inch). What's more, tide tables are rarely wrong! Indeed, it does seem quite remarkable that the tides can be predicted with such accuracy.

Because so many factors and local variables influence the tides along a particular coast, they cannot be accurately predicted using a purely mathematical approach. There is a periodicity to the tides, however, based mostly on an 18.6-year cycle that is related to the orbit of the Moon. Thus, tide tables are based on historical records.

Nevertheless, tide tables can sometimes be wrong because local meteorological conditions that cannot be predicted years in advance may alter the tides. These *meteorological tides* are produced by conditions such as hurricane storm surge (Figure 9D), low atmospheric pressure, or high winds that drive water ashore. Even thermal expansion of warm water associated with El Niño conditions will also create higher sea level than is predicted in tide tables.

How often are conditions right to produce the maximum tide-generating force?

Maximum tides occur when Earth is closest to the Sun (at perihelion), the Moon is closest to Earth (at perigee), and the Earth–Moon–Sun system is aligned (at syzygy) with both the Sun and Moon at zero declination. This rare condition—which creates an absolute *maximum* spring tidal range—occurs once every 1600 years. Fortunately, the next occurrence is predicted for the year 3300.

However, there are other times when conditions produce large tide-generating forces. During early 1983, for example, large, slow-moving low-pressure cells developed in the North Pacific Ocean that caused strong northwest winds (Figure 9E). In late January, the winds produced a near fully developed 3-meter (10-foot) swell that affected the coast from Oregon to Baja California. The large waves would have been trouble enough under normal conditions, but there were also unusually high spring tides of 2.25 meters (7.4 feet) because Earth was near perihelion at the same time that the Moon was at perigee. In addition, a strong El Niño had raised sea level by up to 20 centimeters (8 inches). When the waves hit the coast during these unusual conditions, they caused over $100 million in damages, including the destruction of 25 homes, damage to 3500 others, the collapse of several commercial and municipal piers, and at least a dozen deaths.

Some tide tables I've seen list negative tides. How can there ever be a negative tide?

Negative tides occur because the *datum* (starting point or reference point from which tides are measured) is an average of the tides over many years. Along the west coast of the United States, for instance, the datum is Mean Lower Low Water (MLLW), which is the average of the *lower* of the two low tides that occur daily in a mixed tidal pattern. Because the datum is an average, there will be some days when the tide is

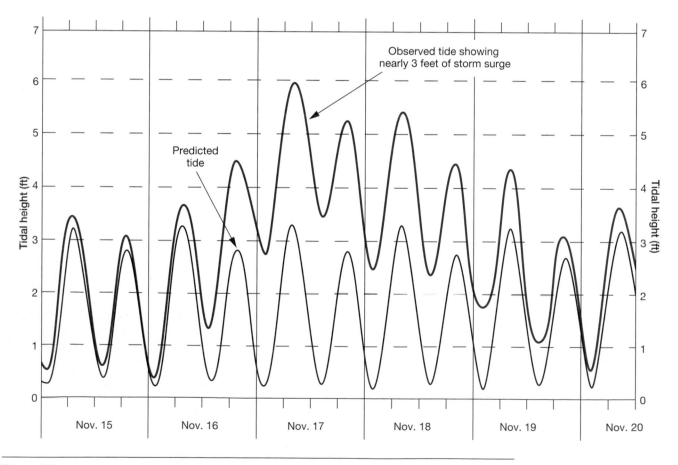

Figure 9D Storm surge effect of Hurricane Gordon at Portsmouth, Virginia, November 16–19, 1994.

less than the average (similar to the distribution of exam scores, some of which will be below the average). These lower-than-average tides are given negative values and are often the best times to visit local tide pool areas.

Do the tides have an effect on Earth's rotation?

Yes. The tides exert friction against the floor of the ocean basins and act as weak brakes that are steadily slowing Earth's rotation. The rate of slowing, however, is not great. Precise measurements by astronomers indicate the length of a day has been increasing by only 0.002 second per century over the past 300 years. Although this may seem inconsequential, over millions of years this small effect becomes very large. Eventually, billions of years into the future, rotation will cease and Earth will no longer have alternating days and nights.

Given that Earth's rotation is slowing, the length of each day in the geologic past must have been shorter—and the corresponding number of days per year must have been greater, because the length of a year (which is determined by Earth's revolution around the Sun) does not change. One method used to investigate this phenomenon involves the examination of shells of well-preserved fossil clams and corals that grow a microscopically thin layer of new shell material each day. Studies using this ingenious technique have revealed that 570 million years ago, there were 424 days in a year, with each day having a length of only 20.7 hours.

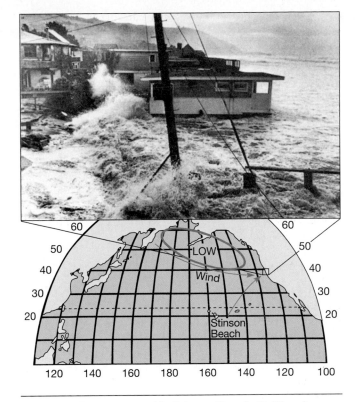

Figure 9E Storm waves hit Stinson Beach, California, during high tides of January 1983.

Chapter in Review

• Gravitational attraction of the Moon and Sun create Earth's tides, which are fundamentally long wavelength waves. According to a simplified model of tides that assumes an ocean of uniform depth and ignores the effects of friction, small horizontal forces (the tide-generating forces) tend to push water into two bulges on opposite sides of Earth. One bulge is directly facing the tide-generating body (the Moon and the Sun), and the other is directly opposite.

• Despite its vastly smaller size, the Moon has about twice the tide-generating effect of the Sun because the Moon is so much closer to Earth. The tidal bulges due to the Moon's gravity (the lunar bulges) dominate, so lunar motions dominate the periods of Earth's tides. However, the changing position of the solar bulges relative to the lunar bulges modifies tides. According to the simplified idealized tide theory, Earth's rotation carries locations on Earth into and out of the various tidal bulges.

• Tides would be easy to predict if Earth were a uniform sphere covered with an ocean of uniform depth. For most places on Earth, the time between successive high tides would be 12 hours 25 minutes (half a lunar day). The $29^1/_2$-day monthly tidal cycle would consist of tides with maximum tidal range (spring tides) and minimum tidal range (neap tides). Spring tides would occur each new moon and full moon, and neap tides would occur each first- and third-quarter phases of the Moon.

• The declination of the Moon varies between 28.5 degrees north or south of the Equator during the lunar month, and the

declination of the Sun varies between 23.5 degrees north or south of the Equator during the year, so the location of tidal bulges usually creates two high tides and two low tides of unequal height per lunar day. Tidal ranges are greatest when Earth is nearest the Sun and Moon.

• When friction and the true shape of ocean basins are considered, the dynamics of tides becomes more complicated. Moreover, the two bulges on opposite sides of Earth cannot exist because they cannot keep up with the rotational speed of Earth. Instead, the bulges are broken up into several tidal patterns that rotate around an amphidromic point, a point of zero tidal range. Rotation is counterclockwise in the Northern Hemisphere and clockwise in the Southern Hemisphere. Many other factors influence tides on Earth, too, such as the placement of the continents and the shapes of the coasts.

• The three types of tidal patterns observed on Earth are diurnal (a single high and low tide each lunar day), semidiurnal (two high and two low tides each lunar day), and mixed (characteristics of both). Mixed tidal patterns usually consist of semidiurnal periods with significant diurnal inequality. Mixed tidal patterns are the most common type in the world.

• The effects of constructive interference and the shoaling and narrowing of coastal bays creates the largest tidal range in the world—17 meters (56 feet)—at the northern end of Nova Scotia's Bay of Fundy. Tidal currents follow a rotary pattern in open-ocean basins but are converted to reversing currents along continental margins. The maximum velocity of reversing currents occurs during flood and ebb currents when the water is halfway between high and low slack waters. Tidal bores are true tidal waves (a wave produced by the tides) that occur in certain rivers and bays due to an incoming high tide.

• The tides are important to many marine organisms. For instance, grunion—small silvery fish that inhabit waters along the west coast of North America—time their spawning cycle to match the pattern of the tides.

• Tides can be used to generate power without need for fossil or nuclear fuel. There are some significant drawbacks, however, to creating a successful power plant using tidal power. Still, many sites worldwide have the potential for tidal power generation.

Key Terms

Amphidromic point (p. 275)

Aphelion (p. 274)

Apogee (p. 274)

Barycenter (p. 266)

Bay of Fundy (p. 277)

Centripetal force (p. 268)

Cotidal line (p. 275)

Declination (p. 273)

Diurnal tidal pattern (p. 276)

Ebb current (p. 279)

Ebb tide (p. 270)

Ecliptic (p. 273)

Flood current (p. 279)

Flood tide (p. 270)

Full moon (p. 271)

Gravitational force (p. 267)

Grunion (*Leuresthes tenuis*) (p. 282)

High slack water (p. 279)

Low slack water (p. 279)

Lunar bulge (p. 270)

Lunar day (p. 270)

Mixed tidal pattern (p. 276)

Nadir (p. 267)

Neap tide (p. 272)

New moon (p. 271)

Newton, Isaac (p. 266)

Perigee (p. 274)

Perihelion (p. 274)

Proxigean (p. 274)

Quadrature (p. 272)

Quarter moon (p. 271)

Resultant force (p. 268)

Reversing current (p. 277)

Rotary current (p. 277)

Semidiurnal tidal pattern (p. 276)

Solar bulge (p. 270)

Solar day (p. 270)

Spring tide (p. 271)

Syzygy (p. 272)

Tidal bore (p. 278)

Tidal period (p. 270)

Tidal range (p. 265)

Tide-generating force (p. 268)

Tides (p. 266)

Waning crescent (p. 273)

Waning gibbous (p. 273)

Waxing crescent (p. 272)

Waxing gibbous (p. 273)

Zenith (p. 267)

Questions And Exercises

1. Explain how a tidal power plant works, using as an example an estuary that has a mixed tidal pattern. Why does potential for usable tidal energy increase with an increase in the tidal range?

2. Explain why the Sun's influence on Earth's tides is only 46% that of the Moon's, even though the Sun is so much more massive than the Moon.

3. Why is a lunar day 24 hours 50 minutes long, while a solar day is 24 hours long?

4. Which is more technically correct: The tide comes in and goes out; or Earth rotates into and out of the tidal bulges. Why?

5. From memory, draw the positions of the Earth–Moon–Sun system during a complete monthly tidal cycle. Indicate the tide conditions experienced on Earth, the phases of the Moon, the time between those phases, and syzygy and quadrature.

6. Explain why the maximum tidal range (spring tide) occurs during new and full moon phases and the minimum tidal range (neap tide) at first-quarter and third-quarter moons.

7. If Earth did not have the Moon orbiting it, would there still be tides? Why or why not?

8. Assume that there are two moons in orbit around Earth that are on the same orbital plane but always on opposite sides of Earth and that each moon is the same size and mass of our Moon. How would this affect the tidal range during spring and neap tide conditions?

9. What is declination? Discuss the degree of declination of the Moon and Sun relative to Earth's Equator. What are the effects of declination of the Moon and Sun on the tides?

10. Diagram the Earth–Moon system's orbit about the Sun. Label the positions on the orbit at which the Moon and Sun are closest to and farthest from Earth, stating the terms used to identify them. Discuss the effects of the Moon's and Earth's positions on Earth's tides.

11. Are tides considered deep-water waves anywhere in the ocean? Why or why not?

12. Describe the number of high and low tides in a lunar day, the period, and any inequality of the following tidal patterns: diurnal, semidiurnal, and mixed.

13. Discuss factors that help produce the world's greatest tidal range in the Bay of Fundy.

14. Discuss the difference between rotary and reversing tidal currents.

15. Of flood current, ebb current, high slack water, and low slack water, when is the best time to enter a bay by boat? When is the best time to navigate in a shallow, rocky harbor?

16. Discuss at least two positive and two negative factors related to tidal power generation.

17. Describe the spawning cycle of grunion, indicating the relationship between tidal phenomena, where grunion lay their eggs, and the movement of sand on the beach.

18. Observe the Moon from a reference location every night at about the same time for two weeks. Keep track of your observations about the shape (phase) of the Moon and its position in the sky. Then, compare these to the reported tides in your area. How do the two compare?

References

Bowditch, N. 1995. *The American practical navigator: An epitome of navigation* (1995 edition). Bethesda, MD: Defense Mapping Agency Hydrographic/Topographic Center.

Clancy, E. P. 1969. *The tides: Pulse of the earth.* Garden City, NY: Doubleday.

Comins, N. F. 1993. *What if the Moon didn't exist? Voyages to Earth that might have been.* New York: HarperCollins.

Defant, A. 1958. *Ebb and flow: The tides of earth, air, and water.* Ann Arbor, MI: University of Michigan Press.

Gill, A. E. 1982. *Atmosphere and ocean dynamics.* International Geophysics Series, Vol. 30. Orlando, FL: Academic Press.

Hauge, C. J. 1972. Tides, currents, and waves. *California Geology* 25:7, 147–160.

Le Provost, C., Bennett, A. F., and Cartwright, D. E. 1995. Ocean tides for and from TOPEX/Poseidon. *Science* 267:5198, 639–642.

Pond, S., and Pickard, G. L. 1978. *Introductory dynamic oceanography.* Oxford: Pergamon Press.

Railsback, L. B. 1991. A model for teaching the dynamic theory of tides. *Journal of Geological Education* 39:1 15–18.

Sverdrup, H. U., Johnson, M. W., and Fleming, R. H. 1942. Renewal 1970. *The oceans: Their physics, chemistry, and biology.* Englewood Cliffs, NJ: Prentice-Hall.

von Arx, W. S. 1962. *An introduction to physical oceanography.* Reading, MA: Addison-Wesley.

***Suggested Reading in* Scientific American**

Goldreich, P. 1972. Tides and the earth-moon system. 226:4, 42–57. The tide-generating force of the Sun and Moon on Earth is discussed, along with the effect of transfer of angular momentum from Earth to the Moon as a result of tidal friction. Also considered are theories of lunar origin.

Greenberg, D. A. 1987. Modeling tidal power. 257:3, 128–131. The effects of using the large tidal ranges experienced in the Bay of Fundy to generate electricity are modeled on a computer, and the idea of building a tidal power plant across the bay is explored.

Lynch, D. K. 1982. Tidal bores. 247:4, 146–157. Tidal bores can be spectacular walls of water rushing up rivers when the tide rises.

Oceanography on the Web

Visit the *Essentials of Oceanography* home page for on-line resources for this chapter. There you will find an on-line study guide with review exercises, and links to oceanography sites to further your exploration of the topics in this chapter. *Essentials of Oceanography* is at: **http://www.prenhall.com/ thurman** (click on the Table of Contents menu and select this chapter).

The Coast: Beaches and Shoreline Processes

- Which coastal areas are parts of the beach?
- What seasonal changes do beaches experience?
- How are longshore currents created, and what is longshore drift?
- Which coastal features are characteristic of erosional and depositional coasts?
- How has changing sea level affected coastal regions?
- What are differences between various U.S. coastal regions?
- What effects do various forms of hard stabilization have on shorelines?

The waves which dash upon the shore are, one by one, broken, but the ocean conquers nevertheless. It overwhelms the Armada, it wears out the rock.

—*Lord Byron (1821)*

THE ULTIMATE PROTECTION: THE NATIONAL FLOOD INSURANCE PROGRAM (NFIP)

In 1965, Hurricane Betsy devastated the east coast of the United States, killing 75 people and causing property damage in excess of $1.4 million, which launched a massive relief effort by the federal government. The government soon realized that it could not lend aid indefinitely in similar situations. So, in 1968, the **National Flood Insurance Program (NFIP)** was established to encourage people to build safely inland of flood-threatened

Figure 10A A house in Leucadia, California, that was built too close to the beach.

areas in return for federally subsidized flood insurance. By reducing fatalities and property damage, it was hoped that fewer and smaller instances of federal disaster relief would be needed. By the time members of Congress had alleviated concerns of the banking, construction, and real estate industries, however, the program had become a vast subsidy that supported risky overdevelopment of the U.S. coastal region (Figure 10A).

During the decade before the program, for instance, hurricanes killed 186 people and damaged $2.2 billion in property. In the first decade of the NFIP, coastal storms caused 411 deaths (a 121% increase) and $4.7 billion in property damage (a 114% increase). Studies conducted by the General Accounting Office indicate that the program operated from 1978 to 1987 at a $652 million deficit, which was made up by federal taxpayers. In spite of the NFIP, the government still pays out billions of dollars with each flood disaster in the form of grants and loan subsidies to repair and replace high-risk structures—in direct opposition to what the program was supposed to prevent!

The problems with the NFIP are twofold: (1) insurance rates are too low, necessitating subsidies by U.S. taxpayers for development that benefits a small number of individuals who choose to live in areas prone to damage, and (2) insurance rates don't adequately assess relative danger. In addition, laws regulating coastal construction do not realistically address the distance buildings need to be set back from the shore. A coastal "setback" of 12 meters (40 feet), for example, is required for all new construction in California. Coastal erosion rates, however, vary widely and a major storm or hurricane can erode the coast 30 meters (100 feet) or more in a single day. Further, the threat of rising sea level may accelerate coastal erosion.

Even after years of research, hearings, and investigations, Congress has failed to pass a bill to reform the program because of special-interest pressure. In 2000, a report by the Federal Emergency Management Agency evaluated the economic impact of coastal erosion and was highly critical of the NFIP, suggesting many improvements.

What can be done to reverse such unsound public policy? What general guidelines should be followed in developing a national program for conservative use of our coastal areas to preserve them for everyone for generations to come? An important first step is to understand the processes operating in the coastal environment that make it one of the most dynamic places on the planet.

Throughout our history, humans have been attracted to the coastal regions of the world for their moderate climate, seafood, transportation, recreational opportunities, and commercial benefits. In the United States, 80% of the population now lives within easy access of the Atlantic, Pacific, and Gulf Coasts, increasing the stress on these important national resources.

The coastal region is constantly changing because waves crash along most shorelines more than 10,000 times a day, releasing their energy from distant storms. Waves cause erosion in some areas and deposition in others, resulting in changes that occur hourly, daily, weekly, monthly, seasonally, and yearly.

In this chapter, we'll examine the major features of the seacoast and shore and the processes that modify them. We'll also discuss ways people interfere with these processes, creating hazards to themselves and to the environment.

The Coastal Region

The **shore** is a zone that lies between the lowest tide level (low tide) and the highest elevation on land that is affected by storm waves. The **coast** extends inland from the shore as far as ocean-related features can be found (Figure 10–1). The width of the shore varies between a few meters and hundreds of meters. The width of the coast may vary from less than a kilometer (0.6 mile) to many tens of kilometers. The **coastline** marks the boundary between the shore and the coast. It is the landward limit of the effect of the highest storm waves on the shore.

Beach Terminology

The beach profile in Figure 10–1 shows features characteristic of a cliffed shoreline. The shore is divided into the **backshore** and the **foreshore**.[1] The backshore is above the high-tide shoreline and is covered with water only during storms. The foreshore is the portion exposed at low tide and submerged at high tide. The **shoreline** migrates back and forth with the tide and is the water's edge. The **nearshore** extends seaward from the low-tide shoreline to the low-tide breaker line. It is never exposed to the atmosphere, but it is affected by waves that touch bottom. Beyond the low-tide breakers is the **offshore** zone, which is deep enough that waves rarely affect the bottom.

A **beach** is a deposit of the shore area. It consists of wave-worked sediment that moves along the **wave-cut bench** (a flat, wave-eroded surface). A beach may continue from the coastline across the nearshore region to the line of breakers. Thus, the beach is the entire active

---

[1]The foreshore is often referred to as the *intertidal* or *littoral* (*litoralis* = the shore) *zone.*

Figure 10–1 Landforms and terminology of coastal regions. The beach is the entire active area affected by waves that extends from the low-tide breaker line to the base of the coastal cliffs.

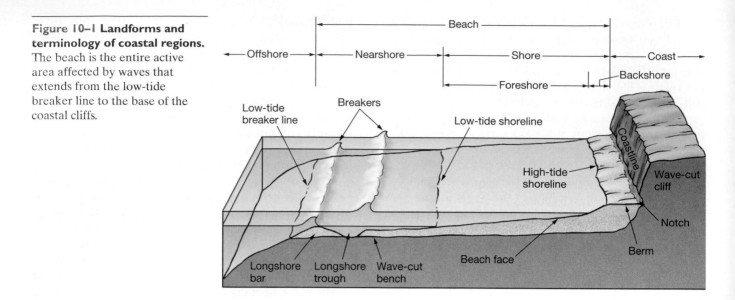

Figure 10–1 Landforms and terminology of coastal regions. The beach is the entire active area affected by waves that extends from the low-tide breaker line to the base of the coastal cliffs.

area of a coast that experiences changes due to breaking waves. The area of the beach above the shoreline is often called the **recreational beach**.

The **berm** is the dry, gently sloping region at the foot of the coastal cliffs or dunes. The berm is often composed of sand, making it a favorite place of beachgoers. The **beach face** is the wet, sloping surface that extends from the berm to the shoreline. It is more fully exposed during low tide, and is also known as the **low tide terrace**. The beach face is a favorite place for runners because the sand is wet and hard-packed. Offshore beyond the beach face is one or more **longshore bars**—sand bars that parallel the coast. A longshore bar may not always be present throughout the year, but when one is, it may be exposed during extremely low tides. Longshore bars can "trip" waves as they approach shore and cause them to begin breaking. Separating the longshore bar from the beach face is a **longshore trough**.

> The beach is the coastal area affected by breaking waves and includes the berm, beach face, longshore trough, and longshore bar.

Beach Composition

Beaches are composed of whatever material is locally available. When this material—sediment—comes from the erosion of beach cliffs or nearby coastal mountains, beaches are composed of mineral particles from these rocks and may be relatively coarse in texture. When the sediment comes primarily from rivers that drain lowland areas, beaches are finer in texture. Often, mud flats develop along the shore because only tiny clay-sized and silt-sized particles are emptied into the ocean. Such is the case for muddy coastlines such as along the coast

of Suriname in South America and the Kerala coast of southwest India.

Other beaches have a significant biologic component. For example, in low-relief, low-latitude areas such as southern Florida, where there are no mountains or other sources of rock-forming minerals nearby, most beaches are composed of shell fragments and the remains of organisms that live in coastal waters. Many beaches on volcanic islands in the open ocean are composed of black or green fragments of the basaltic lava that comprise the islands, or of coarse debris from coral reefs that develop around islands in low latitudes.

Regardless of the composition, though, the material that comprises the beach does not stay in one place. Instead, the waves that crash along the shoreline are constantly moving it. Thus, beaches can be thought of as material in transit along the shoreline.

Movement of Sand on the Beach

The movement of sand on the beach occurs both perpendicular to the shoreline (*toward* and *away from* shore) and parallel to the shoreline (often referred to as *up-coast* and *down-coast*).

Movement Perpendicular to Shoreline Breaking waves move sand perpendicular to the shoreline. As each wave breaks, water rushes up the beach face toward the berm. Some of this **swash** soaks into the beach and eventually returns to the ocean. However, most of the water drains away from shore as **backwash**, though usually not before the next wave breaks and sends its swash over the top of the previous wave's backwash.

While standing in ankle-deep water at the shoreline, you can see that swash and backwash transports sediment up and down the beach face perpendicular to the shoreline. Whether swash or backwash dominates determines whether sand is deposited or eroded from the berm.

In *light wave activity* (characterized by less energetic waves), much of the swash soaks into the beach, so backwash is reduced. The swash dominates the transport system, therefore, causing a net movement of the sand up the beach face toward the berm, making it wide and well developed.

In *heavy wave activity* (characterized by high-energy waves), the beach is saturated with water from previous waves, so very little of the swash soaks into the beach. Backwash dominates the transport system, therefore, causing a net movement of sand down the beach face, which erodes the berm. When a wave breaks, moreover, the incoming swash comes *on top of* the previous wave's backwash, effectively protecting the beach from the swash and adding to the eroding effect of the backwash.

During heavy wave activity, where does the sand from the berm go? The orbital motion in waves is too shallow to move the sand very far offshore. Thus, the sand accumulates just beyond where the waves break and forms one or more offshore sand bars (the longshore bars).

Light and heavy wave activity alternate seasonally at most beaches, so the characteristics of the beaches change, too (Table 10–1). Light wave activity produces a wide sandy berm and an overall steep beach face—a **summertime beach**—at the expense of the longshore bar (Figure 10–2a). Conversely, heavy wave activity produces a narrow rocky berm and an overall flattened beach face—a **wintertime beach**—and builds prominent longshore bars (Figure 10–2b). A wide berm that takes several months to build can be destroyed in just a few hours by high-energy wintertime storm waves.

> Smaller, low-energy waves move sand up the beach face toward the berm and create a summertime beach while larger, high-energy waves scour sand from the berm and create a wintertime beach.

Movement Parallel to Shoreline

At the same time that movement occurs perpendicular to shore, movement parallel to shoreline also occurs. Recall from Chapter 8 that waves refract (bend) and line up *nearly* parallel to the shore. With each breaking wave, the swash moves up onto the exposed beach at a slight angle, then gravity pulls the backwash straight down the beach face. As a result, water moves in a zigzag fashion along the shore, creating a movement of water within the surf zone called a **longshore current** (Figure 10–3).

Longshore currents have speeds up to 4 kilometers (2.5 miles) per hour. Speeds increase as beach slope increases, as the angle at which breakers arrive at the beach increases, as wave height increases, and as wave frequency increases.

Swimmers can be inadvertently carried by longshore currents and find themselves carried far from where they initially entered the water. This demonstrates that longshore currents are strong enough to move people as well as a vast amount of sand in a zigzag fashion along the shore.

Longshore drift or **longshore transport** is the movement of *sediment* in a zigzag fashion by the longshore current (Figure 10–3b). Because a longshore current affects the entire surf zone, the resulting longshore drift works within the entire surf zone as well.

The amount of longshore drift in any coastal region depends on the equilibrium between erosional and depositional forces. Any interference with the movement of sediment along the shore disrupts the equilibrium, forming a new erosional and depositional pattern. Nevertheless, longshore drift moves millions of tons of sediment along coastal regions every year.

Both rivers and coastal zones move water *and* sediment from one area (*upstream*) to another (*downstream*). As a result, the beach has often been referred to as a "river of sand." A longshore current moves in a zigzag fashion, however, and rivers flow mostly in a turbulent, swirling fashion. Additionally, the direction of flow of longshore currents along a shoreline can change, whereas rivers always flow in the same direction (downhill). The longshore current changes direction because the direction with which waves approach the beach changes seasonally. Nevertheless, the longshore current generally flows *southward along both the Atlantic and Pacific shores of the United States.*

> Longshore currents are produced by waves approaching the beach at an angle and create longshore drift, which transports sand along the coast in a zigzag fashion.

Table 10–1   Characteristics of beaches affected by light and heavy wave activity

	Light wave activity	Heavy wave activity
Berm/longshore bars	Berm is built at the expense of the longshore bars	Longshore bars are built at the expense of the berm
Wave energy	Low wave energy (non-storm conditions)	High wave energy (storm conditions)
Time span	Long time span (weeks or months)	Short time span (hours or days)
Characteristics	Creates summertime beach: sandy, wide berm, steep beach face	Creates wintertime beach: rocky, narrow berm, flattened beach face

Figure 10–2 Summertime and wintertime beach conditions. Dramatic differences occur between summertime **(a)** and wintertime **(b)** beach conditions at Boomer Beach in La Jolla, California.

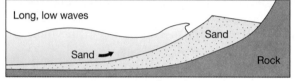

(a) Summertime beach (fair weather)

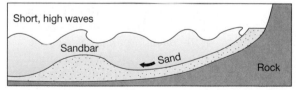

(b) Wintertime beach (storm)

Erosional- and Depositional-Type Shores

Sediment eroded from the beach is transported along the shore and deposited in areas where wave energy is low. Even though all shores experience some degree of both erosion and deposition, shores can often be identi-fied primarily as one type or the other. **Erosional-type shores** typically have well-developed cliffs and are in areas where tectonic uplift of the coast occurs, such as along the U.S. Pacific Coast.

The U.S. southeastern Atlantic Coast and the Gulf Coast, on the other hand, are primarily **depositional-type shores**. Sand deposits and offshore barrier islands

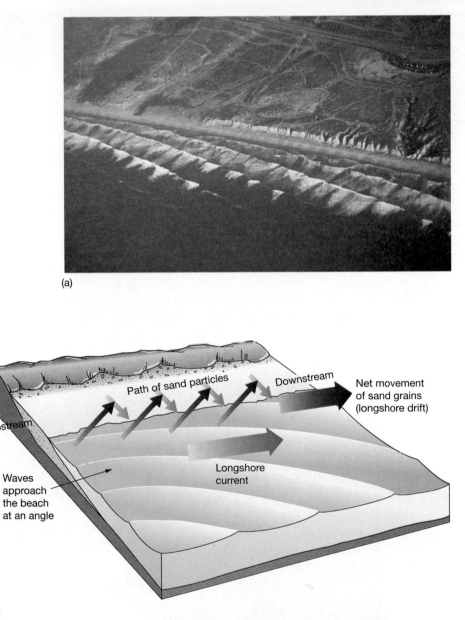

(a)

(b)

Figure 10–3 Longshore current and longshore drift. (a) Waves approaching the beach at a slight angle near Oceanside, California, producing a longshore current moving toward the right of the photo. (b) A longshore current, caused by refracting waves, moves water in a zigzag fashion along the shoreline. This causes a net movement of sand grains (longshore drift) from upstream to downstream ends.

are common there because the shore is gradually subsiding. Erosion can still be a major problem on depositional shores, especially when human development interferes with natural coastal processes.

Features of Erosional-Type Shores

Because of wave refraction, wave energy is concentrated on any **headlands** that jut out from the continent, while the amount of energy reaching the shore in bays is reduced. Headlands, therefore, are eroded and the shoreline retreats. Some of these erosional features are shown in Figure 10–4.

Waves pound relentlessly away at the base of headlands, undermining the upper portions, which eventually collapse to form **wave-cut cliffs**. The waves may form **sea caves** at the base of the cliffs.

As waves continue to pound the headlands, the caves may eventually erode through to the other side,

forming openings called **sea arches** (Figure 10–5). Some sea arches are large enough to allow a boat to maneuver safely through them. With continued erosion, the tops of sea arches eventually crumble to produce **sea stacks** (Figure 10–5). Waves also erode the bedrock of the bench. Uplift of the wave-cut bench creates a gently sloping **marine terrace** above sea level (Figure 10–6).

Rates of coastal erosion are influenced by the degree of exposure to waves, the amount of tidal range, and the composition of the coastal bedrock. Regardless of the erosion rate, all coastal regions follow the same developmental path. As long as there is no change in the elevation of the landmass relative to the ocean surface, the cliffs will continue to erode and retreat until the beaches widen sufficiently to prevent waves from reaching them. The eroded material is carried from the high-energy areas and deposited in the low-energy areas.

Box 10–1
Warning: Rip Currents ... Do You Know What To Do?

The backwash from breaking waves usually returns to the open ocean as a flow of water across the ocean bottom, so it is commonly referred to as "sheet flow." Some of this water, however, flows back in surface **rip currents**. Rip currents typically flow perpendicular to the beach and move away from the shore.

Rip currents are between 15 and 45 meters (15 and 150 feet) wide and can attain velocities of 7 to 8 kilometers (4 to 5 miles) per hour—faster than most people can swim for any length of time. In fact, it is useless to swim for long against a current stronger than about 2 kilometers (1.2 miles) per hour. Rip currents can travel hundreds of meters from shore before they break up. If a light-to-moderate swell is breaking, numerous rip currents may develop, which are moderate in size and velocity. A heavy swell usually produces fewer, more concentrated, and stronger rips. They can

often be recognized by the way they interfere with incoming waves, by their characteristic brown color caused by suspended sediment, or by their foamy and choppy surface (Figure 10B).

The rip currents that occur during heavy swell are a significant hazard to coastal swimmers. In fact, 80% of rescues at beaches by lifeguards involve people who are trapped in rip currents. Swimmers caught in a rip current can escape by swimming parallel to the shore for a short distance (simply swimming out of the narrow rip current) and then riding the waves in toward the beach. However, even excellent swimmers who panic or try to fight the current by swimming directly into it are eventually overcome by exhaustion and may drown. Even though most beaches have warnings posted and are frequently patrolled by lifeguards, many people lose their lives each year because of rip currents.

different color, narrow, only in surf zone

Figure 10B Rip currents. A rip current, which extends outward from shore near the middle of the photo and interferes with incoming waves, and warning sign (*inset*).

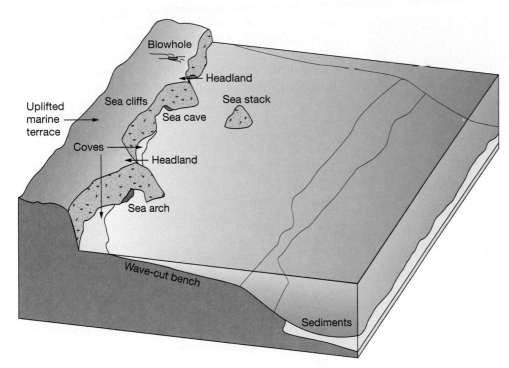

Figure 10–4 Features of erosional coasts.
Diagrammatic view of features characteristic of erosional coasts.

Figure 10–5 Sea arch and sea stack along the coast of Iceland. When the roof of a sea arch (*left*) collapses, a sea stack (*right*) is formed.

Features of Depositional-Type Shores

Coastal erosion of sea cliffs produces large amounts of sediment. Additional sediment, which is carried to the shore by rivers, comes from the erosion of inland rocks. Waves then distribute all of this sediment along the continental margin.

Figure 10–7 shows some of the features of depositional coasts. These features are primarily deposits of sand moved by longshore drift but are also modified by other coastal processes. Some are partially or wholly separated from the shore.

A **spit** (*spit* = spine) is a linear ridge of sediment that extends in the direction of longshore drift from land into the deeper water near the mouth of a bay. The end of the spit normally curves into the bay due to the movement of currents.

Tidal currents or currents from river runoff are usually strong enough to keep the mouth of the bay open. If not, the spit may eventually extend across the bay and connect to the mainland, forming a **bay barrier** or **bay-mouth bar** (Figure 10–8a), which cuts off the bay from the open ocean. Although bay barriers are a buildup of sand usually less than 1 meter (3.3 feet) above sea level, permanent buildings are often constructed on them.

Figure 10–6 **Wave-cut bench and marine terrace.** A wave-cut bench is exposed at low tide along the California coast at Bolinas Point near San Francisco. An elevated wave-cut bench, called a marine terrace, is shown at right.

A **tombolo** (*tombolo* = mound) is a sand ridge that connects an island or sea stack to the mainland (Figure 10–8b). Tombolos can also connect two adjacent islands. Formed in the wave-energy shadow of an island, tombolos are usually perpendicular to the average direction from which waves approach.

Barrier Islands Extremely long offshore deposits of sand lying parallel to the coast are called **barrier islands** (Figure 10–9). They form a barrier to storm waves that otherwise would severely assault the shore. Their origin is complex, but many barrier islands seem to have developed during the worldwide rise in sea level that began with the melting of the most recent major glaciers some 18,000 years ago.

Barrier islands are nearly continuous along the Atlantic Coast of the United States. They extend around Florida and along the Gulf of Mexico coast, where they exist well south of the Mexican border. Barrier islands may exceed 100 kilometers (60 miles) in length and

have widths of several kilometers. They are separated from the mainland by a lagoon. Barrier islands include Fire Island off the New York coast, North Carolina's Outer Banks, and Padre Island off the coast of Texas.

A typical barrier island has the physiographic features shown in Figure 10–10a. From the ocean landward, they are (1) ocean beach, (2) dunes, (3) barrier flat, (4) high salt marsh, (5) low salt marsh, and (6) lagoon between the barrier island and the mainland.

During the summer, gentle waves carry sand to the **ocean beach**, so it widens and becomes steeper. During the winter, higher-energy waves carry sand offshore and produce a narrow, gently sloping beach.

Winds blow sand inland during dry periods to produce coastal **dunes**, which are stabilized by dune grasses. These plants can withstand salt spray and burial by sand. Dunes protect the lagoon against excessive flooding during storm-driven high tides. Numerous passes exist through the dunes, particularly along the southeastern Atlantic Coast, where dunes are less well developed than to the north.

The **barrier flat** forms behind the dunes from sand driven through the passes during storms. Grasses quickly colonize these flats and seawater washes over them during storms. If storms wash over the barrier flat infrequently enough, the plants undergo natural biological succession, with the grasses successively replaced by thickets, woodlands, and eventually forests.

Salt marshes typically lie inland of the barrier flat. They are divided into the *low marsh*, extending from about mean sea level to the high neap-tide line, and the *high marsh*, extending to the highest spring-tide line. The low marsh is by far the most biologically productive part of the salt marsh.

New marshland is formed as overwash carries sediment into the lagoon, filling portions so they become intermittently exposed by the tides. Marshes may be poorly developed on parts of the island that are far

Figure 10–7 **Features of depositional coasts.** Diagrammatic view of features characteristic of depositional coasts.

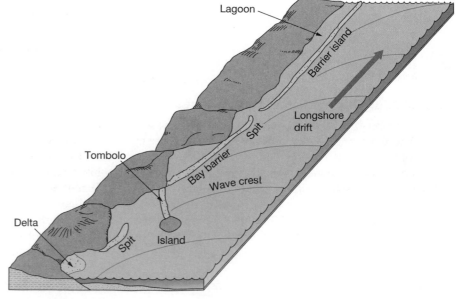

(a)

(b)

Figure 10–8 Coastal depositional features. (a) Barrier coast, spit, and bay barrier along the coast of Martha's Vineyard, Massachusetts. **(b)** Tombolo in the Gulf of California along the coast of Baja California, Mexico.

from floodtide inlets. Their development is greatly restricted on barrier islands where people perform artificial dune enhancement and fill inlets, activities that prevent overwashing and flooding.

The gradual sea level rise experienced along the eastern North American coast is causing barrier islands to migrate landward. The movement of the barrier island is similar to a slowly moving tractor tread, with the entire island rolling over itself, impacting structures built on these islands. **Peat deposits**, which are formed by the accumulation of organic matter in marsh environments, provide further evidence of barrier island migration (Figure 10–10b). As the island slowly rolls over itself and migrates toward land, it buries ancient peat deposits. These peat deposits can be found beneath the island and may even be exposed on the ocean beach when the barrier island has moved far enough.

Deltas Some rivers carry more sediment to the ocean than longshore currents can distribute. These rivers develop a **delta** (*delta* = triangular) deposit at their mouths. The Mississippi River, which empties into the Gulf of Mexico (Figure 10–11a), forms one of the largest deltas on Earth. Deltas are fertile, flat, low-lying areas that are subject to periodic flooding.

Delta formation begins when a river has filled its mouth with sediment. The delta then grows through the formation of **distributaries**, which are branching channels that deposit sediment as they radiate out over the delta in finger-like extensions (Figures 10–11a). When the fingers get too long, they become choked with sediment. At this point, a flood may easily shift the distributary course and provide sediment to low-lying areas between the fingers. When depositional processes exceed coastal erosion and transportation processes, a branching "bird's foot" Mississippi-type delta results.

When erosion and transportation processes exceed deposition, on the other hand, a delta shoreline is smoothed to a gentle curve, like that of the Nile River Delta in Egypt (Figure 10–11b). The Nile Delta is presently eroding because sediment is trapped behind the Aswan High Dam. Prior to completion of the dam in 1964, the Nile carried huge volumes of sediment into the Mediterranean Sea.

Beach Compartments **Beach compartments** consist of three components: a series of rivers that supply sand to a beach; the beach itself where sand is moving due to longshore transport; and offshore submarine canyons where sand is drained away from the beach. The map in Figure 10–12 shows that the coast of southern California contains four separate beach compartments.

Primarily rivers, but also coastal erosion, supply sand to the beach within an individual beach compartment (Figure 10–12, *inset*). The sand moves south with the longshore current, so beaches are wider near the southern (*downstream*) end of each beach compartment. Although some sand is washed offshore along the way,

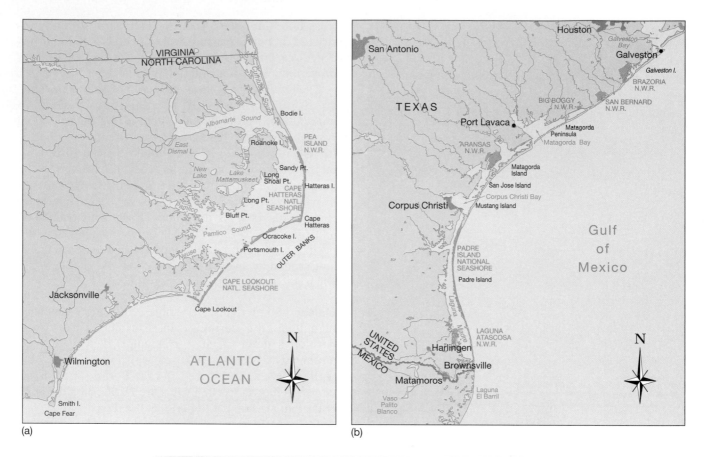

(a)

(b)

(c)

Figure 10–9 Barrier islands. (a) Barrier islands along North Carolina's Outer Banks **(b)** Barrier islands along the south Texas coast. **(c)** A portion of a heavily developed barrier island near Tom's River, New Jersey.

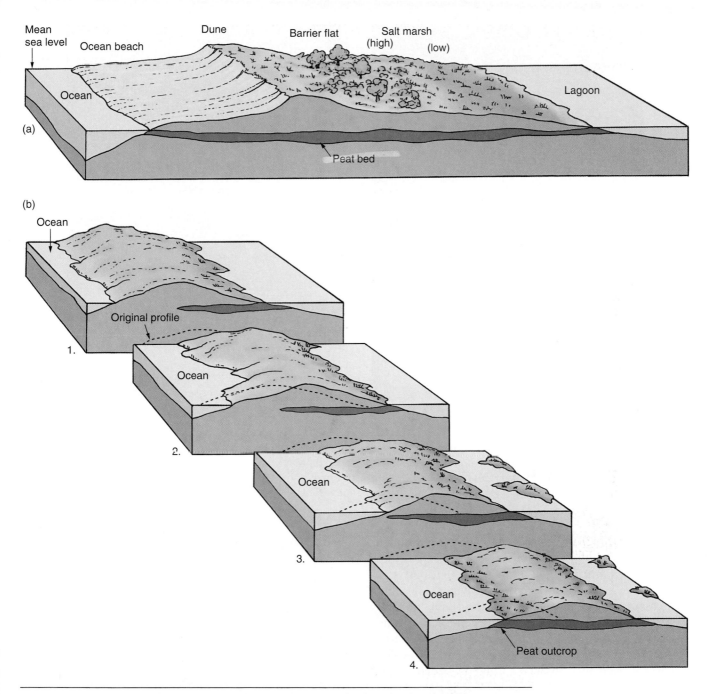

Figure 10–10 Formation of barrier islands. (a) Diagrammatic view showing the major physiographic zones of a barrier island. The peat bed represents ancient marsh environments. **(b)** Sequence (1–4) showing how a barrier island migrates and exposes peat deposits that have been covered by the island as it migrates toward the mainland in response to rising sea level.

most eventually moves near a head of a submarine canyon, where it is diverted away from the beach and onto the ocean floor. When the sand is removed from the coastal environment, it is lost from the beach forever. To the south of this beach compartment, the beaches will be thin and rocky, without much sand. The process begins all over again at the next beach compartment, where rivers add their sediment. Farther downstream, the beach widens and has an abundance of sand until that sand is also moved down a submarine canyon.

Human activities have altered the natural system of beach compartments. When a dam is built along one of the rivers that feeds into the beach compartment, it deprives the beach of sand. Lining rivers with concrete for flood control further reduces the sediment load delivered to coastal regions. Longshore transport continues to sweep the shoreline's sand into the submarine canyons, so the beaches become narrower and experience **beach starvation**. If all the rivers are blocked, the beaches may nearly disappear.

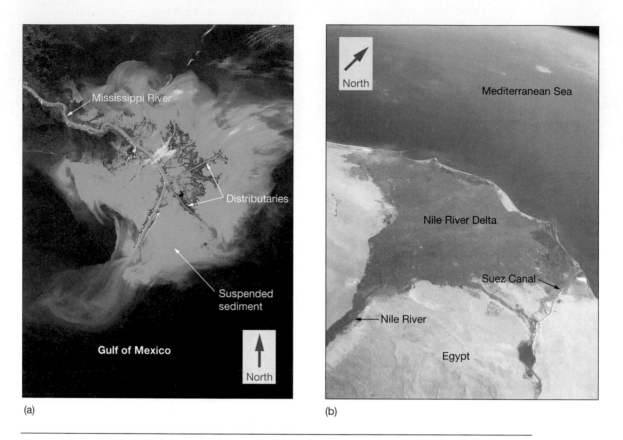

(a)

(b)

Figure 10–11 Deltas. (a) False-color infrared image of the branching "bird's foot" structure of the Mississippi River Delta. Red color is vegetation on land; light blue color is suspended sediment within the water. (b) Photograph from the space shuttle of Egypt's Nile River Delta, which has a smooth, curved shoreline as it extends into the Mediterranean Sea.

What can be done to prevent beach starvation in beach compartments? One obvious solution is to eliminate the dams, which would allow rivers to supply sand to the beach and return beach compartments to a natural balance. However, most dams are built for flood protection, water storage, and the generation of hydropower, so it is unlikely that many will be removed.

Another option is **beach replenishment** (also called **beach nourishment**), in which sand is added to the beach to replace the sediment held back by dams. Beach replenishment is expensive, however, because huge volumes of sand must be continually supplied to the beach. When dams are built, their effect on beaches far downstream is rarely considered. It's not until beach starvation occurs that the rivers are seen as parts of much larger systems that operate along the coast.

> Erosional coasts are characterized by erosional features such as cliffs, sea arches, sea stacks, and marine terraces. Depositional coasts are characterized by depositional features such as spits, tombolos, barrier islands, deltas, and beach compartments.

EIO

For more information and on-line exercises about the Environmental Issue in Oceanography (EIO) "Beaches or Bedrooms? The Dynamic Coastal Environment," visit the EIO Web site at **http://www.prenhall.com/oceanissues** and select Issue #6.

Emerging and Submerging Shorelines

Shorelines can also be classified based on their position relative to sea level. *Sea level, however, has changed throughout time.* It can change because the level of the land changes, the level of the sea changes, or a combination of the two. Shorelines that are rising above sea level are called **emerging shorelines** and those sinking below sea level are called **submerging shorelines**.

Marine terraces (Figures 10–6, 10–13, and 10–18) are one feature characteristic of emerging shorelines. Marine terraces are flat platforms backed by cliffs, which form when a wave-cut bench is exposed above sea level. **Stranded beach deposits** and other evidence of marine processes may exist many meters above the present shoreline, indicating that the former shoreline has risen above sea level.

Features characteristic of submerging shorelines include **drowned beaches** (Figure 10–13), **submerged**

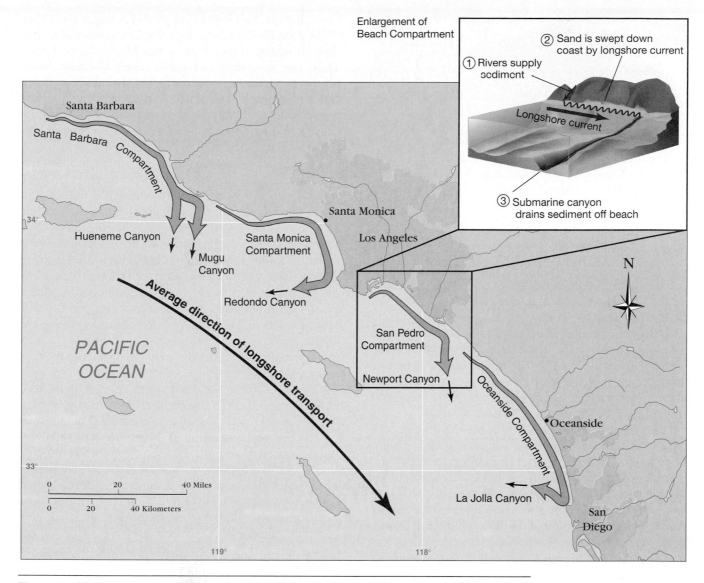

Figure 10–12 Beach compartments. Southern California has several beach compartments, which include rivers that bring sediment to the beach, the beach that experiences longshore transport, and the submarine canyons that remove sand from the beaches. Average longshore transport is toward the south.

dune topography, and **drowned river valleys** along the present shoreline.

What causes the changes in sea level that produce submerging and emerging shorelines? One main cause is tectonic and isostatic movements (as discussed in Chapter 2), which raise or lower the land surface relative to sea level. Another is worldwide changes in sea level, which affect the sea itself.

Tectonic and Isostatic Movements of Earth's Crust

The most dramatic changes in sea level during the past 3000 years have been caused by tectonic processes (movement of the land). These changes include uplift or subsidence of major portions of continents or ocean basins, as well as localized folding, faulting, or tilting of the continental crust.

Earth's crust also undergoes *isostatic adjustment*. It *sinks* under the accumulation of heavy loads of ice, vast piles of sediment, or outpourings of lava, and it *rises* when heavy loads are removed.

Most of the U.S. Pacific Coast is an emerging shoreline because continental margins where plate collisions occur are tectonically active, producing earthquakes, volcanoes, and mountain chains paralleling the coast. Most of the U.S. Atlantic Coast, on the other hand, is a submerging shoreline. When a continent moves away from a spreading center (such as the Mid-Atlantic Ridge), its trailing edge subsides because of cooling and the additional weight of accumulating sediment. Passive margins experience only a low level of

Regional sealevel [handwritten annotation]

Rising [handwritten annotation]

sinking [handwritten annotation]

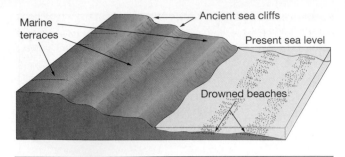

Figure 10–13 Evidence of ancient shorelines. Marine terraces result from exposure of ancient sea cliffs and wave-cut benches above present sea level. Below sea level, drowned beaches indicate the sea level has risen relative to the land.

tectonic deformation, earthquakes, and volcanism, making the Atlantic Coast far more quiet and stable than the Pacific Coast.

At least four major accumulations of glacial ice—and dozens of smaller ones—have occurred in high latitude regions during the last 2.5 to 3 million years. Although Antarctica is still covered by a very large, thick ice cap, much of the ice that once covered northern Asia, Europe, and North America has melted.

The weight of ice sheets as much as 3 kilometers (2 miles) thick caused the crust beneath to sink. Today, these areas are still slowly rebounding, 18,000 years after the ice began to melt. The floor of Hudson Bay, for example, which is now about 150 meters (500 feet) deep, will be close to or above sea level by the time it stops isostatically rebounding. The Gulf of Bothnia, moreover, which lies between Sweden and Finland, has isostatically rebounded 275 meters (900 feet) during the last 18,000 years.

Generally, tectonic and isostatic changes in sea level are confined to a segment of a continent's shoreline. For a *worldwide* change in sea level, there must be a change in seawater volume or ocean basin capacity.

Eustatic Changes in Sea Level

A change in sea level that is experienced worldwide due to changes in seawater volume or ocean basin capacity is called **eustatic** (*eu* = good, *stasis* = standing).[2] The formation or destruction of large inland lakes, for example, causes small eustatic changes in sea level. When lakes form, they trap water that would otherwise run off the land into the ocean, so sea level is lowered worldwide. When lakes are drained and release their water back to the ocean, sea level rises.

Changes in sea-floor spreading rates can change the capacity of the ocean basin, resulting in eustatic sea level changes. Fast spreading produces larger rises, such as the East Pacific Rise, which displace more water than slow-spreading ridges such as the Mid-Atlantic Ridge. Thus, fast spreading raises sea level, whereas slower spreading lowers sea level worldwide. Significant changes in sea level due to changes in spreading rate typically take hundreds of thousands to millions of years and may have changed sea level by 1000 meters (3300 feet) or more.

Ice ages cause eustatic sea level changes, too. As glaciers form, they tie up vast volumes of water on land, eustatically lowering sea level. An analogy to this effect is a sink of water representing an ocean basin. To simulate an ice age, some of the water from the sink is removed and frozen, causing the water level of the sink to be lower. In a similar fashion, worldwide sea level is lower during an ice age. During interglacial stages (such as the one we are in at present), the glaciers melt and release great volumes of water that drain to the sea, eustatically raising sea level. This would be analogous to putting the frozen chunk of ice on the counter near the sink and letting the ice melt, causing the water to drain into the sink and raise "sink level."

Glaciers during the Pleistocene Epoch[3] advanced and retreated many times on land near the poles, causing sea level to fluctuate considerably. The thermal contraction and expansion of the ocean as its temperature decreased and increased, respectively, affected sea level too. The thermal contraction and expansion of seawater works much like a mercury thermometer: As the mercury inside the thermometer warms, it expands and rises into the thermometer; as it cools, it contracts. Similarly, cooler seawater contracts and occupies less volume, thereby eustatically *lowering* sea level. Warmer seawater expands, eustatically *raising* sea level.

For every 1°C (1.8°F) change in the average temperature of the ocean water, sea level changes 2 meters (6.6 feet). Microfossils in Pleistocene ocean sediments suggest that ocean surface temperature may have been as much as 5°C (9°F) lower than at present. Therefore, thermal contraction of the ocean water may have lowered sea level by about 10 meters (33 feet).

Although it is difficult to state definitely the range of shoreline fluctuation during the Pleistocene, evidence suggests that it was at least 120 meters (400 feet) below the present shoreline (Figure 10–14). It is also estimated that if *all* the remaining glacial ice on Earth were to melt, sea level would rise another 60 meters (200 feet). Thus, the *minimum* sea level change during the Pleistocene is on the order of 180 meters (600 feet), most of which was due to the capture and release of Earth's water by land-based glaciers.

---

[2]The term *eustatic* refers to a highly idealized situation in which all of the continents remain static (in *good standing*), while only the sea rises or falls.

[3]The Pleistocene Epoch of geologic time (also called the "Ice Age") is from 1.6 million to 10 thousand years ago (see the Geologic Time Scale, Figure 1–19).

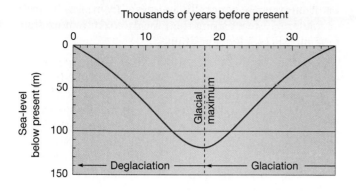

Figure 10–14 Sea level change during the most recent advance and retreat of Pleistocene glaciers. Sea level dropped worldwide by about 120 meters (400 feet) as the last glacial advance removed water from the oceans and transferred it to continental glaciers. About 18,000 years ago, sea level began to rise as the glaciers melted and water was returned to the oceans.

The combination of tectonic and eustatic changes in sea level is very complex, so it is difficult to classify coastal regions as purely emergent or submergent. In fact, most coastal areas show evidence of *both* submergence and emergence in the recent past. Evidence suggest, however, that until recently sea level has experienced only minor changes as a result of melting glacial ice during the last 3000 years.

Sea Level and the Greenhouse Effect

As discussed in Chapter 6, carbon dioxide in the atmosphere has increased 30% over the last 200 years and there has been an increase in global temperature of at least 0.6°C (1.1°F) over the last 130 years. Analy-

sis of worldwide tide records indicate that there has also been a eustatic rise in sea level of between 10 and 25 centimeters (4 and 10 inches) over the last 100 years and that sea level is currently rising 1.8 millimeters (0.07 inch) per year. At certain tide recording stations where data goes back well into the 19th century, there has been an increase in relative sea level of 40 centimeters (16 inches) over the last 150 years (Figure 10–15).

Clearly, sea level is rising. Is this rise the result of increased global warming because of the greenhouse effect or is it part of a long-term natural cycle? At this point, the answer cannot be easily determined, but evidence suggests humans are altering the environment on a global scale with emissions that enhance Earth's greenhouse effect. The rise in sea level most likely represents the combined effect of an increase in ocean volume due to thermal expansion and the observed retreat of small ice caps and glaciers that are adding water to the ocean.

According to some estimates, the rate of sea level rise will increase with increased global warming, which may result in as much as a 1-meter (3.3-foot) rise in sea level by 2100. The recent increase in coastal development puts more homes in danger's path and compounds the problem. The great lesson for humankind in all this is that we cannot dominate nature. Rather, we must learn to live within it.

> Sea level is affected by the movement of land and changes in seawater volume or ocean basin capacity. Sea level has changed dramatically in the past because of changes in Earth's climate.

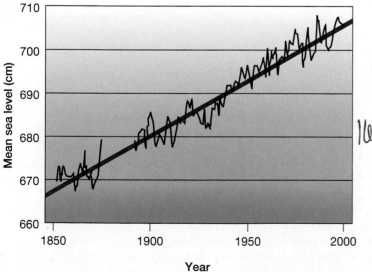

Figure 10–15 Measured relative sea level rise at New York City. Tide-gauge data from New York City shows an increase in sea level of 40 centimeters (16 inches) since 1850. Some of this rise is due to isostatic rebound following Pleistocene glaciation, but the majority is likely caused by thermal expansion of warmer ocean water and the retreat of small ice caps and glaciers.

16 inches

Characteristics of U.S. Coasts

Whether the dominant process along a coast is erosion or deposition depends on the combined effect of many variables, such as composition of coastal bedrock, the degree of exposure to ocean waves, tidal range, tectonic subsidence or emergence, isostatic subsidence or emergence, and eustatic sea level change.

A 1971 study of erosion along the U.S. Coastline by the Army Corps of Engineers found that over 24% to be "seriously eroding." Subsequent studies supported by the U.S. Geological Survey produced the rates of shoreline change presented in Figure 10–16.

The Atlantic Coast

Figure 10–16 shows that the U.S. Atlantic Coast has a variety of complex coastal conditions:

- Most of the Atlantic Coast is exposed to storm waves from the open ocean. Barrier islands from Massachusetts southward, however, protect the mainland from large storm waves.

- Tidal ranges generally increase from less than 1 meter (3.3 feet) along the Florida coast to more than 2 meters (6.5 feet) in Maine.

- Bedrock for most of Florida is a resistant type of sedimentary rock called *limestone*. Most of the bedrock northward through New Jersey, however, consists of nonresistant sedimentary rocks formed in the recent geologic past. As these rocks rapidly erode, they supply sand to barrier islands and other depositional features common along the coast. The bedrock north of New York consists of very resistant rock types.

- From New York northward, continental glaciers affected the coastal region directly. Many coastal features, including Long Island and Cape Cod, are glacial deposits (called *moraines*) left behind when the glaciers melted.

North of Cape Hatteras in North Carolina, the coast is subject to very high-energy waves during fall and winter when storms called "nor'easters" (northeasters) blow in from the North Atlantic. The energy of these

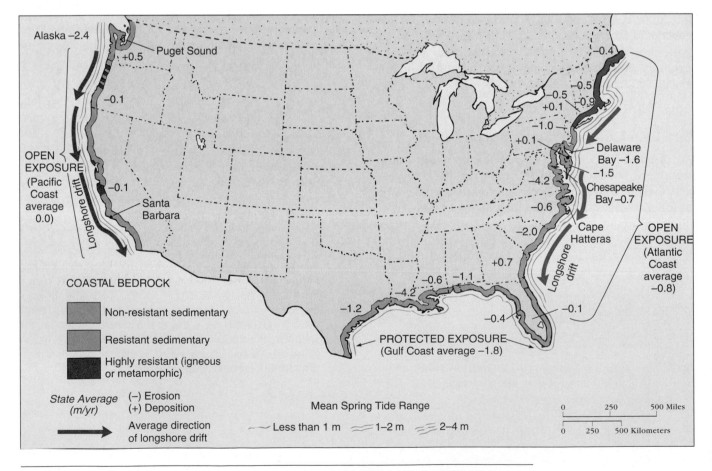

Figure 10–16 U.S. coastal erosion and deposition, 1979–1983. Map showing United States coastal bedrock type (*red, yellow, and blue colors*), the mean spring-tide range (*light blue lines*), degree of exposure, and average direction of longshore drift (*purple arrows*). The map also shows the average rate of erosion (−) or deposition (+) in meters per year for each coastal state and average for each coastal region.

storms generates waves up to 6 meters (20 feet) high, with a 1-meter (3.3-foot) rise in sea level that follows the low pressure as it moves northward. Such high-energy conditions seriously erode coastlines that are predominantly depositional.

Sea level along most of the Atlantic Coast appears to be rising at a rate of about 0.3 meter (1 foot) per century. Drowned river valleys, for instance, are common along the coast and form large bays (Figure 10–17). In northern Maine, however, sea level may be dropping as the continent rebounds isostatically from the melting of the Pleistocene ice sheet.

The Atlantic Coast has an average annual rate of erosion of 0.8 meter (2.6 feet),[4] which means that sea is migrating landward each year by a distance approximately equal to the length of your legs! In Virginia, the loss is over five times that rate at 4.2 meters (13.7 feet) per year, but is confined largely to barrier islands.

Erosion rates for Chesapeake Bay are about average for the Atlantic Coast, but rates for Delaware Bay [1.6 meters (5.2 feet) per year] are about twice the average. Of the observations made along the Atlantic Coast, 79% showed some degree of erosion. Delaware, Georgia, and New York have depositional coasts despite serious erosion problems in these states as well.

The Gulf Coast

The Mississippi River Delta, which is deposited in an area with a tidal range of less than 1 meter (3.3 feet), dominates the Louisiana-Texas portion of the Gulf Coast. Except during the hurricane season (June to November), wave energy is generally low. Tectonic subsidence is common throughout the Gulf Coast, and the average rate of sea level rise is similar to that of the southeast Atlantic Coast, about 0.3 meter (1 foot) per century. Some areas of coastal Louisiana have experienced a 1-meter (3.3-foot) rise during the last century, due to the compaction of Mississippi River sediments by overlying weight.

The average rate of erosion is 1.8 meters (6 feet) per year in the Gulf Coast. The Mississippi River Delta experiences the greatest rate, averaging 4.2 meters (13.7 feet) per year. Erosion is made worse by barge channels dredged through marshlands, and Louisiana has lost more than 1 million acres of delta since 1900. Louisiana is now losing marshland at a rate exceeding 130 square kilometers (50 square miles) per year.

Although all Gulf states show a net loss of land, and the Gulf Coast has a greater erosion rate than the Atlantic Coast, only 63% of the shore is receding because of erosion. The high average rate of erosion reflects the heavy losses in the Mississippi River Delta.

The Pacific Coast

The Pacific Coast is generally experiencing less erosion than the Atlantic and Gulf Coasts. Along the Pacific Coast, relatively young and easily eroded sedimentary rocks dominate the bedrock, with local outcrops of more resistant rock types. Tectonically, the coast is rising, as shown by marine (wave-cut) terraces (Figure 10–18). Sea level still shows at least small rates of rise, except for segments along the coast of Oregon and Alaska. The tidal range is mostly between 1 and 2 meters (3.3 and 6.6 feet).

The Pacific Coast is fully exposed to large storm waves, and is said to have *open exposure*. High-energy waves may strike the coast in winter, with 1-meter (3.3-foot) waves being typical. Frequently, the wave height increases to 2 meters (6.6 feet), and a few times per year 6-meter (20-foot) waves hammer the shore! These high-energy waves erode sand from many beaches. The exposed beaches, which are composed primarily of pebbles and boulders during the winter months, regain their sand during the summer when smaller waves occur.

Many Pacific Coast rivers have been dammed for flood control and hydroelectric power generation. The amount of sediment supplied by rivers to the shoreline for longshore transport is reduced, resulting in beach starvation in some areas.

With an average erosion rate of only 0.005 meter (0.016 foot)[5] per year and only 30% of the coast showing erosion loss, the Pacific Coast is eroded much less than the Atlantic and Gulf Coasts. Nevertheless, high wave energy and relatively soft rocks result in high rates of erosion in some parts of the Pacific Coast. In some parts of Alaska, for example, the average rate of erosion is 2.4 meters (7.9 feet) per year.

Of the Pacific states, only Washington shows a net sediment deposition. The long, protected Washington shoreline within Puget Sound helps skew the Pacific Coast values (Figure 10–16). Although the average erosion rate for California is only 0.1 meter (0.33 foot) per year, over 80% of the California coast is experiencing erosion, with rates up to 0.6 meter (2 feet) per year.

> U.S. coastal regions are affected by many variables, including composition of the coastal bedrock, degree of exposure, and tidal range. Most U.S. coastal regions are experiencing erosion.

Hard Stabilization

People continually modify coastal sediment erosion/deposition in attempts to improve or preserve their property. Structures built to protect a coast from erosion or to prevent the movement of sand along a beach

---

[4]Note that erosion values are shown as *negative* values in Figure 10–16, identifying them as separate from rates of deposition.

[5]0.005 meters is equal to 5 millimeters (0.2 inch).

Figure 10–17 Drowned river valleys. Satellite false-color image of drowned river valleys such as Chesapeake and Delaware Bays along the east coast of the U.S., which were formed by a relative rise in sea level that followed the end of the Ice Age.

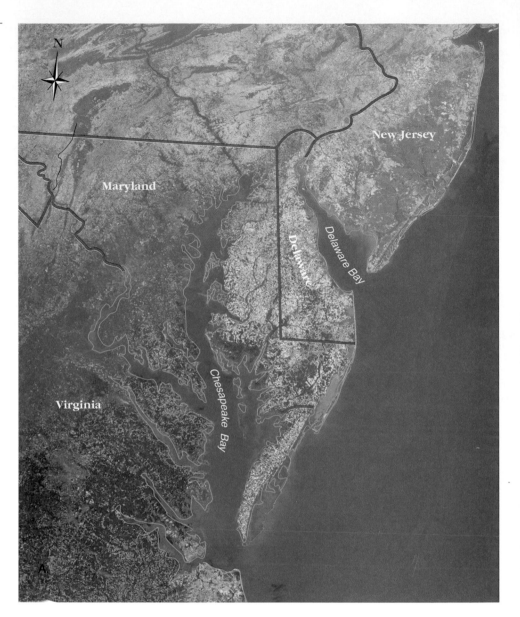

Figure 10–18 Marine (wave-cut) terraces. Each marine terrace on San Clemente Island offshore southern California was created by wave activity at sea level. Subsequently, each terrace has been exposed by tectonic uplift. The highest (oldest) terraces near the top of the photo are now about 400 meters (1320 feet) above sea level.

are known as **hard stabilization**. Hard stabilization can take many forms and often results in predictable yet unwanted outcomes.

Groins and Groin Fields

One type of hard stabilization is a **groin** (*groin* = ground). Groins are built perpendicular to a coastline and are specifically designed to trap sand moving along the coast in longshore transport (Figure 10–19). They are constructed of many types of material, but large blocky material called **rip-rap** is most common. Sometimes, groins are even constructed of sturdy wood pilings (similar to a fence built out into the ocean).

Although a groin traps sand on its *upstream side*, erosion occurs immediately downstream of the groin because the sand that is normally found just downstream of the groin is trapped on the groin's upstream side. To lessen the erosion, another groin can be constructed downstream, which in turn also creates erosion downstream from it. More groins are needed to alleviate the beach erosion, and soon a **groin field** is created (Figure 10–20).

Does a groin (or a groin field) actually retain more sand on the beach? Sand eventually migrates around the end of the groin, so there is no additional sand on the beach; it is only *distributed differently*. With proper engineering, an equilibrium may be reached that allows sufficient sand to move along the coast before excessive erosion occurs downstream from the last groin. However, some serious erosional problems have developed in many areas resulting from attempts to stabilize sand on the beach by the excessive use of groins.

Jetties

Another type of hard stabilization is a **jetty** (*jettee* = to project). A jetty is similar to a groin because it is built perpendicular to the shore and is usually constructed of rip-rap. The purpose of a jetty, however, is to protect harbor entrances from waves and only secondarily does it trap sand (Figure 10–21). Because jetties are usually built in closely spaced pairs and can be quite long, they can cause more pronounced upstream deposition and downstream erosion than groins.

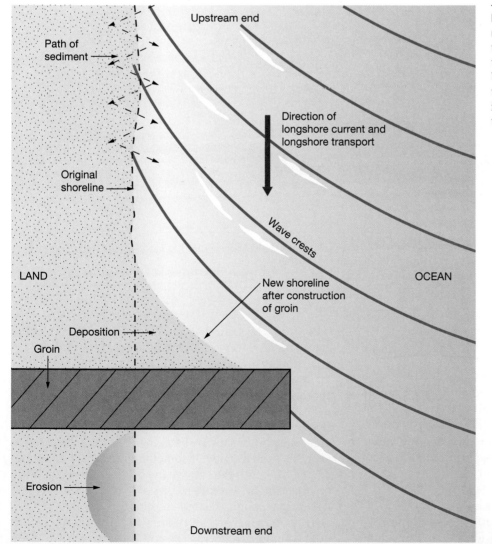

Figure 10–19 Interference of sand movement. Hard stabilization like the groin shown here interferes with the movement of sand along the beach, causing deposition of sand upstream of the groin and erosion immediately downstream, modifying the shape of the beach.

Figure 10–20 Groin field. A series of groins has been built along the shoreline north of Ship Bottom, New Jersey, in an attempt to trap sand, altering the distribution of sand on the beach. The view is toward the north, and the primary direction of longshore current is toward the bottom of the photo (toward the south).

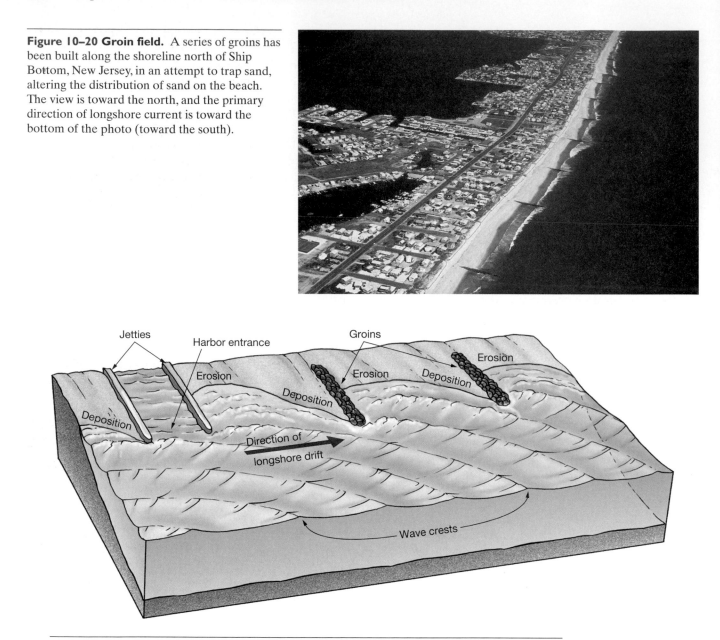

Figure 10–21 Jetties and groins. Jetties protect a harbor entrance and usually occur in pairs. Groins are built specifically to trap sand moving in the longshore transport system and occur individually or as a groin field. Both structures cause deposits of sand on their upstream sides and an equal amount of erosion downstream.

Breakwaters

Figure 10–22 shows a **breakwater**—hard stabilization built parallel to a shoreline—that was constructed to create the harbor at Santa Barbara, California. California's longshore drift is predominantly southward, so the breakwater on the western side of the harbor accumulated sand that had migrated eastward along the coast. The beach to the west of the harbor continued to grow until finally the sand moved around the breakwater and began to fill in the harbor (Figure 10–22).

While abnormal deposition occurred to the west, erosion proceeded at an alarming rate east of the harbor. The waves east of the harbor were no greater than before, but the sand that had formerly moved down the coast was now trapped behind the breakwater.

A similar situation occurred in Santa Monica, California, where a breakwater was built to provide a boat anchorage. A bulge in the beach soon formed upstream of the breakwater and severe erosion occurred downstream (Figure 10–23). The breakwater interfered with the natural transport of sand by blocking the waves that used to keep the sand moving. If something was not done to put energy back into the system, the breakwater would soon be attached by a tombolo of sand, and further erosion downstream might destroy coastal structures.

In Santa Barbara and Santa Monica, dredging was used to compensate for erosion downstream from the

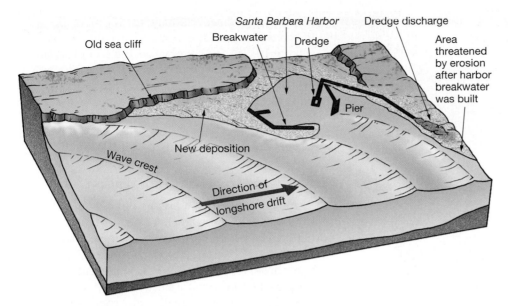

Old sea cliff • Breakwater • *Santa Barbara Harbor* • Dredge • Dredge discharge

Area threatened by erosion after harbor breakwater was built

Pier

New deposition

Wave crest

Direction of longshore drift

Figure 10–22 Santa Barbara Harbor. Construction of a breakwater at Santa Barbara Harbor interfered with the longshore drift, creating a broad beach. As the beach extended around the breakwater into the harbor, the harbor was in danger of being closed off by accumulating sand. As a result, dredging operations were initiated to move sand from the harbor downstream, where it helped reduce coastal erosion.

(a)

(b)

Figure 10–23 Santa Monica breakwater. (a) The shoreline and pier at Santa Monica as it appeared in 1931. **(b)** The same area in 1949, showing that the construction of a breakwater to create a boat anchorage disrupted the longshore transport of sand and caused a bulge of sand in the beach.

breakwater and to keep the harbor or anchorage from filling with sand. Sand dredged from behind the breakwater is pumped down the coast so it can re-enter the longshore drift and replenish the eroded beach.

The dredging operation has stabilized the situation in Santa Barbara, but at a considerable (and ongoing) expense. In Santa Monica, dredging was conducted until the breakwater was largely destroyed during winter storms in 1982–1983. Shortly thereafter, wave energy was able to move sand along the coast again, and the system was restored to near-normal conditions. When

people interfere with natural processes in the coastal region, they must provide the energy needed to replace what they have misdirected through modification of the shore environment.

Seawalls

One of the most destructive types of hard stabilization is the **seawall** (Figure 10–24), which is built parallel to the shore along the landward side of the berm. The purpose of a seawall is to armor the coastline and protect landward developments from ocean waves.

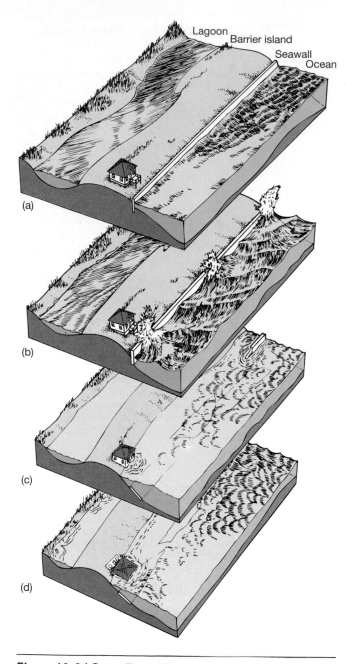

Lagoon
Barrier island
Seawall
Ocean

(a)

(b)

(c)

(d)

Figure 10–24 Seawalls and beaches. When a seawall is built along a beach (such as on this barrier island) to protect beachfront property **(a)**, a large storm can remove the beach from the seaward side of the wall and steepen its seaward slope **(b)**. Eventually, the wall is undermined and falls into the sea **(c)**. The property is lost **(d)** as the oversteepened beach slope advances landward in its effort to reestablish a natural slope angle.

Once waves begin breaking against a seawall, however, turbulence generated by the abrupt release of wave energy quickly erodes the sediment on its seaward side, causing it to collapse into the surf (Figure 10–24). Where seawalls have been used to protect property on barrier islands, the seaward slope of the island beach has steepened and the rate of erosion has increased, causing the destruction of the recreational beach.

A well-designed seawall may last for many decades, but the constant pounding of waves eventually takes its toll (Figure 10–25). In the long run, the cost of repairing or replacing seawalls will be more than the property is worth, and the sea will claim more of the coast through the natural processes of erosion. It's just a matter of time for homeowners who live too close to the coast, many of whom are gambling that their houses won't be destroyed in their lifetimes.

Alternatives to Hard Stabilization

Is it better to preserve the houses of a few people who have built too close to the shore at the expense of armoring the coast with hard stabilization and destroying the recreational beach? If you own coastal property, your response would probably be different from the general beachgoing public. Because hard stabilization has been shown to have negative environmental consequences, alternatives have been sought.

Of course, one alternative to the use of hard stabilization is to restrict construction in areas prone to coastal erosion. Unfortunately, this is becoming less and less an option as coastal regions experience population increases and governments increase the risk of damage and injuries because of programs like the National Flood Insurance Program that encourages construction in unsafe locations (see the chapter-opening feature at the beginning of this chapter). Further, many homeowners spend large amounts of money rebuilding structures and fortifying their property.

However, policy has recently shifted from defending coastal property in high hazard areas to removing structures and letting nature reclaim the beach. This approach is called **relocation**, which involves moving structures to safer locations as they become threatened by erosion. One example of the successful use of this technique is the relocation of the Cape Hatteras Lighthouse in North Carolina (Box 10–2). Relocation, if used wisely, can allow humans to live in balance with the natural processes that continually modify beaches.

Hard stabilization includes groins, jetties, breakwaters, and seawalls, all of which alter the coastal environment, cause erosion and deposition, and result in changes in the shape of the beach.

? Students Sometimes Ask...

What is the difference between a rip current and a rip tide?

Like tidal waves (tsunami), rip tides are a misnomer and have nothing to do with the tides. Rip tides are more correctly called rip currents. Perhaps rip currents have incorrectly been called rip tides because they occur suddenly (like an incoming tide).

Figure 10–25 Seawall damage. A seawall in Solana Beach, California, that has been damaged by waves and needs repair. Although seawalls appear to be sturdy, they can be destroyed by the continual pounding of high-energy storm waves.

What is an undertow? How is it different from a rip current?

An undertow, similar to a rip current, is a flow of water away from shore. An undertow is much wider, however, and is usually more concentrated along the ocean floor. An undertow is really a continuation of backwash that flows down the beach face, and is strongest during heavy wave activity. Undertows can be strong enough to knock people off their feet, but they are confined to the immediate floor of the ocean and only within the surf zone.

Can submarine canyons fill with sediment?

Yes. In many beach compartments, the submarine canyons that drain sand from the beach empty into deep basins offshore. However, given several million years and tons of sediments per year sliding down the submarine canyons, the offshore basins begin to fill up and can eventually be exposed above sea level. In fact, the Los Angeles basin in California was filled in by sediment derived from local mountains in this manner during the geologic past.

I know that the positions of the continents have not always remained the same. Has the movement of the continents changed sea level?

Yes, but in an indirect way. If plate motion is responsible for moving large continental masses into polar regions, then thick continental glaciation can occur (as in Antarctica today). Glacial ice forms from water vapor in the atmosphere (in the form of snow), which is derived originally from the evapora-

tion of seawater. Thus, water is removed from the oceans when continents assume more polar positions, lowering sea level.

Does beach starvation affect coastal cliff erosion?

Most certainly! When beaches narrow, coastal cliffs are more susceptible to attack from waves because there is no longer a wide beach to absorb the energy of storm waves. Instead, storm waves crash directly against the coastal cliffs, increasing erosion.

How much does beach replenishment cost?

The cost of beach replenishment depends on the type and quantity of material placed on the beach, how far the material must be transported, and how it is to be distributed on the beach. Most sand used for replenishment comes from offshore areas, but sand that is dredged from nearby rivers, drained dams, harbors, and lagoons is also used.

The average cost of sand used to replenish beaches is between $5 and $10 per 0.76 cubic meter (1 cubic yard). For comparison, a typical top-loading trash dumpster holds about 2.3 cubic meters (3 cubic yards) of material, and a typical dump truck holds about 45 cubic meters (60 cubic yards) of material. The problem with replenishment projects is that a huge volume of sand is needed, and new sand must be supplied continuously. For example, a small beach replenishment project of several hundred cubic meters can cost around $10,000 per year. Larger projects—several thousand cubic meters of sand—cost several million dollars per year.

Box 10–2 The Move of the Century: Relocating the Cape Hatteras Lighthouse

In spite of efforts to protect structures that are too close to the shore, they can still be in danger of being destroyed by receding shorelines and the destructive power of waves. Such was the case for one of the nation's most prominent landmarks, the candy-striped lighthouse at Cape Hatteras, North Carolina, which is 21 stories tall—the nation's tallest lighthouse and the tallest brick lighthouse in the world.

The lighthouse was built in 1870 on the Cape Hatteras barrier island 457 meters (1500 feet) from the shoreline to guide mariners through the dangerous offshore shoals known as the "Graveyard of the Atlantic." As the barrier island began migrating towards land, its beach narrowed. When the waves began to lap just 37 meters (120 feet) from its brick and granite base, there was concern that even a moderate-strength hurricane could trigger beach erosion sufficient to topple the lighthouse.

In 1970, the U.S. Navy built three groins in front of the lighthouse in an effort to protect the lighthouse from further erosion. The groins initially slowed erosion, but disrupted sand flow in the surf zone, which caused the flattening of nearby dunes and the formation of a bay south of the lighthouse. Attempts to increase the width of the beach in front of the lighthouse included beach nourishment and artificial offshore beds of seaweed, both of which failed to widen the beach substantially. In the 1980s, the Army Corps of Engineers proposed building a massive stone seawall around the lighthouse but decided the eroding coast would eventually move out from under the structure, leaving it stranded at sea on its own island. In 1988, the National Academy of Sciences determined that the shoreline in front of the lighthouse would retreat so far as to destroy the lighthouse and recommended relocation of the tower as had been done with smaller lighthouses. In 1999, the National Park Service, which owns the lighthouse, finally authorized moving the structure to a safer location.

Moving the lighthouse, which weighs 4395 metric tons (4830 short tons), was accomplished by severing it from its foundation and carefully hoisting it onto a platform of steel beams fitted with roller dollies. Once on the platform, it was slowly rolled along a specially designed steel track using a series of hydraulic jacks. A strip of vegetation was cleared to make a runway along which the lighthouse crept 1.5 meters (5 feet) at a time, with the track picked up from behind and reconstructed in front of the tower as it moved. During June and July 1999, the lighthouse was gingerly transported 884 meters (2900 feet) from its original location, making it one of the largest structures ever successfully moved.

After its $12 million move, the lighthouse now resides in a scrub oak and pine woodland 488 meters (1600 feet) from the shore (Figure 10C). Although it now stands further inland, the light's slightly higher elevation makes it visible just as far out to sea, where it continues to warn mariners of the hazardous shoals. At the current rate of shoreline retreat, the lighthouse should be safe from the threat of waves for at least another century.

Figure 10C Relocation of the Cape Hatteras Lighthouse, North Carolina.

Recycled glass that is ground to sand size and spread on the beach has been proposed as a less expensive source for beach replenishment, but the health and safety risks have not been fully explored.

I have the opportunity to live in a house at the edge of a coastal cliff where there is an incredible view along the entire coast. Is it safe from coastal erosion?

Based on what you've described, most certainly not! Geologists have long known that cliffs are naturally unstable. Even if the cliffs appear to be stable (or have been stable for a number of years), one significant storm can make the cliff fail.

The most common cause of coastal erosion is direct wave attack, which undermines the support and causes the cliff to fail.

You might want to check the base of the cliff and examine the local bedrock to determine for yourself if you think it will withstand the pounding of powerful storm waves that can move rocks weighing several tons. Other dangers include drainage runoff, weaknesses in the bedrock, slumps and landslides, seepage of water through the cliff, and burrowing animals.

Even though all states enforce a setback from the edge of the cliff for all new buildings, sometimes that isn't enough because large sections of "stable" cliffs can fail all at once. For instance, several city blocks of real estate have been eroded from the edge of cliffs during the last 100 years in some areas of southern California. Even though the view sounds outstanding, you may find out the hard way that the house is built a little *too* close to the edge of the cliff!

Chapter in Review

- The coastal region changes continuously. The shore is the region of contact between the oceans and the continents, lying between the lowest low tides and the highest elevation on the continents affected by storm waves. The coast extends inland from the shore as far as marine-related features can be found. The coastline marks the boundary between the shore and the coast. The shore is divided into the foreshore, extending from low tide to high tide, and the backshore, extending beyond the high-tide line to the coastline. Seaward of the low tide shoreline are the nearshore zone, extending to the breaker line, and the offshore zone beyond.

- A beach is a deposit of the shore area, consisting of wave-worked sediment that moves along a wave-cut bench. It includes the recreational beach, berm, beach face, low-tide terrace, one or more longshore bars, and longshore trough. Beaches are composed of whatever material is locally available.

- Waves that break at the shore move sand perpendicular to shore (toward and away from shore). In light wave activity, swash dominates the transport system and sand is moved up the beach face toward the berm. In heavy wave activity, backwash dominates the transport system and sand is moved down the beach face away from the berm toward longshore bars. In a natural system, there is a balance between light and heavy wave activity, alternating between sand piled on the berm (summertime beach) and sand stripped from the berm (wintertime beach), respectively.

- Sand is moved parallel to the shore, too. Waves breaking at an angle to the shore create a longshore current that results in a zigzag movement of sediment called longshore drift (longshore transport). Each year, millions of tons of sediment are moved from upstream to downstream ends of beaches. Most of the year, longshore drift moves southward along both the Pacific and Atlantic shores of the United States.

- Erosional-type shores are characterized by headlands, wave-cut cliffs, sea caves, sea arches, sea stacks, and marine terraces (caused by uplift of a wave-cut bench). Wave erosion increases as more of the shore is exposed to the open ocean, tidal range decreases, and bedrock weakens.

- Depositional-type shores are characterized by beaches, spits, bay barriers, tombolos, barrier islands, deltas, and beach compartments. Viewed from ocean side to lagoon side, barrier islands commonly have an ocean beach, dunes, barrier flat, and salt marsh. Deltas form at the mouths of rivers that carry more sediment to the ocean than the longshore current can carry away. Beach starvation occurs when the sand supply is interrupted. Beach replenishment (beach nourishment) is an expensive and temporary way to reduce beach starvation.

- Shorelines can also be classified as emerging or submerging based on their position relative to sea level. Ancient wave-cut cliffs and stranded beaches well above the present shoreline may indicate a drop in sea level relative to land. Old drowned beaches, submerged dunes, wave-cut cliffs, or drowned river valleys may indicate a rise in sea level relative to land. Changes in sea level may result from tectonic processes causing local movement of the landmass or from eustatic processes changing the amount of water in the oceans or the capacity of ocean basins. Melting of continental ice caps during the past 18,000 years has caused a eustatic rise in sea level of about 120 meters (400 feet).

- Sea level is rising along the Atlantic Coast about 0.3 meter (1 foot) per century, and the average erosion rate is −0.8 meter (−2.6 feet) per year. Along the Gulf Coast, sea level is rising 0.3 meter (1 foot) per century, and the average rate of erosion is −1.8 meters (−6 feet) per year. The Mississippi River Delta is eroding at 4.2 meters (13.7 feet) per year, resulting in a large loss of wetlands every year. Along the Pacific Coast, the average erosion rate is only −0.005 meter (−0.016 foot) per year. Different shorelines erode at different rates depending on wave exposure, amount of uplift, and type of bedrock.

- Hard stabilization, such as groins, jetties, breakwaters, and seawalls, is often constructed in an attempt to stabilize a shoreline. Groins (built to trap sand) and jetties (built to protect harbor entrances) widen the beach by trapping sediment on their upstream side, but erosion usually becomes a problem downstream. Similarly, breakwaters (built parallel to a shore) trap sand behind the structure, but cause unwanted erosion downstream. Seawalls (built to armor a coast) often cause loss of the recreational beach. Eventually, the constant pounding of waves destroys all types of hard stabilization. Relocation is a technique that has been successfully used to protect coastal structures.

Key Terms

Backshore (p. 289)

Backwash (p. 290)

Barrier flat (p. 296)

Barrier island (p. 296)

Bay barrier (bay-mouth bar) (p. 295)

Beach (p. 289)

Beach compartment (p. 297)

Beach face (p. 290)

Beach replenishment (beach nourishment) (p. 300)

Beach starvation (p. 299)

Berm (p. 290)

Breakwater (p. 308)

Coast (p. 289)

Coastline (p. 289)

Delta (p. 297)

Depositional-type shore (p. 292)

Distributary (p. 297)

Drowned beach (p. 300)

Drowned river valley (p. 301)

Dune (p. 296)

Emerging shoreline (p. 300)

Erosional-type shore (p. 292)

Eustatic sea level change (p. 302)

Foreshore (p. 289)

Groin (p. 307)

Groin field (p. 307)

Hard stabilization (p. 305)

Headland (p. 293)

Jetty (p. 307)

Longshore bar (p. 290)

Longshore current (p. 291)

Longshore drift (longshore transport) (p. 291)

Longshore trough (p. 290)

Low tide terrace (p. 290)

Marine terrace (p. 293)

National Flood Insurance Program (NFIP) (p. 288)

Nearshore (p. 289)

Ocean beach (p. 296)

Offshore (p. 289)

Peat deposit (p. 297)

Recreational beach (p. 290)

Relocation (p. 310)

Rip current (p. 294)

Rip-rap (p. 307)

Salt marsh (p. 296)

Sea arch (p. 293)

Sea cave (p. 293)

Sea stack (p. 293)

Seawall (p. 309)

Shore (p. 289)

Shoreline (p. 289)

Spit (p. 295)

Stranded beach deposit (p. 300)

Submerged dune topography (p. 300)

Submerging shoreline (p. 300)

Summertime beach (p. 291)

Swash (p. 290)

Tombolo (p. 296)

Wave-cut bench (p. 289)

Wave-cut cliff (p. 293)

Wintertime beach (p. 291)

Questions And Exercises

1. Why has the National Flood Insurance Program (NFIP) had a negative impact on the nation's shores?

2. To help reinforce your knowledge of beach terminology, construct and label your own diagram similar to Figure 10–1 from memory.

3. Describe differences between summertime and wintertime beaches. Explain why these differences occur.

4. What variables affect the speed of longshore currents?

5. What is longshore drift, and how is it related to a longshore current?

6. How is the flow of water in a stream similar to a longshore current? How are the two different?

7. Why does the direction of longshore current sometimes reverse in direction? Along both U.S. coasts, what is the primary direction of annual longshore current?

8. Describe the formation of rip currents. What is the best strategy to ensure that you won't drown if you are caught in a rip current?

9. Discuss the formation of such erosional features as wave-cut cliffs, sea caves, sea arches, sea stacks, and marine terraces.

10. Describe the origin of these depositional features: spit, bay barrier, tombolo, and barrier island.

11. Describe the response of a barrier island to a rise in sea level. Why do some barrier islands develop peat deposits running through them from the ocean beach to the salt marsh?

12. Discuss why some rivers have deltas and others do not. What are the factors that determine whether a "bird's-foot" delta (like the Mississippi Delta) or a smoothly curved delta (like the Nile Delta) will form?

13. Describe all parts of a beach compartment. What will happen when dams are built across all of the rivers that supply sand to the beach?

14. Compare the causes and effects of tectonic versus eustatic changes in sea level.

15. List the two basic processes by which coasts advance seaward, and list their counterparts that lead to coastal retreat.

16. List and discuss four factors that influence the classification of a coast as either erosional or depositional.

17. Describe the tectonic and depositional processes causing subsidence along the Atlantic Coast.

18. Compare the Atlantic Coast, Gulf Coast, and Pacific Coast by describing the conditions and features of emergence-submergence and erosion-deposition that are characteristic of each.

19. List the types of hard stabilization and describe what each is intended to do.

20. Draw an aerial view of a shoreline to show the effect on erosion and deposition caused by constructing a groin, a jetty, a breakwater, and a seawall within the coastal environment.

References

Baltuck, M., Dickey, J., Dixon, T., and Harrison, C. G. A. 1996. New approaches raise questions about future sea level change. *Eos Trans. AGU* 77:40, 385.

Bascom, W. 1980. *Waves and beaches: The dynamics of the ocean surface* (rev. ed.). New York: Anchor Books (Doubleday).

Bird, E. C. F. 1985. *Coastline changes: A global review*. Chichester, U.K.: Wiley.

Burk, K. 1979. The edges of the ocean: An introduction. *Oceanus* 22:3, 2–9.

Coates, R., ed. 1973. *Coastal geomorphology. Publications in Geomorphology*. Binghamton, NY: State University of New York.

Douglas, B. C. 1991. Global sea level rise. *Journal of Geophysical Research* 96:C4 6981–6992.

Federal Emergency Management Agency, 2000. *Evaluation of erosion hazards*. U.S. government publication in association with The Heinz Center for Science, Economics, and the Environment.

Kuhn, G. G., and Shepard, F. P. 1984. *Sea cliffs, beaches, and coastal valleys of San Diego County: Some amazing histories and some horrifying implications*. Berkeley: University of California Press.

Leatherman, S. P. 1983. Barrier dynamics and landward migration with Holocene sea-level rise. *Nature* 301:5899, 415–417.

May, S. K., Kimball, H., Grandy, N., and Dolan, R. 1982. The Coastal Erosion Information System. *Shore Beach* 50, 19–26.

Millemann, B. 1993. The National Flood Insurance Program. *Oceanus* 36:1, 6–8.

Peltier, W. R., and Tushingham, A. M. 1989. Global sea level rise and the greenhouse effect: Might they be connected? *Science* 244:4906, 806–810.

Pilkey, O. H. and Dixon, K. L. 1996. *The corps and the shore*. North Carolina: Island Press.

Pilkey, O. H. et. al. 1998. *The North Carolina shore and its barrier islands: Restless ribbons of sand (living with the shore)*. North Carolina: Island Press.

Rine, J. M. and Ginsburg, R. N. 1985. Depositional facies of a mud shoreface in Suriname, South America—A mud analogue to sandy, shallow-marine deposits. *Journal of Sedimentary Petrology* 55:5, 633–652.

Roemmich, D. 1992. Ocean warming and sea level rise along the southwest U.S. coast. *Science* 257:5068, 373–375.

Sahagian D. L., Schwartz, F. W., and Jacobs, D. K. 1994. Direct anthropogenic contributions to sea levels rise in the twentieth century. *Nature* 367:6458, 54–56.

Shepard, F. P. 1973. *Submarine geology*, 3rd ed. New York: Harper & Row.

———. 1977. *Geological oceanography*. New York: Crane, Russak & Company.

Stanley, D. J., and Warne, A. G. 1993. Nile delta: Recent geological evolution and human impact. *Science* 260:5108, 628–634.

***Suggested Reading in* Scientific American**

Bascom, W. 1960. Beaches. 203:2, 80–97. A comprehensive consideration of the relationship of beach processes, both large and small scale, to release of energy by waves.

Broecker, W. S., and Denton, G. H. 1990. What drives glacial cycles? 262:1, 48–107. Recent research is included in this consideration of the causes of glacial cycles.

Clarke, W. M. 2000. Moving big stuff. *Smithsonian* 30:10, 48–59.

Dolan, R., and Lins, H. 1987. Beaches and barrier islands. 257:1, 68–77. A discussion dealing with the ultimate futility of trying to develop and protect structures on beaches and barrier islands.

Fairbridge, R. W. 1960. The changing level of the sea. 202:5, 70–79. A discussion of what is known of the causes of the changing level of the sea, which appears to be related mostly to the formation and melting of glaciers and changes in the ocean floor.

Phillips, A. Tall order: Cape Hatteras Lighthouse makes tracks. *National Geographic* 197:5, 98–105.

Schneider, D. 1997. The rising seas. 276:3, 112–117. An analysis of the true threat of rising sea level and its relationship to global warming.

Oceanography on the Web

Visit the *Essentials of Oceanography* home page for on-line resources for this chapter. There you will find an on-line study guide with review exercises, and links to oceanography sites to further your exploration of the topics in this chapter. *Essentials of Oceanography* is at: **http://www.prenhall.com/ thurman** (click on the Table of Contents menu and select this chapter).

HTTP://WWW.PRENHALL.COM/THURMAN

11

The Coastal Ocean

●

THE LAW OF THE SEA

Who owns the ocean? Who owns the sea floor? If a company wanted to drill for oil offshore between two different countries, would it have to obtain permission from either country? Extensive exploitation of the ocean floor for minerals and petroleum is well under way, necessitating laws that unambiguously answer these questions. In the future, exploration will occur beyond the jurisdiction of the country with the nearest coastline. Furthermore, overfishing and pollution are worsening. Are these kinds of problems covered by long-established laws? The answer is yes ... and no.

In 1609 Hugo Grotius, a Dutch jurist and statesman whose writings eventually helped formulate international law, urged freedom of the seas to all nations in his treatise *Mare liberum* (*mare* = sea, *liberum* = free), which was premised on the assumption that the sea's major known resource—fish—exists in inexhaustible supply. Nevertheless, controversy continued over whether nations could control a *portion* of an ocean, such as the ocean adjacent to a nation's coastline.

Dutch jurist Cornelius van Bynkershoek attempted to solve this problem in *De dominio maris* (*dominio* = domain, *maris* = sea), published in 1702. It provided for national domain over the sea out to the distance that could be protected by cannons from the shore, an area called the **territorial sea**. Just how far from shore did the territorial sea extend? The British had determined in 1672 that cannon range extended 1 league (3 nautical miles) from shore. Thus, every country with a coastline maintained ownership over a *three-mile territorial limit* from shore.

In response to new technology that facilitated mining the ocean floor, the first **United Nations Conference on the Law of the Sea**, held in 1958 in Geneva, Switzerland, established that prospecting and mining of minerals on the continental shelf was under the control of the country that owned the nearest land. Because the continental shelf is that portion of the sea floor extending from the coastline to where the slope markedly increases, the seaward limit of the shelf is subject to interpretation. Unfortunately, the continental shelf was not well defined in the treaty, which led to disputes. In 1960, the second United Nations Conference on the Law of the Sea was also held in Geneva, but it made little progress toward an unambiguous and fair treaty concerning ownership of the coastal ocean.

Meetings of the third Law of the Sea Conference were held during 1973–1982. A new Law of the Sea treaty was adopted by a vote of 130 to 4, with 17 abstentions. Most developing nations that could benefit significantly from the treaty voted to adopt it. The United States, Turkey, Israel, and Venezuela opposed the new treaty because it made sea floor mining unprofitable. The abstaining countries included the Soviet Union, Great Britain, Belgium, the Netherlands, Italy, and West Germany, all of which were interested in sea floor mining, too. The treaty was ratified by the

key questions

- How does the coastal ocean vary in terms of salinity, temperature, and currents?
- How are estuaries created and what kinds of estuaries exist?
- Why are coastal wetlands important?
- Why is the circulation pattern in the Mediterranean Sea so unusual?
- How is marine pollution defined?
- What are some common types, causes, and environmental effects of coastal pollution?

Sometimes the best, and ironically the most difficult, thing to do in the face of an ecological disaster is to do nothing.

—*Sylvia Earle,
NOAA's chief on-site scientist for
the* Exxon Valdez *oil spill (1989)*

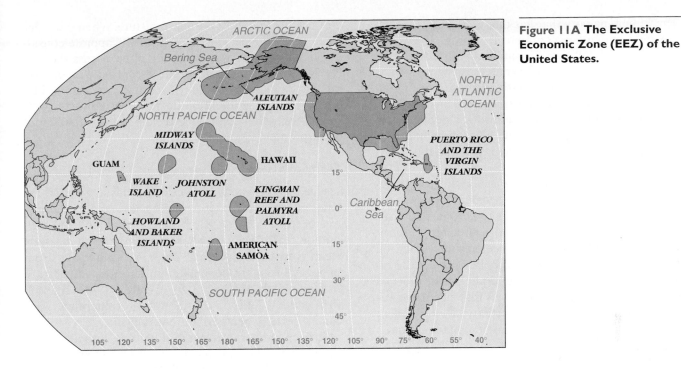

Figure 11A The Exclusive Economic Zone (EEZ) of the United States.

required sixtieth nation in 1993, establishing it as international law. Negotiations removed the objections of nations interested in sea floor mining, and the United States signed the revised treaty in 1994, when it was enacted as law.

The primary components of the treaty are as follows.

1. *Coastal nations jurisdiction.* The treaty established a uniform 12-mile (19-kilometer) territorial sea and a 200-nautical-mile (370-kilometer) **exclusive economic zone (EEZ)** from all land (including islands) within a nation. Each coastal nation has jurisdiction over mineral resources, fishing, and pollution regulation within its EEZ. If the continental shelf (defined geologically) exceeds the 200-mile EEZ, the EEZ is extended to 350 nautical miles (648 kilometers) from shore.

2. *Ship passage.* The right of free passage for all vessels on the high seas is preserved. The right of free passage is also provided within territorial seas and through straits used for international navigation.

3. *Deep-ocean mineral resources.* Private exploitation of sea floor resources may proceed under the regulation of the International Seabed Authority (ISA), within which a mining company will be strictly controlled by the United Nations. This provision, which caused some industrialized nations to oppose ratification, required mining companies to fund two mining operations—their own and one operated by the regulatory United Nations. Recently, this portion of the law was modified to eliminate some of the regulatory components, thus favoring free market principles and development by private companies. Still, this portion of the Law of the Sea has been one of the most contentious issues in international law.

4. *Arbitration of disputes.* A United Nations Law of the Sea tribunal will arbitrate any disputes in the treaty or disputes concerning ownership rights.

Since the passage of the Law of the Sea, 42% of the world's oceans have come under the control of coastal nations. The EEZ of the United States consists of almost 4.4 million square kilometers (1.7 million square miles) (Figure 11A), which is equal to about two-thirds the size of all U.S. lands and has tremendous economic potential.

The coastal ocean is a very busy place, filled with life, commerce, recreation, fisheries, and waste. Of the world fishery,[1] about 95% is obtained within 320 kilometers (200 miles) of shore. Coastal waters also support about 95% of the total mass of life in the oceans. In addition, these waters are the focal point of most shipping routes, oil and gas production, and recreational activities. Unfortunately, they are also the final destination of much of the waste products of those living on the adjacent land.

---

[1]The term *fishery* refers to fish caught from the ocean by commercial fishers.

The human activity that has the most impact on the ocean environment is commercial fishing. The world fishery peaked in 1997 at 93.6 million metric tons (103 million short tons) and has been decreasing ever since because of overfishing by a vastly increased fishing effort. This decreasing catch, combined with an increasing human population that puts additional demands on ocean resources, has put extreme environmental pressure on the ocean and is severely impacting its overall ecology.

Pollution affects the ecology of the oceans, too. Pollution in coastal waters comes from accidental spills of petroleum, the accumulation of sewage, certain chemicals (such as DDT and PCBs), and the element mercury. These pollutants can have severe deleterious effects on organisms living in the ocean. To better predict the effects of pollution on coastal waters, much more must be learned about the physical processes that give these waters their high biological productivity as well as an amazing resiliency to the onslaught of contamination.

Coastal Waters

Coastal waters are those relatively shallow-water areas that adjoin continents or islands. If the continental shelf is broad and shallow, coastal waters can extend several hundred kilometers from land. If it has significant relief or drops rapidly onto the deep-ocean basin, on the other hand, coastal waters will occupy a relatively thin band near the margin of the land. Beyond coastal waters lies the open ocean.

Because of their proximity to land, coastal waters are directly influenced by processes that occur on or near land. River runoff and tidal currents, for example, have a far more significant effect on coastal waters than on the open ocean.

Salinity

Fresh water is less dense than seawater, so river runoff does not mix well with seawater along the coast. Instead, the fresh water forms a wedge at the surface, which creates a well-developed **halocline**[2] (Figure 11–1a). When water is shallow enough, however, tidal mixing causes fresh water to mix with seawater, thus reducing the salinity of the water column (Figure 11–1c). There is no halocline here; instead, the water column is **isohaline** (*iso* = same, *halo* = salt).

Fresh water runoff from the continents generally lowers the salinity of coastal regions compared to the open ocean. Where precipitation on land is mostly rain, river runoff peaks in the rainy season. Where runoff is due mainly to melting snow and ice, on the other hand, runoff always peaks in summer.

Prevailing offshore winds can increase the salinity in some coastal regions. As winds travel over a continent,

they usually lose most of their moisture. When these dry winds reach the ocean, they typically evaporate considerable amounts of water as they move across the surface of the coastal waters. The increased evaporation rate increases surface salinity, creating a halocline (Figure 11–1b). The gradient of the halocline, however, is reversed compared to the one developed from the input of fresh water (Figure 11–1a).

Temperature

Sea ice forms in many high-latitude coastal areas where water temperatures are uniformly cold—generally greater than –2°C (28.4°F) (Figure 11–1d). In low-latitude coastal regions, where circulation with the open ocean is restricted, surface waters are prevented from mixing thoroughly, so maximum surface temperature may approach 45°C (113°F) (Figure 11–1e). In both high- and low-latitude coastal waters, **isothermal** (*iso* = same, *thermo* = heat) conditions prevail.

Surface temperatures in mid-latitude coastal regions are coolest in winter and warmest in late summer. A strong **thermocline**[3] may develop from surface water being warmed during the summer (Figure 11–1f) and cooled during the winter (Figure 11–1g). In summer, very-high-temperature surface water may form a relatively thin layer. Vertical mixing reduces the surface temperature by distributing the heat through a greater volume of water, thus pushing the thermocline deeper and making it less pronounced. In winter, cooling increases the density of surface water, which causes it to sink, thus creating an isothermal water column.

Prevailing offshore winds can significantly affect surface water temperatures. These winds are relatively warm during the summer, so they increase the ocean surface temperature and seawater evaporation. During winter, they are much cooler than the ocean surface, so they absorb heat and cool surface water near shore. Mixing from strong winds may drive the thermoclines in Figures 11–1f and 11–1g deeper and even mix the entire water column, producing isothermal conditions. Tidal currents can also cause considerable vertical mixing in shallow coastal waters.

Coastal Geostrophic Currents

Recall from Chapter 7 that geostrophic (*geo* = earth, *strophio* = turn) currents move in a circular path around the middle of a current gyre. Wind and runoff create geostrophic currents in coastal waters, too, where they are called **coastal geostrophic currents**.

Wind blowing parallel to the coast piles up water along the shore. Gravity eventually pulls this water back toward the open ocean. As it runs downslope away from the shore, the Coriolis effect causes it to curve to the right in the Northern Hemisphere and to the left in the

---

[2]Recall that a *halocline* (*halo* = salt, *cline* = slope) is a layer of rapidly changing salinity, as discussed in Chapter 5.

[3]Recall that a *thermocline* (*themo* = heat, *cline* = slope) is a layer of rapidly changing temperature, as discussed in Chapter 5.

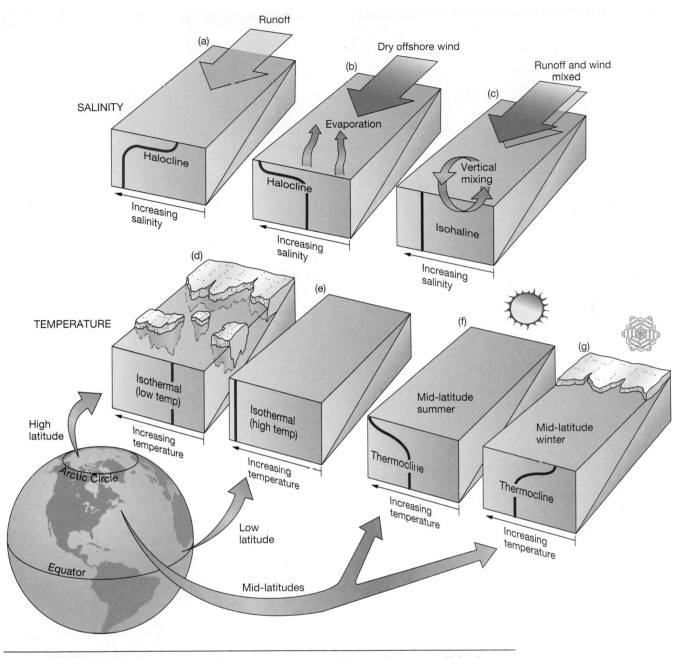

Figure 11–1 Salinity and temperature in the coastal ocean. Changes in coastal salinity (*top row*) can be caused by the input of freshwater runoff **(a)**, by dry offshore winds causing a high rate of evaporation **(b)**, or by both **(c)**. Changes in coastal temperature (*bottom row*) depend on latitude. In high latitudes **(d)**, the temperature of coastal water remains uniformly near freezing. In low latitudes **(e)**, coastal water may become uniformly warm. In the mid-latitudes, coastal surface water is significantly warmed during summer **(f)** and cooled during the winter **(g)**.

Southern Hemisphere. Thus, in the Northern Hemisphere, the coastal geostrophic current curves *northward* on the western coast and *southward* on the eastern coast of continents. These currents are reversed in the Southern Hemisphere.

A high-volume runoff of fresh water produces a surface wedge of fresh water that slopes away from the shore (Figure 11–2). This causes a surface flow of low-salinity water toward the open ocean, which the Coriolis effect curves to the right in the North-

ern Hemisphere and to the left in the Southern Hemisphere.

Coastal geostrophic currents are variable because they depend on the wind and the amount of runoff for their strength. If the wind is strong and the volume of runoff is high, then the currents are relatively strong. They are bounded on the ocean side by the steadier boundary currents comprising the open-ocean gyres.

The **Davidson Current** is a coastal geostrophic current that develops along the coast of Washington and

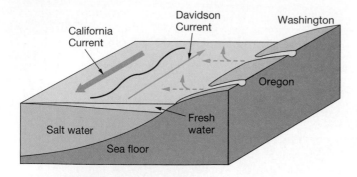

Figure 11–2 **Davidson coastal geostrophic current.** The Davidson Current is a coastal geostrophic current that flows north along the coast of Washington and Oregon. During the winter, runoff produces a freshwater wedge (*light blue*) that thins away from shore. This causes a surface flow of low-salinity water toward the open ocean, which is acted upon by the Coriolis effect, curving to the right.

Oregon during the winter (Figure 11–2). Heavy precipitation (which produces high volumes of runoff) combines with strong southwesterly winds to produce a relatively strong northward-flowing current. It flows between the shore and the southward-flowing California Current.

> The shallow coastal ocean adjoins land and experiences changes in salinity and temperature that are more dramatic than the open ocean. Coastal geostrophic currents can also develop.

Estuaries

An **estuary** (*aestus* = tide) is a partially enclosed coastal body of water in which fresh water runoff dilutes salty ocean water. The most common estuary is a river mouth, where the river empties into the sea. Many bays, inlets, gulfs, and sounds may be considered estuaries, too. All estuaries exhibit large variations in temperature and/or salinity.

The mouths of large rivers form the most economically significant estuaries, because many are seaports, centers of ocean commerce, and important commercial fisheries—Baltimore, New York, San Francisco, Buenos Aires, London, Tokyo, and many others.

Origin of Estuaries

The estuaries of today exist because sea level has risen approximately 120 meters (400 feet) since major continental glaciers began melting 18,000 years ago. As described in Chapter 10, these glaciers covered portions of North America, Europe, and Asia during the Pleis-

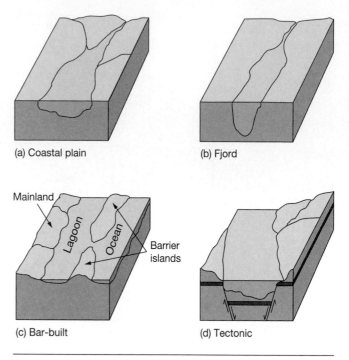

Figure 11–3 **Classifying estuaries by origin.** Diagrammatic views of the four types of estuaries based on origin. **(a)** Coastal plain estuary. **(b)** Glacially carved fjord. **(c)** Bar-built estuary. **(d)** Tectonic estuary.

tocene Epoch, more commonly referred to as the Ice Age. Four major classes of estuaries can be identified based on their origin (Figures 11–3):

1. A **coastal plain estuary** forms as sea level rises and floods existing river valleys. These estuaries, such as Chesapeake Bay in Maryland and Virginia, are called **drowned river valleys** (see Figure 10–17).

2. A **fjord**[4] forms as sea level rises and floods a glaciated valley. Water-carved valleys have V-shaped profiles, but fjords are U-shaped valleys with steep walls. Commonly, a shallowly submerged glacial deposit of debris (called a *moraine*) is located near the ocean entrance, marking the farthest extent of the glacier. Fjords are common along the coasts of Alaska, Canada, New Zealand, Chile, and Norway (Figure 11–4a).

3. A **bar-built estuary** is shallow and is separated from the open ocean by sand bars that are deposited parallel to the coast by wave action. Lagoons that separate **barrier islands** from the mainland are bar-built estuaries. They are common along the U.S. Gulf and East Coasts, including Laguna Madre in Texas and Pamlico Sound in North Carolina (see Figure 10–9).

4. A **tectonic estuary** forms when faulting or folding of rocks creates a restricted downdropped area into

---

[4]The Norwegian word *fjord* is pronounced "FEE-yord."

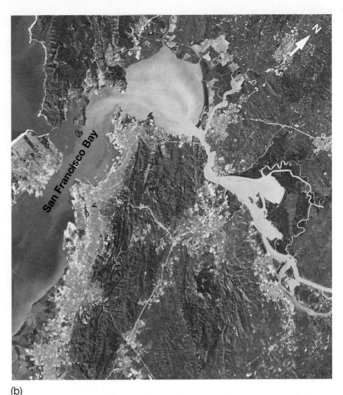

San Francisco Bay

(a)

(b)

Figure 11–4 Estuaries. **(a)** A Norwegian fjord, which is a deep glacially formed estuary that has been flooded by the sea. **(b)** Aerial view of San Francisco Bay in California, which is a tectonic estuary that was created by faulting.

which rivers flow. San Francisco Bay is in part a tectonic estuary (Figure 11–4b), formed by movement along faults including the San Andreas Fault.

Water Mixing in Estuaries

Generally, fresh water runoff moves across the upper layer of the estuary toward the open ocean, whereas denser seawater moves in a layer just below toward the head of the estuary. Mixing takes place at the contact between these water masses.

Estuaries can be classified based on the way fresh water and seawater mix, as shown in Figure 11–5:

1. **Vertically mixed estuary**—a shallow, low-volume estuary where the net flow always proceeds from the head of the estuary toward its mouth. Salinity at any point in the estuary is uniform from surface to bottom because river water mixes evenly with ocean water at all depths. Salinity simply increases from the head to the mouth of the estuary, as shown in Figure 11–5a. Salinity lines curve at the edge of the estuary because the Coriolis effect influences the inflow of seawater.

2. **Slightly stratified estuary**—a somewhat deeper estuary in which salinity increases from the head to

the mouth at any depth, as in a vertically mixed estuary. However, two water layers can be identified. One is the less saline, less dense upper water from the river, and the other is the more saline, more dense deeper water from the ocean. These two layers are separated by a zone of mixing. An **estuarine circulation pattern** begins to develop in this type of estuary, so there is a net surface flow of low-salinity water toward the ocean and a net subsurface flow of seawater toward the head of the estuary (Figure 11–5b).

3. **Highly stratified estuary**—a deep estuary in which upper-layer salinity increases from the head to the mouth, reaching a value close to that of open-ocean water. The deep-water layer has a rather uniform open-ocean salinity at any depth throughout the length of the estuary. An estuarine circulation pattern is well developed in this type of estuary (Figure 11–5c). Mixing at the interface of the upper water and the lower water creates a net movement from the deep-water mass into the upper water. Less-saline surface water simply moves from the head toward the mouth of the estuary, growing more saline as water from the deep mass mixes with it. Relatively strong haloclines develop at the contact between the upper and lower water masses.

Figure 11–5 Classifying estuaries by mixing. The basic flow pattern in an estuary is a surface flow of less dense fresh water toward the ocean and an opposite flow in the subsurface of salty seawater into the estuary. Numbers represent salinity in ‰; arrows indicate flow directions. **(a)** Vertically mixed estuary. **(b)** Slightly stratified estuary. **(c)** Highly stratified estuary. **(d)** Salt wedge estuary.

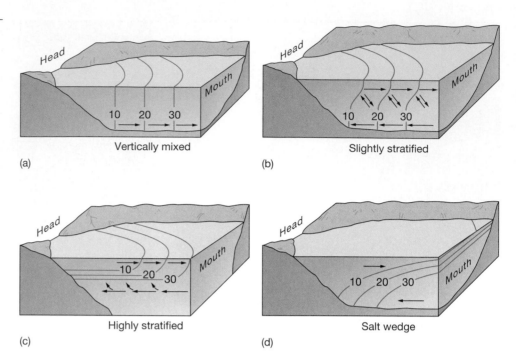

(a) Vertically mixed

(b) Slightly stratified

(c) Highly stratified

(d) Salt wedge

4. **Salt wedge estuary**—an estuary in which a wedge of salty water intrudes from the ocean beneath the river water. This kind of estuary is typical of the mouths of deep, high-volume rivers. No horizontal salinity gradient exists at the surface because surface water is essentially fresh throughout the length of, and even beyond, the estuary (Figure 11–5d). There is, however, a *horizontal* salinity gradient at depth and a very pronounced vertical salinity gradient—a halocline—at any location throughout the length of the estuary. This halocline is shallower and more highly developed near the mouth of the estuary.

The mixing pattern within an estuary may vary with location, season, or tidal conditions.

> Estuaries were formed by the rise in sea level after the last Ice Age. They can be classified based on origin as coastal plain, fjord, bar built, or tectonic. Estuaries can also be classified based on mixing as vertically mixed, slightly stratified, highly stratified, or salt wedge.

Estuaries and Human Activities

Estuaries are important breeding grounds and protective nurseries for many marine animals, so the ecological well-being of estuaries is vital to fisheries and coastal environments worldwide. Nevertheless, estuaries support shipping, logging, manufacturing, waste disposal, and other activities that can potentially damage the environment.

Estuaries are most threatened where human population is large and expanding, but they can be severely damaged where populations are still modest, too. Development in the Columbia River estuary, for example, demonstrates how a relatively small population can damage an estuary.

Columbia River Estuary The Columbia River, which forms most of the border between Washington and Oregon, has a long salt-wedge estuary at its entrance to the Pacific Ocean (Figure 11–6). The strong flow of the river and tides drive a salt wedge as far as 42 kilometers (26 miles) upstream and raise the river's water level over 3.5 meters (12 feet). When the tide falls, the huge flow of fresh water [up to 28,000 cubic meters (1,000,000 cubic feet) per second] creates a fresh water wedge that extends up to hundreds of kilometers into the Pacific Ocean.

Most rivers create floodplains along their lower courses, which have rich soil that can be used for growing crops. In the late 19th century, farmers and dairymen moved onto the floodplains along the Columbia River. Eventually, protective dikes were built to prevent the annual flooding. Flooding brings new nutrients, however, so the dikes deprived the floodplain of the nutrients necessary to sustain agriculture.

The river has been the principal conduit for the logging industry, which dominated the region's economy through most of its modern history. Fortunately, the river's ecosystem has largely survived the additional sediment caused by clear cutting by the logging industry. The construction of over 250 dams along the river and its tributaries, on the other hand, has permanently altered the river's ecosystem. Many of these dams, for

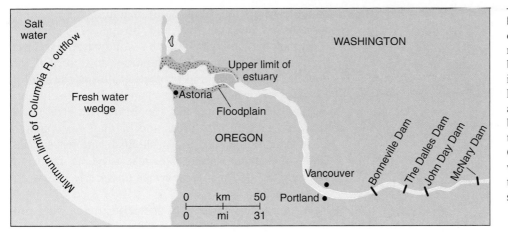

Figure 11–6 Columbia River estuary. The long estuary at the mouth of the Columbia River has been severely affected from interference by floodplains that have been diked, by logging activities, and—most severely—by hydroelectric dams. The tremendous outflow of the Columbia River creates a large wedge of low-density fresh water that remains traceable far out at sea.

example, do not have salmon ladders, which help fish "climb" in short vertical steps around the dams to reach their spawning grounds at the headwaters of their home streams.

Even though the dams have caused a multitude of problems, they do provide flood control, electrical power, and a dependable source of water, all of which have become necessary to the region's economy. To aid shipping operations, the river receives periodic dredging of sediment, which brings an increased risk for pollution. If these kinds of problems have developed in such sparsely populated areas as the Columbia River estuary, then larger environmental effects must exist in more highly populated estuaries, such as Chesapeake Bay.

Chesapeake Bay Estuary   Chesapeake Bay, formed by the drowning of the Susquehanna River (Figure 11–7), is a large coastal plain estuary that is about 320 kilometers (200 miles) long and 50 kilometers (30 miles) wide at its widest point. Most of the fresh water entering the bay comes from its western margin via rivers that drain the slopes of the Appalachian Mountains. About 15 million people live near this estuary.

Chesapeake Bay is a slightly stratified estuary that experiences large seasonal changes in salinity, temperature, and dissolved oxygen. Figure 11–7a shows the estuary's average surface salinity, which increases oceanward. The salinity lines are oriented virtually north-south in the middle of the bay because of the Coriolis effect. The Coriolis effect causes flowing water to curve to the right in the Northern Hemisphere, so seawater entering the bay tends to hug its *eastern* side, and fresh water flowing through the bay toward the ocean tends to hug its *western* side.

With maximum river flow in the spring, a strong halocline (and *pycnocline*[5]) develops, preventing the

fresh surface water and saltier deep water from mixing. Beneath the pycnocline, which can be as shallow as 5 meters (16 feet), waters may become **anoxic** (*a* = without, *oxic* = oxygen) from May through August, as dead organic matter decays in the deep water (Figure 11–7b). Major kills of commercially important blue crab, oysters, and other bottom-dwelling organisms occur during this time.

The degree of stratification and extent of mortality of bottom-dwelling animals have increased since the early 1950s. Increased nutrients from sewage and agricultural fertilizers have been added to the bay during this time, too, which has increased the productivity of microscopic algae (algal blooms). When these organisms die, their remains accumulate as organic matter at the bottom of the bay and promote the development of anoxic conditions. In drier years with less river runoff, however, anoxic conditions aren't as widespread or severe in bottom waters because fewer nutrients are supplied.

Coastal Wetlands

Wetlands are ecosystems in which the water table is close to the surface, so they are typically saturated most of the time. Wetlands can border either fresh water or coastal environments. Coastal wetlands occur along the margins of estuaries and other shore areas that are protected from the open ocean and include swamps, tidal flats, coastal marshes, and bayous.

The two most important types of coastal wetlands are **salt marshes** and **mangrove swamps**. Both are intermittently submerged by ocean water and both have oxygen-poor mud and accumulations of organic matter called *peat deposits*. Marshes support a variety of grasses and are known to occur from the Equator to latitudes as high as 65 degrees (Figures 11–8a and 11–8b). Mangroves are restricted to latitudes below 30 degrees (Figures 11–8a and 11–8c).

Wetlands are some of the most highly productive ecosystems on Earth and provide enormous economic

---

[5]Recall that a *pycnocline* (*pycno* = density, *cline* = slope) is a layer of rapidly changing density, as discussed in Chapter 5. A pycnocline is caused by a change in temperature and/or salinity with depth.

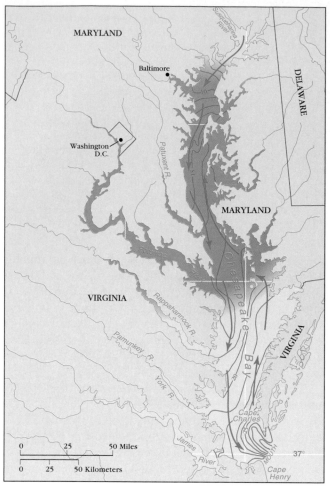

(a)

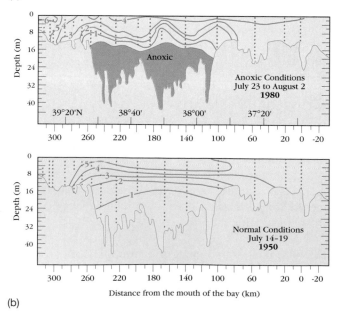

(b)

Figure 11–7 **Chesapeake Bay.** (a) Map of Chesapeake Bay, showing average surface salinity (*blue lines*) in ‰. The purple area in the middle of the bay represents anoxic (oxygen-poor) waters. (b) Profiles of dissolved oxygen levels (in ppm) for July-August 1980 (*top*), which shows deep anoxic waters, and July 1950 (*bottom*), which shows normal conditions.

benefits when left alone. Salt marshes, for example, serve as nurseries for over half the species of commercially important fishes in the southeastern United States. Other fishes, such as flounder and bluefish, use marshes for feeding and protection during the winter. Fisheries of oysters, scallops, clams, eels, and smelt are located directly in marshes, too. Mangrove ecosystems are important nursery areas and habitats for commercially valuable shrimp, prawn, shellfish, and fish species. Both marshes and mangroves also serve as important stopover points for many species of waterfowl and migrating birds.

Wetlands are amazingly efficient at cleansing polluted water. Just 0.4 hectare (1 acre) of wetlands, for example, can filter 730,000 gallons of water each year, cleaning agricultural runoff, toxins, and other pollutants long before they reach the ocean. Wetlands remove inorganic nitrogen compounds (from sewage and fertilizers) and metals (from groundwater polluted by land sources), which become attached to clay-sized particles in the wetland mud. Some nitrogen compounds trapped in sediment are decomposed by bacteria that release the nitrogen to the atmosphere as gas and many of the remaining nitrogen compounds fertilize plants, further increasing the productivity of wetlands. As marsh plants die, their remains either accumulate as peat deposits or are broken up to become food for bacteria, fungi, and fish.

Serious Loss of Valuable Wetlands

Despite all the benefits they provide, over half of the nation's wetlands have vanished. Of the original 87 million hectares (215 million acres) of wetlands that once existed in the conterminous United States (excluding Alaska and Hawaii), only about 43 million hectares (106 million acres) remain. Wetlands have been filled in and developed for housing, industry, and agriculture, because people want to live near the oceans and because they often view wetlands as unproductive, useless land that harbors diseases. Other countries have experienced similar losses of wetlands. The Philippines, for example, has lost 70% of its original mangrove cover.

To help prevent the loss of remaining wetlands, the U.S. Environmental Protection Agency established an Office of Wetlands Protection (OWP) in 1986. At that time, wetlands were being lost to development at a rate of 121,000 hectares (300,000 acres) per year! Currently, the rate of wetland loss has slowed to 24,000 hectares (59,000 acres) per year and the agency's goal is to minimize the loss of wetlands to the point that there is no net loss of wetlands in the U.S. The OWP actively enforces regulations against wetlands pollution and identifies the most valuable wetlands so that they may be protected or restored.

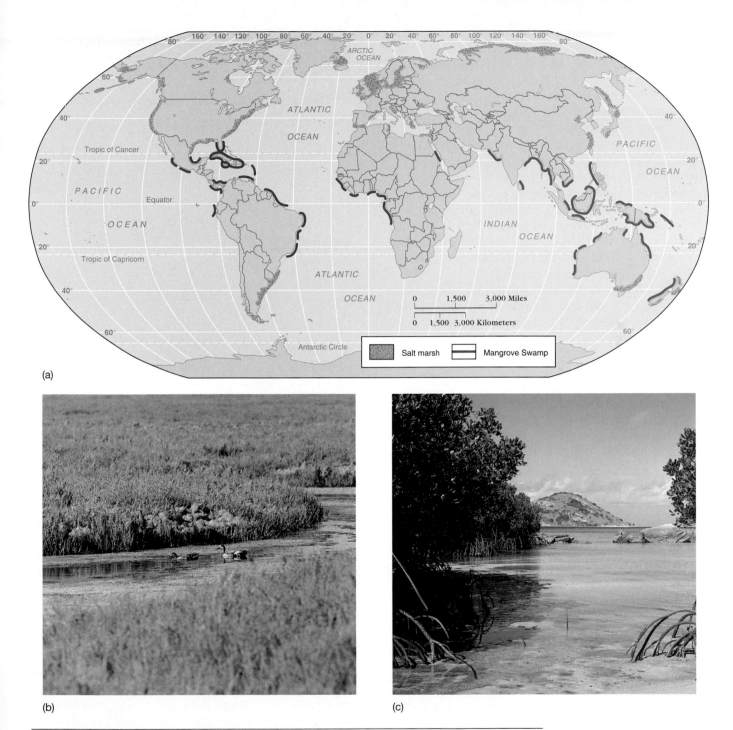

Figure 11–8 Salt marshes and mangrove swamps. (a) Map showing the distribution of salt marshes (higher latitudes) and mangrove swamps (lower latitudes). **(b)** Salt marsh along San Francisco Bay at Shoreline Park, California. **(c)** Mangrove trees on Lizard Island, Great Barrier Reef, Australia.

Coastal wetlands such as salt marshes and mangrove swamps are highly productive areas that serve as important nurseries for many marine organisms and act as a filter for polluted runoff.

Lagoons

Landward of barrier islands lie protected, shallow bodies of water called **lagoons** (see Figure 11–3c). Lagoons form in a bar-built type of estuary. Because of restricted circulation between lagoons and the ocean, three distinct zones can usually be identified within a lagoon. A *freshwater zone* lies near the mouths of rivers that flow into the lagoon. A *transitional zone* of brackish[6] water occurs near the middle of the lagoon. A *saltwater zone* lies close to the entrance (Figure 11–9a).

Salinity within a lagoon is highest near the entrance and lowest near the head (Figure 11–9b). In latitudes that have seasonal variations in temperature and precipitation, ocean water flows through the entrance during a warm, dry summer to compensate for the volume of water lost through evaporation, thus increasing the salinity in the lagoon. Lagoons actually may become hypersaline[7] in arid regions, where the flow of seawater cannot keep pace with the lagoon's surface evaporation. During the rainy season, the lagoon becomes much less saline as fresh water runoff increases.

Tidal effects are greatest near the entrance to the lagoon (Figure 11–9c) and diminish inland from the saltwater zone until they are nearly undetectable in the fresh water zone.

Laguna Madre

Laguna Madre is located along the Texas coast between Corpus Christi and the mouth of the Rio Grande. This long, narrow body of water is protected from the open ocean by Padre Island, a barrier island 160 kilometers (100 miles) long. The lagoon probably formed about 6000 years ago as sea level approached its present height.

The tidal range of the Gulf of Mexico in this area is about 0.5 meter (1.6 feet). The inlets at each end of Padre Island are quite narrow (Figure 11–10), so there is very little tidal interchange between the lagoon and the open sea.

Laguna Madre is a hypersaline lagoon and much of it is less than 1 meter (3.3 feet) deep. As a result, there are large seasonal changes in temperature and salinity. Water temperatures reach 32°C (90°F) in the summer and fall below 5°C (41°F) in winter. Salinities range from 2‰ when infrequent local storms provide large volumes of fresh water to over 100‰ during dry periods. High evaporation generally keeps salinity well above 50‰.[8]

Because even salt-tolerant marsh grasses cannot withstand such high salinities, the marsh has been replaced by an open sand beach on Padre Island. At the inlets, ocean water flows in as a surface wedge *over* the denser water of the lagoon and water from the lagoon flows out as a *subsurface* flow, which is exactly the opposite of the classic estuarine circulation.

---

[6]Brackish water is water with salinity between that of fresh water and seawater.

[7]Hypersaline conditions are created when water becomes excessively salty.

[8]Recall that normal salinity in the open ocean averages 35‰.

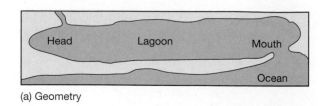

(a) Geometry

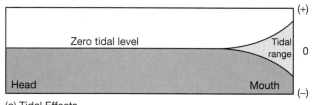

(b) Salinity

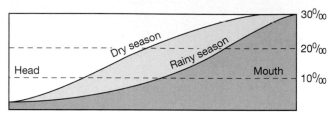

(c) Tidal Effects

Figure 11–9 Lagoons. Typical geometry **(a)** salinity **(b)** and tidal effects **(c)** of a lagoon.

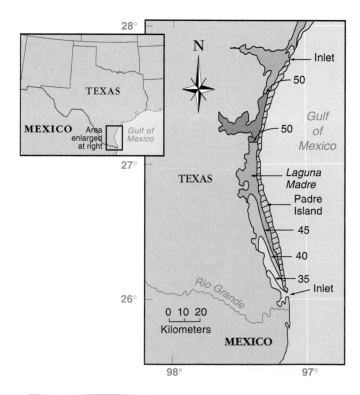

Figure 11–10 Laguna Madre summer surface salinity. Map showing geometry of Laguna Madre, Texas, and typical summer surface salinity (in ‰).

A Case Study: The Mediterranean Sea

The Mediterranean (*medi* = middle, *terra* = land) Sea is actually a number of small seas connected by narrow necks of water into one larger sea. It is the remnant of the ancient Tethys Sea that existed when all the continents were combined about 200 million years ago. It is over 4300 meters (14,100 feet) deep, and is one of the few inland seas in the world underlaid by oceanic crust. Thick salt deposits and other evidence on the floor of the Mediterranean suggest that it nearly dried up about 6 million years ago, only to refill with a large salt water waterfall (see Box 4–2).

The Mediterranean is bounded by Europe and Asia Minor on the north and east and Africa to the south (Figure 11–11a). It is surrounded by land except for very shallow and narrow connections to the Atlantic Ocean through the Strait of Gibraltar [about 14 kilometers (9 miles) wide], and to the Black Sea through the Bosporus [roughly 1.6 kilometers (1 mile) wide]. In addition, the Mediterranean Sea has a man-made passage to the Red Sea via the Suez Canal, a waterway 160 kilometers (100 miles) long that was completed in 1869. The Mediterranean Sea has a very irregular coastline, which divides it into subseas such as the Aegean Sea and Adriatic Sea, each of which has a separate circulation pattern.

An underwater ridge called a **sill**, which extends from Sicily to the coast of Tunisia at a depth of 400 meters (1300 feet), separates the Mediterranean into two major basins. This sill restricts the flow between the two basins, resulting in strong currents that run between Sicily and the Italian mainland through the Strait of Messina (Figure 11–11a).

Mediterranean Circulation

Atlantic Ocean water enters the Mediterranean as a surface flow through the Strait of Gibraltar to replace water that rapidly evaporates in the very arid eastern end of the sea. The water level in the eastern Mediterranean is generally 15 centimeters (6 inches) lower than at the Strait of Gibraltar. The surface flow follows the northern coast of Africa throughout the length of the Mediterranean and spreads northward across the sea (Figure 11–11a).

The remaining Atlantic Ocean water continues eastward to Cyprus. During winter, it sinks to form what is called the *Mediterranean Intermediate Water*, which has a temperature of 15°C (59°F) and a salinity of 39.1‰. This water flows westward at a depth of 200 to 600 meters (650 to 2000 feet) and returns to the North Atlantic as a *subsurface* flow through the Strait of Gibraltar (Figure 11–11b).

By the time Mediterranean Intermediate Water passes through Gibraltar, its temperature has dropped to 13°C (55°F) and its salinity to 37.3‰. It is still denser than even Antarctic bottom water and much denser than water at this depth in the Atlantic Ocean, so it moves down the continental slope. While descending, it mixes with Atlantic Ocean water and becomes less dense. At a depth of about 1000 meters (3300 feet) its density equals that of the surrounding Atlantic Ocean, so it spreads in all directions (Figure 11–11b). It has been detected in deep waters as far north as Iceland.

Circulation between the Mediterranean Sea and the Atlantic Ocean is typical of closed, restricted basins where evaporation exceeds precipitation. Low latitude restricted basins such as this always lose water rapidly to evaporation, so surface flow from the open ocean must replace it. Evaporation of inflowing water from the open ocean increases the sea's salinity to very high values. This denser water eventually sinks and returns to the open ocean as a subsurface flow.

This circulation pattern, which is called **Mediterranean circulation**, is opposite that of estuaries, where fresh water flows at the surface into the open ocean and salty water flows below the surface into the estuary. In estuaries, however, fresh water input exceeds water loss to evaporation, whereas evaporation exceeds input in the Mediterranean.

> High evaporation rates in the Mediterranean Sea cause it to have a shallow inflow of surface seawater and a subsurface high-salinity outflow—a circulation pattern opposite that of most estuaries.

Pollution in Coastal Waters

As the use of coastal areas has increased for residences, recreation, and commerce, pollution of coastal waters has increased as well. Coastal waters are more polluted than the open ocean because more pollution is dumped into coastal waters, and coastal waters are not as well circulated as the open ocean.

What Is Pollution?

Pollution is *any harmful substance*, but how do scientists determine which substances are harmful? For example, a substance may be esthetically unappealing to people yet is not harmful to the environment. Conversely, certain types of pollution cannot be easily detected by humans, yet they can do harm to the environment. A substance may not be immediately harmful, but it may cause harm years, decades, or even centuries later. Also, to whom must this harm be done? For instance, some marine species thrive when exposed to a particular pollutant that is quite toxic to other species. Interestingly, natural conditions in coastal waters, such as dead seaweed on the beach,

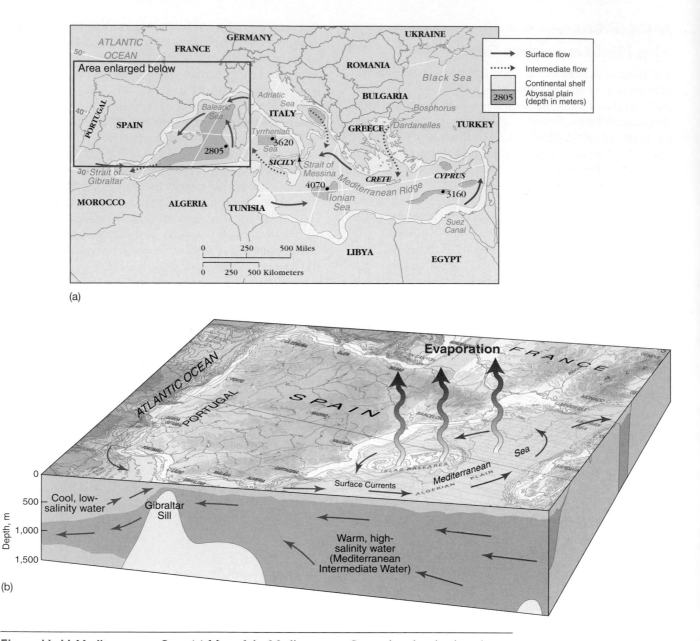

Figure 11–11 Mediterranean Sea. (a) Map of the Mediterranean Sea region showing its subseas, depths, sills (underwater ridges), surface flow, and intermediate flow. **(b)** Diagrammatic view of Mediterranean circulation in the Gibraltar Sill area.

may be considered "pollution" by some people. Nature may produce conditions we dislike, but it does not pollute. The amount of a pollutant is also important: If a substance that causes pollution is present in extremely tiny amounts, can it still be characterized as a pollutant? All of these questions are difficult to answer.

The World Health Organization defines pollution of the marine environment as the following:

The introduction by man, directly or indirectly, of substances or energy into the marine environment, including estuaries, which results or is likely to result in such deleterious effects as

harm to living resources and marine life, hazards to human health, hindrance to marine activities, including fishing and other legitimate uses of the sea, impairment of quality for use of sea water and reduction of amenities.

It is often difficult to determine the degree to which pollution affects the marine environment. Most areas were not studied sufficiently before they were polluted, so scientists do not have an adequate baseline from which to determine how pollutants have altered the marine environment. The marine environment is affected by decade- to century-long cycles, too, so it is difficult to determine whether a change is due to a natural biologic cycle or any number of introduced pollutants, many of which combine to produce new compounds.

To date, the most widely used technique for determining the concentration of pollutants that negatively affect the living resources of the ocean is the **standard laboratory bioassay** (*bio* = biologic, *essaier* = to weigh out). Regulatory agencies such as the Environmental Protection Agency (EPA) use a bioassay that determines the concentration of a pollutant that causes 50% mortality among the test organisms. If a pollutant exceeds a 50% mortality rate, then concentration limits are established for the discharge of the pollutant into coastal waters. One shortcoming of the bioassay is that it does not predict the long-term effect of pollution on marine organisms. Another is that it does not take into account how pollutants may combine with other chemicals, creating new types of pollutants.

Waste disposal facilities on land (such as landfills) have limited capacities that are already being exceeded in many cases. Should additional waste be discarded in the open ocean? Unlike coastal areas, the open ocean has mixing mechanisms (waves, tides, and currents) that distribute pollutants over a wide area—including an entire ocean basin. Diluting pollutants often renders them less harmful. On the other hand, do we really want to distribute a pollutant across an entire ocean without knowing what the long-term effects might be?

Some experts believe that we should not dump *anything* in the ocean, while others believe that the ocean can be a repository for many of society's wastes, as long as proper monitoring is conducted. Unfortunately, there are no easy answers and the issues are complex. What is clear is that more research is needed to assess the impact of pollutants in the ocean.

> Marine pollution is difficult to define but includes any human-induced substance that is harmful to the marine environment.

Petroleum

Major oil (petroleum) spills into the ocean are a fact of our modern oil-powered economy. Some oil spills are the result of loading/unloading accidents, collisions, or tankers running aground, such as the 1989 spill from the *Exxon Valdez* in Prince William Sound, Alaska (Box 11–1). Others are intentionally created, such as the oil that was spilled during the Persian Gulf War in 1991. Still others are caused by the blowout of undersea oil wells during drilling or pumping. The largest such spill occurred in June 1979 in the Gulf of Mexico, when the Petroleus Mexicanos (PEMEX) oil-drilling station

Ixtoc #1 in the Bay of Campeche off the Yucatán peninsula, Mexico, blew out and caught fire. Before it was capped nearly 10 months later, it spewed 530 million liters (140 million gallons) of oil into the Gulf of Mexico, some of which washed up along the coast of Texas (Figure 11–12).

Oil is a **hydrocarbon**, which means it is composed of the elements *hydrogen* and *carbon*. Hydrocarbons are organic substances, so they can be broken down or *biodegraded* by microorganisms. Because hydrocarbons are biodegradable, many experts consider oil to be among the *least* damaging pollutants introduced into the ocean! In fact, natural undersea oil seeps have occurred for millions of years, and the ocean ecosystem seems unaffected or even *enhanced* by them (because oil is a source of energy).

Data from the *Exxon Valdez* oil spill is a case in point. The oil spill released almost 44 million liters (11.6 million gallons) of oil into a pristine wilderness area in Alaska. The affected waters were expected have a long, slow recovery, but the fisheries that closed in 1989 bounced back with record takes in 1990. Ten years after the spill, several key species have rebounded to the point where their numbers are now greater than before the spill (Figure 11–13).

Still, oil is a complex mixture of hydrocarbons and other substances, including the elements oxygen, nitrogen, sulfur, and various trace metals. When this complex chemical mixture combines with seawater another complex chemical mixture, which also contains organisms—the results are usually devastating for marine organisms. Many are killed outright when they are coated

Figure 11–12 Blowout from the Ixtoc #1 oil well, Gulf of Mexico. Location of the 1979 Bay of Campeche blowout and oil slick that affected the Texas coast. The well blew out, caught fire, and flowed for 10 months, spilling 530 million liters (140 million gallons) of oil into the Gulf of Mexico. The accident produced the world's largest oil spill from an oil well.

Box 11–1
The *Exxon Valdez* Oil Spill: Not the Worst Spill Ever

A large percentage of oil enters the oceans from spills by tankers and transportation operations. One of the most publicized oil spills was from the supertanker *Exxon Valdez*, which occurred in Prince William Sound, Alaska, and caused the largest oil spill in U.S. territorial waters.

Crude oil produced from the North Slope of Alaska is carried by pipeline to the southern port of Valdez, Alaska, where it is loaded onto supertankers like the *Exxon Valdez*, which are capable of holding almost 200 million liters (53 million gallons) when full. On March 29, 1989, the tanker left Valdez with a full load of crude oil and was headed toward refineries in California. She was only 40 kilometers (25 miles) out of Valdez when the ship's officers noted icebergs from nearby Columbia Glacier within the shipping channel. While maneuvering around the icebergs, the ship ran aground on a shallowly submerged rocky outcrop known as Bligh Reef (Figure 11B), rupturing eight of the ship's 11 cargo tanks. About 22% of her cargo—almost 44 million liters (about 11.6 million gallons) of oil—spilled into the pristine waters of Prince William Sound, where it subsequently spread into the Gulf of Alaska and fouled over 1775 kilometers (1100 miles) of shoreline.

The U.S. Fish and Wildlife Service reported that at least 994 sea otters and 34,434 birds were killed outright by the spill. By some official estimates, however, the actual kill could have been 10 times that amount, because not all dead organisms are recovered. Due to the remote location and the size of the affected area, the exact total death toll will never be determined.

Immediately after the spill, Exxon spent over $2.5 billion in cleanup efforts and another $900 million in subsequent years for restoration. Absorbent materials and skimming devices were used to remove oil from the water, whereas super hot water (60°C/140°F) sprayed through high-pressure hoses was used to clean oil from the rocky beaches. The hot water removed the oil but also killed most shoreline organisms. Analysis of the cleanup effort, when compared to areas that were left to biodegrade naturally, reveals that the beaches that were left alone recovered more quickly and more completely than the cleaned beaches.

As large and damaging as the *Exxon Valdez* spill was, it ranks as only the *fifty-third* largest oil spill worldwide (Table 11A). The world's largest oil spill occurred because of intentional dumping by the Iraqi army during their invasion of

Figure 11B The 1989 *Exxon Valdez* oil spill, Prince William Sound, Alaska. Map showing location of the oil spill (*left*), the *Exxon Valdez* on Bligh Reef (*upper right*), and spilled oil coating a beach (*lower left*).

Kuwait during the 1991 Persian Gulf War. By the time the Iraqi were driven out of Kuwait and the leaking oil wells and sabotaged production facilities were brought under control, more than 908 million liters (240 million gallons) of oil had spilled into the Persian Gulf (Figure 11C)—more than *20 times* the amount spilled by the *Exxon Valdez*.

Table 11A.   The world's largest oil spills.

Rank	Date	Location	Source of spill	Size of spill	
				million liters	*million gallons*
1	1/1991	Kuwait, Saudi Arabia	Oil terminals, tankers	908	240
2	6/1979	Gulf of Mexico	Ixtoc #1 oil well	530	140
3	3/1992	Uzbekistan	Oil well	333	88
4	2/1983	Iran	Oil well	303	80
5	8/1983	Near coast of South Africa	*Castillo de Bellver* tanker	299	79
6	3/1978	Near coast of France	*Amoco Cadiz* tanker	261	69
53	3/1989	Prince William Sound, Alaska	*Exxon Valdez* tanker	44.0	11.6

Figure 11C Oil pollution from the 1991 Persian Gulf War. Map showing location of the spilled oil, which was confined to the northwest coast of the Persian Gulf by currents and southeasterly winds, and a Saudi Arabian government official examining some of the damage (*inset*).

by oil, rendering their insulating feathers or fur useless (Figure 11–14).

The Florida *Spill in West Falmouth Harbor*

One of the best-studied oil spills in the United States occurred in September 1969 near West Falmouth Harbor in Buzzards Bay, Massachusetts. The barge *Florida* came ashore, ruptured, and spilled about 680,000 liters (180,000 gallons) of No. 2 fuel oil, which is similar to diesel oil and is used in home heating. The oil spread northward into Wild Harbor, where the most severe damage occurred (Figure 11–15).

In the most severely oiled areas, nearly all marsh grasses and most intertidal and subtidal[9] animals were killed. A sharp reduction in *species diversity* (the number of different species present) was accompanied by a rapid increase in the population of polychaete

---

[9]The intertidal zone extends from high to low tide; the subtidal zone is below that.

Figure 11–13 Recovery of organisms affected by the *Exxon Valdez* oil spill. The populations of several key organisms in the Prince William Sound area of Alaska have rebounded after the 1989 *Exxon Valdez* oil spill. The bald eagle is so numerous that it was removed from the endangered species list in 1996. The collapse of the Pacific herring's population is thought to be unrelated to effects of the oil spill.

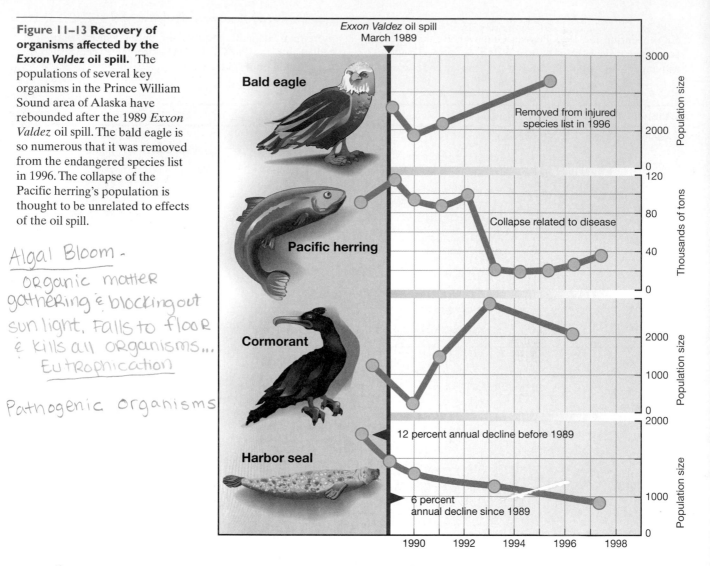

Algal Bloom - organic matter gathering & blocking out sun light. Falls to floor & kills all organisms... Eutrophication

Pathogenic organisms

worms[10] that have a high tolerance of oil. In fact, a single small red polychaete worm species, *Capitella capiata*, accounted for up to 99.9% of the individuals collected in samples of the most severely oiled locations during the first year. Species diversity did not increase appreciably until well into the third year after the spill.

Marsh grasses and animals reentered the area from three to five years after the spill. Remarkably, there was no visible damage after 10 years. After 20 years, there was virtually no oil in the subtidal sediments, and the intertidal marsh sediments were more than 99% oil-free. At the most heavily oiled site in Wild Harbor, however, enough oil was still present in beaches at a depth of 15 centimeters (6 inches) to kill animals that burrow into the sediments.

Despite the extensive damage caused by the oil spill at Wild Harbor in 1969, recovery can occur much more quickly than some researchers thought possible. Evidently, natural processes effectively biodegrade and remove oil from the marine environment, although it can take at least a couple of decades. Studies of other areas affected by oil spills indicate that most return to normal conditions within a few years and seem to experience no long-term damage.

The Argo Merchant *Spill off Nantucket Island*

The *Argo Merchant* sank after running aground on Fishing Rip Shoals 40 kilometers (25 miles) southeast of Nantucket Island, Massachusetts, in December 1976, spilling 29 million liters (7.7 million gallons) of No. 6 fuel oil (Figure 11–16). Fortunately, the winds were such that no oil came ashore. The surface slick moved eastward out to sea and was gone within a month of the spill.

Although the oil was not visible for long, it significantly damaged many organisms, such as planktonic (drifting) fish eggs of pollock and cod (Figure 11–17). Of the 49 pollock eggs collected shortly after the spill, 94% were coated with oil. Of the 60 cod eggs recovered, 60% were also oil fouled. In addition, 20% of the cod eggs and 46% of the pollock eggs were dead or dying. Little is known about the natural mortality rate of these fish eggs, but only 4% of cod eggs spawned in a laboratory under natural conditions were dead or dying at a similar stage of development.

[10]Polychaete worms are close relatives of segmented earthworms found on land.

Figure 11–14 A bird covered by oil from the *Exxon Valdez* oil spill. When marine organisms are covered by oil from an oil spill, their feathers or fur loose their insulation properties, resulting in high fatality rates. Some marine organisms such as this cormorant were rescued and cleaned of oil.

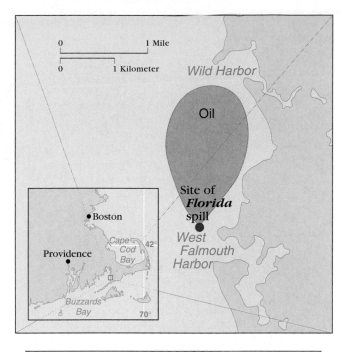

Figure 11–15 *Florida* oil spill at West Falmouth Harbor, Massachusetts. When the barge *Florida* came ashore and ruptured, currents carried its load of No. 2 fuel oil northward into Wild Harbor, where the most severe damage occurred.

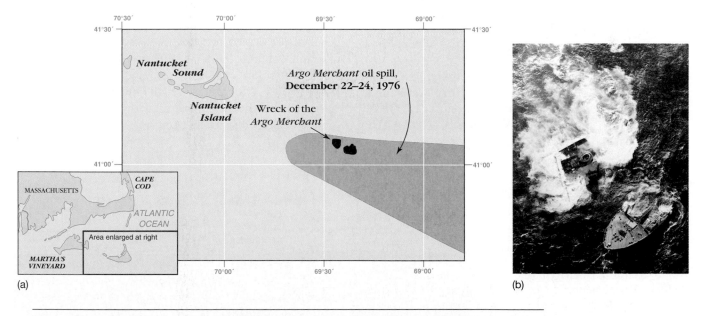

Figure 11–16 *Argo Merchant* oil spill off Nantucket Island, Massachusetts. (a) Map showing the ship's grounding site (*black*) and oiled area of the ocean (*gray*). (b) The *Argo Merchant*, after breaking up and spilling much of its cargo into the Atlantic Ocean southeast of Nantucket Island, Massachusetts.

Most of the oil floated as a surface slick, so contamination in subsurface water samples did not exceed more than 250 parts per billion. Other than the damage to fish eggs and other plankton, little direct evidence of major biological damage was obtained. A large number of dead and oiled birds, however, did wash ashore at Nantucket Island and Martha's Vineyard.

Each season, female pollock and cod spawn about 225,000 and 1 million eggs each, respectively, in an area from New Jersey to Greenland. Because of the huge number of eggs and the large spawning area, it is unlikely that a single oil spill would significantly affect these fisheries. Still, fishing communities of the northwest Atlantic are pleased when oil exploration on

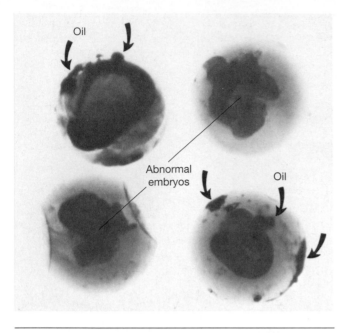

Figure 11–17 Pollock eggs affected by the *Argo Merchant* spill. The pollock eggs at the upper left and lower right show outer membranes contaminated with oil (*arrows*); the other two eggs have abnormal embryos. Each egg is about 1 millimeter (0.04 inch) in diameter.

Georges Bank yielded poor results because the pollock and cod fisheries—both overexploited already—would probably not survive the level of pollution brought about by even a small oil spill.

Cleaning Oil Spills When oil enters the ocean, it initially floats because oil is less dense than water and forms a slick at the surface, where it starts to break down through natural processes (Figure 11–18). The volatile, lighter components of crude oil evaporate over the first few days, leaving behind a more viscous substance that aggregates into tar balls and eventually sinks. The tarry oil also coats suspended particles, which settle to the seafloor, too.

If the floating oil hasn't dispersed, it can be collected with specially designed skimmers or absorbent materials. The collected oil (or oiled materials), however, must still be disposed of elsewhere. Waves, winds, and currents serve to further disperse an oil slick and mix the remaining oil with water to make a frothy emulsion called *mousse*. In addition, bacteria and photooxidation act to break down the oil into compounds that dissolve in water.

Microorganisms such as bacteria and fungi naturally biodegrade oil, so they can be used to help clean oil spills—a method called **bioremediation** (*bio* = biologic, *remedium* = to heal again). Virtually all marine ecosystems harbor naturally occurring bacteria that degrade hydrocarbons. Although certain types of bacteria and fungi can break down particular kinds of hydrocarbons, none is effective against all forms. In 1980, however, microbiologist A. M. Chakrabarty isolated a microorganism capable of breaking down nearly two-thirds of the hydrocarbons in most crude oil spills.

Releasing bacteria directly into the marine environment is one form of bioremediation. For example, a strain of oil-degrading bacteria was released into the Gulf of Mexico to test its effectiveness in cleaning up about 15 million liters (4 million gallons) of crude oil spilled after an explosion disabled the tanker *Mega Borg* in 1990. Preliminary results indicate that the bacteria reduced the amount of oil with no negative effects on the ecology due to the bacteria.

Figure 11–18 Processes acting on oil spills. After an oil spill enters the ocean, it is acted upon by various natural processes that break up the spill. The lighter components evaporate, while the heavier components form tar balls or coat suspended particles and sink. The remaining dispersed oil photo-oxidizes or can mix with water, creating a frothy substance called "mousse."

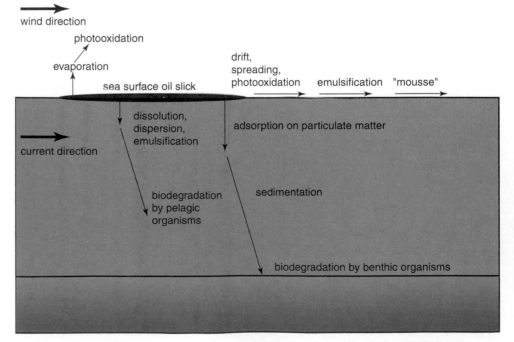

Providing conditions that stimulate the growth of naturally occurring oil-degrading bacteria is another form of bioremediation. Exxon, for example, spent $10 million dollars to spread fertilizers rich in phosphorus and nitrogen on Alaskan shorelines to boost the growth of indigenous oil-eating bacteria after the *Exxon Valdez* spill (Box 11–1). The resulting cleanup rate was more than twice that under natural conditions.

Preventing Oil Spills The best way to protect areas from oil spills is to prevent spills from occurring in the first place. Because our society relies on petroleum products, however, oil spills are a likely occurrence in the future (Figure 11–19), especially as petroleum reserves beneath the continental shelves of the world are increasingly exploited.

In February 1999, the Japanese-owned freighter M/V *New Carissa* ran aground just offshore of Coos Bay, Oregon, with nearly 1,500,000 liters (400,000 gallons) of tar-like fuel oil aboard and began leaking oil through cracks in its hull. When the ship washed into the surf zone and an approaching storm threatened to tear it apart, federal and state authorities decided to ignite the vessel and its fuel rather than risk a larger oil spill (Figure 11–20). This was the first time that oil on a ship in U.S. waters was intentionally burned to prevent an oil spill. Eventually, the ship split in two and about half of its oil burned, limiting the amount of oil spilled into the ocean. Most of the remaining oil was sunk with the wrecked ship a month later when it was towed offshore and sank in water 3 kilometers (1.9 miles) deep by Naval gunfire and a torpedo.

Sewage Sludge   90% Removed

Sewage treated at a facility typically undergoes **primary treatment**, where solids are allowed to settle and dewater, and **secondary treatment**, where it is exposed to bacteria-killing chlorine. **Sewage sludge** is the semisolid material that remains after such treatment. It contains a toxic brew of human waste, oil, zinc, copper, lead, silver, mercury, pesticides, and other chemicals. Since the 1960s, at least 500,000 metric tons (550,000 short tons) of sewage sludge has been dumped into the coastal waters of southern California and more than 8 million metric tons has been dumped in the New York Bight between Long Island and the New Jersey shore.

Although the Clean Water Act of 1972 prohibited the dumping of sewage into the ocean after 1981, the high cost of treating and disposing of sewage sludge on land resulted in extension waivers being granted to many municipalities. In the summer of 1988, however, nonbiodegradable debris including medical waste—probably carried by heavy rains into the ocean through storm drains—washed up on Atlantic coast beaches and adversely affected the tourist business. Although this event was completely unrelated to sewage disposal at sea, it focused public awareness on ocean pollution and

Figure 11–19 Who is at fault?

Figure 11–20 The *New Carissa* on fire off the Oregon coast. When the freighter M/V *New Carissa* ran aground in shallow water offshore Coos Bay, Oregon, in 1999 and began leaking oil, it was intentionally set on fire to prevent further oil from spilling into the ocean.

helped pass new legislation to terminate sewage disposal at sea.

New York's Sewage Sludge Disposal at Sea Sewage sludge from New York and Philadelphia has traditionally been transported offshore by barge and dumped in the ocean at sites totaling 150 square kilometers (58 square miles) within the New York Bight Sludge Site and the Philadelphia Sludge Site (Figure 11–21).

The water depth is about 29 meters (95 feet) at the New York Bight Sludge Site and about 40 meters (130 feet) at the Philadelphia Sludge Site. The water column in such shallow water is relatively uniform, so even the smallest sludge particles reach the bottom without undergoing much horizontal transport, and the ecology of the dump site can be severely affected. At the very least, such a concentration of organic and inorganic matter seriously disrupts the chemical cycling of nutrients. Greatly reduced species diversity results, and in some locations the environment becomes devoid of oxygen (anoxic).

In 1986, the shallow-water sites were abandoned and sewage was subsequently transported to a deep-water site 171 kilometers (106 miles) out to sea (Figure 11–21). The deep-water site is beyond the continental shelf break, so there is usually a well-developed density gradient that separates low-density, warmer surface water from high-density, colder deep water. Internal waves moving along this density gradient can horizontally transport particles at rates 100 times greater than they sink.

Local fishermen reported adverse effects on their fisheries soon after deep-water dumping began. Also, concern was expressed that the sewage could be transported great distances in eddies of the Gulf Stream (see Chapter 7), even as far as the coast of the United Kingdom. This program was terminated in 1993, and municipalities must now dispose of their sewage on land.

Boston Harbor Sewage Project Some 48 different communities that comprise the greater Boston area have, until recently, used an antiquated sewage system to dump sludge and partially treated sewage at the entrance to Boston Harbor. Tidal currents often swept the sewage back into the bay and at other times, the system became overloaded and dumped raw sewage directly

into the bay, making Boston Harbor one of the most polluted bays in the country.

A court-ordered cleanup of Boston Harbor in the 1980s resulted in a new sewage system that came online in 1998. It treats all sewage with bacteria-killing chlorine and carries it through a tunnel 15.3 kilometers (9.5 miles) long into deeper waters offshore (Figure 11–22a), which prevents it from returning to the bay. Since the system started, there has been a dramatic improvement in the bay's water quality. To pay for the $4 billion system, however, the average sewage bill for a Boston-area household is now about $1200.

Some fear that the project will degrade the environment in Cape Cod Bay and Stellwagen Bank, an important whale habitat (Figure 11–22b). The area was recently designated a National Marine Sanctuary, which may affect the feasibility of dumping there.

DDT and PCBs

The pesticide **DDT** (dichlorodiphenyltrichloroethane) and the industrial chemicals called **PCBs** (polychlorinated biphenyls) are now found throughout the marine environment. They are persistent, biologically active chemicals that have been introduced into the oceans entirely as a result of human activities. Because of their toxicity, long life, and propensity for being accumulated in food chains, these and other chemicals have been classified as persistent organic pollutants (POPs) capable of causing cancer, birth defects, and other grave harm.

DDT was widely used in agriculture during the 1950s and improved crop production throughout developing countries for several decades. However, its extreme effectiveness as an insecticide and persistence as a toxin in the environment eventually resulted in a host of environmental problems including devastating effects on marine food chains.

PCBs are industrial chemicals that were once widely used as a liquid coolant and insulation in industrial equipment such as power transformers, where they were released into the environment. They can affect animal reproduction and have been indicated as causes of spontaneous abortions in sea lions and the death of shrimp in Escambia Bay, Florida.

DDT and Eggshells U.S. production of DDT was almost completely banned in 1971. By that time, though, 2 billion kilograms (4.4 billion pounds) had already been manufactured, most by the United States. Since 1972, the use of DDT was banned in the United States by the Environmental Protection Agency. Worldwide, the pesticide is banned from agricultural use, but it continues to be used in limited quantities for public health purposes.

The danger of excessive use of DDT and similar pesticides first became apparent in the marine environment when it affected marine bird populations. During the 1960s, there was a serious decline in the brown pelican population of Anacapa Island off south-

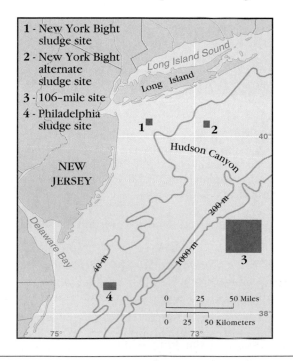

Figure 11–21 Atlantic sewage sludge disposal sites. More than 8 million metric tons of sewage sludge was dumped by barge annually at the New York Bight Sludge Site (*1*) and the Philadelphia Sludge Site (*4*). After 1986, the new dump site is the larger and deeper-water 106-mile site (*3*).

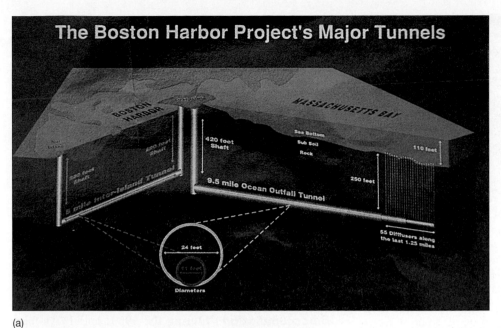

The Boston Harbor Project's Major Tunnels

BOSTON HARBOR

MASSACHUSETTS BAY

Sea Bottom

Sub Soil

Rock

420 feet Shaft

110 feet

250 feet

9.5 mile Ocean Outfall Tunnel

55 Diffusers along the last 1.25 miles

24 feet

11 feet

Diameters

(a)

Figure 11–22 Boston Harbor sewage project.
(a) Diagrammatic view of the Boston Harbor Project tunnels that transport sewage to an outfall 15 kilometers (9.5 miles) offshore at a depth of 76 meters (250 feet) beneath the ocean floor. **(b)** Bathymetric map of the coastal ocean in the Boston–Cape Cod area, showing the proximity of the outfall to Stellwagen Bank. Depths in meters, vertical exaggeration = 100×.

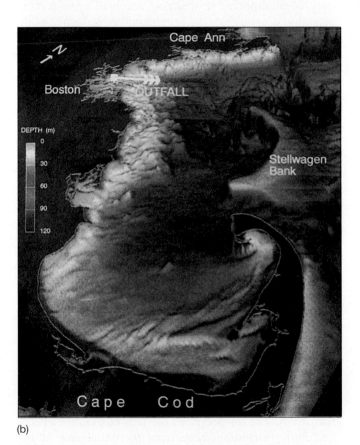

Cape Ann

Boston   OUTFALL

DEPTH (m)

Stellwagen Bank

Cape   Cod

(b)

Persistant Pollutents
• Heavy Metals
- lead (Pb)
- mercury (Hg)
Minimata disease
Chemicals → Pesticides
- DDT

Dredge Spoils

ern California (Figure 11–23). High concentrations of DDT in the fish eaten by the birds had caused them to produce eggs with excessively thin shells.

The osprey is a common bird of prey in coastal waters, similar to a large hawk. The osprey population of Long Island Sound declined in the late 1950s and 1960s because DDT contamination caused them to produce eggs with thin shells, too. Since the ban on DDT, the osprey, brown pelican, and many other species affected by the chemical are making remarkable comebacks.

DDT and PCBs Linger in the Environment DDT and PCBs (which were banned in 1977) generally enter the ocean through the atmosphere and river runoff. They are concentrated initially in the thin slick of organic chemicals at the ocean surface, and then they gradually sink to the bottom, attached to sinking particles. A study off the coast of Scotland indicated that open-ocean concentrations of DDT and PCBs are 10 and 12 times less, respectively, than in coastal waters. Long-term studies have shown that

Figure 11–23 Survival of brown pelicans threatened by DDT. Brown pelicans (*Pelecanus occidentalis*) that breed on Anacapa Island offshore southern California were found to have high levels of DDT, which decreased the thickness of their eggshells. Since DDT has been banned, healthy pelicans have returned to these waters, such as this one in breeding plumage (*inset*).

DDT residue in mollusks along the U.S. coasts peaked in 1968.

DDT and PCBs are so pervasive in the marine environment that even Antarctic marine organisms contain measurable quantities of them. There has been no agriculture or industry in Antarctica to introduce them directly, so they must have been transported from distant sources by winds and ocean currents.

Mercury and Minamata Disease

The metal **mercury**, which is a silvery liquid at room temperature, has many industrial uses. When it enters the ecosystem, however, mercury forms an organic compound that is generally toxic to most living things.

A chemical plant built on Minamata Bay, Japan, in 1938, produced acetaldehyde, which requires mercury in its manufacture. Mercury was discharged into Minamata Bay, where bacteria degraded it into a form that was later ingested and concentrated in the tissues of larger marine organisms. The first ecological changes in Minamata Bay were reported in 1950 and human effects were noted as early as 1953. The mercury poisoning that is now known as Minamata disease became epidemic in 1956, when the plant was only 18 years old. **Minamata disease** is a degenerative neurologic

disorder that affects the human nervous system and causes sensory disturbances including blindness and tremor, brain damage, birth defects, paralysis, and even death. This mercury poisoning was the first major human disaster resulting from ocean pollution. However, the Japanese government did not declare mercury as the cause of the disease until 1968. The plant was immediately shut down, but over 100 people were known to suffer from the disease by 1969 (Figure 11–24), almost half of whom died. A second acetaldehyde plant was closed in 1965 in Niigata, Japan, because it, too, was discharging mercury that poisoned people. Between 1965 and 1970, 47 fishing families contracted Minamata disease. Today the concentration of mercury in Minamata Bay is no longer unusually high, indicating that there has been enough time for the mercury to be widely dispersed within the marine environment.

Bioaccumulation During the 1960s and 1970s mercury contamination in seafood received considerable attention. Certain marine organisms concentrate within their tissues many substances found in minute concentrations in seawater in a process is called **bioaccumulation**. Because the amount of mercury in the ocean has been increasing (mostly from the mercury in disposable batteries), some seafood such as tuna and swordfish were thought to contain unusually high amounts of mercury.

Studies done on the amount of seafood consumed by various human populations helped establish safe levels of mercury in fish to be marketed. To establish these levels, three variables were considered:

1. The rate at which each group of people consumed fish

2. The mercury concentration in the fish consumed by that population

3. The minimum ingestion rate of mercury that induces disease symptoms

These three variables help establish a maximum allowable mercury concentration that will safeguard people from mercury poisoning, as long as they don't exceed the recommended intake of fish.

Figure 11–25 shows the relative risk of contracting Minamata disease for people in the United States, Sweden, and Japan, including those from the Minamata fishing community. The graph shows that the risk increases with increased consumption of fish and that the higher the mercury concentration of the fish, the greater the risk.

▬EIO▬

For more information and on-line exercises about the Environmental Issue in Oceanography (EIO) "Toxic Chemicals in Seawater," visit the EIO Web site at **http://www.prenhall.com/oceanissues** and select Issue #4.

Figure 11–24 A victim of Minamata disease.

Scientists have determined that the minimum level of mercury consumption that causes poisoning symptoms is 0.3 milligrams per day over a 200-day period. Figure 11–25 shows that for people in the United States, which have an average daily consumption of 17 grams of fish per day, mercury poisoning symptoms occur when mercury concentrations in fish exceed 20 parts per million (ppm). Using a safety factor of 10 times, the maximum concentration of mercury in fish that can be safely consumed by people in the U.S. is 2.0 ppm.

The U.S. Food and Drug Administration (FDA) doubled the safety factor and established a limit of mercury concentration for fish at 1 ppm. Based on con-

sumption rates, this limit has adequately protected the health of U.S. citizens because essentially all tuna and most swordfish fall below this concentration. Unless you eat an unusually large amount of tuna or swordfish, there's no need to worry about contracting Minamata disease.

Figure 11–25 shows that for people in Sweden and Japan, the mercury concentration of fish deemed to be at a safe level is lower because these populations eat more fish. The graph also shows the extreme danger to which residents of Minamata were inadvertently subjected when they ate so much of the highly contaminated fish from Minamata Bay.

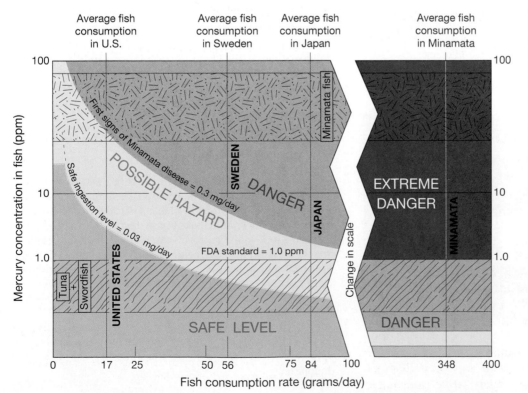

Figure 11–25 Mercury concentrations in fish versus consumption rates for various populations. Graph showing the relative risk of contracting Minamata disease based on the amount of fish consumed and the concentration of mercury in fish for people in the U.S., Sweden, and Japan—including Minamata. It shows that the safe ingestion level of 0.03 milligrams of mercury per day is within the Food and Drug Administration's standard safety level of 1.0 ppm for fish and that tuna and most swordfish are safe to consume.

Non-Point-Source Pollution and Trash

Non-point-source pollution—also called "poison run-off"—is any type of pollution entering the surface water system from sources other than underwater pipelines. Mostly, non-point-source pollution arrives at the ocean via runoff from storm drains, many of which have labels indicating they lead directly to the ocean (Figure 11–26).

Because non-point-source pollution comes from many different locations, it is difficult to pinpoint where it originates, although the *cause* of the pollution may be readily apparent. Trash that is washed down a storm drain to the ocean is one such example. Others include pesticides and fertilizers from agriculture and oil from automobiles that is washed to the ocean whenever it rains. In fact, the amount of road oil and improperly disposed oil regularly discharged each year into U.S. waters as non-point-source pollution is as much as 26 times the amount of the *Exxon Valdez* oil spill!

Trash enters the ocean as a result of ocean dumping, too. According to existing laws (Figure 11–27), certain types of trash such as glass, metal, rags, and food can be legally dumped in the ocean—as long as they are dumped far enough away from shore or ground up small enough. Mostly, this material sinks or biodegrades and does not accumulate at the surface.

Plastic, however, floats and is not biodegradable, so it can remain in the marine environment indefinitely (Box 11–2). Plastic waste has strangled marine organisms and birds that have been caught in plastic netting and packing straps (Figure 11–28). Marine turtles have been killed when they ingested plastic bags, evidently mistaking them for jellyfish or other transparent plankton on which they typically feed. Thus, plastic is one of the few substances that is illegal to dump anywhere in the ocean.

Figure 11–26 A labeled storm drain that leads to the ocean.

Examples of marine pollution include petroleum from oil spills, sewage sludge, chemicals such as PCB's and DDT, mercury, and non-point-source pollution such as road oil and trash.

? Students Sometimes Ask...

Can the influence of highly productive estuaries extend beyond the estuary itself?

Yes. Microscopic plankton (floating algal cells and animals), which form the basis of the food web, abound in nutrient-rich estuary waters. Because plankton float, ocean currents carry them from estuaries to the adjacent continental shelf areas. Estuaries have a powerful influence on the plankton populations and the overall ecology of the adjacent continental shelf.

What is the strategy behind having such large ships for transporting oil? Wouldn't smaller ships be safer?

Theoretically, one larger ship has a lower chance of being damaged and spilling its load than several smaller ships. If smaller ships are used, then more trips are needed to transport the same amount of oil. This increases ship traffic and the exposure of the oil to the ocean, just as frequent trips in your car increase your exposure to being involved in a crash. If only a few trips are made using large ships—and if special precautions are taken during those trips—then less oil will be spilled over time. In addition, larger ships are economically more efficient at transporting large amounts of oil compared to a host of smaller ships.

Legislation has been passed that requires all tankers in U.S. waters to have double hulls by 2015. However, analysis of the *Exxon Valdez* spill indicates that even a double-hulled tanker would not have prevented the disaster. (Currently, only 10% of the tankers operating in Prince William Sound have double hulls.) Tanker designs are also being modified to limit the amount of oil spilled should there be a hull rupture.

I've heard that some organizations want to lift the ban on DDT. Why would they want to do that?

Since production of DDT was banned in 1971, outbreaks of malaria have dramatically increased because DDT was the most effective and readily available pesticide used to kill mosquitoes that transmit malaria. According to the World Health Organization, malaria infects up to 500 million people a year—mostly in tropical regions—and kills as many as 2.7 million, including at least one child every 30 seconds. In addition, drug-resistant strains of malaria have begun to show up worldwide. This resurgence of malaria has caused many health organizations to call for an exception to the ban on DDT— in spite of its well-documented perseverance and negative effects on the environment—so that it can be used selectively to spray houses in malaria-prone areas like tropical Africa and Indonesia.

Under the MARPOL agreement and U.S. federal law, it is illegal for any vessel to discharge plastics or garbage containing plastics into any waters. Additional restrictions on dumping non-plastic waste are outlined below. *All* discharge of garbage is prohibited in the Great Lakes or their connecting or tributary waters. Each knowing violation of these requirements may result in a <u>fine</u> of up to $500,000 and 6 years imprisonment.

3 to 12 nautical miles offshore	12 to 25 nautical miles offshore	Outside 25 nautical miles offshore

Within 3 nautical miles of shore and anywhere in lakes, rivers, bays, and sounds.

ILLEGAL TO DUMP
Plastic
All other trash

ILLEGAL TO DUMP
Plastic
Dunnage, lining & packing materials that float
All other trash if not ground to less than 1"

ILLEGAL TO DUMP
Plastic
Dunnage, lining & packing materials that float

ILLEGAL TO DUMP
Plastic

State and local laws may place further restrictions on the disposal of garbage.

WORKING TOGETHER, WE CAN ALL MAKE A DIFFERENCE!
CENTER FOR MARINE CONSERVATION 1725 DeSales Street, NW Washington, DC 20036 (202)429-5609

Figure 11–27 Current law regulating ocean dumping. Many types of trash can be legally dumped in the ocean, as long as it is ground fine enough and not composed of plastic. In fact, plastic is the only substance that cannot be dumped anywhere in the ocean.

Don't storm drains receive treatment before emptying into the ocean?

Contrary to popular belief, water (and any other material) that goes down a storm drain does *not* receive any treatment before being emptied into a river or directly into the ocean. Sewage treatment plants receive enough waste to process without the additional runoff from storms, so it is important to monitor carefully what is disposed into storm drains. For instance, some people discharge used motor oil into storm drains, thinking that it will be processed by a sewage plant. A good rule of thumb is this: *Don't put anything down a storm drain that you wouldn't put directly into the ocean itself.*

Is dilution the solution to ocean pollution?

That's certainly a catchy phrase, but the implications are controversial. It suggests the oceans can be used to dispose society's wastes, as long as the wastes are diluted to the point they no longer threaten marine organisms (which is difficult to determine). Because the oceans are vast and consist of a good solvent (water), they appear ideally suited to this disposal strategy. In addition, the oceans have good mixing mechanisms (currents, waves, and tides), which would dilute many forms of pollution.

Air pollution was once viewed in a similar manner. Disposal of pollutants into the atmosphere was thought to be acceptable as long as they were dispersed widely and high enough—so, tall smokestacks were constructed. Over time, however, pollutants such as nitric acid and acid sulfates increased in the atmosphere to the point that acid rain is now a problem. The ocean, like the atmosphere, has a finite *holding capacity* for pollutants and even experts disagree on what it is.

As disposal sites on land begin to fill, the ocean is increasingly evaluated as an area for disposal of society's wastes. One thing that we can *all* do is to limit the amount of waste we generate, alleviating some of the problem of where to put waste. It is likely, however, that the ocean will continue to be used as a dumping ground in the foreseeable future. Despite many new disposal techniques, a long-term "solution" to ocean pollution hasn't appeared as yet.

All this ocean pollution is distressing. Is there anything that I can do about it?

Yes, there are many things you can do to help protect the ocean, some of which are listed in this book's afterword. They all involve making intelligent choices and minimizing your impact on the environment. Non-point-source pollution, for example, is something that the general public is directly responsible for, so the best method of prevention may be educating people. Once people understand the impact our choices have on the environment, then the solution is up to *all* of us.

Figure 11–28 Floating plastic strangles marine life. A female northern elephant seal (*Mirounga angustirostris*) with a plastic packing strap around its neck.

Box 11–2
From A to Z in Plastics: The Miracle Substance?

Even though plastic products have been used for over a century, their commercial development occurred during World War II when shortages of rubber and other materials created great demand for alternative products. Plastic products are *lightweight*, *strong*, *durable*, and *inexpensive*, so they have many advantages over other materials. Today, everything from airplane parts to zippers is made of plastic (Figure 11D). We wear plastics, drive in plastics, cook in plastics, and even carry plastic components inside us as artificial parts. The convenience of plastic items intended for one-time use has also contributed to the popularity of plastics.

However, what was once thought of as a miracle substance has several disadvantages. Disposing of plastics has already strained the capacity of land-based solid-waste disposal systems. Plastic waste is now an increasingly abundant component of oceanic flotsam (floating refuse). In fact, plastics constitute the vast majority of floating trash in all oceans worldwide. Unfortunately, the very same properties that make plastics so advantageous make them unusually persistent and damaging when released into the marine environment:

- They are lightweight, so they float and concentrate at the surface.
- They are strong, so they entangle marine organisms.

- They are durable, so they don't biodegrade easily, causing them to last almost indefinitely.
- They are inexpensive, so they are mass-produced and used in almost everything.

Small pellets ranging in size from a BB to a pea are used to produce plastic products (Figure 11E). They are transported in bulk aboard commercial vessels and are found throughout the oceans, probably due to spillage at loading terminals. In coastal waters, plastic products used in fishing are commonly thrown overboard by recreational and commercial vessels. Plastic trash also finds its way into the open-ocean waters from non-point-source pollution by careless people on land.

Even though plastic material does not sink or biodegrade, ocean currents eventually wash it onto the beaches of islands and continents. Thus, even remote beaches are littered with plastic pellets and plastic trash. Studies conducted between 1984 and 1987 showed that the plastic pellet content of beaches throughout the world is increasing. For instance, some Bermuda beaches have up to 10,000 pellets per square meter (10.7 square feet), and some beaches on Martha's Vineyard in Massachusetts yielded 16,000 plastic spherules per square meter (10.7 square feet).

Plastic trash is not limited to beaches. In the northern Sargasso Sea, researchers have counted more than 10,000 plastic pieces and 1500 pellets per square kilometer (0.4 square mile). Between 1972 and 1987, the concentration of plastic pellets doubled, suggesting that the amount of floating plastic trash in the ocean has doubled, too. Indeed, equipment deployed from research vessels often returns entangled in plastic trash such as 6-pack rings, styrofoam, and fishing lines, nets, and floats.

What can be done to limit the amount of plastic in the marine environment? People can limit their use of disposable plastic, recycle plastic material, and dispose of their plastic trash properly, including not dumping any plastic at sea. If these simple guidelines are followed, it would greatly reduce the amount of plastics in the world's oceans.

Figure 11D Plastic products.

Figure 11E Plastic pellets found at a beach.

Chapter in Review

- Coastal waters support about 95% of the total mass of life in the oceans, and they are important areas for commerce, recreation, fisheries, and the disposal of waste. The temperature and salinity of the coastal ocean vary over a greater range than the open ocean because the coastal ocean is shallow and experiences river runoff, tidal currents, and seasonal changes in solar radiation. Coastal geostrophic currents are produced from fresh water runoff and coastal winds.

- Estuaries are semienclosed bodies of water where freshwater runoff from the land mixes with ocean water. Estuaries are classified by their origin as coastal plain, fjord, bar built, or tectonic. Estuaries are also classified by their mixing patterns of fresh and salt water as vertically mixed, slightly stratified, highly stratified, and salt wedge. Typical circulation in an estuary consists of a surface flow of low-salinity water toward its mouth and a subsurface flow of marine water toward its head.

- Estuaries provide important breeding and nursery areas for many marine organisms but often suffer from human population pressures. The Columbia River Estuary, for example, has degraded from agriculture, logging, and the construction of dams upstream. In Chesapeake Bay, an anoxic zone occurs during the summer that kills many commercially important species.

- Wetlands are some of the most biologically productive regions on Earth. Salt marshes and mangrove swamps are important examples of coastal wetlands. Wetlands are ecologically important because they remove land-derived pollutants from water before it reaches the ocean. Nevertheless, human activities continue to destroy wetlands.

- Long offshore deposits called barrier islands protect marshes and lagoons. Some lagoons have restricted circulation with the ocean, so water temperatures and salinity may vary widely with the seasons.

- Circulation in the Mediterranean Sea is characteristic of restricted bodies of water in areas where evaporation greatly exceeds precipitation. Called Mediterranean circulation, it is the reverse of estuarine circulation.

- Although marine pollution seems easily defined, an all-encompassing definition is quite detailed. It is often difficult to establish the degree to which pollution affects ocean areas. The most widely used technique for determining the effects of pollution is the standard laboratory bioassay, which determines the concentration of pollution that causes a 50% mortality rate among marine organisms. The debate continues about whether society's wastes should be dumped in the ocean.

- Oil is a complex mixture of hydrocarbons and other substances, most of which are naturally biodegradable. Thus, many experts consider oil to be among the least damaging of all substances introduced into the marine environment. In areas that have experienced oil spills, recovery can be as rapid as a few years. Still, oil spills can cover large areas and kill many animals.

- Oil pollution reduces species diversity and persists longer on muddy bottoms than on sandy or rocky bottoms. Oil spills that do not come ashore, such as that of the *Argo Merchant*, do considerably less environmental damage than those that do. The greatest damage from the *Argo Merchant* spill was probably to plankton, especially pollock and cod eggs. After the much-publicized *Exxon Valdez* oil spill in Alaska, many novel approaches were used to clean the spilled oil, including oil-eating bacteria (bioremediation).

- Millions of tons of sewage sludge have been dumped offshore in coastal waters. Although 1972 legislation required an end to dumping of sewage in the coastal ocean by 1981, exceptions continue to be made. Increased public concern resulted in new legislation to prohibit sewage dumping in the ocean.

- Plastics are lightweight, strong, durable, and inexpensive. Unfortunately, these same properties make them a relentless source of floating trash in the ocean. The amount of plastic accumulating in the oceans has increased dramatically. Certain forms of plastic are known to be lethal to marine mammals and turtles, and national and international legislation has been enacted to ban the disposal of plastic in the oceans.

- DDT and PCBs are persistent, biologically hazardous chemicals that have been introduced into the ocean by human activities. DDT pollution produced a decline in the Long Island osprey population in the 1950s and the brown pelican population of the California coast in the 1960s. Virtual cessation of DDT use in the Northern Hemisphere in 1972 allowed the recovery of both populations. The DDT thinned the eggshells and reduced the number of successful hatchings. PCBs have been implicated in causing health problems in sea lions and shrimp.

- Mercury poisoning was the first major human disaster resulting from ocean pollution. It is now called Minamata disease after the bay in Japan where it first occurred in 1953. Mercury accumulates in the tissues of many large fish, most notably tuna and swordfish, and works its way up the food web. To prevent mercury poisonings in the United States, stringent mercury contamination levels in fish have been established by the FDA.

Key Terms

Anoxic (p. 323)

Bar-built estuary (p. 320)

Barrier island (p. 320)

Bioaccumulation (p. 338)

Bioremediation (p. 334)

Coastal geostrophic current (p. 318)

Coastal plain estuary (p. 320)

Coastal water (p. 318)

Davidson Current (p. 319)

DDT (p. 336)

Drowned river valley (p. 320)

Estuarine circulation pattern (p. 321)

Estuary (p. 320)

Exclusive economic zone (EEZ) (p. 317)

Fjord (p. 320)

Halocline (p. 318)

Highly stratified estuary (p. 321)

Hydrocarbon (p. 329)

Isohaline (p. 318)

Isothermal (p. 318)

Lagoon (p. 326)

Mangrove swamp (p. 323)

Mediterranean circulation (p. 327)

Mercury (p. 338)

Minamata disease (p. 338)

Non-point-source pollution (p. 340)

PCBs (p. 336)

Pollution (p. 327)

Primary treatment (p. 335)

Salt marsh (p. 323)

Salt wedge estuary (p. 322)

Secondary treatment (p. 335)

Sewage sludge (p. 335)

Sill (p. 327)

Slightly stratified estuary (p. 321)

Standard laboratory bioassay (p. 329)

Tectonic estuary (p. 320)

Territorial sea (p. 316)

Thermocline (p. 318)

United Nations Conference on the Law of the Sea (p. 316)

Vertically mixed estuary (p. 321)

Wetland (p. 323)

Questions And Exercises

1. Discuss possible reasons why less-developed nations believe that the open ocean is the common heritage of all, whereas the more developed nations believe that the open ocean's resources belong to those who recover them.

2. For coastal oceans where deep mixing does not occur, discuss the effect that offshore winds and fresh water runoff will have on salinity distribution. How will the winter and summer seasons affect the temperature distribution in the water column?

3. How does coastal runoff of low-salinity water produce a coastal geostrophic current?

4. Based on their origin, draw and describe the four major classes of estuaries.

5. Describe the difference between vertically mixed and salt wedge estuaries in terms of salinity distribution, depth, and volume of river flow. Which displays the more classical estuarine circulation pattern?

6. Discuss factors that cause the surface salinity of Chesapeake Bay to be greater along its east side, and why periods of summer anoxia in deep water are becoming increasingly severe with time.

7. Name the two types of coastal wetland environments and the latitude ranges where each will likely develop. How do wetlands contribute to the biology of the oceans and the cleansing of polluted river water?

8. What factors lead to a wide seasonal range of salinity in Laguna Madre?

9. Describe the circulation between the Atlantic Ocean and the Mediterranean Sea, and explain how and why it differs from estuarine circulation.

10. Without consulting the textbook, define pollution. Then, consider these items and determine if each one is a pollutant

based on your definition (and refine your definition as necessary):

    a. Dead seaweed on the beach

    b. Natural oil seeps

    c. A small amount of sewage

    d. Warm water dumped into the ocean

11. Why would many marine pollution experts consider oil among the *least* damaging pollutants in the ocean?

12. Describe the effect of oil spills on species diversity and recovery of bottom-dwelling organisms based on the experience at Wild Harbor.

13. Compare oil spills that wash ashore to those that do not, such as the *Argo Merchant*, in terms of destruction to marine life.

14. Discuss techniques used to clean oil spills. Why is it important to begin the cleanup immediately?

15. When and where was the world's largest oil spill? How many times larger was it than the *Exxon Valdez* oil spill?

16. How would dumping sewage in deeper water off the East Coast help reduce the negative effects to the ocean bottom?

17. What properties contributed to plastics being considered a miracle substance? How do those same properties cause them to be unusually persistent and damaging in the marine environment?

18. Discuss the animal populations that clearly suffered from the effects of DDT and the way in which this negative effect was manifested.

19. What causes Minamata disease? What are the symptoms of the disease in humans?

20. What is non-point-source pollution and how does it get to the ocean? What other ways does trash get into the ocean?

References

Boincourt, W. C. 1993. Estuaries: Where the river meets the sea. *Oceanus* 36:2, 29–37.

Borgese, E. M., and Ginsburg, N., eds. *1988 Ocean Yearbook*, Vol. 7. Chicago: University of Chicago Press.

Bragg, J. R. Prince, R. C., Harner, E. J., and Atlas, R. M. 1994. Effectiveness of bioremediation for the *Exxon Valdez* oil spill. *Nature* 368:6470, 413–418.

Cherfas, C. 1990. The fringe of the ocean: Under siege from land. *Science* 248:4952, 163–165.

Clark, W. C. 1989. Managing planet earth. *Scientific American* 261:3, 47–54.

Cronin, T., et. al. 1999. Interdisciplinary environmental project probes Chesapeake Bay down to the core. *Eos Trans. AGU* 80:21, 237–241.

DeCola, E. 2000. *International oil spill statistics: 1999.* Arlington, Massachusetts: Cutter Information Corporation.

Farrington, J. W., Capuzzo, J. M., Leschine, T. M., and Champ, M. A. 1982. Ocean dumping. *Oceanus* 25:4, 39–50.

Fenichelli, S. 1996. *Plastic: The making of a synthetic century.* New York: Harper Business Press.

Fye, P. M. 1982. The law of the sea. *Oceanus* 25:4, 7–12.

Galt, J. A., Lehr, W. J., and Payton, D. L. 1996. Fate and transport of the *Exxon Valdez* oil spill, *in* Pirie, R. G., ed., *Oceanography: Contemporary Readings in Ocean Sciences,* 3rd ed., New York: Oxford University Press.

Hodgson, B. 1990. Alaska's big spill: Can the wilderness heal? *National Geographic* 177:1, 5–43.

Hsü, K. 1983. The Mediterranean was a desert: A voyage of the *Glomar Challenger.* Princeton, NJ: Princeton University Press.

Knauss, J. A. 1974. Marine science and the 1974 Law of the Sea Conference: Science faces a difficult future in changing Law of the Sea. *Science* 184:1335–1341.

Lees, D. C., Houghton, J. P., and Driskell, W. B. 1996. Short-term effects of several types of shoreline treatment on rocky intertidal biota in Prince William Sound. *American Fisheries Society Symposium* 18:329–348.

Manheim, T., and Butman, B. 1994. A crisis in waste management, economic vitality, and a coastal marine environment: Boston Harbor and Massachusetts Bay. *GSA Today* 4:7, 197–199.

Michel, J. 1990. The *Exxon Valdez* oil spill: Status of the shoreline. *Geotimes* 35:5, 20–22.

———. 1991. Prince William Sound, Alaska: The cleanup continues. *Geotimes* 36:3, 16–17.

Mitchell, J. G. 1999. In the wake of the spill: Ten years after the Exxon Valdez. *National Geographic* 195:3, 96–117.

Officer, C. B. 1976. Physical oceanography of estuaries. *Oceanus* 19:5, 3–9.

Officer, C. B., Briggs, R. B., Taft, J. L., Tyler, M. A., and Bovnton, W. R. 1984. Chesapeake Bay anoxia: Origin, development, and significance. *Science* 223:4631, 22–27.

O'Hara, K. J., Iudicello, S., and Bierce, R. 1988. *A citizen's guide to plastics in the ocean: More than a litter problem,* 2nd ed., Washington, DC: Center for Marine Conservation, Inc.

Pickard, G. L. 1964. *Descriptive physical oceanography: An introduction.* New York: Macmillan.

Stanley, D. J. 1990. Med desert theory is drying up. *Oceanus* 33:1, 14–23.

Stommel, H. M., ed. 1950. *Proceedings of the colloquium on "The flushing of estuaries."* Woods Hole, MA: Woods Hole Oceanographic Institution.

Teal, J. M. 1993. A local oil spill revisited. *Oceanus* 36:2, 65–70.

The Open University Course Team. 1991. *Case studies in oceanography and marine affairs.* Oxford: Pergamon Press.

Valiela, I., and Vince, S. 1976. Green borders of the sea. *Oceanus* 19:5, 10–17.

Wilber, R. J. 1987. Plastics in the North Atlantic. *Oceanus* 30:3, 61–68.

Zeder, J. B., and Powell, A. N. 1993. Managing coastal wetlands. *Oceanus* 36:2, 19–28.

Suggested Reading in Scientific American

Borgese, E. M. 1983. The law of the sea. 248:3, 42–49. One hundred nineteen nations sign a convention aimed at establishing a new international scheme of regulating the oceans.

Halloway, M. 1994. Nurturing nature. 270:4, 98–108. Focusing on the Florida Everglades, the potential to restore damaged wetlands is investigated.

———. 1996. Sounding out science. 275:4 106–112. An update on the recovery of Prince William Sound after the *Exxon Valdez* oil spill, with a focus on how science is used to analyze the recovery of affected areas.

HsŸ, K. H. 1972. When the Mediterranean dried up. 227:6, 26–45. Evidence is presented that shows that the Mediterranean Sea was a dry basin 6 million years ago.

McDonald, I. R. 1998. Natural oil spills. 279:5, 56–61. A look at the effects of natural oil seeps in The Gulf of Mexico and the unique biologic communities that consume these hydrocarbons.

Oceanography on the Web

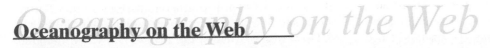

Visit the *Essentials of Oceanography* home page for on-line resources for this chapter. There you will find an on-line study guide with review exercises, and links to oceanography sites to further your exploration of the topics in this chapter. *Essentials of Oceanography* is at: **http://www.prenhall.com/thurman** (click on the Table of Contents menu and select this chapter).

CHAPTER
12
The Marine Habitat

- How are living things classified?
- How are marine organisms classified?
- How many marine species exist?
- What is osmosis, and how does it affect marine organisms?
- What adaptations do marine organisms have that allow them to live in the ocean?
- What are the main divisions of the marine environment?

I never dreamed that islands, about fifty or sixty miles apart, and most of them in sight of each other, formed of precisely the same rocks, placed under a quite similar climate, rising to a nearly equal height, would have been differently tenanted...."
—*Charles Darwin, commenting about species diversity in the Galápagos Islands (1837)*

CHARLES DARWIN AND THE VOYAGE OF HMS *BEAGLE*

In the early 19th century, the English naturalist **Charles Darwin** (Figure 12A, *right inset*), whose interest was investigating the whole of nature, led him to propose the theory of evolution by natural selection. More than any other scientist, Darwin has shaped our understanding of the underlying processes operating in nature. Most of the observations upon which he based his theory were made aboard the vessel HMS *Beagle* during its famous expedition from 1831-1836 that circumnavigated the globe (Figure 12A).

The *Beagle* sailed from Devonport, England, on December 27, 1831, under the command of Captain Robert Fitzroy. The major objective of the voyage was to complete a survey of the coast of Patagonia (Argentina) and Tierra del Fuego, and to make chronometric measurements. The voyage allowed Darwin to study the plants and animals throughout the world, particularly the 14 species of finches in the Galápagos Islands (Figure 12A, *left inset*). These finches differ greatly in the configuration of their beaks, which are suited to their diverse feeding habitats. After his return to England, he noted the adaptations of finches and other organisms living in different environments and concluded that all organisms change slowly over time as a product of their environment.

Darwin also realized that birds and mammals must have evolved from reptiles and that the similar skeletal framework of the human, the bat, the horse, the giraffe, the elephant, the porpoise, and other vertebrates required that they be grouped together. Darwin suggested that the superficial differences between populations were the result of adaptation to different environments and modes of existence.

In 1858, Darwin and Alfred Russel Wallace independently and simultaneously published summaries of their ideas about natural selection. A year later Darwin set forth the structure of his theory and the evidence to support it in *The Origin of Species*, which dealt not so much with the origin of life but with the evolution of living things into their many forms. Darwin's ideas were highly controversial at the time because they stood in stark conflict with what most people believed. Today, most of Darwin's ideas have been so thoroughly embraced by scientists and so widely accepted by the general public that they are now the underpinnings of the modern study of biology.

Darwin's scientific interests were not confined to biology. In fact, before the publication of *The Origin of Species*, Darwin considered himself to be primarily a geologist. During the voyage of the *Beagle*, Darwin formulated a theory on the origin of coral reefs on volcanic islands, which progress through different stages of development (from fringing reef to barrier reef to atoll) as outlined in Chapter 2, "Plate Tectonics and the Ocean Floor." Darwin published his ideas about coral reef development in *The Structure and Distribution of Coral Reefs* in 1842.

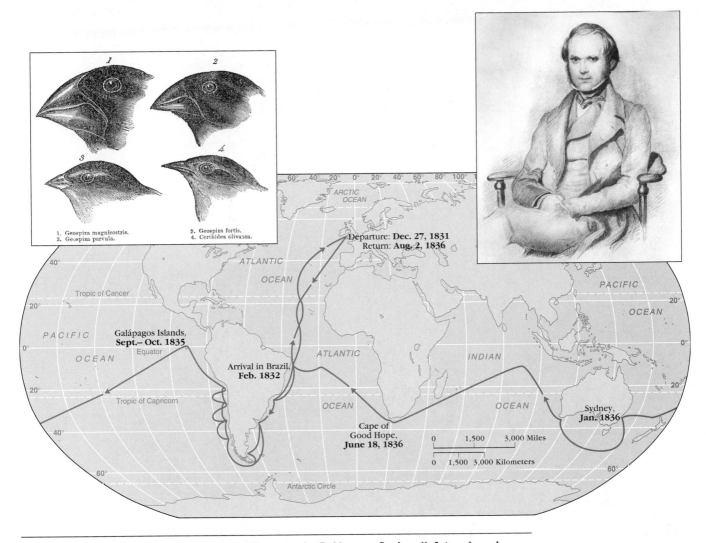

Figure 12A Route of HMS *Beagle*, beak differences in Galápagos finches (*left inset*), and naturalist Charles Darwin (*right inset*).

Awide variety of organisms inhabit the marine environment. These organisms range in size from microscopic bacteria and algae to the blue whale, which is as long as three buses lined up end to end. Marine biologists have identified over 250,000 marine species, and this number is constantly increasing as new organisms are discovered.

Most marine organisms live within the sunlit surface waters of the ocean. Strong sunlight supports photosynthesis by marine algae, which either directly or indirectly provides food for the vast majority of marine organisms. All marine algae live near the surface because they need the sunlight; most marine animals live near the surface because this is where food can be obtained. In shallow water areas close to land, sunlight reaches all the way to the ocean floor, resulting in an abundance of marine life.

There are advantages and disadvantages to living in the marine environment. One advantage is that there is an abundance of water available, which is necessary for supporting all types of life. One disadvantage is that maneuvering in water, which has high density and impedes movement, can be difficult. The individual success of species depends on their ability to find food, avoid predators, and cope with the many physical barriers to their movement.

Classification of Living Things

All living things belong to one of three domains (branches) of life: Archaea, Bacteria, and Eukarya (Figure 12–1). The domain **Archaea** (*archaeo* = ancient) is a group of simple microscopic bacteria-like creatures that includes methane producers and sulfur oxidizers that inhabit deep-sea vents and seeps,[1] as well as other forms—many of which prefer environments with extreme temperatures and/or pressures. The domain **Bacteria** (*bakterion* = rod) includes simple life forms with

---

[1] A *seep* is an area where water trickles out of the sea floor.

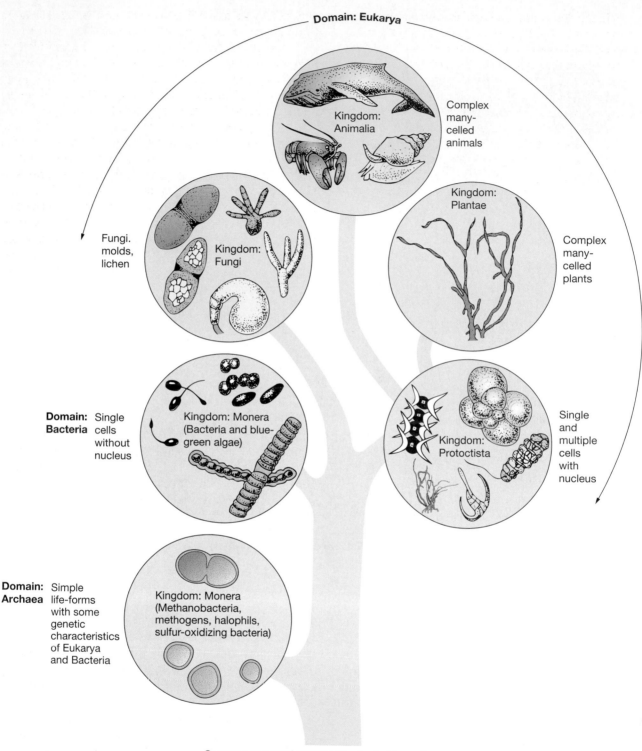

Domain: Eukarya

Kingdom:
Animalia

Complex
many-
celled
animals

Fungi.
molds,
lichen

Kingdom:
Fungi

Kingdom:
Plantae

Complex
many-
celled
plants

Domain: Single
Bacteria cells
without
nucleus

Kingdom: Monera
(Bacteria and blue-
green algae)

Kingdom:
Protoctista

Single
and
multiple
cells
with
nucleus

Domain: Simple
Archaea life-forms
with some
genetic
characteristics
of Eukarya
and Bacteria

Kingdom: Monera
(Methanobacteria,
methogens, halophils,
sulfur-oxidizing bacteria)

Common ancestral community of primitive cells

Figure 12–1 The three domains of life and the five kingdoms of organisms.

cells that usually lack a nucleus, including purple bacteria, green nonsulfur bacteria, and cyanobacteria (blue-green algae). The domain **Eukarya** (*eu* = good, *karuon* = nut) includes complex organisms—protoctists, fungi, and multicellular plants and animals—with cells that usually contain a nucleus.

Within the three domains of life, a system of five kingdoms was first proposed by ecologist and biolo-

gist Robert H. Whittaker in 1959. These five kingdoms of life are Monera, Protoctista, Fungi, Plantae, and Animalia.

Kingdom **Monera** (*monos* = single) includes some of the simplest organisms. These organisms are single-celled but lack a discrete nucleus, so their nuclear material is spread throughout the cell. Included in this kingdom are the cyanobacteria (blue-green algae), heterotrophic bacteria and archaea. Recent discoveries have shown bacteria to be a much more important part of marine ecology than previously believed. These organisms are found throughout the breadth and depth of the oceans.

Kingdom **Protoctista** (*protos* = first, *ktistos* = to establish) includes single- and multicelled organisms that have a nucleus, so they represent a higher stage of evolutionary development than Kingdom Monera. Examples of organisms in Kingdom Protoctista include single-celled algae that produce food for most marine animals, single-celled animals called **protozoa** (*proto* = first, *zoa* = animal), and multicelled marine algae.

Kingdom **Fungi** (*fungus* = [probably from Greek *sp(h)ongos* = sponge]) includes 100,000 species of mold and lichen, though less than one-half of 1% of them are sea-dwellers. Fungi exist throughout the marine environment, but they are much more common in the intertidal zone, where they live with cyanobacteria or green algae to form lichen. Other fungi remineralize organic matter and function primarily as decomposers in the marine ecosystem.

Kingdom **Plantae** (*planta* = plant) comprises the multicelled plants, all of which photosynthesize. Only a few species of true plants—such as surf grass (*Phyllospadix*) and eelgrass (*Zostera*)—inhabit shallow coastal environments. In the ocean, photosynthetic marine algae occupy the ecological niche of land plants. However, certain plants are vital parts of coastal ecosystems, including mangrove swamps and salt marshes.

Kingdom **Animalia** (*anima* = breath) comprises the multicelled animals. Organisms from kingdom Animalia range in complexity from the simple sponges to complex vertebrates (animals with backbones), which also includes humans.

In an effort to determine the relationships of all living things on Earth, Swedish botanist Carl von Linné, who Latinized his name to Carolus Linnaeus, developed a system in 1758 that is the basis of the modern scientific system of classification used today. The systematic classification of organisms—called **taxonomy** (*taxis* = arrangement, *nomia* = a law)—involves placing organisms into the following increasingly specific groupings:

- Kingdom
- Phylum (Division for plants)
- Class
- Order
- Family
- Genus
- Species

All organisms that share a common category (for instance, a family) have certain characteristics and evolutionary similarities (such as the cat or dolphin family). In some cases, subdivisions of these categories are also used, such as subphylum (Table 12–1). The categories

assigned to an individual species must be agreed upon by an international panel of experts.

A species is the fundamental unit of classification. **Species** (*species* = a kind) consist of populations of genetically similar, interbreeding (or potentially interbreeding) individuals that share a collection of inherited characteristics whose combination is unique.

Every type of organism has a unique scientific name that includes its genus and species, which is italicized with the first letter of the generic (genus) name capitalized—for example, *Delphinus delphis*. Most organisms also have one or more common names. *Delphinus delphis*, for instance, is the common dolphin, and *Orcinus orca* is the killer whale or orca. If the scientific name is referred to repeatedly within a document, it is often shortened by abbreviating the genus name to its first letter. Thus, *Delphinus delphis* becomes *D. delphis*.

> Living things can be classified into one of three domains and five kingdoms, each of which is split into increasingly specific groupings of phylum, class, order, family, genus, and species.

Classification of Marine Organisms

Marine organisms can be classified according to where they live (their habitat) and how they move (their mobility). Organisms that inhabit the water column can be classified as either *plankton* (floaters) or *nekton* (swimmers). All other organisms are *benthos* (bottom dwellers).

Plankton (Floaters)

Plankton (*planktos* = wandering) include all organisms—algae, animals, and bacteria—that drift with ocean currents. An individual organism is called a **plankter**. Just because plankters drift does not mean they are unable to swim. Many plankters can swim but either move only weakly or move only vertically. They cannot determine their horizontal position within the ocean.

Among plankton, the algae (microscopic photosynthetic cells) are called **phytoplankton** (*phyto* = plant, *planktos* = wandering) and the animals are called **zooplankton** (*zoo* = animal, *planktos* = wandering). Representative members of each group are shown in Figure 12–2.

Plankton also include bacteria. It has recently been discovered that free-living **bacterioplankton** are much more abundant than previously thought. Having an average diameter of only one-half of a micrometer[2] (0.00002 inch), they were missed in earlier studies because they are so small.

---

[2] One micrometer (also known as one *micron*) is 10^{-9} meter and is designated by the symbol μm.

Table 12–1 Classification of selected organisms.

Category	Human	Common dolphin	Killer whale	Bat star	Giant kelp
Kingdom	Animalia	Animalia	Animalia	Animalia	Protoctista
Phylum	Chordata	Chordata	Chordata	Echinodermata	Phaeophyta
Subphylum	Vertebrata	Vertebrata	Vertebrata		
Class	Mammalia	Mammalia	Mammalia	Asteroidea	Phaeophycae
Order	Primates	Cetacea	Cetacea	Valvatida	Laminariales
Family	Hominidae	Delphinidae	Delphinidae	Oreasteridae	Lessoniaceae
Genus	*Homo*	*Delphinus*	*Orcinus*	*Asterina*	*Macrocystis*
Species	*sapiens*	*delphis*	*orca*	*miniata*	*pyrifera*

Plankton are unbelievably abundant and important within the marine environment. In fact, *most of Earth's biomass—the mass of living organisms—consists of plankton adrift in the oceans*. Even though 98% of marine *species* are bottom dwelling, the vast majority of the ocean's *biomass* is planktonic.

Plankton range greatly in size. They include large floating animals and algae, such as jellyfish and *Sargassum*,[3] which are called **macroplankton** (*macro* = large, *planktos* = wandering) and measure 2 to 20 centimeters (0.8 to 8 inches). Plankton also include bacterioplankton, which are so small that they can be removed from the water only with special microfilters. These very tiny floaters are called **picoplankton** (*pico* = small, *planktos* = wandering) and measure 0.2 to 2 microns (0.000008 to 0.00008 inch).

Plankton can be classified as either phytoplankton, zooplankton, or bacterioplankton. They can also be classified according to the portion of their life cycle spent as plankton. Organisms that spend their entire lives as plankton are **holoplankton** (*holo* = whole, *planktos* = wandering). Many organisms that spend their adult lives as nekton or benthos spend their juvenile and/or larval stages as plankton (Figure 12–3). These organisms are called **meroplankton** (*mero* = a part, *planktos* = wandering).

Nekton (Swimmers)

Nekton (*nektos* = swimming) include all animals capable of moving independently of the ocean currents, by swimming or other means of propulsion. They are capable not only of determining their own positions within the ocean but also, in many cases, of long migrations. Nekton include most adult fish and squid, marine mammals, and marine reptiles (Figure 12–4). When you go ocean swimming, you become nekton, too.

Although nekton move freely, they are unable to move throughout the breadth of the ocean. Gradual changes in temperature, salinity, viscosity, and availability of nutrients effectively limit their lateral range. The deaths of large numbers of fish, for example, can be caused by temporary horizontal shifts of water masses in the ocean. Water pressure normally limits the vertical range of nekton.

Fish may appear to exist everywhere in the oceans, but they are more abundant near continents and islands and in colder waters. Some fish, such as salmon, ascend freshwater rivers to spawn. Many eels do just the reverse, growing to maturity in fresh water and then descending the streams to breed in the great depths of the ocean.

Benthos (Bottom Dwellers)

The term **benthos** (*benthos* = bottom) describes organisms living on or in the ocean bottom. **Epifauna** (*epi* = upon, *fauna* = animal) live on the surface of the sea floor, either attached to rocks or moving along the bottom. **Infauna** (*in* = inside, *fauna* = animal) live buried in the sand, shells, or mud. Some benthos, called **nektobenthos**, live on the bottom but also swim or crawl through the water above the ocean floor. Examples of benthos are shown in Figure 12–5.

The shallow coastal ocean floor contains a wide variety of physical and nutritive conditions, which has allowed a great number of animal species to develop. Moving across the bottom from the shore into deeper water, the *number* of benthos species per square meter may remain relatively constant, but the *biomass* of benthos organisms decreases. In addition, the shallow coastal areas are the only locations where large marine algae (often called "seaweeds") attached to the bottom are found, because these are the only areas of the sea floor that receive sufficient sunlight.

Throughout most of deeper parts of the sea floor, animals live in perpetual darkness, where photosynthetic production cannot occur. They must feed on each other, or on whatever outside nutrients fall from the productive zone near the surface.

The deep-sea bottom is an environment of coldness, stillness, and darkness. Under these conditions, life progresses slowly and organisms that live in the deep sea

---

[3] *Sargassum* is a floating type of brown macro marine algae commonly referred to as a "seaweed" that is particularly abundant in the Sargasso Sea.

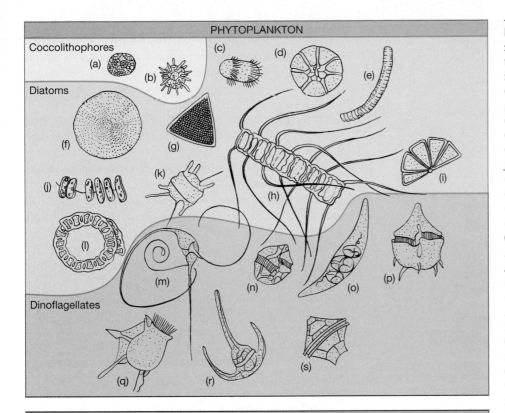

PHYTOPLANKTON

Coccolithophores

Diatoms

Dinoflagellates

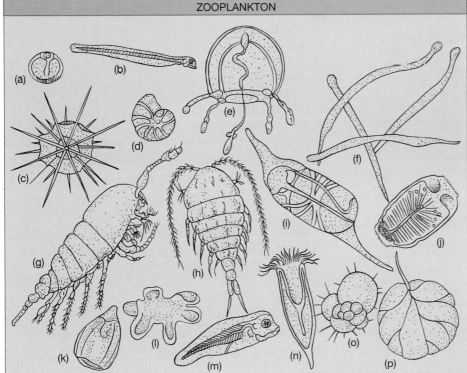

ZOOPLANKTON

Figure 12–2 Phytoplankton and zooplankton (floaters). Not drawn to scale; typical maximum dimension is in parentheses. **Phytoplankton: (a)** and **(b)** Coccolithophoridae (15 μm, or 0.0006 in.). **(c)–(l)** Diatoms (80 μm, or 0.0032 in.): **(c)** *Corethron*; **(d)** *Asteromphalus*; **(e)** *Rhizosolenia*; **(f)** *Coscinodiscus*; **(g)** *Biddulphia favus*; **(h)** *Chaetoceras*; **(i)** *Licmophora*; **(j)** *Thalassiorsira*; **(k)** *Biddulphia mobiliensis*; **(l)** *Eucampia*. **(m)–(s)** Dinoflagellates (100 μm, or 0.004 in.): **(m)** *Ceratium recticulatum*; **(n)** *Goniaulax scrippsae*; **(o)***Gymnodinium*; **(p)** *Goniaulax triacantha*; **(q)** *Dynophysis*; **(r)** *Ceratium bucephalum*; **(s)** *Peridinium*. **Zooplankton: (a)** Fish egg (1 mm, or 0.04 in.). **(b)** Fish larva (5 cm, or 2 in.). **(c)** Radiolaria (0.5 mm, or 0.02 in.). **(d)**Foraminifer (1 mm, or 0.04 in.). **(e)** Jellyfish (30 cm, or 12 in.). **(f)** Arrowworms (3 cm, or 1.2 in.). **(g)** and **(h)** Copepods (5 mm, or 0.2 in.). **(i)** Salp (10 cm, or 4 in.). **(j)** Doliolum (10 cm, or 4 in.). **(k)** Siphonophore (30 cm, or 12 in.). **(l)** Worm larva (1 mm, or 0.04 in.). **(m)** Fish larva (5 cm, or 2 in.). **(n)** Tintinnid (1 mm, or 0.04 in.). **(o)** Foraminifer (1 mm, or 0.04 in.). **(p)** Dinoflagellate (*Noctiluca*) (1 mm, or 0.04 in.).

usually are widely distributed because physical conditions vary little on the deep-ocean floor, even over great distances.

Hydrothermal Vent Biocommunities In 1977, the first biocommunity at a hydrothermal vent was discovered in the Galápagos Rift off South America. This biocommunity demonstrates that high concentrations of deep-ocean benthos are possible. The primary limiting factor for life on the deep-ocean floor is probably the availability of food because food is abundant at these hydrothermal vents. Archaeon (bacteria-like organisms) produce food not by photosynthesis, for no sunlight is available, but by chemosynthesis. The size of individuals and the total biomass in the hydrothermal communities far exceed that previously known

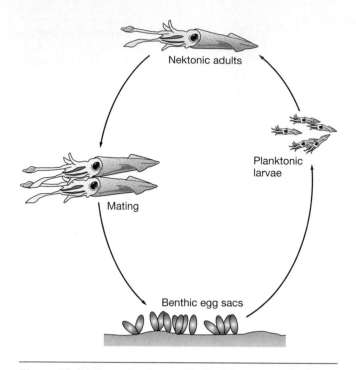

Figure 12–3 Life cycle of a squid. Squid are meroplankton because they are planktonic only during their larval stage. Adult squid are nekton and their egg sacks are benthos.

for the deep-ocean benthos. These biocommunities are discussed in Chapter 15, "Animals of the Benthic Environment."

> Marine organisms can be classified according to their habitat and mobility as plankton (floaters), nekton (swimmers), or benthos (bottom-dwellers).

Distribution of Life in the Oceans

It is difficult to describe the extent to which the marine environment is inhabited because the ocean is immense, so little is known about marine organisms, and some populations fluctuate greatly each season. What is known, however, is that there are over 1,750,000 species[4] in marine and terrestrial environments and only about 14% of all known species live in the ocean (Figure 12–6).

Why Are There So Few Marine Species?

If the ocean is such a prime habitat for life and if life originated there, then why do so few of the world's species live in the ocean? The disparity arises because

---

[4] Many biologists believe there may be from 3 to 10 million additional unnamed and undescribed species living on Earth, of which a large number probably inhabit the oceans.

the marine environment is more stable than the terrestrial environment. The relatively uniform conditions of the open ocean do not pressure organisms to adapt. The need for adaptation is thought to be responsible for the creation of new species. In addition, temperatures are not only stable but also relatively low below the sunlit surface waters. The rate of chemical reaction is slowed, which may further reduce the tendency for variation to occur.

The less stable environment on land, on the other hand, presents many opportunities for natural selection to produce new species to inhabit varied new niches. At least 75% of all land species, for example, are insects that have evolved the capability of inhabiting very restricted environmental niches. If insects are ignored, the sea possesses over 45% of the remaining species living in marine and terrestrial environments.

Figure 12–6 also shows that only 2% or about 5000 of the 250,000 known marine species inhabit the **pelagic** (*pelagios* = of the sea) **environment** and live within the water column. The other 98% inhabit the **benthic** (*benthos* = bottom) **environment** and live either in or on the sea floor. These numbers are minimums, however, because recent discoveries indicate that many more species may inhabit the benthic environment than previously thought.

Why do most marine species inhabit the benthic environment? The ocean floor contains numerous benthic environments (such as rocky, sandy, muddy, flat, sloped, irregular, and mixed bottoms) that create different habitats to which organisms have adapted. On the other hand, most of the pelagic environment—especially that below the sunlit surface waters—is a watery world that is quite uniform from one region to the next and does not experience extreme environmental variability to which organisms need to be adapted in order to survive.

> Marine species represent 14% of the total number of species on Earth. The benthic environment, which has large environmental variability, is home to 98% of the 250,000 known marine species.

Adaptations of Organisms to the Marine Environment

The ocean environment—particularly its temperature—is far more stable than the terrestrial environment. As a result, ocean-dwelling organisms have not developed highly specialized regulatory systems to adjust to sudden changes that might occur within their environment. They can be adversely affected by quite small changes in temperature, salinity, turbidity, pressure, or other environmental variables.

Water constitutes over 80% of the mass of **protoplasm**, the substance of living matter. Over 65% of your

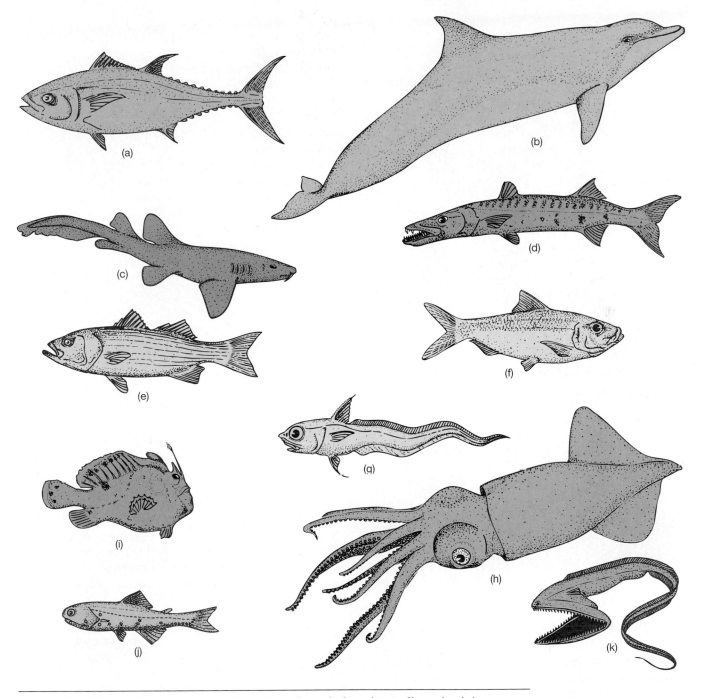

Figure 12–4 Nekton (swimmers). Not drawn to scale; typical maximum dimension is in parentheses. **(a)** Bluefin tuna (2 m, or 6.6 ft.). **(b)** Bottlenose dolphin (4 m, or 13 ft.). **(c)** Nurse shark (3 m, or 10 ft.). **(d)** Barracuda (1 m, or 3.3 ft.). **(e)** Striped bass (0.5 m, or 1.6 ft.). **(f)** Sardine (15 cm, or 6 in.). **(g)** Deep-ocean fish (8 cm, 3 in.). **(h)** Squid (1 m, or 3.3 ft.). **(i)** Anglerfish (5 cm, or 2 in.). **(j)** Lantern fish (8 cm, or 3 in.). **(k)** Gulper (15 cm, or 6 in.).

body's weight, and 95% of a jellyfish's weight, is water (Table 12–2). Water carries dissolved within it the gases and minerals organisms need to survive. Water is also a raw material in the photosynthesis of food by marine phytoplankton.

Land plants and animals have developed complex "plumbing systems" to retain water and to distribute it throughout their bodies. The inhabitants of the open ocean do not risk atmospheric desiccation (drying out), however, because they live in an environment of abundant water.

Need for Physical Support

One basic need of all plants and animals is for simple physical support. Land plants, for example, have vast root systems that anchor the plant securely to the ground. Land animals have skeletons and combinations

Figure 12–5 Benthos (bottom dwellers)—representative intertidal and shallow subtidal forms. Not drawn to scale; typical maximum dimension is in parentheses. **(a)** Sand dollar (8 cm, or 3 in.). **(b)** Clam (30 cm, or 12 in.). **(c)** Crab (30 cm, or 12 in.). **(d)** Abalone (30 cm, or 12 in.). **(e)** Sea urchins (15 cm, or 6 in.). **(f)** Sea anemone (30 cm, or 12 in.). **(g)** Brittle star (20 cm, or 8 in.). **(h)** Sponge (30 cm, or 12 in.). **(i)** Acorn barnacles (2.5 cm, or 1 in.). **(j)** Snail (2 cm, or 0.8 in.). **(k)** Mussels (25 cm, or 10 in.). **(l)** Gooseneck barnacles (8 cm, or 3 in.). **(m)** Sea star (30 cm, or 12 in.). **(n)** Brain coral (50 cm, or 20 in.). **(o)** Sea cucumber (30 cm, or 12 in.). **(p)** Lamp shell (10 cm, or 4 in.). **(q)** Sea lily (10 cm, or 4 in.). **(r)** Sea squirt (10 cm, or 4 in.).

of appendages—legs, arms, fingers, and toes—to support their entire weight.

In the ocean, water physically supports marine plants and animals. Organisms such as photosynthetic phytoplankton, which must live in the upper surface waters of the ocean, depend primarily upon buoyancy and frictional resistance to sinking to maintain their desired position. Still, maintaining position can be difficult, so some organisms have developed special adaptations to increase their efficiency. These adaptations are discussed in this and succeeding chapters.

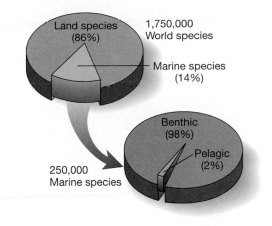

Figure 12–6 Distribution of species on Earth. Of the 1,750,000 known species on Earth, 86% inhabit land environments and 14% inhabit the ocean. Of the 250,000 known marine species, 98% inhabit the benthic environment and live in or on the ocean floor, while only 2% inhabit the pelagic environment and live within the water column as either plankton or nekton.

Table 12–2   Percent water content of selected organisms.

Organism	Percent water content
Human	65
Herring	67
Scallop	78
Lobster	79
Jellyfish	95

Water's Viscosity

Viscosity is a substance's *internal resistance to flow.* As described in Chapter 2, a substance with high resistance to flow (high viscosity)—such as toothpaste—does not flow easily. Conversely, a substance that has low viscosity—such as water—flows more readily. Viscosity is strongly affected by temperature. Tar, for example, must be heated to decrease its viscosity before it can be spread onto roofs or roads.

The viscosity of ocean water increases as salinity increases and temperature decreases. Thus, single-celled organisms that float in colder, higher-viscosity waters have less need for extensions to help them maintain their positions near the surface. Figure 12–7 shows, for example, that a warm-water variety of the same species of floating crustacean has ornate, featherlike appendages, whereas a cold-water variety has no such appendages.

The Importance of Organism Size The major requirements of phytoplankton are that they stay in the upper portion of ocean water where solar radiation is available, that they have available necessary nutrients, that they efficiently take in these nutrients from surrounding waters, and that they expel waste materials. Their ingenious size and shape help single-celled phytoplankton satisfy these requirements without needing specialized multiple cells.

Phytoplankton cannot propel themselves, so they use frictional resistance to maintain their general position near the surface of the water. Frictional resistance to sinking increases as an organism's surface area in contact with surrounding water increases. Figure 12–8 shows the surface area to volume ratio of three different cubes, illustrating that the ratio increases as an organism's size decreases. For instance, cube (a) has twice the surface are per unit of volume of cube (b) and four times the surface area per unit of volume as cube (c). If the cubes were plankton, cube (a) would have four times the resistance to sinking per unit of mass as cube (c), so cube (a) would need to exert far less energy to stay afloat. One-celled organisms, which make up the bulk of photosynthetic marine life, clearly benefit from being as small as possible. They are so small, in fact, that one needs a microscope to see them!

Photosynthetic cells take in nutrients from surrounding water and expel waste through their cell membranes. The efficiency of both functions increases with a higher surface area to volume ratio. Thus, if cubes (a) and (c) in Figure 12–8 were planktonic algae, cube (a) could take in nutrients and dispose of waste four times more efficiently than cube (c). This is why cells in all plants and animals are microscopic, regardless of the overall size of the organism.

Diatoms—one of the most important groups of phytoplankton—often have unusual appendages, needle-like extensions, or even rings (Figure 12–9) to increase their surface area, thus preventing them from sinking below the sunlight surface waters. Other planktonic marine organisms—particularly warm water species—use similar strategies to stay afloat.

Some small organisms produce a tiny droplet of oil, which lowers their overall density and increases buoyancy. Accumulations of vast amounts of these organisms in sediment can produce offshore oil deposits if their droplets of oil are combined, sufficiently matured, and trapped in a reservoir.

Despite adaptations to remain in the upper layers of the ocean, organisms still have a higher density than seawater, so they tend to sink, if ever so slowly. This is not a serious handicap, however, because wind causes considerable mixing and turbulence near the surface. Turbulence, in turn, keeps these organisms positioned to bask in the solar radiation needed to photosynthesize, producing the energy used by essentially all other members of the marine community.

Viscosity and Streamlining As organisms increase in size, viscosity ceases to enhance survival and instead

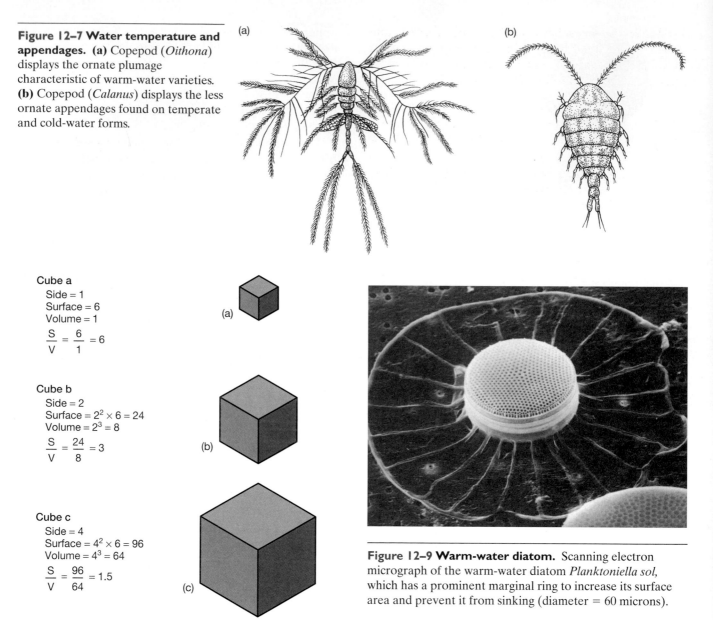

Figure 12–7 Water temperature and appendages. (a) Copepod (*Oithona*) displays the ornate plumage characteristic of warm-water varieties. **(b)** Copepod (*Calanus*) displays the less ornate appendages found on temperate and cold-water forms.

Cube a
Side = 1
Surface = 6
Volume = 1

$$\frac{S}{V} = \frac{6}{1} = 6$$

Cube b
Side = 2
Surface = $2^2 \times 6 = 24$
Volume = $2^3 = 8$

$$\frac{S}{V} = \frac{24}{8} = 3$$

Cube c
Side = 4
Surface = $4^2 \times 6 = 96$
Volume = $4^3 = 64$

$$\frac{S}{V} = \frac{96}{64} = 1.5$$

Figure 12–8 Surface area to volume ratio of cubes of different sizes. As the linear dimension of a cube increases, the ratio of surface area to volume decreases. Thus, smaller bodies have a higher surface area to volume ratio, which allows them to more easily stay afloat and efficiently exchange nutrients and wastes.

Figure 12–9 Warm-water diatom. Scanning electron micrograph of the warm-water diatom *Planktoniella sol*, which has a prominent marginal ring to increase its surface area and prevent it from sinking (diameter = 60 microns).

becomes an obstacle. This is particularly true of large organisms that swim freely in the open ocean. They must pursue prey or flee predators, yet the faster they swim, the more the viscosity of water impedes their progress. Not only must water be displaced ahead of the swimmer, but water also must move in behind it to occupy the space that the animal has vacated.

Figure 12–10 shows the advantage of **streamlining**, which is having a shape that offers the least resistance to fluid flow. Streamlining allows marine organisms to overcome water's viscosity and move more easily through water. A streamlined shape usually consists of a flattened body, which presents a small cross section at the front end, and a gradually tapering back end to reduce the wake created by eddies. It is exemplified in the shape of free-swimming fish (and in mammals such as whales and dolphins).

Temperature

Figure 12–11, which compares extremes in land and ocean surface temperatures, shows that ocean temperatures have a far narrower range than temperatures on land. The minimum surface temperature of the open ocean is seldom much below −2°C (28.4°F) and the maximum surface temperature seldom exceeds 32°C (89.6°F), except in some shallow-water coastal regions, where the temperature may reach 40°C (104°F). On land, however, extremes in temperatures have ranged from −88°C (−127°F) to 58°C (136°F), which represents a temperature range more than four times greater than that experienced by the ocean.

Further, the ocean has a smaller daily, seasonal, and annual temperature range than that experienced on land, which provides a stable environment for marine organisms. The reasons for this are fourfold:

1. Recall from Chapter 5 that the heat capacity of water is much higher than that of land, which causes land to heat up by a greater amount and much more rapidly than the ocean.

2. The warming of the ocean is reduced substantially because of evaporation, a cooling process that stores excess heat as latent heat.

3. Radiation received at the surface of the ocean can penetrate several tens of meters deep and distribute its energy throughout a very large mass. In contrast, solar radiation absorbed by land heats only a very thin surface layer.

4. Unlike solid land surfaces, water has good mixing mechanisms such as currents, waves, and tides that allow heat from one area to be transported to other areas.

In addition, the small daily and seasonal temperature variations are confined to ocean surface waters and decrease with depth, becoming insignificant throughout

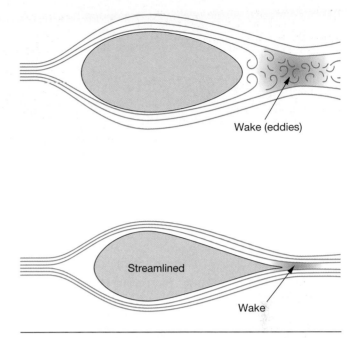

Figure 12–10 Streamlining. To move through water efficiently, a body must produce as little stress as possible as it moves through and displaces the water. After the water has moved past the body, it must flow in behind the body with as little eddy action as possible.

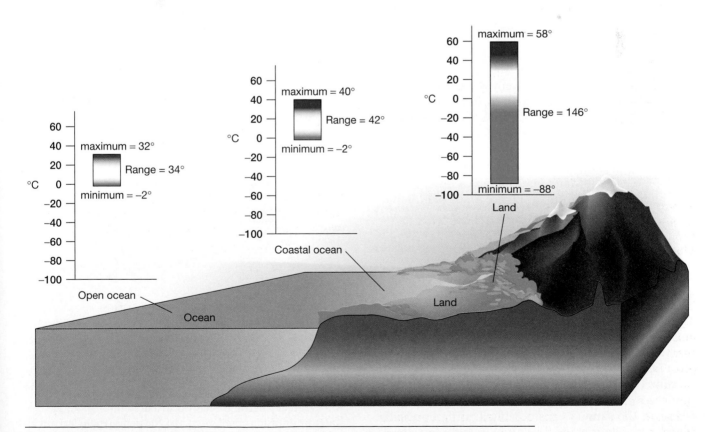

Figure 12–11 A comparison of extremes in ocean and land surface temperatures. In the open ocean, the maximum temperature range is limited to only 34°C (61°F) and increases to 42°C (76°F) in the coastal ocean. On land, the maximum temperature range is 146°C (263°F), over four times the temperature range in the open ocean.

the deeper parts of the ocean. At ocean depths that exceed 1.5 kilometers (0.9 mile), for example, temperatures hover around 3°C (37.4°F) year-round, regardless of latitude.

Comparing Cold- and Warm-Water Species Cold water is denser and has a higher viscosity than warm water. These factors, among others, profoundly influence marine life, resulting in the following differences between warm-water and cold-water species in the marine environment:

- Floating organisms are physically smaller in warm waters than in colder waters. Small organisms expose more surface area per unit of body mass, which helps them more easily maintain their position in the lower viscosity and density of warm seawater.
- Warm-water species often have ornate plumage to increase surface area, which is strikingly absent in the larger cold-water species (see Figures 12–7 and 12–9).
- Warmer temperatures increase the rate of biological activity, which more than doubles with an increase of 10°C (18°F). Tropical organisms apparently grow faster, have a shorter life expectancy, and reproduce earlier and more frequently than those in colder water.
- There are more *species* in warm waters, but the total *biomass* of plankton in colder, high-latitude waters greatly exceeds that of the warmer tropics.

Some animal species can live only in cooler waters, whereas others can live only in warmer waters. Many of these organisms can withstand only very small temperature changes and are called **stenothermal** (*steno* = narrow, *thermo* = temperature). Stenothermal organisms are found predominantly in the open ocean at depths where large temperature ranges do not occur.

Other species are little affected by different temperatures and can withstand large and even rapid changes in temperature. These organisms are called **eurythermal** (*eury* = wide, *thermo* = temperature) and are found predominantly in shallow coastal waters—where the largest temperature ranges are found—and in surface waters of the open ocean.

Salinity

The sensitivity of marine animals to changes in their environment varies from organism to organism. Those that inhabit estuaries, for example, such as oysters, must be able to withstand considerable fluctuations in salinity. The daily rise and fall of the tides forces salty ocean water into river mouths and draws it out again, changing the salinity considerably. During floods, the salinity in estuaries can reach extremely low levels. The coastal organisms that can tolerate large changes in salinity are known as **euryhaline** (*eury* = broad, *halo* = salt).

Marine organisms that inhabit the open ocean, on the other hand, are seldom exposed to a large variation in salinity. They have adapted to a constant salinity and can tolerate only very small changes. These organisms are called **stenohaline** (*steno* = narrow, *halo* = salt).

Extraction of Salinity Components Some organisms extract minerals from ocean water—particularly *silica* (SiO_2) and *calcium carbonate* ($CaCO_3$)—to construct the hard parts of their bodies, which serve as protective coverings. In doing so, they reduce the amount of dissolved material in ocean water. For example, phytoplankton (including diatoms) and microscopic protozoans such as radiolarians and silicoflagellates extract silica from seawater. Coccolithophores, foraminifers, most mollusks, corals, and some algae that secrete a calcium carbonate skeletal structure extract calcium carbonate.

Diffusion Molecules of soluble substances, such as nutrients, move through water from areas of *high* concentration to areas of *low* concentration until the distribution of the substance is uniform (Figure 12–12a). This

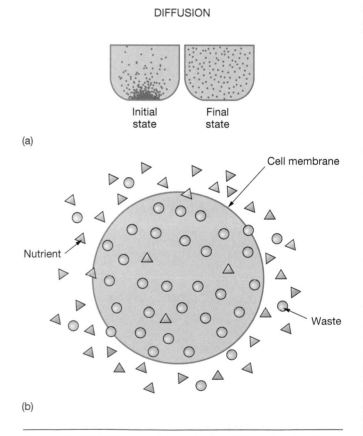

Figure 12–12 Diffusion. **(a)** A water-soluble substance that is added to the bottom of a container of water (*initial state*) will eventually become evenly distributed throughout the water by diffusion (*final state*). **(b)** Nutrients (*triangles*) are in high concentration outside the cell and diffuse into the cell through the cell membrane. Waste particles (*circles*) are in high concentration inside the cell and diffuse out of the cell through the membrane.

process is called **diffusion** (*diffuse* = dispersed) and is caused by random motion of molecules.

The outer membrane of a living cell is permeable to many molecules. Organisms may take in nutrients they need from the surrounding water by diffusion of the nutrients through their cell walls. Because nutrients are usually plentiful in seawater, they pass through the cell wall into the interior, where nutrients are less concentrated (Figure 12–12b).

After a cell uses the energy stored in nutrients, it must dispose of waste. Waste passes out of a cell by diffusion, too. As the concentration of waste materials becomes greater within the cell than in the water surrounding it, these materials pass from within the cell into the surrounding fluid. The waste products are then carried away by circulating fluid that services cells in higher animals, or by the water that surrounds simple one-celled organisms.

Osmosis When water solutions of unequal salinity are separated by a semipermeable membrane (such as skin or the membrane around a living cell), water molecules (but not salt molecules) diffuse through the membrane. Water molecules always move from the *less* concentrated solution into the *more* concentrated solution in a process called **osmosis** (*osmos* = to push) (Figure 12–13a). **Osmotic pressure** is the pressure that must be applied to the more concentrated solution to prevent water molecules from passing into it.

Osmosis causes water to move through an organism's skin (its semipermeable membrane) and affects both marine and freshwater organisms. If the salinity of an organism's body fluid equals that of the ocean, it is **isotonic** (*iso* = same, *tonos* = tension), has equal osmotic pressure, and no net transfer of water will occur through the membrane in either direction (Figure 12–13b).

If seawater has a lower salinity than the fluid within an organism's cells, seawater will pass through the cell walls into the cells (always toward the more concentrated solution). This organism is **hypertonic** (*hyper* = over, *tonos* = tension), which means it is saltier than the surrounding seawater.

If the salinity within an organism's cells is less than that of the surrounding seawater, water from the cells will pass through the cell membranes out into the seawater (toward the more concentrated solution). This organism is **hypotonic** (*hypo* = under, *tonos* = tension) relative to the water outside its body.

In essence, osmosis is diffusion that produces a net transfer of water molecules through a semipermeable membrane from the side with the *greatest concentration* of water molecules to the side with the *lesser concentration* of water molecules.

During osmosis, three things can occur simultaneously across the cell membrane:

1. Water molecules move through the semipermeable membrane toward the side with the lower concentration of water.

2. Nutrient molecules or ions move from where they are more concentrated into the cell, where they are used to maintain the cell.

3. Waste molecules move from within the cell to the surrounding seawater.

Molecules or ions of all the substances in the system are passing through the membrane in both directions. A net transport of molecules of a given substance always occurs from the side on which they are most highly concentrated to the side where the concentration is less, until equilibrium is attained.

The body fluids of marine invertebrates (those without backbones) such as worms, mussels, and octopi, and the seawater in which they live, are nearly isotonic. As a result, these organisms have not had to evolve special mechanisms to maintain their body fluids at a proper concentration. This gives them an advantage over their fresh water relatives, whose body fluids are hypertonic.

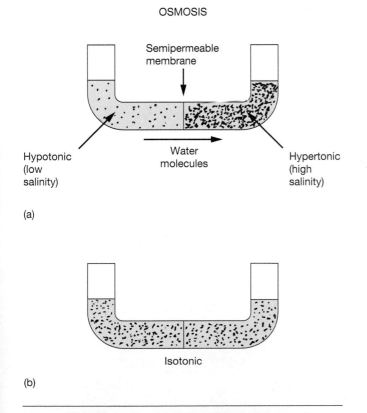

OSMOSIS

Semipermeable membrane

Hypotonic (low salinity)

Water molecules

Hypertonic (high salinity)

(a)

Isotonic

(b)

Figure 12–13 Osmosis. **(a)** The semipermeable membrane that separates two water solutions of different salinities allows water molecules (but not dissolved substances) to pass through it by the process of osmosis. Water molecules will diffuse through the membrane from the less concentrated (hypotonic) solution (*left*) into the more concentrated (hypertonic) solution (*right*). **(b)** If the salinity of the two solutions is the same (isotonic), there is no movement of water.

An Example of Osmosis: Marine Fish Versus Freshwater Fish Marine fish have body fluids that are only slightly more than one-third as saline as ocean water, possibly because they evolved in low-salinity coastal waters. They are, therefore, hypotonic (less salty) compared to the surrounding seawater.

This salinity difference means that saltwater fish, without some means of regulation, would lose water from their body fluids into the surrounding ocean and eventually dehydrate. This loss is counteracted, however, because marine fish drink ocean water and excrete the salts through special chloride-releasing cells located in their gills. These fish also help maintain their body water by discharging a very small amount of very highly concentrated urine (Figure 12–14a).

Freshwater fish are hypertonic (internally more saline) compared to the very dilute water in which they live. The osmotic pressure of the body fluids of such fish may be 20 to 30 times greater than that of the fresh water that surrounds them, so freshwater fish risk rupturing cell walls from excessive quantities of water taken on through osmosis.

To prevent this, freshwater fish do not drink water and their cells have the capacity to absorb salt. They also excrete large volumes of very dilute urine to reduce the amount of water in their cells (Figure 12–14b).

> Osmosis produces a net transfer of water molecules through a semipermeable membrane from the side with the greatest concentration of water molecules to the side with the lesser concentration.

Dissolved Gases

The amount of gases that dissolve in seawater increases as the temperature of seawater decreases, so cold water dissolves more gas than warm water. As a result, vast phytoplankton communities develop in high latitudes during summer, when solar energy becomes available for photosynthesis. These cold waters contain an abundance of dissolved gases, specifically carbon dioxide (which phytoplankton need for photosynthesis) and oxygen (which all organisms need to metabolize their food).

Most animals that live in the ocean—except air-breathing marine mammals and certain fishes—must extract dissolved oxygen from seawater. How do they do it? Most marine animals have specially designed fibrous respiratory organs called **gills** that exchange oxygen and carbon dioxide directly with seawater. Most fish, for instance, take water in through their mouths (which gives them the appearance of "breathing" underwater), pass it through their gills to extract oxygen, and then expel it through the gill slits on the sides of their bodies (Figure 12–15). Most fish need at least 4.0 parts per million (ppm) of dissolved oxygen in seawater to survive for long periods, and even more for activity and rapid growth. That's why aquariums need a pump to continually resupply oxygen to the water in the tank.

During low oxygen conditions, most marine animals with gills cannot simply breathe air at the surface. Their adaptations allow them to use only the oxygen dissolved in water. If dissolved oxygen levels become low enough (such as after an algae bloom when decomposition consumes dissolved oxygen), many marine organisms will suffocate and die.

Gill structure and location vary among animals of different groups. In fishes, gills are located at the rear of the mouth and contain capillaries. In higher aquatic invertebrates, they protrude from the body surface and contain extensions of the vascular system. In mollusks, they are inside the mantle cavity. In aquatic insects, they occur as projections from the walls of the air tubes. In amphibians, gills are usually present only in the larval stage. In higher vertebrates (including humans), they occur as rudimentary, nonfunctional gill slits, which disappear during embryonic development.

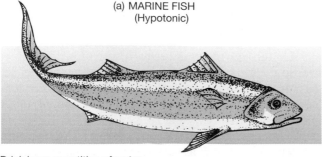

(a) MARINE FISH
(Hypotonic)

Drink large quantities of water
Secrete salt through special cells
Small volume of highly concentrated urine

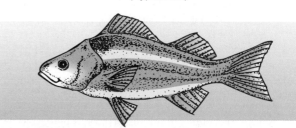

(b) FRESHWATER FISH
(Hypertonic)

Do not drink
Cells absorb salt
Large volume of dilute urine

Figure 12–14 Marine fish versus freshwater fish. Osmotic processes cause these two types of fish to have different adaptations to the environment.

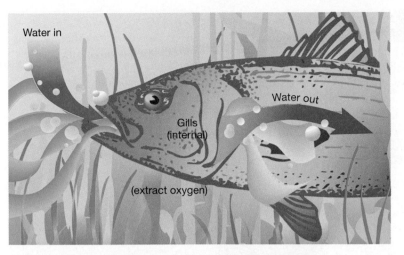

Water in

Water out

Gills
(internal)

(extract oxygen)

Figure 12–15 Gills on fish. Water is taken in through the mouth, passed across the gills to extract dissolved oxygen, and exits through the gill slits.

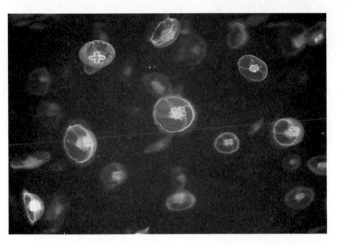

Figure 12–16 Jellyfish. Jellyfish are nearly transparent, which makes them difficult for predators to see.

Water's High Transparency

Water—including seawater—has relatively high transparency compared to many other substances, allowing sunlight to penetrate to a depth of about 1000 meters (3300 feet) in the open ocean. The actual depth depends on the amount of turbidity (suspended sediment) in the water, the amount of plankton in the water, latitude, time of day, and the season.

Because of water's high transparency, many marine organisms have developed keen eyesight, which helps them locate and capture prey. To combat keen-eyed predators, many marine organisms such as jellyfish are themselves nearly transparent, which helps them blend into their environment (Figure 12–16). In fact, almost all open-ocean animals not otherwise protected by teeth, toxins, speed, or small size have some degree of invisibility. Only at depths where sunlight never penetrates is transparency uncommon. Not only can transparency help organisms elude their predators, it can also help organisms stalk their prey.

Other organisms hide by using their coloration pattern as camouflage (Figure 12–17a). Still others use

countershading to blend into their environment, which means they are dark colored on top and light colored on bottom (Figure 12–17b). Many fish—especially flat fish—have countershading so they cannot be easily seen against the dark background of deep water or the ocean floor and they blend into the sunlight when viewed from below. Other organisms undertake a daily vertical migration to deeper, darker parts of the ocean to avoid becoming prey (Box 12–1).

In contrast to species trying to blend into their environment, many species of tropical fish display bright colors (Figure 12–18). Why would they be brightly colored if it makes them easily seen by predators? The bright markings of tropical fish may be an example of **disruptive coloration**, where large, bold patterns of contrasting colors tend to make an object blend in when viewed against an equally variable, contrasting background. Zebras use this principle to evade predators, tigers use it to conceal themselves from their prey, and military uniforms are camouflaged in a similar way.

Even considering disruptive coloration, many tropical fish still don't easily blend into their environment. Perhaps the bright colors and distinctive markings that make tropical fish more apparent make it easier to advertise their identity, to attract mates, or to display weaponry such as spines or poison. Scientists have yet to agree on why tropical fish have such vivid colors, but there must be some biological advantage or the fish wouldn't have them.

Pressure

Water pressure increases about 1 kilogram per square centimeter (1 atmosphere or 14.7 pounds per square inch) with every 10 meters (33 feet) of water depth. Humans are not well adapted to the high pressures that exist below the surface (see the chapter-opening feature in Chapter 1). Even when diving to the bottom of the deep end of a swimming pool, one can feel the dramatic increase in pressure in one's ears.

In the deep ocean, water pressure is on the order of several hundred kilograms per square centimeter

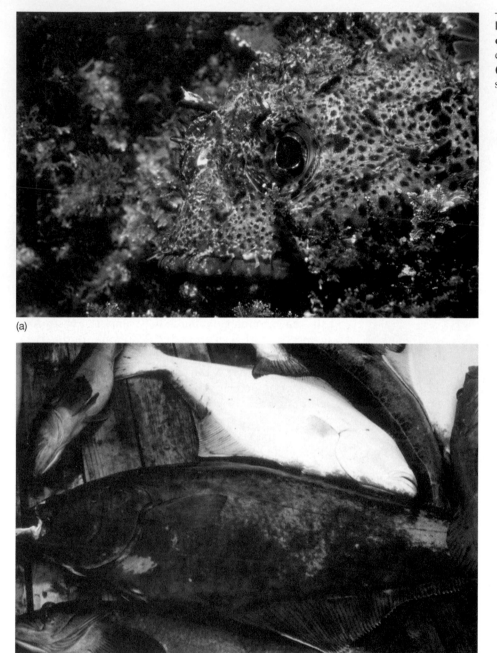

(a)

(b)

Figure 12–17 Camouflage and countershading. (a) The head and eye of a well-camouflaged rock fish. **(b)** Halibut on a dock in Alaska show countershading.

(several hundred atmospheres, or several tons per square inch). How do deep-water marine organisms withstand pressures that can easily kill humans? Most marine organisms lack large compressible air pockets inside their bodies. They do not have lungs, ear canals, or other passageways as we do, so these organisms don't feel the high pressure pushing in on their bodies. Water is nearly incompressible, moreover, so their water-filled bodies have the same amount of pressure pushing outward and they are unaffected by the high pressures found in deep-ocean environments.

A few species appear to be extremely tolerant of pressure changes. In fact, some marine species that are found in nearshore areas can also be found at depths of several kilometers.

The ocean's physical support, viscosity, temperature, salinity, sunlit surface waters, dissolved gases, high transparency, and pressure create conditions to which marine organisms are superbly adapted.

Divisions of the Marine Environment

The oceans can be divided into two main environments. The ocean water itself is the *pelagic environment*, where floaters and swimmers play out their lives in a complex food web. The ocean bottom is the *benthic environment*, where marine algae and animals that do not float or swim (or at least not very well) spend their lives.

Figure 12–18 Use of color by tropical fish. Many tropical fish such as this spotfin butterfly fish have bright colors and bold patterns, which can allow them to blend into the environment by using disruptive coloration. Alternatively, they may also stand out so that they can advertise their identity, sex, or weaponry.

⬛EIO⬛

For more information and on-line exercises about the Environmental Issue in Oceanography (EIO) "Illegal Immigration: Ballast Water and Exotic Species," visit the EIO Web site at **http://www.prenhall.com/oceanissues** and select Issue #7.

Pelagic (Open Sea) Environment

The pelagic environment can be divided into **biozones** that have distinctive biological characteristics, as shown in Figure 12–19.

The pelagic environment is divided into neritic and oceanic provinces (Figure 12–19). The **neritic** (*neritos* = of the coast) **province** extends from the shore seaward, and includes all water less than 200 meters (660 feet) deep. Seaward of the neritic province is the **oceanic province**, where depth increases beyond 200 meters (660 feet). The oceanic province is further subdivided into four biozones:

1. The **epipelagic** (*epi* = top, *pelagios* = of the sea) **zone** from the surface to a depth of 200 meters (660 feet)

2. The **mesopelagic** (*meso* = middle, *pelagios* = of the sea) **zone** from 200 to 1000 meters (660 to 3280 feet)

3. The **bathypelagic** (*bathos* = depth, *pelagios* = of the sea) **zone** from 1000 to 4000 meters (3300 to 13,000 feet)

4. The **abyssopelagic** (*a* = without, *byssus* = bottom, *pelagios* = of the sea) **zone**, which includes all the deepest parts of the ocean below 4000 meters (13,000 feet)

The single most important factor that determines the distribution of life in the oceanic province is the availability of sunlight. Thus, in addition to the four biozones, the distribution of life in the ocean is also divided into zones based on the availability of sunlight as follows.

• The **euphotic** (*eu* = good, *photos* = light) **zone** extends from the surface to a depth where enough light still exists to support photosynthesis. This rarely is deeper than 100 meters (330 feet).

• The **disphotic** (*dis* = apart from, *photos* = light) **zone** has small but measurable quantities of light. It extends from the euphotic zone to a depth where light no longer exists—usually about 1000 meters (3300 feet).

• The **aphotic** (*a* = without, *photos* = light) **zone** has no light, so it exists below about 1000 meters (3300 feet).

Epipelagic Zone The upper half of the epipelagic zone is the only place in the ocean where there is sufficient light to support photosynthesis. The boundary

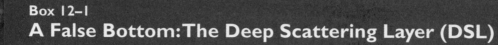

Box 12–1
A False Bottom: The Deep Scattering Layer (DSL)

Within the mesopelagic layer in most parts of the ocean, there exists a curious feature called the **deep scattering layer (DSL)**. It was discovered when the U.S. Navy was testing sonar equipment to detect enemy submarines early in World War II. On many of the sonar recordings, a mysterious sound-reflecting surface appeared that was much too shallow to be the ocean floor. It was often referred to as a "false bottom" (see Figure 3–1). What was even more surprising was that the depth of the deep scattering layer changed with time. It was at a depth of about 100 to 200 meters (330 to 660 feet) during the night, but was as deep at 900 meters (3000 feet) during the day.

With the help of marine biologists, sonar specialists were able to determine that sonar signals were reflecting off densely packed concentrations of marine organisms (Figure 12B). Investigation with plankton nets and submersibles revealed that the DSL contained many different organisms, including copepods (which constitute a large proportion of planktonic animals), euphausids (krill), and lantern fish (family Myctophidae)

The daily movement of the deep scattering layer is caused by the vertical migration of marine organisms that feed in the highly productive surface waters but must protect themselves from being seen by predators. Organisms within the DSL ascend only at night to feed under cover of darkness and then migrate to deeper (darker) water to hide during the day.

Today, fish-finding sonar is still used on commercial and sport fishing trips to track the movement of fish below the surface, often quite successfully.

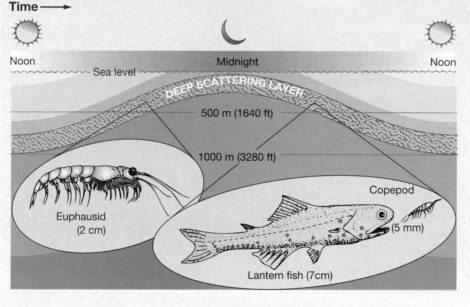

Figure 12B Daily movement and organisms of the deep scattering layer.

between the epipelagic and mesopelagic zones, at 200 meters (660 feet), is also where the level of dissolved oxygen begins to decrease significantly (Figure 12–20, *red curve*).

Oxygen decreases at this depth because no photosynthetic algae live below about 150 meters (500 feet), and dead organic tissue descending from the biologically productive upper waters is decomposing by bacterial oxidation. Nutrient content also increases abruptly below 200 meters (600 feet) (Figure 12–20, *green curve*) and it is the approximate bottom of the mixed layer, seasonal thermocline, and surface water mass.

Mesopelagic Zone A dissolved **oxygen minimum layer (OML)** occurs at a depth of about 700 to 1000 meters (2300 to 3280 feet) (Figure 12–20). The intermediate-water masses that move horizontally in this depth range often possess the highest levels of nutrients in the ocean.

Sunlight from the surface is very, very dim, so fish in the mesopelagic zone have unusually large eyes that are capable of detecting light levels 100 times lower than humans can sense.

Bioluminescent organisms, such as certain species of shrimp, squid, and fish, inhabit the mesopelagic zone, too. These organisms can produce light biologically, causing them to "glow in the dark." Approximately 80% of them have light-producing cells called **photophores**, which are glandular cells containing luminous bacteria surrounded by dark pigments. Some photophores contain lenses to amplify the light radiation.

Bioluminescence is produced when molecules of *luciferin* are excited and emit photons of light in the presence of oxygen, similar to how fireflies and

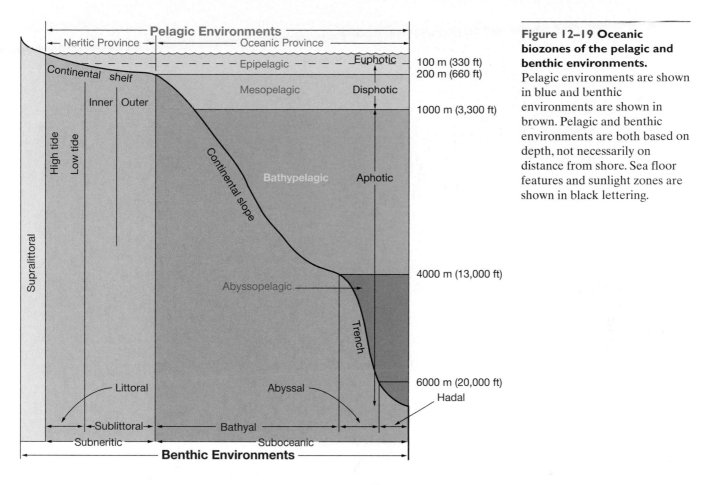

Pelagic Environments
Neritic Province — Oceanic Province
Continental shelf
Inner | Outer
Epipelagic — — Euphotic — 100 m (330 ft) / 200 m (660 ft)
Mesopelagic | Disphotic — 1000 m (3,300 ft)
High tide / Low tide
Continental slope
Bathypelagic | Aphotic
Supralittoral
Abyssopelagic — 4000 m (13,000 ft)
Trench
Littoral | Abyssal — 6000 m (20,000 ft) / Hadal
Sublittoral | Bathyal
Subneritic | Suboceanic
Benthic Environments

Figure 12–19 Oceanic biozones of the pelagic and benthic environments. Pelagic environments are shown in blue and benthic environments are shown in brown. Pelagic and benthic environments are both based on depth, not necessarily on distance from shore. Sea floor features and sunlight zones are shown in black lettering.

glowworms produce light. Only a 1% loss of energy is required to produce this illumination.

Bathypelagic Zone and Abyssopelagic Zone The aphotic (lightless) bathypelagic and abyssopelagic zones represent over 75% of the living space in the oceanic province. Many completely blind fish exist in this region of total darkness and all are small, bizarre-looking, and predaceous.

Many species of shrimp that normally feed on **detritus**[5] become predators at these depths, where the food supply is greatly reduced compared to surface waters. Animals that live in these deep zones feed mostly upon one another. They have evolved impressive warning devices and unusual apparatuses to make them more efficient predators (Figure 12–21). Many also have sharp teeth and extremely large mouths relative to their body size.

Oxygen content increases with depth below the oxygen minimum layer because it is replenished by deep currents originating in polar regions as cold surface water high in oxygen. The abyssopelagic zone is the realm of the bottom-water masses, which commonly move in the direction opposite the deep-water masses in the bathypelagic zone.

[5] *Detritus* is a catchall term for dead and decaying organic matter, including waste products.

Benthic (Sea Bottom) Environment

The benthic, or sea floor, environment is divided into two main units that correspond to the neritic and oceanic provinces of the pelagic environment (see Figure 12–19).

- The **subneritic province** extends from the spring high tide shoreline to a depth of 200 meters (660 feet), approximately encompassing the continental shelf.

- The **suboceanic province** includes the benthic environment below 200 meters (660 feet).

The transitional region from land to sea floor above the spring high-tide line is called the **supralittoral** (*supra* = above, *littoralis* = the shore) **zone**. Commonly called the *spray zone*, it is covered with water only during periods of extremely high tides and when tsunami or large storm waves break on the shore.

The subneritic zone is subdivided into the littoral and sublittoral zones. The *intertidal zone* (the zone between high and low tides) coincides with the **littoral** (*littoralis* = the shore) **zone**. The **sublittoral** (*sub* = below, *littoralis* = the shore) **zone**, or *shallow subtidal zone*, extends from low-tide shore line out to a depth of 200 meters (660 feet).

The sublittoral zone consists of inner and outer regions. The **inner sublittoral zone** extends to the depth at which marine algae no longer grows attached to the ocean bottom [approximately 50 meters (160 feet)], so

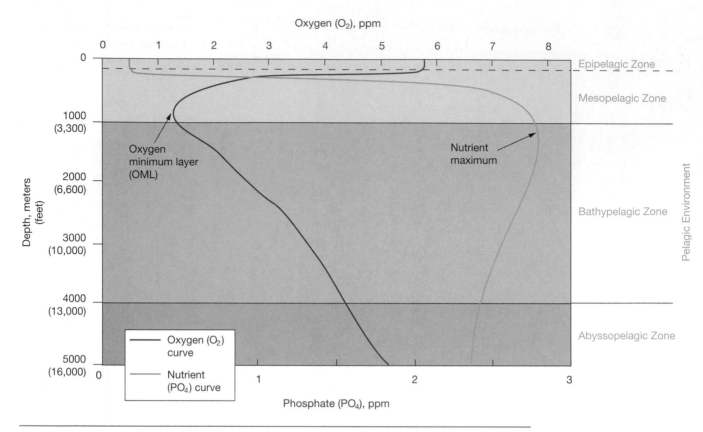

Figure 12–20 Abundance of dissolved oxygen and nutrients with depth. In surface water, oxygen is abundant due to mixing with the atmosphere and plant photosynthesis, and nutrient content (phosphate) is low due to uptake by algae. At deeper depths, oxygen decreases and produces an oxygen minimum layer (OML), which coincides with a nutrient maximum. Below that, nutrient levels remain high and oxygen increases as it is replenished with high-oxygen cold water from polar regions (ppm = parts per million).

Figure 12–21 Deep-sea anglerfish and shrimp. An anglerfish (*Lasiognathus* saccostoma) attracts two deep-water shrimp with its bioluminescent "lure." The fish is about 10 centimeters (4 inches) long.

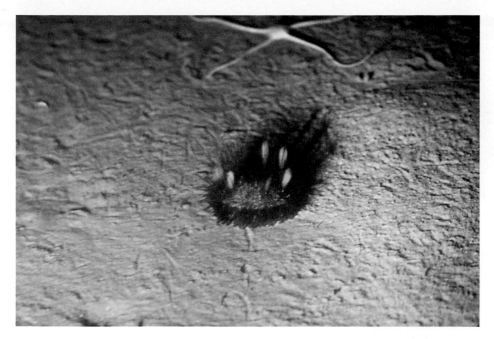

Figure 12–22 Benthic organisms produce tracks on the ocean floor. As benthic organisms—such as this sea urchin (*below*) and brittle star (*above*)—move across or burrow through the ocean bottom, they often leave tracks in the sediment on the ocean floor.

it varies depending on the amount of solar radiation that penetrates the surface water. All photosynthesis seaward of the inner sublittoral zone is carried out by floating microscopic algae.

The **outer sublittoral zone** extends from the inner sublittoral zone out to a depth of 200 meters (660 feet) or the shelf break, which is the seaward edge of the continental shelf.

The suboceanic province is subdivided into bathyal, abyssal, and hadal zones. The **bathyal** (*bathus* = deep) **zone** extends from a depth of 200 to 4000 meters (660 to 13,000 feet) and corresponds generally to the continental slope.

The **abyssal** (*a* = without, *byssus* = bottom) **zone** extends from a depth of 4000 to 6000 meters (13,000 to 20,000 feet) and includes over 80% of the benthic environment. The ocean floor of the abyssal zone is covered by soft oceanic sediment, primarily abyssal clay. Tracks and burrows of animals that live in this sediment can be seen in Figure 12–22.

The **hadal** (*hades* = hell[6]) **zone** extends below 6000 meters (20,000 feet), so it consists only of deep trenches along the margins of continents. Animal communities that are found in these deep environments have been isolated from each other, often resulting in unique adaptations.

> The pelagic environment includes the water column and the benthic environment includes the sea bottom. Subdivisions of pelagic and benthic environments are based on depth, which influences the amount of sunlight.

---

[6] The inhospitable high-pressure environment of the hadal zone is aptly named.

 Students Sometimes Ask...

Why are there scientific names for all organisms? Wouldn't it be easier just to know the common name of an organism?

Each individual species has a *unique* scientific name, so the scientific name identifies one particular species more clearly than the common name. Common names are often used for more than one species of organisms, which can be confusing. "Dolphin," for example, is used to describe dolphins, porpoises, and even a type of fish! Most people would shudder at being served dolphin in a restaurant, but dolphin *fish* (also called mahi-mahi) is often a featured menu item.

Common names can also be confusing because there can be more than one for the same species, and because they vary from language to language. Scientific names, on the other hand, are Latin based, so they are the same in all languages. This allows a Chinese scientist, for example, to communicate effectively with a Greek scientist about a particular organism. Thus, scientific names are useful, descriptive (if you know a bit about Latin terms and word roots), and unambiguous.

What is the difference between kelp, seaweed, and marine algae?

In common usage, they all refer to large, branching, photosynthetic marine organisms that contain various pigments, which give these organisms their color. However, differences between the terms do exist.

Early mariners probably coined the term "seaweed" because they thought these organisms were a nuisance. They clogged harbors, entangled vessels, and washed up on the beach in useless bunches after storms. Although they could be eaten, they could not be consumed in large quantities. Historically, all marine algae (except microscopic species) had similarities to weeds on land, so they have become known collectively as "seaweeds."

It turns out, however, that these organisms are vital to coastal ecosystems, so marine biologists prefer to call them

Figure 12C Sea vegetable dulse (*Palmaria palmata*).

"marine algae." To differentiate them from the microscopic planktonic species of algae, large marine algae are often called "marine macro algae." The branching types of brown algae (phylum Phaeophyta) are called "kelp."

Today marine macro algae have many uses. They are used as a thickener and emulsifier in many products (such as toothpaste) and foods (such as ice cream). Cooking recipes sometimes call for small quantities of "sea vegetable" (Figure 12C), and nori, a type of red alga, is used to wrap sushi. Marine algae are used as fertilizer, and some species have recently been touted as health foods. Marine algae are discussed in Chapter 13, "Biological Productivity and Energy Transfer."

I'm confused. You mentioned that 98% of marine species are benthic, yet the vast majority of the ocean's biomass is planktonic. How can this be true?

The key to this apparent dilemma is understanding the difference between the *number of species* and the *amount of living organisms (biomass)*. A zoo, for instance, has a high number of different species, but a low biomass of each. Conversely, a large cattle farm may have only one species, but a large biomass of that single species.

In benthic regions, there are many different environments, such as sandy bottoms, rocky bottoms, muddy bottoms, upper tide zones, lower tide zones, high-energy shorelines, and low-energy shorelines. The high environmental variability creates ideal conditions for a huge number of species, all with specific adaptations to their particular benthic environment. High species diversity (many different types of species) exists in the benthic environment, but it has a smaller biomass (a smaller overall amount of organisms) than sunlit surface waters. Conversely, the consistency of the pelagic environment has led to low species diversity (many individuals of the same species),

but a large biomass (a large amount of living organisms), provided there are appropriate nutrients and enough sunlight.

I know that some fish living in the ocean return to the same freshwater stream where they were born to spawn. How can fish adapted to saltwater survive in a freshwater environment?

Fish that are adapted to living in saltwater cannot live for long in freshwater because they take on so much water by osmosis, but they do undergo physical changes that allow them to survive for short periods. For example, some species of marine fish—such as the salmon of the Pacific and North Atlantic Oceans—are born in fresh water, spend most of their lives in the sea, then return to the freshwater stream where they were born to spawn. Although it is unknown how these fish accurately return home, most scientists think they use odors and currents as their guides. Marine fish that return to freshwater to spawn are called *anadromous* (*ana* = up, *dromos* = a running). Once Pacific salmon spawn, they die. Atlantic salmon, on the other hand, are able to return to the ocean after as much as a week in freshwater streams.

Why do my fingers get all wrinkly when I stay in the water for a long time?

It's caused by water molecules passing through your skin (a semipermeable membrane) due to osmosis. The water molecules outside your body—contained in either pure water or seawater—flow into your skin cells in an attempt to dilute the dissolved particles inside those cells. After a short time in water, your skin cells contain many more water molecules than before, which hydrates your skin and causes it to look wrinkly. The effect is most obvious on appendages such as fingers and toes because they have a high surface area per unit of volume. It's only a temporary condition, though, because your body absorbs and excretes the excess water molecules after you leave the water.

If animals that have large air pockets inside are affected by the extreme pressures at depth, how are sperm whales able to dive so deeply?

All marine mammals have lungs and breathe air and certain ones have special adaptations that allow them to make extremely deep dives. Sperm whales, for example, can dive to more than 2800 meters (9200 feet) and stay submerged for more than one hour during their search for food. They are able to use small amounts of oxygen very efficiently and have a collapsible rib cage, which forces air out and collapses the lungs, thereby closing the air cavities inside their bodies. Marine mammals and their adaptations are discussed in Chapter 14, "Animals of the Pelagic Environment."

Chapter in Review

• A wide variety of organisms live in the ocean, ranging in size from microscopic bacteria and algae to blue whales. All living things belong to one of three major domains (branches) of life: Archaea, simple microscopic bacteria-like creatures; Bacteria, simple life forms consisting of cells that usually lack

a nucleus; and Eukarya, complex organisms (including plants and animals) consisting of cell that have a nucleus.

• Organisms are further divided into five kingdoms: Monera, single-celled organisms without a nucleus; Protoctista, single-

and multi-celled organisms with a nucleus; Fungi, mold and lichen; Plantae, many-celled plants; and Animalia, many-celled animals. Classification of organisms involves placing individuals within the kingdoms into increasingly specific groupings of phylum, class, order, family, genus, and species, the last two of which denote an organism's scientific name. Many organisms also have one or more common names.

• Marine organisms can be classified into one of three groups based on habitat and mobility. Plankton are free-floating forms with little power of locomotion; nekton are swimmers; and benthos are bottom dwellers. Most of the ocean's biomass is planktonic.

• Only about 14% of all known species inhabit the ocean, and over 98% of marine organisms are benthic. The marine environment—especially the pelagic environment—is much more stable than the terrestrial environment, so there is less pressure on marine organisms to diversify.

• Marine organisms are well adapted to life in the ocean. Those organisms that have established themselves on land have had to develop complex systems for support and for acquiring and retaining water.

• The algae that must stay in surface water to receive sunlight and the small animals that feed on them lack an effective means of locomotion. They depend, therefore, on their small size and other adaptations to increase their ratio of surface area to body mass so they have a greater frictional resistance to sinking. Their small size also allows them to efficiently absorb nutrients and dispose of wastes. Many nektonic organisms have developed streamlined bodies so that they can overcome the viscosity of seawater and more easily move through it.

• Surface temperature of the world ocean does not vary on a daily, seasonally, or yearly basis as much as on land. Organisms living in warm water tend to be individually smaller, have ornate plumage, comprise a greater number of species, and constitute a much smaller total biomass than organisms living in cold water. Warm-water organisms also tend to live shorter lives and reproduce earlier and more frequently than cold-water organisms.

• Osmosis is the passing of water molecules through a semipermeable membrane from a region of higher concentration to a region of lower concentration. If the body fluids of an organism and ocean water are separated by a membrane that allows water molecules to pass through, the organism may become severely dehydrated from osmosis. Many marine invertebrates are essentially isotonic—the salinity of their body fluids is similar to that of ocean water. Most marine vertebrates are hypotonic—the salinity of their body fluids is lower than that of ocean water, so they tend to lose water through osmosis. Freshwater organisms are essentially all hypertonic—the salinity of their body fluids is greater than the water in which they live, so they tend to gain water through osmosis.

• Most marine animals extract oxygen through their gills. Many marine organisms have well-developed eyesight because water is so transparent. To avoid being seen and consumed by predators, many marine organisms are transparent, camouflaged, counter-shaded, or disruptively colored. Unlike humans, most marine organisms are unaffected by the high pressure at depth because they do not have large internal air pockets that can be compressed.

• The marine environment is divided into pelagic (open sea) and benthic (sea bottom) environments. These regions are further divided based on depth and have varying physical conditions to which marine life is adapted.

Key Terms

Abyssal zone (p. 367)	Countershading (p. 361)	Gill (p. 360)
Abyssopelagic zone (p. 363)	Darwin, Charles (p. 346)	Hadal zone (p. 367)
Animalia (p. 349)	Deep scattering layer (DSL) (p. 364)	Holoplankton (p. 350)
Aphotic zone (p. 363)	Detritus (p. 365)	Hypertonic (p. 359)
Archaea (p. 347)	Diffusion (p. 359)	Hypotonic (p. 359)
Bacteria (p. 347)	Disphotic zone (p. 363)	Infauna (p. 350)
Bacterioplankton (p. 349)	Disruptive coloration (p. 361)	Inner sublittoral zone (p. 365)
Bathyal zone (p. 367)	Epifauna (p. 350)	Isotonic (p. 359)
Bathypelagic zone (p. 363)	Epipelagic zone (p. 363)	Littoral zone (p. 365)
Benthic environment (p. 352)	Eukarya (p. 348)	Macroplankton (p. 350)
Benthos (p. 350)	Euphotic zone (p. 363)	Meroplankton (p. 350)
Bioluminescence (p. 364)	Euryhaline (p. 358)	Mesopelagic zone (p. 363)
Biomass (p. 350)	Eurythermal (p. 358)	Monera (p. 348)
Biozone (p. 363)	Fungi (p. 349)	Nektobenthos (p. 350)

Nekton (p. 350)

Neritic province (p. 363)

Oceanic province (p. 363)

Osmosis (p. 359)

Osmotic pressure (p. 359)

Outer sublittoral zone (p. 365)

Oxygen minimum layer (OML) (p. 364)

Pelagic environment (p. 352)

Photophore (p. 364)

Phytoplankton (p. 349)

Picoplankton (p. 350)

Plankter (p. 349)

Plankton (p. 349)

Plantae (p. 349)

Protoctista (p. 349)

Protoplasm (p. 352)

Protozoa (p. 349)

Species (p. 349)

Stenohaline (p. 358)

Stenothermal (p. 358)

Streamlining (p. 356)

Sublittoral zone (p. 365)

Subneritic province (p. 365)

Suboceanic province (p. 365)

Supralittoral zone (p. 365)

Taxonomy (p. 349)

Viscosity (p. 355)

Zooplankton (p. 349)

Questions and Exercises

1. Discuss the accomplishments of the naturalist Charles Darwin.

2. List the three major domains of life and the five kingdoms of organisms. Describe the fundamental criteria used in assigning organisms to these divisions.

3. Describe the lifestyles of plankton, nekton, and benthos. Why is it true that plankton account for a much larger percentage of the ocean's biomass than the benthos and nekton?

4. List the subdivisions of plankton and benthos and the criteria used for assigning individual species to each.

5. List the relative number of species of animals found in the terrestrial, pelagic, and benthic environments, and discuss the factors that may account for this distribution.

6. Describe how higher water temperatures in the tropics may account for the greater number of species in these regions compared with low-temperature, high-latitude areas.

7. Discuss the major differences between marine algae and land plants, and explain the reasons why there is greater complexity in land plants.

8. List several reasons why it is advantageous for organisms in the ocean to be of a very small size.

9. Determine the surface to volume ratio of an organism whose average linear dimension is (a) 1 centimeter (0.4 inch); (b) 3 centimeters (1.1 inches); and (c) 5 centimeters (2 inches). Which one is better able to resist sinking, and why? Discuss some adaptations other than size that are used by organisms to increase their resistance to sinking.

10. Changes in water temperature significantly affect the density, viscosity of water, and ability of water to hold gases in solution. Discuss how decreased water temperature changes these variables and how these changes affect marine life.

11. List differences between cold- and warm-water species in the marine environment.

12. What do the prefixes *eury-* and *steno-* mean? Define the terms *eurythermal/stenothermal* and *euryhaline/stenohaline*. Where in the marine environment will organisms displaying a well-developed degree of each characteristic be found?

13. Describe the process of osmosis. How is it different from diffusion? What three things can occur simultaneously across the cell membrane during osmosis?

14. What is the problem requiring osmotic regulation faced by hypotonic fish in the ocean? How have these animals adapted to meet this problem?

15. How does water temperature affect the water's ability to hold gases? How do marine organisms extract the dissolved oxygen from seawater?

16. What strategies are used by marine organisms to avoid being seen (and perhaps eaten) by predators?

17. How does the depth of the deep scattering layer vary over the course of a day? Why does it do this? Which organisms comprise the DSL?

18. Construct a table listing the subdivisions of the pelagic and benthic environments and the physical factors used in assigning their boundaries.

19. Describe the vertical distribution of oxygen and nutrients in the oceanic province, and discuss the factors that are responsible for this distribution.

20. What is bioluminescence? How are marine organisms able to produce light organically?

References

Borgese, E. M., Ginsburg, N., and Morgan, J. R., eds. 1991. *Ocean Yearbook 9*. Chicago: University of Chicago Press.

Bult, C., et al. 1996. Complete genome sequence of the methanogenic Archaeon, *Methanococcus jannaschii*. *Science* 273:5278, 1058–1072.

Coker, R. E. 1962. *This Great and Wide Sea: An Introduction to Oceanography and Marine Biology*. New York: Harper & Row.

Genin, A., Dayton, P. K., Lonsdale, P. F., and Spiess, F. N. 1986. Corals on seamount peaks provide evidence of current acceleration over deepsea topography. *Nature* 322:6074, 59–61.

Grassle, J. F., and Maciolek, N. J. 1992. Deep-sea species richness: Regional and local diversity estimates from quantitative bottom samples. *American Naturalist* 139:2, 313–341.

Hedpeth, J., and Hinton, S. 1961. *Common Seashore Life of Southern California*. Healdsburg, CA: Naturegraph.

Ishimatsu, A., et al. 1998. Mudskippers store air in their burrows. *Nature* 391:6664, 237–238.

Lalli, C. M., and Parsons, T. R. 1993. *Biological Oceanography: An Introduction*. New York: Pergamon Press.

Levine, J. 1999. In living colors. *Natural History* 108:7, 40–46.

Margulis, L., and Schwartz, K. V. 1988, *Five Kingdoms: An Illustrated Guide to the Phyla of Life on Earth*, 2nd ed. New York: W. H Freeman.

Marshall, J. 1998. Why are reef fish so colorful? *Scientific American Presents: The Oceans* 9:3, 54–57.

May, R. M. 1988. How many species are there on Earth? *Science* 241:4872, 1441–1448.

Morell, V. 1996. Life's last domain. *Science* 273, 1043–1045.

Pimm, S. L., Russell, G. J., Gittlemen, J. L., and Brooks, T. M. 1995. The future of biodiversity. *Science* 269:5222, 347–350.

Sieburth, J. M. N. 1979. *Sea Microbes*. New York: Oxford University Press.

Sumich, J. L. 1976. *An Introduction to the Biology of Marine Life*. Dubuque, IA: Wm. C. Brown.

Sverdrup, H., Johnson, M., and Fleming R. 1942. Renewal 1970. *The Oceans*. Englewood Cliffs, NJ: Prentice-Hall.

Thorson, G. 1971. *Life in the Sea*. New York: McGraw-Hill.

Weiner, J. 1994. *The Beak of the Finch: A Story of Evolution in Our Time*. New York: Random House.

Wilson, E. O. 1992. *The Diversity of Life*. Cambridge, MA: Belknap Press of Harvard University Press.

Wishner, K., Levin, L., Gowing, M., and Mullineaux, L. 1990. Involvement of the oxygen minimum in benthic zonation on a deep seamount. *Nature* 346:6279, 57–59.

Wray, G. A. 2001. A world apart. *Natural History* 110:2, 52–63.

Suggested Reading in Scientific American

Denton, E. 1960. The buoyancy of marine animals. 203:1, 118–129. The means by which some marine animals reduce the energy expenditure required to live in the ocean water far above the ocean floor are discussed.

Doolittle, W. F. 2000. Uprooting the tree of life. 282:2, 90–95. A discussion of how discoveries by molecular biologists have complicated the story of life on Earth.

Eastman, J. T., and DeVries, A. L. 1986. Antarctic fishes. 255:5, 106–114. Explains how one group of fish survived when the Antarctic turned cold.

Herbert, S. 1986. Darwin as a geologist. 254:5, 116–123. Before publication of *The Origin of Species*, Charles Darwin considered himself primarily a geologist. This article outlines his contributions in this field, with special attention to his theory of coral reef formation.

Horn, M. H., and Gibson, R. N. 1988. Intertidal fishes. 258:1, 64–71. Intertidal fishes have undergone remarkable adaptation to survive this physically harsh environment.

Isaacs, J. D. 1969. The nature of oceanic life. 221:3, 146–165. A well-developed survey of the conditions for life in the ocean as they relate to the variety and distribution of marine life forms.

Isaacs, J. D., and Schwartzlose, R. A. 1975. Active animals of the deep-sea floor. 233:4, 84–91. A surprisingly large population of large fishes on the deep-sea floor is suggested by automatic cameras dropped to the ocean bottom.

Johnsen, S. 2000. Transparent animals. 282:2, 80–89. A look at the ingenious adaptations of marine organisms that enable them to be transparent.

Mayr, E. 2000. Darwin's influence on modern thought. 283:1, 78–83. Why Darwin's contributions make him the world's best-known and most influential scientist.

Palmer, J. D. 1975. Biological clock and the tidal zone. 232:2, 70–79. This article investigates the mechanism of biological clocks set to rhythm of the tides, which are found in organisms from diatoms to crabs.

Partridge, B. L. 1982. The structure and function of fish schools. 246:6, 114–123. Schooling benefits and the means by which fish maintain contact with the school are considered.

Vogel, S. 1978. Organisms that capture currents. 239:2, 128–139. The manner in which sponges use ocean currents is an important part of this discussion.

Oceanography on the Web

Visit the *Essentials of Oceanography* home page for on-line resources for this chapter. There you will find an on-line study guide with review exercises, and links to oceanography sites to further your exploration of the topics in this chapter. *Essentials of Oceanography* is at: **http://www.prenhall.com/ thurman** (click on the Table of Contents menu and select this chapter).

CHAPTER

13

Biological Productivity and Energy Transfer

●

BASELINE STUDIES IN THE CALIFORNIA CURRENT: THE CALCOFI PROGRAM

When the colorful description of Monterey, California and its fishing industry was published in John Steinbeck's *Cannery Row* in 1945, California's sardine fishery was the largest single-species fishery in the world. By 1949, however, the sardine fishery had collapsed and scientists could not agree whether it was due to overfishing or natural changes in the environment. To help solve this mystery, the state created the **California Cooperative Oceanic Fisheries Investigations (CalCOFI)**, a consortium of industry, universities, and state and federal agencies appointed to study the

- What is primary productivity?
- What are the key limiting factors in the marine environment that control photosynthesis?
- What types of marine organisms photosynthesize?
- How does productivity differ in polar, tropical, and temperate regions?
- What is the efficiency of energy transfer between various organisms?
- How does a food chain differ from a food web?
- What problems do fisheries have?

We cannot cheat on DNA. We cannot get round photosynthesis. We cannot say I am not going to give a damn about phytoplankton. All these tiny mechanisms provide the preconditions of our planetary life.
—*Barbara Ward,*
Who Speaks for Earth? *(1973)*

Figure 13A CalCOFI sampling bottles.

California current. The program continues today as a partnership among Scripps Institution of Oceanography, the California Department of Fish and Game, and the U.S. National Marine Fisheries Service.

Early CalCOFI investigations proved the futility of studying a single species without examining its ocean environment and the cohabitants of that environment. The focus of the program was expanded so that it was more interdisciplinary and ecosystem-based, including the study of all organisms within the eastern North Pacific Ocean and all source waters of the California Current (from the tip of Baja California, Mexico, to the Columbia River). At the center of the investigations are the CalCOFI surveys, a shipboard monitoring program that makes seasonal measurements using nets, temperature probes, seawater sampling bottles (Figure 13A), and other instruments deployed along precisely plotted pathways off the coast of California. As a result, tens of thousands of measurements of zooplankton abundance, along with temperature, salinity, oxygen, nutrient, and chlorophyll levels, have been collected in the California Current during 300 cruises in the past 50 years.

This record allows oceanographers to define the "normal" patterns of the physical, chemical, and biological components of the California Current. Perhaps more importantly, it provides a baseline against which to identify any effects on organisms from warming trends, such as that predicted from a buildup of greenhouse gases in the environment.

Scientists in the CalCOFI program have developed new methods for estimating the size of adult populations, for determining the growth rates of larvae, and for estimating the number of larvae larger organisms consume. Many of these models and methods are now used worldwide.

Through the CalCOFI investigations, scientists came to understand that many environmental forces that have a minor effect on adult fish can have a tremendous effect on larvae. Young anchovies, for example, perish unless they find dense concentrations of phytoplankton for food. When the ocean is calm, phytoplankton often concentrate in thin layers below the sea surface, providing a source of food. Storms, however, can disrupt the layers of phytoplankton and prevent larvae from finding high concentrations of food needed for survival. Thus, the strength and timing of storms can critically impact the population of adult fish for several years.

In the end, CalCOFI investigators determined that the collapse of the sardine industry had been caused by a combination of natural changes in the environment and overfishing. The CalCOFI program demonstrates the success of long-term cooperative scientific studies between research institutions and state and federal government agencies. Further, it provides a model for today's global oceanographic studies, which must increasingly rely on international cooperation and funding.

Producers are plants and algae that photosynthesize their own food from carbon dioxide, water, and sunlight. Their ability to capture solar energy and bind it into their food sugars is the basis for all nutrition in the ocean (except near hydrothermal vents,[1] where chemosynthesis is the major source of "food" energy). The ocean's producers are the foundation of the ocean food web. The major primary producers of the oceans are marine algae. They capture nearly all of the solar energy used to support the entire marine biological community.

Large species of marine algae play only a minor role in the production of energy for the ocean population as a whole. Instead, marine organisms depend primarily on microscopic marine algae that are scattered throughout the near-surface sunlit waters of the world's oceans. They represent the largest community of **biomass**[2] in the marine environment—the *phytoplankton*.

In the total darkness of the deep sea, where no measurable sunlight penetrates, certain archaeon (bacteria-like organisms) oxidize hydrogen sulfide or methane to synthesize food. This method of food production supports a vast array of unusually large deep-sea benthos.

The chemical energy stored by surface phytoplankton and deep-sea archaeon is passed to the various populations of animals that inhabit the oceans through a series of feeding relationships called food chains and food webs.

Primary Productivity

Primary productivity is the amount of carbon fixed by organisms through the synthesis of organic matter using energy derived from solar radiation (**photosynthesis**) or chemical reactions (**chemosynthesis**). Although chemosynthesis supports hydrothermal vent biocommunities along oceanic spreading centers, it is much less significant than photosynthesis in worldwide marine primary production.

Photosynthetic Productivity

Photosynthesis is a chemical reaction in which energy from the Sun is stored in organic molecules (Figure

---

[1]Hydrothermal vent biocommunities are discussed in more detail in Chapter 15, "Animals of the Benthic Environment."
[2]Biomass is the mass of living organisms.

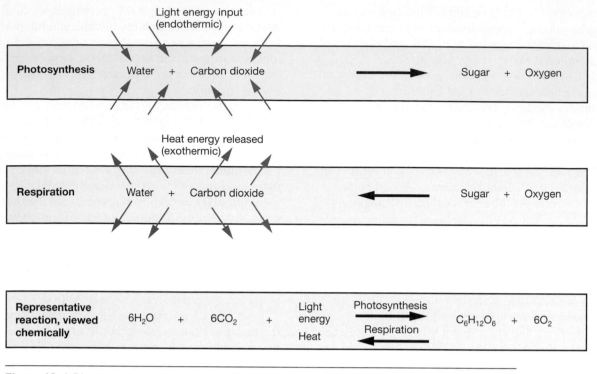

Figure 13–1 Photosynthesis (*top*), respiration (*middle*), and representative reactions viewed chemically (*bottom*).

13–1). The total amount of organic matter produced by photosynthesis per unit of time is the **gross primary production** of the oceans. Algae use some of this organic matter for their own maintenance, through respiration. What remains is **net primary production**,[3] which is manifested as growth and reproduction products. Net primary production supports the rest of the marine population.

Gross primary production has two components, new production and regenerated production. **New production** results from nutrients brought in from outside the local ecosystem by processes such as upwelling. **Regenerated production** results from nutrients that are recycled within the ecosystem. As the ratio of new production to gross primary production increases, so does the ecosystem's ability to support animal populations that are depended on for fisheries, such as pelagic fishes and benthic scallops.

Photosynthetic productivity can be measured by a variety of methods. One of the most direct is to capture plankton in cone-shaped nylon **plankton nets** (Figure 13–2). These fine mesh nets—which resemble windsocks at airports—filter plankton from the ocean as they are towed at a specific depth by research vessels. Analysis of the amounts and types of organisms captured reveals much about the productivity of the area.

Other methods include lowering specially designed bottles into the ocean, laboratory analysis using ra-

dioactive carbon, or using orbiting satellites to determine chlorophyll levels based on seawater surface color. The **SeaWiFS** (<u>Sea</u>-viewing <u>Wi</u>de <u>F</u>ield-of-View <u>S</u>ensor) instrument aboard the SeaStar satellite began operating in 1997, replacing Nimbus-7's Coastal Zone Color Scanner instrument, which operated between 1978 and 1986. SeaWiFS measures the color of the ocean with a radiometer and provides global coverage of ocean chlorophyll levels as well as land productivities every two days.

> Productivity is primarily the amount of carbon (organic matter) produced by plants and algae during photosynthesis. It also includes chemosynthesis.

Availability of Nutrients

The distribution of life throughout the ocean's breadth and depth depends mainly on the availability of nutrients such as nitrate, phosphorous, iron, and silica that are needed by phytoplankton. Where the physical conditions supply large quantities of nutrients, marine populations reach their greatest concentration. The *sources* of nutrients must be considered to understand where these areas are found.

Water in the form of runoff erodes the continents, carrying material to the oceans, and depositing it as sediment at the margins of the continents. Runoff also

[3]The difference between gross and net primary production is similar to the difference between your gross pay (your earnings before taxes) and your net pay (the amount you take home after taxes).

Figure 13–2 Plankton nets. These large, cone-shaped, fine mesh plankton nets being washed during a CalCOFI cruise are lowered into the water and towed behind a research vessel to collect plankton.

dissolves and transports substances such as nitrates and phosphates,[4] which are the basic nutrients for phytoplankton.

Through photosynthesis, phytoplankton combine these nutrients with carbon dioxide and water to produce the carbohydrates, proteins, and fats that the rest of the ocean's biological community depends upon for food.

The continents are the major sources of these nutrients, so the greatest concentrations of marine life are found along the continental margins. The concentration of marine life decreases, however, as the distance from the continental margins into the open sea increases. The vast depth of the world's oceans and the great distance between the open ocean and the coastal regions where nutrients are concentrated account for these differences.

Recent studies in the waters near Antarctica and the Galápagos Islands reveal that photosynthetic produc-

tion is low even though the concentration of all nutrients—except iron—is high.[5] Production is high only in regions of shallow water downcurrent from islands or landmasses where iron from rocks and sediments was dissolved into the water to significant levels.

Availability of Solar Radiation

Photosynthesis cannot proceed unless light energy (solar radiation) is available. Despite the atmosphere's thickness of more than 80 kilometers (50 miles), its high transparency allows sunlight to penetrate it quite readily, so land-based plants almost always have an abundance of solar radiation to conduct photosynthesis.

In the clearest ocean water, however, solar energy may be detected to depths of only about 1 kilometer (0.6 mile) and, even then, the amount reaching these depths is inadequate for photosynthesis. Photosynthesis in the ocean, therefore, is restricted to the uppermost surface waters and those areas of the sea floor where the water is shallow enough to allow light to penetrate.

The depth at which net photosynthesis becomes zero is called the **compensation depth for photosynthesis**. The **euphotic** (*eu* = good, *photos* = light) **zone** extends from the surface down to the compensation depth for photosynthesis, which is approximately 100 meters (330 feet) in the open ocean. Near the coast, the euphotic zone may extend to less than 20 meters (66 feet) because the water contains more suspended inorganic material (turbidity) or microscopic organisms that limit light penetration.

How do the two factors necessary for photosynthesis—the supply of nutrients and the presence of solar radiation—differ between coastal areas and the open ocean? In the open ocean (far from continental margins), solar energy extends deeper into the water column, but concentration of nutrients is low. In coastal regions, on the other hand, light penetration is much less, but the concentration of nutrients is much higher. Because the coastal zone is much more productive, nutrient availability must be the most important factor affecting the distribution of life in the oceans.

Margins of the Oceans

If the stability of the ocean environment is ideal for sustaining life, why are the richest concentrations of marine organisms in the very margins of the oceans, where conditions are the most *un*stable? In the coastal ocean, for instance:

- Water depths are shallow, allowing much greater seasonal variations in temperature and salinity than in the open ocean.

---

[4]Nitrates and phosphates are the basic ingredients in garden and farm fertilizer.

[5]The idea of fertilizing the ocean with iron to stimulate productivity and increase the amount of CO_2 gas absorbed by the ocean is discussed in Chapter 6, "Air–Sea Interaction."

- The thickness of the water column varies in the nearshore region in response to tides that periodically cover and uncover a thin strip of land along the margins of the continents.
- Waves breaking in the surf zone release large amounts of energy that has been carried for great distances across the open ocean.

Each of these conditions stresses organisms. Over the billions of years of geologic time, however, new species have evolved by natural selection to fit every imaginable biological niche—even in environments that pose difficulties for organisms. In fact, many organisms have adapted to live under adverse conditions—such as coastal environments—as long as nutrients are available.

Along continental margins, some areas have more abundant life than others. What characteristics create such an uneven distribution of life? Again, only those basic requirements for the production of food need be considered. For example, areas that have the greatest biomass have the lowest water temperatures, too, because cold water dissolves larger amounts of oxygen and carbon dioxide than warm water. Carbon dioxide, in particular, stimulates phytoplankton growth and phytoplankton growth, in turn, profoundly affects the distribution of all other life in the oceans.

Upwelling and Nutrient Supply

Upwelling is a flow of deep water toward the surface that brings water from depths below the euphotic zone. This deep water is rich in nutrients and dissolved gases because there are no phytoplankton at these depths to consume these compounds. When chilled water from below the surface rises, it hoists nutrients from the depths to the surface, where phytoplankton thrive and make food for larger organisms—copepods, fish, on up to whales.

Highly productive areas of *coastal upwelling* are found along the western margins of continents, where surface currents are moving toward the Equator (Figure 13–3). Ekman transport (see Chapter 7, "Ocean Circulation") moves surface water away from these coasts, so nutrient-rich water from depths of 200 to 1000 meters (660 to 3300 feet) constantly rises to replace it.

> Photosynthetic productivity is limited in the marine environment by the amount of sunlight and the supply of nutrients. Upwelling greatly enhances the conditions for life by lifting cold, nutrient-rich water to the sunlit surface.

Light Transmission in Ocean Water

The graph in Figure 13–4 shows that most solar energy falls in the range of wavelengths called **visible light**. This radiant energy from the Sun powerfully affects three major components of the oceans:

1. The major wind belts of the world, which produce ocean currents and wind-driven ocean waves, ultimately derive their energy from solar radiation. Wind belts and ocean currents strongly influence world climates.

2. A thin layer of warm water at the ocean surface, created by solar heating, overlies the great mass of cold water that fills most of the ocean basins. This is the "life layer" where most sea life exists.

3. Photosynthesis can occur only where sunlight penetrates the ocean water, so phytoplankton and most animals that eat them must live where the light is, in the relatively thin layer of sunlit surface water.

The Electromagnetic Spectrum The Sun radiates a wide range of wavelengths of electromagnetic radiation. Together they comprise the **electromagnetic spectrum**, which is shown in the upper part of Figure 13–4. Only a very narrow portion of the electromagnetic spectrum is visible to humans as visible light. We call it "visible" light because our electromagnetic sensors—our eyes—are adapted to detect only the wavelengths in the visible region. In essence, our eyes "tune into" the visible light wavelengths, just as a radio "tunes into" specific radio waves.

Visible light can be further divided by wavelength into violet, blue, green, yellow, orange, and red energy levels. Together, these different wavelengths produce white light. The shorter wavelengths of energy to the left of visible light (for example, X rays and gamma rays) damage tissue in high enough doses. The longer wavelengths of energy to the right of visible light (for example, infrared, microwaves, and radio waves) are used for heat transfer and communication.

The Color of Objects Light from the Sun includes all the visible colors. Most of the light we see is reflected from objects. All objects absorb and reflect different wavelengths of light, and each wavelength represents a color in the visible spectrum. Vegetation, for example, absorbs most wavelengths except green and yellow, which they reflect, so most plants look green. Similarly, a red jacket absorbs all wavelengths of color except red, which is reflected.

The lower part of Figure 13–4 shows how the ocean selectively absorbs the longer-wavelength colors (red, orange, and yellow) of visible light. The true colors of objects can be observed in natural light only in the surface waters, because only there can all wavelengths of the visible spectrum be found. Red light is absorbed within the upper 10 meters (33 feet) of the ocean, and yellow is completely absorbed before a depth of 100 meters (330 feet). Thus, the shorter-wavelength portion of the visible spectrum (violet, blue, and some green light) is all that can be transmitted to greater depths,

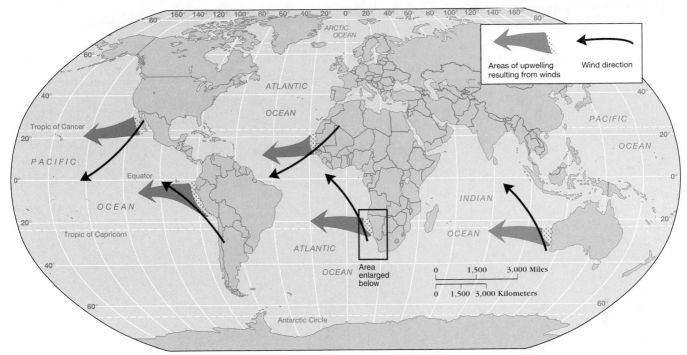

(a)

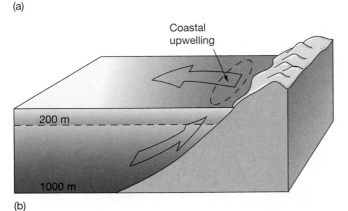

(b)

.01 .02 .03 .05 .1   .2 .3 .5 1   2 3   5 10 15 20 30 50

**Chlorophyll a Concentration
mg/m3**

(c)

Figure 13–3 Coastal upwelling. **(a)** Coastal winds (*black arrows*) cause Ekman transport, which drives surface water away from the west coasts of continents (*blue arrows*). **(b)** Block diagram showing how coastal upwelling is created by surface water that moves away from shore, bringing cold, nutrient-rich water to the surface. **(c)** SeaWiFS image of chlorophyll concentration along the southwest coast of Africa (February 21, 2000). High chlorophyll concentrations indicate high phytoplankton biomass, which is caused by coastal upwelling. Concentration is reported in milligrams of chlorophyll per cubic meter (mg/m^3).

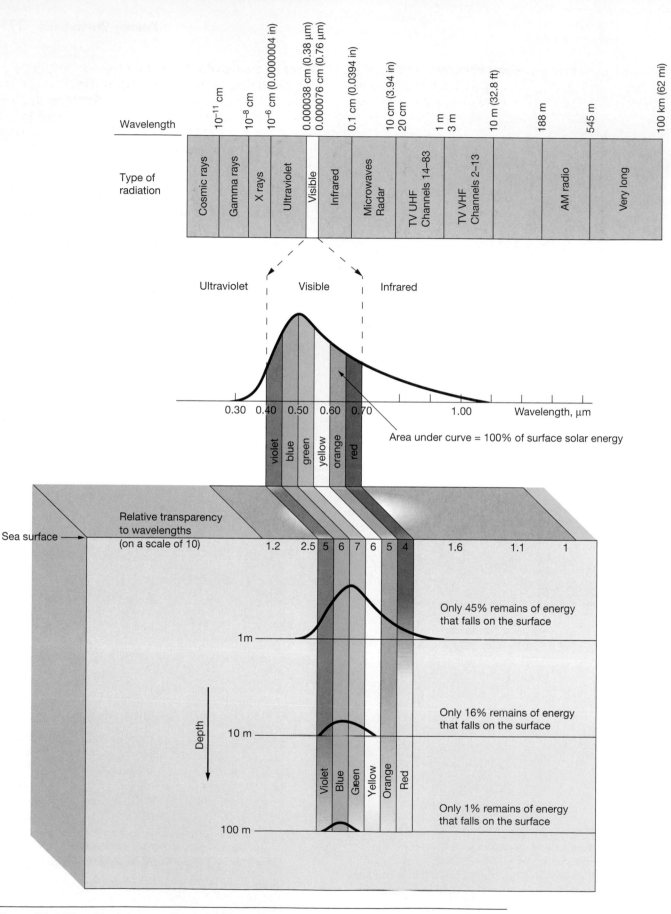

Figure 13–4 The electromagnetic spectrum and transmission of visible light in seawater.
The spectrum (*top*) runs from extremely short cosmic rays (*left side*) with progressively increasing wavelength shown toward the right. The narrow portion of the spectrum that we see as visible light is shown passing through seawater (*bottom*), which absorbs the longer wavelengths (red, orange, and yellow) above a depth of 100 meters.

and even then, their intensity is low. In the open ocean, sunlight strong enough to support photosynthesis occurs only with the euphotic zone to a depth of 100 meters (330 feet) and no sunlight penetrates below a depth of about 1000 meters (3300 feet).

A Secchi disk, like the one shown in Figure 13–5, is used to measure the depth to which visible light penetrates the ocean. The **Secchi** (pronounced "SECK-ee") **disk** is named after its inventor, Angelo Secchi, an Italian astronomer who first used the device to measure water clarity of lakes in 1865. It consists of a round flat disk about 30 centimeters (12 inches) in diameter attached to a line that is marked off at regular intervals. After the disk is lowered into the ocean, the depth at which it can last be seen indicates the water's clarity. Both the amount of microorganisms in the water and the water's turbidity—the amount of suspended material—increase the degree of light absorption, thus decreasing the depth to which visible light can penetrate the ocean.

Figure 13–5 Secchi disk. A Secchi disk is used to measure the depth of penetration of sunlight, and thus indicate the clarity of the water.

Water Color and Life in the Oceans The color of the ocean ranges from deep indigo (blue) to yellow-green. Why are some areas of the ocean blue, whereas others appear green? Ocean color is influenced by (1) the amount of turbidity from runoff and (2) the amount of photosynthetic pigment, which increases with increasing biological production.

Coastal waters and upwelling areas are biologically very productive and almost always yellow-green in color because they contain large amounts of yellow-green microscopic marine algae and suspended particles. These materials disperse solar radiation so that the wavelengths for greenish or yellowish light are scattered most.

Water in the open ocean—particularly in the tropics—lacks productivity (and turbidity), so it is usually a clear, indigo blue color. Water molecules disperse solar radiation so that the wavelength for blue light is scattered most. The atmosphere scatters blue light, too, which is why clear skies are blue.

Satellites can easily detect differences in ocean color, which is related to the water's productivity. Figure 13–6 is a SeaStar satellite/SeaWiFS instrument view of worldwide productivity, showing that highly productive areas, which are called **eutrophic** (*eu* = good, *tropho* = nourishment), are found in shallow-water coastal regions, areas of upwelling, and high-latitude regions. Areas of low productivity, on the other hand, which are called **oligotrophic** (*oligo* = few, *tropho* = nourishment), are found in the open oceans of the tropics.

Photosynthetic Marine Organisms

Many types of marine organisms photosynthesize. Mostly, they are represented by microscopic marine algae, but also include larger forms of algae and seed-bearing plants.

Seed-Bearing Plants (Spermatophyta)

The only members of kingdom Plantae that exist in the marine environment belong to the highest group of plants, the seed-bearing Spermatophyta, and they occur exclusively in shallow coastal areas. Eelgrass (*Zostera*), for example, is a grasslike plant with true roots that exists primarily in the quiet waters of bays and estuaries from the low-tide zone to a depth of 6 meters (20 feet). Surf grass (*Phyllospadix*) (Figure 13–7), which is also a seed-bearing plant with true roots, is typically found in the high-energy environment of exposed rocky coasts from the intertidal zones down to a depth of 15 meters (50 feet).

Other seed-bearing plants are found in salt marshes and include grasses (mostly of the genus *Spartina*), whereas mangrove swamps contain primarily mangroves (genera *Rhizophora* and *Avicennia*). All of

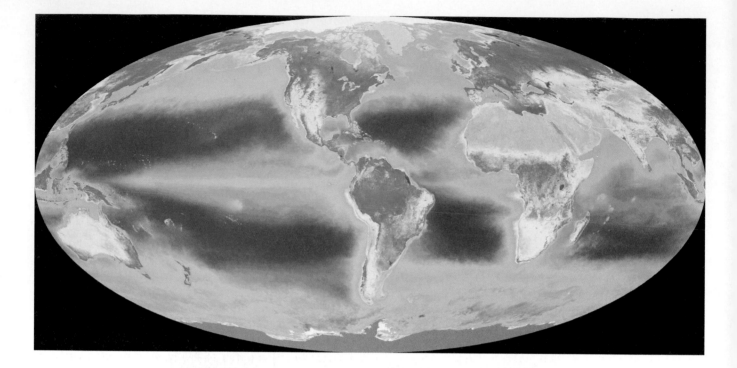

Figure 13–6 **World productivity.** SeaWiFS image of world productivity (September 1997–August 2000). The SeaWiFS instrument aboard the SeaStar satellite detects changes in seawater color caused by changing concentrations of chlorophyll, which varies with the photosynthetic productivity. Ocean concentrations are reported in milligrams of chlorophyll per cubic meter (mg/m³).

Figure 13–7 **Surf grass (***Phyllospadix***).**

these plants are important sources of food and protection for the marine animals that inhabit nearshore environments.

Macroscopic (Large) Algae

Various types of marine macro algae (the "seaweeds") are typically found in shallow waters along the ocean margins. These algae are usually attached to the bottom, but a few species float. Algae are classified in part on the color of the pigment they contain (Figure 13–8).

Brown Algae The brown algae of the phylum Phaeophyta (*phaeo* = dusky, *phytum* = a plant) include the largest members of the attached (not free-floating) species of marine algae. Their color ranges from very

(a)

(b)

(c)

(d)

Figure 13–8 Macroscopic algae. (a) Brown algae, *Sargassum*. This attached form is similar to the floating form that is the namesake of the Sargasso Sea. **(b)** Brown algae, *Macrocystis*, a major component of kelp beds. **(c)** Green algae, *Codium fragile*, also known as sponge weed or dead man's fingers. **(d)** Red algae, *Lithothamnion*, an encrusting form that produces a pink coating on these rounded cobbles.

light brown to black. Brown algae occur primarily in temperate and cold-water areas. Their sizes range widely. The smallest is a black encrusting patch of *Ralfsia* of the upper and middle intertidal zones. The largest is bull kelp (*Pelagophycus*), which may grow in water deeper than 30 meters (100 feet) and extend to the surface. Other types of brown algae include *Sargassum* (Figure 13–8a) and *Macrocystis* (Figure 13–8b).

Green Algae Although green algae of phylum Chlorophyta (*chloro* = green, *phytum* = a plant) are common in freshwater environments, they are not well represented in the ocean. Most marine species are intertidal or grow in shallow bay waters. They contain the pigment chlorophyll, which gives them their green color. They grow only to moderate size, seldom exceeding 30 centimeters (12 inches) in the largest dimension. Forms range from finely branched filaments to thin sheets.

Various species of sea lettuce (*Ulva*), a thin membranous sheet only two cell layers thick, are widely scattered throughout colder-water areas. Sponge weed (*Codium*), a two-branched form more common in warm waters, can exceed 6 meters (20 feet) in length (Figure 13–8c).

Red Algae Red algae of phylum Rhodophyta (*rhodo* = red, *phytum* = a plant) are the most abundant and widespread of marine macroscopic algae. Over 4000 species occur from the very highest intertidal levels to the outer edge of the inner sublittoral zone. Many are attached to the bottom, either as branching forms or as forms that encrust surfaces (Figure 13–8d). They are very rare in fresh water. Red algae range from just barely visible to the unaided eye to 3 meters (10 feet) long. While found in both warm and cold waters, the warm-water varieties are relatively small.

The color of red algae varies considerably depending on its depth in the intertidal or inner sublittoral zones. In upper, well-lighted areas, it may be green to black or purplish. In deeper-water zones, where less light is available, it may be brown to pinkish red.

The bulk of marine photosynthetic productivity occurs within the surface layer of the ocean to a depth of 100 meters (330 feet), which corresponds to the depth of the euphotic zone. At this depth, the amount of light is reduced to 1% of that available at the surface. A red alga, however, has been documented growing at a depth of 268 meters (880 feet) on a seamount near San Salvadore, Bahamas. Available light at this depth was only 0.05% of that available at the surface.

Microscopic (Small) Algae

Microscopic algae produce food either directly or indirectly for over 99% of marine animals. Most microscopic algae are phytoplankton—photosynthetic organisms that live in the upper surface waters and drift with currents—although some live on the bottom in the nearshore environment, where sunlight reaches the shallow ocean floor.

Golden Algae   The golden algae of phylum Chrysophyta (*chrysus* = golden, *phytum* = a plant) contain the orange-yellow pigment **carotin**. They consist of diatoms and coccolithophores, both of which are described in Chapter 4, "Marine Sediments," and they store food as carbohydrates and oils.

Diatoms   The **diatoms** (*diatoma* = cut in half) are a class of algae that are contained in a microscopic shell called a **test** (*testa* = shell). The tests are composed of opaline silica ($SiO_2 \cdot nH_2O$) and are important geologically because they accumulate on the ocean bottom, producing **diatomaceous earth**. Some deposits of diatomaceous earth that have been elevated above the water surface by tectonic forces are mined and used in filtering devices and numerous other applications (see Box 4–1). Diatoms are the most productive group of marine algae.

The tests of diatoms have a variety of shapes, but all have a top and bottom half that fit together (Figure 13–9a). The single cell is contained within this test, and it exchanges nutrients and waste with the surrounding water through holes in its test.

Coccolithophores   **Coccolithophores** (*coccus* = berry; *lithos* = stone; *phorid* = carrying) are covered with small calcareous plates called **coccoliths**, made of calcium carbonate ($CaCO_3$) (Figure 13–9b). The individual plates are about the size of a bacterium, and the entire organism is too small to be captured in plankton nets. Coccolithophores contribute significantly to calcareous deposits in temperate and warmer oceans.

Dinoflagellates   The **dinoflagellates** (*dino* = whirling, *flagellum* = a whip) belong to the phylum Pyrrophyta (*pyrro* = red, *phytum* = a plant) (Figures 13–9c and d). They are the second most productive group of marine algae. They possess **flagella** (small, whip-like structures) for locomotion, giving them a slight capacity to move into areas that are more favorable for photosynthetic

productivity. Dinoflagellates are rarely important geologically because their tests are made of cellulose, which is biodegradable and not preserved as deposits on the sea floor. Many dinoflagellates can bioluminesce and they sometimes exist in great abundance, coloring surface waters red and producing **red tides** (Box 13–1)—which have nothing to do with tidal phenomena—or **harmful algal blooms (HABs)**. In addition, many of the 1100 species undergo structural changes in response to changes in their environment (Box 13–2).

> Marine photosynthetic organisms include seed-bearing plants (such as surf grass), macroscopic algae (seaweeds), and microscopic algae (diatoms, coccolithophores, and dinoflagellates).

═EIO═

For more information and on-line exercises about the Environmental Issue in Oceanography (EIO) "Coastal Population Growth," visit the EIO Web site at **http://www.prenhall.com/oceanissues** and select Issue #1.

Regional Productivity

Primary photosynthetic production in the oceans varies dramatically from place to place (see Figure 13–6). Typical units of photosynthetic production are in weight of carbon (*grams of carbon*) per unit of area (*square meter*) per unit of time (*year*), which is abbreviated as gC/m^2/yr. Values range from as low as 1 gC/m^2/yr in some areas of the open ocean to as much as 4000 gC/m^2/yr in some highly productive coastal estuaries (Table 13–1). This variability is the result of the uneven distribution of nutrients throughout the photosynthetic zone and seasonal changes in the availability of solar energy.[6]

About 90% of the biomass generated in the euphotic (sunlit) zone of the open ocean is decomposed into inorganic nutrients before descending below this zone. The remaining 10% of this organic matter sinks into deeper water, where all but about 1% of it is decomposed. The 1% that reaches the deep-ocean floor accumulates there. The process of removing material from the euphotic zone to the sea floor is called a **biological pump**, because it "pumps" carbon dioxide and nutrients from the upper ocean and concentrates them in deep-sea waters and sea floor sediments.

Throughout much of the subtropical gyres, a permanent **thermocline** (and resulting **pycnocline**[7]) develops. It forms a barrier to vertical mixing, so it prevents the resupply of nutrients to the sunlit surface layer. In the mid-latitudes, a thermocline develops only during

---

[6]For a review of Earth's seasons, see Chapter 6, "Air–Sea Interaction."
[7]Recall that a *thermocline* is a layer of rapidly changing temperature, and a *pycnocline* is a layer of rapidly changing density. Development of ocean thermoclines and pycnoclines is discussed in Chapter 5, "Water and Seawater."

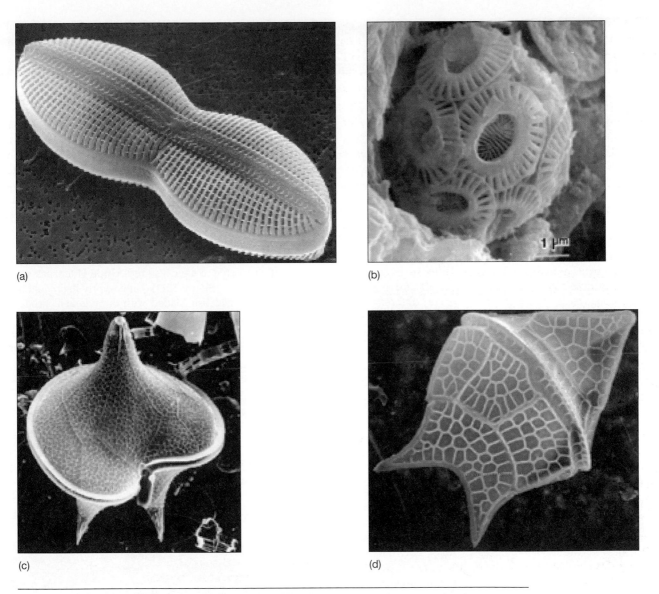

(a)

(b)

(c)

(d)

Figure 13–9 Microscopic algae. **(a)** Peanut-shaped diatom *Diploneis* (length = 50 microns, or 0.002 inch). **(b)** Coccolithophore *Emiliania huxleyi*, showing disk-shaped calcium carbonate ($CaCO_3$) plates—called coccoliths—that cover the organism (bar scale = 1 micron, or 0.00004 inch). **(c)** Dinoflagellate *Protoperidinium divergnes* (length = 70 microns, or 0.003 inch). **(d)** Leaf-like tropical dinoflagellate *Heterodinium whittingae* (length = 100 microns, or 0.004 inch).

the summer season and, in polar regions, a thermocline does not usually develop. The degree to which waters develop a thermocline profoundly affects the patterns of biological production observed at different latitudes.

Productivity in Polar Oceans

Polar regions such as the Arctic Ocean's Barents Sea, which is off the northern coast of Europe, experiences continuous darkness for about three months of winter and continuous illumination for about three months during summer. Diatom productivity peaks there during May (Figure 13–10a), when the Sun rises high enough in the sky so that there is deep penetration of sunlight into the water. As soon as the diatoms develop, zooplankton—mostly small crustaceans (Figure 13–10d)—begins feeding on them. The zooplankton biomass peaks in June and continues at a relatively high level until winter darkness begins in October.

In the Antarctic region—particularly at the southern end of the Atlantic Ocean—productivity is somewhat greater. This is caused by the upwelling of North Atlantic Deep Water, which forms on the opposite side of the ocean basin where it sinks and moves southward below the surface. Hundreds of years later, it rises to the surface near Antarctica, carrying with it high concentrations of nutrients (Figure 13–10b). When the summer Sun provides sufficient radiation, there is an explosion of biological productivity.

Blue whales, the largest of all whales (see Figure 14–19), eat mostly zooplankton and time their migration through temperate and polar oceans to coincide

Box 13–1
Red Tides: Was Alfred Hitchcock's *The Birds* Based on Fact?

Conditions in the oceans sometimes stimulate the productivity of certain dinoflagellates. During these times, up to 2 million dinoflagellates may be found in 1 liter (about 1 quart) of water, giving the water a reddish color and causing what is known as a red tide (Figure 13B). Red tides are by no means a new phenomenon. In fact, the Old Testament makes reference to waters turning blood red, which is most likely how the Red Sea got its name.

Although many red tides are harmless to marine animals and humans, they can still be responsible for mass die-offs of marine organisms. When huge numbers of dinoflagellates die, oxygen is removed from seawater during decomposition and many types of marine life literally suffocate to death. In other cases, dinoflagellates that are responsible for many red tides produce toxins that can spread to many different types of organisms—including humans (Figure 13C). *Ptychodiscus* and *Gonyaulax*, for example, are two common genera of dinoflagellates in red tides that produce water-soluble toxins. Certain filter-feeding shellfish called bivalves—various clams, mussels, and oysters—then strain the dinoflagellates from the water for food. *Ptychodiscus* toxin kills fish and shellfish. *Gonyaulax* toxin is not poisonous to shellfish, but it concentrates in their tissues and is poisonous to humans who eat the shellfish, even after the shellfish are cooked. This malady is called **paralytic shellfish poisoning (PSP)**.

The symptoms of PSP in humans are similar to those of drunkenness—incoherent speech, uncoordinated movement, dizziness, and nausea—and can occur only 30 minutes after ingesting contaminated shellfish. There is no known antidote for the toxin, which attacks the human nervous system, but the critical period usually passes within 24 hours. At least 300 fatal and 1750 nonfatal cases of PSP have been documented worldwide.

April through September are particularly dangerous months for red tides in the Northern Hemisphere. In most areas, quarantines exist to prohibit harvesting those shellfish that feed on toxic microscopic organisms.

Worldwide, increasing numbers of mysterious poisonings are implicating species of toxic dinoflagellates that can spread throughout the marine food web and to humans. For instance, domoic acid—a toxin produced by a diatom (*Pseudonitzschia*)—was first recognized in 1987 as the poison that infected over 100 people who ate contaminated mussels from Prince Edward Island, Canada. Four of the victims died and 10 suffered permanent memory loss, which lead researchers to call poisoning by domoic acid **amnesic shellfish poisoning**.

Domoic acid poisoning may also be responsible for birds attacking humans in California's Monterey Bay in 1961. The incident—reported to have inspired Alfred Hitchcock's 1963 movie, *The Birds* (Figure 13D)—apparently happened after seagulls consumed fish loaded with domoic acid produced by a bloom of diatoms. Infected by the poison, the birds smashed into structures during flight and pecked eight people. In the same area in September 1991, brown pelicans and

Figure 13B Red tide.

with maximum zooplankton productivity. This enables the whales to develop and support calves that can exceed 7 meters (23 feet) in length at birth. The mother blue whale suckles the calf with rich, high-fat milk for six months. By the time the calf is weaned, it is over 16 meters (50 feet) long. In two years, it will be 23 meters (75 feet) long, and after about three years, it will weigh 60 tons! This phenomenal growth rate gives some indication of the enormous biomass of small copepods and krill upon which these large mammals feed.[8]

Density and temperature change very little with depth in polar regions (Figure 13–10c), so these waters are **isothermal** (*iso* = same, *thermo* = temperature) and there is no barrier to mixing between surface waters and

deeper, nutrient-rich waters. In the summer, however, melting ice creates a thin, low-salinity layer that does not readily mix with the deeper waters. This stratification is crucial to summer production, because it helps prevent phytoplankton from being carried into deeper, darker waters. Instead, they are concentrated in the sunlit surface waters where they reproduce continuously.

Nutrient concentrations (phosphates and nitrates) are usually adequate in high-latitude surface waters, so the availability of solar energy limits photosynthetic productivity in these areas more than the availability of nutrients.

Productivity in Tropical Oceans

Perhaps surprisingly, productivity is low in tropical regions of the open ocean. Because the Sun is more

---

[8]As an analogy, consider how many ants you would have to eat as a child to grow to adult size.

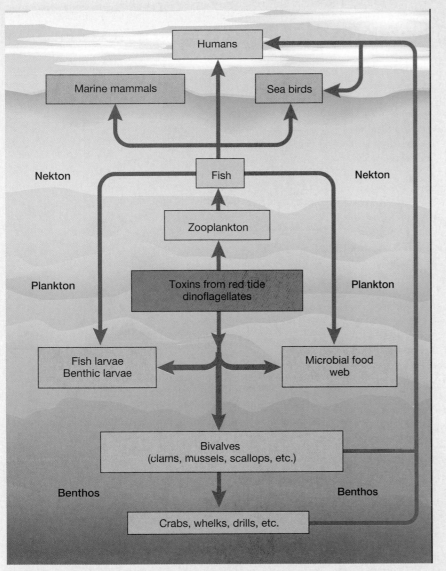

Figure 13C Routes through which dinoflagellate toxins spread to marine organisms and humans.

Brandt's cormorants exhibited similar behavior—they acted drunk, swam in circles, and made loud squawking sounds. Research revealed that toxic diatoms (*Pseudonitzschia australis*) were ingested by unaffected anchovies that were subsequently eaten by birds, causing more than 100 birds to wash up dead at the shore. In 1998, another toxic diatom bloom in Monterey Bay caused the death of 400 California sea lions that ate fish contaminated with domoic acid.

Red tides are affecting larger areas and more marine species (including fish, dolphins, and humpback whales) are succumbing to the toxins. In fact, toxic species of dinoflagellates have been implicated in several group strandings of marine mammals. Human activities have been shown to contribute to red tides when excess nutrients in the form of fertilizer, sewage, and animal waste make their way into coastal waters, producing ocean **eutrophication** (*eu* = good, *tropho* = nourishment, *ation* = action). Eutrophication is the enrichment of waters by a previously scarce nutrient and can result in dangerous phytoplankton blooms. In addition, red tides appear to be occurring more frequently around the globe. It is also possible, however, that the system for reporting them has simply improved.

Figure 13D Actress Tippi Hedren fights off a seagull in Hitchcock's 1963 classic, *The Birds*.

Box 13–2
Pfiesteria: A Morphing Peril to Fish and Humans

In 1995, the death of 15 million fish—mostly Atlantic menhaden (*Brevoortia tyrannus*)—in North Carolina's Neuse Estuary (Figure 13E) was traced to an outbreak of ***Pfiesteria piscicida***, a toxic dinoflagellate that caused a similar fish kill in nearby Pamlico Sound in 1991. This and related dinoflagellate species kill millions of fish yearly in coastal waters off North Carolina and in Chesapeake Bay by undermining the ability of fish to reproduce and resist disease.

P. piscicida is known to change into as many as 24 distinct forms (Figure 13F) within a life cycle that can be initiated by the presence of fish. In water that is at least 26°C (79°F), cysts lie dormant in sediments and produce nontoxic zoospores that feed on other microorganisms present in the water. When fish are present, however, the nontoxic zoospores become toxic and attach to fish, where they drug the fish, destroy their skin, and weaken their resistance to disease-causing bacteria and fungi. The spores then feed on the material oozing from sores that develop on the skin of infected fish. Toxic zoospores reproduce asexually and also produce gametes that merge to produce planozygotes that resemble the zoospores. When the fish die, the zoospores and planozygotes change into amoebae that gorge on the dead fish. In colder waters, amoebae rise from the sediment when fish are sensed, attack the fish with toxins and return to the bottom to feed on the fish when it dies. Unable to perform photosynthesis on their own, zoospores can save chloroplasts from algae they ingest in their nontoxic form and use them for weeks to generate food.

Pfiesteria outbreaks are initiated by an overabundance of nutrients such as nitrogen that allow a proliferation of algae that serve as a food sources for zoospores. These zoospores

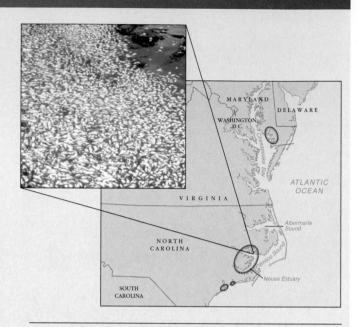

Figure 13E Fish kills related to *Pfiesteria* outbreaks. As many as 100,000 fish died in Pamlico Sound (*inset*) during an outbreak of the dinoflagellate *Pfiesteria* in 1991. Since then, similar fish kills related to *Pfiesteria* have occurred in Chesapeake Bay and along the North Carolina coast (*red circled areas*).

---

directly overhead, light penetrates much deeper into tropical oceans than temperate and polar waters, and solar energy is available year round, but productivity is low in tropical regions of the open ocean because a permanent thermocline produces a stratification (layering) of water masses. This prevents mixing between surface waters and nutrient-rich deeper waters, effectively eliminating any supply of nutrients from deeper waters below (Figure 13–11).

At about 20 degrees north and south latitude, phosphate and nitrate concentrations are commonly less than $1/100$ of their concentrations in temperate oceans during winter. In fact, nutrient-rich waters in the tropics lie below 150 meters (500 feet), with the highest concentrations between 500 and 1000 meters (1640 and 3300 feet). So, productivity in tropical regions is limited by the lack of nutrients (unlike polar regions, where productivity is limited by the lack of sunlight).

Generally, primary production in tropical oceans occurs at a steady but rather low rate. The total annual production of tropical oceans is only about half of that found in temperate oceans.

Exceptions to the general pattern of low productivity in topical oceans include:

- *Equatorial upwelling.* Where trade winds drive westerly equatorial currents on either side of the Equator, Ekman transport causes surface water to diverge toward higher latitudes (see Figure 7–9). This surface water is replaced by nutrient-rich water from depths of up to 200 meters (660 feet). Equatorial upwelling is best developed in the eastern Pacific Ocean.

- *Coastal upwelling.* Where the prevailing winds blow toward the Equator and along western continental margins, surface waters are driven away from the coast. They are replaced by nutrient-rich waters from depths of 200 to 900 meters (660 to 2950 feet). This upwelling promotes high primary production along the west coasts of continents (see Figure 13–3), which can support large fisheries.

- *Coral reefs.* Organisms that comprise and live among coral reefs are superbly adapted to low-nutrient conditions, similar to the way certain organisms are then reproduce rapidly and are ready to attack fish when they arrive in coastal waters.

Humans can suffer from *Pfiesteria*, too, but not from eating *Pfiesteria*-contaminated fish. Harm comes to humans from getting toxin-laden water on their skin or even breathing air over the toxic waters. Researchers and others exposed to the *Pfiesteria*-infested water often reported symptoms such as shortness of breath, itchy or burning eyes, headaches, and forgetfulness.

The *Pfiesteria* outbreaks are part of a generally worsening pattern of increasing harmful algae blooms and ciguat-

era, which is caused by dinoflagellate toxins accumulating in tropical fish including barracuda, red snapper and grouper and causes more human illness than any other form of seafood poisoning. Factors that lead to harmful algae blooms include poor water quality in streams that enter the ocean due to runoff of nutrients and other chemicals, destruction of wetlands that naturally filter water from these streams, and the increasing human population in coastal areas. Sickness in our coastal waters must be taken seriously, as it affects the quality of life for fish in coastal waters, which is linked to the quality of life for humans.

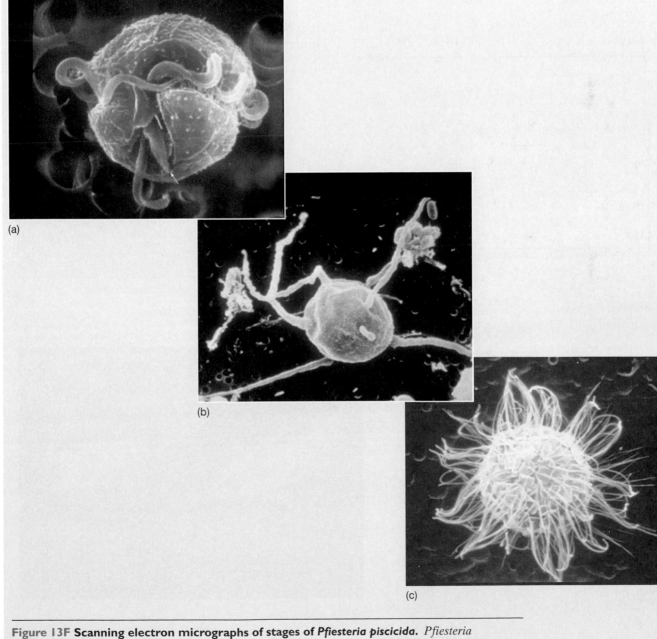

(a)

(b)

(c)

Figure 13F Scanning electron micrographs of stages of *Pfiesteria piscicida*. *Pfiesteria* undergoes numerous transformations during the stages of its life cycle, some of which are shown here. **(a)** The toxic zoospore stage, when *Pfiesteria* releases toxins that kill fish. **(b)** The amoeba stage, when *Pfiesteria* feed on dead fish. **(c)** The cyst stage, when *Pfiesteria* is covered with bristles and settles to the bottom after a fish kill.

Table 13–1   Values of net primary productivity for various ecosystems.

	Ecosystem	Range (gC/m²/yr)	Average (gC/m²/yr)
Oceanic	Algae beds and coral reefs	1000–3000	2000
	Estuaries	500–4000	1800
	Upwelling zone	400–1000	500
	Continental shelf	300–600	360
	Open ocean	1–400	125
Land	Freshwater swamp and marsh	800–4000	2500
	Tropical rainforest	1000–5000	2000
	Mid-latitude forest	600–2500	1300
	Cultivated land	100–4000	650

adapted to desert life on land. Symbiotic algae living within the tissues of coral and other species allow coral reefs to be highly productive ecosystems. Coral reefs also tend to retain and concentrate what little nutrients exist. Coral reef ecosystems are discussed further in Chapter 15, "Animals of the Benthic Environment."

Productivity in Temperate Oceans

Productivity is limited by available sunlight in polar regions and by nutrient supply in the low-latitude tropics. In temperate (mid-latitude) regions, a combination of these two limiting factors controls productivity as shown in Figure 13–12a (which shows the pattern for the Northern Hemisphere; in the Southern Hemisphere, the seasons are reversed).

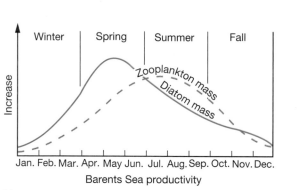

(a)

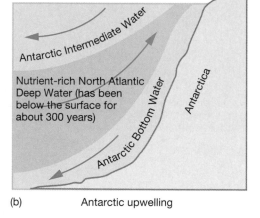

(b)   Antarctic upwelling

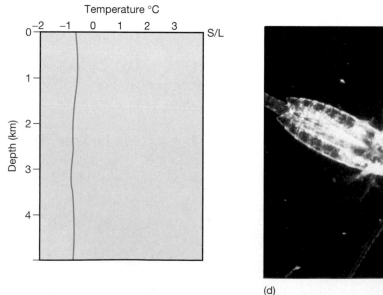

(c)

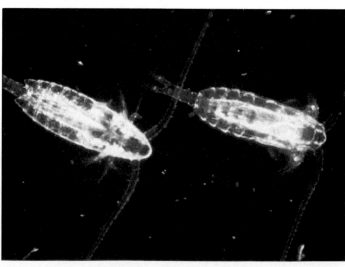

(d)

Figure 13–10 Productivity in polar oceans. (a) A springtime increase of diatom mass is followed closely by an increase in zooplankton abundance. **(b).** The continuous upwelling of North Atlantic Deep Water keeps Antarctic waters rich in nutrients **(c)** Polar water shows nearly uniform temperature with depth (an isothermal water column). **(d)** Copepods of the genus *Calanus*, each about 8 millimeters (0.3 inch) in length.

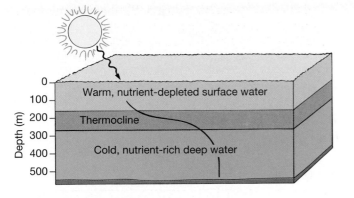

Figure 13–11 Productivity in tropical oceans. Although tropical regions receive adequate sunlight year-round, a permanent thermocline prevents the mixing of surface and deep water. As phytoplankton consume nutrients in the surface layer, productivity is limited because the thermocline prevents replenishment of nutrients from deeper water. Thus, productivity remains at a steady, low level.

Winter Productivity in temperate oceans is very low during winter, even though nutrient concentration is *highest* at this time (Figure 13–12a). The water column is isothermal, too, similar to polar regions, so nutrients are well distributed throughout the water column. Figure 13–12b (*winter*) shows, however, that the Sun is at its lowest position above the horizon during winter, so a high percentage of the available solar energy is reflected, leaving only a small percentage to be absorbed into surface waters. As a result, the compensation depth for photosynthesis is so shallow that phytoplankton do not grow much. The absence of a thermocline, moreover, allows algal cells to be carried down beneath the euphotic zone for extended periods by turbulence associated with winter waves.

Spring The Sun rises higher in the sky during spring [Figure 13–12b (*spring*)], so the compensation depth for photosynthesis deepens. A **spring bloom** of phytoplankton occurs because solar energy and nutrients are available, and a seasonal thermocline develops (due to increased solar heating) that traps algae in the euphotic zone (Figure 13–12a). This creates a tremendous demand for nutrients in the euphotic zone, so the supply becomes limited, causing productivity to decrease sharply. Even though the days are lengthening and sunlight is increasing, productivity during the spring bloom is limited by the lack of nutrients. In most areas of the Northern Hemisphere, therefore, phytoplankton populations decrease in April due to insufficient nutrients and because their population is being consumed by zooplankton (grazers).

Summer The Sun rises even higher in the summer [Figure 13–12b (*summer*)], so surface waters in temperate parts of the ocean continue to warm. A strong seasonal thermocline is created at a depth of about 15 meters (50 feet). The thermocline, in turn, prevents vertical mixing, so nutrients depleted from surface waters cannot be replaced by those from deeper waters. Throughout summer, the phytoplankton population remains relatively low (Figure 13–12a). Even though the compensation depth for photosynthesis is at its maximum, phytoplankton can actually become scarce in late summer.

Fall Solar radiation diminishes in the fall as the Sun drops lower in the sky [Figure 13–12b (*fall*)], so surface temperatures drop and the summer thermocline breaks down. Nutrients return to the surface layer as increased wind strength mixes surface waters with deeper waters. These conditions create a **fall bloom** of phytoplankton, which is much less dramatic than the spring bloom (Figure 13–12a). The fall bloom is very short-lived because sunlight (not nutrient supply, as in the spring bloom) becomes the limiting factor as winter approaches to repeat the seasonal cycle.

Figure 13–13 compares the seasonal variation in phytoplankton biomass of tropical, north polar, and north temperate regions, where the total area under each curve represents photosynthetic productivity. The figure shows that the highest overall productivity occurs in the temperate region.

> In polar regions, productivity peaks during the summer and is limited by sunlight. In tropical regions, productivity is low year-round and is limited by nutrients. In temperate regions, productivity varies seasonally due to a combination of sunlight and nutrients and produces a spring and fall phytoplankton bloom.

Energy Flow

Energy flow is not a *cycle* but a *unidirectional flow* that begins with a constant input of solar energy and ends with a high level of **entropy** (*en* = in, *trope* = transformation), a state when energy is so randomly distributed it can no longer do work.

Energy Flow in Marine Ecosystems

A **biotic community** is the assemblage of organisms that live together within some definable area. An **ecosystem** includes the biotic community plus the environment with which it exchanges energy and chemical substances. A kelp forest biotic community, for instance, includes all organisms living within or near the kelp and receiving some benefit from it. A kelp forest ecosystem, on the other hand, includes all those organisms plus the surrounding seawater, the hard substrate onto which the kelp is attached, and the atmosphere where gases are exchanged.

In an algae-supported biotic community (Figure 13–14), energy enters the system when algae absorb solar radiation. Photosynthesis converts this solar energy into chemical energy, which is used for the algae's

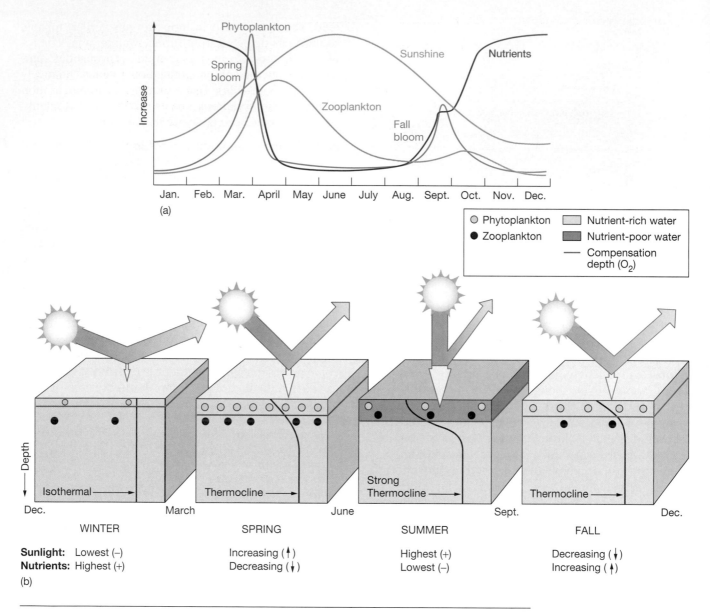

Figure 13–12 Productivity in temperate oceans (Northern Hemisphere). (a) Relationship among phytoplankton, zooplankton, amount of sunshine, and nutrient levels for surface waters in northern temperate latitudes. **(b)** The seasonal cycle of sunlight affects the presence and depth of the thermocline, which affects the availability of nutrients. This, in turn, affects the abundance of phytoplankton and other organisms such as zooplankton that rely on phytoplankton for food.

respiration. This chemical energy is also passed on to the animals that consume the algae for their growth and other life functions. The animals expend mechanical and heat energy, which are progressively less recoverable forms of energy. Finally, the residual energy becomes biologically useless as entropy increases.

Generally, three basic categories of organisms exist within an ecosystem: **producers**, **consumers**, and **decomposers** (Figure 13–14). Algae, archaeons, and certain bacteria are called **autotrophic** (*auto* = self, *tropho* = nourishment) producers because they can nourish themselves through chemosynthesis or photosynthesis. Consumers and decomposers are called **heterotrophic** (*hetero* = different, *tropho* = nourishment) organisms because they depend on the organic compounds produced by the autotrophs for their food supply.

Consumers may be categorized as **herbivores** (*herba* = grass, *vora* = eat), which feed directly on plants or algae; **carnivores** (*carni* = meat, *vora* = eat), which feed only on other animals; **omnivores** (*omni* = all, *vora* = eat), which feed on both; and **bacteriovores** (*bacterio* = bacteria, *vora* = eat), which feed only on bacteria.

Decomposers such as bacteria break down organic compounds that comprise **detritus** (*detritus* = to lessen)—dead and decaying remains and waste products of organisms—for their own energy requirements. In the decomposition process, compounds are released that are again available for use by algae and plants as nutrients.

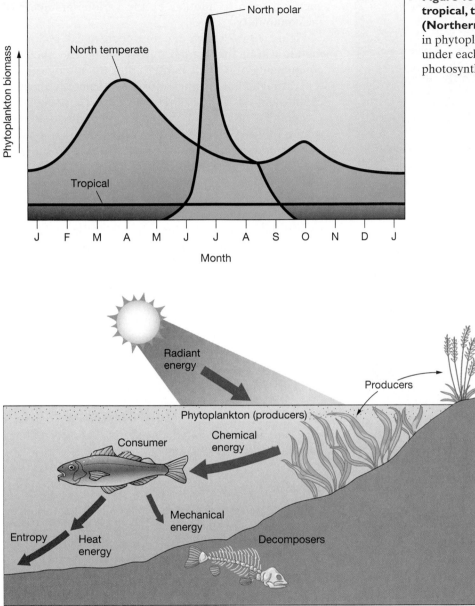

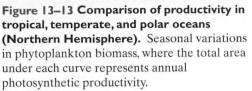

Figure 13–13 Comparison of productivity in tropical, temperate, and polar oceans (Northern Hemisphere). Seasonal variations in phytoplankton biomass, where the total area under each curve represents annual photosynthetic productivity.

Figure 13–14 Energy flow through a photosynthetic marine ecosystem. Energy enters a marine ecosystem as radiant solar energy and is converted to chemical energy through photosynthesis by producers. Metabolism in the fish (a consumer) then releases the chemical energy for conversion to mechanical energy. Energy also is lost from the biotic community (the algae and fish) as heat, which increases the entropy of the ecosystem. Decomposers work to break down the remaining energy after an organism dies.

Symbiosis

Symbiosis (*sym* = together, *bios* = life) occurs when two or more organisms associate in a way that benefits at least one of them. Symbiotic relationships are classified as commensalism, mutualism, or parasitism.

In **commensalism** (*commensal* = sharing a meal, *ism* = process), a smaller or less dominant participant benefits without harming its host, which affords subsistence or protection to the other. Remoras, for example, attach themselves to a shark or other fish to obtain food and transportation, generally without harming its host (Figure 13–15a).

In **mutualism** (*mutuus* = borrowed, *ism* = process), both participants benefit. For example, the stinging tentacles of the sea anemone protect the clownfish and the clown fish serves as bait to draw other fish within reach of the anemone's tentacles (Figure 13–15b).

In **parasitism** (*parasitos* = person who eats at some else's table, *ism* = process), one participant (the parasite) benefits at the expense of the other (the host). Many fish are hosts to isopods, which attach to the fish and derive their nutrition from the body fluids of the fish, thereby robbing the host of some of its energy supply (Figure 13–15c). Usually, the parasite does not rob enough energy to kill the host because if the host dies, so does the parasite.

Biogeochemical Cycling

Unlike the noncyclic, unidirectional flow of energy through a biotic community, the flow of nutrients depends on **biogeochemical cycles**.[9] That is, matter does

[9]Biogeochemical cycles are so named because they involve biological, geological (Earth processes), and chemical components.

(a)

(b)

(c)

Figure 13–15 Symbiosis. (a) Commensalism, when an organism benefits without harming its host, such as this remora attached to a Caribbean grouper. **(b)** Mutualism, when both participants benefit, such as this clown fish and sea anemone. **(c)** Parasitism, when one participant benefits at the expense of the other, such as this isopod that has attached itself to the head of a blackbar soldierfish.

not dissipate (as energy does) but is *cycled* from one chemical form to another by the various members of the community.

Figure 13–16 shows the biogeochemical cycling of matter within the marine environment. The chemical components of organic matter enter the biological system through photosynthesis (or, less commonly, through chemosynthesis at hydrothermal vents). These chemical components are passed on to animal populations (consumers) through feeding. When organisms die, some of the material is used and reused within the euphotic zone, while some sinks as detritus. Some of this detritus feeds organisms living in deep water or on the sea floor, while some undergoes bacterial or other decomposition processes that convert organic remains into useable nutrients (nitrates and phosphates). When upwelling hoists these nutrients to the surface again, they can be used by algae and plants to begin the cycle anew.

Trophic Levels and Biomass Pyramids

As producers make food (organic matter) available to the consuming animals of the ocean, it passes from one feeding population to the next. Only a small percentage of the energy taken in at any level is passed on to the next because energy is consumed and lost at each level. As a result, the producers' biomass in the ocean is many times greater than the mass of the top consumers, such as sharks or whales.

Trophic Levels

Chemical energy stored in the mass of the ocean's algae (the "grass of the sea") is transferred to the animal community mostly through feeding. Zooplankton are *herbivores*, like cows,[10] so they eat diatoms and other microscopic marine algae. Larger herbivores feed on the larger algae and marine plants that grow attached to the ocean bottom near shore.

The herbivores are then eaten by larger animals, the *carnivores*. They, in turn, are eaten by another population of larger carnivores, and so on. Each of these feeding stages is called a **trophic** (*tropho* = nourishment) **level**.

Generally, individual members of a feeding population are larger—but not too much larger—than the organisms they eat. There are conspicuous exceptions, however, such as the blue whale. At 30 meters (100 feet) long, it is possibly the largest animal that has ever existed on Earth, yet it feeds on krill, which have a maximum length of only 6 centimeters (2.4 inches).

The transfer of energy from one population to another is a continuous *flow* of energy. Small-scale recycling and storage interrupts the flow, which slows the

---

[10]In fact, zooplankton might be considered the miniature drifting "cows of the sea."

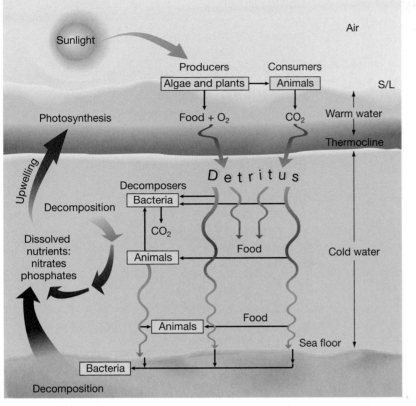

Figure 13–16 Biogeochemical cycling of matter. The chemical components of organic matter enter the biological system through photosynthesis and are passed on to consumers through feeding. Detritus sinks and feeds organisms living below the surface or undergoes decomposition, which returns nutrients to the water that can be hoisted to the surface by upwelling.

conversion of potential (chemical) energy to kinetic energy, then to heat energy, and finally to entropy.

Transfer Efficiency

The transfer of energy between trophic levels is very inefficient. The efficiencies of different algal species vary, but the average is only about *2%*, which means that 2% of the light energy absorbed by algae is ultimately synthesized into food and made available to herbivores.

The **gross ecological efficiency** at any trophic level is *the ratio of energy passed on to the next higher trophic level divided by the energy received from the trophic level below.* The ecological efficiency of herbivorous anchovies, for example, would be the energy consumed by carnivorous tuna that feed on the anchovies divided by the energy contained in the phytoplankton that the anchovies consumed.

Figure 13–17 shows that some of the chemical energy taken in as food by herbivores is excreted as feces and the rest is assimilated. Of the assimilated chemical energy, much is converted through respiration to kinetic energy for maintaining life, and what remains is available for growth and reproduction. Thus, only about 10% of the food mass consumed by herbivores is available to the next trophic level.

Figure 13–18 shows the passage of energy between trophic levels through an entire ecosystem, from the solar energy assimilated by phytoplankton through all trophic levels to the ultimate carnivore—humans. Be-

cause energy is lost at each trophic level, it takes thousands of smaller marine organisms to produce a *single* fish that is so easily consumed during dinner!

The efficiency of energy transfer between trophic levels depends on many variables. Young animals, for example, have a higher growth efficiency than older animals. In addition, when food is plentiful, animals expend more energy in digestion and assimilation than when food is scarce.

Most ecological efficiencies in natural ecosystems range between 6% and 15%, and average about 10%. There is some evidence, however, that ecological efficiencies in populations important to present fisheries may run as high as 20%. The true value of this efficiency is of practical importance because it determines the size of the fish harvest that can be safely taken from the oceans without damaging the ecosystem.

> The transfer of energy between various trophic levels operates at low efficiencies, averaging only 2% for marine algae and 10% for most other levels.

Biomass Pyramid

The loss of energy between each feeding population limits the number of feeding populations in an ecosystem. If there were too many levels, there would not be

Figure 13–17 Passage of energy through a trophic level. As food mass initially produced by phytoplankton passes from herbivores to carnivores, a large percentage is excreted by feces, used during respiration, or dies uneaten. Thus, only about 10% of the food mass consumed by herbivores is available for consumption by carnivores.

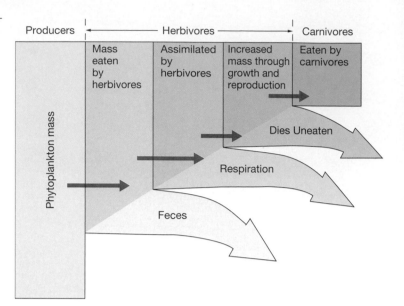

enough energy to support the organisms in higher and higher trophic levels. In addition, each feeding population necessarily must have less mass than the population it eats. As a result, individual members of a feeding population are generally *larger in size* and *less numerous* than their prey.

Food Chains   A **food chain** is a sequence of organisms through which energy is transferred, starting with an organism that is the primary producer, then a herbivore, then one or more carnivores, finally culminating with the "top carnivore," which is not usually preyed upon by any other organism.

Because energy transfer between trophic levels is inefficient, it is advantageous for fishers to choose a population that feeds as close to the primary producing population as possible. This increases the biomass available for food and the number of individuals available to be taken by the fishery. Newfoundland herring, for example, are an important fishery that usually represents the third trophic level in a food chain. They feed primarily on small crustaceans (copepods) that feed, in turn, upon diatoms (Figure 13–19a).

Food Webs   Feeding relationships are rarely as simple as that of the Newfoundland herring. More often, top carnivores in a food chain feed on a number of different animals, each of which has its own simple or complex feeding relationships. This constitutes a **food web**, as shown in Figure 13–19b for North Sea herring.

Animals that feed through a food web rather than a food chain are more likely to survive because they have alternative foods to eat should one of their food sources diminish in quantity or even disappear. Newfoundland herring, on the other hand, eat only copepods, so the disappearance of copepods would catastrophically effect their population.

Conversely, Newfoundland herring are more likely to have a larger biomass to eat, because they are only

two steps removed from the producers, whereas North Sea herring are three steps removed in some of the food chains within their web.

The ultimate effect of energy transfer between trophic levels can be seen in the **biomass pyramid** in Figure 13–20. The *number of individuals* and *total biomass* decrease at successive trophic levels because the amount of available energy decreases. The figure also shows that organisms *increase in size* at successive tropic levels up the food web.

A food chain is a linear feeding relationship among producers and one or more consumers. A food web is a branching network of feeding relationships among many different organisms.

◄EIO►

For more information and on-line exercises about the Environmental Issue in Oceanography (EIO) " Lifestyles of the Large and Blubbery: How to Grow a Blue Whale," visit the EIO Web site at **http://www. prenhall.com/oceanissues** and select Issue #8.

Ecosystems and Fisheries

Since well before the beginning of recorded history, humans have used the sea as a source of food. Over the last several decades, **fisheries** (fish caught from the ocean by commercial fishers) have provided about 16% of the protein consumed by the world's population.

Figure 13–21 shows that the world marine fishery is drawn from five ecosystems. In descending order, they are: (1) nontropical shelves, (2) tropical shelves, (3) upwellings, (4) coastal and coral systems, and (5) open ocean. The largest proportion of the marine fishery is found in highly productive shallow shelf and coastal waters, whereas low productivity open ocean areas

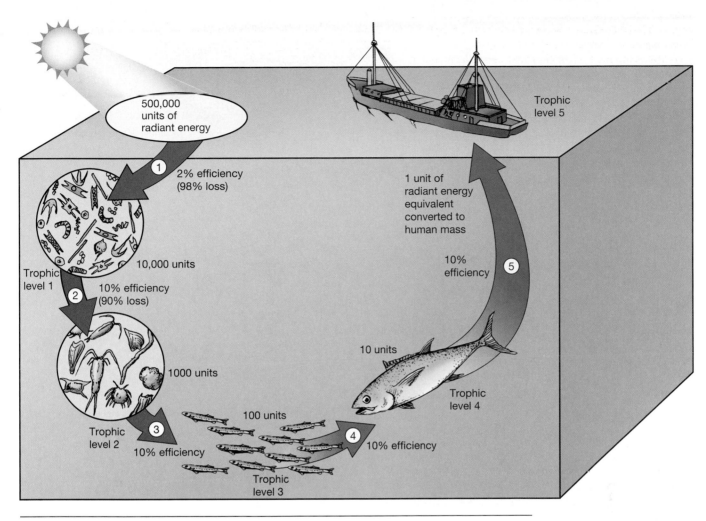

Figure 13–18 Ecosystem energy flow and efficiency. For every 500,000 units of radiant energy input available to the producers (phytoplankton), only one unit of equivalent mass is added to the fifth trophic level (humans). Average phytoplankton transfer efficiency is 2% (98% loss) and all other trophic levels average 10% efficiency (90% loss).

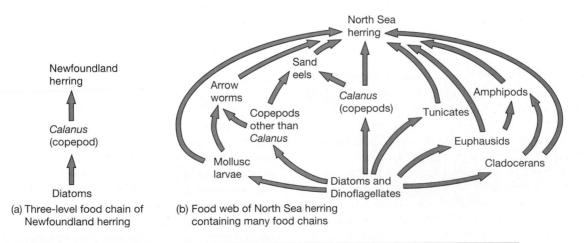

Figure 13–19 Comparison between a food chain and a food web. (a) A food chain, showing the passage of energy along a single path, such as from diatoms to copepods to Newfoundland herring in three trophic levels. **(b)** A food web, showing multiple paths for food sources of the North Sea herring, which may be at the third or fourth trophic level.

Figure 13–20 Biomass pyramid. A huge mass of phytoplankton constitutes the base of the biomass pyramid. At each step up the pyramid, there are larger organisms but fewer individuals and a smaller total biomass because transfer efficiency between steps averages only 10%.

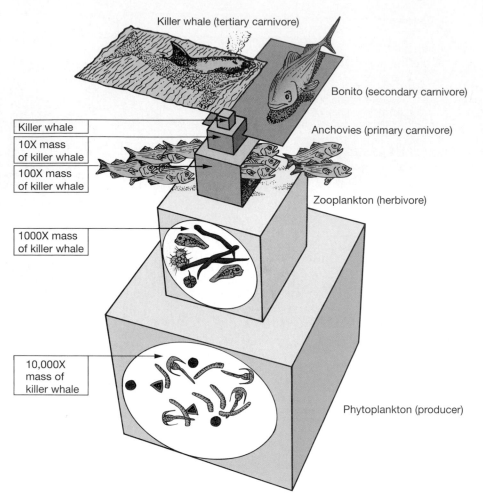

Killer whale (tertiary carnivore)

Bonito (secondary carnivore)

Anchovies (primary carnivore)

Zooplankton (herbivore)

Phytoplankton (producer)

Killer whale

10X mass of killer whale

100X mass of killer whale

1000X mass of killer whale

10,000X mass of killer whale

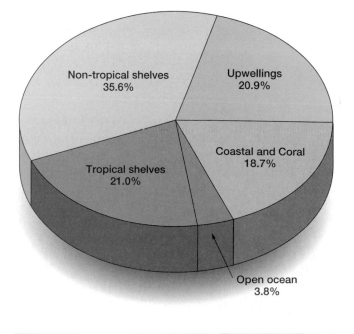

Non-tropical shelves 35.6%

Upwellings 20.9%

Coastal and Coral 18.7%

Tropical shelves 21.0%

Open ocean 3.8%

Figure 13–21 Marine fishery ecosystems and their contribution to the total world fishery.

comprise only 3.8% of the total. Nearly 21% of the world's total catch is from very highly productive upwelling areas, which represent only about 0.1% of the ocean surface area.

Fisheries harvest from the **standing stock** of a population, which is the mass present in an ecosystem at a given time. Successful fisheries leave enough individuals from the standing stock to repopulate the ecosystem after fisheries have made their harvest. **Overfishing** occurs when adult fish are harvested faster than their natural rate of reproduction, resulting in the decline of marine fish populations and the reduction of a fishery's **maximum sustainable yield** (the maximum fishery biomass that can be removed yearly and still be sustained by the fishery ecosystem). Worldwide, about 30% of fish stocks are now depleted, while another 40% of fish stocks are being fished at their biological limit. In U.S. waters, 80% of 191 commercial stocks are fully exploited or overfished. Overfishing can only be reversed by curtailing fish harvests.

Figure 13–22 shows the world total fish production in marine waters since 1950. After increasing steadily for more than 30 years, the world catch of ocean fish declined in 1990. After that decline, it peaked again in 1997 at 93.6 million metric tons (103 million short

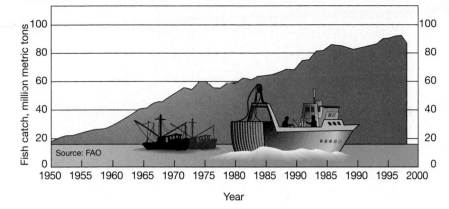

Figure 13–22 World total fish production in marine waters since 1950.

tons). The National Research Council estimated in 1999 that the maximum catch that might be expected from the ocean is about 100 million metric tons (110 million short tons). This amount might be exceeded if new species—such as Antarctic krill, squid, and pelagic crabs—become significant new fisheries in the future.

Incidental Catch

Incidental catch or **bycatch** includes any marine organisms that are caught incidentally by fishers seeking commercial species. On average, close to one-fourth of the catch is discarded, although for some fisheries, such as shrimp, the incidental catch may be up to eight times larger than the catch of the target species. Incidental catch includes birds, turtles, sharks, and dolphins, as well as many species of non-commercial fish. In most cases, these animals die before they are thrown back overboard, even though some of them are protected by United States and international law. Each year, an estimated 25 million metric tons (27.5 short tons) of bycatch is produced by the fishing industry.

Tuna and Dolphins Schools of yellowfin tuna are commonly found swimming beneath spotted and spinner dolphins in the eastern Pacific Ocean (Figure 13–23). Fishers commonly used these dolphins to locate tuna and set a **purse seine net** around the entire school (Figure 13–24). When an underwater line is drawn tight, the net traps the tuna underwater as well as the dolphins at the surface.

The problem of dolphin deaths caused by tuna fishing was graphically presented to the world in March 1988 through video footage of dolphins struggling in tuna fishing nets taken by biologist Samuel F. La Budde. In 1990, under intense public outcry and a boycott on tuna, the United States tuna canning industry declared that it would not buy or sell tuna caught by methods that kill or injure dolphins. In 1992, a special addendum was added to the **Marine Mammal Protection Act**,

further protecting dolphins. The policy of the U.S. tuna canning industry has greatly reduced the practice of using dolphins to locate tuna. In addition, purse seine nets have been modified so that dolphins can be released safely.

Driftnets Another means of netting tuna and other species is by use of **driftnets** or **gill nets**, which are made of monofilament fishing line that is virtually invisible and cannot be detected by most marine organisms. Depending on the size of the holes in the net, it is highly effective at catching anything large enough to become entangled in it. As a result, driftnets often have high amounts of bycatch.

Up until 1993, Japan, Korea, and Taiwan had the largest driftnet fleets, deploying as many as 1500 fishing vessels into the North Pacific and setting over 48,000 kilometers (30,000 miles) of driftnets in one day. Although driftnetting was supposed to be restricted to specific fisheries, some fishers who claimed to be fishing for squid were involved in illegally taking large quantities of salmon and steelhead trout. Driftnetters were also targeting immature tuna in the South Pacific, which could result in the reduced abundance of South

Figure 13–23 Spotted dolphin (*Stenella attnuata*), which are commonly associated with yellowfin tuna.

Figure 13–24 Purse seiner. Purse seine nets are set around schools of yellowfin tuna, often trapping dolphins at the surface as bycatch.

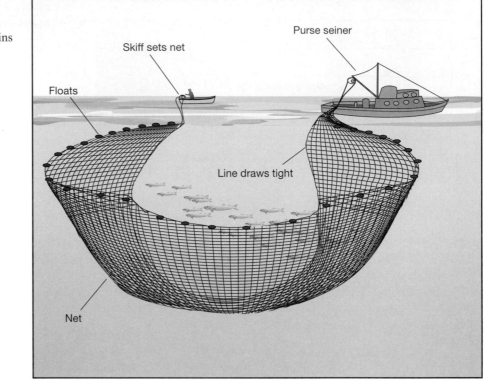

Pacific tuna. In addition, tens of thousands of birds, turtles, dolphins, and other species were killed annually in these nets.

A prohibition against the importation of fish caught in driftnets was included in the International Convention of Pacific Long Driftnet Fishing, signed in 1989. Although driftnets are now banned by international law, some fishers continue to illegally use them.

Fisheries Management

Fisheries management is the organized effort directed at regulating fishing activity with the goal of maintaining a long-term fishery. It includes assessing ecosystem health, determining fish stocks, analyzing fishing practices, and setting and enforcing catch limits. Unfortunately, however, fisheries management has historically been more concerned with maintaining human employment than preserving a self-sustaining marine ecosystem. For example, fisheries such as the anchovy, cod, flounder, haddock, herring, and sardine are suffering from overfishing in spite of being managed (Box 13–3).

One of the problems with fisheries management results from the fact that some fisheries encompass the waters of many different countries, including a variety of ecosystems. Some species of commercial fish, for instance, are raised in coastal estuaries and migrate long distances across international waters to their preferred environment. Fishing limits are difficult to enforce internationally, and if human interference occurs at any location where the fish exist, these species can be severely reduced in number.

A major regulatory failure has been the absence of restrictions on the number of fishing vessels. Figure 13–25 shows that the number of decked (large) fishing vessels in the world more than doubled between 1970 and 1995. The slight decrease in 2000 is more a reflection of decreasing fish stocks than any limitation on fishing vessels. Many of these larger vessels use nets that can hold up to 27,000 kilograms (60,000 pounds) of fish in one haul. In addition, there are more than 1.6 million smaller nondecked fishing vessels, mostly in Asia and Africa.

The increase in fishing vessels has resulted in an increased fishing effort, which often leads to overfishing. In some locations, the fish are becoming so scarce that the fishing effort costs more than what the catch is worth! In 1989, for instance, the world fishing fleet spent $92 billion to catch $70 billion worth of fish. To make up for the shortfall, many governments have given subsidies to fishers, which compounds the problem by maintaining (or even *increasing*) the number of fishing vessels.

If fisheries are to remain viable in the future, marine ecosystems must be better studied to document the natural relationships among organisms in marine food webs, the critical environmental factors for the health of the fishery, and the effects of removing so many organisms as fishery catch. In addition, critical fish habitats must be protected and fishing limits must be upheld despite political factors. Only after such obstacles are overcome can a fishery be successfully managed.

Box 13–3 A Case Study in Fisheries Mismanagement: The Peruvian Anchoveta Fishery

The waters off the coast of Peru have historically been filled with life, particularly a small silvery fish called the *anchoveta* or **anchovy** (*Engraulis ringens*), which swims through the water with its mouth open to catch its food (Figure 13G, *inset*). Human disruption of the anchovy ecosystem combined with periodic oceanographic changes resulted in depletion of anchovy stocks in 1972.

Off the coast of Peru, winds parallel to the coast from the southeast produce Ekman transport, which moves surface water away from shore and causes the upwelling of cold, nutrient-rich water from below (see Figure 7–11) that creates high productivity and an abundance of phytoplankton. Phytoplankton support a large population of zooplankton, which in turn are the food source for the anchovies. Anchovies are a vital link in the food web for larger marine animals such as birds, large fishes, marine mammals, and squid.

Recall from Chapter 7 that the waters offshore Peru periodically experience warm surface water, a deeper thermocline, and a reduction of upwelling during El Niño years. This reduction of upwelling severely affects productivity, limiting the number of anchovies and the abundance of many other marine organisms that feed on them. In 1972, a severe El Niño affected the area, decreasing the total biomass of anchovies to about 10% of its former level. Normally, anchovy populations had rebounded in the years following El Niño events, but failed to do so after the 1972 El Niño. What had prevented the repopulation of anchovies?

The sharp reduction of anchovies after 1972 may have resulted from a combination of the effects of El Niño and the establishment of the anchoveta fishing industry, which began in 1957. In 1960, the Intituto del Mar del Peru was created with the aid of the United Nations to study the anchoveta fishery and initiate a management program to avoid problems like those experienced by the Japan herring and California sardine fisheries, which had collapsed as a result of overexploitation. Still, the production of anchovies increased through the 1960s (Figure 13G).

Biologists had estimated that 10 million metric tons (11 million short tons) was the maximum sustainable yield of the anchoveta (Figure 13G, *green line*). However, this amount was exceeded in 1968, 1970, and 1971. Anchovy production peaked in 1970 at 12.3 million metric tons (13.5 million short tons), representing nearly one-quarter of the *entire* catch of fish from the sea worldwide! The reported amounts of anchovy catch may in fact be higher if processing losses, spoilage, and underreporting are considered.

In addition, there was a lack of understanding of the anchoveta fishery with respect to natural cycles. In January 1972, for example, the anchovy catch was set at 1.2 million metric tons (1.3 million short tons), which was reached easily in waters close to shore. The abundance of anchovies in nearshore waters raised expectations of a record harvest—in spite of the strengthening El Niño. As a result, the allotment for February 1972 was raised to 1.8 million metric tons (2.0 million short tons). The warm water of the 1972 El Niño, however, began to influence surface water temperatures. The combination of warm water plus an increased fishing effort severely reduced the abundance of anchovies, and in March,

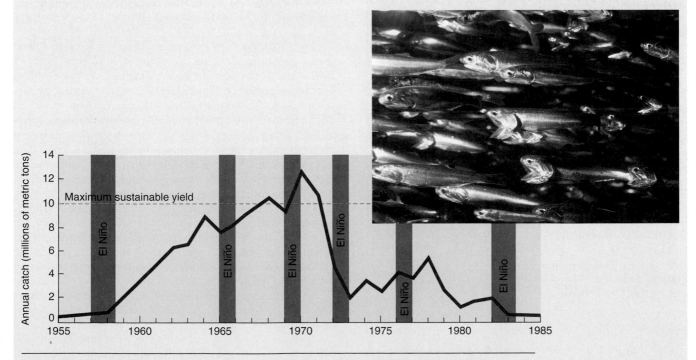

Figure 13G Peruvian anchovies (*inset*) and annual catch, 1955–1985. Graph showing the annual catch of Peruvian anchovies, which severely declined after a peak production in 1970. Also shown are El Niño years (*red bars*) and estimated maximum sustainable yield (*green line*).

(continued)

(continued)

the fishery collapsed, leading to the loss of many fishers' livelihood.

Further, it wasn't realized that anchovies play a vital role in the health of the entire ecosystem. Studies have revealed that even with upwelling, ammonia (NH_3) from decomposition of large quantities of anchovy excrement is required to support enough phytoplankton to allow the anchovy population to expand. When anchovy populations are small and scattered, they do not provide much nutrient enrichment. When the population is larger, the whole region receives additional enrichment as part of a natural cycle. It will require a gradual increase in the anchovy population over many years to supply enough nutrients throughout the entire region for the ecosystem's health to be restored.

In the meantime, anchovies are gradually being replaced by sardines, which are a more commercially valuable variety of fish because sardines are used for direct human consumption (unlike anchovies, most of which are ground up and converted to fish meal that is used to feed poultry and hogs). Currently, Peru's fishing activity emphasizes species conversion, so the anchovy population may recover and be viable in the future if it is managed properly. Because the area was severely overfished immediately prior to an El Niño, however, it may never reach previous levels of anchovy abundance.

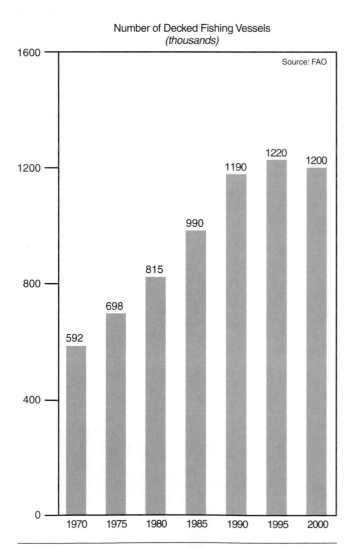

Number of Decked Fishing Vessels
(thousands)

Source: FAO

Figure 13–25 **Number of decked (large) fishing vessels in the world (thousands).**

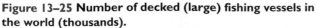

The fishing industry suffers from overfishing, practices that produce a large amount of unwanted bycatch, and a lack of adequate fisheries management.

? Students Sometimes Ask...

The number and variety of tropical species on land is astounding. I don't understand how the tropical oceans can have such low productivity.

Life on land does not necessarily correspond to life in the ocean! Tropical rain forests support an amazing diversity of species and an enormous biomass. In the tropical ocean, however, a strong, permanent thermocline limits the availability of nutrients that are necessary for the growth of phytoplankton. Without abundant phytoplankton, not much else can live in the ocean. In fact, these areas are often considered biological deserts. It is ironic that the clear blue water of the tropics so prominently displayed in tourist brochures indicates seawater that is biologically quite sterile!

I've heard of the "dead zone" that occurs periodically along the coast of the Gulf of Mexico. What is it?

A large hypoxic (*hypo* = under, *oxid* = oxygen) "dead zone"—which reached the size of the state of New Jersey in 1999—usually develops during the summer near the mouth of the Mississippi River (Figure 13H). A smaller "dead zone" has occurred each summer in the region for decades, but increased dramatically in size after the record-breaking Midwest floods in 1993. Oxygen levels within the zone drop from above 5.0 parts per million (ppm) to below 2.0 ppm, which is lower than most marine animals can tolerate. Some of the more mobile marine organisms can flee the area, but it kills many bottom-dwelling organisms that cannot swim or crawl.

These "dead zones" appear to be related to runoff of nutrients from agricultural activities that are washed down the Mississippi River, eventually reaching the Gulf, where they stimulate algal blooms. Once these algae die and sink to the bottom, bacteria feed on them and on fecal matter, depleting the water of oxygen along the bottom.

Although many other coastal waters experience summer-time hypoxic conditions, the Gulf's "dead zone" is the largest in the Western Hemisphere. Proposals to combat the spread of the "dead zone" include controlling nutrient runoff from agriculture, preserving and utilizing wetlands that filter runoff before it enters the Gulf, planting buffer strips of trees and grasses between farm fields and streams, altering the times when fertilizers are applied, improving crop rotation, and enforcing existing clean water regulations.

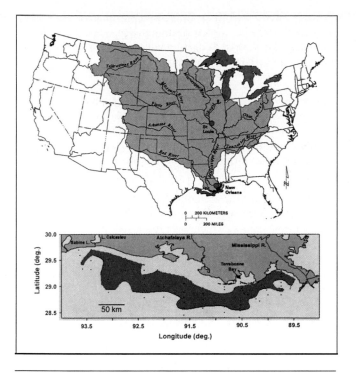

Figure 13H Mississippi River drainage basin *(yellow)* and extent of the 1999 Gulf of Mexico dead zone *(red)*.

Why does a red tide glow blue-green at night?

Many of the species of dinoflagellates that produce red tides (most notably those of genus *Gonyaulax*) also have bioluminescent capabilities—that is, they can produce light organically. When the organisms are disturbed, they emit a faint blue-green glow. When waves break during a red tide at night, the waves are often spectacularly illuminated by millions of bioluminescent dinoflagellates. During these times, one can easily observe marine animals moving through the water because their bodies are silhouetted by bioluminescent dinoflagellates that light up as they pass over the animal's body.

How likely am I to see a whale during a whale-watching trip?

It depends on the area and the time of the year, but generally your chances are quite low because large marine organisms comprise such a small proportion of marine life. In fact, it has been estimated that large nektonic organisms like whales comprise only *one-tenth of 1%* of all biomass in the sea! With your knowledge of food pyramids, it should be no surprise that the majority of the ocean's biomass is comprised of phytoplankton. Perhaps commercial boat operators should conduct *plankton*-watching trips! Everyone would be guaranteed to see dozens of different species—all that would be required is a plankton net and a microscope.

My friends are planning an ocean-fishing trip. To get to good fishing grounds that haven't been fished out, one of my friends has suggested that we go far offshore. Should we expect to find more fish there?

Generally, no. Most people fish coastal waters because they have high nutrient levels and are highly productive, whereas few people fish the open ocean because it has a low nutrient level and correspondingly low productivity. Far offshore, there just isn't enough food for many fish to live (see, for example, the distribution of phytoplankton shown in Figure 13–6). Thus, the farther you go from land, the less likely there would be good fishing, especially past the continental shelf. However, there are some exceptions to this, such as shallowly submerged banks and areas of upwelling far from shore.

Does aquaculture relieve some of the demand for wild fish?

Even though global production of farmed fish and shellfish has more than doubled in the past 15 years and now provides over one-quarter of all fish directly consumed by humans, some types of aquaculture have actually *increased* the demand for wild fish. This is because the farming of carnivorous species such as salmon, tuna, cod, and seabass require large inputs of wild fish for feed. Some aquaculture systems further reduce wild fish supplies through habitat modification, the collection of wild fish for initially stocking aquaculture operations, and other negative ecological impacts including waste disposal, exotic species introduction, and pathogen invasions, all of which probably have contributed to the collapse of fishery stocks worldwide. If the aquaculture industry is to sustain its contribution to world fish supplies, it must reduce wild fish inputs in feed and adopt more ecologically sound management practices.

Chapter in Review

• Microscopic planktonic algae that photosynthesize represent the largest biomass in the ocean. They are the ocean's primary producers—the foundation of the ocean's food web. Organic biomass is also produced near deep-sea hydrothermal springs through chemosynthesis, in which bacteria-like organisms trap chemical energy by the oxidation of hydrogen sulfide.

• The availability of nutrients and the amount of solar radiation limits the photosynthetic productivity of the oceans. Nutrients—such as nitrate, phosphorous, iron, and silica—are most abundant in coastal areas, due to runoff and upwelling. The depth at which net photosynthesis is zero is the compensation depth for photosynthesis. Algae cannot live successfully below this depth, which may be less than 20 meters (65 feet) in turbid coastal waters or as much as 100 meters (330 feet) in the open ocean.

• Marine life is most abundant along continental margins, where nutrients and sunlight are optimal. It decreases with distance from the continents and with increased depth. In addition, cool water typically supports more abundant life than warm water because cool water can dissolve more of the gases necessary for life (oxygen and carbon dioxide). Areas of upwelling bring cold, nutrient-rich water to the surface and have some of the highest productivities.

• Ocean water selectively absorbs the colors of the visible spectrum. Red and yellow light are absorbed at relatively shallow depths, whereas blue and green light are the last to be removed. Ocean water of low biologic productivity scatters the short wavelengths of visible light, producing a blue color. Turbidity and photosynthetic algae in more-productive ocean water scatter more green light wavelengths, which produces a green color.

• There are many different types of photosynthetic marine organisms. The seed-bearing Spermatophyta are represented by a few genera of nearshore plants such as eelgrass (*Zostera*), surf grass (*Phyllospadix*), marsh grass (*Spartina*), and mangrove trees (genera *Rhizophora* and *Avicennia*). Macroscopic algae include brown algae (Phaetophyta), green algae (Chlorophyta), and red algae (Rhodophyta). Microscopic algae include diatoms and coccolithophores (Chrysophyta), and dinoflagellates (Pyrrophyta, which are responsible for red tides).

• A thermocline is generally absent in high-latitude (polar) areas, so upwelling can readily occur. The availability of solar radiation limits productivity more than the availability of nutrients. In low-latitude (tropical) regions, however, a strong thermocline may exist year-round, so the lack of nutrients limits its productivity, except in areas of upwelling or near coral reefs. In temperate regions, productivity peaks in the spring and fall and is limited by lack of solar radiation in the winter and lack of nutrients in the summer.

• Radiant energy captured by algae is converted to chemical energy and passed through the different trophic levels of the biotic community. It is expended as mechanical and heat energy and ultimately reaches a high state of entropy, where it is biologically useless. Upon death, organisms are decomposed to an inorganic form that algae can use again for nutrients.

• Marine ecosystems are composed of populations of organisms called producers (which photosynthesize or chemosynthesize), consumers (which eat producers), and decomposers (which break down detritus). Animals can be categorized as herbivores (eat plants), carnivores (eat animals), omnivores (eat both), or bacteriovores (eat bacteria). Some organisms live closely together in symbiotic relationships. The organisms of a biotic community cycle nutrients and other chemicals from one form to another.

• On average, only about 10% of the mass taken in at one feeding level is passed on to the next. As a result, the size of individuals increases but the number of individuals decreases with each trophic level of the food chain or food web. Overall, the total biomass of populations decreases the higher they are in the biomass pyramid.

• Marine fisheries harvest standing stocks of populations from various ecosystems. Overfishing occurs when adult fish are harvested faster than they can reproduce and results in the decline of fish populations as well as a reduction of a fishery's maximum sustainable yield. Many fishing practices capture unwanted bycatch. Despite the management of fisheries, many fish stocks worldwide are declining.

Key Terms

Amnesic shellfish poisoning (p. 384)

Anchovy (anchoveta) (p. 399)

Autotrophic (p. 390)

Bacteriovore (p. 390)

Biogeochemical cycle (p. 391)

Biological pump (p. 382)

Biomass (p. 373)

Biomass pyramid (p. 394)

Biotic community (p. 389)

Bycatch (p. 397)

California Cooperative Oceanic Fisheries Investigations (CalCOFI) (p. 372)

Carnivore (p. 390)

Carotin (p. 382)

Chemosynthesis (p. 373)

Coccolith (p. 382)

Coccolithophore (p. 382)

Commensalism (p. 390)

Compensation depth for photosynthesis (p. 375)

Consumer (p. 390)

Decomposer (p. 390)

Detritus (p. 390)

Diatom (p. 382)

Diatomaceous earth (p. 382)

Dinoflagellate (p. 382)

Driftnet (p. 397)

Ecosystem (p. 389)

Electromagnetic spectrum (p. 376)

Entropy (p. 389)

Euphotic zone (p. 375)

Eutrophic (p. 379)

Eutrophication (p. 385)

Fall bloom (p. 389)

Flagella (p. 382)

Fisheries (p. 394)

Fisheries management (p. 398)

Food chain (p. 394)

Food web (p. 394)

Gill net (p. 397)

Gross ecological efficiency (p. 393)

Gross primary production (p. 374)

Harmful algae bloom (HAB) (p. 382)

Herbivore (p. 390)

Heterotrophic (p. 390)

Incidental catch (p. 397)

Isothermal (p. 384)

Marine Mammal Protection Act (p. 397)

Maximum sustainable yield (p. 396)

Mutualism (p. 391)

Net primary production (p. 374)

New production (p. 374)

Oligotrophic (p. 379)

Omnivore (p. 390)

Overfishing (p. 396)

Paralytic shellfish poisoning (PSP) (p. 384)

Parasitism (p. 391)

Pfiesteria piscicida (p. 386)

Purse seine net (p. 397)

Photosynthesis (p. 373)

Plankton net (p. 374)

Primary productivity (p. 373)

Producer (p. 390)

Pycnocline (p. 382)

Red tide (p. 382)

Regenerated production (p. 374)

SeaWiFS (p. 374)

Secchi disk (p. 379)

Spring bloom (p. 389)

Standing stock (p. 396)

Symbiosis (p. 391)

Test (p. 382)

Thermocline (p. 382)

Trophic level (p. 392)

Upwelling (p. 376)

Visible light (p. 376)

Questions And Exercises

1. Describe the mission of the CalCOFI program. How did the CalCOFI program help solve the mystery it intended to answer?

2. Discuss chemosynthesis as a method of primary productivity. How does it differ from photosynthesis?

3. How does gross primary production differ from net primary production? What are the two components of gross primary production, and how do they differ?

4. An important variable in determining the distribution of life in the oceans is the availability of nutrients. How are the following variables related: proximity to the continents, availability of nutrients, and the concentration of life in the oceans?

5. Another important determinant of productivity is the availability of solar radiation. Why is biological productivity relatively low in the tropical open ocean, where the penetration of sunlight is greatest?

6. Discuss the characteristics of the coastal ocean where unusually high concentrations of marine life are found.

7. Why does everything in the ocean at depths below the shallowest surface water take on a blue-green appearance?

8. What factors create the color difference between coastal waters and the less-productive open-ocean water? What color is each?

9. Compare the macroscopic algae in terms of color, maximum depth in which they grow, common species, and size.

10. The golden algae include two classes of important phytoplankton. Compare their composition and the structure of their tests and explain their importance in the geologic fossil record.

11. Discuss and compare the contributions of the Pyrrophyta genera *Ptychodiscus* and *Gonyaulax* to red tide development.

12. How does paralytic shellfish poisoning (PSP) differ from amnesic shellfish poisoning? What types of microorganisms create each?

13. Describe how a biological pump works. What percentage of organic material from the euphotic zone accumulates on the sea floor?

14. Compare the biological productivity of polar, temperate, and tropical regions of the oceans. Consider seasonal changes, the development of a thermocline, the availability of nutrients, and solar radiation.

15. Generally, the productivity in tropical oceans is rather low. What are three environments that are exceptions to this, and what factors contribute to their higher productivity?

16. Describe the flow of energy through the biotic community and include the forms into which solar radiation is converted. How does this flow differ from the manner in which matter is moved through the ecosystem?

17. What are the three types of symbiosis, and how do they differ?

18. What is the average efficiency of energy transfer between trophic levels? Use this efficiency to determine how much phytoplankton mass is required to add *1 gram* of new mass to a killer whale, which is a third-level carnivore. Include a diagram that shows the different trophic levels and the relative size and abundance of organisms at different levels. How would your answer change if the efficiency were half the average rate, or twice the average rate?

19. Describe the advantage that a top carnivore gains by eating from a food web as compared to a single food chain.

20. When a species is overfished, what changes are there in the standing stock and the maximum sustainable yield? What are some problems with fisheries management?

References

Alexander, R. B., Smith, R. A., and Schwarz, G. E. 2000. Effect of stream channel size on the delivery of nitrogen to the Gulf of Mexico. *Nature* 403:6771, 758–761.

Chung, J. E. 1999. Surveying changes in the sea. *Scripps Institution of Oceanography Explorations* 6:1, 20–26, Scripps Institution of Oceanography, University of California, San Diego.

Culotta, E. 1996. Red menace in the world's oceans, *in* Pirie, R. G., ed., *Oceanography: Contemporary Readings in Ocean Sciences*, 3rd ed., New York: Oxford University Press.

Ducklow, H. W. 1983. Production and fate of bacteria in the oceans. *Bioscience* 33:8, 494–501.

Dugdale, R. C., and Wilkerson, F. P. 1998. Silicate regulation of new production in the equatorial Pacific upwelling. *Nature* 391:6664, 270–273.

Falkowski, P. G., Barber, R. T., and Smetacek, V. 1998. Biogeochemical controls and feedbacks on ocean primary production. *Science* 281:5374, 200–206.

Ferber, D. 2001. Keeping the stygian waters at bay. *Science* 291:5506, 968–973.

Food and Agriculture Organization, 1999. *The state of world fisheries and aquaculture 1998*. Rome: Food and Agriculture Organization of the United Nations.

George D., and George, J. 1979. *Marine life: An Illustrated Encyclopedia of Invertebrates in the Sea*. New York: Wiley-Interscience.

Goolsby, D. A. 2000. Mississippi basin nitrogen flux believed to cause Gulf hypoxia. *Eos Trans. AGU* 81:29, 321–327.

Grassle, J. F., et al. 1979. Galápagos '79: Initial findings of a deep-sea biological quest. *Oceanus* 22:2, 2–10.

Howard, J. 1995. Vanishing act: Critical link in marine food chain may be at risk. *Scripps Institution of Oceanography Explorations* 2:2, 11–17, Scripps Institution of Oceanography, University of California, San Diego.

———. 1996. Red tides rising: Local plankton bloom possible sign of growing global threat. *Scripps Institution of Oceanography Explorations* 2:3, 2–9, Scripps Institution of Oceanography, University of California, San Diego.

Jenkins, W. J. 1982. Oxygen utilization rates in North Atlantic subtropical gyre and primary production in oligotrophic systems. *Nature* 300, 246–248.

Lalli, C. M., and Parsons, T. R. 1993. *Biological Oceanography: An Introduction*. New York: Pergamon Press.

Little, M. M., Little, D. S., Blair, S. M., and Norris, J. N. 1985. Deepest known plant life discovered on an uncharted seamount. *Science* 227:4683, 57–59.

Malakoff, D. 1998. Death by suffocation in the Gulf of Mexico. *Science* 281:5374, 190–192.

National Research Council, 1999. *Sustaining marine fisheries*. Washington D.C.: National Academy Press.

Naylor, et al., 2000. Effect of aquaculture on world fish supplies. Nature 405:6790, 1017–1024.

Parsons, T. R., Takahashi, M., and Hargrave, B. 1974. *Biological Oceanographic Processes*, 3rd ed. New York: Pergamon Press.

Pauly, D., and Christensen. 1995. Primary production required to sustain global fisheries. *Nature* 374:6519, 255–257.

Pauly, D., et al., 1998. Fishing down marine food webs. *Science* 279: 5352, 860–863.

Pimm, S. L., Lawton, J. H., and Cohen, J. E. 1991. Food web patterns and their consequences. *Nature* 350:6320, 669–674.

Platt, T., and Sathyendranath, S. 1988. Oceanic primary production: Estimation by remote sensing at local and regional scales. *Science* 241:4873, 1613–1619.

Platt, T., Subba Rao, D. V., and Irwin, B. 1983. Photosynthesis of picoplankton in the oligotrophic ocean. *Nature* 310:5902, 702–704.

Russell-Hunter, W. D. 1970. *Aquatic Productivity*. New York: Macmillan.

Sathyendranath, S., et al., 1991. Estimation of new production in the ocean by compound remote sensing. *Nature* 353:6340, 129–133.

Seliger, H. H. 1996. Bioluminescence: Excited states under cover of darkness, *in* Pirie, R. G., ed., *Oceanography: Contemporary Readings in Ocean Sciences*, 3rd ed. New York: Oxford University Press.

Sherr, B. F., Sherr, E. B., and Hopkinson, C. S. 1988. Trophic interactions within pelagic microbial communities: Indications of feedback regulation of carbon flow. *Hydrobiologia* 159:1, 19–26.

Shulenberger, E., and Reid, J. L. 1981. The Pacific shallow oxygen maximum, deep chlorophyll maximum, and primary productivity reconsidered. *Deep Sea Research* 28A:9, 901–919.

Strahler, A. H., and Strahler, A. N. 1992. *Modern Physical Geography*, 4th ed. New York: Wiley.

Sullivan, C. W., Arrigo, K. R., McClain, C. R., Comiso, J. C., and Firestone, J. 1993. Distributions of phytoplankton blooms in the Southern Ocean. *Science* 262:5141, 1832–1836.

Suggested Reading in Scientific American

Anderson, D. M. 1994. Red tides. Dense blooms of algae are becoming more frequent in coastal waters. They are also involving species of algae and animals not previously known to be associated with red tide occurrences.

Benson, A. A. 1975. Role of wax in oceanic food chains. 232:3, 76–89. A report on the findings from observations made of the content of wax in the bodies of many marine animals from copepods to small deep-water fishes and their implications.

Burkholder, J. M. 1999. The lurking perils of *Pfiesteria*. 281:2, 42–49. Examines the life cycle of the dinoflagellate *Pfiesteria*, which causes fish kills, harm to people, and other negative environmental effects.

Childress, J. J., Feldback, H., and Somero, G. N. 1987. Symbiosis in the deep sea. 256:5, 114–121. Deep-sea hydrothermal vent animals have a symbiotic relationship with sulfur-oxidizing bacteria that allows them to live in the darkness of the deep ocean.

Coleman, G., and Coleman, W. J. 1990. How plants make oxygen. 262:2, 50–67. The process of oxygen production by plants is explained.

King, M. D. and Herring, D. D. 2000. Monitoring Earth's vital signs. 282:4, 92–97. Information about Terra, the flagship of NASA's Earth Observing System fleet of satellites that monitor Earth's vital signs including ocean productivity.

Levine, R. P. 1969. The mechanism of photosynthesis. 221:6, 58–71. Reveals what is known of the process by which energy is captured by plants and converted to useful forms of chemical energy while freeing oxygen to the atmosphere.

Pettit, J., Drucker, S., and Knox, B. 1981. Submarine pollination. 244:3, 134–144. Discusses the pollination of sea grasses by wave action.

Oceanography on the Web

Visit the *Essentials of Oceanography* home page for on-line resources for this chapter. There you will find an on-line study guide with review exercises, and links to oceanography sites to further your exploration of the topics in this chapter. *Essentials of Oceanography* is at: **http://www.prenhall.com/ thurman** (click on the Table of Contents menu and select this chapter).

CHAPTER
14

Animals of the Pelagic Environment

* How are marine organisms able to stay above the ocean floor?

* What types of fins do fish have, and how are they used?

* What kinds of adaptations do fish have that live in the deep ocean?

* How are pelagic organisms adapted for seeking prey?

* What are the advantages of schooling?

* Which marine mammals live in the ocean?

* Why do gray whales make the longest yearly migration of any mammal?

If you were to make little fishes talk, they would talk like whales.
—Oliver Goldsmith (1773)

ALEXANDER AGASSIZ: ADVANCEMENTS IN OCEAN SAMPLING

Biological oceanography was strongly supported in America by the activities of Alexander Agassiz (Figure 14A). The son of the great Swiss scientist Louis Agassiz, he immigrated to the United States in 1846 with his parents. He graduated from Harvard University in 1855, became superintendent of a large copper mining operation in 1866, and was a multimillionaire by age 40. Agassiz was a friend of C. Wyville Thompson of the *Challenger* Expedition, and his successor Sir John Murray, who helped prepare the reports from the voyage of the *Challenger*. In fact, Agassiz helped process the organisms collected during the *Challenger* Expedition.

Agassiz contributed substantial money and effort to the development of U.S. ocean study and was strongly opinionated—but not always right. For example, he disagreed with Darwin's theory of coral reef development (see the chapter-opening feature in Chapter 12), and with the idea that there was an extensive midwater plankton community, both of which became well accepted. These controversies, however, did not stop Agassiz from financing scientific voyages and supporting the Museum of Comparative Zoology at Harvard, where valuable analysis of marine organisms has been conducted ever since.

Agassiz is widely acknowledged as the driving force that brought oceanography recognition as a science. He initiated the system that first brought oceanographic research strong financial and institutional support. He was also noted for his ability to organize efficient and effective research voyages, many of which occurred in the Caribbean Sea, the Gulf of Mexico, and off the Atlantic Coast of Florida.

In addition, Agassiz's training as a mining engineer led him to develop ingenious oceanographic sampling devices—the prototypes for many devices in use on research vessels today—that improved the quantitative value of the biological samples recovered. One such device, a dredge used to collect organisms living on the bottom, was designed to work equally well no matter which way it landed on the ocean floor. It worked so well, in fact, that one haul from 3219 meters (10,560 feet) brought up more specimens of deep-sea fishes than the *Challenger* had collected in its entire three-and-a-half-year expedition!

Figure 14A Alexander Agassiz (*standing*) and associates collecting specimens.

Pelagic organisms live suspended in seawater (not on the ocean floor) and comprise the vast majority of the ocean's **biomass**.[1] Phytoplankton live within the sunlit surface waters of the ocean and are the food source for nearly all other marine life, so most marine animals live in surface waters too so they can be close to their food supply. The immense depth of the oceans makes it challenging for these organisms to remain afloat.

Phytoplankton depend primarily on their small size to provide a high degree of frictional resistance to sinking. Most animals, however, are more dense than ocean water, and have less surface area per unit of body mass. Therefore, they tend to sink more rapidly than phytoplankton.

To remain in surface waters where the food supply is greatest, pelagic marine animals must increase their buoyancy or continually swim. Animals apply one or both of these strategies in a wonderful variety of adaptations and lifestyles.

Staying Above the Ocean Floor

Some animals increase their buoyancy to remain in near-surface waters. They may have containers of gas, which significantly reduce their average density, or they may have soft bodies void of hard, high-density parts. Larger animals with bodies denser than seawater must exert more energy to propel themselves through the water.

Gas Containers

Air is approximately 1000 times less dense than water at sea level, so even a small amount of air inside an organism can dramatically increase its buoyancy. Some animals, such as the cephalopods (*cephalo* = the head, *podium* = a foot), have rigid gas containers in their bod-

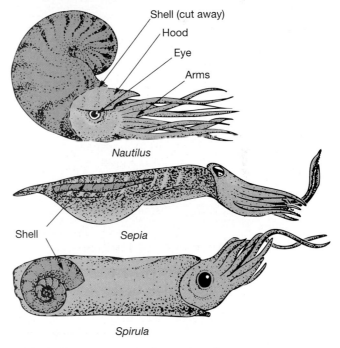

Figure 14–1 Gas containers in cephalopods. The *Nautilus* has an external chambered shell, while *Sepia* and *Spirula* have rigid internal chambered structures that can be filled with gas to provide buoyancy.

ies. For instance, the genus *Nautilus* have an external shell, whereas the cuttlefish *Sepia*[2] and deep-water squid *Spirula* have an internal chambered structure (Figure 14–1). These animals are neutrally buoyant, which means the amount of air in their bodies regulates their density, so they can remain at a particular depth.

Because the pressure in their air chambers is always 1 kilogram per square centimeter (1 atmosphere, or 14.7

---

[1]Remember that *biomass* is the mass of living organisms.

[2]Many species of cephalopods have an inking response. The ink of the cuttlefish *Sepia* was used as writing ink (with the brand name Sepia) before alternatives were developed.

pounds per square inch), the *Nautilus* must stay above a depth of approximately 500 meters (1640 feet) to prevent collapse of its chambered shell. The *Nautilus*, therefore, rarely ventures below about 250 meters (800 feet).

Some slow-moving fish use an internal organ called a **swim bladder** (Figure 14–2) to achieve neutral buoyancy. Very active swimmers (such as tuna) or fish that live on the bottom do not usually have a swim bladder because they don't have a problem maintaining their positions in the water column.

A change in depth either expands or contracts the swim bladder, so fish must remove or add gas to maintain a constant volume. In some fish, a pneumatic duct connects the swim bladder to the esophagus, so these fish can quickly add or remove gases through the duct. In other fish without the pneumatic duct, the gases of the swim bladder must be added or removed more slowly by an interchange with the blood, so they cannot withstand rapid changes in depth.

The composition of gases in the swim bladders of shallow-water fishes is similar to that of the atmosphere. At the surface, the concentration of oxygen in the swim bladder is about 20% and as depth increases, the oxygen concentration increases to more than 90%. Fish with swim bladders have been captured from as deep as 7000 meters (23,000 feet), where the pressure is 700 kilograms per square centimeter (700 atmospheres, or 10,300 pounds per square inch). Pressure this high compresses the gas to a density of 0.7 gram per cubic centimeter.[3] This is approximately the same density as fat, so many deep-water fish have special organs for buoyancy that are filled with fat instead of compressed gas.

Floating Organisms (Zooplankton)

Organisms floating at the surface range in size from microscopic to relatively large, such as the familiar jellyfish. These floating organisms—collectively called zooplankton—comprise the second largest biomass in the ocean after the phytoplankton. Microscopic forms have a hard shell or test. Many larger forms have soft, gelatinous bodies with little if any hard tissue, which reduces their density and allows them to stay afloat.

Microscopic Zooplankton Microscopic zooplankton are incredibly abundant in the ocean. They are *primary consumers* because they eat microscopic phytoplankton, the primary producers. Thus, many zooplankton are herbivores. Others are omnivores because they eat other zooplankton in addition to phytoplankton. Most types of microscopic zooplankton have adaptations to increase the surface area of their bodies (or shells) so they can remain in the sunlit surface waters near their food source.

Three of the most important groups of zooplankton are the radiolarians, foraminifers (both of which are discussed in more detail in Chapter 4, "Marine Sediments"), and copepods.

Radiolarians (*radio* = a spoke or ray) are single-celled, microscopic organisms that build their hard shells (*tests*) out of silica (Figure 14–3). Their tests have intricate ornamentation including long projections. Although the spikes and spines appear to be a defense mechanism against predators, they increase the test's surface area so the organism won't sink through the water column.

Foraminifers (*foramen* = an opening) are microscopic to (barely) macroscopic single-celled animals. While the most abundant types of foraminifers are planktonic, the most diverse (in terms of number of species) are benthic. Foraminifers produce a hard test made of calcium carbonate (Figure 14–4) that is segmented or chambered with a prominent opening in one end. The tests of both radiolarians and foraminifers are common components of deep-sea sediment.

Copepods (*kope* = oar, *pod* = a foot) are microscopic shrimp-like animals of the subphylum Crustacea, which also includes shrimps, crabs, and lobsters. Like other crustaceans, copepods have a hard exoskeleton (*exo* = outside) and a segmented body with jointed legs (Figure 14–5). Copepods probably represent the majority of the ocean's zooplankton biomass and are an important link in many marine food webs. They have special adaptations for filtering their tiny floating food from seawater.

Macroscopic Zooplankton Many types of zooplankton are large enough to be seen without the aid of a microscope. One important group is **krill**, which means "young fry of fish" in Norwegian. Krill, however, are in the subphylum Crustacea (genus *Euphausia*) and

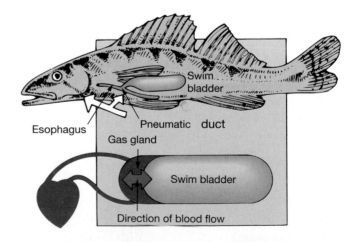

Figure 14–2 Swim bladder. Some bony fishes have a swim bladder, which is connected to the esophagus by the pneumatic duct, allowing air to be added or removed rapidly. In fish with no pneumatic duct, all gas must be added or removed through the blood, which requires more time.

---

[3]For comparison, note that the density of water is 1.0 gram per cubic centimeter.

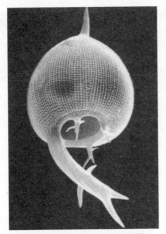

(a) *Euphysetta elegans, x280*

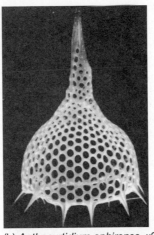

(b) *Anthocyrtidium ophirense, x230*

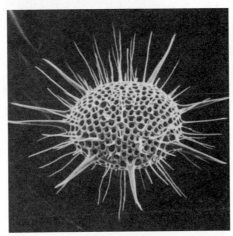

(c) *Larcospira quadrangula, x190*

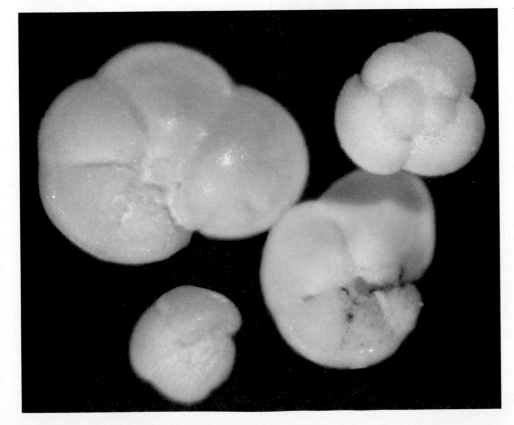

(d) *Heliodiscus asteriscus, x200*

Figure 14–3 Radiolarians.
Scanning electron micrographs of various radiolarians.
(a) *Euphysetta elegans* (magnified 280 times). **(b)** *Anthocyrtidium ophirense* (magnified 230 times).
(c) *Larcospira quadrangula* (magnified 190 times).
(d) *Heliodiscus asteriscus* (magnified 200 times).

Figure 14–4 Foraminifers.
Photomicrograph of various species of foraminifers, the largest of which is 1 millimeter (0.04 inch) long. These pelagic foraminifers were collected from the Ontong Java Plateau in the western Pacific Ocean.

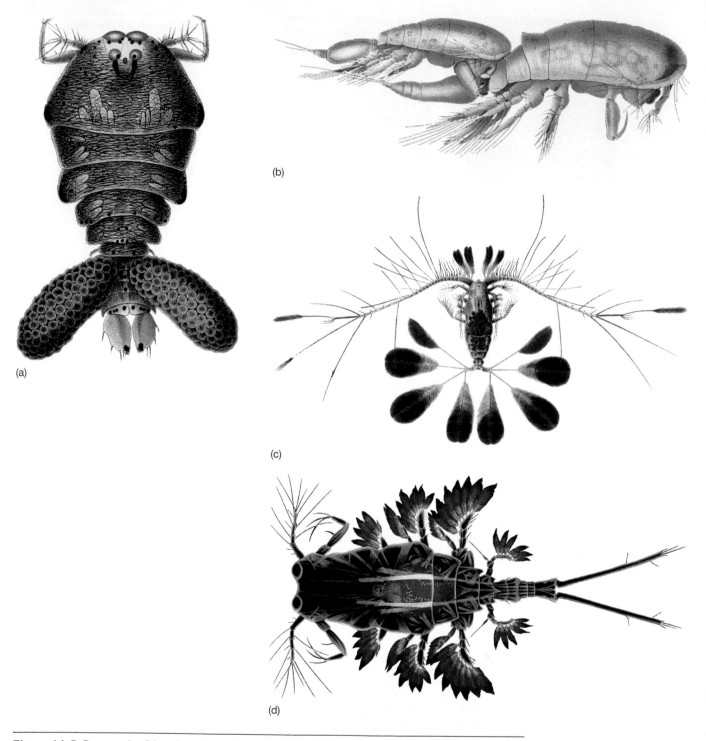

Figure 14–5 Copepods. Line drawings of various copepods (shown many times their actual size) from Wilhelm Giesbrecht's 1892 book on the flora and fauna of the Gulf of Naples. **(a)** The adult female *Sapphirina auronitens* carries a pair of lobe-like egg sacks. **(b)** Copulating pair of *Oncaea conifera*. **(c)** *Calocalanus pavo* showing elaborate feathery appendages that are characteristic of warm-water species. **(d)** *Copilia vitrea* uses its appendages to cling to large particles in the water column or to larger zooplankton.

resemble mini-shrimp or large copepods (Figure 14–6). There are over 1500 species of krill, most of which achieve a length no longer than 5 centimeters (2 inches). They are abundant near Antarctica and form a critical link in the food web there, supplying food for many organisms from sea birds to the largest whales in the world.

Another important group of macroscopic zooplankton is the **coelenterates** (*coel* = hollow, *entreon* = the intestines), which have soft bodies that are more than

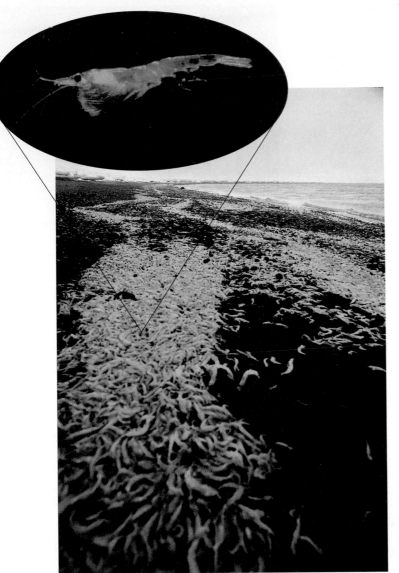

Figure 14–6 Krill. Krill that has washed up on a beach in Antarctica and close-up view of *Meganyctiphanes norvegica*, which is about 3.8 centimeters (1.5 inches) long (*inset*).

95% water. There are two basic types, the siphonophores and the scyphozoans (jellyfish).

Siphonophore (*siphon* = tube, *phoros* = bearing) coelenterates are represented in all oceans by the Portuguese man-of-war (genus *Physalia*) and "by-the-wind sailor" (genus *Velella*). Their gas chambers, called pneumatophores, serve as floats and sails that allow the wind to push them across the ocean surface (Figure 14–7a, *left*). Sometimes the wind pushes large numbers of these organisms toward a beach, where they can wash ashore and die. In the living organism, a colony of tiny individuals is suspended beneath the float. Portuguese man-of-war tentacles may be many meters long and possess nematocysts long enough to penetrate the skin of humans. They have been known to inflict a painful and occasionally dangerous neurotoxin poisoning.

Jellyfish, or **scyphozoan** (*skuphos* = cup, *zoa* = animal) coelenterates, have a bell-shaped body with a fringe of tentacles and a mouth at the end of a clapper-like extension hanging beneath the bell-shaped float (Figure 14–7). Ranging in size from nearly microscopic to 2 meters (6.6 feet) in diameter, most jellyfish are less than 0.5 meter (1.6 feet) in diameter. The largest jellyfish can have tentacles as long as 60 meters (200 feet).

Jellyfish move by muscular contraction. Water enters the cavity under the bell and is forced out when muscles that circle the bell contract, jetting the animals ahead in short spurts. To allow the animal to swim generally in an upward direction, sensory organs that are light sensitive or gravity sensitive are spaced around the outer edge of the bell. The ability to orient is important because jellyfish feed by swimming to the surface and sinking slowly through the rich surface waters.

Marine organisms use a variety of adaptations to stay within the sunlit surface waters, such as rigid gas containers, swim bladders, spines to increase their surface area, or soft bodies.

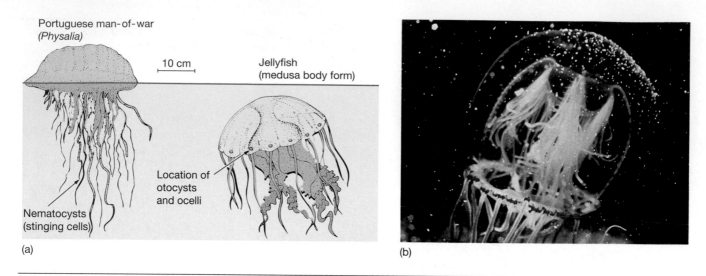

Portuguese man-of-war
(*Physalia*)

10 cm

Jellyfish
(medusa body form)

Location of
otocysts
and ocelli

Nematocysts
(stinging cells)

(a)

(b)

Figure 14–7 Planktonic cnidarians. **(a)** Portuguese man-of-war (*Physalia*) and jellyfish.
(b) A medusa jellyfish.

Swimming Organisms (Nekton)

Many larger pelagic animals can swim easily against currents. These organisms include invertebrate squids, fish, and marine mammals. Because they can swim, some of these organisms undertake long migrations.

Swimming squid include the common squid (genus *Loligo*), flying squid (*Ommastrephes*), and giant squid (*Architeuthis*). Active predators of small fish, the smaller squid varieties have long, slender bodies with paired fins (Figure 14–8). Unlike the less active *Sepia* and *Spirula* shown in Figure 14–1, they have no hollow chambers in their bodies and therefore require more energy to remain in the upper water of the oceans without sinking.

Squid can swim about as fast as any fish their size, and do so by trapping water in a cavity between their soft body and pen-like shell and forcing it out through a siphon. To capture prey, they use two long arms with pads containing suction cups at the ends (Figure 14–8). Eight shorter arms with suckers convey the prey to the mouth, where it is crushed by a mouthpiece that resembles a parrot's beak.

Locomotion in fish occurs when a wave of lateral body curvature passes from the front of the fish to the back. This is achieved by the alternate contraction and relaxation of muscle segments, called **myomeres** (*myo* = muscle, *merous* = parted), along the sides of the body. The backward pressure of the fish's body and fins produced by the movement of this wave provides the forward thrust (Figure 14–9).

Fin Designs in Fish   Most active swimming fish use two sets of paired fins—*pelvic fins* and *pectoral* (*pectoralis* = breast) *fins*—to turn, brake, and balance (Figure 14–9). When not in use, these fins can be folded against the body. Vertical fins, both *dorsal* (*dorsum* = back) and *anal*, serve primarily as stabilizers.

The fin that is most important in propelling the high-speed fish is the tail fin, or *caudal* (*cauda* = tail) *fin*. Caudal fins flare vertically to increase the surface area available to develop thrust against the water.[4] The increased surface area also increases frictional drag. The efficiency of the design of a caudal fin depends on its shape, which is one of the five basic forms illustrated in Figure 14–10 and keyed by letter to the following descriptions:

a.   The *rounded fin* is flexible and useful in accelerating and maneuvering at slow speeds.

b. & c.   The somewhat flexible *truncate fin* (b) and *forked fin* (c) are found on faster fish and still may be used for maneuvering.

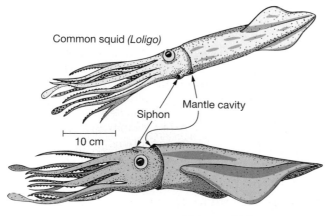

Common squid (*Loligo*)

Mantle cavity

Siphon

10 cm

Flying squid (*Ommastrephes*)

Figure 14–8 Squid. Squid move by trapping water in their mantle cavity between their soft body and pen-like shell. They then jettison the water through their siphon for rapid propulsion.

[4]This is equivalent to humans donning swimming fins on their feet to enable them to swim more efficiently.

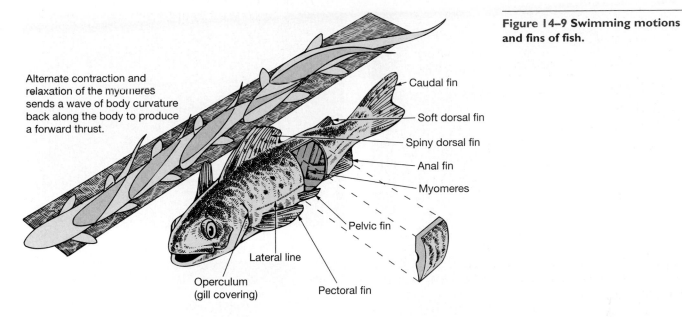

Figure 14–9 Swimming motions and fins of fish.

Alternate contraction and relaxation of the myomeres sends a wave of body curvature back along the body to produce a forward thrust.

Caudal fin
Soft dorsal fin
Spiny dorsal fin
Anal fin
Myomeres
Pelvic fin
Lateral line
Operculum (gill covering)
Pectoral fin

d. The *lunate fin* is found on the fast-cruising fishes such as tuna, marlin, and swordfish; it is very rigid and useless for maneuverability, but very efficient for propulsion.

e. The *heterocercal* (*hetero* = uneven, *cercal* = tail) *fin* is asymmetrical, with most of its mass and surface area in the upper lobe. The heterocercal fin produces a significant lift to sharks as it is moved from side to side. Sharks need this lift because they have no swim bladder and tend to sink when they stop moving. The pectoral (chest) fins, moreover, are large and flat and positioned so they function like airplane wings to lift the front of the shark's body, balancing the rear lift supplied by the caudal fin. Although the shark gains tremendous lift from this adaptation of its pectoral fins, it sacrifices maneuverability. This is why sharks tend to swim in broad circles—like a circling airplane—and do not make sharp turns while swimming.

> Fish use their fins for swimming and staying afloat within the water column. The fin that provides the most thrust is the caudal (tail) fin, which can have a variety of shapes depending on its use.

Deep-Water Nekton Living below the surface water but still above the ocean floor are deep-water nektonic species—mostly various species of fish—that are specially adapted to the deep-water environment, where it is very still and completely dark. Their food source is either **detritus**—dead and decaying organic matter and waste products that slowly settle through the water column from the surface waters above—or each other. The lack of abundant food limits the *number* of organisms (total biomass) and the *size* of these organisms. As a result, small populations of these organisms exist and most individuals are less than 30 centimeters (1 foot) long. Many have low metabolic rates to conserve energy, too.

These **deep-sea fish** (Figure 14–11) have special adaptations to efficiently find and collect food. They have good sensory devices, for example, such as long antennae or sensitive lateral lines that can detect movement of other organisms within the water column.

Many species have large and sensitive eyes—perhaps 100 times more sensitive to light than our own—that enable them to see potential prey. To avoid being prey, most species are dark so they blend into the environment. Other species are blind and rely on senses such as smell to track down prey.

Well over half of deep-sea fish can **bioluminesce**, which means they can produce light organically. Bioluminescent fish produce light by using specially designed structures or cells called **photophores**, some of which contain symbiotic luminescent bacteria. In a world of darkness, the ability to produce light has the following uses:

- Attracting prey (the deep-sea anglerfish in Figure 14–11f uses its specially modified dorsal fin as a bioluminescent lure)
- Staking out territory by constantly patrolling an area
- Communicating or seeking a mate by sending signals
- Escaping from predators by using a flash of light to temporarily blind them

Other adaptations to the deep-sea environment include large sharp teeth, expandable bodies to

(a)

(b)

(d)

(c)

(e)

Figure 14–10 Caudal fin shapes. **(a)** Rounded fin on a queen angel (other examples: sculpin, flounder). **(b)** Truncate fin on a gray angelfish (also on salmon, bass). **(c)** Forked fin on a goatfish (also on herring, yellowtail). **(d)** Lunate fin on a blue marlin (also on bluefish, tuna). **(e)** Heterocercal fin on a silvertip shark (also on many other types of sharks).

accommodate large food items, hinged jaws that can unlock to open widely, and mouths that are huge in proportion to their bodies (Figure 14–12). These adaptations allow deep-sea fish to ingest species that are larger than they are and to process food efficiently whenever it is captured.

Deep-sea fish have special adaptations such as good sensory devices and bioluminescence that allow them to survive in this still and completely dark environment.

Box 14–1
Some Myths (and Facts) About Sharks

Sharks (Figure 14B) are the fish humans most fear. The strength, large size, sharp teeth, and unpredictable nature of sharks are enough to keep some people from *ever* entering the ocean. The publicity generated by the occasional shark attack on humans has led to many myths about sharks.

- *Myth #1: All sharks are dangerous.* Of the more than 350 shark species, about 80% are unable to hurt people or rarely encounter people. The largest shark—also the largest fish in the world—is the whale shark (*Rhincodon typus*), which reaches lengths of up to 15 meters (50 feet) but eats only plankton and so is not considered dangerous.

- *Myth #2: Sharks are voracious eaters that must eat continuously.* Like other large animals, sharks eat periodically depending upon their metabolism and the availability of food. Humans are not a primary food source of any shark, and many large sharks prefer the higher fat content of seals and sea lions.

- *Myth #3: Most people attacked by a shark are killed.* Of every 100 people attacked by sharks, 85 survive. Many large sharks commonly attack by biting their prey to immobilize it before trying to eat it. Consequently, many potential prey escape and survive.

- *Myth #4: Many people are killed by sharks each year.* The chances of a person being killed by a shark are quite low (Table 14A). Sharks kill an average of only 5-15 people worldwide each year. Humans, on the other hand, kill as many as 100 million sharks a year (mostly as bycatch from fishing activities). Compared to many other fishes, sharks have low reproduction rates and grow slowly, which may result in many sharks being designated as endangered species in the future.

- *Myth #5: The great white shark is a common, abundant species found off most beaches.* Great white sharks are relatively uncommon predators that prefer cooler waters. At most beaches, great whites are rarely encountered.

- *Myth #6: Sharks are not found in fresh water.* A specialized osmoregulatory system enables some species (such as the bull shark) to cope with dramatic changes in salinity, from the high salinity of seawater to the low salinity of freshwater rivers and lakes.

- *Myth #7: All sharks need to swim constantly.* Some sharks can remain at rest for long periods on the bottom and obtain enough oxygen by opening and closing their mouths to pump oxygen through their gills. Typically, sharks swim very slowly—cruising speeds are less than 9 kilometers (6 miles) per hour—but they can swim at bursts of over 37 kilometers (23 miles) per hour.

- *Myth #8: Sharks have poor vision.* The lens of a shark's eye is up to seven times more powerful than that of a human's. Sharks can even distinguish color.

- *Myth #9: Eating shark meat makes one aggressive.* There is no indication that eating shark meat will alter a person's temperament. The firm texture, white flesh, low fat content, and mild taste of shark meat have made it a favorite seafood in many countries.

- *Myth #10: No one would ever want to enter water filled with sharks.* Long regarded with fear and suspicion, sharks are more recently being viewed with fascination and wonder. Diving tours specializing in close encounters with sharks are becoming increasingly popular.

Figure 14B **Great white shark (*Carcharodon carcharias*).**

Table 14A **Types and number of occurrences in the United States.**

Occurrence to people in the U.S.	Average number per year
Highway fatalities	2611
Struck by lightning	352
Killed by lightning	94
Bit by a squirrel in New York City	88
Killed by a snake bite	15
Bit by a shark	10
Killed by a shark	0.4

Figure 14–11 Deep-sea fish.
(a) A hatchet fish. **(b)** A lantern fish.
(c) A stomiatoid. **(d)** A hatchet fish.
(e) A gulper eel. **(f)** A female deep-sea anglerfish, with attached parasitic male **(g)**.

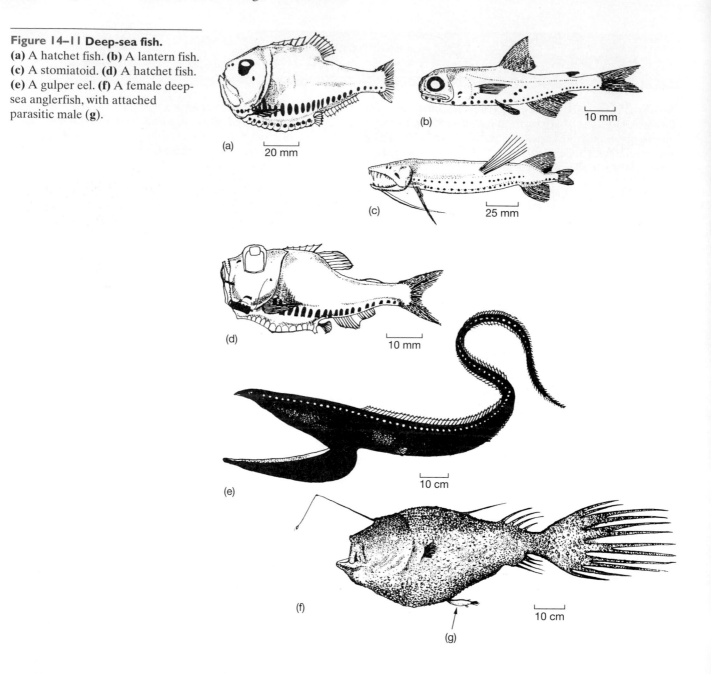

Adaptations for Seeking Prey

Several factors affect each species' adaptation for capturing food, including mobility (lunging versus cruising), speed, body length, body temperature, and circulatory system.

Lungers Versus Cruisers

Some fish wait patiently for prey and exert themselves only in short bursts as they lunge at the prey. Others cruise relentlessly through the water, seeking prey. A marked difference occurs in the musculature of fishes that use these different styles of obtaining food.

Lungers, such as the grouper in Figure 14–13a, sit and wait for prey to come close by. Lungers have truncate caudal fins for speed and maneuverability, and almost all their muscle tissue is white.

Cruisers, such as the tuna in Figure 14–13b, actively seek prey. Less than half of a cruiser's muscle tissue is white; most is red.

What is the significance of red versus white muscle tissue? *Red muscle fibers* are 25 to 50 microns in diameter (0.01 to 0.02 inch), whereas *white muscle fibers* are 135 microns (0.05 inch) and contain lower concentrations of **myoglobin** (*myo* = muscle, *globus* = sphere), a red pigment with an affinity for oxygen. Red fibers, therefore, get a much greater oxygen supply than white fibers. They supports a metabolic rate six times that of white fibers, which cruisers need for endurance.

Lungers need very little red tissue because they do not move continually. Instead, they need white tissue, which fatigues much more rapidly than red tissue, for quick bursts of speed to capture prey. Cruisers use white tissue, too, for short periods of acceleration while on the attack.

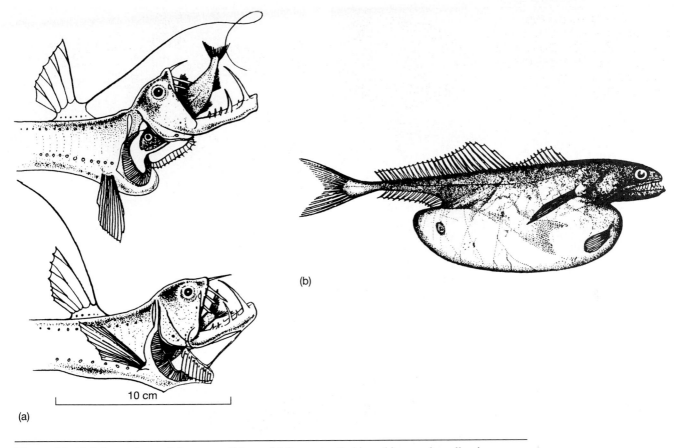

Figure 14–12 Adaptations of deep-sea fishes. (a) Large teeth, hinged jaw, and swallowing mechanism of the deep-sea viper fish *Chauliodus sloani*. **(b)** Ingestion capability of *Chiasmodon niger*, with a curled-up fish in its stomach that is longer than it is.

Speed and Body Size

Fish swim slowly when cruising, fast when hunting for prey, and fastest of all when trying to escape from predators. Generally, the larger the fish, the faster it can swim. For tuna, well adapted for sustained cruising and short, high-speed bursts, cruising speed averages about three body lengths per second. They can maintain a maximum speed of about 10 body lengths per second, but only for one second. A yellowfin tuna (*Thunnus albacares*) has been clocked at 74.6 kilometers (46 miles) per hour! This speed is more than 20 body lengths per second, but the tuna could maintain it for only a *fraction* of a second.

Theoretically, a 4-meter (13-foot) bluefin tuna (*Thunnus thynnus*) can reach speeds up to about 144 kilometers (90 miles) per hour.[5] Like fish, many of the toothed whales can swim fast, too. For instance, dolphins of the genus *Stenella* have been clocked at 40 kilometers (25 miles) per hour, and it is believed that the top speed of killer whales may exceed 55 kilometers (34 miles) per hour during short bursts.

Cold-Blooded Versus Warm-Blooded

Fish are mostly **cold-blooded** or **poikilothermic** (*poikilos* = spotted, *thermos* = heat), so their body temperatures are nearly the same as their environment. Usually, these fish are not fast swimmers. The mackerel (*Scomber*), yellowtail (*Seriola*), and bonito (*Sarda*), on the other hand, are fast swimmers. Their body temperatures are 1.3, 1.4, and 1.8°C (2.3, 2.5, and 3.2°F), respectively, above the surrounding seawater.

The mackerel shark (*Lamna* and *Isurus*) and tuna (*Thunnus*) genera have body temperatures much higher than their environment. Bluefin tuna can maintain a body temperature of 30 to 32°C (86 to 90°F) regardless of the water temperature, which is characteristic of **warm-blooded** or **homeothermic** (*homeo* = alike, *thermos* = heat) organisms. Although these tuna are more commonly found in warmer water, where the temperature difference between fish and water is no more than 5°C (9°F), body temperatures of 30°C (86°F) have been measured in bluefin tuna swimming in 7°C (45°F) water.

Why do these fish exert so much energy to maintain their body temperatures at high levels, when other fish do quite well with ambient body temperatures? It may be that for cruisers, any adaptation (high temperature

---

[5]Imagine how difficult it would be to clock a bluefin tuna accurately in the ocean at this speed!

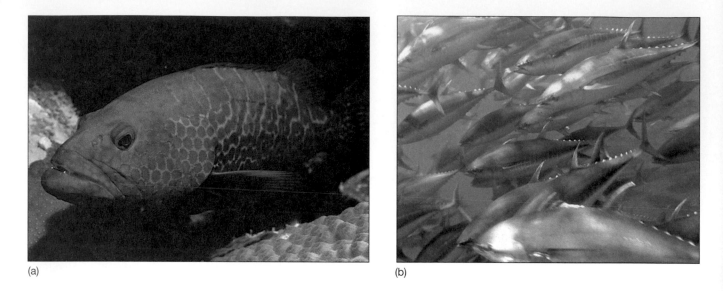

(a)   (b)

Figure 14–13 Feeding styles—lungers and cruisers. (a) Lungers, such as this tiger grouper, sit patiently on the bottom and capture prey with quick, short lunges. **(b)** Cruisers, such as these yellowfin tuna, swim constantly in search of prey and capture it with short periods of high-speed swimming.

and high metabolic rate) that increases the power output of their muscle tissue helps them seek and capture prey.

Circulatory System Modifications

A modified circulatory system (Figure 14–14) helps mackerel sharks and tuna maintain their high body temperatures. Most fish have a **dorsal aorta** located just beneath the vertebral column that provides blood to the swimming muscles. Some fish, such as mackerel sharks and tuna, have additional **cutaneous** (*cutane* = skin) **arteries** just beneath the skin on either side of the body. As cool blood flows into red muscle tissue, muscle contractions generate heat that increases the temperature of the blood. A fine network of tiny blood vessels within the muscle tissue is designed to minimize heat loss.

The vessels that return the blood to the **cutaneous vein**, parallel to the cutaneous artery along the side of the fish, are all paired with small vessels carrying blood into the muscle tissue. In this way, the warm blood leaving the tissue helps to heat the cooler blood entering from the cutaneous artery.

> Adaptations of pelagic organisms for seeking prey include mobility (lunging versus cruising), high swimming speed, body length, high body temperature, and a modified circulatory system.

Adaptations to Avoid Being Prey

Obtaining food occupies most of the time of many inhabitants of the open ocean. Some animals are fast and

agile, and obtain food through active predation. Other animals, called **filter feeders**, move more leisurely as they filter their small prey from seawater.

Many animals have unique adaptations to avoid being captured and eaten. Some fish use speed to avoid predators. Others use transparency, camouflage, or countershading, as discussed in Chapter 12 ("The Marine Habitat"). Still others use schooling.

Schooling

Although vast populations of phytoplankton and zooplankton may be highly concentrated in certain areas of the ocean, they are not usually referred to as schools. The term **school** is usually reserved for large numbers of fish, squid, or shrimp that form well-defined social groupings.

The number of individuals in a school can vary from a few larger predaceous fish (such as bluefin tuna) to hundreds of thousands of small filter feeders (such as anchovies). Within the school, individuals move in the same direction and are evenly spaced. Spacing is probably maintained through visual contact and, in the case of fish, by use of the lateral line system (see Figure 14–9) that detects vibrations of swimming neighbors. The school can turn abruptly or reverse direction as individuals at the head or rear of the school assume leadership positions (Figure 14–15).

What is the advantage of schooling? During spawning, schooling ensures that there will be males to release sperm to fertilize the eggs released into the water or deposited on the bottom by females. The most important function of schooling in small fish, however, is protection from predators.

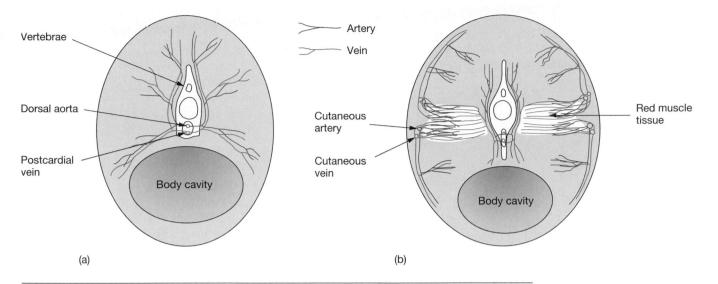

Figure 14–14 Circulatory modifications in fishes. (a) Most cold-blooded (poikilothermic) fishes have major blood vessels arranged so that blood flows to muscle tissue from the dorsal aorta and returns to the postcardial vein beneath the vertebral column. **(b)** Warm-blooded (homeothermic) fishes such as the bluefin tuna have cutaneous arteries and veins that help maintain high blood temperature by using heat energy generated by contracting muscle tissue.

Figure 14–15 Schooling. A school of soldier fish near a reef in the Maldives, Indian Ocean.

It may seem illogical that schooling would be protective. For instance, schooling creates tighter groupings of organisms so that any predator lunging into a school would surely catch something, just as land predators run a herd of grazing animals until one weakens and becomes a meal. So, aren't the smaller fish making it easier for the predators by forming a large target?

Scientists who study fish behavior suggest that schooling does indeed serve to protect a group of organisms based on the following strategies that give them "safety in numbers":

1. If members of a species form schools, they reduce the percentage of ocean volume in which a cruising predator might find one of their kind.

2. Should a predator encounter a large school, it is less likely to consume the entire unit than if it encounters a small school or an individual.

3. The school may appear as a single large and dangerous opponent to the potential predator and prevent some attacks.

4. Predators may find the continually changing position and direction of movement of fish within the school confusing, making attack particularly difficult for predators, who can attack only one fish at a time.

In addition, the fact that over 2000 fish species are known to form schools and half of all fishes join schools during a portion of their lives suggests that schooling enhances survival of species, especially for those with no other means of defense. Schooling may also help fish swim greater distances than individuals because each schooling fish gets a boost from the vortex created by the fish swimming in front of it.

> Many pelagic species (especially fish) school to increase their chances of avoiding predators.

Marine Mammals

All organisms in class Mammalia (including marine mammals) share the following characteristics:

- They are warm-blooded
- They breathe air
- They have hair (or fur) in at least some stage of their development
- They bear live young[6]
- The females of each species have mammary glands that produce milk for their young

Recent discoveries of ancient whale fossils in Egypt, Pakistan, and China provide strong evidence that marine mammals evolved from mammals on land about 50 million years ago. Some have small, unusable hind legs, suggesting that the land mammal predecessor had no need for its hind legs when it developed a large paddle-

---

[6]This is true except for a few egg-laying mammals of Australia from the subclass Prototheria, which include the duck-billed platypus and the spiny anteater (echidna).

shaped structure for a tail used to swim through water. Many anatomical similarities between land mammals and marine mammals are observed, too. Mammals, which originally evolved from organisms that inhabited the sea millions of years ago, may have returned to the sea because of more abundant food sources.

Marine mammals include at least 116 species within the orders Carnivora, Sirenia, and Cetacea.

Order Carnivora

Animals within order **Carnivora** (*carni* = meat, *vora* = eat) have prominent canine teeth, including the cat and dog families on land. Marine representatives or order Carnivora include sea otters, polar bears, and the **pinnipeds** (*pinni* = feather, *ped* = a foot). The name pinniped describes these organisms' prominent skin-covered flippers, well adapted for propelling them through water and include walruses, seals, sea lions, and fur seals.

Sea otters (Figure 14–16a) lack an insulating layer of blubber but have extremely dense fur. This fur was highly sought for pelts, causing sea otters to be hunted to the brink of extinction in the late 1800s. Fortunately, they have made a remarkable comeback and now inhabit most areas where they were formerly hunted. Sea otters are some of the smallest marine mammals, and seem particularly playful because they continually scratch themselves, which serves to clean their fur and adds an insulating layer of air. They eat various shellfish and crustaceans, and often use a rock to break open the shells of their food while floating at the surface on their backs. Sea otters commonly inhabit kelp beds.

Polar bears (Figure 14–16b) have massive webbed paws that make them excellent swimmers. The polar bear's fur is thick and each hair is hollow for better insulation. Polar bears also have large teeth and sharp

(a)

(b)

Figure 14–16 Marine mammals of order Carnivora. (a) Sea otter. (b) Polar bear.

claws, which they use for prying and killing. Their diet consists mainly of seals, which they often capture at holes in the Arctic ice when the seals come up for a breath of air.

Walruses have large bodies and adults (both male and female) have ivory tusks up to 1 meter (3 feet) long (Figure 14–17a), which are used for territorial fighting, for hauling themselves onto icebergs, and sometimes for stabbing their prey.

Seals—also called the *earless seals* or *true seals*—differ from the **sea lions** and **fur seals**—also called the *eared seals*—in the following ways:

- Seals lack prominent ear flaps that are specific to sea lions and fur seals (compare Figures 14–17b and 14–17c).
- Seals have smaller and less-prominent front flippers (called *fore flippers*) than sea lions and fur seals.
- Seals have prominent claws that extend from their fore flippers that sea lions and fur seals lack (Figure 14–18).
- Seals have a different hip structure than sea lions and fur seals. Thus, seals cannot move their rear flippers underneath their bodies as sea lions and fur seals can (Figure 14–18).
- Seals with their smaller front flippers and different hip structure do not move around on land very well, and can only slither along like a caterpillar. Sea lions and fur seals, on the other hand, use their large front flippers and their rear flippers, which they can turn under their bodies, to walk easily on land, and can even ascend steep slopes, climb stairs, and do other acrobatic tricks.
- Seals propel themselves through the water using a back-and-forth motion of their rear flippers (similar to a wagging tail), whereas sea lions and fur seals flap their large front flippers.

Order Sirenia

Animals of order **Sirenia** (*siren* = a mythical mermaid-like creature with an enticing voice) include the manatees and dugongs, collectively known as "sea cows." Manatees are concentrated in coastal areas of the tropical Atlantic Ocean, while the dugongs populate the tropical regions of the Indian and Western Pacific Oceans. Both of these animals have a paddle-like tail and rounded front flippers (Figure 14–19). The land-dwelling ancestors of sirenians were elephant-like, and sirenians today have large bodies and manatees have nails on their front flippers. The sirenians have sparse hairs covering their bodies, concentrated around the mouth.

Sirenians eat only shallow-water grasses and are thus the only vegetarian marine mammals. They spend most of their lives in heavily traveled shallow coastal waters, resulting in much competition with humans for space. In addition, there have been many accidents, with boats

(a)

(b)

(c)

Figure 14–17 Marine mammals of order Carnivora (pinnipeds). (a) Walruses. **(b)** Harbor seals. **(c)** California sea lions.

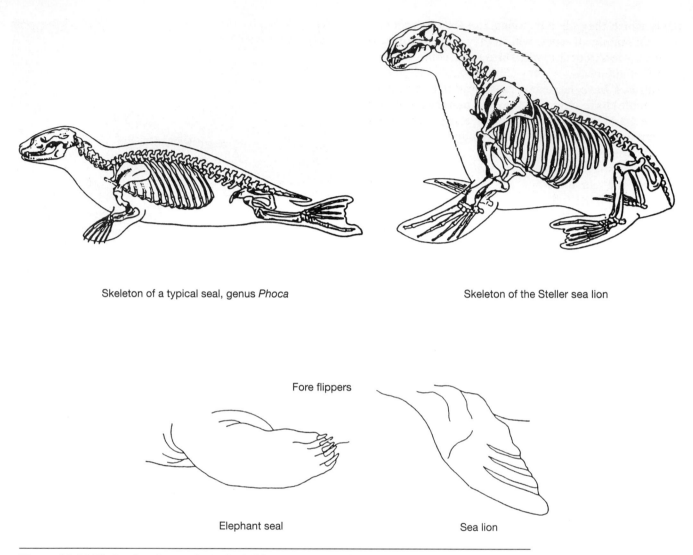

Skeleton of a typical seal, genus *Phoca*

Skeleton of the Steller sea lion

Fore flippers

Elephant seal

Sea lion

Figure 14–18 Skeletal and morphological differences between seals and sea lions.

running over these large animals that move slowly and cannot be easily seen. The populations of manatees and dugongs have been decreasing, and both are considered endangered.

Order Cetacea

The order **Cetacea** (*cetus* = a whale) includes the whales, dolphins, and porpoises (Figure 14–20). The cetacean body is more or less cigar shaped and insulated with a thick layer of blubber. Cetacean forelimbs are modified into flippers that move only at the "shoulder" joint. The hind limbs are vestigial (rudimentary), not attached to the rest of the skeleton, and are usually not visible externally. All cetaceans share the following characteristics:

- An elongated (telescoped) skull
- Blowholes on top of the skull
- Very few hairs
- A horizontal tail fin called a *fluke* that is used for propulsion by vertical movements

These characteristics make cetacean's bodies very streamlined, allowing them to be excellent swimmers.

Modifications to Increase Swimming Speed
Cetaceans' muscles are not a great deal more powerful than those of other mammals, so their ability to swim fast must result from modifications that reduce frictional drag. The muscles of a small dolphin, for example, would need to be five times stronger than they are to swim at 40 kilometers (25 miles) per hour in turbulent flow.

In addition to a streamlined body, cetaceans improve the flow of water around their bodies with a specialized skin structure. Their skin consists of a soft outer layer that is 80% water and has narrow canals filled with spongy material, and a stiffer inner layer composed mostly of tough connective tissue. The soft layer decreases the pressure differences at the skin–water interface by compressing when pressure is high and expanding when pressure is low, reducing turbulence and drag.

Modifications to Allow Deep Diving Humans can free-dive to a maximum depth of 130 meters (428 feet) and hold their breath in rare instances for up to six minutes. In contrast, sperm whales (*Physeter macrocephalus*) dive deeper than 2800 meters (9200 feet), and northern bottlenose whales (*Hyperoodon ampullatus*) can stay submerged for up to two hours. These remarkable feats require special adaptations, such as special structures that allow them to use oxygen efficiently and an ability to resist nitrogen narcosis.

Cetacean Breathing Figure 14–21 shows the internal structures that allow cetaceans to remain submerged for extended periods. Inhaled air finds its way to tiny terminal chambers, the **alveoli** (*alveus* = a small hollow). Alveoli are lined with a thin alveolar membrane that is in contact with a dense bed of capillaries. The exchange of gases between the inhaled air and the blood (oxygen in, carbon dioxide out) occurs across the alveolar membrane. Some cetaceans have an exceptionally large concentration of capillaries surrounding the alveoli (Figure 14–21b), which have muscles that move air against the membrane by repeatedly contracting and expanding.

Cetaceans take from one to three breaths per minute while resting, compared with about 15 in humans. Because they hold the inhaled breath much longer, and because of the large capillary mass in contact with the alveolar membrane and the circulation of the air by muscular action, cetaceans can extract almost 90% of the oxygen in each breath, whereas terrestrial mammals extract only 4 to 20%.

To use oxygen efficiently during long dives, cetaceans store it and reduce their use of it. The storage of so much oxygen is possible because prolonged divers have such a large blood volume per unit of body mass.

Some cetaceans have twice as many red blood cells per unit of blood volume and up to nine times as much myoglobin in their muscle tissue as terrestrial animals. Large supplies of oxygen can be chemically stored in the **hemoglobin** (*hemo* = blood, *globus* = sphere) of the red blood cells and the myoglobin of the muscles.

Additionally, muscles at the start of a dive that have a significant oxygen supply can continue to function through anaerobic respiration when oxygen is used up. The muscle tissue is relatively insensitive to high levels of carbon dioxide.

Research has shown that cetaceans' swimming muscles can function without oxygen during a dive. This suggests that these muscles and other organs, such as the digestive tract and kidneys, may be sealed off from the circulatory system by constriction of key arteries. The circulatory system would then service only essential components, such as the heart and brain. Because of the decreased circulatory requirements, the heart rate can be reduced by 20 to 50% of normal. Other research has shown, however, that no such reduction in heart rate occurs during dives by the common dolphin (*Delphinus delphis*), the white whale (*Delphinapterus leucas*), or the bottlenose dolphin (*Tursiops truncatus*).

(a)

(b)

Figure 14–19 Marine mammals of order Sirenia.
(a) Manatees. **(b)** Dugong.

Nitrogen Narcosis Another difficulty with deep and prolonged dives is the absorption of compressed gases into the blood. When humans dive using compressed air—which includes nitrogen and oxygen—they can experience nitrogen narcosis or decompression sickness (the "bends").[7] The effect of nitrogen narcosis is similar to drunkenness, and it can occur when a diver either goes too deep or stays too long at depths greater than 30 meters (100 feet).

If a diver surfaces too rapidly, the lungs cannot remove the excess gases fast enough, and the reduced

---

[7]The effects on human physiology in response to diving into the marine environment are discussed in the chapter-opening feature in Chapter 1, "Introduction to Planet 'Earth'."

NORTHERN RIGHT WHALE (*Eubalaena glacialis*)
Baleen in northern right whales can reach 2.8 m (9 ft) in length and
is used to strain copepods and krill from surface waters. A similar
species, the southern right whale, is found in high southern latitudes.
Length is to 18 m (60 ft).

SPERM WHALE (*Physeter macrocephalus*)
Found mostly in tropical waters, these deep-diving toothed whales
have a huge snout that contains a large amount of oil. Length of male
is to 18 m (59 ft); length of female is to 10.5 m (35 ft).

Figure 14–20 Marine mammals of order Cetacea. A composite drawing of representatives of
the two whale suborders, drawn to relative scale. The toothed whales (Odontoceti) form complex
social communities and include the bottlenose dolphin, killer whale, narwhal, and sperm whale.
The baleen whales (Mysticeti) are the largest of all whales and include the right whale, gray whale,
humpback whale, and blue whale.

BOTTLENOSE DOLPHIN (*Tursiops truncatus*)
Found in all oceans, this is one of the many species of oceanic dolphins. Length is to 3 m (10 ft).

KILLER WHALE (*Orcinus orca*)
Cosmopolitan in distribution, this whale is unique in that it not only feeds on fish but also on seals, birds, and other whales. The male has a larger dorsal fin than the female. Length is to 9 m (31 ft).

NARWHAL (*Monodon monoceros*)
The scientific name, which means "one tooth, one horn" is accurate for the male of this Arctic whale. Length is to 6 m (20 ft).

GRAY WHALE (*Eschrichtius robustus*)
This bottom-feeding baleen whale has a small head and very reduced dorsal fin. It stays close to shore, and even enters the surf zone. Length is to 14 m (45 ft).

HUMPBACK WHALE (*Megaptera novaeangliae*)
These baleen whales produce some of the most complex songs of any marine mammal and are known for their energetic and athletic behaviors, such as breaching. The genus name is derived from their extraordinary large wing-like flippers. Length is to 16 m (52 ft).

BLUE WHALE (*Balaenoptera musculus*)
Probably the largest animal to ever inhabit the Earth, these baleen whales weigh up to 150 tons and can consume over 5 tons of krill per day. Length is to 30 m (100 ft).

Figure 14–21 Cetacean modifications to allow prolonged submergence. (a) Basic lung design. Air enters the lung through the trachea, and oxygen is absorbed into the blood through the walls of the alveoli. **(b)** Oxygen exchange in the alveolus. A dense mat of capillaries receives oxygen through the alveolar membrane, allowing whales to extract as much as 90% of oxygen from each breath.

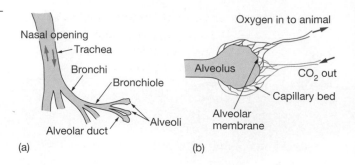

(a)   (b)

pressure may cause small bubbles to form in the blood and tissue. The bubbles interfere with blood circulation, and the resulting decompression sickness can cause excruciating pain, severe physical debilitation, or even death.

Cetaceans and other marine mammals do not suffer from these difficulties. By the time a cetacean has reached a depth of 70 meters (230 feet), its rib cage has collapsed under the 8 kilograms per square centimeter (8 atmospheres, or 118 pounds per square inch) of pressure. The lungs within the rib cage also collapse, removing all air from the alveoli. This, in turn, prevents the blood from absorbing additional gases across the alveolar membrane, so nitrogen narcosis cannot occur.

It is possible, however, that a collapsible rib cage is not the main defense against nitrogen narcosis. When enough nitrogen was put into the tissue of a dolphin to give a human a severe case of the "bends," for instance, the dolphin suffered no ill effects. This dolphin, as well as other marine mammal species, may have simply evolved an insensitivity to nitrogen gas.

Suborder Odontoceti   Members of order Cetacea can be divided into the suborders Odontoceti (the toothed whales) and Mysticeti (the baleen whales). Suborder **Odontoceti** (*odonto* = a tooth, *cetus* = a whale) includes the killer whale, sperm whale,[8] porpoises, and dolphins.

Differences between Dolphins and Porpoises   Dolphins and porpoises are small toothed whales of suborder Odontoceti. They have similarities in appearance, behavior, and range, so are easily confused. For instance, both dolphins and porpoises (as well as seals, sea lions, and fur seals) can exhibit a behavior known as "porpoising," which is leaping out of the water while swimming. However, there are several morphological differences between dolphins and porpoises.

Porpoises are somewhat smaller and have a more "stout" (bulky and robust) body shape compared to the more elongated and streamlined dolphins. Generally, porpoises have a blunt snout (rostrum) while dolphins have a longer rostrum. Porpoises have a smaller and more triangular (or, on one species, no) dorsal fin,

whereas a dolphin's dorsal fin is sickle-shaped or **falcate** (*falcatus* = sickle) and appears hooked and curved backwards in profile view.

They also have differences in the shape of their teeth, although it is often difficult to get close enough to see them. The teeth of dolphins end in points, while the teeth of porpoises are blunt or flat (shovel shaped) and resemble our incisors (front teeth). Killer whales have teeth that end in points (Figure 14–22) and thus are members of the dolphin family.

Characteristics of Suborder Odontoceti   All toothed whales have prominent teeth, which are used to hold and orient fish and squid, although the killer whale is known to feed on a variety of larger animals, including other whales. The toothed whales form complex and long-lived social groups. Toothed whales have one external nasal opening (blowhole), while baleen whales have two. Although both toothed and baleen whales can emit and receive sounds, the ability to use sound is best developed in toothed whales (particularly sperm whales, which are the most vocal cetaceans).

Despite their lack of vocal cords, toothed whales can produce a variety of sounds—some of which are within the range of human hearing. Sounds are emitted from the blowhole or, in sperm whales, near a special

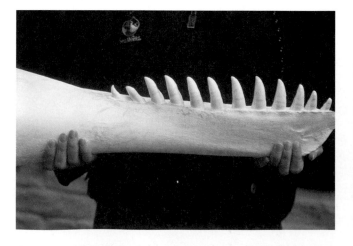

Figure 14–22 Jawbone of a killer whale. The lower jawbone of a killer whale (*Orcinus orca*), showing large teeth that end in points. Thus, killer whales are in the dolphin family.

---

[8]Recent DNA studies suggest that the sperm whale is more closely related to baleen whales than to other toothed whales.

structure called the *museau du singe* ("monkey's muzzle") (Figure 14–23). Contractions of muscles in these structures produce sound that reflects off the front of the skull, which is bowl shaped and resembles a radar dish. The sound is concentrated as it passes through an organ called the **melon** (or the **spermaceti organ** in sperm whales). This organ can form various shapes and sizes of lenses (Figure 14–23a, *inset*) and acts as an acoustical lens, focusing the sound. Speculations about the sounds' purpose range from **echolocation**—using sound to determine the direction and distance of ob-

jects—(clearly true) to a highly developed language (doubtful). In fact, what marine biologists know about cetaceans' use of sound is limited.

All marine mammals have good vision, but conditions often limit its effectiveness. In coastal waters (where suspended sediment and dense plankton blooms make the water turbid) and in deeper waters (where light is limited or absent), echolocation surely assists the pursuit of prey or location of objects.

Using lower-frequency clicks at great distance and higher frequency at closer range, the bottlenose dolphin

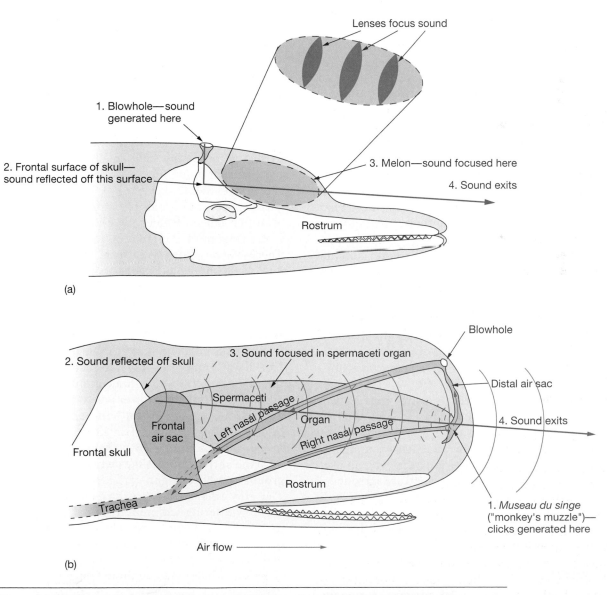

Figure 14–23 Generation of echolocation clicks in small toothed and sperm whales.
(a) In small toothed whales, clicks are generated within the blowhole, reflected off the frontal surface of the skull, and then focused by the melon, which can form lenses to focus the sound (*inset*). **(b)** In the sperm whale, air passes from the trachea through the right nasal passage across the *museau du singe* ("monkey's muzzle"), where clicks are generated. The air passes along the distal air sac, past the closed blowhole, and returns to the lungs along the left nasal passage. The sounds are reflected off the frontal surface of the skull and are focused by the spermaceti organ before leaving the whale.

Figure 14–24 Echolocation. Sounds are generated by toothed whales and bounced off objects in the ocean to determine their size, shape, distance, movement, density—and even internal structure.

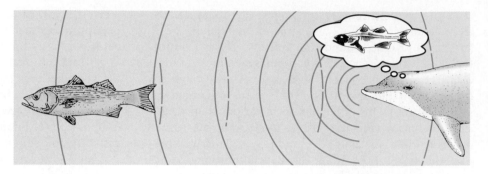

(*Tursiops truncatus*) can detect a school of fish at distances exceeding 100 meters (330 feet). It can pick out an individual fish 13.5 centimeters (5.3 inches) long at a distance of 9 meters (30 feet). Sperm whales, moreover, can detect their main prey— squid—from distances of up to 400 meters (1300 feet).

To locate an object and determine its distance, toothed whales send sound signals through the water, some of which are reflected from various objects and are returned to the animal and interpreted (Figure 14–24). Because sound penetrates objects, echolocation can produce a three-dimensional image of the object's internal structure and density (which is more than eyesight alone can do). Recent research indicates that the cetaceans may also use a sharp burst of sound to stun their prey before they close in for the kill.

How reflected sound is received is not yet fully understood. In most cetaceans, the bony housing of the inner ear is fused to the skull. When submerged, sounds transmitted through the water are picked up by the skull and travel to the hearing structure from many directions, which makes it impossible to accurately locate the source of the sounds. This kind of hearing structure would not work for an animal that depends on echolocation to find objects in water.

All cetaceans have evolved structures that insulate the inner ear housing from the rest of the skull. In toothed whales, the inner ear is separated from the rest of the skull and surrounded by an extensive system of air sinuses (cavities). The sinuses are filled with an insulating emulsion of oil, mucus, and air and are surrounded by fibrous connective tissue and venous networks. In many toothed whales, it is believed that sound is picked up by the thin, flaring jawbone and passed to the inner ear via the connecting oil-filled body.[9]

◼EIO◼

For more information and on-line exercises about the Environmental Issue in Oceanography (EIO) "Sonar and Whales," visit the EIO Web site at **http://www.prenhall. com/oceanissues** and select Issue #2.

Suborder Mysticeti   Suborder **Mysticeti** (*mystic* = a moustache, *cetus* = a whale)—also known as the baleen whales—includes the world's largest whales (the blue whale, finback whale, and humpback whale) and the gray whale (a bottom feeder).

Baleen whales are generally much larger than toothed whales because of their different food sources. Baleen whales eat lower on the food web (including zooplankton such as krill and small nektonic organisms), which are relatively abundant in the marine environment. How are the largest whales in the world able to survive on eating such small prey, especially when these smaller organisms are widely dispersed in the marine environment?

To concentrate small prey and separate them from seawater, baleen whales have parallel rows of **baleen** plates (Figure 14–25a) in their mouths instead of teeth. These baleen plates hang from the whale's upper jaw and, when the whale opens its mouth, the baleen resembles a moustache (except that it is on the *inside* of their mouths), which is why these whales are sometimes called the moustached whales (Figure 14–25b). Baleen is made of flexible keratin—the same as human nails and hair—and can be up to 4.3 meters (14 feet) long[10] (Figure 14–25c). To feed, baleen whales fill their mouths with water that contains their prey items, allowing their pleated lower jaw to balloon in size. The whales force the water out between the fibrous plates of baleen, trapping small fish, krill, and other plankton inside their mouths. Mostly, baleen whales feed at or near the surface, sometimes working together in large groups and surfacing in vertical lunges (Figure 14–26). The gray whale, however, has short baleen slats and feeds by filtering sediment from the shallow bottom of its North Pacific and Arctic feeding grounds, straining out a diet of benthic organisms including amphipods and shellfish.

Baleen whales include three families:

1. The **gray whale**, which has short, coarse baleen, no dorsal fin, and only two to five ventral grooves on its lower jaw

---

[9]To simulate this, try pushing the end of a vibrating tuning fork into your chin. The sound is transmitted through your jaw directly to your ear.

[10]Baleen (also called whalebone) was used for such items as buggy whips and corset stays before synthetic materials were substituted.

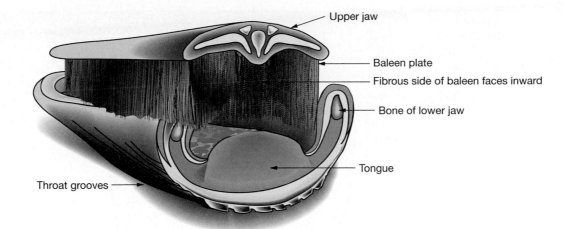

Upper jaw

Baleen plate

Fibrous side of baleen faces inward

Bone of lower jaw

Tongue

Throat grooves

(a)

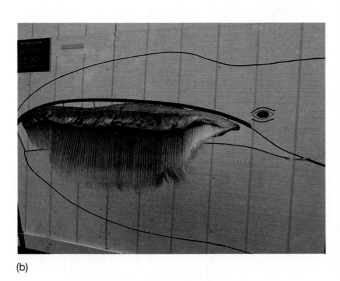

(b)

(c)

Figure 14–25 Baleen. (a) A diagrammatic cross section through the head of a typical baleen whale. The baleen plates, hanging from the upper jaw, form a sieve that allows these whales to concentrate and eat large quantities of smaller organisms. **(b)** A rack of baleen from a bottom-feeding gray whale. **(c)** Individual slat of baleen from a surface-skimming northern right whale.

2. The **rorqual**[11] **whales**, which have short baleen, many ventral grooves, and are divided into two subfamilies:

   a. The *balaenopterids*, which have long, slender bodies, small sickle-shaped dorsal fins, and flukes with smooth edges (minke, Bryde's, sei, fin, and blue whales)

   b. The *megapterids*, or humpback whales, which have a more robust body, long flippers, flukes with uneven trailing edges, tiny dorsal fins, and tubercles (a row of large bumps) on the head

3. The **right whales**[12], which have long, fine baleen, broad triangular flukes, no dorsal fin, and no ventral grooves. The northern right whale is most threatened with extinction. The other members of

---

[11]The term "rorqual" refers to the longitudinal grooves on the lower jaw, called rorqual folds. The term rorqual is from the Old Norse term *raudhr*, meaning "red."

[12]The right whales are so named because, from a whaler's point of view, they were the "right" whales to take.

Figure 14–26 Cooperative feeding by humpback whales. A humpback whale (*Megaptera novaeangliae*) surfaces in a vertical lunge with mouth open during cooperative feeding, where groups of humpback whales work together as a team to capture prey. Baleen plates occur as parallel rows on the upper jaw that are separated by the roof of the mouth (*pink*). The expanded throat grooves can also be seen near the water line.

this family are the Southern Hemisphere's southern right whale and the bowhead whale, which remains near the edge of the Arctic pack-ice.

Baleen whales also produce sound, but at much lower frequencies than toothed whales. Gray whales produce pulses, (possibly for echolocation) and moans that may maintain contact with other gray whales. Rorqual whales produce moans that last from one to many seconds. These sounds are extremely low in frequency, and are probably used to communicate over distances of up to 50 kilometers (31 miles). Blue whales produce sounds that may travel along the SOFAR channel across entire ocean basins (see Box 6–3 for information about the SOFAR channel). Songs of humpback whales are thought to be a form of sexual display, but it is unclear whether its main purpose is to repel other males or to attract females.

> Marine mammals include the order Carnivora (sea otters, polar bears, and pinnipeds—walrus, seals, sea lions, and fur seals); order Sirenia (manatees and dugongs); and Cetacea (whales, dolphins, and porpoises).

An Example of Migration: Gray Whales

Fish, sea turtles, and marine mammals migrate seasonally. Some of the longest migrations known in the open ocean are those of baleen whales, and one of the best-studied is that of the gray whale (*Eschrichtius robustus*, often called the "California" gray whale).

The migratory routes of commercially important baleen whales (including gray whales) have been well known since the mid–1800s. Marine mammals are rela-

tively easy to track because they must surface periodically for air. Gray whales are particularly easy to track because they spend their entire lives in nearshore areas. More recently, radio tracking of individual gray whales has delineated the route and timing of the migration.

Why Gray Whales Migrate Gray whales undertake a 22,000-kilometer (13,700-mile) round-trip journey every year, the longest known migration of any mammal. Gray whales feed in cold high-latitude waters in the coastal Arctic Ocean and the far northern Pacific Ocean near Alaska; they breed and give birth in warm tropical lagoons along the west coast of Baja California and mainland Mexico (Figure 14–27). Feeding occurs during summer when the long hours of sunlight produce a vast feast of crustaceans and other bottom-dwelling organisms in the shallow Bering, Chukchi, and Beaufort Seas. Without this bountiful food, the whales could not sustain themselves during the long migrations and the mating and calving season, when feeding is minimal.

Initially, gray whales were thought to migrate so far because the physical environment of their cold-water feeding grounds does not meet the needs of young gray whales. Recent research on the physiology of newborn gray whales indicates, however, that gray whale calves can survive in much colder water. So why do they migrate? Perhaps the migration is a relic from the Ice Age, when sea level was lower. During that time, the feeding grounds that are so productive today were above sea level. Hence, gray whales could not feast on the abundant food and probably gave birth to smaller calves that could not survive in the cold water. This necessitated the migration to warmer-water regions, which continues to this day despite the abundant food supply.

Alternatively, gray whales may have left the colder waters to avoid killer whales, which are more numerous there and are a major threat to young whales. This may also explain why only those lagoons in Mexico with shallow entrances are used for calving. Killer whales have been seen near the lagoons, however, and have also been observed to feed in extremely shallow water (see Figure 14C).

Timing of Migration   The timing of the gray whale migration is closely linked to physical oceanographic conditions. The migration usually begins in September,

after the high latitude summer bloom in productivity has peaked. By this time, the whales have stored enough fat to last them until they return to high-latitude waters. Gray whales have been observed to feed during their migration, however, when the opportunity presents itself. When pack ice begins to form over their feeding grounds along the continental shelf, the whales move south.

First to leave are the pregnant females. They are followed by the nonpregnant mature females, immature females, mature males, and then immature males. After navigating through passages between the Aleutian

Box 14–2
Killer Whales: A Reputation Deserved?

Killer whales (*Orcinus orca*) are sleek and powerful predatory animals that inhabit all oceans of the world. Once considered ferocious man-eaters, it turns out they are social animals that don't harm people. Why, then, are they called *killer* whales?

Killer whales prey mainly on fish. When fish are not plentiful, some feed on other dolphins, large whales, penguins, birds, squid, turtles, seals, or sea lions (Figure 14C). Killer whales often hunt in groups, employing various strategies, including deception, to trap their prey. Their distinctive black-and-white coloration—a type of **disruptive coloration**—and white eye patch often confuse their prey as they close in for the kill. Killer whales have often been called "wolves of the sea" because of their group hunting tactics.

Killer whales often play with their food before killing it. Similar to the way a cat plays with a mouse, killer whales have been known to hone their hunting skills by releasing their prey only to catch it over and over again. Sometimes,

they throw their prey completely out of the water. These behaviors appear cruel, so they may have contributed to the whale's common name.

Killer whales have rammed and sunk boats many times and have even been known to tug on divers' fins, but there is only one documented case of a killer whale ever killing a human. This occurred in 1991 when trainer Keltie Byrne of the marine park Sealand of the Pacific in Victoria, Canada, accidentally slipped into the killer whale pool. As she attempted to climb out, a killer whale pulled her back into the water, where she was tossed about by killer whales in the pool for 10 minutes. Finally, one whale carried her underwater in its mouth for longer than she could hold her breath, causing her to drown. Evidently, the killer whales saw her as a new play object and inadvertently killed her. In the wild, however, there has never been a documented case where killer whales have maliciously attacked and killed a human being.

Figure 14C Killer whale (*Orcinus orca*) hunting sea lions near Valdes, Argentina.

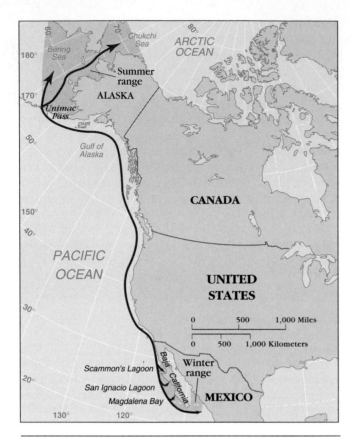

Figure 14–27 Gray whale migration route. Gray whales (*Eschrichtius robustus*) undertake as much as a 22,000 kilometers (13,700 miles) annual migration, the longest of any mammal. They migrate from Arctic summer feeding areas in the Bering and Chukchi Seas to warmer winter breeding and calving lagoons offshore Mexico.

Islands, they follow the coast throughout their southern journey. Traveling about 200 kilometers (125 miles) per day, most reach the lagoons of Baja California by the end of January.

In these warm-water lagoons, the pregnant females give birth to 2-ton calves. The calves nurse on milk that is almost half butterfat and has the consistency of cheese, allowing the calves to put on weight quickly during the next two months. While the calves are nursing, the mature males breed with the mature females that did not bear calves. Producing large offspring (the gestation period is up to one year) and providing them with fat-rich milk for several months requires enormous amounts of energy, so it is not uncommon for females to mate only once every two or three years.

Late in March, they return north in reverse order (beginning, that is, with the immature males). Most of the whales are back in their high-latitude feeding grounds by the end of June, which coincides with the beginning of the summer bloom in productivity. The whales feed on prodigious quantities of bottom-dwelling organisms to replenish their depleted stores of fat and blubber before their next trip south.

Gray whales undertake the longest migration of any mammal, traveling from high latitude Arctic summer feeding areas to low latitude winter birthing and breeding lagoons in Mexico.

Students Sometimes Ask...

Those deep-sea fish look frightening. Do they ever come to the surface? Are they related to piranhas?

Deep-sea fish live in a world of darkness and never come to the surface. That is fortunate for us, because they are vicious predators. They are only distantly related to piranhas (they are within the same group of bony fishes), but their similarly adapted large, sharp teeth may be a good example of *convergent evolution*: the evolution of similar characteristics on different organisms independent of one another yet adapted to the same problem (in this case, a low food supply).

In a battle between a killer whale and a great white shark, which one would win?

Although many people who are fascinated with large and powerful wild animals have often wondered which of the two would win such a fight, there was little evidence to settle the dispute until recently. A remarkable video was taken in waters off northern California in 1997, documenting a battle between a 6-meter (20-foot) juvenile killer whale and a 3.6-meter (12-foot) adult great white shark. The video clearly shows that the killer whale completely severed the shark's head! If this is representative of the way these two animals interact in the wild, then the killer whale is the top carnivore in the ocean. It is believed that the killer whale's superior maneuverability and use of echolocation helped it conquer the shark.

Do sharks ever get cancer?

Legend has it that sharks don't get cancer, making the creature's cartilage popular in the alternative health market as a cure for the disease. However, researchers who study tumors in animals have recently reported that sharks (and their close relatives, skates and rays) can and do get cancer.

I've seen dolphins swimming and playing around just in front of the bow of a moving ship. Are they in danger of being run over by the ship?

Not really. Dolphins or porpoises that swim in front of a moving ship are actually surfing the "bow wave" that forms as the ship plies through the water. Studies on trained dolphins conducted during this activity reveal that their respiratory rates are just barely above resting. In essence, they are being pushed along and catching a free ride from the ship! The dolphins or porpoises will stay with the ship as long as the ship travels in the direction they're going—or until a food source is located—which may be for an hour or more. Not only do dolphins and porpoises ride the bow waves of ships, they have also been observed to ride the bow waves of large baleen whales, which is a spectacular sight!

Dolphins and porpoises can swim much faster than ships, so they are in no real danger of being run over. However, there have been a few recorded instances where they have been inadvertently hit by a ship, probably because they became either too brave or too careless.

If the ocean is such a good transmitter of sound, can marine organisms hear noises?

Yes, many marine organisms can hear sounds—and the ocean is a very noisy place! The constant sound of waves breaking at the shoreline, fish making cracking noises, whales clicking and singing, crabs snapping their claws, boats motoring around on the surface can all be heard, sometimes for hundreds or thousands of kilometers in all directions. Sound travels as a compressive wave, so sounds in the ocean are felt by an organism's entire body. The effect is similar to being able to "feel" the music from a loud rock concert even if you are wearing earplugs. If you ever go whale watching, don't think for a minute that you are sneaking up on the whales. They have heard you coming toward them since your boat left the harbor! However, seeing whales in their natural environment can be enhanced by knowing that the whales are aware of your presence and that they are *allowing* you to observe them at close range.

How intelligent are toothed whales?

There may not be a definitive answer to that question, but the following facts about toothed whales (suborder Odontoceti) imply a certain level of intelligence:

- They communicate with each other by using sound.
- They have large brains relative to their body size.
- Their brains are highly convoluted—a characteristic shared by many organisms that are considered to have highly developed intelligence (such as humans and other primates).
- Some dolphins have been reported to assist drowning humans in the wild.
- Some dolphins have been trained to respond to hand signals and do tricks on command (such as retrieve objects).

Although Odontocetes have remarkable abilities, this does not necessarily imply intelligence. Pigeons, for instance, which are not known for being highly intelligent, have also been trained to retrieve objects by using hand signals. Perhaps many of us would like to think that whales and dolphins are more intelligent than they really are because we feel an attachment to these playful, seemingly ever-smiling, air-breathing creatures. It is interesting to note that even experts in the field of animal intelligence disagree on how to assess *human* intelligence accurately, let alone that of a marine mammal.

If the large brain that odontocetes have is not an indication of intelligence, then why is their brain so large? Leading whale researches don't exactly know, but it might be because odontocetes need a large brain to process the wealth of information they receive from the sound echoes they transmit. Because intelligence is difficult to measure, perhaps it is best to say that animals of suborder Odontoceti are tremendously well adapted to the marine environment.

Are gray whales still an endangered species?

Although many other whales exist in smaller populations than before whaling (Figure 14D) and are endangered species, the North Pacific gray whale was removed from the endangered species list in 1993 when their numbers exceeded 20,000, which surpassed the estimated size of their population prior to whaling. Other populations of gray whales were not so fortunate. For instance, the gray whales that used to inhabit the North Atlantic Ocean were hunted to extinction several centuries ago, and the gray whales that live in the waters near Japan may also have recently gone extinct.

These slow-moving whales of suborder Mysticeti spend most of their lives in coastal waters, which made them easy targets for whalers. In the mid–1800s, they were traced to their birthing and breeding lagoons and were hunted to the brink of extinction. A common strategy was to harpoon a calf and then its mother when she came to its rescue. At this time, gray whales were known as "devilfish" because they would often capsize small whaling boats when the adults came to the aid of their young. By the late 1800s, the number of gray whales had diminished to the point that they were difficult to find during their annual migration. Fortunately, these low numbers also made it difficult for whalers to hunt them successfully.

In 1938, the International Whaling Treaty banned the taking of gray whales, which were thought to be nearly extinct. This

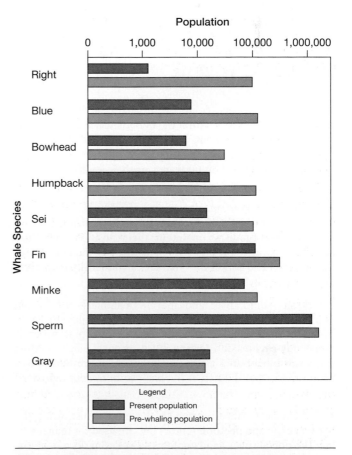

Figure 14D World whale populations (present and pre-whaling).

Figure 14E A gray whale (*Eschrichtius robustus*) exhibits friendly behavior and approaches a boat in Scammon's Lagoon, Baja California, Mexico.

protection has allowed them to steadily increase in number to this day and to become the first marine creature to be removed from endangered status. Their repopulation is truly one of the most impressive success stories of how protecting animals can ensure their continued survival. What is perhaps most surprising is how friendly they are now toward people in boats (Figure 14E) in the same lagoons where they were hunted to near extinction over 150 years ago. Devilfish, indeed!

Chapter in Review

• Pelagic animals that comprise the majority of the ocean's biomass remain mostly within the upper surface waters of the ocean, where their primary food source exists. Those animals that are not planktonic (floating forms such as microscopic zooplankton) depend on buoyancy or their ability to swim to help them remain in food-rich surface waters.

• The rigid gas containers in some cephalopods and the expandable swim bladders in some fishes help increase buoyancy. Other organisms maintain their positions near the surface with gas-filled floats (such as those of the Portuguese man-of-war) and soft bodies that lack high-density hard parts (such as the jellyfish).

• Nekton—squid, fish, and marine mammals—are strong swimmers that depend on their swimming ability to avoid predators and obtain food. Squid swim by trapping water in their body cavities and forcing it out through a siphon. Most fish swim by creating a wave of body curvature that passes from the front of the fish to the back and provides a forward thrust.

• The caudal (rear) fin provides the most thrust, while the paired pelvic and pectoral (chest) fins are used for maneuvering. The dorsal (back) and anal fins serve primarily as stabilizers. A rounded caudal fin is flexible and can be used for maneuvering at slow speeds. The lunate fin is rigid and is of lit-

tle use in maneuvering, but produces thrust efficiently for fast swimmers such as tuna.

• Deep-water nekton feed on detritus and each other. They have special adaptations—such as good sensory devices and bioluminescence—that allow them to survive in this still and completely dark environment.

• Fish can be categorized as lungers (such as groupers) or cruisers (such as tuna). Lungers sit motionless and lunge at passing prey. They have mostly white muscle tissue, which fatigues more quickly than red muscle tissue. Cruisers swim constantly in search of prey and possess mostly red, myoglobin- rich muscle tissue.

• Fish swim slowly when cruising, fast when hunting for prey, and fastest when trying to escape from predators. Although most fish are cold blooded, the fast-swimming tuna, *Thunnus*, is homeothermic, meaning it maintains its body temperature well above water temperature.

• Many marine organisms such as fish, squid, and crustaceans exhibit schooling, probably because it increases their chances of avoiding predation than swimming alone and serves to preserve the species.

• Marine mammals are warm blooded; breathe air; have hair or fur; bear live young; and the females have mammary glands.

Good fossil evidence exists that marine mammals evolved from land-dwelling animals about 50 million years ago. Marine mammals belong to orders Carnivora, Sirenia, and Cetacea.

• Marine mammals within order Carnivora have prominent canine teeth and include sea otters, polar bears, and the pinnipeds (walruses, seals, sea lions, and fur seals). Marine mammals of order Sirenia, which include manatees and dugongs ("sea cows"), have toenails (manatees only), sparse hairs covering their bodies, and are vegetarians.

• The mammals best-adapted to life in the open ocean are those of the order Cetacea, which includes whales, dolphins, and porpoises. Cetaceans have highly streamlined bodies so that they are fast swimmers. Other adaptations—such as being able to absorb 90% of the oxygen they inhale, store large quantities of oxygen, reduce the use of oxygen by noncritical organs, and collapse their lungs below depths of 100 meters (330 feet)—allow them to dive deeply without suffering the effects of nitrogen narcosis.

• Cetaceans are divided into suborder Odontoceti (the toothed whales) and suborder Mysticeti (the baleen whales). Odontocetes use echolocation to find their way through the ocean and locate prey. They emit clicking sounds, and can determine the size, shape, internal structure, and distance of the objects from the nature of the returning signals and the time elapsed.

• Mysticetes, which include the largest whales in the world, separate their small prey from seawater using their baleen plates as a strainer. Baleen whales include the gray whale, the rorqual whales, and the right whales.

• Gray whales migrate from their cold-water summer feeding grounds in the Arctic to warm, low-latitude lagoons in Mexico during winter for breeding and birthing purposes. This behavior may have evolved to allow their young to be born into warm water during the Ice Age, when lower sea level eliminated today's highly productive Arctic feeding areas.

Key Terms

Alveoli (p. 423)
Baleen (p. 428)
Bioluminesce (p. 413)
Biomass (p. 407)
Carnivora (p. 420)
Cetacea (p. 422)
Coelenterate (p. 410)
Cold-blooded (p. 416)
Copepod (p. 408)
Cruiser (p. 415)
Cutaneous artery (p. 418)
Cutaneous vein (p. 418)
Deep-sea fish (p. 413)
Detritus (p. 413)
Disruptive coloration (p. 431)
Dorsal aorta (p. 418)
Echolocation (p. 427)

Falcate (p. 426)
Filter feeder (p. 418)
Foraminifer (p. 408)
Fur seal (p. 421)
Gray whale (p. 429)
Hemoglobin (p. 423)
Homeothermic (p. 416)
Krill (p. 408)
Lunger (p. 415)
Melon (p. 427)
Myoglobin (p. 415)
Myomere (p. 412)
Mysticeti (p. 428)
Odontoceti (p. 426)
Photophore (p. 413)
Pinniped (p. 420)
Poikilothermic (p. 416)

Polar bear (p. 420)
Radiolarian (p. 408)
Right whale (p. 430)
Rorqual whale (p. 429)
School (p. 418)
Scyphozoan (p. 411)
Sea lion (p. 421)
Sea otter (p. 420)
Seal (p. 421)
Siphonophore (p. 411)
Sirenia (p. 421)
Spermaceti organ (p. 427)
Swim bladder (p. 408)
Walrus (p. 421)
Warm-blooded (p. 416)

Questions And Exercises

1. What contributions did Alexander Agassiz make to advancing the study of the marine environment?

2. Discuss why the rigid gas chamber in cephalopods limits the depth to which they can descend. Why do fish with a swim bladder not have this limitation?

3. Draw and describe several different types of microscopic zooplankton and macroscopic zooplankton.

4. Name and describe the different types of fins that fish exhibit. What are the five basic shapes of caudal fins, and what are their uses?

5. What are the two food sources of deep-water nekton? List several adaptations of deep-water nekton that allow them to survive in their environment.

6. Do humans have more to fear about being attacked by a shark, or do sharks have more to fear about being killed by a human?

7. What are the major structural and physiological differences between the fast-swimming cruisers and lungers that patiently lie in wait for their prey?

8. Are most fast swimming fish cold-blooded or warm-blooded? What circulatory system modifications do these fish have to minimize heat loss?

9. What are several benefits of schooling?

10. What common characteristics do all organisms in class Mammalia share?

11. Describe marine mammals within the order Carnivora, including their adaptations for living in the marine environment.

12. How can true seals be differentiated from the eared seals (sea lions and fur seals)?

13. Describe the marine mammals within the order Sirenia, including their distinguishing characteristics.

14. How can dolphins be differentiated from porpoises?

15. List the modifications that are thought to give some cetaceans the ability to (a) increase their swimming speed; (b) dive to great depths without suffering the "bends"; and (c) stay submerged for long periods.

16. Describe differences between cetaceans of the suborder Odontoceti (toothed whales) with those of the suborder Mysticeti (baleen whales). Be sure to include examples from each suborder.

17. Describe the process by which the sperm whale produces echolocation clicks.

18. Discuss how sound reaches the inner ear of toothed whales.

19. Describe the mechanism by which baleen whales feed.

20. Discuss reasons why gray whales leave their cold-water feeding grounds during the winter season.

References

Bonner, N. 1989. *Whales of the world*. New York: Facts-on-File.

Carey, F. G. 1973. Fishes with warm bodies. *Scientific American* 228:2, 36–44.

Carwardine, M. 1995. *Whales, dolphins and porpoises: A visual guide to all the world's cetaceans*. New York: Dorling Kindersley.

Denton, E. J., and Shaw, T. I. 1962. The buoyancy of gelatinous marine animals. *Journal of Physiology* 161:14P–15P.

Doak, W. 1989. *Encounters with whales and dolphins*. Dobbs Ferry, NY: Sheridan House.

Ellis, R. 1996. *Deep Atlantic*. New York: Alfred A. Knopf.

George, D., and George, J. 1979. *Marine life: An illustrated encyclopedia of invertebrates in the sea*. New York: Wiley-Interscience.

Giesbrecht, W. 1892. *Fauna und flora des Golfes von Neapel und der Angrenzenden Meeres—Abschnitte*. Berlin: Verlag Von R. Friedländer and Sohn.

Gingerich, P. D. 1994. The whales of Tethys. *Natural History* 103:4, 86–88.

Idyll, C. P., Ed. 1969. *The science of the sea: A History of Oceanography*. London: Thomas Y. Crowell.

Kanwisher, J. W., and Ridgway, S. H. 1983. The physiological ecology of whales and porpoises. *Scientific American* 248:6, 110–121.

Klimley, P A. 1996. The predatory behavior of the white shark, in Pirie, R. G., ed., *Oceanography: Contemporary readings in ocean sciences*, 3rd ed. New York: Oxford University Press.

Kunzig, R. 1990. Invisible garden. *Discover* 11:4, 66–74.

Lalli, C. M., and Parsons, T. R. 1993. *Biological oceanography: An introduction*. New York: Pergamon Press.

Leatherwood, S., and Reeves, R. R. 1983 *The Sierra Club handbook of whales and dolphins*. San Francisco: Sierra Club Books.

MacGinitie, G. E., and MacGinitie, N. 1968. *Natural history of marine animals*. 2nd ed. New York: McGraw-Hill.

Mrosovsky, N., Hopkins-Murphy, S. R., and Richardson, J. I. 1984. Sex ratio of sea turtles: Seasonal changes. *Science* 225:4663, 739–740.

Noad, M. J., et al. 2000. Cultural revolution in whale songs. *Nature* 408:6812, 537.

Norris, K. S., and Harvey, G. W. 1972. A theory for the function of the spermaceti organ of the sperm whale (*Physeter catodon* L), in *Animal Orientation and Navigation*, 397–417. Washington, DC: National Aeronautics and Space Administration.

Obee, B. 1992. The great killer whale debate: Should captive orcas be set free? *Canadian Geographic* 112:1, 20–31.

Pike, G. C. 1962. Migration and feeding of the gray whale (*Eschrichtius gibbosus*). *Journal of the Fisheries Research Board of Canada* 19:815–838.

Reeves, R. R., Stewart, B. S., and Leatherwood, S. 1992. *The Sierra Club handbook of seals and Sirenians*. San Francisco: Sierra Club Books.

Thewissen, J. G. M. 1998. *The emergence of whales: Evolutionary patterns in the origin of Cetacea*. New York: Plenum Press.

Thewissen, J. G. M., Hussain, S. T., and Arif, M. 1994. Fossil evidence for the origin of aquatic locomotion in archaeocete whales. *Science* 263:5144, 210–212.

Thorson, G. 1971. *Life in the sea*. New York: McGraw-Hill.

Waller, G., ed. 1996. *SeaLife: A complete guide to the marine environment.* Washington, DC: Smithsonian Institution Press.

Würsig, B. 1989. Cetaceans. *Science* 244:4912, 1550–1557.

***Suggested Reading in* Scientific American**

Denton, E. 1960. The buoyancy of marine animals. 303:11, 118–128. How various marine animals use buoyancy to maintain their position in the water column.

Gosline, J. M., and DeMont, M. E. 1985. Jet propelled swimming in squids. 252:1, 96–103. By using jet propulsion resulting from expelling water through their siphons, squids can move as fast as the speediest fishes.

Gray, J. 1957. How fishes swim. 197:2, 48–54. The roles of musculature and fins in the swimming of fishes.

Horn, M. H., and Gibson, R. N. 1988. Intertidal fishes. 258:1, 64–71. A discussion of a wide variety of fishes that inhabit tide pools.

Kooyman, G. L. 1969. The Weddell seal. 221:2, 100–107. The lifestyle and problems related to this mammal's living in water permanently covered with ice.

Martin, K., Shariff, K., Psarakos, S., and White, D. J. 1996. Ring bubbles of dolphins. 275:2, 82–87. Shows how dolphins emit bubble-shaped rings from their blowholes and use them as play toys.

O'Shea, T. J. 1994. Manatees. 271:1, 66–73. Presents information about how manatees are thought to have evolved from the same ancestors as elephants and aardvarks and discusses threats to their survival.

Rudd, J. T. 1956. The blue whale. 195:6, 46–65. A description of the ecology of the largest animal that ever lived.

Sanderson, S. L., and Wassersug, R. 1990. Suspension-feeding vertebrates. 263:3, 96–102. A discussion of the ecology of filter-feeding fishes and cetaceans.

Shaw, E. 1962. The schooling of fishes. 206:6, 128–136. A consideration of theories seeking to explain the schooling behavior of fishes.

Whitehead, H. 1985. Why whales leap. 252:3, 84–93. An attempt to answer the enigmatic question, "Why do whales breach?"

Würsig, B. 1988. The behavior of baleen whales. 258:4, 102–107. A summary of the facts concerning the behavior of baleen whales.

Zapol, W. M. 1987. Diving adaptations of the Weddell seal. 256:6, 100–107. The physiological adaptations that enable the Weddell seal to make deep, long dives are discussed.

Oceanography on the Web

Visit the *Essentials of Oceanography* home page for on-line resources for this chapter. There you will find an on-line study guide with review exercises, and links to oceanography sites to further your exploration of the topics in this chapter. *Essentials of Oceanography* is at: **http://www. prenhall.com/thurman** (click on the Table of Contents menu and select this chapter).

15

Animals of the Benthic Environment

key questions

- What types of organisms occur along rocky shores?
- What types of organisms occur within sediment covered shores?
- What are the conditions necessary for coral growth?
- How are corals able to survive in nutrient-depleted warm water?
- What are the environmental conditions of the deep ocean floor?
- What types of biocommunities exist on the deep ocean floor?

THE GREAT DEBATE ON LIFE IN THE DEEP OCEAN: THE ROSSES AND EDWARD FORBES

One of the first major controversies in marine science developed around the work of Sir John Ross (1777–1856), his nephew Sir James Clark Ross (1800–1862), and Edward Forbes (1815–1854)—all British naturalists and scientists. Sir John Ross explored Baffin Bay in Canada in 1817 and 1818, where he made depth measurements. He also collected samples of bottom-dwelling organisms and sediments with a device of his own design called a *deep-sea clamm*, which resembled a pair of metal jaws that retrieved "bites" of the bottom mud. He obtained worms in mud samples from a depth of 1.8 kilometers (1.1 miles) and even retrieved a starfish from this depth, which had become entangled in the line that was used to lower the clamm to the bottom.

Sir James Clark Ross extended the soundings to greater depths on voyages to the Antarctic during 1839–1843 (Figure 15A). Despite using a 7-kilometer (4.3-mile) sounding line, it was insufficient at times to reach the bottom of the ocean. The animals recovered from the cold waters of the Antarctic were the same species his uncle had recovered from the Arctic, and they were found to be very sensitive to temperature increase. Sir James concluded that the waters of the deep ocean must have a uniformly low temperature and that life exists at all levels in the ocean, including the deep ocean floor.

During this same time, Edward Forbes was conducting oceanographic shipboard experiments to study the vertical distribution of life in the

From however great a depth we may be able to bring the mud and stones of the bed of the ocean, we shall find them teeming with animal life.
—Sir James Clark Ross (circa 1840)

Figure 15A The Ross Expedition in Antarctica (circa 1840).

ocean. He observed that plant life (marine algae) is limited to the sunlit surface waters. Forbes also found that the concentration of animals was greater near the surface and decreased with increasing depth until only a small trace of life—if any—remained in the deepest waters.

Forbes's followers may have misinterpreted his findings because they concluded that no life whatsoever existed in the deep ocean. They reasoned that life would be impossible in the deep ocean because the pressure is so high and there is no light and oxygen. Their conclusions, moreover, contradicted the findings of the Rosses, who had discovered many types of life in the deep ocean. Today, it is well established that life exists at all levels in the ocean, and those organisms that are found in the deep ocean are well adapted to the dark, cold, high pressure environment.

Of the 250,000 known species that inhabit the marine environment, more than 98% (about 245,000) live in or on the ocean floor. Ranging from the rocky, sandy, and muddy intertidal zone to the muddy deposits of the deepest ocean trenches, the ocean floor provides a tremendously varied environment that is home to a diverse group of specially adapted organisms.

Living at or near the interface of the ocean floor and seawater, an organism's success is closely related to its ability to cope with the physical conditions of the water, the ocean floor, and other members of the biological community. The vast majority of known benthic species live on the continental shelf, where the water is often shallow enough to allow sunlight to penetrate to the ocean bottom.

The number of benthic species found at similar latitudes on opposite sides of an ocean basin depends on how ocean surface currents affect coastal water temperature—one of the most important variables affecting species diversity. The Gulf Stream, for example, warms the European coast from Spain to the northern tip of Norway, giving rise to more three times the number of benthic species than are found in similar latitudes along the Atlantic coast of North America, where the Labrador Current cools the water as far south as Cape Cod, Massachusetts.

The distribution of benthic **biomass**[1] (Figure 15–1) closely matches the distribution of photosynthetic productivity in surface waters (compare with Figure 13–6). Mostly, life on the ocean floor depends on the productivity of the ocean's surface waters and great abundances of benthic life are found beneath areas of high productivity and in shallow water.

Rocky Shores

Rocky shorelines teem with organisms that live on the surface of the ocean floor. These organisms are called **epifauna** (*epi* = upon, *fauna* = animal) and are either permanently attached to the bottom (such as marine algae) or they move over it (such as crabs). Table 15–1 lists some of the special adaptations these organisms have to withstand the rigors of life on rocky shores.

A typical rocky shore (Figure 15–2a) can be divided into a **spray zone**, which is above the spring high-tide line and is covered by water only during storms, and an **intertidal zone**, which lies between the high and low tidal extremes. Along most shores, the intertidal zone can be clearly separated into the following subzones (Figure 15–2):

- The **high tide zone**, which is relatively dry and is covered only by the highest high tides
- The **middle tide zone**, which is alternately covered by all high tides and exposed during all low tides
- The **low tide zone**, which is usually wet but is exposed during the lowest low tides

The subzones of the intertidal zone can also be delineated based on the populations of organisms that attach themselves to the bottom. Each centimeter of the rocky shore has a significantly different character than the centimeter above and below it, so organisms have evolved to withstand very specific degrees of exposure to the atmosphere. Consequently, the most finely delineated biozones in the marine environment can be found along rocky shores.

Diversity of species that inhabit rocky shores varies widely. Overall, rocky intertidal ecosystems have a moderate diversity of species compared to other benthic environments. The greatest animal diversity is at lower (tropical) latitudes, while the diversity of algae is greater in the mid-latitudes, probably because of better availability of nutrients.[2]

Spray (Supralittoral) Zone

Organisms that live within the spray zone [also known as the **supralittoral** (*supra* = above, *littora* = the seashore) **zone**] must avoid drying out during the long periods when they are above water level. Consequently, many animals, such as the periwinkle snail (genus *Littorina*; Figure 15–2b), have shells, and few species of marine algae are found.

---

[1]Remember that *biomass* is the mass of living organisms.

[2]As discussed in Chapter 13, "Biological Productivity and Energy Transfer," this increased nutrient supply is a result of the lack of a permanent thermocline in mid-latitude regions.

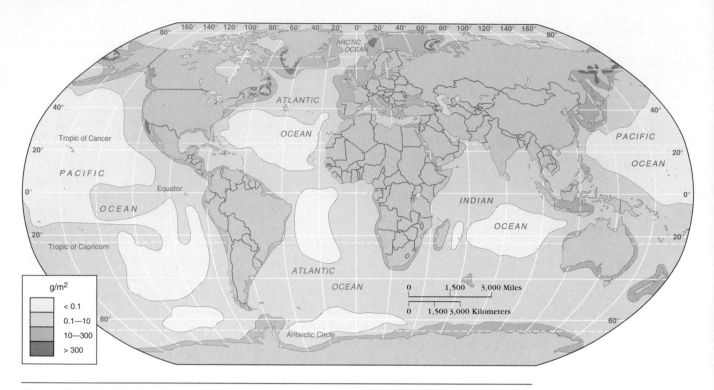

Figure 15–1 Distribution of benthic biomass. The distribution of benthic biomass (in grams of biomass per square meter) shows that the lowest benthic biomass is beneath the centers of subtropical gyres, and the highest values are in high-latitude continental shelf areas. The pattern is similar to that of surface productivity (see Figure 13–6), suggesting that most of the benthic community receives their food from surface waters.

Found among the cobbles and boulders that typically cover the floors of sea caves well above the high tide line are rock lice or sea roaches (isopods of the genus *Ligia*). These scavengers reach lengths of 3 centimeters (1.2 inches) and scurry about at night feeding on organic debris (Figure 15–2c). During the day, they hide in crevices.

A distant relative of the periwinkle snail, the limpet is also found in the spray zone (genus *Acmaea*; Figure 15–2d). Both limpets and periwinkle snails feed on marine algae. The limpet has a flattened conical shell and a muscular foot with which it clings tightly to rocks.

High Tide Zone

Like animals within the spray zone, most animals that inhabit the high tide zone have a protective covering to prevent them from drying out. Periwinkles, for example, have a protective shell and can move between the spray zone and the high tide zone. Buckshot barnacles (Figure 15–2e) have a protective shell, too, but they cannot live above the high tide shoreline because they filter-feed from seawater, and their larval form is planktonic.

The most conspicuous algae in the high tide zone are rock weeds, members of the genus *Fucus* that live in colder latitudes (Figure 15–2f) and *Pelvetia* that live in warmer latitudes. Both have thick cell walls to reduce water loss during periods of low tide.

The rock weeds are among the first organisms to colonize a rocky shore. Later, **sessile** (*sessilis* = sitting on) animal forms—those that are attached to the bottom, such as barnacles and mussels—begin to establish themselves, competing for attachment sites with the rock weeds.

Middle Tide Zone

Seawater constantly bathes the middle tide zone, so more types of marine algae and soft-bodied animals can live there. The total biomass is much greater than in the high tide zone so there is much greater competition for rock space among sessile forms.

Shelled organisms inhabiting the middle tide zone include acorn barnacles (Figure 15–2g), goose-necked barnacles (*Pollicipes*) (Figure 15–2h), which attach themselves to rocks with a long, muscular neck, and various mussels (genera *Mytilus* and *Modiolus*). Mussels attach to bare rock, algae, or barnacles during their planktonic stage and remain in place by means of strong *byssus* threads.

Carnivorous snails and sea stars (such as genera *Pisaster* and *Asterias*) feed upon the mussels. To pry open the mussel shell, sea stars pull on either side with hundreds of tube-like feet. The mussel eventually becomes fatigued and can no longer hold its shell halves closed. When the shell opens ever so slightly, the sea star turns its stomach inside out, slips it through the crack in

Table 15–1   Adverse conditions of rocky intertidal zones and adaptations for adverse conditions

Adverse conditions of rocky intertidal zones	Adaptations for adverse conditions	Examples
Drying out during low tide	Ability to seek shelter or withdraw into shells	Sea slugs, snails, crabs
	Thick exterior or exoskeleton to prevent water loss	
Strong wave activity	Strong holdfasts (in algae) to prevent being washed away	Kelp, snails, sea stars, sea urchins
	Strong attachment threads, a muscular foot, multiple legs, or hundreds of tube feet (in animals) to allow them to attach firmly to bottom	
	Hard structures adapted to withstand wave energy	
Predators occupy area during low tide	Firm attachment	Mussels, sea anemones, sea slugs, octopi, sea stars
	Stinging cells	
	Camouflage	
	Inking response	
	Ability to break off body parts and regrow them later (regenerative capability)	
Difficulty finding mates for attached species	Release of large numbers of egg/sperm into the water column during reproduction	Abalones, sea urchins
Rapid changes in temperature, salinity, pH, and oxygen content	Ability to withdraw into shells to minimize exposure to rapid changes in environment	Snails, barnacles
	Ability to exist in varied temperature, salinity, pH, and low-oxygen environments for extended periods	
Lack of abundant attachment sites	Organisms attach to others	Bryozoans, coral

the mussel shell, and digests the edible tissue inside (Figure 15–3).

The most striking feature of the middle tidal zone along most rocky coasts is a mussel bed that thickens toward the bottom until it reaches an abrupt bottom limit. This is where the physical conditions restrict barnacle growth and in most areas, the boundary is quite pronounced. Protruding from the mussel bed will be numerous goose-necked barnacles, and sea stars browsing on the mussels are concentrated in the lower levels of the bed. Less-conspicuous forms common to mussel beds are varieties of algae, worms, clams, and crustaceans.

Where the rock surface flattens out within the middle tidal zone, tide pools trap water as the tide ebbs. These pools support microecosystems containing a wide variety of organisms. The most conspicuous member of this community is often the sea anemone (see Figure 15–2i), which is a relative of the jellyfish.

Shaped like a sack, anemones have a flat foot disk that provides a suction attachment to the rock surface. Directed upward, the open end of the sack is the mouth, which leads directly to the gut cavity and is surrounded by rows of tentacles (Figure 15–4). The tentacles are covered with stinging needle-like cells called **nematocysts** (*nemato* = thread, *cystis* = bladder) (Figure 15–4, inset), which inject the victim with a potent neurotoxin. Nematocysts are automatically released when any organism brushes against a sea anemone's tentacles.

Hermit crabs (*Pagurus*) inhabit tide pools, too. They have a well-armored pair of claws and upper body, but a soft, unprotected abdomen, which they protect by inhabiting an abandoned snail shell (Figure 15–5a). They can often be seen scurrying around the tide pool area or fighting with other hermit crabs for new shells. Their abdomen has even evolved a curl to the right to make it fit properly into snail shells. Once in the snail shell, the crab can protect itself by closing off the shell's opening with its large claws.

In tide pools near the lower limit of the middle tide zone, sea urchins may be found feeding on algae (Figure 15–5b). Sea urchins have a five-toothed mouth centered on the bottom side of their hard spherical shell, consisting of fused calcium carbonate plates perforated to allow tube feet and water to pass through. Resembling a pincushion, the shell of a sea urchin has numerous spines for protection and to scrape out protective holes in rocks.

Low Tide Zone

The low tide zone is almost always submerged, so algae, not animals, dominate. A diverse community of animals exists, too, but they are hidden by the great variety of marine algae and surf grass (*Phylospadix*) (Figure 15–6). The encrusting red algae (*Lithothamnion*), which are also seen in middle-zone tide pools, becomes very abundant in the lower tide pools (see Figure 15–2d). In temperate latitudes, moderate-sized red and brown algae provide a drooping canopy beneath which much of the animal life can hide during low tide.

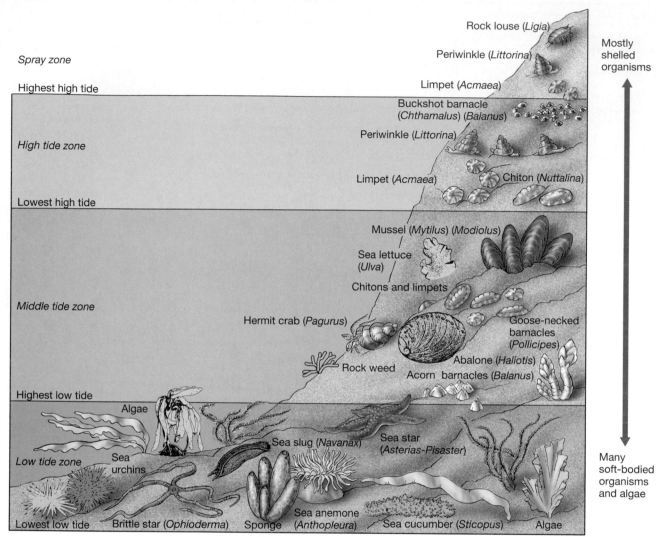

Rock louse (*Ligia*)

Periwinkle (*Littorina*)

Limpet (*Acmaea*)

Buckshot barnacle
(*Chthamalus*) (*Balanus*)

Periwinkle (*Littorina*)

Limpet (*Acmaea*)   Chiton (*Nuttalina*)

Mussel (*Mytilus*) (*Modiolus*)

Sea lettuce
(*Ulva*)

Chitons and limpets

Hermit crab (*Pagurus*)

Goose-necked
barnacles
(*Pollicipes*)

Abalone (*Haliotis*)

Rock weed   Acorn barnacles (*Balanus*)

Algae

Sea slug (*Navanax*)   Sea star
(*Asterias-Pisacter*)

Sea
urchins

Brittle star (*Ophioderma*)   Sponge   Sea anemone
(*Anthopleura*)   Sea cucumber (*Sticopus*)   Algae

Spray zone

Highest high tide

High tide zone

Lowest high tide

Middle tide zone

Highest low tide

Low tide zone

Lowest low tide

Mostly
shelled
organisms

Many
soft-bodied
organisms
and algae

(a)

(b)

(c)

(d)

(e)

(f)

(g)

(h)

(i)

Figure 15–2 The rocky shore. (a) Zonations and typical organisms of a rocky shore (not to scale). **(b)** Periwinkles (*Littorina*), spray zone to upper high-tide zone. **(c)** Rock louse (*Ligia*), spray zone. **(d)** Rough keyhole limpet (*Diodora aspera*) with encrusting red algae (*Lithothamnion*), high tide zone. **(e)** Buckshot barnacles (*Chthamalus*), high tide zone. **(f)** Rock weed (*Fucus filiformes*), middle tide zone. **(g)** Acorn barnacles (*Balanus*), middle tide zone. **(h)** Goose-necked barnacles (*Pollicipes*) and blue mussels (*Mytilus*), middle tide zone. **(i)** Sea anemone (*Anthopleura*), low tide zone.

Figure 15–3 Sea star feeding on a mussel. An ochre sea star (*Pisaster*) pulls apart the two halves of a mussel's (*Mytilus*) shell with its tube feet. The star then turns its own stomach inside out and forces it through the opening between the mussel's shells, where it digests the mussel's soft tissue within the mussel's own shell.

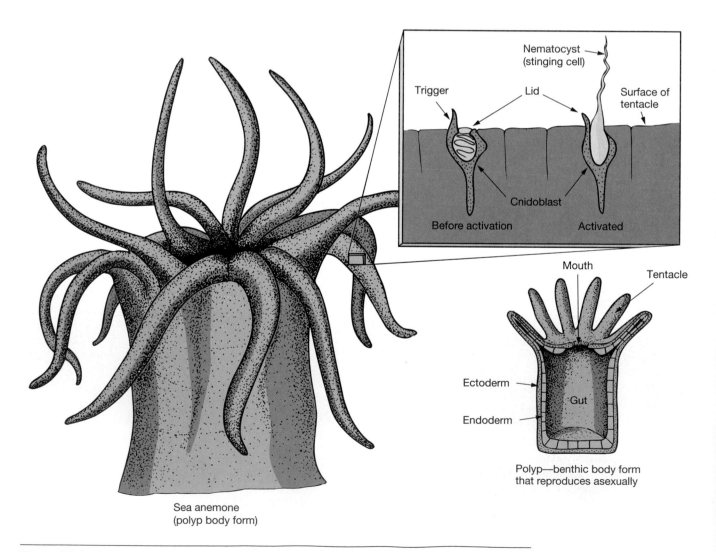

Sea anemone
(polyp body form)

Polyp—benthic body form
that reproduces asexually

Figure 15–4 Sea anemone. Sea anemone morphology and detail of its stinging nematocysts (*inset*).

(a)

(b)

Figure 15–5 Hermit crab and sea urchins.
(a) Hermit crab (*Pagurus*) that has taken up residence in a *Maxwellia gemma* shell. **(b)** Sea urchins (*Echinus*) that have burrowed into the bottom of a tide pool within the middle tide zone.

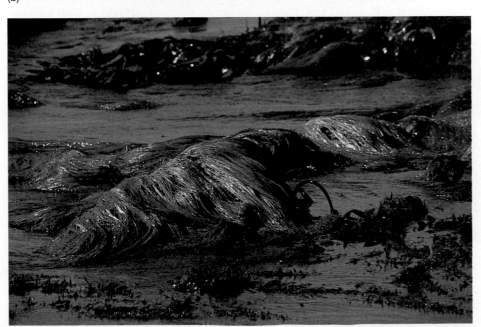

Figure 15–6 Marine algae and surf grass. Dark-colored sea palms (a brown alga) and green surf grass (*Phylospadix*) are exposed during an extremely low tide in a California low tide zone but provide protection for many organisms.

Scampering from crevice to crevice and in and out of tide pools across the full range of the intertidal zone are various species of shore crabs (Figure 15–7). These scavengers help keep the shore clean. Shore crabs spend most of the day hiding in cracks or beneath overhangs. At night, they eat algae as rapidly as they can tear them from the rock surface with their large front claws, called *chelae* (*khele* = claw). Their hard exoskeleton prevents them from drying out too quickly so they can spend long periods of time out of water.

> Rocky shores are divided into the spray zone and high, middle, and low tide (intertidal) zones. Many shelled organisms inhabit the upper zones while more soft-bodied organisms and algae inhabit the lower zones.

Sediment-Covered Shores

The sediment-covered shore ranges from steep boulder beaches, where wave energy is high, to the mud flats of quiet, protected embayments, and includes sandy beaches where wave energy is usually moderate.

Nearly all organisms that inhabit sediment-covered shores are called **infauna** (*in* = inside, *fauna* = animal) because they can burrow into the sediment. Most sediment-covered shores have intertidal zones similar to rocky shores. Also, there is much less species diversity in sediment-covered shores, but the organisms are usually found in great numbers.

The Sediment

Sediment-covered shores include *beaches*, *salt marshes*, and *mud flats*, which represent progressively lower-energy environments and are consequently composed of progressively finer sediment. The energy level that a shore experiences is related to the strength of waves and longshore currents. Along shores that experience low energy levels, particle size becomes smaller, the sediment slope decreases, and overall sediment stability increases. Thus, the sediment in a fine-grained mud flat is more stable than that of a high-energy sandy beach.

A large quantity of water from breaking waves rapidly sinks into the sand and brings a continual supply of nutrients and oxygen-rich water for the animals that live there. This supply of oxygen also enhances bacterial decomposition of dead tissue. The sediment in salt marshes and mud flats is not nearly so rich in oxygen, however, so decomposition occurs more slowly.

Intertidal Zonation

The intertidal zone of the sediment-covered shore consists of supralittoral, high tide, middle tide, and low tide zones, as shown in Figure 15–8. These zones are best developed on steeply sloping, coarse-sand beaches and are less distinct on the more gentle sloping, fine-sand beaches. On mud flats, the tiny clay-sized particles form a deposit with essentially no slope, so zonation is not possible in this protected, low-energy environment.

The species of animals differ from zone to zone. As in intertidal rocky shores, however, the maximum *number* of species and the greatest *biomass* in intertidal sediment-covered shores are found near the low-tide shoreline, and both diversity and biomass decrease toward the high-tide shoreline.

Life in the Sediment

Life on and in the sediment requires very different adaptations than on the rocky shore. The sandy beach supports fewer species than the rocky shore, and mud flats fewer still, but the total *number* of individuals may be as high. In the low tide zone of some beaches and on mud flats, for example, as many as 5000 to 8000 burrow-

(a)

(b)

Figure 15–7 Shore crabs. **(a)** Coral crab, or queen crab (Bonaire Island, Netherlands Antilles). **(b)** Shore crab, *Pachygrapsis crassipes*. This female is carrying eggs (dark oval structure curled under her abdomen).

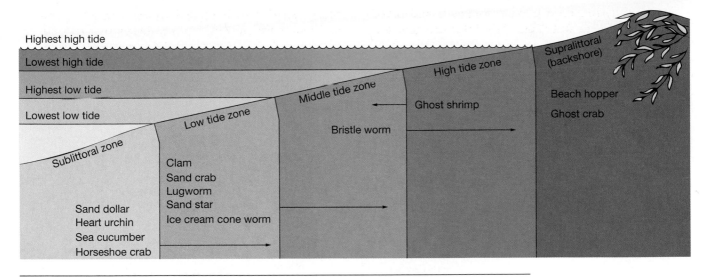

Figure 15–8 Intertidal zonation and typical organisms on a sediment-covered shore.
Intertidal zonation is related to the amount of exposure during low tide. The zonation is best
displayed on coarse sand beaches with steep slopes; as the sediment becomes finer and the beach
slope decreases, zonation becomes less distinct and disappears entirely on mud flats.

ing clams have been counted in only 1 square meter
(10.8 square feet).

Burrowing is the most successful adaptation for life
in sediment-covered shores, so organisms are much less
obvious than in other environments. By burrowing only
a few centimeters beneath the surface, they encounter a
much more stable environment where they are not
bothered by fluctuations of temperature and salinity or
the threat of drying out.

Suspension feeding—also called filter feeding—and
deposit feeding are two techniques animals of sediment-covered shores commonly use to obtain food
from the clear water above or from the sediment itself.
In **suspension feeding**, organisms that are buried in sediment use specially designed structures to filter plankton from seawater (Figure 15–9a). Clams, for example,
bury themselves in sediment and extend siphons
through the surface. They pump in overlying water and
filter suspended plankton and other organic matter
from it.

In **deposit feeding**, organisms feed on food items that
occur as deposits. These deposits include detritus—dead
and decaying organic matter and waste products—and
the sediment itself, which is coated with organic matter.
Some deposit feeders, such as the segmented worm
Arenicola (Figure 15–9b), feed by ingesting sediment
and extracting organic matter from it. Others, such as
the amphipod *Orchestoidea* (Figure 15–9c), feed on
more concentrated deposits of organic matter (detritus)
on the sediment surface.

Another less common method is **carnivorous feeding**. The sand star *Astropecten* (Figure 15–9d), for example, cannot climb rocks the way its sea star relatives can,
but it can burrow rapidly into the sand, where it feeds
voraciously on crustaceans, mollusks, worms, and other
echinoderms.

Sandy Beaches

Most animals at the beach burrow into the sand and are
safely hidden from view because there is no stable, fixed
surface (as on rocky shores) to which they can attach.

Bivalve Mollusks A **bivalve** (*bi* = two, *valva* = a
valve) is an animal having two hinged shells, such as a
clam or a mussel. A **mollusk** is a member of the phylum
Mollusca (*molluscus* = soft), characterized by a soft
body and either an internal or external hard calcium
carbonate shell.

Bivalve mollusks are well adapted to life in the sediment. A single foot digs into the sediment to pull the
creature down into the sand. Their siphon extends
vertically through the sediment for feeding (Figure
15–9a). The way clams bury themselves is shown in Figure 15–10.

How deeply a bivalve can bury itself depends on the
length of its siphons, which must reach above the sediment surface to pull in water for food (plankton) and
oxygen. Indigestible matter is forced back out the
siphon periodically by quick muscular contractions. The
greatest biomass of clams is burrowed into the low-tide
region of sandy beaches and it decreases where the sediment becomes muddier.

Annelid Worms A variety of **annelids** (*annelus* = a
ring)—segmented worms—are also well adapted to life
in the sediment. The lugworm (*Arenicola* species), for
example, lives in a U-shaped burrow (see Figure 15–9b),
the walls of which are strengthened with mucus.
The worm moves forward to feed and extends its proboscis (snout) up into the head shaft of the burrow to
loosen sand with quick pulsing movements. A cone-
shaped depression forms at the surface over the head
end of the burrow as sand continually slides into the

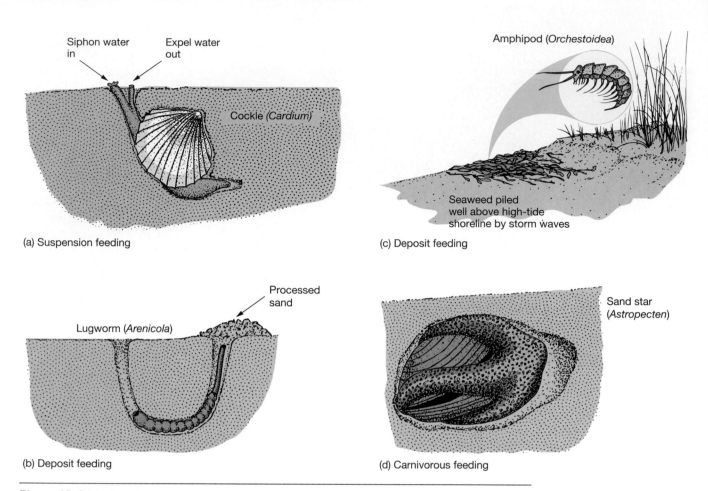

Figure 15–9 Modes of feeding along sediment-covered shores. (a) Suspension feeding by the clam (cockle) *Cardium*, which uses its siphon to filter plankton and other organic matter that is suspended in the water. **(b)** Deposit feeding by the segmented worm *Arenicola*, which feeds by ingesting sediment and extracting organic matter from it. **(c)** Deposit feeding by the amphipod *Orchestoidea*, which feeds on more concentrated deposits of organic matter (detritus) on the sediment surface. **(d)** Carnivorous feeding of a clam by the sand star *Astropecten*.

burrow and is ingested by the worm. As the sand passes through its digestive tract, the organic content is digested, and the processed sand is deposited at the surface.

Crustaceans Crustaceans (*crusta* = shell)—such as crabs, lobsters, shrimps, and barnacles—include predominately aquatic animals that are characterized by a segmented body, a hard exoskeleton, and paired, jointed limbs. On most sandy beaches, numerous crustaceans called *beach hoppers* feed on kelp cast up by storm waves or high tides. A common genus is *Orchestoidea*, which is only 2 to 3 centimeters (0.8 to 1.2 inches) long (see Figure 15–9c), but can jump more than 2 meters (6.6 feet) high. Laterally flattened, beach hoppers usually spend the day buried in the sand or hidden in kelp. They become particularly active at night, when large groups many hop at the same time and form clouds above the piles of kelp on which they feed.

Sand crabs (*Emerita*) (Figure 15–11) are a type of crustacean common to many sandy beaches. Ranging in length from 2.5 to 8 centimeters (1 to 3 inches), they move up and down the beach near the shoreline. They bury their bodies in the sand and leave their long, curved, V-shaped antennae pointing up the beach slope. These little crabs filter food particles from the water, and can be located by looking in the lower intertidal zone for a V-shaped pattern in the swash as it runs down the beach face.

Echinoderms Echinoderms (*echino* = spiny, *derma* = skin) found in beach deposits include the sand star (*Astropecten*) and heart urchins (*Echinocardium*). Sand stars prey on invertebrates that burrow into the low-tide region of sandy beaches. The sand star (see Figure 15–9d) is well designed for moving through sediment, with five tapered legs with spines and a smooth back.

More flattened and elongated than the sea urchins of the rocky shore, heart urchins live buried in the sand near the low-tide line. They gather sand grains into their mouths, where the coating of organic matter is scraped off and ingested (Figure 15–12).

Meiofauna Meiofauna (*meio* = lesser, *fauna* = animal) are small organisms that live in the spaces between sediment particles. These organisms, only 0.1 to 2 mil-

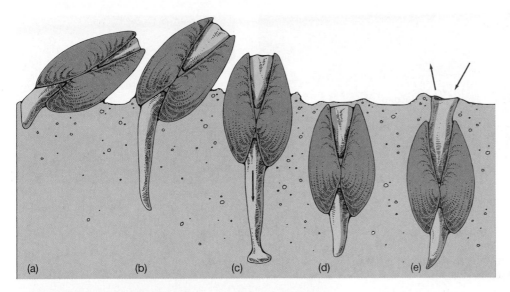

Figure 15–10 How a clam burrows. A clam burrows into the sediment by **(a)** extending its pointed foot into the sediment and **(b)** forcing the foot deeper into the sediment and using this increasing leverage to bring the exposed, shell-clad body toward vertical. When the foot has penetrated deeply enough, a bulbous anchor forms at the bottom **(c),** and a quick muscular contraction pulls the entire animal into the sediment **(d).** The siphons are then pushed up above the sediment to pump in water, from which the clam extracts food and oxygen **(e).**

Figure 15–11 Sand crab. A sand crab (*Emerita*) emerges from the sand. They can often be found just beneath the surface within the lower intertidal zone.

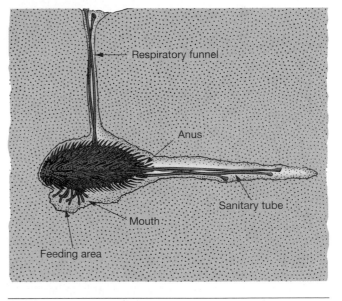

Figure 15–12 Heart urchin. Feeding and respiratory structures of a heart urchin (*Echinocardium*), which feeds on the film of organic matter that covers sand grains.

limeters (0.004 to 0.08 inch) long, feed primarily on bacteria removed from the surface of sediment particles. Meiofauna include polychaetes, mollusks, arthropods, and nematodes (Figure 15–13), and are found in sediment from the intertidal zone to the deep-ocean trenches.

Mud Flats

Eelgrass (*Zostera*) and turtle grass (*Thalassia*) are widely distributed in the low tide zone of mud flats and the adjacent shallow coastal regions. Numerous openings at the surface of mud flats attest to a large population of bivalve mollusks and other invertebrates.

Fiddler crabs (*Uca*) live in burrows that may exceed 1 meter (3 feet) deep in the mud flats. Relatives of the shore crabs, they usually measure no more than 2 centimeters (0.8 inch) across the body. Male fiddler crabs have one small claw and one oversized claw, which is up to 4 centimeters (1.6 inches) long (Figure 15–14). Fiddler crabs get their name because this large claw is waved around as if they were playing an imaginary fiddle. The females have two normal-sized claws. The large claw of the male is used to court females and to fight competing males.

Sediment covered shores—including sandy beaches and mudflats—have similar intertidal zonations as rocky shores but contain many organisms that live within the sediment (infauna).

(a) (b) (c)

Figure 15-13 Scanning electron micrographs of meiofauna. (a) Nematode head, magnified 804 times. The projections and pit on the right side are sensory structures. **(b)** Amphipod, magnified 20 times. This 3-millimeter-long organism builds a burrow of cemented sand grains. **(c)** A 1-millimeter long polychaete worm (magnified 55 times) with its proboscis (mouth) extended.

Shallow Offshore Ocean Floor

An environment that is mainly sediment covered extends from the spring low-tide shoreline to the seaward edge of the continental shelf. Rocky exposures may occur locally near shore. On rocky exposures, many types of marine algae exist, which have adaptations (such as gas-filled floats) for reaching from the shallow sea floor to near the sunlit surface waters.

The sediment-covered shelf has moderate to low species diversity. The diversity of benthic organisms is *lowest* beneath upwelling regions because upwelling carries nutrients to the surface, making pelagic production so great that an excess of dead organic matter is produced. When this matter rains down on the bottom and decomposes, it consumes oxygen, so the oxygen supply can be locally depleted. The giant kelp bed associated with rocky bottoms is a specialized shallow-water community with higher diversity.

Rocky Bottoms (Sublittoral)

A rocky bottom within the shallow inner **sublittoral** (*sub* = under, *littora* = the seashore) **zone** is usually covered with various types of marine macro algae. Along the North American Pacific coast, the giant brown bladder kelp (*Macrocystis*) attaches to rocks as deep as 30 meters (100 feet), and is attached to rocky bottoms with a root-like anchor called a *holdfast* (Figure 15–15a) so strong only large storm waves can break the algae free. The *stipes* and *blades* of the algae are supported by gas-filled floats called *pneumatocysts* (*pneumato* = breath, *cystis* = bladder), which allow the algae to grow upward and extend for another 30 meters (100 feet) along the surface to allow for good exposure to sunlight. Under ideal conditions, *Macrocystis* can grow up to 0.6 meter (2 feet) per day.

The giant brown bladder kelp and bull kelp (*Nereocystis*), another fast-growing kelp, often form beds

Figure 15-14 Fiddler crab. A male fiddler crab (*Uca*) among eelgrass (*Zostera*) at Cape Hatteras, North Carolina. The fiddler crab uses its large claw for protection and for attracting mates.

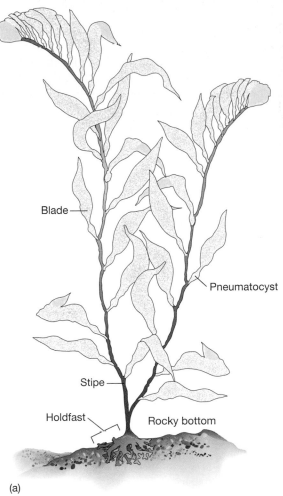

Figure 15–15 *Macrocystis* and other kelp forest inhabitants. (a) Structure of the giant brown bladder kelp (*Macrocystis*). **(b)** A kelp forest. **(c)** Sea hares (*Aplysia californica*) and sea urchins (*Echinus*) in a kelp forest.

called **kelp forests** along the Pacific coast (Figure 15–15b). Smaller tufts of red and brown algae are found on the bottom and also live on the kelp blades.

Kelp forests are highly productive ecosystems that provide shelter for a wide variety of organisms living within or directly upon the kelp as epifauna. These organisms are an important food source for many of the animals living in and near the kelp forest, including mollusks, sea stars, fishes, octopus, lobsters, and marine mammals. Surprisingly, very few animals feed directly on the living kelp plant. Among those that do are the large sea hare (*Aplysia*) and sea urchins (Figure 15–15c).

Lobsters Large crustaceans—including lobsters and crabs—are common along rocky bottoms. The spiny lobsters are named for their spiny covering and have two very large, spiny antennae (Figure 15–16a). These antennae serve as feelers and are equipped with noise making devices near their base that are used in protection. The genus *Panuliris*, considered a delicacy, lives deeper than 20 meters (65 feet) along the European coast and reaches lengths to 50 centimeters (20 inches). For unknown reasons, the Caribbean species *Palinuris argus* sometimes migrates single-file across the sea floor in lines that are several kilometers long.

Panulirus interruptus is the spiny lobster of the American West Coast. All spiny lobsters are taken for food, but none are as highly regarded as the so-called true lobsters (genus *Homarus*), which include the American lobster, *Homarus americanus* (Figure 15–16b). Although they are scavengers like their spiny relatives, the true lobsters also feed on live animals, including mollusks, crustaceans, and other lobsters.

Oysters Oysters are thick-shelled sessile (anchored) bivalve mollusks found in estuaries. They grow best where there is a steady flow of clean water to provide plankton and oxygen.

Oysters are food for sea stars, fishes, crabs, and snails that bore through the shell and rasp away the soft tissue inside (Figure 15–17). In fact, this may be one of the main reasons that oysters have such a thick shell.[3] Oysters also have great commercial importance to humans throughout the world as a food source.

Oyster beds are composed of empty shells of many previous generations that are cemented to a hard substrate or to one another, with the living generation on top. Each female produces many millions of eggs each year, which become planktonic larvae when fertilized. After a few weeks as plankton, the larvae attach themselves to the bottom. As a material upon which to anchor, the oyster larvae prefer (in order) live oyster shells, dead oyster shells, and rock.

Coral Reefs

Individual corals—called **polyps** (*poly* = many, *pous* = foot)—are small benthic marine animals that feed with stinging tentacles and are related to jellyfish. Most species of corals are about the size of an ant, live in large colonies, and construct hard calcium carbonate structures for protection. Coral species are found throughout the ocean, but coral accumulations that are

---

[3]This is an example of a pattern in nature where a type of armor or defense possessed by one species creates evolutionary pressure in another species for a weapon to defeat it.

(a)

(b)

Figure 15–16 Spiny and American lobsters. (a) Spiny lobster (*Panulirus interruptus*).
(b) American lobster (*Homarus americanus*).

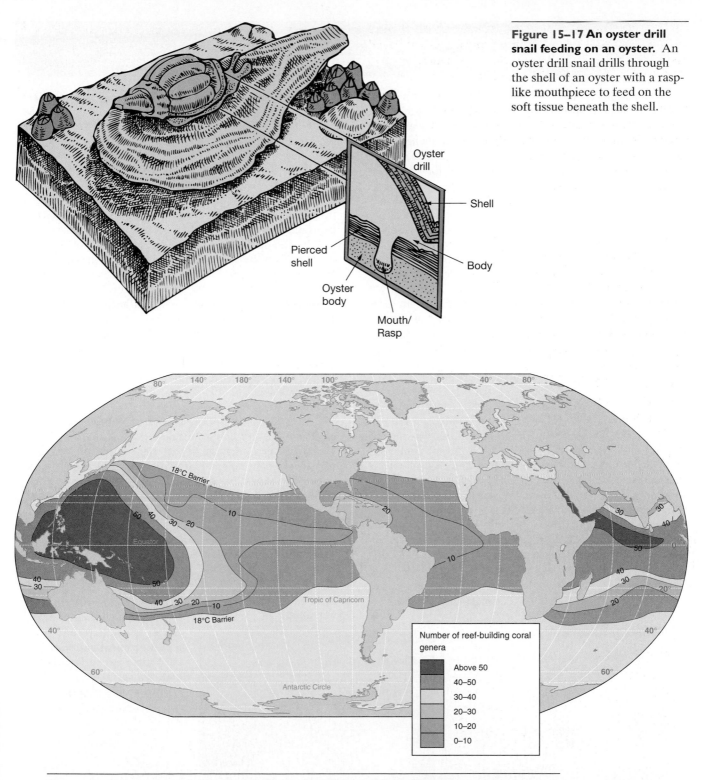

Figure 15–17 An oyster drill snail feeding on an oyster. An oyster drill snail drills through the shell of an oyster with a rasp-like mouthpiece to feed on the soft tissue beneath the shell.

Figure 15–18 Coral reef distribution and diversity. Coral reef development is restricted to warm tropical waters between the two 18°C (64°F) temperature lines. On the western side of each ocean basin, the coral reef belt is wider and the diversity of coral genera is greater, which is most likely related to surface circulation patterns and the presence of numerous tropical islands that favor speciation.

classified as **coral reefs** are restricted to shallow warmer-water regions.

Corals are very temperature sensitive. To survive, they need water where the average monthly temperature exceeds 18°C (64°F) throughout the year (Figure 15–18). If the water exceeds 30°C (86°F), however, many corals die, too. Warmer-than-normal sea-surface temperatures occur during El Niño events, which appear to be related to outbreaks of coral bleaching (Box 15–1).

Box 15–1
How White I Am: Coral Bleaching and Other Diseases

Coral bleaching is the loss of color in coral reef organisms—often in response to an increase in water temperature—that causes them to turn white (Figure 15B). Bleaching occurs when the coral's symbiotic partner, the zooxanthellae algae, is removed or expelled. Once bleached, the coral no longer received nourishment from the algae and if the coral does not regain its symbiotic algae, it will eventually die. Bleaching often occurs in surface waters—the top two or three meters (7 to 10 feet)—but has recently been observed at depths of 30 meters (100 feet) and can occur as quickly as overnight. The coral does not die immediately, but is weakened and does not grow. Recovery from a minor bout of bleaching can take as little as four weeks, but severe bouts can take as long as four years.

Florida's coral reefs have experienced at least eight widespread bleachings since the early 1900s, and at least 70% of the corals along the Pacific Central American coast died due to bleaching associated with the severe El Niño event of 1982–1983. Coral reefs around the Galápagos Islands thrive in ocean water at or below 27°C (81°F). If the water is even 1 or 2°C (2 or 4°F) warmer for an extended period, however, the coral may expel the algae, in effect "bleaching" itself. The warming during 1982–1983 was so severe and long-lived that two species of Panamanian coral became extinct during this El Niño. The bleaching episode of 1987 affected coral reefs worldwide, especially those in Florida and throughout the Caribbean. Since then, widespread bleachings have occurred with increasing frequency and intensity. For instance, the El Niño event of 1997–1998 raised water temperatures several degrees higher than normal and has been blamed for the most geographically widespread bleaching ever recorded, including the equatorial eastern Pacific Ocean, the Yucatán coast, the Florida Keys, and the Netherlands Antilles.

Besides abnormally high surface water temperatures (such as El Niño events), the algae may leave their host due to pollution, elevated ultraviolet radiation levels, changes in salinity, invasion of disease, or a combination of factors. Some researchers believe that excess oxygen builds up in the coral's tissues and becomes toxic when temperatures are excessively warm. Then the algae is expelled or, perhaps, the algae leave with dead tissue. The strong correlation between coral bleaching and elevated water temperature concerns scientists, some of whom believe that coral bleaching may be one of the first oceanic indications of global warming. Whatever the reason, experts agree that bleaching indicates that the coral is experiencing severe environmental stress.

John Porter—a coral reef ecologist at the University of Georgia—and his colleagues study diseases that affect corals. They have been monitoring the health of corals in the Florida Keys since 1995 and have discovered the reappearance of *white plague disease* as well as a dozen new diseases such as *white band disease*, *white pox*, *black band disease*, *yellow-band* (*yellow-blotch*) *disease*, *patchy necrosis*, and *rapid wasting disease*.

The cause of most of these diseases is still being investigated, and it is not know if the new diseases are from the invasion of microorganisms—bacteria, viruses, or fungi—or related to environmental stress like coral bleaching is. As human population has increased along the Florida Keys, the coral reefs of the Keys have begun to show signs of stress, thus making them more susceptible to a host of diseases. The increased nutrient levels and water turbidity resulting from soil runoff and improper sewage disposal in the Keys may contribute to the problem, too.

Easter Island

March 1999   March 2000

Figure 15B Normal coral (*Pocillopora verrucosa*) near Easter Island (*left*) and bleached coral one year later after exceptionally high sea surface temperatures (*right*).

Water warm enough to support coral growth is found primarily within the tropics. Reefs also grow as far north and south as 35 degrees latitude on the western margins of ocean basins, however, where warm-water currents raise average sea surface temperatures (Figure 15–18).

The map in Figure 15–18 also shows the greater diversity of reef-building corals on the western side of ocean basins. More than 50 genera of corals thrive in a broad area of the western Pacific Ocean and a narrow belt of the western Indian Ocean. Fewer than 30 genera occur in the Atlantic Ocean, with the greatest diversity occurring in the Caribbean Sea. The greatest diversity is most likely related to present (or past) surface current circulation patterns and the presence of numerous tropical islands that favor speciation.

Besides warm water, other environmental conditions that allow for coral growth include the following:

- Strong sunlight (not for the corals themselves, which are animals and can exist in deeper water, but for a symbiotic photosynthetic microscopic algae called **zooxanthellae** that lives within the coral's tissues[4])

- Strong wave or current action (to bring nutrients and oxygen)

- Lack of turbidity (suspended particles in the water tend to interfere with the coral's filter-feeding capability and absorb radiant energy, so corals are not usually found close to areas where major rivers drain into the sea)

- Salt water (corals die if the water is too fresh, which is another reason coral reefs do not form near the mouths of freshwater rivers)

- A hard substrate for attachment (corals cannot attach to a muddy bottom, so they often build upon the hard skeletons of their ancestors, creating coral reefs that are several kilometers thick)

> Coral are small colonial animals with stinging cells that are found primarily in shallow tropical water and need strong sunlight, wave or current action, lack of turbidity, normal salinity seawater, and a hard substrate for attachment.

Because of changes in wave energy, salinity, water depth, temperature, and other less obvious factors, there is a well-developed vertical and horizontal zonation of the reef slope (Figure 15–19). These zones can be readily identified by the types of coral present and the assemblages of other organisms found in and near the reef.

Symbiosis of Coral and Algae "Coral reefs" are more than just coral. Algae, mollusks, and foraminifers contribute to the reef structure, too. Individual reef-building corals are **hermatypic** (*herma* = secret, *typi* = type) because they have a mutualistic relationship[5] with microscopic algae (zooxanthellae) that live within the tissue of the coral polyp. The algae provide their coral host with a continual supply of food, and the corals provide the zooxanthellae with nutrients. Although coral polyps capture tiny planktonic food with their stinging tentacles, most reef-building corals receive up to 90% of their nutrition from symbiotic zooxanthellae algae. In this way, corals are able to survive in the nutrient-poor waters characteristic of the tropical oceans.

Other reef animals also have a symbiotic relationship with various types of marine algae. Those that derive part of their nutrition from their algae partners are called **mixotrophs** (*mixo* = mix, *tropho* = nourishment), and include coral, foraminifers, sponges, and mollusks (Figure 15–20). The algae not only nourish the coral but also may contribute to their calcification by extracting carbon dioxide from the coral's body fluids.

Coral reefs actually contain up to three times as much algal biomass as animal biomass. Zooxanthellae, for example, account for up to 75% of the biomass of reef-building coral. Nevertheless, zooxanthellae account for less than 5% of the reef's overall algal mass (most of the rest is filamentous green algae).

Because the algae need sunlight for photosynthesis, the greatest depth to which active coral growth extends is 150 meters (500 feet). Water motion is less at these depths, so relatively delicate plate corals can live on the outer slope of the reef from 150 meters (500 feet) up to about 50 meters (165 feet), where light intensity is as low as 4% of the surface intensity (see Figure 15–19).

From 50 meters (164 feet) to about 20 meters (66 feet), water motion from breaking waves increases on the side of the reef facing into the prevailing current flow. Correspondingly, the mass of coral growth and the strength of the coral structure supporting it increase toward the top of this zone, where light intensity is as low as 20% of the surface value.

The reef flat may have a water depth of a few centimeters to a few meters at low tide, so it has at least 60% of the surface light intensity. Many species of colorful reef fish inhabit this shallow water, as well as sea cucumbers, worms, and mollusks. In the protected water of the reef lagoon live gorgonian coral, anemones, crustaceans, mollusks, and echinoderms (Figure 15–21).

Coral Reefs and Nutrient Levels When human populations increase on land adjacent to coral reefs, the

---

[4]It is zooxanthellae (*zoo* = animal, *xanthos* = yellow, *ella* = small) algae that give corals their distinctive bright coloration (which can be many colors besides yellow).

[5]See Chapter 13, "Biological Productivity and Energy Transfer" for a discussion of the types of symbiosis, including mutualism.

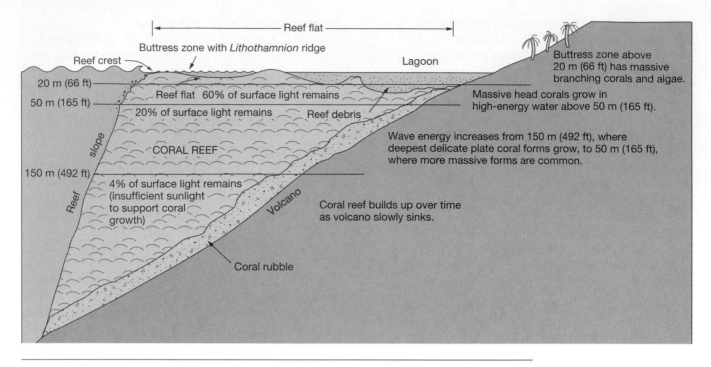

Figure 15–19 Coral reef zonation. As depth increases, wave energy decreases and light intensity decreases. Massive branching corals occur above 20 meters (66 feet), where wave energy is great. Corals become more delicate with increasing depth, until a depth of around 150 meters (500 feet), where too little solar radiation is available to allow the survival of their symbiotic zooxanthellae algae.

reefs deteriorate. Fishing, trampling, boat collisions with the reef, sediment increase due to development, and removal of reef inhabitants by visitors all damage the reef. One of the more subtle effects is the inevitable increase in the nutrient levels of the reef waters from sewage discharge and farm fertilizers.

As nutrient levels increase in reef waters, the dominant benthic community changes:

- At low nutrient levels, hermatypic corals and other reef animals that contain algal symbiotic partners thrive.
- Moderate nutrient levels favor fleshy benthic plants, and high nutrient levels favor suspension feeders such as clams.
- At high nutrient levels, the phytoplankton mass exceeds the benthic algal mass, so benthic populations tied to the phytoplankton food web dominate.

Increased phytoplankton biomass reduces the clarity of the water, too, which interferes with the coral's filter-feeding capability. The fast-growing members of the phytoplankton-based ecosystem destroy the reef structure by overgrowing the slow-growing coral and through **bioerosion**, which is erosion of the reef by organisms. Bioerosion by sea urchins and sponges is particularly damaging to many coral reefs.

Corals are able to survive in nutrient-depleted warm water by living symbiotically with zooxanthellae algae, which live within the coral's tissues, provide it with food, and give the coral its color.

The Deep-Ocean Floor

The vast majority of the ocean floor lies submerged below several kilometers of water. Less is known about life in the deep ocean than about life in any of the shallower nearshore environments because it is difficult and expensive to investigate the deep sea. Just to obtain samples from the deep-ocean floor requires a specially designed submersible or a properly equipped research vessel that has a spool of high-strength cable at least 12 kilometers (7.5 miles) long.

Collecting samples by submersible or with a biological dredge is a time-consuming process. Because the supply of oxygen is limited, manned submersibles can stay down for only 12 hours, and it may take eight of those hours to descend and ascend. To send a dredge to the deep-ocean floor and retrieve it from the depths takes about 24 hours.

Robotics and remotely operated vehicles (ROVs) are making it possible to observe and sample even the deepest reaches of the ocean more easily. Because they

(a)

(b)

(c)

Figure 15–20 Coral reef inhabitants that rely on symbiotic algae. **(a)** Coral polyps, which are nourished by internal zooxanthellae algae and also by extending their tentacles to capture tiny planktonic organisms from the surrounding water. **(b)** The blue-gray sponge *Niphates digitalis* (*left*), and the brown sponge *Angelas* (*right*), which contain symbiotic algae or bacteria. **(c)** A giant clam (*Tridacna gigas*), which depends on symbiotic algae living within its mantle tissue.

are unmanned, ROVs are cheaper to operate and can stay beneath the surface for months if necessary. These developments should lead to further discoveries in one of Earth's least-known habitats.

The Physical Environment

The deep-ocean floor includes bathyal, abyssal, and hadal zones.[6] Here, the physical environment is much different than at the surface, but it is quite stable and homogeneous. Light is present in only the lowest concentrations down to a maximum of 1000 meters (3300 feet), and absent below this depth. Everywhere the temperature rarely exceeding 3°C (37°F) and falls as low as −1.8°C (28.8°F) in the high latitudes. Salinity remains at slightly less than 35‰.[7] Oxygen content is constant and relatively high. Pressure exceeds 200 kilograms per square centimeter (200 atmospheres or 2940 pounds per square inch) on the oceanic ridges, exceeds 300–500 kilograms per square centimeter (300–500 atmospheres or 4410–7350 pounds per square inch) on the deep-ocean abyssal plains, and exceeds 1000 kilograms per square centimeter (1000 atmospheres or 14,700 pounds per square inch) in the deepest trenches.[8] Bottom currents are generally slow but more variable than once believed. **Abyssal storms** created by warm- and cold-core eddies of surface currents affect certain areas, lasting several weeks and causing bottom currents to reverse and/or increase in speed.

A thin layer of sediment covers much of the deep-ocean floor. On abyssal plains and in deep trenches, sediment is composed of mud-like abyssal clay deposits. The accumulation of oozes—composed of dead planktonic organisms that have sunk through the water column—occurs on the flanks of oceanic ridges and rises. On the continental rise, there may be some coarse sediment from nearby land sources. Sediment may be absent on steep areas of the continental slope. It may also be absent near the crest of the mid-ocean ridge and along the slopes of seamounts and oceanic islands, where it had not had enough time to accumulate on newly formed ocean floor.

Food Sources and Species Diversity

Because of the lack of light, photosynthetic primary production cannot occur. Except for the chemosynthetic productivity that occurs around hydrothermal vents, all benthic organisms receive their food from the surface waters above. Only about 1% to 3% of the food produced in the euphotic zone reaches the deep-ocean floor, so the scarcity of food that drifts down from the sunlit surface waters—not low temperature or high pressure—limits deep-sea benthic biomass. However, some variability in the supply of food is caused by seasonal phytoplankton blooms at the surface. Figure 15–22 shows the food sources for deep-sea organisms.

Many of the organisms that inhabit the deep sea have special adaptations to help them detect food using chemical clues. Once food is found, these organisms are efficient at consuming it (Box 15–2).

For many years, it was believed that the species diversity of the deep-ocean floor was quite low compared with shallow-water communities. Researchers studying sediment-dwelling animals in the North Atlantic, however, discovered an unexpectedly large diversity of species. An area of 21 square meters (225 square feet) contained 898 species, of which 460 were new to science. After analyzing 200 samples, new species were being discovered at a rate that suggested millions of deep-sea species!

It turns out that deep sea species diversity—especially for small infaunal deposit feeders—rivals tropical rain forests. It also appears, however, that the distribution of deep-sea life is patchy and depends to a large degree on the presence of certain microenvironments.

> The deep ocean floor is a stable environment of darkness, cold water, and high pressure but still supports life. The food source for most deep-sea organisms is from the sunlit surface waters.

Deep-Sea Hydrothermal Vent Biocommunities

An active hydrothermal (*hydro* = water, *thermo* = heat) vent field on the ocean floor was visited for the first time in 1977 during a dive of the submersible *Alvin*. The field exists in complete darkness in water below 2500 meters (8200 feet) in the Galápagos Rift, near the Equator in the eastern Pacific Ocean (Figures 15–23 and 15–24). Water temperature near the vents was 8 to 12°C (46 to 54°F), whereas normal water temperature at these depths is about 2°C (36°F).

These vents supported the first known **hydrothermal vent biocommunities**, consisting of organisms that were unknown to science and unusually large for those depths. The most prominent were tube worms over 1 meter (3.3 feet) long, giant clams up to 25 centimeters (10 inches) long, large mussels, two varieties of white crabs, and extensive microbial mats. These biocommunities had up to 1000 times more biomass than the rest of the deep-ocean floor. In a region of scarce nutrients and

---

[6]The bathyal, abyssal, and hadal zones are described in Chapter 12, "The Marine Habitat."

[7]Remember that average surface seawater salinity is 35 parts per thousand (‰).

[8]The pressure is 1 atmosphere (1 kilogram per square centimeter) at the ocean surface and increases by 1 atmosphere for each 10 meters (33 feet) of depth. Thus, a pressure of 1000 atmospheres is 1000 times that at the ocean's surface.

(a)

(b)

Figure 15–21 Non-reef building inhabitants. **(a)** Coral reefs supply habitat and protection for many fishes, including this puffer (*Arothron*). Puffers usually aren't quick enough to escape a predator, but they can expand their bodies to produce a large, spherical shape that cannot be easily eaten. **(b)** Unlike reef-building corals, some corals do not secrete a hard calcium carbonate structure, such as this soft gorgonian coral, which has feeding polyps (*purple*) extending from its branches.

small populations of organisms, these hydrothermal vents are truly the oases of the deep ocean.

At 21 degrees north latitude on the East Pacific Rise, south of the tip of Baja California, tall underwater chimneys were found to belch hot vent water (350°C or 662°F) so rich in metal sulfides that it colored the water black. These chimney vents, first observed in 1979, were composed primarily of sulfides of copper, zinc, and silver and came to be called **black smokers**.

The most important members of these hydrothermal vent biocommunities are the microscopic **archaea** (*archaeo* = ancient), which are simple bacteria-like life forms. Through **chemosynthesis** (*chemo* = chemistry, *syn* = with, *thesis* = an arranging) of hydrogen sulfide, archaea manufacture carbohydrates from carbon dioxide and water (Figure 15–25) and form the base of the food web for vent organisms. Although some animals feed directly on archaea and larger prey, many of them depend primarily on a symbiotic relationship with archaea. The tube worms and giant clams, for instance, depend entirely on sulfur-oxidizing archaea that live symbiotically within their tissues (Figure 15–26).

In 1981, humans in a submersible first visited the Juan de Fuca Ridge biocommunity offshore of Oregon. Although vent fauna at this site are less abundant than at the Galápagos Rift and on the East Pacific Rise, the metallic sulfide deposits from the vents aroused much interest because they are the only active hydrothermal vent deposits in U.S. waters.

In 1982, the first hydrothermal vents beneath a thick layer of sediment were discovered during a submersible dive in the Guaymas Basin of the Gulf of California. In this region, a spreading center is actively working to rift apart the sea floor as it is being covered with sediment. Sediment samples recovered in this region were high in sulfide and saturated with hydrocarbons, which may have entered the food chain through bacteria. The abundance and diversity of life discovered here may exceed that of the Galápagos Rift and along the East Pacific Rise.

Like the Guaymas Basin, the Mariana Basin of the western Pacific has a small spreading center beneath a sediment-filled basin. A research dive in a submersible in 1987 revealed many new species of hydrothermal vent organisms (Figure 15–26c). Subsequent exploration has revealed numerous hydrothermal vent biocommunities in other parts of the Pacific Ocean (see Figure 15–24) and additional new species.

In 1985, the first active hydrothermal vents with associated biocommunities in the Atlantic Ocean were discovered at depths below 3600 meters (11,800 feet) near the axis of the Mid-Atlantic Ridge between 23 and 26 degrees north latitude. The predominant fauna of these vents consists of shrimp that have no eye lens but can detect levels of light emitted by the black smoker chimneys that are invisible to the human eye (Figure 15–27).

In 1993, a hydrothermal vent community was discovered on a flat-topped volcano rising to 1525 meters

Box 15–2
How Long Would Your Remains Remain on the Sea Floor?

What happens to people who are buried at sea? How long do their remains remain on the sea floor? How long do the remains of a large organism such as a whale remain on the sea bottom? Oceanographers who study deep-sea biocommunities have conducted experiments in the deep sea to help answer these questions.

One experiment was conducted along the sea floor in the Philippine Trench at a depth of 9600 meters (31,500 feet) in 1975. Several whole fish were placed on the sea floor and an underwater camera positioned above them took a picture every few minutes to observe how long they remained there (Figure 15C). *Hirondellea gigas*, a scavenging benthic amphipod (a shrimp-like animal), discovered the bait after only a few hours. The bait was swarming with amphipods after nine hours and it was stripped of flesh in as little as 16 hours! Other studies obtained similar results, suggesting that organisms the size of humans would have their soft tissue devoured within a day on the deep ocean floor.

For deep-sea organisms, large food falls represent an unpredictable but intense nutrient supply. Deep-sea scavengers such as amphipods, hagfish, and sleeper sharks use special chemoreceptive sensory devices to identify and quickly locate food on the sea floor. Whale carcasses, for example, can support a thriving ecosystem of benthic organisms, including some species that inhabit hydrothermal vents.

To test how long a whale's remains lasted on the sea floor, researchers used two dead juvenile gray whales that washed up on the beach in southern California in 1996 and 1997. With permission from the National Marine Fisheries Service, the 5000-kilogram (5.5-ton) whales were intentionally weighted and sunk in the San Diego Trough. Researchers in a deep-diving submersible visited the carcasses at regular intervals and found that the gray whales were completely stripped of flesh in four months! Other studies indicate that even blue whales, which can weigh 25 times more than a gray whale, are stripped of their flesh in as little as six months.

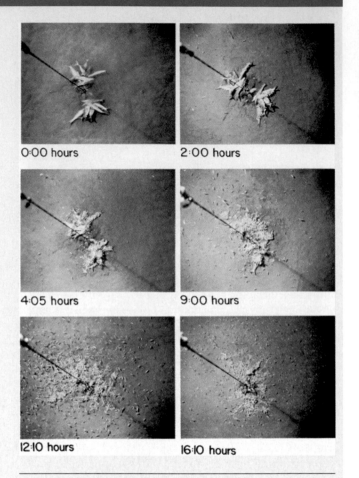

0:00 hours 2:00 hours
4:05 hours 9:00 hours
12:10 hours 16:10 hours

Figure 15C Time-sequence photography of fish remains on the deep-ocean floor.

(5000 feet)—well above the walls of the Mid-Atlantic Ridge rift valley. Called the "Lucky Strike" vent field, it is about 1000 meters (3300 feet) shallower than most other sites. It is the only Mid-Atlantic Ridge site to possess the mussels common at many other vent sites and is the only location where a new species of pink sea urchin has been found.

Life Span of Hydrothermal Vents
Because the hot-water plumbing of the sea floor is controlled by unpredictable volcanic activity associated with mid-ocean ridge spreading centers, a vent may remain active for only limited periods—years or sometimes decades. For instance, a hydrothermal vent field called the Coaxial Site along the Juan de Fuca Ridge offshore of Washington that had been active was revisited a few years later and found to be inactive. Inactive sites such as this one are identified

by an accumulation of large numbers of dead hydrothermal vent organisms. When the vent becomes inactive and the hydrogen sulfide that serves as the source of energy for the community is no longer available, organisms of the community die if they cannot move elsewhere.

Other sites indicate an increase in volcanic activity. For example, at a site along the East Pacific Rise known as Nine-Degrees North, a large number of tube worms were cooked by lava flowing into their midst in what has been described as a "tube-worm barbecue." The discovery of newly formed and ancient vent areas along spreading centers indicates that hydrothermal vents can suddenly appear or cease to operate. Moreover, areas of active venting may lie hundreds of kilometers apart.

Hydrothermal vent organisms are well adapted to the temporary nature of hydrothermal vents. Most have

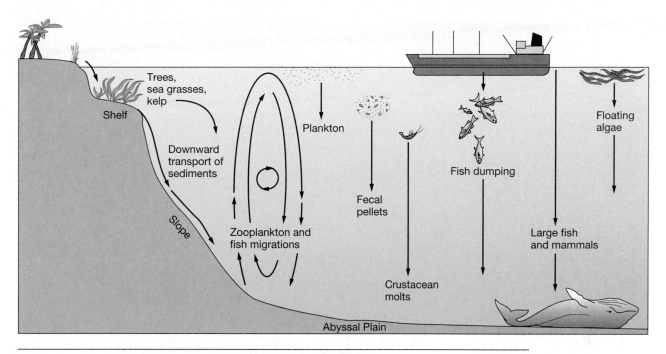

Figure 15–22 Food sources for deep-sea organisms. Most deep-sea organisms obtain their food from surface water after it slowly settles through the water column to the sea floor. The food supply is usually limited, except when large fish or mammals (such as whales) sink to the bottom.

Figure 15–23 *Alvin* **approaches a hydrothermal vent community.** Schematic view of a hydrothermal vent area, showing lava pillows and a black smoker that spews hot (350°C, or 662°F), sulfide-rich water from a chimney. Organisms (counterclockwise from *Alvin*) include the grenadier fish (or rattail fish), octacoras, a sea anemone, white brachyuran crabs, large clams (*Calypotogena*), and tube worms (*Riftia*).

high metabolic rates, for example, which cause them to mature rapidly so they can reproduce while the vent is still active.

Studies of several hydrothermal vent sites suggest that species diversity is low. In fact, only a little more than 300 animal species have been identified to date. Many species, however, are common to widely sepa-

rated hydrothermal vent fields. Although hydrothermal vent animals typically release drifting larvae into the water, it is not clear how the larvae are able to survive the journey to hydrothermal vents that lie at such great distances from one another.

One idea, called the "*dead whale hypothesis*," suggests that when large animals die they may sink to the

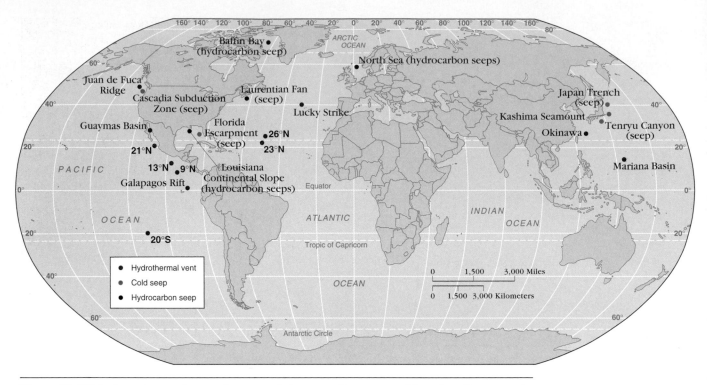

Figure 15–24 Vents and seeps known to support deep-sea biocommunities. Location map of selected hydrothermal vents (*red*), cold seeps (*blue*), and hydrocarbon seeps (*black*).

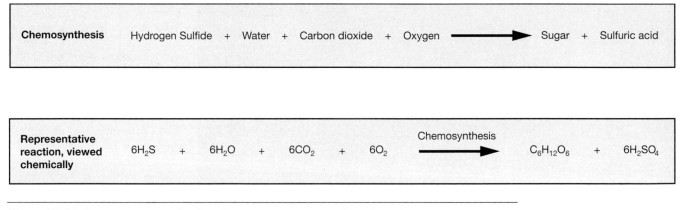

Figure 15–25 Chemosynthesis (*top*) and representative reaction viewed chemically (*bottom*).

deep-ocean floor, decompose, and provide an energy source in stepping stone fashion for the larvae of hydrothermal vent organisms. The organisms settle and grow here, then breed and release their own larvae, some of which make it to the next hydrothermal vent field. Other researchers believe that deep-ocean currents are strong enough to transport drifting larvae to new sites. Still others have suggested that the rift valleys of mid-ocean ridges act as passageways along which drifting larvae traverse to inhabit new vent fields. By whatever means they travel, they colonize new hydrothermal vents soon after the vents are created. In 1989, for example, a newly formed hydrothermal vent along the Juan de Fuca Ridge had no life forms. By

1993, however, tube worms and other life forms had already established themselves.

Hydrothermal Vents and the Origin of Life Life is thought to have begun in the oceans, and environments similar to those of the hydrothermal vents must have been present in the early history of the planet. The uniformity of conditions and abundant energy of the vents, therefore, has led some scientists to propose that hydrothermal vents would have provided an ideal habitat for the origin of life. In fact, hydrothermal vents may represent one of the oldest life-sustaining environments, because hydrothermal activity occurs wherever there are both volcanoes and water. The presence of

(a)

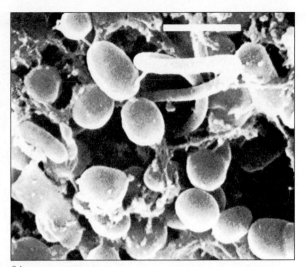

(b)

(c)

Figure 15–26 Chemosynthetic life. (a) Tube worms up to 1 meter (3.3 feet) long are found at the Galápagos Rift and other deep-sea hydrothermal vents. **(b)** Sulfur-oxidizing archaea (enlarged 20,000 times; white bar at top is 1 micron long) that live symbiotically within the tissue of tube worms, clams, and mussels found at hydrothermal vents. **(c)** A hydrothermal vent biocommunity from the Mariana Back-Arc Basin, which included a new genus and species of sea anemone (*Marianactis bythios*), a new family, genus, and species of gastropod (*Alviniconcha hessleri*, the first known conch to contain chemosynthetic bacteria), and the galatheid crab (*Munidopsis marianica*).

archaea, which have ancient genetic makeup, helps support this idea.

Low-Temperature Seep Biocommunities

Three additional submarine seep environments—locations where water trickles out of the sea floor—have been found that chemosynthetically support biocommunities similar to hydrothermal vent communities.

Hypersaline Seeps In 1984, a hypersaline seep was studied in water depths below 3000 meters (9800 feet) at the base of the Florida Escarpment in the Gulf of Mexico (Figure 15–28a). The water from this seep had a salinity of 46.2‰ but its temperature was not warmer than normal. Researchers discovered a **hypersaline seep biocommunity** similar in many respects to the hydrothermal vent communities. The seeping water appears to flow from fractures at the base of a limestone escarpment (Figure 15–28b) and move out across the clay deposits of the abyssal plain at a depth of about 3200 meters (10,500 feet).

The hydrogen sulfide-rich waters support a number of white microbial growths called mats, which conduct chemosynthesis in a fashion similar to archaea at hydrothermal vents. These and other chemosynthetic archaea may provide most of the sustenance for a diverse community of animals that includes sea stars, shrimp, snails, limpets, brittle stars, anemones, tube worms, crabs, clams, mussels, and a few species of fish (Figure 15–28c).

Figure 15–27 Atlantic Ocean hydrothermal vent organisms. Swarm of particulate-feeding shrimp, the predominant animals observed at hydrothermal vents near 26 degrees north latitude on the Mid-Atlantic Ridge.

Hydrocarbon Seeps Also observed in 1984 were dense biological communities associated with oil and gas seeps on the Gulf of Mexico continental slope (Figure 15–29). Trawls at depths of between 600 and 700 meters (2000 and 2300 feet) recovered fauna similar to those observed at hydrothermal vents and at the hypersaline seep in the Gulf of Mexico. Subsequent investigations identified seeps with associated communities down to depths of 2200 meters (7300 feet) on the continental slope.

Carbon-isotope analysis indicates that these **hydrocarbon seep biocommunities** are based on chemosynthesis that derives its energy from hydrogen sulfide and/or methane. Microbial oxidation of methane produces calcium carbonate slabs found here and at other hydrocarbon seeps (see Figure 15–24).

Subduction Zone Seeps In 1984, a **subduction zone seep biocommunity** was discovered during one of *Alvin's* dives to study folding of the sea floor in a subduction zone. The seep is located near the Cascadia subduction zone of the Juan de Fuca Plate at the base of the continental slope off the coast of Oregon (Figure 15–30a). The trench is filled with sediments, which are folded into a ridge at the seaward edge of the slope. At the crest of this ridge, water slowly flows from the 2-million-year-old folded sedimentary rocks into a thin overlying layer of soft sediment on the sea floor. Even-

tually, the water is released from the sediment through seeps on the ocean bottom.

At a depth of 2036 meters (6678 feet), the seeps produce water that is only about 0.3°C (0.5°F) warmer than seawater at that depth. The vent water contains methane that is probably produced by decomposition of organic material in the sedimentary rocks. Microbes oxidize the methane, chemosynthetically producing food for themselves and the rest of the community, which contains many of the same genera found at other vent and seep sites (Figure 15–30b).

Since the detection of subduction zone seeps, similar communities have been discovered in other subduction zones, including the Japan Trench and the Peru-Chile Trench. All these subduction zone seeps are located on the landward side of the trenches at depths from 1300 to 5640 meters (4265 to 18,500 feet).

> Hydrothermal vent biocommunities occur near black smokers and rely on chemosynthetic archaea for food. Other deep-sea biocommunities exist along hypersaline, hydrocarbon, and subduction zone seeps.

 ## Students Sometimes Ask...

What is an urchin barren?

An urchin barren is created when the population of sea urchins goes unchecked and they devour entire areas of the giant brown bladder kelp (*Macrocystis*), one of the main types of algae in kelp forests. The sea urchins crawl across the ocean floor to new areas in search of food, and set the kelp adrift by chewing through the holdfast structure that the kelp needs to remain attached to the ocean floor. In California, the elimination of species that prey upon sea urchins (such as the wolf eel and the sea otter) has upset the natural balance in oceanic food webs. Consequently, sea urchins have proliferated and urchin barrens now exist where there were once lush kelp beds.

I've been at a tide pool and seen sea anemones. When I put my finger on one, it tends to gently grab my finger. Why does it do that?

The sea anemone is trying to kill you and wants to eat you (seriously!). Disguised as a harmless flower, the sea anemone is actually a vicious predator that will attack any unsuspecting animal (even a human) that its stinging tentacles entrap. Fortunately, the skin on our hands is thick enough to resist the stinging nematocyst and its neurotoxin. A couple of people, however, were interested in finding out if the sea anemone grabbed other things with its tentacles, so they put their *tongue* into a sea anemone. After a short time, their throats swelled almost completely closed, and they had to be rushed to a hospital. They lived, but the moral of this story is: *NEVER* put your tongue into a sea anemone!

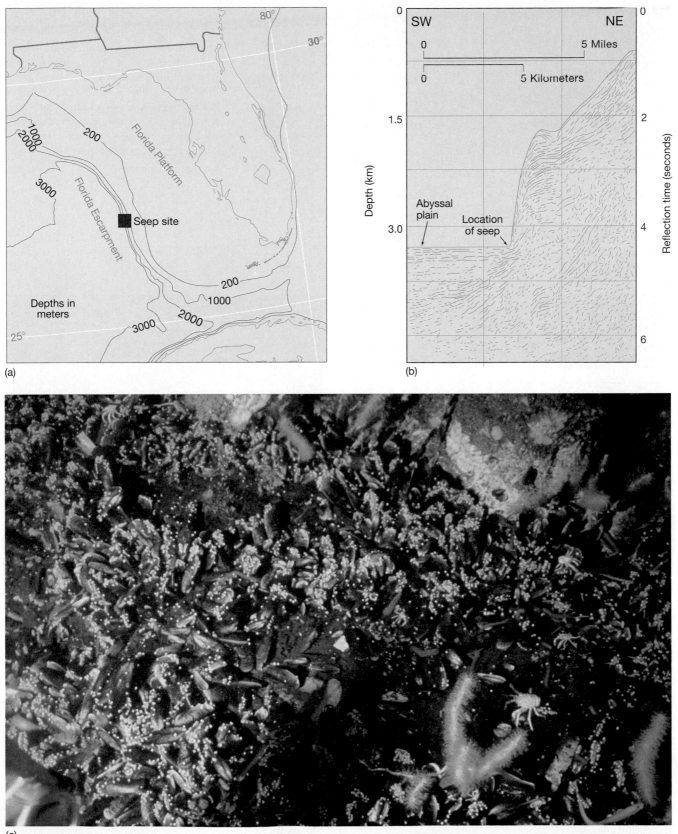

(a)

(b)

(c)

Figure 15–28 Hypersaline seep biocommunity at the base of the Florida Escarpment.
(a) Location of seep and biocommunity. **(b)** Seismic reflection profile of Florida Escarpment and abyssal sediments at its base. Arrow marks location of seep. **(c)** Florida Escarpment seep biocommunity of dense mussel beds. White dots are small gastropods on mussel shells. Tube worms (*lower right*) are covered with hydrozoans and galatheid crabs.

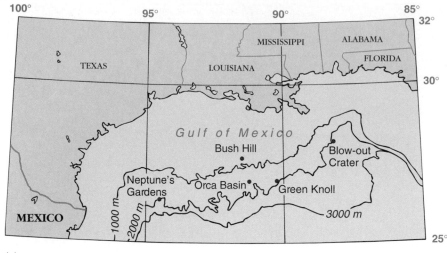

(a)

Figure 15–29 Hydrocarbon seeps on the continental slope in the Gulf of Mexico. (a) Locations of known hydrocarbon seeps with associated biocommunities. (b) Chemosynthetic mussels and tube worms from the Bush Hill seep. (c) Alaminos Canyon site (Neptune's Gardens) discovered in 1990, contains a new species of mussel and two new species of tube worms.

(b)

(c)

I know that coral reefs are pretty, but do they have any practical uses?

Many. Coral reefs are some of the largest structures created by living creatures on Earth [the Great Barrier Reef, for example, is over 2000 kilometers (1250 miles) long. Coral reefs foster a diversity of species that surpasses even that of tropical rain forests. Reefs provide shelter, food, and breeding grounds for an estimated 35,000 to 60,000 species worldwide, including almost a third of the world's estimated 20,000 species of marine fishes. Other species that inhabit reefs include creatures as diverse as anemones, sea stars, crabs, eels, sea slugs, clams, sharks, and sponges.

Coral reefs are also important to the economies of countries that have them. Many of these tropical countries receive over 50% of their gross national product as tourism related to reefs, which provides a much-needed incentive for these countries to protect them. Fisheries associated with

reefs supply more than one-sixth of all fish from the sea. Recently, pharmacologists and marine chemists have discovered a storehouse of new medical compounds that fight maladies such as cancer and infections. In addition, reefs help to prevent shoreline erosion and protect coastal communities from waves and storms. The hard calcium carbonate skeletons of coral have even been used in some human bone grafts!

I've heard that the Great Barrier Reef is plagued by an organism called the crown-of-thorns. What is it?

The crown-of-thorns (*Acanthaster planci*) is a sea star (Figure 15D). Since 1962, its proliferation has destroyed living coral on many reefs throughout the western Pacific Ocean. The sea star moves across reefs and eats the coral polyps. Normally, the coral can grow back if it has enough time to do so. Vast numbers of crown-of-thorns sea stars upset the natural

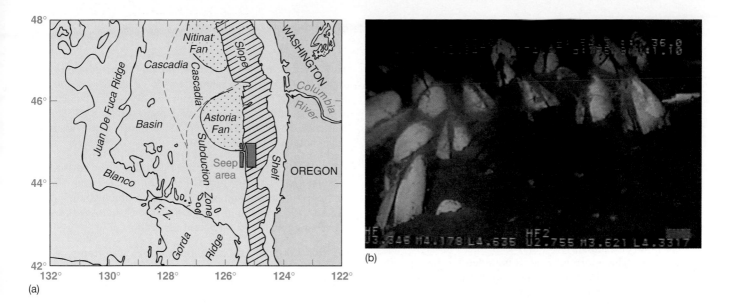

(b)

Figure 15–30 Locations of vent communities off the coast of Oregon. (a) Map showing sea floor features and the location of vent communities off the coast of Oregon. These communities are associated with the Cascadia subduction zone, where sediment filling the trench is folded into a ridge with vents at its crest. (b) Giant white clams (*Calyptogena soyoae*) half buried in methane-rich mud at 1100 meters (3600 feet) near the Japan Trench. The clams host sulfide-oxidizing microorganisms, which oxidize methane and supply the clams with food.

balance, however, decimating coral communities. Initially, divers were employed to smash the crown-of-thorns, but sea stars (which have tremendous regenerating capabilities) can easily produce new individuals from various body parts, so it only made the problem worse.

Some investigators believe the proliferation of the crown-of-thorns sea star is a modern phenomenon brought about by the activities of humans, although there is little supporting evidence. Studies suggest that during the past 80,000 years, the crown-of-thorns sea star has been even more abundant on the reefs studied than it is today. Thus, the sea star may be an integral part of the reef ecology in this region, and their increase may be part of a long-term natural cycle rather than a destructive event triggered by human actions.

What happens to the zooxanthellae algae that are released by corals that become bleached?

This has not been thoroughly researched, but it is believed that zooxanthellae algae that leave or are expelled from corals enter the stream of phytoplankton, which is largely consumed by primary consumers such as copepods and other zooplankton. It is not believed that they can exist outside of their host, so those that are not eaten probably die.

The mussel beds near hydrothermal vents are certainly extensive. Are those organisms edible?

Not for humans. The microbes that form the base of the food web use hydrogen sulfide gas (which has a characteristic "rot-

ten egg" odor) as a source of energy. Sulfide, therefore, which is poisonous to most organisms even at low levels, tends to concentrate within the tissues of these organisms. Although organisms within the hydrothermal biocommunities can ingest sulfide and have unique adaptations for getting rid of it, hydrothermal vent organisms such as mussels and clams are toxic to humans. Even if they were edible, they would be difficult to harvest economically because they live in such deep water.

Figure 15D Crown-of-thorns sea star.

Chapter in Review

- Over 98% of the 250,000 known marine species live in diverse environments within or on the ocean floor. Species diversity of these benthic organisms depends on their ability to adapt to the conditions of their environment, particularly temperature. With few exceptions, the biomass of benthic organisms closely matches that of photosynthetic productivity in surface waters above.

- Many adverse conditions exist in the intertidal zone of rocky shores, but organisms have adapted so they can densely populate these environments. Influenced by the tides, rocky shores can be divided into a high tide zone (mostly dry), a middle tide zone (equally wet and dry), and a low tide zone (mostly wet). The intertidal zone is bounded by the supralittoral (covered only by storm waves) and the sublittoral, which extends below the low-tide shoreline.

- Each of these zones contains characteristic types of life. Periwinkle snails, rock louses, and limpets can be found in the sublittoral zone. Sessile organisms can be found in the high-tide zone, especially buckshot barnacles. Algae become more abundant in the middle tide zone, and the diversity and abundance of the flora and fauna increase toward the lower intertidal zone. Acorn barnacles, goose-necked barnacles, mussels, and sea stars are commonly found in the middle tide zone, as well as sea anemones, fishes, hermit crabs, and sea urchins. The low tide zone in temperate latitudes has a variety of moderately sized red and brown algae that provide a drooping canopy for the animal life.

- Many varieties of burrowing infauna are common along sediment-covered shores (that is, beaches, salt marshes, and mud flats). Compared with rocky shores, however, the diversity of species in sediment-covered shores is less. As with the rocky shore, the diversity of species and abundance of life on the sediment-covered shore increases toward the low-tide shoreline.

- In more protected segments of the shore, wave energy is lower, so sand and mud are deposited. Sand deposits are usually well oxygenated compared to mud deposits. The intertidal region of sediment-covered shores has high-, middle-, and low-tide zones, similar to rocky shores.

- Common methods for feeding on sediment-covered shores include suspension feeding (filtering planktonic organisms from the water), deposit feeding (ingesting sediment and detritus), and carnivorous feeding (preying directly upon other organisms). Organisms characteristic of sandy beaches include bivalve (two-shelled) mollusks, lugworms, beach hoppers, sand crabs, sand stars, and heart urchins. Organisms characteristic of mud flats include eelgrass, turtle grass, bivalve mollusks, and fiddler crabs.

- Attached to the rocky sublittoral bottom just beyond the shoreline is a band of algae that often creates kelp forests. Kelp forests are the home of many organisms, including other varieties of algae, mollusks, sea stars, fishes, octopus, lobsters, marine mammals, sea hares, and sea urchins.

- Spiny lobsters are common to rocky bottoms in the Caribbean and along the West Coast, and the American lobster is found from Labrador to Cape Hatteras. Oyster beds found in estuarine environments consist of individuals that attach themselves to the bottom or to the empty shells of previous generations.

- Coral reefs consist of large colonies of coral polyps and many other species that need warm water and strong sunlight to live. Coral reefs are usually found in nutrient-poor tropical waters. Reef-building corals and other mixotrophs are hermatypic, containing symbiotic algae (zooxanthellae) in their tissues. Delicate varieties are found at 150 meters, and they become more massive near the surface, where wave energy is higher. The potentially lethal "bleaching" of coral reefs is caused by the removal or expulsion of symbiotic algae, probably under stress of elevated temperature.

- The physical conditions of the deep ocean floor are much different from those of shallow water. There is no light and the water is uniformly cold. The primary food source is from the surface waters above, which limits biomass. Species diversity in the deep ocean, however, is much higher than was previously thought.

- Primary production in hydrothermal vent communities near black smokers is due to chemosynthesis. Some evidence suggests that hydrothermal vents may have been some of the first regions where life became established on Earth, despite the short life span of individual vents. Chemosynthesis has also been identified in low-temperature seep biocommunities near hypersaline, hydrocarbon, and subduction zone seeps.

Key Terms

Abyssal storm (p. 458)

Annelid (p. 447)

Archaea (p. 460)

Bioerosion (p. 458)

Biomass (p. 439)

Bivalve (p. 447)

Black smoker (p. 460)

Carnivorous feeding (p. 447)

Chemosynthesis (p. 460)

Coral bleaching (p. 454)

Coral reef (p. 452)

Crustacean (p. 448)

Deposit feeding (p. 447)

Echinoderm (p. 448)

Epifauna (p. 439)

Hermatypic (p. 456)

High tide zone (p. 439)

Hydrocarbon seep biocommunity (p. 464)

Hydrothermal vent biocommunity (p. 459)

Hypersaline seep biocommunity (p. 464)

Infauna (p. 446)

Intertidal zone (p. 439)

Kelp forest (p. 452)

Low tide zone (p.439)

Meiofauna (p. 448)

Middle tide zone (p. 439)

Mixotroph (p. 456)

Mollusk (p. 447)

Nematocyst (p. 441)

Polyp (p. 452)

Sessile (p. 440)

Spray zone (p. 439)

Subduction zone seep biocommunity (p. 464)

Sublittoral zone (p. 450)

Supralittoral zone (p. 439)

Suspension feeding (p. 447)

Zooxanthellae (p. 455)

Questions And Exercises

1. What controversy developed over the observations of the Rosses and Edward Forbes? How has the controversy been resolved?

2. What are some adverse conditions of rocky intertidal zones? What are some organism's adaptations for those adverse conditions? Which conditions seem to be most important in controlling the distribution of life?

3. Draw a diagram of the zones within the rocky-shore intertidal region and list characteristic organisms of each zone.

4. One of the most dominant features of the middle tide zone along rocky coasts is a mussel bed. Describe general characteristics of mussels, and include a discussion of other organisms that are associated with mussels.

5. Describe how sandy and muddy shores differ in terms of energy level, particle size, sediment stability, and oxygen content.

6. Describe the two types of feeding styles (other than predation) that are characteristic of rocky, sandy, and muddy shores. One of these feeding styles is rather well represented in all three types of shores; name it and for each type of shore, give an example of an organism that uses it.

7. How does the diversity of species on sediment-covered shores compare with that of the rocky shore? Suggest at least one reason why this occurs.

8. In which intertidal zone of a steeply sloping, coarse sand beach would you find each of the following organisms: clams; beach hoppers; ghost shrimp; sand crabs; and heart urchins?

9. Discuss the dominant species of kelp, their epifauna, and animals that feed on kelp in Pacific coast kelp forests.

10. Describe the environmental conditions required for development of coral reefs.

11. Describe the zones of the reef slope, the characteristic coral types, and the physical factors related to zonation.

12. What is coral bleaching? How does it occur? What other diseases affect corals?

13. What are some problems that limit the study of the deep-ocean floor? How are advances in technology overcoming these problems?

14. As one moves from the shoreline to the deep-ocean floor, what changes in the physical environment are experienced?

15. How long would human remains remain on the sea floor? What happens to the flesh? How long would a whale's remains remain on the sea floor?

16. Where does the food come from to supply organisms living on the deep-ocean floor? How does this affect benthic biomass?

17. Is species diversity on the deep-ocean floor relatively high or low? Explain.

18. Describe the characteristics of hydrothermal vents. What evidence suggests that hydrothermal vents have short life spans?

19. What is the "dead whale hypothesis"? What other ideas have been suggested to help explain how organisms from hydrothermal vent biocommunities populate new vent sites?

20. What are the major differences between the conditions and biocommunities of the hydrothermal vents and the cold seeps? How are they similar?

References

Brandon, J. L., and Rokip, F. J. 1985. *Life between the tides: The natural history of the common seashore life of southern California.* San Diego, CA: American Southwest Publishing Company of San Diego.

Broad, W. J. 1997. *The universe below: Discovering the secrets of the deep sea.* New York: Simon & Schuster.

Bruckner, A. W., and Bruckner, R. J., 1997. Emerging infections on the reefs. *Natural History* 106:11, 48.

Chadwick, W. W., Embley, R. W., and Fox, C. G. 1991. Evidence for volcanic eruption on the southern Juan de Fuca ridge between 1981 and 1987. *Nature* 350, 416–418.

Childress, J. J., Fisher, C. R., Brooks, J. M., Kennicutt, M. C., II, Bidigare, R., and Anderson, A. E. 1986. A methanotrophic marine molluscan (Bivalvia, Mytilidae) symbiosis: Mussels fueled by gas. *Science* 233:4770, 1306–1308.

Dawson, E. Y., and Foster, M. S. 1982. *Seashore plants of southern California.* Berkeley: University of California Press.

Dybas, C. L. 1999. Undertakers of the deep. *Natural History* 108:9, 40–47.

Fagoonee, I., et al. 1999. The dynamics of zooxanthellae populations: A long-term study in the field. *Science* 283:5403, 843–845.

Fisher, C. R. 1990. Chemoautrophic and methanotrophic symbioses in marine invertebrates. *Reviews in Aquatic Sciences* 2:3 and 4, 399–436.

George, D., and George, J. 1979. *Marine life: An illustrated encyclopedia of invertebrates in the sea.* New York: Wiley-Interscience.

Grassle, F. J., and Maciolek, N. J. 1992. Deep-sea species richness: Regional and local diversity estimates from quantitative bottom samples. *American Naturalist* 139:2, 313–341.

Hallock, P., and Schlager, W. 1986. Nutrient excess and the demise of coral reefs and carbonate platforms. *Palaios* 1:389–398.

Hessler, R. R., Ingram, C. L., Yayanos, A. A., and Burnett, B. R. 1978. Scavenging amphipods from the floor of the Philippine Trench. *Deep-Sea Research* 25:1029–1047.

Hessler, R. R., and Lonsdale, P. F. 1991. Biogeography of Mariana Trough hydrothermal vent communities. *Deep-Sea Research* 38:2, 185–199.

Huyghe, P. 1990. The storm down below. *Discover* 11:11, 70–76.

Kennicutt, M. C., II, Brooks, J. M., Bidigare, R. R., Fay, R. R., Wade, T. L., and McDonald, T. J. 1985. Vent-type taxa in a hydrocarbon seep region on the Louisiana slope. *Nature* 317:6035, 351–353.

Klum, L. D., et al. 1986. Oregon subduction zone: Venting fauna and carbonates. *Science* 231:4738, 561–566.

Lalli, C. M., and Parsons, T. R. 1993. *Biological oceanography: An introduction.* New York: Pergamon Press.

Lutz, R. A., and Kennish, M. J. 1996. Ecology of deep-sea hydrothermal vent communities: A review, *in* Pirie, R. G., ed., *Oceanography: Contemporary readings in ocean sciences,* 3rd ed., New York: Oxford University Press.

MacGinitie, G. E., and MacGinitie, N. 1968. *Natural history of marine animals,* 2nd ed. New York: McGraw-Hill.

Milliman, J. D., ed. 1998. Deep sea biodiversity: A compilation of recent advances in honor of Robert R. Hessler. *Deep Sea Research* 45:1–2.

Ricketts, E. F., Calvin, J., and Hedgpeth, J. 1968. *Between Pacific tides.* Stanford, CA: Stanford University Press.

Rona, P. A., Klinkhammer, G., Nelson, T. A., Trefry, J. H., and Elderfield, H. 1986. Black smokers, massive sulphides and vent biota at the Mid-Atlantic Ridge. *Nature* 321:6065, 33–37.

Smith, S. 1996. Life after death on the seafloor: Aerobic oases of whales' bones. *Pacific Discovery* 49:1, 32–35.

Thorne-Miller, B., and Catena, J. 1991. *The living ocean: Understanding and protecting marine biodiversity.* Washington, DC: Island Press.

Walbran, P. D., Henderson, R. A., Jull, A. J. T., and Head, M. J. 1989. Evidence from sediments of long-term *Acanthaster planci* predation on corals of the Great Barrier Reef. *Science* 245:4920, 847–850.

Waller, G., ed. 1996. *SeaLife: A complete guide to the marine environment.* Washington, DC: Smithsonian Institution Press.

Wellington, G. M., et al. 2001. Crisis on coral reefs linked to climate change. *Eos Trans. AGU* 82:1, 1–5.

Williams, E. H., Jr., and Bunkley-Williams, L. 1990. The worldwide coral reef bleaching cycle and related sources of coral mortality. *Smithsonian Atoll Research Bulletin* 335.

Williams, E. H., Jr., Goenaga, C., and Vicente, V. 1987. Mass bleaching on Atlantic coral reefs. *Science* 238:4830, 877–878.

Yonge, C. M. 1963. *The sea shore.* New York: Atheneum.

Zenevitch, L. A., Filatove, A., Belyaev, G. M., Lukanove, T. S., and Suetove, I. A. 1971. Quantitative distribution of zoobenthos in the world ocean. *Bulletin der Moskauer Gen der Naturforscher, Abt. Biol.* 76, 27–33.

ZoBell, C. E. 1968. Bacterial life in the deep sea, *in* Proceedings of the U.S.–Japan Seminar on Marine Microbiology, August 1966, Tokyo. *Bulletin Misaki Marine Biology, Kyoto Inst. Univ.* 12, 77–96.

***Suggested Reading in* Scientific American**

Caldwell, R. L., and Dingle, H. 1976. Stomatopods. 234:1, 80–89. Presents the ecology of these interesting crustaceans that have appendages specialized for spearing and smashing prey.

Feder, H. A. 1972. Escape responses in marine invertebrates. 227:1, 92–100. Discusses the surprisingly rapid movements and other responses made by invertebrates to the presence of predators. Some interesting photographs accompany the text, which describes the escape responses of limpets, snails, clams, scallops, sea urchins, and sea anemones.

Martini, F. H. 1998. Secrets of the slime hag. 279:4, 70-75. An examination of the biology and life history of hagfishes, scavengers of the deep sea.

Wicksten, M. K. 1980. Decorator crabs. 242:2, 146–157. Describes how species of spider crabs use materials from their environment to camouflage themselves.

Younge, C. M. 1975. Giant clams. 232:4, 96–105. The distribution and general ecology of the tridacnid clams, some of which grow to lengths well over 1 meter (3.3 feet), are investigated.

Oceanography on the Web

Visit the *Essentials of Oceanography* home page for on-line resources for this chapter. There you will find an on-line study guide with review exercises, and links to oceanography sites to further your exploration of the topics in this chapter. *Essentials of Oceanography* is at: **http://www.prenhall.com/ thurman** (click on the Table of Contents menu and select this chapter).

Afterword

At the end of our journey through this book together, it seems appropriate to examine people's perceptions of the ocean. Many people describe the ocean as "powerful," "awe inspiring," "moving," "serene," "abundant," and "majestic." Others call it "vast," "infinite," or "boundless." These are all appropriate descriptions, because even though humans have been exploring and studying the ocean for centuries, the ocean still holds many secrets. Newly discovered ocean life—such as a new species of bottlenose whale discovered in 2000—demonstrate how limited our knowledge of the ocean really is.

Despite the ocean's impressive size, it is beginning to feel the effects of human activities. Every ocean contains floating plastic trash, for instance, and even remote beaches are littered with trash. Organisms living in the ocean also feel the effect of humankind's use of the ocean. Whaling in the 19th and 20th centuries, for example, pushed many great whale populations to near extinction. Restrictions on whaling and the development of substitutes for whale products helped whales survive this threat, but now they face destruction of their feeding and breeding grounds. Overfishing, however, more than any other human activity, has altered the marine ecosystem, suggesting that the ocean is not quite as "vast," "infinite," or "boundless" as most people believe.

Whether it is coral bleaching, increased concentrations of greenhouse gases in the lower atmosphere, depletion of upper-atmosphere ozone, the increased frequency of red tides and "dead zones," or polluting the ocean with petroleum, plastics, sewage, chemicals, or toxin-laden sediment, human-induced changes in the environment are broad and far-reaching. All of these problems may be symptoms of a worldwide pathology (sickness) that will require major changes in human behavior before it can be cured.

Marine Sanctuaries and Marine Reserves

In 1972, the U.S. Congress began establishing national **marine sanctuaries** to protect vital pockets of the ocean from further degradation. Today, 13 national marine sanctuaries cover more than 6200 square kilometers (2400 square miles) in areas such as the Florida Keys, the Stellwagen Bank off Massachusetts, the Channel Islands off southern California, the Flower Garden Banks in the Gulf of Mexico, and the Hawaiian Islands. Many activities that degrade the marine environment, however, are still allowed in marine sanctuaries, such as fishing, recreational boating, and even mining of some resources.

Many scientists who recognize the importance of preserving vital marine habitats are calling for governments to set aside large "no-take" refuges called **marine reserves** that would prohibit fishing and other activities. Such action would allow the recovery of heavily overfished stocks and protect sea floor communities decimated by trawling the sea floor with nets. Although the fishing industry has opposed creating marine reserves, research indicates that marine reserves provide benefits to surrounding regions by boosting populations of fish outside their borders. Other countries are also realizing the economic benefit of fully protecting ocean resources. A healthy coral reef, for example, can be worth more as a tourist draw than it might be as a source of seafood.

Although the ocean is feeling the effects of humans, it is very resilient and has a tremendous ability to withstand change. Studies reveal that once human impact is reduced, the ocean often returns to a near pristine state because natural processes in the ocean tend to disperse and eventually remove many types of pollutants. The vast majority of ocean water, moreover, is still relatively unpolluted, except in shallow-water coastal areas near large population centers and near the mouths of major rivers.

What Can I Do?

Because the ocean is vast and capable of absorbing many substances, it has been used as a dumping ground for many of society's wastes. Even today, humans are adding pollutants to the ocean at staggering rates. What can each of us do to help? Some ways to help the environment in general and the ocean in particular include:

- Minimize your impact on the environment. Reduce the amount of waste you generate by making wise consumer choices. Avoid products with excessive packaging and support companies that have good environmental records. Reuse and recycle items, then help close the recycling loop by buying goods made of recycled materials. Do simple things that make a positive impact on the environment (Box Aft–1).

- Become politically aware. Many ocean-related issues come before the public and require a majority of voters to approve a proposal before they are enacted into law. This "Majority rules" is true for local as well as national and international issues. Within our lifetimes, for instance, we may very well decide whether

Box Aft–1
Ten Simple Things You Can Do to Prevent Marine Pollution[1]

There are many simple things you can do every day to help prevent marine pollution such as:

1. **Snip six-pack rings.** Six-pack rings entangle many marine organisms, so snip each circle with a scissors before you toss them into the garbage. If you find any at a beach, pick them up, snip them, and discard them properly.

2. **Don't overfertilize your lawn and use a clean laundry detergent.** Lawn fertilizers contain nitrates and phosphates, which cause algae blooms when runoff from land enters the ocean. Detergents also contain phosphates, so read the detergent label to find one that is low in phosphate or phosphate free. Using a smaller amount of detergent than recommended also helps.

3. **Clean up after your pet.** Dog and cat feces that wash into a stream and eventually to the ocean also provide nitrates and phosphates, which create algae blooms.

4. **Make sure that your car doesn't leak oil.** Oil that leaks from automobiles is responsible for a large percentage of the oil that gets into the ocean. Annually, the amount of oil that enters the ocean as runoff from road sources is greater than a major oil spill. If you change your car's oil yourself, be sure to recycle the used oil at an appropriate recycling center.

5. **Drive less and carpool more.** Reducing the amount of gasoline you use reduces the amount of oil that must be transported across the ocean, which minimizes the potential for oil spills.

6. **Take your own bags to the grocery store.** Although paper bags are biodegradable, plastic bags are not and are becoming an increasing problem in the ocean, especially for animals like sea turtles that eat plastic bags when they mistake them for jellyfish.

7. **Don't release balloons.** Balloons that are released far from the ocean can still wind up there, where they deflate, quickly loose their color, and resemble drifting jellyfish that marine animals can ingest.

8. **Don't litter.** Any material that is carelessly discarded on land can become non-point-source pollution when it washes down a storm drain, into a stream, and eventually into the ocean.

9. **Pick up trash at the beach or volunteer for an organized beach clean-up.** It is important to remove trash that washes up at the beach so that it can't endanger marine organisms. Trash arrives at the beach from non-point-source pollution, ships, recreational boaters, beachgoers, and other sources. It is truly surprising (and somewhat horrifying) to discover what winds up at the beach.

10. **Inform and educate others.** Many people are unaware that their actions have a negative influence on the environment—especially the marine environment.

Nobody made a greater mistake than he who did nothing because he could only do a little.
— *Edmund Burke (circa 1790)*

---

[1]With thanks to The Earthworks Group, 1989, *50 Simple Things You Can Do to Save the Earth*, Berkeley, California: Earthworks Press.

to spend large amounts of money to add finely ground iron to the ocean to reduce the amount of carbon dioxide in the atmosphere. Many political issues in the future will involve the ocean.

• Educate yourself about how the ocean works. A recent poll indicated that over 90% of the American public consider themselves to be scientifically illiterate. With our society becoming more scientifically advanced, people need to understand how science operates. Science is not meant to be comprehended by an elite few. Rather, science is for everyone. By studying oceanography, you have begun to understand how the ocean works. It is our hope that you will be a life-long student of the ocean.

In the end, we will conserve only what we love. We love only what we understand. We will understand only what we are taught.

— *Baba Dioum, Senegalese conservationist (1968)*

Figure Aft–1 Sunset at the ocean.

APPENDIX I

Metric and English Units Compared

How many inches are there in a mile? How many cups in a gallon? How many pounds in a ton? In our daily lives, we often need to convert between units. Worldwide, the metric system of measurement is the most widely used system. Besides the United States, only *two other countries in the world*—Liberia and Myanmar (formerly Burma)—still use English units as their primary system of measurement. The metric system has many advantages over the English system. It is simple, logical, and easy to convert between units. For those of us in the United States, it is only a matter of time before the change to the metric system occurs.

On December 23, 1975, U.S. President Gerald R. Ford signed the Metric Conversion Act of 1975. It defined the metric system as the International System of Units (officially called the *Système International d'Unités*, or SI) as interpreted in the United States by the secretary of commerce. The Trade Act of 1988 and other legislation declared the metric system the preferred system of weights and measures for U.S. trade and commerce, called for the federal government to adopt metric specifications, and mandated the Commerce Department to oversee the program. All of this legislation comes 200 years after Benjamin Franklin first proposed that the United States adopt the metric system.

Although the metric system has not become the system of choice for most Americans' daily use, and there is great resistance to using it, the United States must change over in order to remain competitive in world markets. Many of the sciences are leading the way in this changeover. Oceanographers all over the world have been using the metric system for years.

The English System

The English system actually consists of two related systems—the U.S. Customary System[1] (used in the United States and dependencies), and the British Imperial System (used in Great Britain). Great Britain has now largely converted to the metric system. The basic unit of length in the English system is the *yard*; the basic unit of weight (not mass[2]) is the *pound*.

In the English system, the units of length were initially based arbitrarily on dimensions of the body. The yard as a measure of length, for example, can be traced to the early Saxon kings. They wore a sash around their waists that could be used as a convenient measuring device. Thus, the word *yard* comes from the Saxon word *gird* (like a girdle), in reference to the circumference of a person's waist. The circumference of a person's waist varies from person to person, however, and even varies from time to time on the *same* person, so it is of limited use. It had to be standardized to be useful, so King Henry I of England decreed that the yard should be the distance from the tip of his nose to the end of his thumb!

Romans initially defined the mile (*mille passuum* = 1000 paces) as an even 5000 feet. The early Tudor rulers in England, however, defined a *furlong* ("furrow-long") as 220 yards, based on the length of agricultural fields. To facilitate the conversion between miles and furlongs, Queen Elizabeth I declared in the 16th century that the traditional Roman mile of 5000 feet would be forever replaced by one of 5280 feet, making the mile exactly 8 furlongs. Today, a furlong is an archaic unit of measurement (except in horse races), but a mile is still 5280 feet.

The Metric System

The metric system of weights and measures was devised in France and adopted there in 1799. It is based on a unit of length called the *meter* and a unit of mass called the *kilogram*. Originally defined as one ten-millionth the distance from the North Pole to the Equator, the meter is now defined as the distance light travels through a vacuum in $1/_{299,792,458}$ of a second. The kilogram was originally related to the volume of one cubic meter of water. It is now defined as the mass of the International Prototype Kilogram, a platinum-iridium cylinder kept at Sèvres (near Paris), France. Other metric units can be defined in terms of the meter and the kilogram.

Fractions and multiples of the metric units are related to each other by powers of 10, allowing conversion from one unit to a multiple of it simply by shifting the decimal point. The lengthy arithmetical operations required with English units can be avoided. Even the names of the metric units indicate how many are in a larger unit. For instance, how many cents are there in a dollar? It is the same as the number of *centi*grams in a gram. The prefixes listed in the following table, "Common Prefixes for Basic Metric Units" have been

---

[1] The names of the units and the relationships between them are generally the same in the U.S. customary System and the British Imperial System, but the sizes of the units differ, sometimes considerably.

[2] The basic unit of mass in the English system is the *slug*, although it is rarely used.

accepted for designating multiples and fractions of the meter, the gram (which equals $1/_{1000}$ of a kilogram), and other units.

Common Prefixes for Basic Metric Units

Factor	Name	Prefix Name
$10^{12} = 1,000,000,000,000$	trillion	tera
$10^9 = 1,000,000,000$	billion	giga
$10^6 = 1,000,000$	million	mega
$10^3 = 1,000$	thousand	kilo
$10^2 = 100$	hundred	hecto
$10^1 = 10$	ten	deka
$10^{-1} = 0.1$	tenth	deci
$10^{-2} = 0.01$	hundredth	centi
$10^{-3} = 0.001$	thousandth	milli
$10^{-6} = 0.000001$	millionth	micro
$10^{-9} = 0.000000001$	billionth	nano
$10^{-12} = 0.000000000001$	trillionth	pico

In spite of many fears that changing to the metric system will be an economic burden on businesses in the United States, the conversion has already begun. Examples include 35-millimeter film, 2-liter soft drink bottles, 750-milliliter wine bottles, computers with gigabytes of memory, and 10-kilometer runs. Cars are now manufactured metrically in order to compete with other countries. Most people in the United States will experience the increasing use of the metric system in their daily lives. Some ways to ease yourself into this unfamiliar (yet sensible) system of units include noticing distances on road signs in kilometers, using a meterstick instead of a yardstick, and charting your weight in kilograms.

Temperature

The Celsius (centigrade) temperature scale was invented in 1742 by a Swedish astronomer named Anders Celsius. The scale was designed so that the freezing point of pure water—the temperature at which it changes state from a liquid to a solid—is set as 0 degrees. Similarly, the boiling point of pure water—the temperature at which it changes state from a liquid to a gas—is set at 100 degrees. Doing so made the liquid range of water—0 degrees to 100 degrees—an even 100 units, giving the scale its name: centigrade (*centi* = 100, *grade* = graduations).

The Fahrenheit temperature scale, on the other hand, sets water's freezing and boiling points at 32 degrees

and 212 degrees, respectively, for a spread of 180 units. This scale was devised by a German-born physicist named Gabriel Daniel Fahrenheit, who invented the mercury thermometer in 1714. He initially designed the scale so that normal body temperature was set at 100°F. However, because of inaccuracies in early thermometers or how the scale was initially devised, normal human body temperature is now known to be 98.6°F (37°C).

Conversion Tables

Length	
1 micrometer (μm)	0.001 millimeter 0.0000394 inch
1 millimeter (mm)	1,000 micrometers 0.1 centimeter 0.001 meter 0.0394 inch
1 centimeter (cm)	10 millimeters 0.01 meter 0.394 inch
1 meter (m)	100 centimeters 39.4 inches 3.28 feet 1.09 yards 0.547 fathom
1 kilometer (km)	1,000 meters 1,093 yards 3,280 feet 0.62 statute mile 0.54 nautical mile
1 inch (in)	25.4 millimeters 2.54 centimeters
1 foot (ft)	12 inches 30.5 centimeters 0.305 meter
1 yard (yd)	3 feet 0.91 meter
1 fathom (fm)	6 feet 2 yards 1.83 meters
1 statute mile (mi)	5,280 feet 1,760 yards 1,609 meters 1.609 kilometers 0.87 nautical mile
1 nautical mile (nm)	1 minute of latitude 6,076 feet 2,025 yards 1,852 meters 1.15 statute miles
1 league (lea)	5,280 yards 15,840 feet 4,805 meters 3 statute miles 2.61 nautical miles

Area

1 square centimeter (cm^2)	0.155 square inch 100 square millimeters
1 square meter (m^2)	10,000 square centimeters 10.8 square feet
1 square kilometer (km^2)	100 hectares 247.1 acres 0.386 square mile 0.292 square nautical mile
1 square inch (in^2)	6.45 square centimeters
1 square foot (ft^2)	144 square inches 929 square centimeters

Volume

1 cubic centimeter (cc; cm^3)	1 milliliter 0.061 cubic inch
1 liter (l)	1,000 cubic centimeters 61 cubic inches 1.06 quarts 0.264 gallon
1 cubic meter (m^3)	1,000,000 cubic centimeters 1,000 liters 264.2 gallons 35.3 cubic feet
1 cubic kilometer (km^3)	0.24 cubic mile 0.157 cubic nautical mile
1 cubic inch (in^3)	16.4 cubic centimeters
1 cubic foot (ft^3)	1,728 cubic inches 28.32 liters 7.48 gallons

Mass

1 gram (g)	0.035 ounce
1 kilogram (kg)	2.2 pounds[a] 1,000 grams
1 metric ton (mt)	2,205 pounds 1,000 kilograms 1.1 U.S. short tons
1 pound[a] (lb)	16 ounces 454 grams 0.454 kilogram
1 U.S. short ton (ton; t)	2,000 pounds 907.2 kilograms 0.91 metric ton

[a]The pound is a weight unit, not a mass unit, but is often used as such.

Pressure

1 atmosphere (atm) (at sea level)	760 millimeters of mercury 14.7 pounds per square inch 29.9 inches of mercury 33.9 feet of fresh water 33 feet of seawater

Speed

1 centimeter per second (cm/s)	0.0328 foot per second
1 meter per second (m/s)	2.24 statute miles per hour 1.94 knots 3.28 feet per second 3.60 kilometers per hour
1 kilometer per hour (kph)	27.8 centimeters per second 0.62 mile per hour 0.909 foot per second 0.55 knot
1 statute mile per hour (mph)	1.61 kilometers per hour 0.87 knot
1 knot (kt)	1 nautical mile per hour 51.5 centimeters per second 1.15 miles per hour 1.85 kilometers per hour

Oceanographic data

Velocity of sound in 34.85‰ seawater	4,945 feet per second 1,507 meters per second 824 fathoms per second
Seawater with 35 grams of dissolved substances per kilogram of seawater	3.5 percent 35 parts per thousand 35,000 parts per million 35,000,000 parts per billion

Temperature

Exact Formula	Approximation (easy way)
$°C = \dfrac{(°F - 32)}{1.8}$	$°C = \dfrac{(°F - 30)}{2}$
$°F = (1.8 \times °C) + 32$	$°F = (2 \times °C) + 30$

Some useful equivalent temperatures:

100°C = 212°F	(boiling point of pure water)
40°C = 104°F	(heat wave conditions)
37°C = 98.6°F	(normal body temperature)
30°C = 86°F	(very warm—almost hot)
20°C = 68°F	(room temperature)
10°C = 50°F	(a warm winter day)
3°C = 37°F	(average temperature of deep water)
0°C = 32°F	(freezing point of pure water)

In degrees centigrade:

Thirty is hot, twenty is pleasing;

Ten is not, and zero is freezing.

APPENDIX II

Geographic Locations

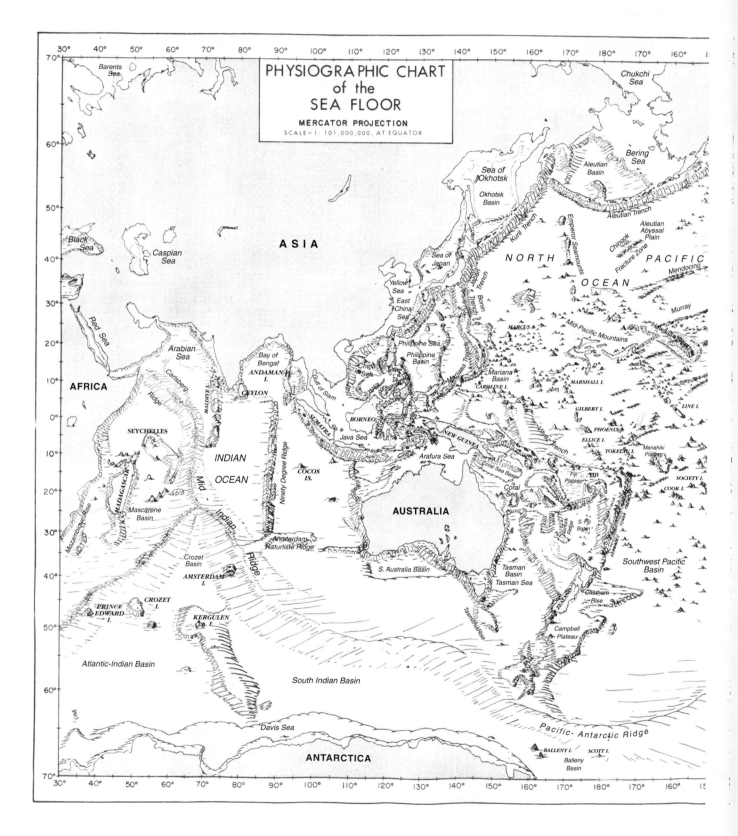

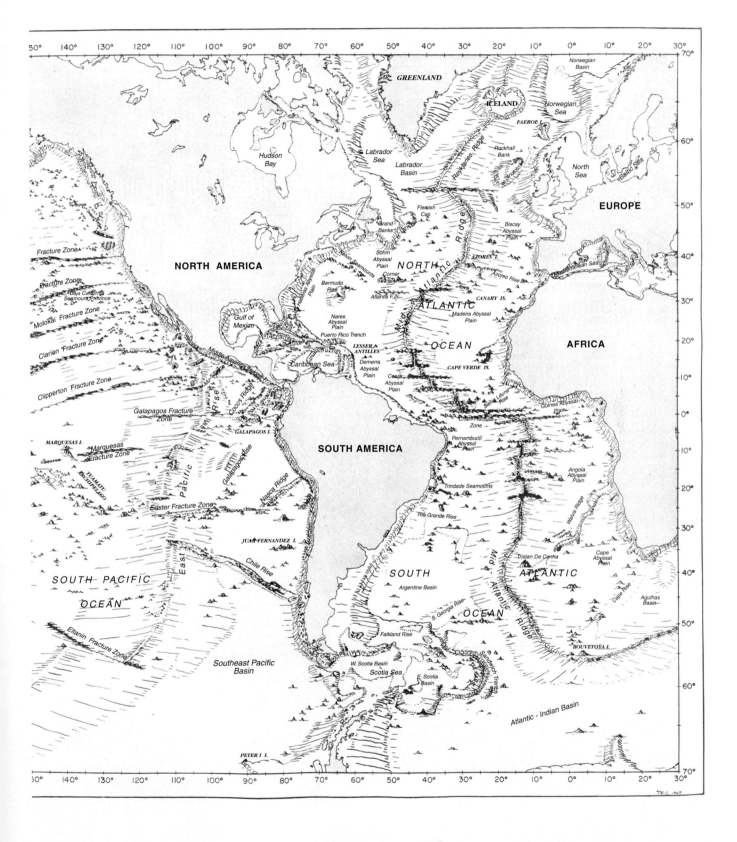

APPENDIX III

Latitude and Longitude on Earth

Suppose that you find a great fishing spot or a shipwreck on the sea floor. How would you remember where it is—far from any sight of land—so that you might return? How do sailors navigate a ship when they are at sea? Even using sophisticated equipment such as the global positioning system (GPS), which uses satellites to determine location, how is this information reported?

To solve problems like these, a navigational grid—a series of intersecting lines across a globe—is used. Once starting places are fixed for a grid, a location can be identified based on its position within the grid. Some cities, for example, use the regular numbering of streets as a grid. Even if you have never been to the intersection of 5th Avenue and 42nd Street in New York City, you would be able to locate it on a map, or know which way to travel if you were at 10th Avenue and 42nd Street. Although many grid (or coordinate) systems are used today, the one most universally accepted uses *latitude* and *longitude*.

A series of north–south and east–west lines that together comprise a grid system can be used to locate points on Earth's surface whether at sea or on land. The north–south lines of the grid are called *meridians* and extend from pole to pole (Figure A3–1). The meridian lines converge at the poles and are spaced farthest apart at the Equator. The east–west lines of the grid are called *parallels* because they are parallel to one another. The longest parallel is the *Equator* (so called because it divides the globe into two equal hemispheres), and the parallels at the poles are a single point (Figure A3–1).

Latitude and Longitude

Latitude is the angular distance (in degrees of arc) measured north or south of the starting line (the Equator) from the center of Earth. All points that lie along the same parallel are an identical distance from the Equator, so they all have the same latitude. The latitude of the Equator is 0 degrees, and the North and South Poles lie at 90 degrees north and 90 degrees south, respectively. Figure A3–2 shows that the latitude of New Orleans is an angular distance of 30 degrees north from the Equator.

Longitude is the angular distance (in degrees of arc) measured east or west of the starting line from the center of Earth. All meridians are identical (none is longer or shorter than the others, and they all pass through both poles), so the starting line of longitude has been chosen arbitrarily. Before it was agreed upon by all nations, different countries used different starting points for 0 degrees longitude. Examples include

Figure A3–1 Earth's grid system.

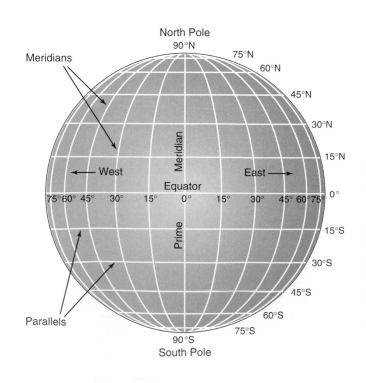

478

the Canary Islands, the Azores, Rome, Copenhagen, Jerusalem, St. Petersburg, Pisa, Paris, and Philadelphia. At the 1884 International Meridian Conference, it was agreed that the meridian that passes through the Royal Observatory at Greenwich, England, would be used as the zero or starting point of longitude. This zero degree line of longitude is also called the *prime meridian,* and the longitude for any place on Earth is measured east or west from this line. Longitude can vary from 0 degrees along the prime meridian to 180 degrees (either east or west), which is halfway around the globe and is known as the *International Date Line.* In Figure A3–2, New Orleans is 90 degrees west of the prime meridian.

A particular latitude and longitude defines a unique location on Earth, provided the direction from the starting point is given, too. For instance, 42 degrees north and 120 degrees west defines a unique spot on Earth because it is different from 42 degrees south and 120 degrees west and from 42 degrees north and 120 degrees east.

A degree of latitude or longitude can be divided into smaller units. One degree (°) of arc (angular distance) is equal to 60 minutes (') of arc. One minute of arc is equal to 60 seconds (") of arc. When using a small-scale map or a globe, it may be difficult to estimate latitude and longitude to the nearest degree or two. When using a large-scale map, however, it is often possible to determine the latitude and longitude to the nearest fraction of a minute.

Determination of Latitude and Longitude

Today, latitude and longitude can be determined very precisely by using satellites that remain in orbit around Earth in fixed positions. How did navigators determine their position before this technology was available?

Latitude was determined from the positions of particular stars. Initially, navigators in the Northern Hemisphere measured the angle between the horizon and the North Star (Polaris), which is directly above the North Pole. The latitude north of the Equator is the angle between the two sightings (Figure A3–3). In the Southern Hemisphere, the angle between the horizon and the Southern Cross was used because the Southern Cross is directly overhead at the South Pole. Later, the angle of the sun above the horizon, corrected for the date, was also used.

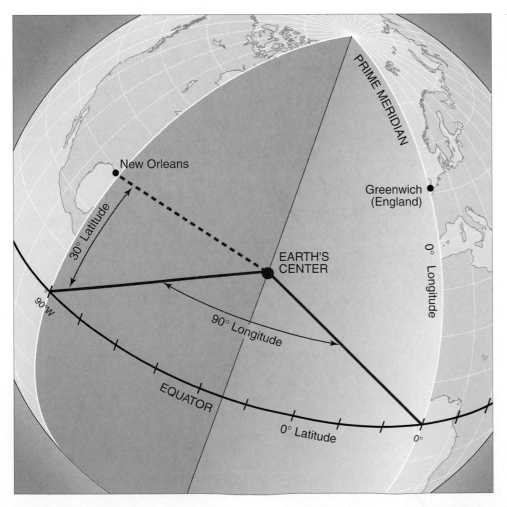

Figure A3–2 New Orleans is located at 30 degrees north latitude and 90 degrees west longitude.

There was no method of determining longitude until one based on time was developed at the end of the 18th century. As Earth turns on its rotational axis, it moves through 360 degrees of arc every 24 hours (one complete rotation on its axis) (Figure A3–4a). Earth, therefore, rotates through 15 degrees of longitude per hour (360° ÷ 24 hours = 15 degrees/hour). As a result, a navigator needed only to know the time at the prime or Greenwich meridian (Figure A3–4b) at the exact same time the sun was at its highest point (local noon) at the ship's location. In this way, a navigator aboard a ship could calculate the ship's longitude each day at noon. That's why the development of John Harrison's chronometer (see Box 1–1 in Chapter 1) was so crucial for navigation.

Suppose a ship sets sail west across the Atlantic Ocean from Europe, checking its longitude each day at noon local time. One day when the sun is at the noon position (noon local time), the chronometer reads 16:18 hours (0:00 is midnight and 12:00 is noon). What is the ship's longitude (Figure A3–4c)?

If the clock is keeping good time (which Harrison's chronometer did), then we know that the ship is 4 hours and 18 minutes behind (west of) Greenwich time. To determine the longitude, we must convert time into longitude. Each hour represents 15 degrees of longitude based on the rotation of Earth. Thus, the four hours represents 60 degrees of longitude (4 hours × 15 degrees of longitude = 60 degrees). One degree is divided into 60 minutes of arc, so Earth rotates through $\frac{1}{4}$ degree (15') of arc per minute of time. Thus, 18 minutes of time multiplied by 15 minutes of arc per minute of time equals 270 minutes of arc. To convert 270 minutes of arc into degrees, it must be divided by 60 minutes per degree, which gives 4.5 degrees of longitude. Therefore, the answer is 60 degrees plus 4.5 degrees, or 64.5 degrees west longitude.

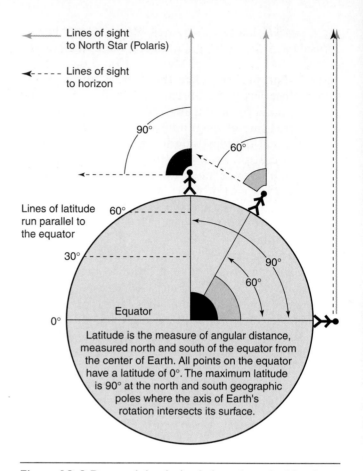

Figure A3–3 Determining latitude based on the North Star (Polaris). Latitude can be determined by noting the angular difference between the horizon and the North Star, which is directly over the North Pole of Earth.

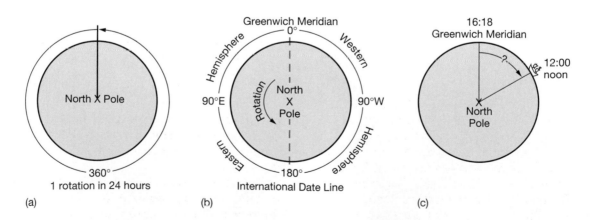

Figure A3–4 Determining longitude based on time. View of Earth as seen from above the North Pole. **(a)** Earth rotates 360 degrees of arc every 24 hours. **(b)** The Greenwich Meridian is set as 0 degree longitude, which divides the globe into Eastern and Western Hemispheres. The International Date Line is 180 degrees from the Greenwich Meridian. **(c)** An example of how a ship at sea can determine its longitude using time.

APPENDIX IV

A Chemical Background: Why Water has 2 H's and 1 O

An **element** (*elementum* = a first principle) is a substance comprised entirely of like particles that cannot be broken into smaller particles by chemical means. The **atom** (*a* = not, *tomos* = cut) is the smallest particle of an element that can combine with similar particles of other elements to produce compounds. The periodic table of elements shown in Figure A4–1 lists the elements and describes their atoms. A **compound** (*compondre* = to put together) is a substance containing two or more elements combined in fixed proportions. A **molecule** (*molecula* = a mass) is the smallest particle of an element or compound that, in the free state, retains the characteristics of the substance.

As an illustration of these terms, consider **Sir Humphrey Davey**'s use of electrical dissociation to break the compound water into its component elements, hydrogen and oxygen. Atoms of the elements hydrogen (H) and oxygen (O) combine in the proportion 2 to 1, respectively, to produce molecules of water (H_2O). As an electric current is passed through the water, the molecules dissociate into hydrogen atoms that collect near the cathode (negatively charged electrode) and oxygen atoms that collect near the anode (positively charged electrode). Here they combine to form the diatomic gaseous molecules of the elements hydrogen (H_2) and oxygen (O_2). Because there are twice as many hydrogen atoms as oxygen atoms in a given volume of water, twice as many molecules of hydrogen gas (H_2) as oxygen gas (O_2) are formed. Further, the volume of gas under identical conditions of temperature and pressure is proportional to the number of gas particles (molecules) present, so two volumes of hydrogen gas are produced for each volume of oxygen gas.

A Look at the Atom

Building upon earlier discoveries, the Danish physicist **Niels Bohr** (1884–1962) developed his theory of the atom as a small solar system in which a positively charged nucleus takes the place of the sun and the planets that orbit around it are represented by negatively charged electrons. Although this theory has since been diagrammatically altered, it is still commonly used to demonstrate the arrangement of electrons and nuclear particles in the atom.

Bohr's earliest concern was with the atom of hydrogen, which he considered to consist of a single positively charged **proton** (*protos* = first) in its nucleus orbited by a single negatively charged **electron** (*electro* = electricity). Since the mass of the electron is only about $1/1840$ the mass of the proton, it will be considered negligible in our discussion of atomic masses.

According to Bohr, the number of protons—or units of positive charge in the nucleus—coincides with the **atomic number** of the element. Thus hydrogen, with an atomic number of 1, has a **nucleus** (*nucleos* = a little nut) containing a single proton. Helium, the next heavier element, having an atomic number of 2, contains two nuclear protons, and so forth. The atomic number also indicates the number of electrons in a normal atom of any element, because this number is equal to the number of protons in the nucleus (Figure A4–2).

An **isotope** (*isos* = equal, *topos* = place) is an atom of an element that has a different **atomic mass** than other atoms of the same element. As we will see, chemical properties of atoms are determined by the electron arrangement that surrounds the nucleus. This arrangement, in turn, is determined by the number of protons in the nucleus. Because some isotopes have different atomic masses but identical chemical characteristics, there must be a nuclear particle in the nucleus other than the proton that can influence the atomic mass of an atom but does not affect the electron structure surrounding the nucleus (Figure A4–3).

Nuclear research has discovered the additional particle, the **neutron** (*neutr* = neutral), which was postulated by **Ernest Rutherford** in 1920 and was first detected by his associate **James Chadwick** in 1932. It has a mass very similar to that of the proton but no electrical charge. This characteristic of being electrically neutral has made it one of the particles most utilized by nuclear physicists, who continue to explore the atomic nucleus and to make new discoveries of nuclear particles. We will not consider these particles, as knowledge of them is not necessary for our understanding of the chemical nature of atoms.

An atom of a given element can be changed to an ion of that element by adding or taking away one or more electrons, and it is changed to an isotope of the element by adding or taking away one or more neutrons. Adding

Periodic Table of the Elements

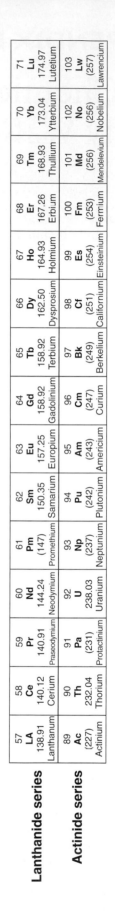

Figure A4–I The periodic table of the elements.

Figure A4–2 Bohr-Stoner orbital models for atoms. Each atom is composed of a positively charged nucleus with negatively charged electrons around it. The nucleus, which occupies very little space, contains most of the mass of the atom. The atomic number is equal to the number of protons (positively charged nuclear particles) in the nucleus. Note that the first shell holds only two electrons and other shells can hold eight (or more) electrons.

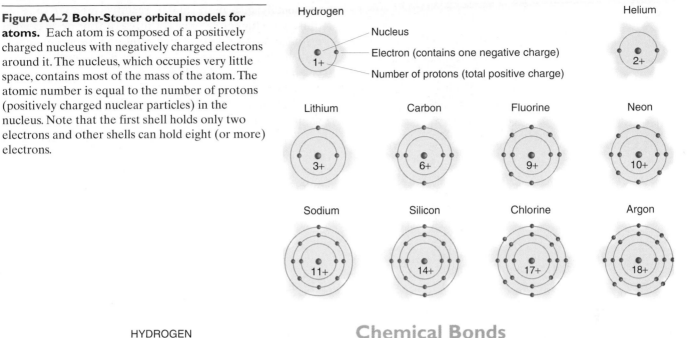

HYDROGEN

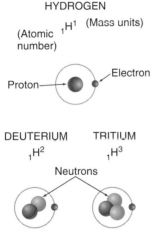

Figure A4–3 Hydrogen isotopes. Isotopes are atoms of an element that have different atomic masses. The hydrogen atom ($_1H^1$) accounts for 99.98% of the hydrogen atoms on Earth and has a nucleus that contains only one proton. Deuterium ($_1H^2$) contains one neutron and one proton in its nucleus and combines with oxygen to form "heavy water" with a molecular mass of 20. Tritium ($_1H^3$) is a very rare radioactive isotope of hydrogen that has a nucleus containing one proton and two neutrons.

or taking away one or more protons will change the atom to an atom of a different element.

The atom can be divided into two parts—the nucleus, which contains neutrons and protons, and the **electron cloud** surrounding the nucleus that is involved in chemical reactions. The random motion of electrons makes it impossible to determine the precise location of electrons in this cloud at any instant, but it is possible to estimate the most probable position of an electron in this cloud. The regions in which there would more likely be particular electrons can be viewed as concentric spheres, or shells, that surround the nucleus (see Figure A4–2).

Chemical Bonds

In considering the chemical reactions in which atoms are involved, we will be concerned primarily with the distribution of electrons in the outer shell. When atoms combine to form compounds, they usually do so by forming one of two bonds:

1. **Ionic** (*ienai* = to go) **bonds,** where electrons are either gained or lost.

2. **Covalent** (*co* = with, *valere* = to be strong) **bonds,** where electrons are shared between atoms.

Ionic bonding produces an ion, an electrically charged atom that no longer has the properties of a neutral atom of the element it represents. A positively charged ion, a **cation** (*kation* = something going down), is produced by the loss of electrons from the outer shell, the positive charge being equal to the number of electrons lost. A negatively charged ion, an **anion** (*anienai* = to go up), is produced by the gain of electrons in the outer shell of an atom, and its charge is equal to the number of electrons gained.

An example of a compound formed by ionic bonding is common table salt, which is sodium chloride (Figure A4–4). In the formation of this compound, the sodium atom, which has one electron in its outer shell, loses this electron and forms a sodium ion with a positive electrical charge of one. The chlorine atom, which contains seven electrons in its outer shell, completes its shell by gaining an electron and becoming a chloride ion with a negative charge of one. These two ions are held in close proximity by an electrostatic attraction between the two ions of equal and opposite charge.

Moreover, it is the tendency of an individual atom to assume the outer-shell electron content of the inert gases such as helium (2), neon (8), and argon (8) that make it chemically reactive (see Figure A4–2).

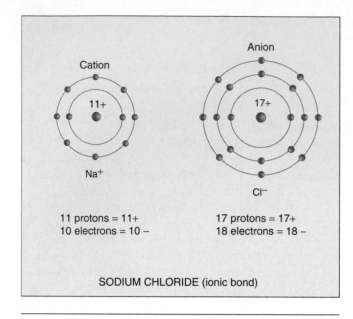

SODIUM CHLORIDE (ionic bond)

Cation

11+

Na⁺

11 protons = 11+
10 electrons = 10 −

Anion

17+

Cl⁻

17 protons = 17+
18 electrons = 18 −

Figure A4–4 Ionic Bonds in sodium chloride (table salt).
Ions of sodium (*left*) and chlorine (*right*) combine to form
sodium chloride, producing common table salt. Notice the
contrast in the outer shells of the ions of sodium (Na⁺)
and chlorine (Cl⁻) with the atoms of these elements shown
in Figure A4–2.

Normally, if an atom can assume this configuration by
either sharing one or two electrons with an atom of
another element or by losing or gaining one or two
electrons, the elements are highly reactive. By contrast,
the elements are less reactive if three or four electrons
must be shared, gained or lost to achieve the desired
configuration.

Valence (*valentia* = capacity) is the number of hydrogen atoms with which an atom of a given element can
combine with either ionic or covalent bonds. Elements
with lower valences of one or two, for example, combine
chemically in a more highly reactive manner than do
those with higher valences of three or four or, in certain
instances, more than four. Although those elements
with higher valences do not react as violently as low-
valence elements, they have a greater combining power
and can gather about them larger numbers of atoms of
other elements than can the atoms with lower valence
values.

An example of a covalent bond is the sharing of
electrons by hydrogen and oxygen atoms in the water
molecule (Figure A4–5). In the formation of this molecule, both hydrogen atoms and the oxygen atom assume
the inert gas configuration they seek by sharing
electrons.

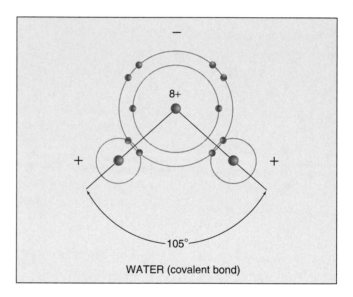

−

8+

+   +

105°

WATER (covalent bond)

Figure A4–5 Covalent bonds in water. Water (H_2O) is
composed of 2 atoms of hydrogen and 1 atom of oxygen, held
together by covalent bonds.

APPENDIX V

Careers in Oceanography

Many people think a career in oceanography consists of swimming with marine animals at a marine life park or snorkeling in crystal-clear tropical waters studying coral reefs. In reality, these kinds of jobs are extremely rare and there is intense competition for them. Most oceanographers use science to answer questions about the ocean, such as the following:

- What is the role of the ocean in limiting the greenhouse effect?
- What kinds of pharmaceuticals can be found naturally in marine organisms?
- How does sea floor spreading relate to the movement of tectonic plates?
- What economic deposits are there on the sea floor?
- Can rogue waves be predicted?
- What is the role of longshore transport in the distribution of sand on the beach?
- How does a particular pollutant affect organisms in the marine environment?

Preparation for a Career in Oceanography

Preparing yourself for a career in oceanography is probably one of the most interesting and rewarding (yet difficult) paths to travel. The study of oceanography is typically divided into four different academic disciplines (or subfields) of study:

- *Geological oceanography* is the study of the structure of the sea floor and how the sea floor has changed through time; the creation of sea floor features; and the history of sediments deposited on it.
- *Chemical oceanography* is the study of the chemical composition and properties of seawater; how to extract certain chemicals from seawater; and the effects of pollutants.
- *Physical oceanography* is the study of waves, tides, and currents; the ocean–atmosphere relationship that influences weather and climate; and the transmission of light and sound in the oceans.
- *Biological oceanography* is the study of oceanic life forms and their relationships to one another; adaptations to the marine environment, and developing ecologically sound methods of harvesting seafood.

Other disciplines include ocean engineering, marine archaeology, and marine policy. Oceanography is an *interdisciplinary* science because it utilizes all the disciplines of science as they apply to the oceans. Some of the most exciting work and best employment opportunities combine two or more of these disciplines.

Individuals working in oceanography and marine-related fields need a strong background (typically an undergraduate degree) in at least one area of basic science (for example, geology, physics, chemistry, or biology) or engineering. In almost all cases, mathematics is required as well. Marine archaeology requires a background in archaeology or anthropology; marine policy studies require a background in at least one of the social sciences (such as law, economics, or political science).

The ability to speak and write clearly—as well as critical thinking skills—are prerequisites for any career. Fluency in computers—preferably PC systems, not Macintosh—is rapidly becoming a necessity. Because many job opportunities in oceanography require trips on research vessels (Figure A5–1), any shipboard experience is also desirable. Mechanical ability (the ability to fix equipment while on board a vessel without having to return to port) is a plus. Depending on the type of work that is required, other traits that may be desirable include the ability to speak one or more foreign languages; certification as a scuba diver; the ability to work for long periods of time in cramped conditions; physical stamina; physical strength; and a high tolerance to motion sickness.

Oceanography is a relatively new science (with much room left for discovery), so most people enter the field with an advanced degree (master's or doctorate). Work as a marine technician, however, usually requires only a bachelor's degree or applicable experience. It does take a large commitment to achieve an advanced degree but, in the end, the journey itself is what makes all the hard work worthwhile.

Job Duties of Oceanographers

There has been and will continue to be enormous expansion in the number of ocean-related jobs. Job opportunities for oceanographers exist with scientific research institutions (universities), various government agencies, and private companies that are engaged in searching for economic sea floor deposits, investigating areas for sea farming, and evaluating natural energy

Figure A5–I Oceanographers at work on a research vessel. Oceanographers deploy a sonar device on a cable using the A-frame on Scripps Institution of Oceanography's R/V *Melville*.

production from waves, currents, and tides. The duties of oceanographers vary from job to job, but can be generally described as follows:

- *Geological oceanographers* and *geophysicists* explore the ocean floor and map submarine geological structures. Studies of the physical and chemical properties of rocks and sediments give us valuable information about Earth's history. The results of their work help us understand the processes that created the ocean basins and the interactions between the ocean and the sea floor.

- *Chemical oceanographers* and *marine geochemists* investigate the chemical composition of seawater and its interaction with the atmosphere and the sea floor. Their work may include analysis of seawater components, desalination of seawater, and studying the effects of pollutants. They also examine chemical processes operating within the marine environment and work with biological oceanographers to study

living systems. Their study of trace chemicals in seawater helps us understand how ocean currents move seawater around the globe, and how the ocean affects climate.

- *Physical oceanographers* investigate ocean properties such as temperature, density, wave motions, tides, and currents. They study ocean–atmosphere relationships that influence weather and climate, the transmission of light and sound through water, and the ocean's interactions with its boundaries at the sea floor and the coast.

- *Biological oceanographers, marine biologists,* and *fisheries scientists* study marine plants and animals. They want to understand how marine organisms develop, relate to one another, adapt to their environment, and interact with their environment. They develop ecologically sound methods of harvesting seafood and study biological responses to pollution. New fields associated with biological oceanography include marine biotechnology (the use of natural marine resources in the development of new industrial and biomedical products) and molecular biology (the study of the structure and function of bioinformational molecules—such as DNA, RNA, and proteins—and the regulation of cellular processes at the molecular level). Because marine biology is the most well known field of oceanography (and because the larger marine animals have such wide appeal), it is currently the most competitive, too.

- *Marine* and *ocean engineers* apply scientific and technical knowledge to practical uses. Their work ranges from designing sensitive instruments for measuring ocean processes to building marine structures that can withstand ocean currents, waves, tides, and severe storms. Subfields include acoustics, robotics, electrical, mechanical, civil, and chemical engineering as well as naval architecture. They often use highly specialized computer techniques.

- *Marine archaeologists* systematically recover and study material evidence, such as shipwrecks, graves, buildings, tools, and pottery remaining from past human life and culture that is now covered by the sea. Marine archaeologists use state-of-the-art technology to locate underwater archaeological sites.

- *Marine policy experts* combine their knowledge of oceanography and social sciences, law, or business to develop guidelines and policies for the wise use of the ocean and coastal resources.

Other job opportunities for oceanographers include work as science journalists specializing in marine science, teachers at various grade levels, and aquarium and museum curators.

Box A5–1
Report from a Student/Oceanographer

One of the true pleasures of teaching is that some of your students become so interested in the subject matter that they pursue a career in it. One of Al Trujillo's former students, Joe Cooney (Figure A5A), works as a facilities manager for the Hubbs Sea World Research Institute's Sea Bass Fish Hatchery in Carlsbad, California. His job is to maintain the equipment used to raise sea bass for release to the ocean to restore the former abundance of this sportfish in southern California waters. Joe writes:

> After completing coursework in oceanography at a community college, I wanted more information about what types of jobs were available and what degree would be most valuable. I was not certain if I wanted to be involved in work like marine biologists I had seen, or look more towards physical oceanography as a career. Fortunately, I had the opportunity to get a summer job at a marine fish hatchery, which led to an offer of a permanent position. I am now involved daily in a marine sciences working environment where my primary responsibility is to monitor and maintain the seawater systems for the hatchery.
>
> Unlike a traditional fish hatchery that uses water diverted from a river to provide large amounts of clean water at virtually no cost, a saltwater hatchery must pump in its entire water supply from the sea. Also, conditions must be maintained so that open-ocean fish will live and reproduce, and that their larvae will survive and grow.
>
> Water is pumped to recirculating and experimental systems, as well as to traditional flow-through raceways where juvenile fish are briefly held before being released into the ocean. Incoming water is diverted throughout a hatchery building via computer-controlled valves, pumps, filters and other automated systems. Some water is temperature regulated and its gas content manip-
ulated for experimental use. Controlled light and temperature conditions are maintained in larger tanks to achieve regulated spawning for year-round fish production. Much like the automated system that runs a car engine, sensors throughout the hatchery's system allow a computer to maintain desired conditions, and signal an alarm if conditions vary from prescribed limits. The system contains many components, and maintenance of it requires knowledge of everything from computer hardware to large industrial pumps. In addition, the saltwater is very corrosive and constant maintenance is needed to keep everything running smoothly. Much of what is done is experimental, which presents daily challenges to design, build, and run systems that will achieve the desired results. Both the animal caretakers and the machines must work correctly to successfully produce fish.

> Lacking an advanced college degree, I was able to draw on past work experience to get the position that I have. In order to be considered for most jobs in marine science, a degree is a virtual necessity. However, experience in any marine-related field is good not only to make you more valuable to an employer, but also to allow you to better choose your goals. My position has exposed me to current developments, which in turn reinforces my desire to continue working in this field and obtain further schooling.

> As humans place increasing demands on the ocean, I believe there will be an even larger need for more people to work in many different marine-related areas. Even in such specialized operations such as aquaculture, public aquariums, and research facilities, there are many jobs in animal care, research, lab work, and system operation. The ocean is only recently being studied extensively, and I am looking forward to the investigation into what is found, which should continue for a long time to come.

Figure A5A Joe Cooney at work in the Hubbs Sea World Sea Bass Fish Hatchery in Carlsbad, California.

Sources of Information

- Consult the catalog of any college or university that offers a curriculum in oceanography or marine science.

- The Oceanography Society publishes an excellent brochure entitled *Careers in Oceanography and Marine-Related Fields.* The Oceanography Society can be contacted at 4052 Timber Ridge Drive, Virginia Beach, VA 23455.

- The National Sea Grant College Program of NOAA publishes a comprehensive brochure entitled *Marine Science Careers: A Sea Grant Guide to Ocean Opportunities,* which includes interviews with working oceanographers. The Sea Grant College can be reached c/o NOAA, SSMC3 Room 11606, 1315 East-West Highway, Silver Springs, MD 20910.

- The Scripps Institution of Oceanography at the University of California, San Diego, publishes an informative brochure aimed at perspective students entitled *Preparing for a Career in Oceanography.* General information about Scripps can be obtained by contacting the Scripps Communication Office at the Scripps Institution of Oceanography, University of California San Diego, 9500 Gilman Drive, Department 0233, La Jolla, CA 92093-0233.

- Sea World manages marine and adventure parks and sometimes hires educators and animal trainers—especially those interested in training marine mammals. Their Education Department publishes a booklet on working in a marine life park entitled *The Sea World/Bush Gardens Guide to Zoological Park Careers,* available from Sea World of California Education Department, 1720 South Shores Road, San Diego, CA. 92109-7995.

- For more information about graduate school, The National Academy of Sciences has published a book entitled *Careers in Science and Engineering: A Student Planning Guide to Graduate School and Beyond* (National Academy Press, 1996). The National Acad-

emy of Sciences can be contacted at 2101 Constitution Ave., NW Washington, DC. 20418. Their toll-free telephone number is (800) 624-6242.

Some WWW sites that contain oceanography career information on-line include:

- The International Oceanographic Foundation at: **http://www.rsmas.miami.edu/iof/**

- The U.S. Navy's web site on careers in oceanography at: **http://www.cnmoc.navy.mil/educate/career-o.html** or **http://www.oc.nps.navy.mil/careers.html**

- The Office of Naval Research's web site, which includes The Oceanography Society's brochure entitled *Careers in Oceanography and Marine-Related Fields* at: **http://www.onr.navy.mil/onr/careers/default.htm**

- A comprehensive list of information about careers in oceanography, marine science, and marine biology is available through the Scripps Institution of Oceanography Science Library's web site at: **http://scilib.ucsd.edu/sio/guide/career.html**

- The popular "So You Want to Become A Marine Biologist" web site at: **http://www-siograddept.ucsd.edu/Web/To_Be_A_Marine_Biologist.html**

- A listing of marine laboratories and institutions is available at: **http://life.bio.sunysb.edu/marinebio/mblabs.html**

- Woods Hole Oceanographic Institution has developed a web site devoted to the achievements of women oceanographers, including biographies and unique perspectives of women scientists at: **http://www.womenoceanographers.org/**

Glossary

Abiotic environment   The nonliving components of an ecosystem.

Abyssal clay   Deep-ocean (oceanic) deposits containing less than 30% biogenous sediment. Often oxidized and red in color, thus commonly termed red clay.

Abyssal hill   Volcanic peaks rising less than 1 kilometer (0.6 mile) above the ocean floor.

Abyssal hill province   Deep-ocean regions, particularly in the Pacific Ocean, where oceanic sedimentation rates are so low that abyssal plains do not form and the ocean floor is covered with abyssal hills.

Abyssal plain   A flat depositional surface extending seaward from the continental rise or oceanic trenches.

Abyssal storm   Storm-like occurrences of rapid current movement affecting the deep ocean floor. They are believed to be caused by warm- and cold-core eddies of surface currents.

Abyssal zone   The benthic environment between 4000 and 6000 meters (13,000 and 20,000 feet).

Abyssopelagic   Open-ocean (oceanic) environment below 4000 meters (13,000 feet) in depth.

Acid   A substance that releases hydrogen ions (H^+) in solution.

Acoustic Thermometry of Ocean Climate (ATOC)   The measurement of ocean-wide changes in water properties such as temperature by transmitting and receiving low-frequency sound signals.

Active margin   A continental margin marked by a high degree of tectonic activity, such as those typical of the Pacific Rim. Types of active margins include convergent active margins (marked by plate convergence) and transform active margins (marked by transform faulting).

Adiabatic   Pertaining to a change in the temperature of a mass resulting from compression or expansion. It requires no addition of heat to or loss of heat from the substance.

Agulhas Current   A warm current that carries Indian Ocean water around the southern tip of Africa and into the Atlantic Ocean.

Air mass   A large area of air that has a definite area of origin and distinctive characteristics.

Albedo   The fraction of incident electromagnetic radiation reflected by a surface.

Algae   Primarily aquatic, eukaryotic, photosynthetic organisms that have no root, stem, or leaf systems. Can be microscopic or macroscopic.

Alkaline   A substance that releases hydroxide ions (OH^-) in solution. Also called basic.

Alveoli   A tiny, thin-walled, capillary-rich sac in the lungs where the exchange of oxygen and carbon dioxide takes place.

Amino acid   One of more than 20 naturally occurring compounds that contain NH_2 and COOH groups. They combine to form proteins.

Amnesic shellfish poisoning   Partial or total loss of memory resulting from poisoning caused by eating shellfish with high levels of domoic acid, a toxin produced by a diatom.

Amphidromic point   A nodal or "no-tide" point in the ocean or sea around which the crest of the tide wave rotates during one tidal period.

Amphipoda   Crustacean order containing laterally compressed members such as the "beach hoppers."

Anadromous   Pertaining to a species of fish that spawns in fresh water and then migrates into the ocean to grow to maturity.

Anaerobic respiration   Respiration carried on in the absence of free oxygen (O_2). Some bacteria and protozoans carry on respiration this way.

Anchovy   A small silvery fish (*Engraulis ringens*) that swims through the water with its mouth open to catch its food. Also called anchoveta.

Andesite   A gray, fine-grained volcanic rock composed chiefly of plagioclase and feldspar.

Anhydrite   A colorless, white, gray, blue, or lilac evaporite mineral (anhydrous calcium sulfate, $CaSO_4$) that usually occurs as layers associated with gypsum deposits.

Animalia   Kingdom of many-celled animals.

Anion   An atom that has gained one or more electrons and has an electrical negative charge.

Annelida   Phylum of elongated segmented worms.

Anomalistic month   The time required for the Moon to go from perigee to perigee, 27.5 days.

Anoxic   Without oxygen.

Antarctic Bottom Water   A water mass that forms in the Weddell Sea, sinks to the ocean floor, and spreads across the bottom of all oceans.

Antarctic Circle   The latitude of 66.5° south.

Antarctic Convergence   The zone of convergence along the northern boundary of the West Wind Drift where the southward-flowing boundary currents of the subtropical gyres converge on the cold Antarctic waters.

Antarctic Divergence   The zone of divergence separating the westward-flowing East Wind Drift and the eastward-flowing West Wind Drift.

Antarctic Intermediate Water   Antarctic zone surface water that sinks at the Antarctic convergence and flows north at a depth of about 900 meters (2950 feet) beneath the warmer upper-water mass of the South Atlantic subtropical gyre.

Anticyclonic flow   The flow of air around a region of high pressure clockwise in the Northern Hemisphere.

Antilles Current   A warm current that flows north seaward of the Lesser Antilles from the north equatorial current of the Atlantic Ocean to join the Florida Current.

Antinode   Zone of maximum vertical particle movement in standing waves where crest and trough formation alternate.

Aphelion   The point in the orbit of a planet or comet where it is farthest from the Sun.

Aphotic zone   Without light. The ocean is generally in this state below 1000 meters (3280 feet).

Apogee   The point in the orbit of the Moon or an artificial satellite that is farthest from Earth.

Aragonite   A form of $CaCO_3$ that is less common and less stable than calcite. Pteropod shells are usually composed of aragonite.

Archaea   One of the three major domains of life. The domain consists of simple microscopic bacteria-like creatures (including methane producers and sulfur oxidizers that inhabit deep-sea vents and seeps) and other microscopic life forms that prefer environments of extreme conditions of temperature and/or pressure.

Archipelago   A large group of islands.

Arctic Circle   The latitude of 66.5° north.

Arctic Convergence   A zone of converging currents similar to the Antarctic Convergence but located in the Arctic.

Aspect ratio   The index of propulsive efficiency obtained by dividing the square of the height by fin area.

Asthenosphere   A plastic layer in the upper mantle 80 to 200 kilometers (50 to 124 miles) deep that may allow lateral movement of lithospheric plates and isostatic adjustments.

Atlantic-type margin   See *passive margin*.

Atoll   A ring-shaped coral reef growing upward from a submerged volcanic peak. It may have low-lying islands composed of coral debris.

Atom   A unit of matter, the smallest unit of an element, having all the characteristics of that element and consisting of a dense, central, positively charged nucleus surrounded by a system of electrons.

Atomic mass   The mass of an atom, usually expressed in atomic mass units.

Atomic number   The number of protons in an atomic nucleus.

Autotroph   Algae, plants, and bacteria that can synthesize organic compounds from inorganic nutrients.

Autumnal equinox   The passage of the Sun across the Equator as it moves from the Northern Hemisphere into the Southern Hemisphere, approximately September 23. During this time, all places in the world experience equal lengths of night and day. Also called fall equinox.

Backshore   The inner portion of the shore, lying landward of the mean spring tide high water line. Acted upon by the ocean only during exceptionally high tides and storms.

Backwash   The flow of water down the beach face toward the ocean from a previously broken wave.

Bacteria   One of the three major domains of life. The domain includes unicellular, prokaryotic microorganisms that vary in terms of morphology, oxygen and nutritional requirements, and motility.

Bacterioplankton   Bacteria that live as plankton.

Bacteriovore   Organisms that feed on bacteria.

Bar-built estuary   A shallow estuary (lagoon) separated from the open ocean by a bar deposit such as a barrier island. The water in these estuaries usually exhibits vertical mixing.

Barrier flat   Lying between the salt marsh and dunes of a barrier island, it is usually covered with grasses and even forests if protected from overwash for sufficient time.

Barrier island   A long, narrow, wave-built island separated from the mainland by a lagoon.

Barrier reef   A coral reef separated from the nearby landmass by open water.

Barycenter   The center of mass of a system.

Basalt   A dark-colored volcanic rock characteristic of the ocean crust. Contains minerals with relatively high iron and magnesium content.

Base   See *alkaline.*

Bathyal zone   The benthic environment between the depths of 200 and 4000 meters (660 and 13,000 feet). It includes mainly the continental slope and the oceanic ridges and rises.

Bathymetry   The measurement of ocean depth.

Bathypelagic zone   The pelagic environment between the depths of 1000 and 4000 meters (3300 and 13,000 feet).

Bathyscaphe   A specially-designed deep-diving submersible.

Bathysphere   A specially-designed deep-diving submersible that resembles a sphere.

Bay barrier   A marine deposit attached to the mainland at both ends and extending entirely across the mouth of a bay, separating the bay from the open water. Also known as a bay-mouth bar.

Beach   Sediment seaward of the coastline through the surf zone that is in transport along the shore and within the surf zone.

Beach compartment   A series of rivers, beaches, and submarine canyons involved in the movement of sediment to the coast, along the coast, and down one or more submarine canyons.

Beach face   The wet, sloping surface that extends from the berm to the shoreline. Also known as the low tide terrace.

Beach replenishment   The addition of beach sediment to replace lost or missing material. Also called beach nourishment.

Beach starvation   The interruption of sediment supply and resulting narrowing of beaches.

Benguela Current   The cold eastern boundary current of the South Atlantic subtropical gyre.

Benthic   Pertaining to the ocean bottom.

Benthos   The forms of marine life that live on the ocean bottom.

Berm   The dry, gently-sloping region on the backshore of a beach at the foot of the coastal cliffs or dunes.

Berm crest   The area of a beach that separates the berm from the beach face. The berm crest is often the highest portion of a berm.

Bicarbonate ion (HCO_3^-)   An ion that contains the radical group HCO_3^-.

Bioaccumulation   The accumulation of a substance, such as a toxic chemical, in various tissues of a living organism.

Bioerosion   Erosion of reef or other solid bottom material by the activities of organisms.

Biogenous sediment   Sediment containing material produced by plants or animals, such as coral reefs, shell fragments, and housings of diatoms, radiolarians, Foraminifera, and coccolithophores.

Biogeochemical cycle   The natural cycling of compounds among the living and nonliving components of an ecosystem.

Biological pump   The movement of CO_2 that enters the ocean from the atmosphere through the water column to the sediment on the ocean floor by biological processes—photosynthesis, secretion of shells, feeding, and dying.

Bioluminescence   Light organically produced by a chemical reaction. Found in bacteria, phytoplankton, and various fishes (especially deep-sea fish).

Biomass   The total mass of a defined organism or group of organisms in a particular community or in the ocean as a whole.

Biomass pyramid   A representation of trophic levels that illustrates the progressive decrease in total biomass at successive higher levels of the pyramid.

Bioremediation   The technique of using microbes to assist in cleaning toxic spills.

Biotic community   The living organisms that inhabit an ecosystem.

Biozone   A region of the environment that has distinctive biological characteristics.

Bivalve   A mollusk, such as an oyster or a clam, that has a shell consisting of two hinged valves.

Black smoker   A hydrothermal vent on the ocean floor that emits a black cloud of hot water filled with dissolved metal particles.

Body wave   A longitudinal or transverse wave that transmits energy through a body of matter.

Boiling point   The temperature at which a substance changes state from a liquid to a gas at a given pressure.

Bore   A steep-fronted tide crest that moves up a river in association with an incoming high tide.

Boundary current   The northward- or southward-flowing currents that form the western and eastern boundaries of the subtropical circulation gyres.

Brackish   Low salinity water caused by the mixing of fresh water and saltwater.

Brazil Current   The warm western boundary current of the South Atlantic subtropical gyre.

Breaker zone   Region where waves break at the seaward margin of the surf zone.

Breakwater   Any artificial structure constructed roughly parallel to shore and designed to protect a coastal region from the force of ocean waves.

Brittle   Descriptive term for a substance that is likely to fracture when force is applied to it.

Bryozoa   Phylum of colonial animals that often share one coelomic cavity. Encrusting and branching forms secrete a protective housing (zooecium) of calcium carbonate or chitinous material. Possess lophophore feeding structure.

Buffering   The process by which a substance minimizes a change in the acidity of a solution when an acid or base is added to the solution.

Buoyancy   The ability or tendency to float or rise in a liquid.

Bycatch   Any marine organisms that are caught incidentally by fishers seeking commercial species.

Calcareous   Containing calcium carbonate.

Calcite   The mineral of the chemical formula CaCO₃.

Calcite compensation depth (CCD)   The depth at which the amount of calcite (CaCO₃) produced by the organisms in the overlying water column is equal to the amount of calcite the water column can dissolve. No calcite deposition occurs below this depth, which, in most parts of the ocean, is at a depth of 4500 meters (15,000 feet).

Calcium carbonate (CaCO₃)   A chalk-like substance secreted by many organisms in the form of coverings or skeletal structures.

California Current   The cold eastern boundary current of the North Pacific subtropical gyre.

Calorie   Unit of heat, defined as the amount of heat required to raise the temperature of 1 gram of water 1°C.

Calving   The process by which a glacier breaks at an edge, so that a portion of the ice separates and falls from the glacier.

Canary Current   The cold eastern boundary current of the North Atlantic subtropical gyre.

Capillarity   The action by which a fluid, such as water, is drawn up in small tubes as a result of surface tension.

Capillary wave   An ocean wave whose wavelength is less than 1.74 centimeters (0.7 inch). The dominant restoring force for such waves is surface tension.

Carapace   Chitinous or calcareous shield that covers the cephalothorax of some crustaceans. Dorsal portion of a turtle shell.

Carbohydrate   An organic compound containing the elements carbon, hydrogen, and oxygen with the general formula $(CH_2O)_n$.

Carbonate ion (CO₃⁻²)   An ion that contains the radical group CO_3^{-2}.

Caribbean Current   The warm current that carries equatorial water across the Caribbean Sea into the Gulf of Mexico.

Carnivora   An order of marine mammals that includes the sea otter, polar bear, and pinnipeds.

Carnivore   An animal that depends on other animals solely or chiefly for its food supply.

Carotin   An orange-yellow pigment found in plants.

Cation   An atom that has lost one or more electrons and has an electrical positive charge.

Centigrade temperature scale   A temperature scale based on the freezing point (0°C = 32°F) and boiling point (100°C = 212°F) of pure water. Also known as the Celsius scale after its founder.

Centripetal force   A center-seeking force that tends to make rotating bodies move toward the center of rotation.

Cephalopoda   A class of the phylum Mollusca with a well-developed pair of eyes and a ring of tentacles surrounding the mouth. The shell is absent or internal on most members. The class includes the squid, octopus, and nautilus.

Cetacea   An order of marine mammals that includes the whales, dolphins, and porpoises.

Chalk   A soft, compact form of calcite, generally gray-white or yellow-white in color and derived chiefly from microscopic fossils.

Chemical energy   A form of potential energy stored in the chemical bonds of compounds.

Chemosynthesis   A process by which bacteria or archaea synthesize organic molecules from inorganic nutrients using chemical energy released from the bonds of a chemical compound (such as hydrogen sulfide) by oxidation.

Chloride ion (Cl⁻)   A chlorine atom that has become negatively charged by gaining one electron.

Chlorinity   The amount of chloride ion and ions of other halogens in ocean water expressed in parts per thousand (‰) by weight.

Chlorophyll   A group of green pigments that make it possible for plants to carry on photosynthesis.

Chlorophyta   Green algae. Characterized by the presence of chlorophyll and other pigments.

Chondrite   A stony meteorite composed primarily of silicate rock material and containing chondrules (spheroidal granules). They are the most commonly found meteorites.

Chronometer   An exceptionally precise timepiece.

Chrysophyta   An important phylum of planktonic algae that includes the diatoms. The presence of chlorophyll is masked by the pigment carotin that gives the plants a golden color.

Cilium   A short, hairlike structure common on lower animals. Beating in unison, cilia may create water currents that carry food toward the mouth of an animal or may be used for locomotion.

Circadian rhythm   Behavioral and physiological rhythms of organisms related to the 24-hour day. Sleeping and waking patterns are an example.

Circular orbital motion   The motion of water particles caused by a wave as the wave is transmitted through water.

Circumpolar Current   Eastward-flowing current that extends from the surface to the ocean floor and encircles Antarctica.

Clay   **1.** A particle size between silt and colloid. **2.** Any of various hydrous aluminum silicate minerals that are plastic, expansive, and have ion exchange capabilities.

Climate   The meteorological conditions, including temperature, precipitation, and wind, that characteristically prevail in a particular region; the long-term average of weather.

Cnidoblast   Stinging cell of phylum Coelenterata that contains the stinging mechanism (nematocyst) used in defense and capturing prey.

Coast   A strip of land that extends inland from the coastline as far as marine influence is evidenced in the landforms.

Coastal geostrophic current   See *geostrophic current.*

Coastal plain estuary   An estuary formed by rising sea level flooding a coastal river valley.

Coastal upwelling   The movement of deeper nutrient-rich water into the surface water mass as a result of windblown surface water moving offshore.

Coastal waters   Those relatively shallow water areas that adjoin continents or islands.

Coastline   Landward limit of the effect of the highest storm waves on the shore.

Coccolith   Tiny calcareous discs averaging about 3 micrometers (0.00012 inch) in diameter that form the cell wall of coccolithophores.

Coccolithophore   A microscopic planktonic form of algae encased by a covering composed of calcareous discs (coccoliths).

Coelenterata   Phylum of radially symmetrical animals that includes two basic body forms, the medusa and the polyp. Includes jellyfish (medusoid) and sea anemones (polypoid). Preferred name is now Cnidaria.

Cohesion   The intermolecular attraction by which the elements of a body are held together.

Cold core ring   A circular eddy of a surface current that contains cold water in its center and rotates counterclockwise.

Cold front   A weather front in which a cold air mass moves into and under a warm air mass. It creates a narrow band of intense precipitation.

Cold-blooded   See *ectothermic.*

Colonial animals   Animals that live in groups of attached or separate individuals. Groups of individuals may serve special functions.

Columbia River estuary   An estuary at the border between the states of Washington and Oregon that has been most adversely affected by the construction of hydroelectric dams.

Comb jelly   Common name for members of the phylum Ctenophora. (See *Ctenophora.*)

Commensalism   A symbiotic relationship in which one party benefits and the other is unaffected.

Compensation depth for photosynthesis   The depth at which net photosynthesis becomes zero and photosynthetic organisms can no longer survive. This depth is greater in the open ocean (up to 100 meters or 330 feet) than near the shore due to factors that limit light penetration in coastal regions.

Compound   A substance containing two or more elements combined in fixed proportions.

Condensation   The conversion of water from the vapor to the liquid state. When it occurs, the energy required to vaporize the water is released into the atmosphere. This is about 585 calories per gram of water at 20°C.

Conduction   The transmission of heat by the passage of energy from particle to particle.

Conjunction   The apparent closeness of two heavenly bodies. During the new moon phase, Earth and the Moon are in conjunction on the same side of the Sun.

Constant Proportions, Principle of   A principle stating that the major constituents of ocean-water salinity are found in the same relative proportions throughout the ocean-water volume, independent of salinity.

Constructive interference   A form of wave interference in which two waves come together in phase, for example, crest to crest, to produce a greater displacement from the still-water line than that produced by either of the waves alone.

Consumer   An animal within an ecosystem that consumes the organic mass produced by the producers.

Continent   About one-third of Earth's surface that rises above the deep-ocean floor to be exposed above sea level. Continents are composed primarily of granite, an igneous rock of lower density than the basaltic oceanic crust.

Continental accretion   Growth or increase in size of a continent by gradual external addition of crustal material.

Continental arc   An arc-shaped row of active volcanoes produced by subduction that occurs along convergent active continental margins.

Continental borderland   A highly irregular portion of the continental margin that is submerged beneath the ocean and is characterized by depths greater than those characteristic of the continental shelf.

Continental drift   A term applied to early theories supporting the possibility the continents are in motion over Earth's surface.

Continental margin   The submerged area next to a continent comprising the continental shelf, continental slope, and continental rise.

Continental rise   A gently sloping depositional surface at the base of the continental slope.

Continental shelf   A gently sloping depositional surface extending from the low water line to the depth of a marked increase in slope around the margin of a continent or island.

Continental slope   A relatively steeply sloping surface lying seaward of the continental shelf.

Convection   Heat transfer in a gas or liquid by the circulation of currents from one region to another.

Convection cell   A circular-moving loop of matter involved in convective movement.

Convergence   The act of coming together from different directions. There are polar, tropical, and subtropical regions of the oceans where water masses with different characteristics come together. Along these lines of convergence, the denser mass will sink beneath the others.

Convergent plate boundary   A lithospheric plate boundary where adjacent plates converge, producing ocean trench-island arc systems, ocean trench-continental volcanic arcs, or folded mountain ranges.

Copepoda   An order of microscopic to nearly microscopic crustaceans that are important members of zooplankton in temperate and subpolar waters.

Coral   A group of benthic anthozoans that exist as individuals or in colonies and secrete CaCO₃ external skeletons. Under the proper conditions, corals may produce reefs composed of their external skeletons and the CaCO₃ material secreted by varieties of algae associated with the reefs.

Coral bleaching   The loss of color in coral reef organisms that causes them to turn white. Coral bleaching is caused by the removal or expulsion of the coral's symbiotic zooxanthellae algae in response to high water temperatures or other adverse conditions.

Coral reef   A calcareous organic reef composed significantly of solid coral and coral sand. Algae may be responsible for more than half of the $CaCO_3$ reef material. Found in waters where the minimum average monthly temperature is 18°C or higher.

Core   **1.** The core of Earth is composed primarily of iron and nickel. It has a liquid outer portion 2270 kilometers (1410 miles) thick and a solid inner core with a radius of 1216 kilometers (756 miles). **2.** A cylinder of sediment and/or rock material.

Coriolis effect   An apparent force resulting from Earth's rotation causes particles in motion to be deflected to the right in the Northern Hemisphere and to the left in the Southern Hemisphere.

Cosmogenous sediment   All sediment derived from outer space.

Cotidal lines   Lines connecting points where high tide occurs simultaneously.

Countershading   Protective coloration in an animal or insect, characterized by darker coloring of areas exposed to light and lighter coloring of areas that are normally shaded.

Covalent bond   A chemical bond formed by the sharing of one or more electrons, especially pairs of electrons, between atoms.

Crest (wave)   The portion of an ocean wave that is displaced above the still-water level.

Cruiser   Fish, such as the bluefin tuna, that constantly cruise the pelagic waters in search of food.

Crust   **1.** Unit of Earth's structure that is composed of basaltic ocean crust and granitic continental crust. The total thickness of the crustal units may range from 5 kilometers (3 miles) beneath the ocean to 50 kilometers (30 miles) beneath the continents. **2.** A hard covering or surface layer of hydrogenous sediment.

Crustacea   A class of phylum Arthropoda that includes barnacles, copepods, lobsters, crabs, and shrimp.

Crystalline rock   Igneous or metamorphic rocks. These rocks are made up of crystalline particles with orderly molecular structures.

Current   A horizontal movement of water.

Cutaneous artery   The artery that runs down both sides of some cruiser-type fish to help maintain a constant elevated temperature in the myomere musculature used for swimming.

Cutaneous vein   The vein that runs down both sides of some cruiser-type fish to help maintain a constant elevated temperature in the myomere musculature used for swimming.

Cyclone   An atmospheric system characterized by the rapid, inward circulation of air masses about a low-pressure center, usually accompanied by stormy, often destructive, weather. Cyclones circulate counterclockwise in the Northern Hemisphere and clockwise in the Southern Hemisphere.

Cyclonic flow   The flow of air around a region of low pressure counterclockwise in the Northern Hemisphere.

Davidson Current   A northward-flowing current along the Washington-Oregon coast that is driven by geostrophic effects on a large freshwater runoff.

DDT   An insecticide that caused damage to marine bird populations in the 1950s and 1960s. Its use is now banned throughout most of the world.

Decay distance   The distance over which waves change from a choppy "sea" to uniform swell.

Declination   The angular distance of the Sun or Moon above or below the plane of Earth's Equator.

Decomposers   Primarily bacteria that break down nonliving organic material, extract some of the products of decomposition for their own needs, and make available the compounds needed for primary production.

Deep boundary current   Relatively strong deep currents flowing across the continental rise along the western margin of ocean basins.

Deep current   Density-driven circulation that is initiated at the ocean surface by temperature and salinity conditions that produce a high-density water mass, which sinks and spreads slowly beneath surface waters.

Deep ocean basin   Areas of the ocean floor that have deep water, are far from land, and are underlain by basaltic crust.

Deep scattering layer (DSL)   A layer of marine organisms in the open ocean that scatter signals from an echo sounder. It migrates daily from depths of slightly over 100 meters (330 feet) at night to more than 800 meters (2600 feet) during the day.

Deep-sea fan   A large fan-shaped deposit commonly found on the continental rise seaward of such sediment-laden rivers as the Amazon, Indus, or Ganges-Brahmaputra. Also known as a submarine fan.

Deep-sea fish   Any of a large group of fishes that lives within the aphotic zone and has special adaptations for finding food and avoiding predators in darkness.

Deep-sea system   Includes the bathyal, abyssal, and hadal benthic environments.

Deep-water   The water beneath the permanent thermocline (and resulting pycnocline) that has a uniformly low temperature.

Deep-water wave   Ocean wave traveling in water that has a depth greater than one-half the average wavelength. Its speed is independent of water depth.

Delta   A low-lying deposit at the mouth of a river, usually having a triangular shape as viewed from above.

Density   Mass per unit volume of a substance. Usually expressed as grams per cubic centimeter (g/cm^3). For ocean water with a salinity of 35‰ at 0°C, the density is 1.028 g/cm^3.

Density stratification   Layering based on density, where the highest density material occupies the lowest space.

Deposit feeder   An organism that feeds on food items that occur as deposits, including detritus and various detritus-coated sediment.

Depositional-type shores   Shorelines dominated by processes that form deposits (such as sand bars and barrier islands) along the shore.

Desalination   The removal of salt ions from ocean water to produce pure water.

Destructive interference   A form of wave interference in which two waves come together out of phase, for example, crest to trough, and produce a wave with less displacement than the larger of the two waves would have produced alone.

Detritus   1. Any loose material produced directly from rock disintegration. 2. Material resulting from the disintegration of dead organic remains.

Diatom   Member of the class Bacillariophyceae of algae that possesses a wall of overlapping silica valves.

Diatomaceous earth   A deposit composed primarily of the tests of diatoms mixed with clay. Also called diatomite.

Dipolar   Having two poles. The water molecule possesses a polarity of electrical charge with one pole being more positive and the other more negative in electrical charge.

Diffraction   Any change in the direction or intensity of a wave after passing an obstacle that cannot be interpreted as refraction or reflection.

Diffusion   A process by which fluids move through other fluids by random molecular movement from areas of high concentration to areas in which they are in lower concentrations.

Dinoflagellates   Single-celled microscopic organisms that may possess chlorophyll and belong to the plant phylum Pyrrophyta (autotrophic) or may ingest food and belong to the class Mastigophora of the animal phylum Protozoa (heterotrophic).

Discontinuity   An abrupt change in a property such as temperature or salinity at a line or surface.

Disphotic zone   The dimly lit zone, corresponding approximately to the mesopelagic, in which there is not enough light to support photosynthetic organisms; sometimes called the twilight zone.

Disruptive coloration   A marking or color pattern that confuses prey.

Dissolved oxygen   Oxygen that is dissolved in ocean water.

Distillation   A method of purifying liquids by heating them to their boiling point and condensing the vapor.

Distributary   A small stream flowing away from a main stream. Such streams are characteristic of deltas.

Diurnal inequality   The difference in the heights of two successive high tides or two successive low tides during a lunar (tidal) day.

Diurnal tidal pattern   A tidal pattern exhibiting one high tide and one low tide during a tidal day; a daily tide.

Divergence   A horizontal flow of water from a central region, as occurs in upwelling.

Divergent plate boundary   A lithospheric plate boundary where adjacent plates diverge, producing an oceanic ridge or rise (spreading center).

Doldrums   A belt of light variable winds within 10° to 15° of the Equator, resulting from the vertical flow of low-density air within this equatorial belt.

Dolphin   1. A brilliantly colored fish of the genus *Coryphaena*. 2. The name applied to the small, beaked members of the cetacean family Delphinidae.

Dorsal   Pertaining to the back or upper surface of most animals.

Dorsal aorta   For most fish, this is the only major artery that runs the length of the fish through openings in the vertebrae and supplies blood. Some pelagic cruisers also have cutaneous arteries.

Downwelling   In the open or coastal ocean where Ekman transport causes surface waters to converge or impinge on the coast, surface water will be carried down beneath the surface.

Drift bottle   Any equipment used to study current movement by drifting with currents.

Driftnet   A fishing net made of monofilament fishing line that catches organisms by entanglement.

Drifts   Thick sediment deposits on the continental rise produced where the deep boundary current slows and loses sediment when it changes direction to follow the base of the continental slope.

Drowned beach   An ancient beach now beneath the coastal ocean because of rising sea level.

Drowned river valley   The lower part of a river valley that has been submerged by rising sea level.

Dunes   Coastal deposits of sand lying landward of the beach and deriving their sand from onshore winds that transport beach sand inland.

Dynamic topography   A surface configuration resulting from the geopotential difference between a given surface and a reference surface of no motion. A contour map of this surface is useful in estimating the nature of geostrophic currents.

Earthquake   A sudden motion or trembling in Earth, caused by the sudden release of slowly accumulated strain by faulting (movement along a fracture in Earth's crust) or volcanic activity.

East Australian Current   The warm western boundary current of the South Pacific subtropical gyre.

East Pacific Rise   A fast-spreading divergent plate boundary extending southward from the Gulf of California through the eastern South Pacific Ocean.

East Wind Drift   The coastal current driven in a westerly direction by the polar easterly winds blowing off Antarctica.

Eastern boundary current   Equatorward-flowing cold drifts of water on the eastern side of all subtropical gyres.

Ebb current   The flow of water seaward during a decrease in the height of the tide.

Echinodermata   Phylum of animals that have bilateral symmetry in larval forms and usually a five-sided radial symmetry as adults. Benthic and possessing rigid or articulating exoskeletons of calcium carbonate with spines, this phylum includes sea stars, brittle stars, sea urchins, sand dollars, sea cucumbers, and sea lilies.

Echolocation   A sensory system in odontocete cetaceans in which usually high-pitched sounds are emitted and their echoes interpreted to determine the direction and distance of objects.

Echosounder   A device that transmits sound from a ship's hull to the ocean floor where it is reflected back to receivers. The speed of sound in the water is known, so the depth can be determined from the travel time of the sound signal.

Ecliptic   The plane of the center of the Earth-Moon system as it orbits around the Sun.

Ecosystem   All the organisms in a biotic community and the abiotic environmental factors with which they interact.

Ectothermic   Of or relating to an organism that regulates its body temperature largely by exchanging heat with its surroundings; cold-blooded.

Eddy   A current of any fluid forming on the side of a main current. It usually moves in a circular path and develops where currents encounter obstacles or flow past one another.

Ekman spiral   A theoretical consideration of the effect of a steady wind blowing over an ocean of unlimited depth and breadth and of uniform viscosity. The result is a surface flow at 45° to the right of the wind in the Northern Hemisphere. Water at increasing depth below the surface will drift in directions increasingly more slowly and to the right until at about 100 meters (330 feet) depth it may move in a direction opposite to that of the wind.

Ekman transport   The net transport of surface water set in motion by wind. Due to the Ekman spiral phenomenon, it is theoretically in a direction 90° to the right and 90° to the left of the wind direction in the Northern Hemisphere and Southern Hemisphere, respectively.

El Niño   A southerly flowing warm current that generally develops off the coast of Ecuador around Christmastime. Occasionally it will move farther south into Peruvian coastal waters and cause the widespread death of plankton, fish, and other organisms dependent upon fish for food.

El Niño-Southern Oscillation (ENSO)   The correlation of El Niño events with an oscillatory pattern of pressure change in a persistent high-pressure cell in the southeastern Pacific Ocean and a persistent low-pressure cell over the East Indies.

Electrical conductivity   The ability or power to conduct or transmit electricity.

Electrolysis   A separation process by which salt ions are removed from saltwater through water-impermeable membranes toward oppositely charged electrodes.

Electromagnetic energy   Energy that travels as waves or particles with the speed of light. Different kinds possess different properties based on wavelength. The longest wavelengths belong to radio waves, up to 100 kilometers (60 miles) in length. At the other end of the spectrum are cosmic rays with greater penetrating power and wavelengths of less than 0.000001 micrometer.

Electromagnetic spectrum   The spectrum of radiant energy emitted from stars and ranging between cosmic rays with wavelengths of less than 10 to 11 centimeters (4 to 4.3 inches) and very long waves with wavelengths in excess of 100 kilometers (60 miles).

Electron   A subatomic particle that orbits the nucleus of an atom and has a negative electric charge.

Electron cloud   The diffuse area surrounding the nucleus of an atom where electrons are found.

Electrostatic force   A force caused by electric charges at rest.

Element   One of a number of substances, each of which is composed entirely of like particles—atoms—that cannot be broken into smaller particles by chemical means.

Emerging shoreline   A shoreline resulting from the emergence of the ocean floor relative to the ocean surface. It is usually rather straight and characterized by marine features usually found at some depth.

Endothermic reaction   A chemical reaction that absorbs energy. For example, energy is stored in the organic products of the chemical reaction photosynthesis.

ENSO   See *El Niño–Southern Oscillation.*

ENSO index   An index showing the relative strength of El Niño and La Niña conditions.

Entropy   A thermodynamic quantity that reflects the degree of randomness in a system. It increases in all natural systems.

Environment   The sum of all physical, chemical, and biological factors to which an organism or community is subjected.

Epicenter   The point on Earth's surface that is directly above the focus of an earthquake.

Epifauna   Animals that live on the ocean bottom, either attached or moving freely over it.

Epipelagic zone   A subdivision of the oceanic province that extends from the surface to a depth of 200 meters (660 feet).

Equator   The imaginary great circle around Earth's surface, equidistant from the poles and perpendicular to Earth's axis of rotation. It divides Earth into the Northern Hemisphere and the Southern Hemisphere.

Equatorial   Pertaining to the equatorial region.

Equatorial countercurrent   Eastward-flowing currents found between the north and south equatorial currents in all oceans, but particularly well developed in the Pacific Ocean.

Equatorial current   Westward-flowing currents that travel along the Equator in all ocean basins, caused by the trade winds. They are called North or South Equatorial Currents depending on their position north or south of the Equator.

Equatorial low   A band of low atmospheric pressure that encircles the globe along the Equator.

Equatorial upwelling   The movement of deeper nutrient-rich water into the surface water mass as a result of divergence of currents along the Equator.

Erosion   The group of natural processes, including weathering, dissolution, abrasion, corrosion, and transportation, by which material is worn away from Earth's surface.

Erosional-type shores   Shorelines dominated by processes that form erosional features (such as cliffs and sea stacks) along the shore.

Estuarine circulation pattern   A flow pattern in an estuary characterized by a net surface flow of low-salinity water toward the ocean and an opposite net subsurface flow of seawater toward the head of the estuary.

Estuary   A partially enclosed coastal body of water in which salty ocean water is significantly diluted by fresh water from land runoff. Examples of estuaries include river mouths, bays, inlets, gulfs, and sounds.

Eukarya   One of the three major domains of life. The domain includes single-celled or multicellular organisms whose cells usually contain a distinct membrane-bound nucleus.

Euphotic zone   A layer that extends from the surface of the ocean to a depth where enough light exists to support photosynthesis, rarely deeper than 100 meters (330 feet).

Euryhaline   Descriptive term for organisms with a high tolerance for a wide range of salinity conditions.

Eurythermal   Descriptive term for organisms with a high tolerance for a wide range of temperature conditions.

Eustatic sea level change   A worldwide raising or lowering of sea level.

Eutrophic   A region of high productivity.

Eutrophication   The enrichment of waters by a previously scarce nutrient.

Evaporation   To change from a liquid to the vapor state.

Evaporite   A sedimentary deposit that is left behind when water evaporates. Evaporite minerals include gypsum, calcite, and halite.

Evolution   The change of groups of organisms with the passage of time, mainly as a result of natural selection, so that descendants differ morphologically and physiologically from their ancestors.

Exclusive Economic Zone (EEZ)   A coastal zone 200 nautical miles wide over which the coastal nation has jurisdiction over mineral resources, fishing, and pollution. If the continental shelf extends beyond 200 miles, the EEZ may be up to 350 miles (about 560 kilometers) in width.

Exothermic reaction   A chemical reaction that liberates energy. For example, the energy stored in the products of photosynthesis is released by the chemical reaction respiration.

Extrusive rocks   Igneous rocks that flow out onto Earth's surface before cooling and solidifying (lavas).

Eye **1.** An organ of vision or of light sensitivity. **2.** The circular area of relative calm at the center of a hurricane.

Fact Something having real, demonstrable existence. A scientific fact is an occurrence that has been repeatedly confirmed.

Fahrenheit temperature scale (°F) A temperature scale whereby the freezing point of water is 32° and the boiling point of water is 212°.

Falcate Curved and tapering to a point; sickle-shaped.

Falkland Current A northward-flowing cold current found off the southeastern coast of South America.

Fall bloom A mid-latitude bloom of phytoplankton that occurs during the fall and is limited by the availability of sunlight.

Fall equinox See *autumnal equinox*.

Fan A gently sloping, fan-shaped feature normally located near the lower end of a canyon. Also known as a submarine fan.

Fat An organic compound formed from alcohol, glycerol and one or more fatty acids; a lipid, it is a solid at atmospheric temperatures.

Fathom (fm) A unit of depth in the ocean, commonly used in countries using the English system of units. It is equal to 1.83 meters (6 feet).

Fault A fracture or fracture zone in Earth's crust along which displacement has occurred.

Fault block A crustal block bounded on at least two sides by faults. Usually elongate; if it is down-dropped, it produces a graben; if uplifted, it is a horst.

Fauna The animal life of any particular area or of any particular time.

Fecal pellet Excrement of planktonic crustaceans that assist in speeding up the descent rate of sedimentary particles by combining them into larger packages.

Ferrel Cell The large atmospheric circulation cell that occurs between 30° and 60° latitude in each hemisphere.

Ferromagnesium Minerals rich in iron and magnesium.

Fetch **1.** Pertaining to the area of the open ocean over which the wind blows with constant speed and direction, thereby creating a wave system. **2.** The distance across the fetch (wave-generating area) measured in a direction parallel to the direction of the wind.

Filter feeder An organism that obtains its food by filtering seawater for other organisms. Also known as suspension feeder.

Fishery Fish caught from the ocean by commercial fishers.

Fishery management The organized effort directed at regulating fishing activity with the goal of maintaining a long-term fishery.

Fissure A long, narrow opening; a crack or cleft.

Fjord A long, narrow, deep, U-shaped inlet that usually represents the seaward end of a glacial valley that has become partially submerged after the melting of the glacier.

Flagellum A whiplike living structure used by some cells for locomotion.

Flood current A tidal current associated with increasing height of the tide, generally moving toward the shore.

Flora The plant life of any particular area or of any particular time.

Florida Current A warm current flowing north along the coast of Florida that merges into the Gulf Stream.

Folded mountain range Mountain ranges formed as a result of the convergence of lithospheric plates. They are characterized by masses of folded sedimentary rocks that formed from sediments deposited in the ocean basin that was destroyed by the convergence.

Food chain The passage of energy materials from producers through a sequence of a herbivore and a number of carnivores.

Food web A group of interrelated food chains.

Foraminifera An order of planktonic and benthic protozoans that possess protective coverings, usually composed of calcium carbonate.

Forced wave A wave that is generated and maintained by a continuous force such as the gravitational attraction of the Moon.

Foreshore The portion of the shore lying between the normal high and low water marks; the intertidal zone.

Fossil Any remains, print, or trace of an organism that has been preserved in Earth's crust.

Fracture zone An extensive linear zone of unusually irregular topography of the ocean floor, characterized by large seamounts, steep-sided or asymmetrical ridges, troughs, or long, steep slopes. Usually represents ancient, inactive transform fault zones.

Free wave A wave created by a sudden rather than a continuous impulse that continues to exist after the generating force is gone.

Freeze separation The desalination of seawater by multiple episodes of freezing, rinsing, and thawing.

Freezing The process by which a liquid is converted to a solid at its freezing point.

Freezing point The temperature at which a liquid becomes a solid under any given set of conditions. The freezing point of water is 0°C at one atmosphere pressure.

Fringing reef A reef that is directly attached to the shore of an island or continent. It may extend more than 1 kilometer (0.6 mile) from shore. The outer margin is submerged and often consists of algal limestone, coral rock, and living coral.

Fucoxanthin The reddish-brown pigment that gives brown algae its characteristic color.

Full Moon The phase of the Moon that occurs when the Sun and Moon are in opposition; that is, they are on opposite sides of Earth. During this time, the lit side of the Moon faces Earth.

Fully developed sea The maximum average size of waves that can be developed for a given wind speed when it has blown in the same direction for a minimum duration over a minimum fetch.

Fungi Any of numerous eukaryotic organisms of the kingdom Fungi, which lack chlorophyll and vascular tissue and range in form from a single cell to a body mass of branched filamentous structures that often produce specialized fruiting bodies. The kingdom includes the yeasts, molds, lichens, and mushrooms.

Fur seal Any of several eared seals of the genera *Callorhinus* or *Arctocephalus*, having thick, soft underfur.

Fusion reaction A type of nuclear reaction where hydrogen atoms are converted to helium atoms, thereby releasing large amounts of energy.

Galápagos Rift A divergent plate boundary extending eastward from the Galápagos Islands toward South America. The first deep-sea hydrothermal vent biocommunity was discovered here in 1977.

Galaxy One of the billions of large systems of stars that make up the universe.

Gas hydrate A lattice-like compound composed of water and natural gas (usually methane) formed in high pressure and low temperature environments such as those found in deep ocean sediments. Also known as clathrates because of their cage-like chemical structure.

Gaseous state A state of matter in which molecules move by translation and only interact through chance collisions.

Gastropoda A class of mollusks, most of which possess an asymmetrical, spiral one-piece shell and a well-developed flattened foot. A well-developed head will usually have two eyes and one or two pairs of tentacles. Includes snails, limpets, abalone, cowries, sea hares, and sea slugs.

Geostrophic current A current that grows out of Earth's rotation and is the result of a near balance between gravitational force and the Coriolis effect.

Gill A thin-walled projection from some part of the external body or the digestive tract used for respiration in a water environment.

Gill net See *driftnet*.

Glacial deposit A sedimentary deposit formed by a glacier and characterized by poor sorting.

Glacier A large mass of ice formed on land by the recrystallization of old, compacted snow. It flows from an area of accumulation to an area of wasting where ice is removed from the glacier by melting.

Glauconite A group of green hydrogenous minerals consisting of hydrous silicates of potassium and iron.

Global Positioning System (GPS)   A system of satellites that transmit microwave signals to Earth, allowing people at the surface to accurately locate themselves.

Globigerina ooze   An ooze that contains a large percentage of the calcareous tests of the foraminiferan Globigerina.

Goiter   An enlargement of the thyroid gland, visible as a swelling at the front of the neck, often associated with iodine deficiency.

Gondwanaland   A hypothetical protocontinent of the Southern Hemisphere named for the Gondwana region of India. It included the present continental masses Africa, Antarctica, Australia, India, and South America.

Graded bedding   Stratification in which each layer displays a decrease in grain size from bottom to top.

Gradient   The rate of increase or decrease of one quantity or characteristic relative to a unit change in another. For example, the slope of the ocean floor is a change in elevation (a vertical linear measurement) per unit of horizontal distance covered. Commonly measured in meters per kilometer.

Grain size   The average size of the grains of material in a sample. Also known as fragment or particle size.

Granite   A light-colored igneous rock characteristic of the continental crust. Rich in nonferromagnesian minerals such as feldspar and quartz.

Gravitational force   The force of attraction that exists between any two bodies in the universe that is proportional to the product of their masses and inversely proportional to the square of the distance between the centers of their masses.

Gravity wave   A wave for which the dominant restoring force is gravity. Such waves have a wavelength of more than 1.74 centimeters (0.7 inch), and their speed of propagation is controlled mainly by gravity.

Gray whale   A baleen whale (*Eschrichtius robustus*) of northern Pacific waters, having grayish-black coloring with white blotches.

Greenhouse effect   The heating of Earth's atmosphere that results from the absorption by components of the atmosphere such as water vapor and carbon dioxide of infrared radiation from Earth's surface.

Groin   A low artificial structure built perpendicular to the shore and designed to interfere with longshore transportation of sediment so that it traps sand and widens the beach on its upstream side.

Groin field   A series of closely-spaced groins.

Gross ecological efficiency   The amount of energy passed on from a trophic level to the one above it divided by the amount it received from the one below it.

Gross primary production   The total carbon fixed into organic molecules through photosynthesis or chemosynthesis by a discrete autotrophic community.

Grunion   A small fish (*Leuresthes tenuis*) of coastal waters of California and Mexico that spawns at night along beaches during the high tides of spring and summer. Grunion time their reproductive activities to coincide with tidal phenomena and are the only fish that come completely out of water to spawn.

Gulf Stream   The high-intensity western boundary current of the North Atlantic Ocean subtropical gyre that flows north off the east coast of the United States.

Guyot   See *tablemount*.

Gypsum   A colorless, white, or yellowish evaporite mineral, $CaSO_4 \cdot 2H_2O$.

Gyre   A large horizontal circular-moving loop of water. Used mainly in reference to the circular motion of water in each of the major ocean basins centered in subtropical high pressure regions.

Habitat   A place where a particular plant or animal lives. Generally refers to a smaller area than environment.

Hadal   Pertaining to the deepest ocean environment, specifically that of ocean trenches deeper than 6 kilometers (3.7 miles).

Hadal zone   Pertaining to the deepest ocean benthic environment, specifically that of ocean trenches deeper than 6 kilometers (3.7 miles).

Hadley Cell   The large atmospheric circulation cell that occurs between the Equator and 30° latitude in each hemisphere.

Half-life   The time required for half the atoms of a sample of a radioactive isotope to decay to an atom of another element.

Halite   A colorless or white evaporite mineral, NaCl, which occurs as cubic crystals and is used as table salt.

Halocline   A layer of water in which a high rate of change in salinity in the vertical dimension is present.

Hard stabilization   Any form of artificial structure built to protect a coast or to prevent the movement of sand along a beach. Examples include groins, breakwaters, and seawalls.

Harmful algae bloom (HAB)   See *red tide*.

Headland   A steep-faced irregularity of the coast that extends out into the ocean.

Heat   Energy moving from a high temperature system to a lower temperature system. The heat gained by the one system may be used to raise its temperature or to do work.

Heat capacity   Usually defined as the amount of heat required to raise the temperature of 1 gram of a substance 1°C.

Heat energy   Energy of molecular motion. The conversion of higher forms of energy such as radiant or mechanical energy to heat energy within a system increases the heat energy within the system and the temperature of the system.

Heat flow (flux)   The quantity of heat flow to Earth's surface per unit of time.

Hemoglobin   A red pigment found in red blood corpuscles that carries oxygen from the lungs to tissue and carbon dioxide from tissue to lungs.

Herbivore   An animal that relies chiefly or solely on plants for its food.

Hermatypic coral   Reef-building corals that have symbiotic algae in their ectodermal tissue. They cannot produce a reef structure below the euphotic zone.

Heterotroph   Animals and bacteria that depend on the organic compounds produced by other organisms as food. Organisms not capable of producing their own food by photosynthesis.

High slack water   The period of time associated with the peak of high tide when there is no visible flow of water into or out of bays and rivers.

High tide zone   That portion of the littoral zone that lies between the lowest high tides and highest high tides that occur in an area. It is, on the average, exposed to desiccation for longer periods each day than it is covered by water.

High water (HW)   The highest level reached by the rising tide before it begins to recede.

Higher high water (HHW)   The higher of two high waters occurring during a tidal day where tides exhibit a mixed tidal pattern.

Higher low water (HLW)   The higher of two low waters occurring during a tidal day where tides exhibit a mixed tidal pattern.

Highly stratified estuary   A relatively deep estuary in which a significant volume of marine water enters as a subsurface flow. A large volume of freshwater stream input produces a widespread low surface-salinity condition that produces a well-developed halocline throughout most of the estuary.

Holoplankton   Organisms that spend their entire life as members of the plankton.

Homeothermic   Of or relating to an animal that maintains a precisely controlled internal body temperature using its own internal heating and cooling mechanisms; warm-blooded.

Horse latitudes   The latitude belts between 30° and 35° north and south latitude where winds are light and variable, since the principal movement of air masses at these latitudes is one of vertical descent. The climate is hot and dry, resulting in the creation of the major continental and maritime deserts of the world.

Hotspot   The relatively stationary surface expression of a persistent column of molten mantle material rising to the surface.

Hurricane   A tropical cyclone in which winds reach speeds in excess of 120 kilometers (74 miles) per hour. Generally applied to such storms in the North Atlantic Ocean, eastern North Pacific Ocean, Caribbean Sea, and Gulf of Mexico. Such storms in the western Pacific Ocean are called typhoons and those in the Indian ocean are known as cyclones.

Hydration   The condition of being surrounded by water molecules, as when sodium chloride dissolves in water.

Hydrocarbon   Any organic compound consisting only of hydrogen and carbon. Crude oil is a mixture of hydrocarbons.

Hydrocarbon seep biocommunity   Deep bottom-dwelling community of organisms associated with a hydrocarbon seep from the ocean floor. The community depends on methane and sulfur-oxidizing bacteria as producers. The bacteria may live free in the water, on the bottom, or symbiotically in the tissues of some of the animals.

Hydrogen bond   An intermolecular bond that forms within water because of the dipolar nature of water molecules.

Hydrogenous sediment   Sediment that forms from precipitation from ocean water or ion exchange between existing sediment and ocean water. Examples are manganese nodules, phosphorite, glauconite, phillipsite, and montmorillonite.

Hydrologic cycle   The cycle of water exchange among the atmosphere, land, and ocean through the processes of evaporation, precipitation, runoff, and subsurface percolation. Also called the water cycle.

Hydrothermal spring   Vents of hot water found primarily along the spreading axes of oceanic ridges and rises.

Hydrothermal vent   Ocean water that percolates down through fractures in recently formed ocean floor is heated by underlying magma and surfaces again through these vents. They are usually located near the axis of spreading along mid-ocean ridges.

Hydrothermal vent biocommunity   Deep bottom-dwelling community of organisms associated with a hydrothermal vent. The hot water vent is usually associated with the axis of a spreading center, and the community is dependent on sulfur-oxidizing bacteria that may live free in the water, on the bottom, or symbiotically in the tissue of some of the animals of the community.

Hypersaline   Waters that are highly or excessively saline.

Hypersaline lagoon   Shallow lagoons such as Laguna Madre, which may become hypersaline due to little tidal flushing and seasonal variability in freshwater input. High evaporation rates and low freshwater input can result in very high salinities.

Hypersaline seep   At the base of the Florida Escarpment there is a biocommunity that depends on methane and sulfur-oxidizing bacteria as producers. The bacteria may live free in the water, on the bottom, or symbiotically in the tissue of some of the animals within the biocommunity.

Hypersaline seep biocommunity   Bottom-dwelling community of organisms associated with a hypersaline seep.

Hypertonic   Pertaining to the property of an aqueous solution having a higher osmotic pressure (salinity) than another aqueous solution from which it is separated by a semipermeable membrane that will allow osmosis to occur. The hypertonic fluid will gain water molecules through the membrane from the other fluid.

Hypothesis   A tentative, testable statement about the general nature of the phenomenon observed.

Hypotonic   Pertaining to the property of an aqueous solution having a lower osmotic pressure (salinity) than another aqueous solution from which it is separated by a semipermeable membrane that will allow osmosis to occur. The hypotonic fluid will lose water molecules through the membrane to the other fluid.

Hypsographic curve   A curve that displays the relative elevations of the land surface and depths of the ocean.

Ice Age   The most recent glacial period, which occurred during the Pleistocene epoch.

Ice rafting   The movement of trapped sediment within or on top of ice by flotation.

Ice sheet   An extensive, relatively flat accumulation of ice.

Iceberg   A massive piece of glacier ice that has broken from the front of the glacier (calved) into a body of water. It floats with its tip at least 5 meters (16 feet) above the water's surface and at least four-fifths of its mass submerged.

Igneous rock   One of the three main classes into which all rocks are divided (igneous, metamorphic, and sedimentary). Rock that forms from the solidification of molten or partly molten material (magma).

In situ   In place; in situ density of a sample of water is its density at its original depth.

Incidental catch   See *bycatch.*

Inertia   Newton's first law of motion. It states that a body at rest will stay at rest and a body in motion will remain in uniform motion in a straight line unless acted on by some external force.

Infauna   Animals that live buried in the soft substrate (sand or mud).

Infrared radiation   Electromagnetic radiation lying between the wavelengths of 0.8 micrometers (0.000003 inch) and about 1000 micrometers (0.04 inch). It is bounded on the shorter wavelength side by the visible spectrum and on the long side by microwave radiation.

Inner sublittoral   Pertaining to the inner continental shelf, above the intersection with the euphotic zone, where attached plants grow.

Insolation   The rate at which solar radiation is received per unit of surface area at any point at or above Earth's surface.

Integrated Ocean Drilling Program (IODP)   A new drilling program that is scheduled to replace the Ocean Drilling Program in 2003 with a new drill ship that has riser technology, enabling cores to be collected from deeper within Earth's interior.

Interface   A surface separating two substances of different properties, such as density, salinity, or temperature. In oceanography, it usually refers to a separation of two layers of water with different densities caused by significant differences in temperature and/or salinity.

Interface wave   An orbital wave that moves along an interface between fluids of different density. An example is ocean surface waves moving along the interface between the atmosphere and the ocean, which is about one thousand times more dense than air.

Interference (wave)   The overlapping of different wave groups, either in phase (constructive interference, which results in larger waves), out of phase (destructive interference, which results in smaller waves), or some combination of the two (mixed interference).

Intermolecular bond   A relatively weak bond that forms between molecules of a given substance. The hydrogen bond and the van der Waals bonds are intermolecular bonds.

Internal wave   A wave that develops below the surface of a fluid, the density of which changes with increased depth. This change may be gradual or occur abruptly at an interface.

Intertidal zone   The ocean floor within the foreshore region that is covered by the highest normal tides and exposed by the lowest normal tides, including the water environment of tide pools within this region.

Intertropical Convergence Zone (ITCZ)   Zone where northeast trade winds and southeast trade winds converge. Averages about 5° north in the Pacific and Atlantic Oceans and 7° south in the Indian Ocean.

Intraplate feature   Any feature that occurs within a tectonic plate and not along a plate boundary.

Intrusive rocks   Igneous rocks such as granite that cool slowly beneath Earth's surface.

Invertebrate   Animal without a backbone.

Ion   An atom that becomes electrically charged by gaining or losing one or more electrons. The loss of electrons produces a positively charged cation, and the gain of electrons produces a negatively charged anion.

Ionic bond A chemical bond formed as a result of the electrical attraction.

Irminger Current A warm current that branches off from the Gulf Stream and moves up along the west coast of Iceland.

Iron hypothesis, the A hypothesis that states that an effective way of increasing productivity in the ocean is to fertilize the ocean by adding the only nutrient that appears to be lacking—iron. Adding iron to the ocean also increases the amount of carbon dioxide removed from the atmosphere.

Irons A meteorite consisting essentially of iron but may also contain up to 30% nickel.

Island arc A linear arrangement of islands, many of which are volcanic, usually curved so that the concave side faces a sea separating the islands from a continent. The convex side faces the open ocean and is bounded by a deep-ocean trench.

Island mass effect As surface current flows past an island, surface water is carried away from the island on the downcurrent side. This water is replaced in part by upwelling of water on the downcurrent side of the island.

Isohaline Of the same salinity.

Isopoda An order of dorsoventrally flattened crustaceans that are mostly scavengers or parasites on other crustaceans or fish.

Isostasy A condition of equilibrium, comparable to buoyancy, by which Earth's brittle crust floats on the plastic mantle.

Isostatic adjustment The adjustment of crustal material due to isostasy.

Isostatic rebound The upward movement of crustal material due to isostasy.

Isotherm A line connecting points of equal temperature.

Isopycnal Of the same density.

Isothermal Of the same temperature.

Isotonic Pertaining to the property of having equal osmotic pressure. If two such fluids were separated by a semipermeable membrane that will allow osmosis to occur, there would be no net transfer of water molecules across the membrane.

Isotope One of several atoms of an element that has a different number of neutrons, and therefore a different atomic mass, than the other atoms, or isotopes, of the element.

Jellyfish **1.** Free-swimming, umbrella-shaped medusoid members of the coelenterate class Scyphozoa. **2.** Also frequently applied to the medusoid forms of other coelenterates.

Jet stream An easterly moving air mass at an elevation of about 10 kilometers (6 miles). Moving at speeds that can exceed 300 kilometers (185 miles) per hour, the jet stream follows a wavy path in the midlatitudes and influences how far polar air masses may extend into the lower latitudes.

Jetty A structure built from the shore into a body of water to protect a harbor or a navigable passage from being closed off by the deposition of longshore drift material.

Juan de Fuca Ridge A divergent plate boundary off the Oregon-Washington coast.

K–T event An extinction event marked by the disappearance of the dinosaurs that occurred 65 million years ago at the boundary between the Cretaceous (K) and Tertiary (T) Periods of geologic time.

Kelp Large varieties of Phaeophyta (brown algae).

Kelp forest A extensive bed of various species of macroscopic brown algae that provides a habitat for many other types of marine organisms.

Key A low, flat island composed of sand or coral debris that accumulates on a reef flat.

Kinetic energy Energy of motion. It increases as the mass or speed of the object in motion increases.

Knot (kt) Unit of speed equal to 1 nautical mile per hour, approximately 1.15 statute (land) miles per hour.

Krill A common name frequently applied to members of crustacean order Euphausiacea (euphausids).

Kuroshio Current The warm western boundary current of the North Pacific subtropical gyre.

Kyoto Protocol An agreement among 60 nations to voluntarily limit greenhouse gas emissions, signed in 1997 in Kyoto, Japan.

La Niña An event where the surface temperature in the waters of the eastern South Pacific fall below average values. It usually occurs at the end of an El Niño-Southern Oscillation event.

Labrador Current A cold current flowing south along the coast of Labrador in the northwest Atlantic Ocean.

Lagoon A shallow stretch of seawater partly or completely separated from the open ocean by an elongate narrow strip of land such as a reef or barrier island.

Laguna Madre A hypersaline lagoon located landward of Padre Island along the south Texas coast.

Laminar flow Manner in which a fluid flows in parallel layers or sheets such that the direction of flow at any point does not change with time. Also known as nonturbulent flow.

Land breeze The seaward flow of air from the land caused by differential cooling of Earth's surface.

Langmuir circulation A cellular circulation set up by winds that blow consistently in one direction with speeds in excess of 12 kilometers (7.5 miles) per hour. Helical spirals running parallel to the wind direction are alternately clockwise and counterclockwise.

Larva An embryo that has a different form before it assumes the characteristics of the adult of the species.

Latent heat The quantity of heat gained or lost per unit of mass as a substance undergoes a change of state (such as liquid to solid) at a given temperature and pressure.

Latent heat of condensation The heat energy that must be removed from one gram of a substance to convert it from a vapor at a given temperature below its boiling point. For water, it is 585 calories at 20°C.

Latent heat of freezing The heat energy that must be removed from one gram of a substance at its melting point to convert it to a solid. For water, it is 80 calories.

Latent heat of evaporation The heat energy that must be added to one gram of a liquid substance to convert it to a vapor at a given temperature below its boiling point. For water, it is 585 calories at 20°C.

Latent heat of melting The heat energy that must be added to one gram of a substance at its melting point to convert it to a liquid. For water, it is 80 calories.

Latent heat of vaporization The heat energy that must be added to one gram of a substance at its boiling point to convert it to a vapor. For water, it is 540 calories.

Lateral line system A sensory system running down both sides of fishes to sense subsonic pressure waves transmitted through ocean water.

Latitude Location on Earth's surface based on angular distance north or south of the Equator. Equator = 0°; North Pole = 90° north; South Pole = 90° south.

Laurasia An ancient landmass of the Northern Hemisphere. The name is derived from Laurentia, pertaining to the Canadian Shield of North America, and Eurasia, of which it was composed.

Lava Fluid magma coming from an opening in Earth's surface, or the same material after it solidifies.

Law of gravitation See *gravitational force*.

Leeuwin Current A warm current that flows south out of the East Indies along the western coast of Australia.

Leeward Direction toward which the wind is blowing or waves are moving.

Levee **1.** Natural levees are low ridges on either side of river channels that result from deposition during flooding. **2.** Artificial levees are built by human beings.

Light Electromagnetic radiation that has a wavelength in the range from about 4,000 (violet) to about 7,700 (red) angstroms and may be perceived by the normal unaided human eye.

Light-year The distance traveled by light during one year at a speed of 300,000 kilometers (186,000 miles) per second. It equals 9.8 trillion kilometers (6.2 trillion miles).

Limestone A class of sedimentary rocks composed of at least 80% carbonates of calcium or magnesium. Limestones may be either biogenous or hydrogenous.

Limpet A mollusk of the class Gastropoda that possesses a low conical shell that exhibits no spiraling in the adult form.

LIMPET 500 The world's first commercial wave power plant that can generate up to 500 kilowatts of power. It is located on Islay, a small island off the west coast of Scotland, and began generating electricity in November 2000.

Liquid state A state of matter in which a substance has a fixed volume but no fixed shape.

Lithify Process by which sediment becomes hardened into sedimentary rock.

Lithogenous sediment Sediment composed of mineral grains derived from the weathering of rock material and transported to the ocean by various mechanisms of transport, including running water, gravity, the movement of ice, and wind.

Lithosphere The outer layer of Earth's structure, including the crust and the upper mantle to a depth of about 200 kilometers (124 miles). Lithospheric plates are the major components involved in plate tectonic movement.

Lithothamnion ridge A feature common to the windward edge of a reef structure, characterized by the presence of the red algae, *Lithothamnion*.

Littoral zone The benthic zone between the highest and lowest spring tide shorelines; the intertidal zone.

Lobster Large marine crustacean considered a delicacy. *Homarus americanus* (American lobster) possesses two large chelae (pincers) and is found off the New England coast. *Panulirus sp.* (spiny lobsters or rock lobsters) have no chelae but possess long, spiny antennae effective in warding off predators. *P. argue* is found off the coast of Florida and in the West Indies, whereas *P. interruptus* is common along the coast of southern California.

Longitude Location on Earth's surface based on angular distance east or west of the Prime (Greenwich) Meridian (0° longitude). 180° longitude is the International Date Line.

Longitudinal wave A wave phenomenon when particle vibration is parallel to the direction of energy propagation.

Longshore bar A deposit of sediment that forms parallel to the coast within or just beyond the surf zone.

Longshore current A current located in the surf zone and running parallel to the shore as a result of waves breaking at an angle to the shore.

Longshore drift The load of sediment transported along the beach from the breaker zone to the top of the swash line in association with the longshore current. Also called longshore transport or littoral drift.

Longshore trough A low area of the beach that separates the beach face from the longshore bar.

Lophophore Horseshoe-shaped feeding structure bearing ciliated tentacles characteristic of the phyla Bryozoa, Brachiopoda, and Phoronidea.

Low slack water The period of time associated with the peak of low tide when there is no visible flow of water into or out of bays and rivers.

Low tide terrace See *beach face*.

Low tide zone That portion of the intertidal zone that lies between the lowest low-tide shoreline and the highest low-tide shoreline.

Low water (LW) The lowest level reached by the water surface at low tide before the rise toward high tide begins.

Lower high water (LHW) The lower of two high waters occurring during a tidal day where tides exhibit a mixed tidal pattern.

Lower low water (LLW) The lower of two low waters occurring during a tidal day where tides exhibit a mixed tidal pattern.

Lunar day The time interval between two successive transits of the Moon over a meridian, approximately 24 hours and 50 minutes of solar time.

Lunar hour One-twenty-fourth of a lunar day; about 62.1 minutes.

Lunar tide The part of the tide caused solely by the tide-producing force of the Moon.

Lunger Fish such as groupers that sit motionless on the ocean floor waiting for prey to appear. A quick burst of speed over a short distance is used to capture prey.

Lysocline The level in the ocean at which calcium carbonate begins to dissolve, typically at a depth of about 4000 meters (13,100 feet). Below the lysocline, calcium carbonate dissolves at an increasing rate with increasing depth until the calcite compensation depth (CCD) is reached.

Macroplankton Plankton larger than 2 centimeters (0.8 inch) in their smallest dimension.

Magma Fluid rock material from which igneous rock is derived through solidification.

Magnetic anomaly Distortion of the regular pattern of Earth's magnetic field resulting from the various magnetic properties of local concentrations of ferromagnetic minerals in Earth's crust.

Magnetic dip The dip of magnetite particles in rock units of Earth's crust relative to sea level. It is approximately equivalent to latitude. Also called magnetic inclination.

Magnetic field A condition found in the region around a magnet or an electric current, characterized by the existence of a detectable magnetic force at every point in the region and by the existence of magnetic poles.

Magnetic inclination See *magnetic dip*.

Magnetite The mineral form of black iron oxide, Fe_3O_4, that often occurs with magnesium, zinc, and manganese and is an important ore of iron.

Magnetometer A device used for measuring the magnetic field of Earth.

Manganese nodules Concretionary lumps containing oxides of manganese, iron, copper, cobalt, and nickel found scattered over the ocean floor.

Mangrove swamp A marshlike environment dominated by mangrove trees. They are restricted to latitudes below 30°.

Mantle The relatively plastic zone rich in ferromagnesian minerals between the core and crust of Earth.

Mantle plume A rising column of molten magma from Earth's mantle.

Marginal sea A semienclosed body of water adjacent to a continent and floored by submerged continental crust.

Mariculture The application of the principles of agriculture to the production of marine organisms.

Marine Mammals Protection Act An act by U.S. Congress in 1972 that specifies rules to protect marine mammals in U.S. waters.

Marine terrace Wave-cut benches that have been exposed above sea level by a drop in sea level.

Marsh An area of soft, wet, flat land that is periodically flooded by salt water and common in portions of lagoons.

Maturity A texture of lithogenous sediment, where increasing maturity (caused by increased time of transport) is indicated by decreased clay content, increased sorting, and increased rounding of the grains within the deposit.

Maximum sustainable yield The maximum fishery biomass that can be removed yearly and still be sustained by the fishery ecosystem.

Mean high water (MHW) The average height of all the high waters occurring over a 19-year-period.

Mean low water (MLW) The average height of the low waters occurring over a 19-year period.

Mean sea level (MSL) The mean surface water level determined by averaging all stages of the tide over a 19-year period, usually determined from hourly height observations along an open coast.

Mean tidal range The difference between mean high water and mean low water.

Meander A sinuous curve, bend, or turn in the course of a current.

Mechanical energy Energy manifested as work being done; the movement of a mass some distance.

Mediterranean circulation Circulation characteristic of bodies of water with restricted circulation with the ocean that results from an excess of evaporation as compared to precipitation and runoff similar to the Mediterranean Sea. Surface flow is into the restricted body of water with a subsurface counterflow as exists between the Mediterranean Sea and the Atlantic Ocean.

Medusa Free-swimming, bell-shaped coelenterate body form with a mouth at the end of a central projection and tentacles around the periphery.

Meiofauna Small species of animals that live in the spaces among particles in a marine sediment.

Melon A fatty organ located forward of the blowhole on certain odontocete cetaceans that is used to focus echolocation sounds.

Melting point The temperature at which a solid substance changes to the liquid state.

Mercury A silvery-white poisonous metallic element, liquid at room temperature and used in thermometers, barometers, vapor lamps, and batteries and in the preparation of chemical pesticides.

Meridian of longitude Great circles running through the North and South Poles.

Meroplankton Planktonic larval forms of organisms that are members of the benthos or nekton as adults.

Mesopelagic zone That portion of the oceanic province 200 to 1000 meters (660 to 3300 feet) deep. Corresponds approximately with the disphotic (twilight) zone.

Mesosaurus An extinct presumably aquatic reptile that lived about 250 million years ago. The distribution of its fossil remains helped support plate tectonic theory.

Mesosphere The middle region of Earth below the asthenosphere and above the core.

Metal sulfide A compound containing one or more metals and sulfur.

Metamorphic rock Rock that has undergone recrystallization while in the solid state in response to changes of temperature, pressure, and chemical environment.

Meteor A bright trail or streak that appears in the sky when a meteoroid is heated to incandescence by friction with Earth's atmosphere. Also called a falling or shooting star.

Meteorite A stony or metallic mass of matter that has fallen to Earth's surface from outer space.

Methane hydrate A white compact icy solid made of water and methane. The most common type of gas hydrate.

Microplankton Net plankton. Plankton not easily seen by the unaided eye, but easily recovered from the ocean with the aid of a silk-mesh plankton net.

Mid-Atlantic Ridge A slow-spreading divergent plate boundary running north-south and bisecting the Atlantic Ocean.

Mid-ocean ridge A linear, volcanic mountain range that extends through all the major oceans, rising 1 to 3 kilometers (0.6 to 2 miles) above the deep-ocean basins. Averaging 1500 kilometers (930 miles) in width, rift valleys are common along the central axis. Source of new oceanic crustal material.

Middle tide zone That portion of the intertidal zone that lies between the highest low-tide shoreline and the lowest high-tide shoreline.

Migration Long journeys undertaken by many marine species for the purpose of successful feeding and reproduction.

Minamata Bay, Japan The site of the occurrence of human poisoning in the 1950s by mercury contained in marine organisms that were consumed by victims.

Minamata disease A degenerative neurological disorder caused by poisoning with a mercury compound found in seafood obtained from waters contaminated with mercury-containing industrial waste.

Mineral An inorganic substance occurring naturally on Earth and having distinctive physical properties and a chemical composition that can be expressed by a chemical formula. The term is also sometimes applied to organic substances such as coal and petroleum.

Mixed interference A pattern of wave interference in which there is a combination of constructive and destructive interference.

Mixed surface layer The surface layer of the ocean water mixed by wave and tide motions to produce relatively isothermal and isohaline conditions.

Mixed tidal pattern A tidal pattern exhibiting two high tides and two low tides per tidal day with a marked diurnal inequality. Coastal locations that experience such a tidal pattern may also show alternating periods of diurnal and semidiurnal patterns. Also called mixed semidiurnal.

Mixotroph An organism that depends on a combination of autotrophic and heterotrophic behavior to meet its energy requirements. Many coral reef species exhibit such behavior.

Mohorovičić discontinuity (Moho) A sharp compositional discontinuity between the crust and mantle of Earth. It may be as shallow as 5 kilometers (3 miles) below the ocean floor or as deep as 60 kilometers (37 miles) beneath some continental mountain ranges.

Molecular motion Molecules move in three ways: vibration, rotation, and translation.

Molecule A group of two or more atoms bound together by ionic or covalent bonds.

Mollusca Phylum of soft, unsegmented animals usually protected by a calcareous shell and having a muscular foot for locomotion. Includes snails, clams, chitons, and octopi.

Monera Kingdom of organisms that do not have nuclear material confined within a sheath but spread throughout the cell. Includes bacteria, blue-green algae, and Archaea.

Mononodal Pertaining to a standing wave with only one nodal point or nodal line.

Monsoon A name for seasonal winds derived from the Arabic word for season, *mausim*. The term was originally applied to winds over the Arabian Sea that blow from the southwest during summer and the northeast during winter.

Moraine A deposit of unsorted material deposited at the margins of glaciers. Many such deposits have become important economically as fishing banks after being submerged by the rising level of the ocean.

Mud Sediment consisting primarily of silt and clay-sized particles smaller than 0.06 millimeters (0.002 inch).

Mutualism A symbiotic relationship in which both participants benefit.

Myoglobin A red, oxygen-storing pigment found in muscle tissue.

Myomere A muscle fiber.

Mysticeti The baleen whales.

Nadir The point on the celestial sphere directly opposite the zenith and directly beneath the observer.

Nanoplankton Plankton less than 50 micrometers (0.002 inch) in length that cannot be captured in a plankton net and must be removed from the water by centrifuge or special microfilters.

Nansen bottle A device used by oceanographers to obtain samples of ocean water from beneath the surface.

National Flood Insurance Program (NFIP) A program financed by the U.S. government that was intended to prevent costly federal aid after a natural disaster but instead encourages building in risk-prone areas.

Natural selection The process in nature by which only the organisms best adapted to their environment tend to survive and transmit their genetic characters in increasing numbers to succeeding generations while those less adapted tend to be eliminated.

Nauplius   A microscopic free-swimming larval stage of crustaceans such as copepods, ostracodes, and decapods. Typically has three pairs of appendages.

Neap tide   Tides of minimal range occurring about every two weeks when the Moon is in either first or third quarter moon phase.

Nearshore   The zone of a beach that extends from the low tide shoreline seaward to where breakers begin forming.

Nebula   A diffuse mass of interstellar dust and/or gas.

Nebular hypothesis   A model that describes the formation of the solar system by contraction of a nebula.

Nektobenthos   Those members of the benthos that can actively swim and spend much time off the bottom.

Nekton   Pelagic animals such as adult squids, fish, and mammals that are active swimmers to the extent that they can determine their position in the ocean by swimming.

Nematath   A linear chain of islands and/or seamounts that are progressively older in one direction. It is created by the passage of a lithospheric plate over a hotspot.

Nematocyst   The stinging mechanism found within the cnidoblast of members of the phylum Cnidaria (Coelenterata).

Neritic province   That portion of the pelagic environment from the shoreline to where the depth reaches 200 meters (660 feet).

Neritic sediment   That sediment composed primarily of lithogenous particles and deposited relatively rapidly on the continental shelf, continental slope, and continental rise.

Net primary production   The primary production of producers after they have removed what is needed for their metabolism.

Neutral   A state in which there is no excess of either the hydrogen or the hydroxide ion.

Neutron   An electrically neutral subatomic particle found in the nucleus of atoms that has a mass approximately equivalent to that of a proton.

New moon   The phase of the Moon that occurs when the Sun and the Moon are in conjunction; that is, they are both on the same side of Earth. During this time, the dark side of the Moon faces Earth.

New production   Primary production supported by nutrients supplied from outside the immediate ecosystem by upwelling or other physical transport.

Niche   The ecological role of an organism and its position in the ecosystem.

Niigata, Japan   The site of mercury poisoning of humans in the 1960s by ingestion of contaminated seafood.

Node   The point on a standing wave where vertical motion is lacking or minimal. If this condition extends across the surface of an oscillating body of water, the line of no vertical motion is a nodal line.

Non-point-source pollution   Any type of pollution entering the surface water system from sources other than underwater pipelines; also called "poison runoff."

North Atlantic Current   The northernmost current of the North Atlantic current gyre.

North Atlantic Deep Water   A deep-water mass that forms primarily at the surface of the Norwegian Sea and moves south along the floor of the North Atlantic Ocean.

North Pacific Current   The northernmost current of the North Pacific current gyre.

Northeast Monsoon   A northeast wind that blows off the Asian mainland onto the Indian Ocean during the winter season.

Norwegian Current   A warm current that branches off from the Gulf Stream and flows into the Norwegian Sea between Iceland and the British Isles.

Nucleus   The positively charged central region of an atom, composed of protons and neutrons and containing almost all of the mass of the atom.

Nudibranch   Sea slug. A member of the mollusk class Gastropoda that has no protective covering as an adult. Respiration is carried on by gills or other projections on the dorsal surface.

Nutrients   Any number of organic or inorganic compounds used by primary producers. Nitrogen and phosphorus compounds are important examples.

Observation   Occurrences that can be measured with one's senses.

Ocean   The entire body of salt water that covers more than 71% of Earth's surface.

Ocean acoustical tomography   A method by which changes in water temperature may be determined by changes in the speed of transmission of sound. It has the potential to help map ocean circulation patterns over large ocean areas.

Ocean beach   The beach on the open-ocean side of a barrier island.

Ocean current   A mass of oceanic water that flows from one place to another.

Ocean Drilling Program (ODP)   In 1983, this program replaced the Deep Sea Drilling Project and focused on drilling the continental margins using the drill ship JOIDES *Resolution*.

Ocean thermal energy conversion (OTEC)   A technique of generating energy by using the difference in temperature between surface waters and deep waters in low latitude regions.

Oceanic common water   Deep water found in Pacific and Indian Oceans as a result of mixing of Antarctic Bottom Water and North Atlantic Deep Water.

Oceanic crust   A mass of rock with a basaltic composition that is about 5 kilometers (3 miles) thick.

Oceanic province   That division of the pelagic environment where the water depth is greater than 200 meters (660 feet).

Oceanic ridge   A portion of the global mid-ocean ridge system that is characterized by slow spreading and steep slopes.

Oceanic rise   A portion of the global mid-ocean ridge system that is characterized by fast spreading and gentle slopes

Oceanic sediment   The inorganic abyssal clays and the organic oozes that accumulate slowly on the deep-ocean floor.

Oceanic spreading center   The axes of oceanic ridges and rises that are the locations at which new lithosphere is added to lithospheric plates. The plates move away from these axes in the process of seafloor spreading.

Ocelli   Light-sensitive organ around the base of many medusoid bells.

Odontoceti   Toothed whales.

Offset   Separation that occurs due to movement along a fault.

Offshore   The comparatively flat submerged zone of variable width extending from the breaker line to the edge of the continental shelf.

Oligotrophic   Areas such as the midsubtropical gyres where there are low levels of biological production.

Omnivore   An animal that feeds on both plants and animals.

Oolite   A deposit formed of small spheres from 0.25 to 2 millimeters (0.01 to 0.08 inch) in diameter. They are usually composed of concentric layers of calcite.

Ooze   A pelagic sediment containing at least 30% skeletal remains of pelagic organisms, the balance being clay minerals. Oozes are further defined by the chemical composition of the organic remains (siliceous or calcareous) and by their characteristic organisms (ex: diatomaceous ooze, foraminiferan ooze).

Opal   An amorphous form of silica ($SiO_2 \cdot nH_2O$) that usually contains from 3 to 9% water. It forms the shells of radiolarians and diatoms.

Opposition   The separation of two heavenly bodies by 180° relative to Earth. The Sun and Moon are in opposition during the full moon phase.

Orbital wave   A wave phenomenon in which energy is moved along the interface between fluids of different densities. The wave form is propagated by the movement of fluid particles in orbital paths.

Orthogonal lines   Lines constructed perpendicular to wave fronts and spaced so that the energy between lines is equal at all times. Orthogonals are used to help determine how energy is distributed along the shoreline by breaking waves.

Orthophosphate Phosphoric oxide (P_2O_5) can combine with water to produce orthophosphates ($3H_2O \cdot P_2O_5$ or H_3PO_4) that may be used as nutrients by photosynthetic organisms.

Osmosis Passage of water molecules through a semipermeable membrane separating two aqueous solutions of different solute concentration. The water molecules pass from the solution of lower solute concentration into the other.

Osmotic pressure A measure of the tendency for osmosis to occur. It is the pressure that must be applied to the more concentrated solution to prevent the passage of water molecules into it from the less concentrated solution.

Osmotic regulation Physical and biological processes used by organisms to counteract the osmotic effects of differences in osmotic pressures of their body fluids and the water in which they live.

Ostracoda An order of crustaceans that are minute and compressed within a bivalve shell.

Otocyst Gravity-sensitive organs around the bell of a medusa.

Outer sublittoral zone The continental shelf below the intersection with the euphotic zone where no plants grow attached to the bottom.

Outgassing The process by which gases are removed from within Earth's interior.

Overfishing Occurs when adult fish in a fishery are harvested faster than their natural rate of reproduction.

Oxygen compensation depth The depth in the ocean at which marine plants receive just enough solar radiation to meet their basic metabolic needs. It marks the base of the euphotic zone.

Oxygen minimum layer (OML) A zone of low dissolved oxygen concentration that occurs at a depth of about 700 to 1000 meters (2300 to 3280 feet).

Pacific Decadal Oscillation (PDO) A natural oscillation in the Pacific Ocean that lasts 20 to 30 years and appears to influence sea surface temperatures.

Pacific Ring of Fire An extensive zone of volcanic and seismic activity that coincides roughly with the borders of the Pacific Ocean.

Pacific-type margin See *active margin*.

Paleoceanography The study of the physical and biological changes of the oceans brought about by the changing shapes and positions of the continents.

Paleogeography The study of the historical changes of shapes and positions of the continents and oceans.

Paleomagnetism The study of Earth's ancient magnetic field.

Pangaea An ancient supercontinent of the geologic past that contained all Earth's continents.

Panthalassa A large, ancient ocean that surrounded Pangaea.

Paralytic shellfish poisoning (PSP) Paralysis resulting from poisoning caused by eating shellfish contaminated with the toxic dinoflagellate *Gonyaulax*.

Parasitism A symbiotic relationship between two organisms in which one benefits at the expense of the other.

Parts per thousand (‰) A unit of measurement used in reporting salinity of water equal to the number of grams of dissolved substances in 1000 grams of water. One ‰ is equivalent to .1% or 1000 ppm.

Passive margin A continental margin that lacks a plate boundary and is marked by a low degree of tectonic activity, such as those typical of the Atlantic Ocean.

PCBs A group of industrial chemicals used in a variety of products; responsible for several episodes of ecological damage in coastal waters.

Peat deposit Partially carbonized organic matter found in bogs and marshes that can be used as fertilizer and fuel.

Pelagic environment The open ocean environment, which is divided into the neritic province (water depth 0 to 200 meters or 656 feet) and the oceanic province (water depth greater than 200 meters or 656 feet).

Pelecypoda A class of mollusks characterized by two more or less symmetrical lateral valves with a dorsal hinge. These filter feeders pump water through the filter system and over gills through posterior siphons. Many possess a hatchet-shaped foot used for locomotion and burrowing. Includes clams, oysters, mussels, and scallops.

Perigee The point on the orbit of an Earth satellite (Moon) that is nearest Earth.

Perihelion That point on the orbit of a planet or comet around the Sun that is closest to the Sun.

Permeability Capacity of a porous rock or sediment for transmitting fluid.

Peru Current The cold eastern boundary current of the South Pacific subtropical gyre.

Petroleum A naturally occurring liquid hydrocarbon.

Pfiesteria piscicida A species of toxic dinoflagellate that has been known to cause fish kills.

pH scale A measure of the acidity or alkalinity of a solution, numerically equal to 7 for neutral solutions, increasing with increasing alkalinity and decreasing with increasing acidity. The pH scale commonly in use ranges from 0 to 14.

Phaeophyta Brown algae characterized by the carotinoid pigment fucoxanthin. Contains the largest members of the marine algal community.

Phosphate Any of a number of phosphorous-bearing compounds.

Phosphorite A sedimentary rock composed primarily of phosphate minerals.

Photic zone The upper ocean in which the presence of solar radiation is detectable. It includes the euphotic and disphotic zones.

Photophore One of several types of light-producing organs found primarily on fishes and squids inhabiting the mesopelagic and upper bathypelagic zones.

Photosynthesis The process by which plants and algae produce carbohydrates from carbon dioxide and water in the presence of chlorophyll, using light energy and releasing oxygen.

Phycoerythrin A red pigment characteristic of the Rhodophyta (red algae).

Phytoplankton Algal plankton. The most important community of primary producers in the ocean.

Picoplankton Small plankton within the size range of 0.2 to 2.0 micrometers (0.000008 to 0.00008 inch) in size. Composed primarily of bacteria.

Pillow basalt A basalt exhibiting pillow structure. See *pillow lava*.

Pillow lava A general term for those lavas displaying discontinuous pillow-shaped masses (pillow structure) caused by the rapid cooling of lava as a result of underwater eruption of lava or lava flowing into water.

Ping A sharp, high-pitched sound made by the transmitting device of many sonar systems.

Pinniped A group of marine mammals that have prominent flippers; includes the sea lions/fur seals, seals, and walruses.

Plankter Informal term for plankton.

Plankton Passively drifting or weakly swimming organisms that are not independent of currents. Includes mostly microscopic algae, protozoa, and larval forms of higher animals.

Plankton bloom A very high concentration of phytoplankton, resulting from a rapid rate of reproduction as conditions become optimal during the spring in high latitude areas. Less obvious causes produce blooms that may be destructive in other areas.

Plankton net Plankton-extracting device that is cone-shaped and typically of a silk material. It is towed through the water or lifted vertically to extract plankton down to a size of 50 micrometers (0.0002 inch).

Plantae Kingdom of many-celled plants.

Plastic 1. Capable of being shaped or formed. 2. Composed of plastic or plastics.

Plate tectonics Global dynamics having to do with the movement of a small number of semirigid sections of Earth's crust, with seismic activity and volcanism occurring primarily at the margins of these sections. This movement has resulted in changes in the geographic positions of continents and the shape and size of ocean basins and continents.

Plume　A rising column of molten mantle material that is associated with a hotspot when it penetrates Earth's crust.

Plunging breaker　Impressive curling breakers that form on moderately sloping beaches.

Pneumatic duct　An opening into the swim bladder of some fishes that allows rapid release of air into the esophagus.

Poikilotherm　An organism whose body temperature varies with and is largely controlled by its environment.

Polar　Pertaining to the polar regions.

Polar Cell　The large atmospheric circulation cell that occurs between 60° and 90° latitude in each hemisphere.

Polar easterly winds　Cold air masses that move away from the polar regions toward lower latitudes.

Polar emergence　The emergence of low- and mid-latitude temperature-sensitive deep-ocean benthos onto the shallow shelves of the polar regions where temperatures similar to that of their deep-ocean habitat exist.

Polar front　The boundary between the prevailing westerly and polar easterly wind belts that is centered at about 60° latitude in each hemisphere.

Polar high　The region of high atmospheric pressure that surrounds the poles.

Polar wandering curve　A curve that shows the change in the position of a pole through time.

Polarity　Intrinsic polar separation, alignment, or orientation, especially of a physical property (such as magnetic or electrical polarity).

Pollution (marine)　The introduction of substances that result in harm to the living resources of the ocean or humans who use these resources.

Polychaeta　Class of annelid worms that includes most of the marine segmented worms.

Polyp　A single individual of a colony or a solitary attached coelenterate.

Population　A group of individuals of one species living in an area.

Porifera　Phylum of sponges. Supporting structure composed of $CaCO_3$ or SiO_2 spicules or fibrous spongin. Water currents created by flagella-waving choanocytes enter tiny pores, pass through canals, and exit through a larger osculum.

Porosity　The ratio of the volume of all the empty spaces in a material to the volume of the whole.

Precession　Describes the change in the attitude of the Moon's orbit around Earth as it slowly changes its direction. The cycle is completed every 18.6 years and is accompanied by a clockwise rotation of the plane of the Moon's orbit that is completed in the same time interval.

Precipitate　To cause a solid substance to be separated from a solution, usually due to a change in physical or chemical conditions.

Precipitation　In a meteorological sense, the discharge of water in the form of rain, snow, hail, or sleet from the atmosphere onto Earth's surface.

Prevailing westerlies　The air masses moving from subtropical high pressure belts toward the polar front. They are southwesterly in the Northern Hemisphere and northwesterly in the Southern Hemisphere.

Primary productivity　The amount of organic matter synthesized by organisms from inorganic substances within a given volume of water or habitat in a unit of time.

Prime meridian　The meridian of longitude 0° used as a reference for measuring longitude that passes through the Royal Observatory at Greenwich, England. Also known as the Greenwich Meridian.

Producer　The autotrophic component of an ecosystem that produces the food that supports the biocommunity.

Productivity　See *primary productivity.*

Progressive wave　A wave in which the waveform progressively moves.

Propagation　The transmission of energy through a medium.

Protein　A very complex organic compound made up of large numbers of amino acids. Proteins make up a large percentage of the dry weight of all living organisms.

Protoctista　A kingdom of organisms that includes any of the unicellular eukaryotic organisms and their descendant multicellular organisms. Includes single- and multi-celled marine algae as well as single-celled animals called protozoa.

Protoearth　The young, early-developing Earth.

Proton　A positively charged subatomic particle found in the nucleus of atoms that has a mass approximately equivalent to that of a neutron.

Protoplanet　Any planet that is in its early stages of development.

Protoplasm　The self-perpetuating living material making up all organisms, mostly consisting of the elements carbon, hydrogen, and oxygen combined into various chemical forms.

Protozoa　Phylum of one-celled animals with nuclear material confined within a nuclear sheath.

Proxigean　A tidal condition of extremely large tidal range that occurs when spring tides coincide with perigee. Also called "closest of the close moon" tides.

Pseudopodia　An extension of protoplasm in a broad, flat, or long needlelike projection used for locomotion or feeding. Typical of amoeboid forms such as Foraminiferans and Radiolarians.

Pteropoda　An order of pelagic gastropods in which the foot is modified for swimming and the shell may be present or absent.

Purse seine net　A style of large fishing net that resembles a purse. It is set around a grouping of organisms (such as tuna) and the bottom is drawn tight to capture the organisms.

Pycnocline　A layer of water in which a high rate of change in density in the vertical dimension is present.

Pyrrophyta　A phyla of dinoflagellates that possess flagella for locomotion.

Quadrature　The state of the Moon during the first and third quarter moon phases (at right angles to one another relative to Earth).

Quarter moon　First and third quarter moon phases, which occur when the Moon is in quadrature about one week after the new moon and full moon phases, respectively. The third quarter moon phase is also known as the last quarter moon phase.

Quartz　A very hard mineral composed of silica, SiO_2.

Radiata　A grouping of phyla with primary radial symmetry—phyla Coelenterata and Ctenophora.

Radioactivity　The spontaneous breakdown of the nucleus of an atom resulting in the emission of radiant energy in the form of particles or waves.

Radiolaria　An order of planktonic and benthic protozoans that possess protective coverings usually made of silica.

Radiometric age dating　The use of half-lives to determine the age of rock units in years before present, usually accurate to within two or three percent.

Ray　A cartilaginous fish in which the body is dorsoventrally flattened, eyes and spiracles are on the upper surface, and gill slits are on the bottom. The tail is reduced to a whiplike appendage. Includes electric rays, manta rays, and stingrays.

Recreational beach　The area of a beach above shoreline, including the berm, berm crest, and the exposed part of the beach face.

Red clay　See *abyssal clay.*

Red muscle fiber　Fine muscle fibers rich in myoglobin that are abundant in cruiser-type fishes.

Red tide　A reddish-brown discoloration of surface water, usually in coastal areas, caused by high concentrations of microscopic organisms, usually dinoflagellates. It probably results from increased availability of certain nutrients. Toxins produced by the dinoflagellates may kill fish directly; decaying plant and animal remains or large populations of animals that migrate to the area of abundant plants may also deplete the surface waters of oxygen and cause asphyxiation of many animals.

Reef A strip or ridge of rocks, sand, coral, or man-made objects that rises to or near the surface of the ocean and creates a navigational hazard.

Reef flat A platform of coral fragments and sand on the lagoonal side of a reef that is relatively exposed at low tide.

Reef front The upper seaward face of a reef from the reef edge (seaward margin of reef flat) to the depth at which living coral and coralline algae become rare, 16 to 30 meters (50 to 100 feet).

Reflection The process in which a wave has part of its energy returned seaward by a reflecting surface.

Refraction The process by which the part of a wave in shallow water is slowed down, causing the wave to bend and align itself nearly parallel to underwater contours.

Regenerated production The portion of gross primary production that is supported by nutrients recycled within an ecosystem.

Relict beach A beach deposit laid down and submerged by a rise in sea level. It is still identifiable on the continental shelf, indicating that no deposition is presently taking place at that location on the shelf.

Relict sediment A sediment deposited under a set of environmental conditions that still remains unchanged although the environment has changed, and it remains unburied by later sediment. An example is a beach deposited near the edge of the continental shelf when sea level was lower.

Relocation The strategy of moving a structure that is threatened by being claimed by the sea.

Residence time The average length of time a particle of any substance spends in the ocean. It is calculated by dividing the total amount of the substance in the ocean by the rate of its introduction into the ocean or the rate at which it leaves the ocean.

Respiration The process by which organisms use organic materials (food) as a source of energy. As the energy is released, oxygen is used and carbon dioxide and water are produced.

Restoring force A force such as surface tension or gravity that tends to restore the ocean surface displaced by a wave to that of a still water level.

Resultant force The difference between the provided gravitational force of various bodies and the required centripetal force on Earth. The horizontal component of the resultant force is the tide-producing force.

Reverse osmosis A method of desalinating ocean water that involves forcing water molecules through a water-permeable membrane under pressure.

Reversing current The tide current as it occurs at the margins of landmasses. The water flows in and out for approximately equal periods of time separated by slack water where the water is still at high and low tidal extremes.

Rhodophyta Phylum of algae composed primarily of small encrusting, branching, or filamentous plants that receive their characteristic red color from the presence of the pigment phycoerythrin. With a worldwide distribution, they are found at greater depths than other algae.

Rift valley A deep fracture or break, about 25 to 50 kilometers (15 to 30 miles) wide, extending along the crest of a mid-ocean ridge.

Rifting The movement of two plates in opposite directions such as along a divergent boundary.

Right whales Surface-feeding baleen whales of the family Balaenidae that were the favorite targets of early whalers.

Rip current A strong narrow surface or near-surface current of short duration and high speed flowing seaward through the breaker zone at nearly right angles to the shore. It represents the return to the ocean of water that has been piled up on the shore by incoming waves.

Rip-rap Any large blocky material used to armor coastal structures.

Rogue wave An unusually large wave that usually occurs unexpectedly amid other waves of smaller size. Also known as a superwave.

Rorqual whales Large baleen whales with prominent ventral groves (rorqual folds) of the family Balaenopteridae: the minke, Baird's, Bryde's, sei, fin, blue, and humpback whales.

Rotary current Tidal current as observed in the open ocean. The tidal crest makes one complete rotation during a tidal period.

Salinity A measure of the quantity of dissolved solids in ocean water. Formally, it is the total amount of dissolved solids in ocean water in parts per thousand (‰) by weight after all carbonate has been converted to oxide, the bromide and iodide to chloride, and all the organic matter oxidized. It is normally computed from conductivity, refractive index, or chlorinity.

Salinometer An instrument that is used to determine the salinity of seawater by measuring its electrical conductivity.

Salpa Genus of pelagic tunicates that are cylindrical, transparent, and found in all oceans.

Salt Any substance that yields ions other than hydrogen or hydroxyl. Salts are produced from acids by replacing the hydrogen with a metal.

Salt marsh A relatively flat area of the shore where fine sediment is deposited and salt-tolerant grasses grow. One of the most biologically productive regions of Earth.

Salt wedge estuary A very deep river mouth with a very large volume of freshwater flow beneath which a wedge of saltwater from the ocean invades. The Mississippi River is an example.

Saffir–Simpson scale A scale of hurricane intensity that divides tropical cyclones into categories based on wind speed and damage.

San Andreas Fault A transform fault that cuts across the state of California from the northern end of the Gulf of California to Point Arena north of San Francisco.

Sand Particle size of 0.0625 to 2 millimeters (0.002 to 0.08 inch). It pertains to particles that lie between silt and granules on the Wentworth scale of grain size.

Sargasso Sea A region of convergence in the North Atlantic lying south and east of Bermuda where the water is a very clear, deep blue color, and contains large quantities of floating *Sargassum*.

Sargassum A brown alga characterized by a bushy form, substantial holdfast when attached, and a yellow-brown, green-yellow, or orange color. Two species, *S. fluitans* and *S. natans,* make up most of the macroscopic vegetation in the Sargasso Sea.

Scarp A linear steep slope on the ocean floor separating gently sloping or flat surfaces.

Scavenger An animal that feeds on dead organisms.

Schooling Well-defined large groups of fish, squid, and crustaceans that apparently aid them in survival.

Scientific method The principles and empirical processes of discovery and demonstration considered characteristic of or necessary for scientific investigation, generally involving the observation of phenomena, the formulation of a hypothesis concerning the phenomena, experimentation to demonstrate the truth or falseness of the hypothesis, and a conclusion that validates or modifies the hypothesis.

Scyphozoa A class of coelenterates that includes the true jellyfish in which the medusoid body form predominates and the polyp is reduced or absent.

Scuba An acronym for self-contained underwater breathing apparatus, it is a portable device containing compressed air that is used for breathing under water.

Sea 1. A subdivision of an ocean, generally enclosed by land and usually composed of salt water. Two types of seas are identifiable and defined. They are the Mediterranean seas, where a number of seas are grouped together collectively as one sea, and adjacent seas, which are connected individually to the ocean. 2. A portion of the ocean where waves are being generated by wind.

Sea anemone A member of the class Anthozoa whose bright color, tentacles, and general appearance make it resemble flowers.

Sea arch An opening through a headland caused by wave erosion. Usually develops as sea caves are extended from one or both sides of the headland.

Sea breeze   The landward flow of air from the sea caused by differential heating of Earth's surface.

Sea cave   A cavity at the base of a sea cliff formed by wave erosion.

Sea cow   See *Sirenia*.

Sea cucumber   A common name given to members of the echinoderm class Holotheuroidea.

Sea-floor spreading   A process producing the lithosphere when convective upwelling of magma along the oceanic ridges moves the ocean floor away from the ridge axes at rates between 2 to 12 centimeters (0.8 to 5 inches) per year.

Sea ice   Any form of ice originating from the freezing of ocean water.

Sea lion   Any of several eared seals with a relatively long neck and large front flippers, especially the California sea lion *Zalophus californianus* of the northern Pacific. Along with the fur seals, these marine mammals are known as eared seals.

Sea otter   A seagoing otter that has recovered from near extinction along the North Pacific coasts. It feeds primarily on abalone, sea urchins, and crustaceans.

Sea snake   A reptile belonging to the family Hydrophiidae with venom similar to that of cobras. They are found primarily in the coastal waters of the Indian Ocean and the western Pacific Ocean.

Sea stack   An isolated, pillarlike rocky island that is detached from a headland by wave erosion.

Sea turtle   Any of the reptilian order Testudinata found widely in warm water.

Sea urchin   An echinoderm belonging to the class Echinoidea possessing a fused test (external covering) and well-developed spines.

Seaknoll   See *abyssal hill*.

Seal   **1.** Any of several earless seals with a relatively short neck and small front flippers. Also known as true seals. **2.** A general term that describes any of the various aquatic, carnivorous marine mammals of the families *Phocidae* and *Otariidae* (true seals and eared seals), found chiefly in the Northern Hemisphere and having a sleek, torpedo-shaped body and limbs that are modified into paddlelike flippers.

Seamount   An individual volcanic peak extending over 1000 meters (3300 feet) above the surrounding ocean floor.

Seasonal thermocline   A thermocline that develops due to surface heating of the oceans in mid- to high latitudes. The base of the seasonal thermocline is usually above 200 meters (656 feet).

Seawall   A wall built parallel to the shore to protect coastal property from the waves.

SeaWiFS   An instrument aboard the SeaStar satellite launched in 1997 that measures the color of the ocean with a radiometer and provides global coverage of ocean chlorophyll levels as well as land productivities every two days.

Secchi disk   A light colored disk-shaped device that is lowered into water in order to measure the water's clarity.

Sediment   Particles of organic or inorganic origin that accumulate in loose form.

Sediment maturity   A condition in which the roundness and degree of sorting increase and clay content decreases within a sedimentary deposit.

Sedimentary rock   A rock resulting from the consolidation of loose sediment, or a rock resulting from chemical precipitation, such as sandstone and limestone.

Seep   An area where water of various temperature trickles out of the sea floor.

Seiche   A standing wave of an enclosed or semi-enclosed body of water that may have a period ranging from a few minutes to a few hours, depending on the dimensions of the basin. The wave motion continues after the initiating force has ceased.

Seismic   Pertaining to an earthquake or Earth vibration, including those that are artificially induced.

Seismic moment-magnitude   A scale used for measuring earthquake intensity based on energy released in creating very long-period seismic waves.

Seismic sea wave   See *tsunami*.

Seismic surveying   The use of sound-generating techniques to identify features on or beneath the ocean floor.

Semidiurnal tidal pattern   A tidal pattern exhibiting two high tides and two low tides per tidal day with small inequalities between successive highs and successive lows; a semidaily tide.

Sessile   Permanently attached to the substrate and not free to move about.

Sewage sludge   Semisolid material precipitated by sewage treatment.

Shallow-water wave   A wave on the surface having a wavelength of at least 20 times water depth. The bottom affects the orbit of water particles and speed is determined by water depth.

Shelf break   The depth at which the gentle slope of the continental shelf steepens appreciably. It marks the boundary between the continental shelf and continental rise.

Shelf ice   Thick shelves of glacial ice that push out into Antarctic seas from Antarctica. Large tabular icebergs calve at the edge of these vast shelves.

Shoaling   To become shallow.

Shore   Seaward of the coast, extends from highest level of wave action during storms to the low water line.

Shoreline   The line marking the intersection of water surface with the shore. Migrates up and down as the tide rises and falls.

Shoreline of emergence   Shorelines that indicate a lowering of sea level by the presence of stranded beach deposits and marine terraces above it.

Shoreline of submergence   Shorelines that indicate a rise in sea level by the presence of drowned beaches or submerged dune topography.

Side-scan sonar   A method of mapping the topography of the ocean floor along a strip up to 60 kilometers (37 miles) wide using computers and sonar signals that are directed away from both sides of the survey ship.

Silica   Silicon dioxide (SiO_2).

Silicate   A mineral whose crystal structure contains SiO_4 tetrahedra.

Siliceous   A condition of containing abundant silica (SiO_2).

Sill   A submarine ridge partially separating bodies of water such as fjords and seas from one another or from the open ocean.

Silt   A particle size of 0.008 to 0.0625 millimeter (0.0003 to 0.002 inch). It is intermediate in size between sand and clay.

Siphonophora   An order of hydrozoan coelenterates that forms pelagic colonies containing both polyps and medusae. An example is *Physalia*.

Sirenia   An order of large, vegetarian marine mammals that includes dugongs and manatees, which are also know as sea cows.

Slack water   Occurs when a reversing tidal current changes direction at high or low water. Current speed is zero.

Slick   A smooth patch on an otherwise rippled surface caused by a monomolecular film of organic material that reduces surface tension.

Slightly stratified estuary   An estuary of moderate depth in which marine water invades beneath the freshwater runoff. The two water masses mix so the bottom water is slightly saltier than the surface water at most places in the estuary.

SOFAR channel   Sound fixing and ranging channel, which is a low-velocity sound travel zone that coincides with the permanent thermocline in low and mid-latitudes.

Solar day   The 24-hour period during which Earth completes one rotation on its axis.

Solar distillation   A process by which ocean water can be desalinated by evaporation and the condensation of the vapor on the cover of a container. The condensate then runs into a separate container and is collected as fresh water. Also called solar humidification.

Solar humidification   See *solar distillation*.

Solar system   The Sun and the celestial bodies, asteroids, planets, and comets that orbit around it.

Solar tide   The partial tide caused by the tide-producing forces of the Sun.

Solid state   A state of matter in which the substance has a fixed volume and shape. A crystalline state of matter.

Solstice   The time during which the Sun is directly over one of the tropics. In the Northern Hemisphere the summer solstice occurs on June 21 or 22 as the Sun is over the Tropic of Cancer, and the winter solstice occurs on December 21 or 22 when the Sun is over the Tropic of Capricorn.

Solute   A substance dissolved in a solution. Salts are the solute in saltwater.

Solution   A state in which a solute is homogeneously mixed with a liquid solvent. Water is the solvent for the solution that is ocean water.

Solvent   A liquid that has one or more solutes dissolved in it.

Somali Current   A current that flows north along the Somali coast of Africa during the southwest monsoon season.

Sonar   An acronym for sound navigation and ranging, it is a method by which objects may be located in the ocean.

Sorting   A texture of sediments, where a well-sorted sediment is characterized by having great uniformity of grain sizes.

Sounding   A measured depth of water beneath a ship.

Southwest Monsoon   A southwest wind that develops during the summer season. It blows off the Indian Ocean onto the Asian mainland.

Southwest Monsoon Current   During the southwest monsoon season, this eastward-flowing current replaces the west-flowing North Equatorial Current in the Indian Ocean.

Space dust   Micrometeoroid space debris.

Species   A fundamental category of taxonomic classification, ranking below a genus or subgenus and consisting of related organisms capable of interbreeding.

Species diversity   The number or variety of species found in a subdivision of the marine environment.

Specific gravity   The ratio of density of a given substance to that of pure water at 4°C and at one atmosphere pressure.

Specific heat   The quantity of heat required to raise the temperature of 1 gram of a given substance 1°C. For water it is 1 calorie.

Spermaceti organ   A large fatty organ located within the head region of sperm whales (*Physeter catodon*) that is used to focus echolocation sounds.

Spermatophyta   Seed-bearing plants.

Spherule   A cosmogenous microscopic globular mass composed of silicate rock material (tektites) or of iron and nickel.

Spicule   A minute, needlelike calcareous or siliceous projection found in sponges, radiolarians, chitons, and echinoderms that acts to support the tissue or provide a protective covering.

Spilling breaker   A type of breaking wave that forms on a gently sloping beach, which gradually extracts the energy from the wave to produce a turbulent mass of air and water that runs down the front slope of the wave.

Spit   A small point, low tongue, or narrow embankment of land commonly consisting of sand deposited by longshore currents and having one end attached to the mainland and the other terminating in open water.

Splash wave   A long-wavelength wave created by a massive object or series of objects falling into water; a type of tsunami.

Sponge   See *Porifera*.

Spray zone   The shore zone lying between the high-tide shoreline and the coastline. It is covered by water only during storms.

Spreading center   A divergent plate boundary.

Spreading rate   The rate of divergence of plates at a spreading center.

Spring bloom   A mid-latitude bloom of phytoplankton that occurs during the spring and is limited by the availability of nutrients.

Spring equinox   See *vernal equinox*.

Spring tide   Tide of maximum range occurring about every two weeks when the Moon is in either new or full moon phase.

Stack   An isolated mass of rock projecting from the ocean off the end of a headland from which it has been separated by wave erosion.

Standard laboratory bioassay   An standard assessment technique that determines the concentration of a pollutant that causes 50% mortality among selected test organisms.

Standard seawater   Ampules of ocean water for which the chlorinity has been determined by the Institute of Oceanographic Services in Wormly, England. The ampules are sent to laboratories all over the world so that equipment and reagents used to determine the salinity of ocean water samples can be calibrated by adjustment until they give the same chlorinity as is shown on the ampule label.

Standing stock   The mass of fishery organisms present in an ecosystem at a given time

Standing wave   A wave, the form of which oscillates vertically without progressive movement. The region of maximum vertical motion is an antinode. On either side are nodes, where there is no vertical motion but maximum horizontal motion.

Stenohaline   Pertaining to organisms that can withstand only a small range of salinity change.

Stenothermal   Pertaining to organisms that can withstand only a small range of temperature change.

Stick chart   A device made of sticks or pieces of bamboo that was used by early navigators at sea.

Still water level   The horizontal surface halfway between crest and trough of a wave. If there were no waves, the water surface would exist at this level. Also known as zero energy level.

Storm   An atmospheric disturbance characterized by strong winds accompanied by precipitation and often by thunder and lightning.

Storm surge   A rise above normal water level resulting from wind stress and reduced atmospheric pressure during storms. Consequences can be more severe if it occurs in association with high tide.

Strait of Gibraltar   The narrow opening between Europe and Africa through which the waters of the Atlantic Ocean and Mediterranean Sea mix.

Stranded beach   An ancient beach deposit found above present sea level because of a lowering of sea level.

Streamlining   The shaping of an object so it produces the minimum of turbulence while moving through a fluid medium. The teardrop shape displays a high degree of streamlining.

Subduction   The process by which one lithospheric plate descends beneath another as they converge.

Subduction zone   A long, narrow region beneath Earth's surface in which subduction takes place.

Subduction zone seep biocommunity   Animals that live in association with seeps of pore water squeezed out of deeper sediments. They depend on sulfur-oxidizing bacteria that act as producers for the ecosystem.

Sublittoral zone   That portion of the benthic environment extending from low tide to a depth of 200 meters (660 feet); considered by some to be the surface of the continental shelf.

Submarine canyon   A steep, V-shaped canyon cut into the continental shelf or slope.

Submarine fan   See *deep-sea fan*.

Submerged dune topography   Ancient coastal dune deposits found submerged beneath the present shoreline because of a rise in sea level.

Submerging shoreline   A shoreline formed by the relative submergence of a landmass in which the shoreline is on landforms developed under subaerial processes. It is characterized by bays and promontories and is more irregular than a shoreline of emergence.

Subneritic province   The benthic environment extending from the shoreline across the continental shelf to the shelf break. It underlies the neritic province of the pelagic environment.

Suboceanic province   Benthic environments seaward of the continental shelf.

Subpolar   Pertaining to the oceanic region that is covered by sea ice in winter. The ice melts away in summer.

Subpolar low   A region of low atmospheric pressure located at about 60° latitude.

Substrate   The base on which an organism lives and grows.

Subsurface current   A current usually flowing below the pycnocline, generally at slower speed and in a different direction from the surface current.

Subtropical   Pertaining to the oceanic region poleward of the tropics (about 30° latitude).

Subtropical Convergence   The zone of convergence that occurs within all subtropical gyres as a result of Ekman transport driving water toward the interior of the gyres.

Subtropical gyre   The trade winds and westerly winds initiated in the subtropical regions of all ocean basins, with the influence of the Coriolis effect, set large regions of ocean water in motion. They rotate clockwise in the Northern Hemisphere and counterclockwise in the Southern Hemisphere, and they are centered in the subtropics.

Subtropical high   A region of high atmospheric pressure located at about 30° latitude.

Sulfur   A yellow mineral composed of the element sulfur. It is commonly found in association with hydrocarbons and salt deposits.

Sulfur-oxidizing bacteria   Bacteria that support many deep-sea hydrothermal vent and cold-water seep biocommunities by using energy released by oxidation to synthesize organic matter chemosynthetically.

Summer solstice   In the Northern Hemisphere, it is the instant when that Sun moves north to the Tropic of Cancer before changing direction and moving southward toward the Equator, approximately June 21.

Summertime beach   A beach that is characteristic during summer months. It typically has a wide sandy berm and a steep beach face.

Superwave   See *rogue wave*.

Supralittoral zone   The splash or spray zone above the spring high-tide shoreline.

Surf beat   An irregular wave pattern caused by mixed interference that results in a varied sequence of larger and smaller waves.

Surf zone   The region between the shoreline and the line of breakers where most wave energy is released.

Surface tension   The tendency for the surface of a liquid to contract owing to intermolecular bond attraction.

Surging breaker   A compressed breaking wave that builds up over a short distance and surges forward as it breaks. It is characteristic of abrupt beach slopes.

Suspension feeder   See *filter feeder*.

Suspension settling   The process by which fine-grained material that is being suspended in the water column slowly accumulates on the sea floor.

Sverdrup (Sv)   A unit of flow rate equal to one million cubic meters per second. Named after Norwegian explorer Otto Sverdrup.

Swash   A thin layer of water that washes up over exposed beach as waves break at the shore.

Swell   A free ocean wave by which energy put into ocean waves by wind in the sea is transported with little energy loss across great stretches of ocean to the margins of continents where the energy is released in the surf zone.

Swim bladder   A gas-containing, flexible, cigar-shaped organ that aids many fishes in attaining neutral buoyancy.

Symbiosis   A relationship between two species in which one or both benefit or neither or one is harmed. Examples are commensalism, mutualism, and parasitism.

Syzygy   Either of two points in the orbit of the Moon (full or new moon phase) when the Moon lies in a straight line with the Sun and Earth.

Tablemount   A conical volcanic feature on the ocean floor resembling a seamount except that it has had its top truncated to a relatively flat surface.

Taxonomy   The classification of organisms in an ordered system that indicates natural relationships.

Tectonic estuary   An estuary, the origin of which is related to tectonic deformation of the coastal region.

Tectonics   Deformation of Earth's surface by forces generated by heat flow from Earth's interior.

Tektite   See *spherule*.

Temperate   Pertaining to the oceanic region where pronounced seasonal change occurs (about 40° to 60° latitude). Also known as the mid-latitudes.

Temperature   A direct measure of the average kinetic energy of the molecules of a substance.

Temperture–salinity (T–S) diagram   A diagram with axes representing water temperature and salinity, whereby the density of the water can be determined.

Temperature of maximum density   The temperature at which a substance reaches its highest density. For water, it is 4°C.

Terrigenous sediment   Sediment produced from or of the Earth. Also called lithogenous sediment.

Territorial sea   A strip of ocean, 12 nautical miles wide, adjacent to land over which the coastal nation has control over the passage of ships.

Test   The supporting skeleton or shell (usually microscopic) of many invertebrates.

Tethys Sea   An ancient body of water that separated Laurasia to the north and Gondwanaland to the south. Its location was approximately that of the present Alpine-Himalayan mountain system.

Texture   The general physical appearance of an object.

Theory   A well-substantiated explanation of some aspect of the natural world that can incorporate facts, laws (descriptive generalizations about the behavior of an aspect of the natural world), logical inferences, and tested hypotheses.

Thermal contraction   To reduce in size during times when temperature is lowered.

Thermocline   A layer of water beneath the mixed layer in which a rapid change in temperature can be measured in the vertical dimension.

Thermohaline circulation   The vertical movement of ocean water driven by density differences resulting from the combined effects of variations in temperature and salinity; produces deep currents.

Tidal bore   A steep-fronted wave that moves up some rivers when the tide rises in the coastal ocean.

Tidal bulges   The theoretical mounds of water on both sides of Earth caused by the relative positions of the Moon (lunar tidal bulges) and the Sun (solar tidal bulges).

Tidal period   The time that elapses between successive high tides. In most parts of the world, it is 12 hours and 25 minutes.

Tidal range   The difference between high tide and low tide water levels over any designated time interval, usually one lunar day.

Tide   Periodic rise and fall of the surface of the ocean and connected bodies of water resulting from the gravitational attraction of the Moon and Sun acting unequally on different parts of Earth.

Tide-generating force   The magnitude of the centripetal force required to keep all particles of Earth having identical mass moving in identical circular paths required by the movements of the Earth-Moon system is identical. This required force is provided by the gravitational attraction between the particles and the Moon. This gravitational force is identical to the required centripetal force only at the center of Earth. For ocean tides, the horizontal component of the small force that results from the difference between the required and provided forces is the tide-generating force on that individual particle. These forces are such that they tend to push the ocean water into bulges toward the tide-generating body on one side of Earth and away from the tide-generating body on the opposite side of Earth.

Tide wave   The long-period gravity wave generated by tide-generating forces described above and manifested in the rise and fall of the tide.

Tissue   An aggregate of cells and their products developed by organisms for the performance of a particular function.

Tombolo   A sand or gravel bar that connects an island with another island or the mainland.

Topography   The configuration of a surface. In oceanography it refers to the ocean bottom or the surface of a mass of water with given characteristics.

Trade winds   The air masses moving from subtropical high-pressure belts toward the Equator. They are northeasterly in the Northern Hemisphere and southeasterly in the Southern Hemisphere.

Transform fault   A fault characteristic of mid-ocean ridges along which they are offset. Oceanic transform faults occur wholly on the ocean floor, while continental transform faults occur on land.

Transform plate boundary   The boundary between two lithospheric plates formed by a transform fault.

Transitional wave   A wave moving from deep water to shallow water that has a wavelength more than twice the water depth but less than 20 times the water depth. Particle orbits are beginning to be influenced by the bottom.

Transverse wave   A wave in which particle motion is at right angles to energy propagation.

Trench   A long, narrow, and deep depression on the ocean floor with relatively steep sides that is caused by plate convergence.

Trophic level   A nourishment level in a food chain. Plant producers constitute the lowest level, followed by herbivores and a series of carnivores at the higher levels.

Tropic of Cancer   The latitude of 23.5° north, which is the furthest location north that receives vertical rays of the Sun.

Tropic of Capricorn   The latitude of 23.5° south, which is the furthest location south that receives vertical rays of the Sun.

Tropical   Pertaining to, characteristic of, occurring in, or inhabiting the tropics.

Tropical cyclone   See *hurricane*.

Tropical tide   A tide occurring twice monthly when the Moon is at its maximum declination near the tropics of Cancer and Capricorn.

Tropics   The region of Earth's surface lying between the Tropic of Cancer and the Tropic of Capricorn. Also known as the Torrid Zone.

Troposphere   The lowermost portion of the atmosphere, which extends from Earth's surface to 12 kilometer (7 miles). It is where all weather is produced.

Trough (wave)   The part of an ocean wave that is displaced below the still-water level.

Tsunami   Seismic sea wave. A long-period gravity wave generated by a submarine earthquake or volcanic event. Not noticeable on the open ocean but builds up to great heights in shallow water.

Turbidite deposit   A sediment or rock formed from sediment deposited by turbidity currents characterized by both horizontally and vertically graded bedding.

Turbidity   A state of reduced clarity in a fluid caused by the presence of suspended matter.

Turbidity current   A gravity current resulting from a density increase brought about by increased water turbidity. Possibly initiated by some sudden force such as an earthquake, the turbid mass continues under the force of gravity down a submarine slope.

Turbulent flow   Flow in which the flow lines are confused heterogeneously due to random velocity fluctuations.

Typhoon   See *hurricane*.

Ultraplankton   Plankton for which the greatest dimension is less than 5 micrometers (0.0002 inch). They are very difficult to separate from the water.

Ultrasonic   Sound frequencies above those that can be heard by humans (above 20,000 Hertz).

Ultraviolet radiation   Electromagnetic radiation shorter than visible radiation and longer than X rays. The approximate range is 1 to 400 nanometers.

Upper water   Includes the mixed layer and the permanent thermocline. It is approximately the top 1000 meters (3300 feet) of the ocean.

Upwelling   The process by which deep, cold, nutrient-laden water is brought to the surface, usually by diverging equatorial currents or coastal currents that pull water away from a coast.

Valence   The combining capacity of an element measured by the number of hydrogen atoms with which it will combine.

van der Waals force   Weak attractive force between molecules resulting from the interaction of one molecule and the electrons of another.

Vapor   The gaseous state of a substance that is liquid or solid under ordinary conditions.

Vent   An opening on the ocean floor that emits hot water and dissolved minerals.

Ventral   Pertaining to the lower or under surface.

Vernal equinox   The passage of the Sun across the Equator as it moves from the Southern Hemisphere into the Northern Hemisphere, approximately March 21. During this time, all places in the world experience equal lengths of night and day. Also known as the spring equinox.

Vertebrata   Subphylum of chordates that includes those animals with a well-developed brain and a skeleton of bone or cartilage; includes fish, amphibians, reptiles, birds, and land animals.

Vertically mixed estuary   Very shallow estuaries such as lagoons in which freshwater and marine water are totally mixed from top to bottom so that the salinity at the surface and the bottom is the same at most places within the estuary.

Viscosity   A property of a substance to offer resistance to flow caused by internal friction.

Volcanic arc   An arc-shaped row of active volcanoes directly above a subduction zone. Can occur as a row of islands (island arc) or mountains on land (continental arc).

Walker Circulation Cell   The pattern of atmospheric circulation that involves the rising of warm air over the East Indies low-pressure cell and its descent over the high-pressure cell in the southeastern Pacific Ocean off the coast of Chile. It is the weakening of this circulation that accompanies an El Niño event, which has led to the development of the term El Niño-Southern Oscillation event.

Walrus   A large marine mammal (*Odobenus rosmarus*) of Arctic regions belonging to the order Pinnipedia and having two long tusks, tough wrinkled skin, and four flippers.

Waning crescent   When the Moon is between third quarter and new moon phases.

Waning gibbous   When the Moon is between full moon and third quarter phases.

Warm-blooded   See *homeothermic*.

Warm front   A weather front in which a warm air mass moves into and over a cold air mass producing a broad band of gentle precipitation.

Water mass   A body of water identifiable from its temperature, salinity, or chemical content.

Wave   A disturbance that moves over the surface or through a medium with a speed determined by the properties of the medium.

Wave base   The depth at which circular orbital motion becomes negligible. It exists at a depth of one-half wavelength, measured vertically from still water level.

Wave-cut beach   A gently sloping surface produced by wave erosion and extending from the base of the wave-cut cliff out under the offshore region.

Wave-cut cliff   A cliff produced by landward cutting by wave erosion.

Wave dispersion   The separation of waves as they leave the sea area by wave size. Larger waves travel faster than smaller waves and thus leave the sea area first, to be followed by progressively smaller waves.

Wave frequency (*f*)   The number of waves that pass a fixed point in a unit of time (usually one second). A wave's frequency is the inverse of its period.

Wave height (*H*)   The vertical distance between a crest and the adjoining trough.

Wave period (*T*)   The elapsed time between the passage of two successive wave crests (or troughs) past a fixed point. A wave's period is the inverse of its frequency.

Wave speed (*S*)   The rate at which a wave travels. It can be calculated by dividing a wave's wavelength (*L*) by its period (*T*).

Wave steepness   Ratio of wave height (*H*) to wavelength (*L*). If a 1:7 ratio is ever exceeded by the wave, then the wave breaks.

Wave train   A series of waves from the same direction. Informally known as a wave set.

Wavelength (*L*)   The horizontal distance between two corresponding points on successive waves, such as from crest to crest.

Waxing crescent   When the Moon is between new moon and first quarter phases.

Waxing gibbous   When the Moon is between first quarter and full moon phases.

Weather   The state of the atmosphere at a given time and place, with respect to variables such as temperature, moisture, wind velocity, and barometric pressure.

Weathering   A process by which rocks are broken down by chemical and mechanical means.

Wentworth scale of grain size   A logarithmic scale for size classification of sediment particles.

West Australian Current   A cold current that forms the eastern boundary current of the Indian Ocean subtropical gyre. It is separated from the coast by the warm Leeuwin Current except during El Niño-Southern Oscillation events when the Leeuwin Current weakens.

West Wind Drift   A surface current driven in an easterly direction around Antarctica by the strong prevailing westerly wind belt. Also called the Antarctic Circumpolar Current.

Westerly winds   The air masses moving away from the subtropical high pressure belts toward higher latitudes. They are southwesterly in the Northern Hemisphere and northwesterly in the Southern Hemisphere.

Western boundary current   Poleward-flowing warm currents on the western side of all subtropical gyres.

Western boundary undercurrent (WBUC)   A bottom current that flows along the base of the continental slope eroding sediment from it and redepositing the sediment on the continental rise. It is confined to the western boundary of deep-ocean basins.

Westward intensification   Pertaining to the intensification of the warm western portion of the subtropical gyre currents that is manifested in higher velocity, faster flow, and deeper flow compared with the cold eastern boundary currents that drift leisurely toward the Equator.

Wetlands   Biologically productive regions bordering estuaries and other protected coastal areas. They are usually salt marshes at latitudes greater than 30° and mangrove swamps in lower latitudes.

White muscle fiber   Thick muscle fibers with relatively low concentrations of myoglobin that make up a large percentage of the muscle fiber in lunger-type fishes.

White smoker   Similar to a black smoker, but emits water of a lower temperature that is white in color.

Wind-driven circulation   Any movement of ocean water that is driven by winds. This includes most horizontal movements in the surface waters of the world's oceans.

Windward   The direction from which the wind is blowing.

Winter solstice   The instant the southward-moving Sun reaches the Tropic of Cancer before changing direction and moving north back toward the Equator, approximately December 21.

Wintertime beach   A beach that is characteristic during winter months. It typically has a narrow rocky berm and a flat beach face.

Zenith   That point on the celestial sphere directly over the observer.

Zooplankton   Animal plankton.

Zooxanthellae   A form of algae that lives as a symbiont in the tissue of corals and other coral reef animals and provides varying amounts of their required food supply.

Credits and Acknowledgments

Frontspiece
By Bruce C. Heezen and Marie Tharp, 1977. Copyright © 1977 Marie Tharp. Reproduced by permission of Marie Tharp.

Introduction
Figure I–1 APT photo. **Figure I–3** Courtesy of Johnson Space Center, NASA. **Figure I–4** F. Stuart Westmorland/Photo Researchers, Inc.

Chapter 1
Figure 1A Courtesy of Scripps Institution of Oceanography Archives, University of California, San Diego. **Figure 1B** APT photo; Polynesian stick chart courtesy of Glen Foss. **Figure 1C** APT photo. **Figure 1D** National Maritime Museum Picture Library, London, England. **Figure 1E** Courtesy of U.S. Department of Defense. **Figure 1–1** From the *Challenger* Report, Great Britain, 1895. **Figure 1–3** Data from Kennish (1994) and the National Geographic Society. **Figure 1–5** Official photograph U.S. Navy. **Figure 1–9** Courtesy of U.S. Navy. **Figure 1–11** Reprinted by permission from Tarbuck, E. J., and Lutgens, F. K., *Earth Science,* 6th Ed. (Fig. 19.2), Macmillan Publishing Company, 1991. **Figure 1–12** After Tarbuck, E. J., and Lutgens, F. K., *The Earth: An Introduction to Physical Geology,* 5th Ed. (Fig. 1.10), Prentice-Hall, 1996. **Figure 1–13** Reprinted by permission from Tarbuck, E. J., and Lutgens, F. K., *The Earth: An Introduction to Physical Geology,* 4th Ed. (Fig. 1.11), Macmillan Publishing Company, 1993. **Figure 1–19** After Tarbuck, E. J., and Lutgens, F. K., *The Earth: An Introduction to Physical Geology,* 5th Ed. (Fig. 1.9), Prentice-Hall, 1997. Data from the Geological Society of America.

Chapter 2
Figure 2A Courtesy of Woods Hole Oceanographic Institution. **Figure 2B** Courtesy of Deeanne Edwards, Scripps Institution of Oceanography, University of California, San Diego. **Figure 2–1** Courtesy of Bildarchiv Preussischer Kulturbesitz, West Berlin. **Figure 2–2** After Dietz, R. S., and Holden, J. C. 1970. Reconstruction of Pangaea: Breakup and dispersion of continents, Permian to present. *Journal of Geophysical Research* 75:26, 4939–4956. **Figure 2–3** From *Continental Drift* by Don and Maureen Tarling, © 1971 by G. Bell & Sons, Ltd. Reprinted by permission of Doubleday & Co., Inc. **Figure 2–4** After Tarbuck, E. J., and Lutgens, F. K., *Earth Science,* 8th Ed. (Fig. 7.6), Prentice-Hall, 1997. **Figure 2–5** After Tarbuck, E. J., and Lutgens, F. K., *Earth Science,* 8th Ed. (Fig. 7.7), Prentice-Hall, 1997. **Figure 2–6** After Tarbuck, F. J., and Lutgens, F. K., *Earth Science,* 8th Ed. (Fig. 7.4), Prentice-Hall, 1997. **Figure 2–7a** Reprinted by permission from Tarbuck, E. J., and Lutgens, F. K., *The Earth: An Introduction to Physical Geology,* 3rd Ed. (Fig. 18.8), Merrill Publishing Company, 1990. **Figure 2–7b** Reprinted by permission from Tarbuck, E. J., and Lutgens, F. K., *The Earth: An Introduction to Physical Geology,* 3rd Ed. (Fig. 18.), Merrill Publishing Company, 1990. **Figure 2–8** After Tarbuck, E. J., and Lutgens, F. K., *Earth Science,* 8th Ed. (Fig. 7.17), Prentice-Hall, 1997. **Figure 2–9** After Tarbuck, E. J., and Lutgens, F. K., *Earth Science,* 8th Ed. (Fig. 7.18), Prentice-Hall, 1997. **Figure 2–10** Reprinted by permission from Tarbuck, E. J., and Lutgens, F. K., *The Earth: An Introduction to Physical Geology,* 3rd Ed. (Fig. 18.11), Merrill Publishing Company, 1990. **Figure 2–11** After Tarbuck, E. J., and Lutgens, F. K., *Earth Science,* 8th Ed. (Fig. 7.20), Prentice-Hall, 1997. **Figure 2–12** Reprinted by permission from Tarbuck, E. J., and Lutgens, F. K., *The Earth: An Introduction to Physical Geology,* 3rd Ed. (Fig. 19.16), Merrill Publishing Company, 1990. After *The*

Bedrock Geology of the World, by R. L. Larson et al., Copyright © 1985 by W. H. Freeman. **Figure 2–13a** Reprinted by permission from Tarbuck, E. J., and Lutgens, F. K., *The Earth: An Introduction to Physical Geology,* 3rd Ed. (Fig. 5.11), Merrill Publishing Company, 1990. Data from National Geophysical Data Center/NOAA. **Figure 2–13b** After Tarbuck, E. J., and Lutgens, F. K., *The Earth: An Introduction to Physical Geology,* 3rd Ed. (Fig. 5.11), Merrill Publishing Company, 1990; modified with data from W. B. Hamilton, U.S. Geological Survey. **Figure 2–15** Reprinted by permission from Tarbuck, E. J., and Lutgens, F. K., *The Earth: An Introduction to Physical Geology,* 3rd Ed. (Fig. 5.19), Merrill Publishing Company, 1990. **Figure 2–17** Reprinted by permission from Tarbuck, E. J., and Lutgens, F. K., *The Earth: An Introduction to Physical Geology,* 3rd Ed. (Fig. 6.7), Merrill Publishing Company, 1990. **Figure 2–18** After Tarbuck, E. J., and Lutgens, F. K., *Earth Science,* 8th Ed. (Fig. 7.10), Prentice-Hall, 1997. **Figure 2–19** Courtesy of Patricia Deen, Palomar College. **Figure 2–20** After Tarbuck, E. J., and Lutgens, F. K., *Earth Science,* 8th Ed. (Fig. 7.12), Prentice-Hall, 1997. **Figure 2–21** After Tarbuck, E. J., and Lutgens, F. K., *Earth Science,* 8th Ed. (Fig. 7.11), Prentice-Hall, 1997. **Figure 2–22** After Schmidt, V. A. 1986. *Planet Earth and the New Geoscience* (Fig. V–4), Dubuque, Iowa: Kendall/Hunt. Sea floor images courtesy of NOAA National Geophysical Data Center and Peter W. Sloss. **Figure 2–23** Reprinted by permission from Tarbuck, E. J., and Lutgens, F. K., *Earth Science,* 5th Ed. (Fig. 6.11), Merrill Publishing Company, 1988. **Figure 2–24c** Photo courtesy of U.S. Geologic Survey, Cascades Volcano Observatory. **Figure 2–25** After Tarbuck, E. J., and Lutgens, F. K., *Earth Science,* 8th Ed. (Fig. 7.15), Prentice-Hall, 1997. **Figure 2–26** After Tarbuck, E. J., and Lutgens, F. K., *The Earth: An Introduction to Physical Geology,* 3rd Ed. (Figs. 18.22, 18.23), Merrill Publishing Company, 1990. **Figure 2–27** Data from Robert A. Duncan. **Figure 2–28** After Tarbuck, E. J., and Lutgens, F. K., *The Earth: An Introduction to Physical Geology,* 6th Ed. (Fig. 19.30), Prentice-Hall, 1999. **Figure 2–32** Photo courtesy of NASA. **Figure 2–33** Courtesy of Paul C. Lowman, Jr., and others, Goddard Space Flight Center, Greenbelt, Maryland. **Figure 2–34** Plate tectonic reconstructions by Christopher R. Scotese, PALEOMAP Project, University of Texas at Arlington. **Figure 2–35** After Dietz, R. S., and Holden, J. C. 1970. The breakup of Pangaea, *Scientific American* 223:4, 30–41.

Chapter 3
Figure 3A Official photograph U.S. Navy. **Figure 3B** After Gross, M. G., *Oceanography,* 6th Ed. (Figure 16.10), Prentice-Hall, 1993. **Figure 3C, Part A** Bathymetric chart reproduced with the permission of the Canadian Hydrographic Service. **Figure 3C, Part B** Courtesy of David Sandwell, Scripps Institution of Oceanography, University of California, San Diego. **Figure 3D** Courtesy of David Sandwell, Scripps Institution of Oceanography, University of California, San Diego. **Figure 3E** After Heezen, B. C., and Ewing, M. 1952, Turbidity currents and submarine slumps, and the 1929 Grand Banks earthquake, *American Journal of Science* 250:867. **Figure 3–1** Courtesy of Peter A. Rona, Hudson Laboratories of Columbia University. **Figure 3–2, inset** Courtesy of Daniel J. Fornari, Lamont-Doherty Geological Observatory, Columbia University. Reprinted with permission of the American Geophysical Union. **Figure 3–4** After Tarbuck, E. J., and Lutgens, F. K., *Earth Science,* 6th Ed. (Fig. 10.2), Macmillan Publishing Company, 1991. **Figure 3–5** After Tarbuck, E. J., and Lutgens, F. K., *The Earth: An Introduction to Physical Geology,* 5th Ed. (Fig. 19.3), Prentice-Hall, 1996. **Figure 3–6** After Plummer, C. C. and

McGeary, D., *Physical Geology,* 5th Ed. (Fig. 18.6), Wm. C. Brown, 1991. **Figure 3–7** After Tarbuck, E. J., and Lutgens, F. K., *The Earth: An Introduction to Physical Geology,* 5th Ed. (Fig. 19.4), Prentice-Hall, 1996. **Figure 3–8a** After Tarbuck, E. J., and Lutgens, F. K., *The Earth: An Introduction to Physical Geology,* 5th Ed. (Fig. 19.6), Prentice-Hall, 1996. **Figure 3–8b** Courtesy of Francis P. Shepard Photographic Archives/Collections; with thanks to G. G. Kuhn. **Figure 3–9** Courtesy of Charles Hollister, Woods Hole Oceanographic Institution. **Figure 3–12** Courtesy of Aluminum Company of America (Alcoa). **Figure 3–13a** Courtesy of Woods Hole Oceanographic Institution. **Figure 3–13b** Courtesy of Br. Robert McDermott, S. J. **Figure 3–14** Courtesy of A. E. J. Engel and Scripps Geological Collections, Scripps Institution of Oceanography, University of California, San Diego. **Figure 3–15a** After Tarbuck, E. J., and Lutgens, F. K., *The Earth: An Introduction to Physical Geology,* 5th Ed. (Fig. 21.23), Prentice-Hall, 1996. Inset photo by Fred N. Spiess, Scripps Institution of Oceanography, University of California, San Diego. **Figure 3–15b** Photo Ifremer, from Manaute Cruise, courtesy of Jean-Marie Auzende.

Chapter 4
Figure 4A Courtesy of the Ocean Drilling Program. **Figure 4B** Courtesy of the Ocean Drilling Program. **Figure 4C** Photo courtesy of World Minerals, Inc., Lompoc, California. **Figure 4C, inset** Reprinted by permission from Hallegraeff, G. M., *Plankton: A Microscopic World,* 1988 (p. 20). Courtesy of E. J. Brill, Inc. **Figure 4F** Courtesy of the Ocean Drilling Program, Texas A&M University. **Figure 4–1** Courtesy of the RISE Project Group, F. N. Spiess et al., Scripps Institution of Oceanography, University of California, San Diego. **Figure 4–2** APT photo. **Figure 4–3** After Tarbuck, E. J., and Lutgens, F. K., *The Earth: An Introduction to Physical Geology,* 5th Ed. (Fig. 5.11), Prentice-Hall, 1996. **Figure 4–4a** APT photo. **Figure 4–4b** Annie Griffiths Belt/CORBIS. **Figure 4–4c** APT photo. **Figure 4–4d** APT photo. **Figure 4–5** Courtesy of Walter N. Mack, Michigan State University. **Figure 4–6** After Leinen, M. et al., 1986, Distribution of Biogenic Silica and Quartz in Recent Deep-sea Sediments. *Geology* 14:3, 199–203. **Figure 4–6 inset** Provided by the SeaWiFS Project, NASA/Goddard Space Flight Center and ORBIMAGE. **Figure 4–8a** Reprinted by permission from Hallegraeff, G. M., *Plankton: A Microscopic World,* 1988 (p. 46). Courtesy of E. J. Brill, Inc. **Figure 4–8b** Courtesy of Warren Smith, Scripps Institution of Oceanography, University of California, San Diego. **Figure 4–8c** Photo courtesy of World Minerals, Inc., Lompoc, California (sample from Celite Corporations Diatomite Mine in Lompoc, California). **Figure 4–9a** Reprinted by permission from Hallegraeff, G. M., *Plankton: A Microscopic World,* 1988 (p. 8). Courtesy of E. J. Brill, Inc. **Figure 4–9b** Reprinted by permission from Hallegraeff, G. M., *Plankton: A Microscopic World,* 1988 (p. 16). Courtesy of E. J. Brill, Inc. **Figure 4–9c** Courtesy of Memorie Yasuda, Scripps Institution of Oceanography, University of California, San Diego. **Figure 4–9d** Courtesy of the Deep Sea Drilling Project, Scripps Institution of Oceanography, University of California, San Diego. **Figure 4–10** APT photo. **Figure 4–13** After Biscaye, P. E. et al., 1976; Berger, W. H. et al., 1976; and Kolla V. and Biscaye, P.E., 1976. **Figure 4–14** Courtesy of Scripps Institution of Oceanography, University of California, San Diego. **Figure 4–15** APT photo. **Figure 4–16** Reprinted by permission of The Open University Course Team, *Ocean Chemistry and Deep-Sea Sediments,* Butterworth-Heinemann, 1989. **Figure 4–18** After image by Joseph Holliday, El Camino College. **Figure 4–19** After Sverdrup, H. U. et al., 1942. **Figure 4–20** Courtesy of Susumu Honjo, Woods Hole Oceanographic Institution. **Figure 4–21** Photo by Earl Roberge, courtesy Photo Researchers, Inc. **Figure 4–22** Courtesy of GEOMAR Research Center, Kiel, Germany. **Figure 4–24** Courtesy of Patricia Deen, Palomar College. **Figure 4–25** Courtesy of Deep Sea Ventures, Inc. **Figure 4–26** After Cronan, D. S. 1977. Deep sea nodules: Distribution and geochemistry, *in* Glasby, G. P., ed., *Marine Manganese Deposits,* Elsevier Scientific Publishing Co. **Table 4–1** After Patricia Deen,

Palomar College. **Table 4–2** Source: Wentworth, 1992; After Udden, 1898.

Chapter 5
Figure 5A inset From C. W. Thompson and Sir John Murray, Report on the Scientific Results of the Voyage of H.M.S. *Challenger,* Vol. 1. Great Britain: *Challenger* Office, 1895, Plate 1. **Figure 5B** Peter Arnold, Inc. **Figure 5–1** After Tarbuck, E. J., and Lutgens, F. K., *The Earth: An Introduction to Physical Geology,* 5th Ed. (Fig. 2.4), Prentice-Hall, 1996. **Figure 5–3** Used with permission from R. W. Christopherson, *Geosystems,* 2nd Ed., Macmillan Publishing Company, 1994 (Fig. 7–7, p. 186). **Figure 5–12** Photo by Electron Microscopy Laboratory, ARS, USDA. **Figure 5–19** After Tarbuck, E. J., and Lutgens, F. K., *The Earth: An Introduction to Physical Geology,* 4th Ed. (Fig. 10.2), Macmillan Publishing Company, 1993. **Figure 5–20** After G. L. Pickard, *Descriptive Physical Oceanography,* © 1963. By permission of Pergamon Press Ltd., Oxford, England. **Figure 5–21** After Sverdrup, H. U. et al., 1942. **Figure 5–23** After G. L. Pickard, *Descriptive Physical Oceanography,* © 1963. By permission of Pergamon Press Ltd., Oxford, England. **Figure 5–24** After G. L. Pickard, *Descriptive Physical Oceanography,* © 1963. By permission of Pergamon Press Ltd., Oxford, England.

Chapter 6
Figure 6A left HuttonGetty/Liaison Agency, Inc. **Figure 6A right** Courtesy of Woods Hole Oceanographic Institution, Emory Kristof/National Geographic Society. **Figure 6B** Photo courtesy of Morgan P. Sanger, The Columbus Foundation. **Figure 6C** Courtesy National Oceanic and Atmospheric Administration/Department of Commerce. **Figure 6D inset** Photo courtesy of Walter Munk, Scripps Institution of Oceanography, University of California, San Diego. **Figure 6E** Courtesy of J. H. Golden, NOAA/OAR. **Figure 6–4** After Gross, M. G., *Oceanography,* 6th Ed. (Fig. 5–2), Prentice-Hall, 1993. **Figure 6–13** After Gross, M. G., *Oceanography,* 6th Ed. (Fig. 5–17), Prentice-Hall, 1993. **Figure 6–14** Reprinted by permission from Lutgens, F. K., and Tarbuck, E. J., *The Atmosphere,* 6th Ed. (Fig. 8.2), Prentice-Hall, 1995. **Figure 6–18 inset** Photo by Bob Stovall, Bruce Coleman, Inc. **Figure 6–19** Courtesy of National Hurricane Center, NOAA. **Figure 6–23** After Tarbuck. E. J., and Lutgens, F. K. *The Earth: An Introduction to Physical Geology,* 6th Edition (Fig. 21.9), Prentice-Hall, 1999. **Figure 6–24** Data for 1958-1998 from laboratory of Charles Keeling, Scripps Institution of Oceanography, University of San Diego, California. Carbon dioxide concentrations prior to 1958 estimated from air bubbles in polar ice cores (adapted from *Our Changing Climate,* Reports to the Nation on Our Changing Planet, Fall 1997, No. 4, University Corporation for Atmospheric Research). **Figure 6–25** Adapted from *Our Changing Climate,* Reports to the Nation on Our Changing Planet, Fall 1997, No. 4, University Corporation for Atmospheric Research. **Table 6–6** Source: After Rodhe, H., 1990.

Chapter 7
Figure 7A Courtesy of U.S. Navy. **Figure 7B** Map courtesy of *Eos Transactions* AGU 73:34, 361 (1992); inset courtesy of *Eos Transactions* AGU 75:37, 425 (1994). **Figure 7C** Photo courtesy of The Fram Museum, Oslo, Norway. **Figure 7D** Photo courtesy of Martin Wikelski, University of Illinois. **Figure 7–1a** Courtesy of Douglas Alden, Scripps Institution of Oceanography, University of California, San Diego. **Figure 7–1b** Courtesy of Aanderaa Instruments. **Figure 7–2** Courtesy of NASA. **Figure 7–8a** Courtesy of NASA and NOAA. **Figure 7–16a** Image courtesy of Charles McLain at the Rosenstiel School of Marine and Atmospheric Science, University of Miami. **Figure 7–19** Maps courtesy of the International Research Institute for Climate Prediction, Lamont-Doherty Earth Observatory, Columbia University. **Figure 7–20** Courtesy of Klaus Wolter and NOAA. **Figure 7–22** Courtesy Paul C. Fiedler at the National Marine Fisheries Service. **Figure 7–27** Courtesy of Woods Hole Oceanographic Institution.

Chapter 8

Figure 8A Courtesy of California Geology. **Figure 8D** Courtesy of Oregon Department of Geology. **Figure 8–1b** Photo courtesy of NASA. **Figure 8–2** After Kinsman, B., 1965. **Figure 8–3b** From *The Tasa Collection: Shorelines.* Published by Macmillan Publishing Co., New York. Copyright © 1986 by Tasa Graphic Arts, Inc. All rights reserved. **Figure 8–10** Courtesy of Jet Propulsion Laboratory, NASA. **Figure 8–12** Official photograph U.S. Navy. **Figure 8–14** After Gross, M. G. *Oceanography.* 6th Ed. (Fig. 8–4), Prentice–Hall, 1993. **Figure 8–15** From *The Tasa Collection: Shorelines.* Published by Macmillan Publishing, New York. Copyright © 1986 by Tasa Graphic Arts, Inc. All rights reserved. **Figure 8–16a** © Peter Arnold, Inc. **Figure 8–16b** Vince Cavataio/Allsport/ Agency Vandystadt/Photo Researchers, Inc. **Figure 8–16c** Photo © Woody Woodworth, Creation Captured. All rights reserved. **Figure 8–18** Adapted from Tarbuck, E. J., and Lutgens, F. K., *Earth Science,* 6th Ed. New York: Macmillan Publishing Company, 1991. **Figure 8–19** Photo by Hal Thurman. **Figure 8–21b** Photos courtesy Kyodo News Agency. **Figure 8–22** CORBIS. **Figure 8–23** After González, F. I., Tsunami!, *Scientific American* 280:5, 59, 1999. **Figure 8–24** Map constructed from data in U.S. Navy Summary of Synoptic Meteorological Observations (SSMO). Adapted from Sea Secrets, *Sea Frontiers* 33: 4, 260–261, International Oceanographic Foundation, 1987.

Chapter 9

Figure 9A Photo courtesy of Phototeque/Electricite de France. **Figure 9B** Photo courtesy of New Brunswick Department of Tourism. **Figure 9C** Photo by Eda Rogers. **Figure 9E** Tom Strickland, Courtesy of Wide World Photos. **Figure 9–8** From *The Tasa Collection: Shorelines.* Published by Macmillan Publishing Co., New York. Copyright © 1986 by Tasa Graphic Arts, Inc. All rights reserved. **Figure 9–14** After von Arx, W. S., 1962; original by H. Poincaré 1910, Leçons de Mécanique Céleste, a Gauther-Crofts, Vol. 3. **Figure 9–15** After C. Hauge, Tides, currents, and waves. *California Geology,* July 1972. **Figure 9–17** Photos courtesy of Nova Scotia Department of Tourism.

Chapter 10

Figure 10A Courtesy of Patricia Deen, Palomar College. **Figure 10B** APT photo. **Figure 10C** Courtesy of Drew Wilson, Virginian–Pilot © 1999. **Figure 10–2** APT photo. **Figure 10–3a** Photo by John S. Shelton. **Figure 10–3b** After Tarbuck, E. J., and Lutgens, F. K., *The Earth: An Introduction to Physical Geology,* 4th Ed. (Fig. 14.8), Macmillan Publishing Company, 1993. **Figure 10–5** Photo by Bruce F. Molnia, Terra-photographics/BPS. **Figure 10–6** Photo by John S. Shelton. **Figure 10–8a** Photo by USDA-ASCS. **Figure 10–8b** APT photo. **Figure 10–9a and b** After Tarbuck, E. J., and Lutgens, F. K., *The Earth: An Introduction to Physical Geology,* 4th Ed. (Fig. 14.12), Macmillan Publishing Company, 1993. **Figure 10–9c** Photo by USDA-ASCS. **Figure 10–11a** Photo by GEOPIC®, Earth Satellite Corporation. **Figure 10–11b** Photo courtesy of NASA. **Figure 10–15** After Neumann, J. E., et at., *Sea-level Rise and Global Climate Change: A Review of Impacts to U.S. Coasts,* PEW Center on Global Climate Change, 2000. **Figure 10–18** Photo by John S. Shelton. **Figure 10–20** Photo by John S. Shelton. **Figure 10–21** After Tarbuck, E. J., and Lutgens, F. K., *Earth Science,* 5th Ed., Merrill Publishing Company, 1988. **Figure 10–23a** Courtesy of The Fairchild Aerial Photography Collection at Whittier College, California. Flight C–1670, Frame 4 (9/18/31). **Figure 10–23b** Courtesy of The Fairchild Aerial Photography Collection at Whittier College, California. Flight C–14180, Frame 3:61 (10/21/49). **Figure 10–25** APT photo.

Chapter 11

Figure 11B upper Natalie Forbes/NGS Image Collection. **Figure 11B lower** Photo courtesy of NOAA. **Figure 11C** Bob Jordan/AP/Wide World Photos. **Figure 11D** © Tony Freeman/ Photo Edit. **Figure 11E** APT photo. **Figure 11–4a** Photo courtesy of The Stock Market. **Figure 11–4b** Courtesy NASA. **Figure 11–7** After Officer et al., 1984. **Figure 11–8b** Photo © Eda Rogers. **Figure 11–8c** Photo © R. N. Mariscal/Bruce Coleman, Inc. **Figure 11–11a** Adapted from *Encyclopedia of Oceanography,* edited by Rhodes Fairbridge, © 1966. Reprinted by permission of Dowden, Hutchinson, & Ross, Inc., Stroudsburg, PA. **Figure 11–11b** After Judson, S. et al., *Physical Geology,* 7th Ed. Prentice-Hall, 1987. **Figure 11–12** Photo courtesy of NOAA. **Figure 11–13** After Mitchell, J. G., In the Wake of the Spill: Ten Years After the Exxon Valdez, *National Geographic* 195:3, 96–117, 1999. **Figure 11–14** Gary Braasch/CORBIS. **Figure 11–16b** Photo courtesy of U.S. EPA/EMSL. **Figure 11–17** Courtesy of A. Crosby Longwell. **Figure 11–19** Courtesy of John Trever, Albuquerque Journal. **Figure 11–20** AP/World Wide Photos. **Figure 11–22a** Courtesy of Massachusetts Water Resources Authority. **Figure 11–22b** Courtesy of U.S. Geological Survey, Woods Hole, Massachusetts. **Figure 11–23** Photo by Thomas D. Mangelsen, courtesy Peter Arnold, Inc. **Figure 11–24** Minamata, Tokomo is bathed by her mother, ca 1972. Photograph by W. Eugene Smith. Collection, Center for Creative Photography, The University of Arizona. © Aileen M. Smith, Courtesy Black Star, Inc., New York. **Figure 11–26** APT photo. **Figure 11–27** Reprinted with permission of the Ocean Conservancy (formerly the Center for Marine Conservation). **Figure 11–28** Photo by Wayne Perryman, National Marine Fisheries Service. **Table 11A** Data from The Oil Spill Intelligence Report at **http://www.cutter.com/osir/biglist.htm.**

Chapter 12

Figure 12A left Smithsonian Institution Libraries © 2001 Smithsonian Institution. **Figure 12A right** Photo courtesy of Corbis–The Bettmann Archive. **Figure 12C** APT photo. **Figure 12–7** After Sverdrup, H. U. et al., 1942. **Figure 12–9** Reprinted by permission from Hallegraeff, G. M., *Plankton: A Microscopic World,* 1988 (p. 21). Courtesy of E. J. Brill, Inc. **Figure 12–16** Photo © Ralph Lee Hopkins/Wilderland Images. **Figure 12–17a** Courtesy of Deeanne Edwards, Scripps Institution of Oceanography, University of California, San Diego. **Figure 12–17b** APT photo. **Figure 12–18** Courtesy National Oceanic and Atmospheric Administration/Department of Commerce. **Figure 12–21** Photo by Peter Arnold, Inc. 12-22-Woods Hole Oceanograhic Institution.

Chapter 13

Chapter Opening Feature Reprinted with permission and light editing from Howard, J., Vanishing Act: Critical Link in Marine Food Chain May Be at Risk. *Scripps Institution of Oceanography Explorations* 2:2, p. 13, Scripps Institution of Oceanography, University of California, San Diego, 1995. **Figure 13A** Courtesy of Elizabeth Venrick, Scripps Institution of Oceanography, University of California, San Diego. **Figure 13B** Carleton Ray/Photo Researchers, Inc. **Figure 13D** Copyright © 1999 by Universal City Studios, Inc. Courtesy of Universal Studios Publishing Rights, a division of Universal Studios Licensing, Inc. All Rights Reserved. Photo Courtesy of the Academy of Motion Picture Arts and Sciences. **Figure 13E inset** Courtesy of North Carolina Divison of Water Quality. **Figure 13F upper** Courtesy of JoAnn Burkholder. **Figure 13F middle** Courtesy of JoAnn Burkholder and Howard Glasgow. **Figure 13F lower** Courtesy of JoAnn Burkholder et al., with permission from *Nature.* **Figure 13G inset** Photo courtesy of National Marine Fisheries Service. **Figure 13H** Courtesy of the American Geophysical Union. **Figure 13–2** Courtesy of Elizabeth Venrick, Scripps Institution of Oceanography, University of California, San Diego. **Figure 13–3c** Provided by the SeaWiFS Project, NASA/Goddard Space Flight Center and ORBIMAGE. **Figure 13–5** Courtesy of Patricia Deen, Palomar College. **Figure 13–6** Provided by the SeaWiFS Project, NASA/Goddard Space Flight Center and ORBIMAGE. **Figure 13–7** APT photo. **Figure 13–8 a, b, and d** APT photos. **Figure 13–8c** Photo by Hal Thurman. **Figure 13–9a** Reprinted by permission from Hallegraeff, G. M., *Plankton: A Microscopic World,* 1988 (p. 43).

Courtesy of E. J. Brill, Inc. **Figure 13–9b** Courtesy of Wuchang Wei, Scripps Institution of Oceanography, University of California, San Diego. **Figure 13–9c** Reprinted by permission from Hallegraeff, G. M., *Plankton: A Microscopic World,* 1988 (p. 91). Courtesy of E. J. Brill, Inc. **Figure 13–9d** Reprinted by permission from Hallegraeff, G. M., *Plankton: A Microscopic World,* 1988 (p. 81). Courtesy of E. J. Brill, Inc. **Figure 13–10d** Photo courtesy of Scripps Institution of Oceanography, University of California, San Diego. **Figure 13–15a** Photo © Marty Snyderman. **Figure 13–15b** Photo by Christopher Newbert. **Figure 13–15c** Photo © Marty Snyderman. **Figure 13–21** After Pauley and Christensen, 1995. **Figure 13–22** Data from the Food and Agriculture Organization of the United Nations. **Figure 13–23** Photo by W. High, courtesy of National Marine Fisheries Service. **Figure 13–25** Data from the Food and Agriculture Organization of the United Nations. **Table 13–1** Data after Strahler, A. H. and Strahler, A. N., *Modern Physical Geograph,* 4th Ed., Wiley, 1992 (Table 25.1).

Chapter 14

Figure 14A Photo courtesy of Corbis–The Bettmann Archive. **Figure 14B** Photo by Kelvin Aitken, courtesy Peter Arnold, Inc. **Figure 14C** Photo by Francois Gohier, courtesy of Photo Researchers, Inc. **Figure 14D** Data from National Marine Fisheries Service. **Figure 14E** Photo © Ralph Lee Hopkins, Wilderland Images. **Figure 14–3** Courtesy of Kozo Takahashi, Kyushu University, Fukuoka, Japan. **Figure 14–4** APT photo. **Figure 14–5** From Wilhelm Giesbrecht's 1892 book *Fauna und Flora des Golfes von Neapel und der Angrenzenden Meeres–Abschnitte, Berlin:* Verlag Von R. Friedländer and Sohn. Courtesy of Scripps Institution of Oceanography Explorations, Scripps Institution of Oceanography, University of California, San Diego. **Figure 14–6** Courtesy of Scripps Institution of Oceanography, University of California, San Diego. Photographer: Paige Jennings. **Figure 14–6 inset** Photo by Pam Blades-Eckelbarger, Harbor Branch Oceanographic Institution, Inc. **Figure 14–7b** Photo by Larry Ford. **Figure 14–10a** Photo © Marty Snyderman. **Figure 14–10b** Photo © Wayne and Karen Brown. **Figure 14–10c** Photo © Greg Ochocki/The Stock Market. **Figure 14–10d** Photo © Bob Gomel/The Stock Market. **Figure 14–10e** Photo © Valerie Taylor/Peter Arnold, Inc. **Figure 14–11** After Lalli and Parsons, 1993 (Fig. 6.5). **Figure 14–12** After Lalli and Parsons, 1993 (Fig. 6.6b and 6.7a). **Figure 14–13a** Photo © Fred Bavendam/Peter Arnold, Inc. **Figure 14–13b** Photo courtesy of National Marine Fisheries Service. **Figure 14–15** Photo © Thomas Ives/The Stock Market. **Figure 14–16a** © Jeff Foott/Bruce Coleman, Inc. **Figure 14–16b** Photo © Ralph Lee Hopkins, Wilderland Images. **Figure 14–17a** Photo © Ralph Lee Hopkins/Wilderland Images. **Figure 14–17b** Photo © C. Allan Morgan/Peter Arnold, Inc. **Figure 14–17c** APT photo. **Figure 14–19a** Photo © Fred Bavendam/ Peter Arnold, Inc. **Figure 14–19b** D. Fleetham/Animals Animals/ Earth Scenes. **Figure 14–22** APT photo, courtesy Sea World of California. **Figure 14–25b and c** APT photos, courtesy Sea World of California. **Figure 14–26** Photo by Kennan Ward Photography (© 1990). **Table 14A** Adapted from the International Shark Attack File at http://www.flmnh.ufl.edu/fish/Sharks/isaf/isaf.htm. **Box 14–1** Adapted from the International Shark Attack File at http://www.flmnh.ufl.edu/fish/Sharks/isaf/isaf.htm.

Chapter 15

Figure 15A Reprinted by permission of The Bettman Archive. **Figure 15B** Courtesy of Jerry Wellington, University of Houston. **Figure 15C** Photos courtesy of Robert Hessler, Scripps Institution of Oceanography, University of California, San Diego. Reprinted from *Deep-Sea Research,* Vol. 25, Hessler, R. R., Ingram, C. L., Yayanos, A. A., and Burnett, B. R., Scavenging amphipods from the floor of the Philippine Trench, Fig. 1, Copyright 1978, with permission from Elsevier Science. **Figure 15D** Courtesy of Deeanne Edwards, Scripps Institution of Oceanography, University of California, San Diego. **Figure 15–1** After Zenkevitch, L. A. et al., 1971. **Figure 15–2b, c, e, g, h** Photos by Hal Thurman. **Figure 15–2d** Photo © Norbet Wu/Peter Arnold, Inc. **Figure 15–2f** Photo © Breck P. Kent. **Figures 15–2i** APT photo. **Figure 15–3** Photo © Joy Sparr/Bruce Coleman, Inc. **Figure 15–5a** Photo by James McCullagh. **Figure 15–5b** Photo by Hal Thurman. **Figure 15–6** Photo by Hal Thurman. **Figure 15–7a** Photo © Fred Bavendam/ Peter Arnold, Inc. **Figure 15–7b** Photo © Eda Rogers. **Figure 15–11** Photo by Hal Thurman. **Figure 15–13** Courtesy of Howard J. Spero, University of South Carolina. **Figure 15–14** Photo by Stephen J. Kraseman © Peter Arnold, Inc. **Figure 15–15b** Photo by Mia Tegner. **Figure 15–15c** Photo by James McCullagh. **Figure 15–16a** Photo by B. Kiwala. **Figure 15–16b** Photo by Harold W. Pratt/Biological Photo Service. **Figure 15–18** After Stehli, F. G. and Wells, J. H., Diversity and Age Patterns in Hermatypic Corals, *Systematic Zoology* 20, 115–126, 1971. **Figure 15–20a** Photo by Christopher Newbert. **Figure 15–20b** Photo by C. R. Wilkerson, Australian Institute of Marine Science. **Figure 15–20c** Photo by Ken Lucas/Biological Photo Service. **Figure 15–21** Photos by Christopher Newbert. **Figure 15–22** After Lalli and Parsons, 1993 (Flg. 8.16). **Figure 15–25** After Ron Johnson, Old Dominion University. **Figure 15–26a** Photo courtesy of Woods Hole Oceanographic Institution. **Figure 15–26b** SEM photomicrograph courtesy of Woods Hole Oceanographic Institution. **Figure 15–26c** Photo courtesy of Robert Hessler, Scripps Institution of Oceanography, University of California, San Diego. **Figure 15–27** Courtesy of Peter A. Rona, NOAA. **Figure 15–28c** Courtesy of C. K. Paull, Scripps Institution of Oceanography, University of California, San Diego. **Figure 15–29b and c** Courtesy of Charles R. Fisher, Pennsylvania State University. **Figure 15–30b** Photo courtesy of JAMSTEC.

Afterword

Figure Aft–1 APT photo.

Appendices

All Reprinted with permission of Hubbard Scientific, Inc. Physiographic Chart of the Sea Floor © Hubbard Scientific. **Figure A5A** APT photo. **Figure A5–1** APT photo.

Index

Abalone, 354
Abyssal clay, 104–105
Abyssal hill provinces, 85
Abyssal hills (seaknolls), 85
Abyssal plains, 84–85
  formation, 87
  volcanic peaks, 85
Abyssal storms, 458
Abyssal zone, 367
Abyssopelagic zone, 363–365
Acanthaster planci, 467
Acid, 146
Acorn barnacles, 354, 440, 443
Acoustic Thermometry of Ocean
  Climate (ATOC)
  experiment, 192–193
Active margins, 80–83
Actual geostrophic flow, 208
Adventure, 17
Aegean Sea, 327
African Plate, 54
Agassiz, Alexander, 406
Age of Discovery, 13–17
Agulhas Current, 225
Air flow, high- and low-pressure
  regions and, 177
Air masses, 176. *See also*
  Atmosphere
  U.S. weather and, 178
Alaska
  Lituya Bay, 236–237
  Prince William Sound, 329–332
  Scotch Cap, 256
Alaskan Current, 217
Albedo, 164
Aleutian Trench, 62
Alexander the Great, 34
Algae
  in biogenous sediment, 106
  blue-green (cyanobacteria),
    348
  calcium carbonate secretion,
    358
  macroscopic
    brown (Phaeophyta),
      380–381, 445
      green (Chlorophyta), 381
      red (Rhodophyta), 381–382
  marine, 367–368
  microscopic
    golden (Chrysophyta), 382
    photos of, 383
  red (*Lithothamnion*), 441, 443
  stipes and blades, 450
  symbiosis with coral, 456–457
  zooxanthellae, 455–457, 468
Alkaline, 146–147
Alvarez, Luis, 118
Alvarez, Walter, 118
Alveoli, 423, 426
Alvin, 34–35, 163, 461, 464
Alviniconcha hessleri, 463
Amazon River, 278
Amnesic shellfish poisoning, 385
Amphidromic point, 275
Ampules, 144
Anacapa Island, brown pelicans
  on, 336–337
Anadromous fish, 368
Anaerobic bacteria, 23

Anal fins, 412
Anchovies (anchovetas), 217,
  399–400
  fishing industry, 230
  schools, 418
Andes Mountains, 55, 82, 86, 88
Andesite, 55
Anemones, 441, 443–444, 461,
  463, 466
Angelas, 457
Anglerfish, 353, 366
  deep-sea, 413, 415
Animalia, 349
Animals. *See also* Marine
  organisms
  evolution of, 23–25
  multicellular, 348
  benthic environment. *See*
    Benthic organisms
  pelagic environment. *See*
    Pelagic organisms
Annapolis River, 265
Annelid worms, 447–449
Anoxic, 323
Antarctic Bottom Water, 226
Antarctic Circle, seasonal
  changes and, 165
Antarctic Circumpolar Current,
  211
Antarctic Convergence, 211, 213,
  226
Antarctic Divergence, 212
Antarctic Intermediate Water,
  226
Antarctic Ocean, 8–10
Antarctic region, productivity in,
  383–384
Antarctica
  Cook's exploration, 17
  Ross expedition, 438
  separation from Australia, 66
  shelf break around, 82
Anthocyrtidium ophirense, 409
Anthopleura, 443
Anti-cyclonic flow, 175
Antilles Current, 214
Antinodes, 253
Anydrite, 114
Aphelion, 274
Aplysia californica, 451
Apogee, 274
Appalachian Mountains, 323
Aquaculture, 401
Aqualung, 7
Aquanauts, 74–75
Aragonite, 112
Archaea, 347, 349, 460
Archaeons, 90
Architeuthis, 412
Arctic Circle, seasonal changes
  and, 165
Arctic Convergence, 226
Arctic Ocean, 9, 10
  exploration of, 207
  gray whale migration, 430
  shelf break, 82
Arenicola, 447–449
Argo, 163
Argo Merchant oil spill, 332–334

Aristotle, 6
Arothron, 459
Arrowworms, 351
Asaro, Frank, 118
Asia, collision with India, 60
Asterias, 440
Asteromphalus, 351
Asthenosphere, 49, 50
Astroblemes, 119
Astropecten, 447, 448
Aswan High Dam, 297
Atlantic Coast of U.S., 304–305
Atlantic Equatorial
  Countercurrent, 214
Atlantic Ocean, 8
  abyssal plains, 85, 87
  Bay of Fundy, 277. *See also* Bay
    of Fundy
  "Bermuda Triangle", 29
  coral, 455
  early navigators, 13
  hills of water, 207
  hydrothermal vent organisms,
    463
  Madeira abyssal plain, 87
  North
    circulation, 215
    floor, 89
    Franklin's interest, 197
    major regions, 82
    mountain ranges across, 37
  ooze, 119
  passive margins, 83
  pattern of age distribution, 46
  pelagic sediment, 120
  Strait of Gibraltar and, 115
  subsurface water masses, 228
  surface circulation pattern, 201
  surface currents, 213–217
Atmosphere
  composition, 166
  density, 166–167
  greenhouse effect, 185–190
  heating of Earth's, 187
  movement in, 168
  origin, 21–22
  pressure, 168–169
  temperature, 166–167
  water vapor content, 167
Atmospheric circulation cells,
  172–175
  boundaries, 173
  pressure, 172
  wind belts, 172–173
Atmospheric tides, 281
Atmospheric waves, 238
ATOC. *See* Acoustic
  Thermometry of Ocean
  Climate (ATOC)
  experiment
Atoll, 65–66
Atomic structure, 131–132
Atoms, 20, 131–132
Australia
  Great Barrier Reef, 107–108
  separation from Antarctica, 66
Autotrophic, 389–390
Autotrophs, 23
Autumnal equinox, 164
Avicennia, 379

Backshore, 289
Backwash, 290
Bacon, Francis (Sir), 35
Bacteria, 347–348
  heterotrophic, 349
  oil-degrading, 334
Bacterioplankton, 349–350
Bacteriovores, 390
Badger, 236
Baffin Bay, 438
Bahama Banks, 107–108
Balaenoptera musculus, 425
Balaenopterids, 429
Balanus, 443
Balboa, Vasco Núñez de, 14
Baleen whales, 428–429
Ballard, Robert, 34, 162–163
Baltic Sea, salinity of, 142
Bar-built estuary, 320
Barnacles, 440, 443
Barotrauma, 7
Barracuda, 353
Barrier flat, 296
Barrier islands, 296, 298
  formation of, 299
  lagoons separating, 320
Barrier reef, 64–66
Barton, Otis, 34
Barycenter, 266
Basalt, 50
Bascom, Willard, 1
Base, 146–147
Bathyal zone, 367
Bathymetry, 75–77
Bathypelagic zone, 363–365
Bathyscaphe, 10, 12
Bathysphere, 34
Bay barrier, 295
Bay of Fundy, 265, 277, 280
Bay-mouth bar, 295
Beach. *See also* Sand
  black sand, 124–125
  composition, 290
  defined, 289
  drowned, 300–301
  movement of sand, 290–292
  ocean, 296
  recreational, 290
  sandy, benthic organisms on,
    447–449
  seawalls and, 310
  as sediment-covered shores,
    446
  summertime, 291–292
  terminology, 289–290
  wave activity on, 291
  wintertime, 291–292
Beach compartments, 297–298,
  301
Beach face, 290
Beach hoppers, 448
Beach nourishment, 300
Beach replenishment, 300, 311
Beach sand, 103
Beach starvation, 299, 311
Beagle, 63–65, 346–347
Beaufort Sea, gray whale
  migration in, 430
Beebe, William, 34
"Bends," 7, 423–424

Benguela Current, 214
Bennington, 247
Benthic (sea bottom) environment, 365–367
Benthic organisms, 367
　in coral reefs, 452–458
　on deep-ocean floor, 458–466
　rocky shorelines and, 439–446
　on sandy beaches, 447–449
　on sediment-covered shores, 446–450
　on shallow offshore ocean floor, 450–458
Benthos, 350–352, 354
Bering Sea, 430
Bering Strait, 206
Berm, 290
"Bermuda Triangle," 29
Biddulphia favus, 351
Biddulphia mobiliensis, 351
Bioaccumulation, 338–339
Bioassay, 329
Bioerosion, 458
Biogenous sediment, 100
　composition, 106
　defined, 105
　distribution, 106–110
　macroscopic, 105
　microscopic, 105
　origin, 105–106
Biogeochemical cycling, 391–392
Biological oceanography, 2
Biological productivity
　polar oceans, 383–384, 388
　primary productivity, 373–379
　regional productivity, 382–389
　temperate oceans, 388–390
　tropical oceans, 384–389
　See also Productivity
Biological pump, 382
Bioluminesce, 413
Bioluminescent organisms, 364
Biomass
　distribution of benthic, 440
　pelagic organisms in, 407
　phytoplankton, 373
　plankton, 350, 358
Biomass pyramid, 393–394, 396
Bioremediation, 334
Biotic community, 389
Biozones, 363–367
Birds, The (Hitchcock), 384
Bivalve mollusks, 447
Black band disease, 454
Black Sea, 327
Black smokers, 89, 92, 93, 460. *See also* Hydrothermal vents
Blackbar soldierfish, 392
Bladder kelp, 450–451
Blake Nose, 118
Blue marlin, 414
"Blue moon," 281
Blue mussels, 443
Blue whale, 383–384, 425, 428
Bluefin tuna, 353, 416
Blue-gray sponge, 457
Boiling point, 135
Bonds, 132
Bonito, 416
Boston Harbor sewage project, 335–337
Bottlenose dolphin, 353, 423, 425, 428
Bottlenose whales, 423
Bottom dwellers, 350–352, 354
Boundary currents, 200

Brachyuran crabs, 461
Brackish water, 142, 326
Brain coral, 354
Brazil Current, 200, 214
Breakers, 250–251
Breakwaters, 308–309
Brevoortia tyrannus, 386
Brittle, 49
Brittle star, 354
Brown algae, 380–381
Brown sponge, 457
Bull kelp, 450
Bullard, Edward (Sir), 35–36
Burrowing, 447
　clams, 449
Bycatch, 397–398

Caglioti, Luciano, 130
Calanus, 356. *See also* Copepod
Calcareous ooze, 106, 110, 114–115
Calcite, 106, 146. *See also* Calcium carbonate
Calcite compensation depth (CCD), 110
Calcium carbonate, 106–108, 358
　distribution, 113
　shells and, 146
California
　Anacapa Island, brown pelicans, 336–337
　grunion along beaches, 282–283
　Jalama Beach, 2
　Leucadia, house too close to beach, 288
　Monterey Bay, contaminated fish, 385
　Newport Harbor, wave reflection, 253
　San Andreas Fault, 60–61, 82, 321
　San Clemente Island, 306
　Santa Monica, breakwater, 308–309
　sea-surface temperatures off southern, 223
　Solana Beach, seawall, 311
　temperature range comparisons, 136
California Cooperative Oceanic Fisheries Investigations (CalCOFI), 372–373, 375
California Current, 217
Calocalanus pavo, 410
Calorie, 134
Calypotogena, 461
Calypso, 91
Calyptogena soyoae, 467
Camoflage, 361–362
Canada, Baffin Bay, 438
Canary Current, 215
Cannery Row (Steinbeck), 372
Cannon, Berry, 75
Caño, Juan Sebastian del, 15
Cape Cod, Massachusetts, 336, 439
Cape Hatteras, North Carolina, 304, 310, 312
Capillary waves, 244–245
Carbon dioxide. *See* Atmosphere; Greenhouse effect; Photosynthesis
Carbonate buffering system, 146–147
Carbonate deposits, 107
Carbonates, 112, 114

Carcharodon carcharias, 417
Cardium, 448
Caribbean Current, 214
Caribbean grouper, 392
Caribbean Sea
　barrier reefs, 64
　coral, 455
Carnivora, 420–421
Carnivores, 390, 392
Carnivorous feeding, 447
Carotin, 382
Carpenter, M. Scott, 75
Carson, Rachel, 97
Cascade Mountains, 55, 59
Cascadia subduction zone, 257, 467
Caspian Sea, 10
Caudal fins, 412, 414
CCD. *See* Calcite compensation depth
Celerity, 242
Centripetal force, 267–269
Cephalopods, gas containers in, 407
Ceratium bucephalum, 351
Ceratium recticulatum, 351
Cetacea, 422–432
　breathing, 423
　deep diving and, 422–423
　nitrogen narcosis and, 423–424
　suborder Mysticeti, 428–430
　suborder Odontoceti, 426–428
　swimming speed, 422
Chaetoceras, 351
Chakrabarty, A. M., 334
Chalk, 106
Challenger Deep, 10
Challenger Deep area, 86
Chauliodus sloani, 416
Chelonia mydas, 44
Chemical oceanography, 2
Chemical precipitate, 89–90
Chemosynthesis, 23, 373, 460, 462
Chemosynthetic life, 463
Chesapeake Bay estuary, 323–324
Chiasmodon niger, 416
Chicxulub Crater, 118
Chignecto Bay, 277–278
Chlorinity, 144
Chlorofluorocarbons, 199
Chlorophyll, 24
Chlorophyta, 381
Chondrites, 114
Chronometer, 14
Chukchi Sea, gray whale migration in, 430
Circular orbital motion, 241
Circulation, ocean. *See* Ocean currents
Circulation cells, 172–175
Clam, 354, 461, 448
　giant, 457
　burrowing, 449
　giant white, 467
Clathrates, 120–121
Clean Water Act of 1972, 335
Climate. *See also* Weather
　continental drift and, 37–38
　defined, 175
　North Atlantic Currents and, 215–216
　ocean currents and, 208–209
　patterns in oceans, 182–185
　thermohaline circulation and, 228–229
Cline, Isaac, 184

Cnidarians, planktonic, 412
Coast, 289. *See also* Coastal regions; Coastline; Shore; Shoreline
　characteristics of U.S., 304–305
Coastal cliff erosion, 311, 313
Coastal downwelling, 211, 212
Coastal geostrophic currents, 318–320
Coastal ocean, 316–345
Coastal plain estuary, 320
Coastal regions. *See also* Beach; Coast; Shore; Shoreline
　Atlantic Coast, 304–305
　beach composition, 290
　beach terminology, 289–290
　estuaries. *See* Estuaries
　in general, 289
　geostrophic currents, 318–319
　Gulf Coast, 305
　movement of sand, 290–292
　Pacific Coast, 305
　pollution affecting, 327–336
　primary productivity, 375–376
　rocky bottom, 450–452
　rocky shores, 439–446
　salinity, 318–319
　sediment-covered shores, 446–450
　shallow offshore ocean floor, 450–452
　temperature, 318–319
　wetlands, 323
Coastal tidal currents, 277–280
Coastal upwelling, 209, 212, 376–377, 386
Coastal waters, 318–320
　pollution, 327–336
　salinity, 318–319
　temperature, 318–319
Coastal wetlands, 323–327
Coastal Zone Color Scanner (CZCS), 80
Coastline, 289. *See also* Coast; Coastal regions; Shore
Cobalt, 123
Coccolith ooze, 110
Coccolithophores, 106, 108, 358, 382, 383
Coccolithophoridae, 351
Coccoliths, 106, 382, 383
Cockle, 448
Codium fragile, 381
Coelenterates, 410
Cohesion, 133
Cold core rings, 214
Cold front, 176, 179
Cold-blooded fish, 416–418
Color
　of objects, 376–377
　water, ocean life and, 379
Coloration, disruptive, 431
Columbia River estuary, 322–323
Columbus, Christopher, 13–14, 16, 174–175
Comet Shoemaker-Levy, 119
Comets, icy, 26
Commensalism, 390–392
Compensation depth for photosynthesis, 375
Compressed-air diving, 27–28
Condensation, latent heat of, 137
Condensation point, 135
Conshelf project, 74
Constant Proportions, Principle of, 131

Constructive interference, 246, 248, 271
Consumers, 389
    primary, 408
Continental accretion, 65–66
Continental arc, 55, 86
Continental crust, 21, 50
Continental drift
    climate evidence, 37–38
    defined, 35
    evidence for, 35–40
    fit of continents, 35–36
    objections to model, 39–40
    organism distribution and, 38–39
    sequences of rocks and mountain chains, 36–37
Continental margin
    active, 80–83
    continental rise, 84
    continental shelf, 82
    continental slope, 82–83
    defined, 77
    distribution of sediment across, 117
    passive, 77, 80, 82, 84
    submarine canyons, 83
    turbidity currents, 83
Continental rise, 84
Continental shelf, 82
    deposits, 103
Continental slope, 82–83
Continental sulfide ore deposits, 93
Continental transform fault, 59
Continental-continental convergence, 56
Continents. See also Plate boundaries
    comparing to oceans, 10
    crust, 21, 50
    Pangaea, 35–37
Convection cells, 43, 167
Convergent active margins, 80–81
Convergent boundaries, 52, 55–59
    earthquakes and, 56–57
    three sub-types, 58
Convergent evolution, 432
Converging surface water, 209
Conveyer-belt circulation, 227–229
Cook, James (Captain), 17
Coos Bay, Oregon, 335
Copepods, 351, 356, 408, 410
Copilia vitrea, 410
Coral
    calcium carbonate, 358
    individual, 452
    normal vs. bleached, 454
    soft gorgonian, 459
    symbiosis with algae, 456–457
Coral bleaching, 454
Coral crab, 446
Coral reef
    benthic organisms, 452–458
    defined, 453
    development, 63–65
    distribution and diversity, 453
    nutrient levels, 458
    practical uses, 466–467
    productivity, 386
    zonation, 456
Core of Earth, 21, 49
Cores, 97
Corethron, 351
Coriolis effect

changes in, with latitude, 171
coastal geostrophic currents and, 318–319
Ekman spiral and, 203–204
estuary salinity and, 323
in general, 168–169
merry-go-round example, 169–170
missile example, 170–171
summary of, 173
Coscinodiscus, 351
Cosmogenous sediment, 100, 114
Cosmogenous spherule, 116
Cotidal lines, 275–276
Countershading, 361–362
Cousteau, Jacques-Yves, 6–7, 74, 91
Covalent, 132
Crab, 354
    fiddler, 449, 450
    galatheid, 463
    sand, 448–449
    shore, 446
Crests, 240
Cretaceous–tertiary (K–T) Event, 118–119
Crown-of-thorns, 467
Cruisers, 415, 418
Crust
    continental, 50
    Earth's, 21, 49, 301–302
    oceanic, 50
    oceanic vs. continental, 51
    as sediment, 123–124
Crustacea, 408–409
Crustaceans, 448
Currents. See also Ocean currents
    longshore, 291
    rip, 294
    tidal, 277–280
    turbidity, 83, 86, 103
Curves, monthly, 279
Cutaneous arteries, 418
Cutaneous vein, 418
Cyanobacteria, 348
Cyclones, 178
Cyclonic flow, 175
CZCS. See Coastal Zone Color Scanner

da Gama, Vasco, 13
Darwin, Charles, 63, 108, 346
Davidson Current, 319–320
DDT, 336–338, 340
Dead man's fingers, 381
Dead Sea, 143
Dead whale hypothesis, 463
"Dead zone," 400
Decay distance, 245
Deccan Traps, 118
Declination, 164, 273–274
Decomposers, 389
Decompression illness, 7
Deep current, 198
    conveyer-belt circulation, 227–229
    model, 229
    thermohaline circulation, 225–226
Deep scattering layer (DSL), 364
Deep Sea Drilling Project (DSDP), 97
Deep water, 153
    circulation model, 229
    dissolved oxygen, 228
    sources, 226–227

Deep-ocean basin, 77
    abyssal plains, 84–85
    sediment, 101
Deep-ocean fish, 353
Deep-ocean floor
    benthic organisms, 458–466
    food sources, 459, 461
    historical record, 97–129
    hydrothermal vent biocommunities, 459–464
    low-temperature sccp biocommunities, 464–468
    physical environment, 458–459
    species diversity, 459
Deep-sea anglerfish, 415
Deep-sea clam, 438
Deep-sea fans, 84–85
Deep-sea fish, 413, 415
Deep-water mapping, 42
Deep-water waves, 241–243
    shoaling, 248
    speed, 243
DeLong, George Washington, 206
Delphinapterus leucas, 423
Delphinus delphis, 423
Delta, 297, 300
Density stratification, 20
Deposit feeding, 447
Deposition, U.S. coastal, 304
Depositional-type shores, 292, 295–300
Desalination, 155–157
Destruction, 107
Destructive interference, 246, 248
Detritus, 390, 413, 447
"Devil's Triangle," 29
Diatom
    Diploneis, 383
    in general, 106–108
    illustration, 351
    surface area, 355
    test of, 382
    warm-water, 356
Diatomaceous earth, 106, 382
Diatomaceous ooze, 108
Diaz, Bartholomeu, 13
Diffusion, 358–359
Dilution, 107, 341
Dinoflagellate, 351, 382, 383
    toxic, 385–386
Dinosaurs, extinction of, 118
Diodora aspera, 443
Diploneis, 383
Dipolar molecules, 133
Disphotic zone, 363
Disruptive coloration, 361, 431
Distillation, 155–156
Distributaries, 297
Dittmar, William, 131, 144
Diurnal inequalities, 276
Diurnal tidal pattern, 276
Divergent boundaries, 52–55
Diving
    compressed-air, 27–28
    history of, 6–7
Doldrums, 173
Doliolum, 351
Dolphins
    bottlenose, 423, 425, 428
    diving by, 423
    locating tuna with, 397
    in order Cetacea, 422
    porpoises vs., 426
    speed of, 416
    spotted, 397
    swimming ability of, 432–433

Tuffy, 75
Doppler flow meter, 199
Dorsal aorta, 418
Down-coast, 290
Downstream, 291
Downwelling, 209, 211, 212, 219
Drake Passage, 212
Dredge, 97, 308
Drift bottles, 203
Driftnets, 397
Drilling rig, 121
Drowned beaches, 300–301
Drowned river valleys, 300–301, 306, 320
DSDP. See Deep Sea Drilling Project
DSL. See Deep scattering layer
Dugongs, 421
Dunes, 296
Dynamic topography, 199
Dynophysis, 351

Eared seals, 421
Earle, Sylvia, 316
Earth
    atmospheric circulation cells, 172–175
    changes to environment, 25
    crust, 301–302
    distance from Moon, 274
    distribution of species, 355
    elliptical orbit, 274
    energy radiated by, 186
    explanation for surface features, 35
    gravity, 267
    heating of atmosphere, 187
    if nonspinning, 168–169
    latent heat exchange, 137–138
    magnetic field, 40–43
    organic compound reservoirs, 122
    origin, 19–21
    paleogeographic reconstructions of, 68
    paleomagnetism, 40
    as planet, 6–33
    rocks, 26
    rotatio, 270–271, 285
    satellite positioning of locations, 67
    seasons, 164–165
    size, 271
    solar heating, 164–168
    solar radiation, 165
    structure, 47–52. See also Earth structure
    tilt of axis, 165
    timeline of major events, 28
Earth structure
    chemical composition, 47, 49
    internal, 21, 50–51
    isostatic adjustment, 50–53
    layers close to surface, 50
    physical properties, 47, 49
Earth tides, 281
Earth–moon system, 267–269
Earthquakes, 47–48
    convergent boundaries and, 56–57
    Grand Banks, 86
    Mid-Atlantic Ridge, 54
    predicting, 69
    tsunami and, 254
East African rift valleys, 57
East Australian Current, 217

East Pacific Rise, 53, 57
East Wind Drift, 211
Easter Island, 11, 454
Eastern boundary currents, 200
Ebb current, 279
Ebb tide, 270
Echinocardium, 448
Echinoderms, 448
Echinus, 445, 451
Echolocation, 427–428
Echosounder, 76
Ecliptic, 273
Ecliptic orbit, 164
Ecosystems
    defined, 389
    energy flow and efficiency in, 395
    fisheries and, 394–398
    marine fishery, 396
Eddies, 214
Edrie, 236
Eelgrass, 379, 449, 450
Ekman, V. Walfrid, 202, 207
Ekman spiral, 202, 204–205, 230
Ekman transport, 204–205, 209, 230, 376, 386, 399
El Niño
    anchovy populations and, 399
    catastrophic weather and, 163
    conditions, 218
    coral bleaching and, 453–454
    defined, 219
    effects of, 221–222
    equatorial countercurrents and, 207
    examples from recent, 221–222
    marine iguanas and, 224–225
    predicting events, 223
    unusual weather and, 230
El Niño–Southern Oscillation (ENSO) conditions, 162, 219
    See also El Niño; La Niña
Electrical conductivity, 144
Electrolysis, 156
Electromagnetic spectrum, 376–377
Electrons, 132
Electrostatic attraction, 133
Elephant seal, 341
Ellesmere Island, glaciers on, 162
Emerging shorelines, 300–301
Emerita, 448–449
Emiliania huxley, 383
Endeavour, 17
Endothermic, 24
Energy flow, 389–392
    biogeochemical cycling, 391–392
    marine ecosystems, 389–390
    symbiosis and, 390–391
    ecosystem, 395
    trophic levels, 394
England
    Royal Society of, 130
    White Cliffs of, 106, 109
English Channel, tidal power plant near, 265–266
ENSO. *See* El Niño–Southern Oscillation (ENSO) conditions
Epifauna, 350, 439
Epipelagic zone, 363–364
Equator
    air movement near, 183
    geographical vs. meteorological, 209

Moon's orbit and, 273
    temperatures near, 149
    water salinity near, 148
Equatorial countercurrents, 207
Equatorial currents, 200
Equatorial low, 172
Equatorial regions, 183
Equatorial tides, 283
Equatorial upwelling, 209, 211, 386
Eratosthenes, 12
Erik "the Red," 13
Eriksson, Leif, 13
Eroded, 100
Erosion, U.S. coastal, 304
Erosional-type shores, 292–295
Eschrichtius robustus, 425
Estuaries
    circulation pattern, 321
    classifying, 322
    defined, 320
    human activities and, 322–323
    influence of productive, 340
    origin of, 320–321
    tidal power plants in, 265
    water mixing in, 321–322
Eubalaena glacialis, 424
Eucampia, 351
Eukarya, 348
Euphotic zone, 363, 375
Euphysetta elegans, 409
Europa, ocean on, 2–3
Europe
    Age of Discovery in, 13–17
    early navigators, 12–13
European Space Agency, ERS-1 satellite, 79
Euryhaline, 358
Eurythermal, 358
Eustatic, 302–303
Eutrophic, 379
Eutrophication, 385
Evaporation
    defined, 137
    latent heat of, 137
    Mediterranean circulation and, 327
Evaporation latitudes, 138
Evaporative salts, 122
Evaporite minerals, 114
Evaporite salts, 116
Evaporites, 145
Evolution, 23–25
    convergent, 432
    natural selection and, 24–25
Exclusive economic zone (EEZ), 317
Exothermic, 24
Exploration, ocean, 10–17
Exxon Valdez oil spill, 329–333
Eye of the hurricane, 179–180

Fact, scientific, 18
Fairweather Fault, 236
Falcate, 426
Falco, Albert, 74
Falkland Current, 214
Fall bloom, 389
Fathom, 75
Fault movement, tsunami and, 253
Fecal pellets, 119, 121
Ferrel, William, 172–173
Ferrel cell, 172
Ferreras, Francisco, 27–28
Fetch, 244

Fiddler crab, 449, 450
Filter feeders, 418
Fin designs, 412–413
Finback whale, 428
Fire Island, 296
Fish. *See also* Nekton; Pelagic organisms; *and specific names of fish*
    anadromous, 368
    circulatory modifications in, 419
    cold- vs. warm-blooded, 416–418
    deep-sea, 413, 415–, 416
    fin designs, 412–413
    gills, 360–361
    mercury concentrations, 339
    osmosis in marine vs. freshwater, 360
    saltwater vs. freshwater, 368
    speed and body size, 416
    tropical, use of color by, 363
Fish egg, 351
Fish larva, 351
Fisheries
    defined, 317n1
    ecosystems and, 394–398
    incidental catch, 397–398
    management, 398–399
Fissures, 87, 90
Fjord, 320
Flagella, 382
Flippers, fore, 421
Floaters, 349–350. *See also* Plankton
Floating objects, transport of, 204
Flood current, 279
Flood tide, 270
Floodplains, along rivers, 322
Florida
    bedrock in, 304–305
    continental margin off, 118
    Escambia Bay, PCBs in, 336
    Miami Beach, 5, 143
Florida Current, 214
Florida Escarpment, 464–465
Florida oil spill, 331, 333
Flotsam, 342
Fluke, 397
Folger, Timothy, 197
Food chain, 394–395
Food webs, 394–395
Foraminifer ooze, 110
Foraminifers, 106, 108, 351, 358, 408–409
Forbes, Edward, 438
Forced waves, 275
Fore flippers, 421
Foreshore, 289
Forked fins, 412
Forsal fins, 412
Fossil fuels, 25
Fossils
    Mesosaurus, 39
    record on Earth, 22
Fracture zones, 91, 93
Fram, 202, 206–207
France, La Rance tidal power plant, 265–266
Franklin, Benjamin, 197–198
Freeze separation, 157
Freezing, latent heat of, 137
Freezing point, 135
Freons, 199
Frequency, wave, 240
Freshwater fish, osmosis in, 360

Freshwater zone, 326
Fringing reefs, 63–64
Fucus filiformes, 443
Full moon, 271
Fully developed sea, 244
Fungi, 349
Fur seals, 421
Fusion reaction, 20

Gagnan, Émile, 7
Galápagos Islands
    coral reefs, 454
    El Niño and, 222
    finches, 346–347
    hotspots, 61
    marine iguana, 224–225
    photosynthetic production, 375
Galápagos Rift, 351, 460–461
Galatheid crab, 463
Galveston, Texas, tidal pattern in, 279
Gas containers, 407–408
Gas hydrates, 120–122
Gaseous state, 135
Geography of oceans, 7–10
Geologic time scale, 25–30
Geological oceanography, 2
Georges Bank, 334
Geosat, 79, 80
Geostrophic currents, 205, 208
    coastal, 318–319
Giant brown bladder kelp, 450, 466
Giant clam, 457
Giant white clams, 467
Giesbrecht, Wilhelm, 410
Gill nets, 397
Gills, 360–361
Glacial ages, continental drift and, 37–38
Glacial deposits, 103–104, 304
Glaciers, 302
Global Positioning system (GPS), 16
Global warming, 186–189
Globigerina ooze, 110
Glomar Challenger, 97
GLORIA, 76–77
Goatfish, 414
Goiters, 143
Golden algae, 382
Goniaulax scrippsae, 351
Goniaulax triacantha, 351
Gonyaulax, 384
Goose-necked barnacles, 354, 440, 443
GPS. *See* Global Positioning system
Graded bedding, 84
Grain size, 101
Grand Banks, 279
Grand Banks Earthquake, 86
Granite, 50
Gravitational force, 267
Gravity corer, 97
Gravity waves, 244–245
Gray angelfish, 414
Gray whale, 425, 428–429
    endangered status and, 433–434
    friendly behavior of, 434
    migration of, 430–432
Great Barrier Reef, 64–65, 107–108, 466–467
Great Lakes, 3
Great Salt Lake, 143

Great white shark, 417, 432
Green algae, 381
Green sea turtles, 44
Greenhouse effect
    air–sea interactions and,
        163–164
    in general, 25, 185–190
    sea level and, 303
Greenhouse gases, 186–190
Greenland, 13, 162, 206
Greenwich Meridian, 275
Grenadier fish, 461
Groin, 307–308
Groin field, 307–308
Gross ecological efficiency, 393
Gross primary productivity, 374
Grotius, Hugo, 316
Grunion, 282–283
Guaymas Basic, 460
Gulf of Bothnia, 302
Gulf of California, 460
Gulf Coast, 305
Gulf of Mexico
    barrier islands, 296
    Chicxulub Crater, 118
    Florida Escarpment, 464–465
    hydrocarbon seeps on, 466
    oil spill, 329
    oil-degrading bacteria, 334
    tidal range, 326
Gulf Stream, 197–198, 200,
        214–217
    benthic species in, 439
    life in, 231
Gulper, 353
Gulper eel, 415
Guyots, 62–63
Gymnodinium, 351
Gypsum, 114
Gyres
    Atlantic Ocean, 213–214
    in general, 201–202
    Indian Ocean, 225
    Pacific Ocean, 217
    subtropical, 209

HABs. *See* Harmful algae blooms
Hadal zone, 367
Hadley cells, 172
Half-life, 26
Halite, 114, 122
Halocline, 149–150, 318, 323
Hansa Carrier, 203
Hard stabilization, 305
    alternatives, 310
    breakwaters, 308–309
    growns and groin fields, 307
    seawalls, 309–310
Harmful algae blooms (HABs),
        382
Harrison, John, 14–15
Hatchet fish, 415
Hawaii
    black sand beach, 124–125
    Cook's exploration of, 17
    early exploration, 11
    Hilo, tsunami damage, 256
    hotspots, 61–62
    Pacific Tsunami Warning
        Center (PTWC), 257–258
Hawaiian Islands–Emperor
        Seamount chain, 61–63
Headlands, 293
Heart urchins, 448
Heat
    in general, 134

latent, 136–139
Heat capacity, 135–136
Heat flow
    in general, 46–47
    oceanic, 165
Heat transfer, principles of, 139
Heavy wave activity, 291
Heliocentric theory, 19
Heliodiscus asteriscus, 409
Hemoglobin, 423
Henry the Navigator, 13
Herbivores, 390, 392
Herjolfsson, Bjarni, 13
Hermatypic, 456
Hermit crabs, 441, 445
Herodotus, 8
Herring, Newfoundland, 394
Hess, Harry, 43
Heterocercal fins, 413
Heterodinium whittingae, 383
Heterotrophic, 390
Heterotrophs, 23
Heyerdahl, Thor, 11–12
High marsh, 296
High slack water, 279
High tide zone, 439, 440
Highly stratified estuary, 321
Himalayan Mountains, 56, 60
Hirondellea gigas, 460
HMS *Beagle,* 63–65, 346–347
HMS *Challenger,* 75, 125, 130–131
Holoplankton, 350
Homarus, 452
Homeothermic, 416
Homogenous, 20
"Hoodoo Sea," 29
Horse latitudes, 173
Hotspots, 61
Hudson Bay, 302
Humboldt Current, 217
Humpback whale, 425, 428, 430
Hurricane Andrew, 180, 184
Hurricane Betsy, 288
Hurricane Camille, 180
Hurricane Dot, 182
Hurricane Gorky, 180
Hurricane Iniki, 182
Hurricane Iwa, 182
Hurricane Mitch, 180
Hurricanes, 178–182. *See also*
        Tropical cyclones
    eye of, 179–180
    in Galveston, Texas, 184–185
    internal structure, 181
    typical storm track, 181
    historic, 180, 182
Hydration, 133
Hydrocarbon, 329
Hydrocarbon seep
        biocommunities, 464
Hydrogen bonds, 133, 138
Hydrogenous sediment, 100,
        111–114
Hydrologic cycle, 148–149
Hydrothermal vents. *See also*
        Black smokers
    biocommunities, 459–464,
        351–352
    in general, 87, 89, 92–93
    origin of life and, 464
Hypersaline, 143
Hypersaline seep biocommunity,
        464–465
Hypertonic, 359
Hypothesis, 18
Hypotonic, 359

Hypoxic zone, 400
Hypsographic curve, 77
Hysographic curve, 81

Ice, formation of, 140
Ice Age, 37, 51, 103, 302n3
    migration as relic from, 431
Ice rafting, 104
Ice sheets, 37
Icebergs, 147, 162, 190
Iceland, 13, 61, 87, 295, 327
Icy comets, 26
Idealized tide prediction, 274–275
Igneous rocks, 40
Iguanas, marine, 224–225
In phase, 246
Incidental catch, 397–398
India, collision with Asia, 60
Indian Ocean, 8, 10
    abyssal plains, 85
    coral, 455
    early navigators, 12–13
    ooze, 119
    passive margins, 83
    pelagic sediment, 120
    South, hills of water, 207
    surface currents, 223, 225
Indian Ocean Gyre, 225
Infauna, 350, 446
Inner core, 49
"Inner space," 34
Inner sublittoral zone, 365
Integrated Ocean Drilling
        Program (IODP), 98
Interface waves, 240
Interference patterns, 245, 248
Intergovernmental Panel on
        Climate Change (IPCC), 189
Internal waves, 238–239, 259
International Convention of
        Pacific Long Driftnet
        Fishing, 398
International Ice Patrol, 162, 190
Intertidal zone
    in benthic environment, 365
    of rocky shore, 439, 441
    of sediment-covered shores,
        446–447
Intertropical Convergence Zone
        (ITCZ), 173
IODP. *See* Integrated Ocean
        Drilling Program
Ionic bond, 133
Ions, 132
IPCC. *See* Intergovernmental
        Panel on Climate Change
Iridium, 118
Irminger Current, 215
"Iron hypothesis," 190
Irons, 114
Island, barrier, 296, 298
Island arc, 55, 86
Isohaline, 318
Isopycnal water column, 155
Isostatic adjustment
    in general, 50–52
    shoreline movement and,
        301–302
Isostatic rebound, 51
Isothermal, 318, 384
Isothermal water column, 155
Isotonic, 359
Isurus, 416
ITCZ. *See* Intertropical
        Convergence Zone
Ixtoc #1 oil well, 329

Japan
    as island arc, 86
    tsunami hits, 257
Japan Current, 217
Japan Trench, 464
Jason Jr., 279
Jason-1 mission, 80
Jeanette, 206
Jellyfish, 351, 361, 411
    medusa, 412
Jet stream, 178
Jetty, 307–308
JOIDES Resolution, 97–98, 125
Joint Oceanographic Institutions
        for Deep Earth Sampling
        (JOIDES), 97
Juan de Fuca Plate, 59, 257, 464
Juan de Fuca Ridge, 61, 87, 460,
        462
Jules' Undersea Lodge, 75
Junger, Sebastian, 236
Jurassic Period, 108

K–T Event, 118–119
Kaiko, 34
Keller, Hannes, 7
Kelp, 368, 381
    bladder, 450–451
    bull, 450
    giant brown bladder, 450, 466
Kelp forests, 451–452
Killer whale, 425, 431, 432
    jawbone of, 426
Kinetic energy, 134
Kingdoms of organisms, 348
Kon Tiki, 11–12
Krakatau volcano, 256
Krill, 408–409, 411
Kuroshio Current, 205, 217, 231
Kyoto Protocol, 189

La Niña, 218–219, 221
Labrador Current, 215
Lagoons, 320, 326
Laguna Madre, 326
Lamna, 416
Lamp shell, 354
Land breezes, 175, 177
Landslides, 102
Lantern fish, 353, 415
Lapita people, 11, 12
Larcospira quadrangula, 409
Lasiognathus saccostoma, 366
Latent heat, 136–139
Latent heat of condensation, 137
Latent heat of evaporation, 137
Latent heat of freezing, 137
Latent heat of melting, 136–137
Latent heat of vaporization, 137
Latitude
    changes in coriolis effect with,
        171
    climate and, 38
    determining at sea, 14
    evaporation, 138
    horse, 173
    precipitation, 138
Lava, 40
Law of the sea, 316
Lederman, Leon, 5
Leeuwin Current, 225
LicmophoraI, 351
Life
    beginning of, 22–25
    distribution of, in oceans, 352
    importance of oxygen to, 22–23

origin of, hydrothermal vents and, 464
Light
    on deep-ocean floor, 458
    transmission of, in ocean water, 376–379
Light wave activity, 291
Ligia, 440, 443
Limestone, 107, 304–305
LIMPET 500, 259
Linnaeus, Carolus, 349
Liquid air, 28–29
Liquid heat, 28–29
Liquid state, 134
Lithified, 99
Lithogenous sediment
    beach sand, 103
    composition of, 101
    distribution of, 103–105
    map, 104
    origin of, 100–101
    texture of, 101–103
Lithosphere, 35, 50
Lithospheric plates, 48
Lithothamnion, 381, 443
Littoral zone, 365
Littorina, 443
Lituya Bay, 236–237
Living things, classification of, 347–349
Lobsters, 452
Loihi, 62
Loligo, 412
Long Island Sound, 337
Long waves, 242
Longitude, determining, 14
Longitudinal waves, 238–239
Longshore bars, 290
Longshore current, 291
Longshore drift, 291
Longshore transport, 291
Longshore trough, 290
Loudon, 256
Low marsh, 296
Low slack water, 279
Low tide, 275–276
Low tide terrace, 290
Low tide zone, 439, 441–446, 449
Low-temperature seep biocommunities, 464–468
Luciferin, 364
"Lucky Strike" vent field, 462
Lugworm, 447–449
Lunar bulges, 270
Lunar cycle, 271
Lunar day, 270–271
Lunate fins, 413
Lungers, 415, 418
Lysocline, 110

Mackerel shark, 416
Macrocystis, 381, 450–451, 466
Macroplankton, 350
Macroscopic algae, 380–382
Macroscopic biogenous sediment, 105
Madeira abyssal plain, 87
Magellan, Ferdinand, 8, 14, 16
Magma, 40
Magnetic anomalies, 43
Magnetic dip, 40
Magnetic field, 40–44, 69
Magnetic inclination, 40
Magnetite, 40
Magnetometer, 42
Magnetoreception, 44
Malvinas Current, 214

Manatees, 421
Manganese nodules, 112, 123–125
Mangrove swamps, 323–325
Man-of-war, Portuguese, 412
Mantle, 21, 49
Mantle plumes, 61
Mapping
    deep-water, 42
    sea floor, 78–79, 93
Maps
    atmospheric transport of surplus heat, 138
    calcium carbonate distribution, 111
    climate regions of ocean, 183
    cotidal, 276
    East African rift valleys, 57
    four principal oceans, 8
    global sea surface elevation, 80
    Herodotus' (450 b.c.), 8
    lithogenous sediments, 104
    Mediterranean Sea, 115, 328
    Mesosaurus fossils, 39
    Pacific islands, peopling of, 12
    Pangaea, 36–38
    sea-surface temperature anomalies, 220
    sea-surface temperature off southern California, 223
    tropical cyclone paths, 180
    U.S. coastal erosion and deposition, 304
    Viking colonies, 13
    voyages of Columbus and Magellan, 16
Mariana Basin, 461
Mariana Trench, 131
    Challenger Deep area, 86
    Challenger Deep region, 10
Marianactis bythios, 463
Marine algae, 367–368, 445
"Marine dust," 12
Marine ecosystems, energy flow in, 389–390
Marine environment
    adaptations of species to, 352–363
    divisions of, 363–367
Marine fish. *See also* Fish
    osmosis in, 360
Marine habitat, 346–371
Marine iguanas, 224–225
Marine Mammal Protection Act, 397
Marine mammals, 420–432
    order Carnivora, 420–421
    order Cetacea, 422–432
    order Sirenia, 421–422
Marine organisms
    benthic. *See* Benthic organisms
    bioluminescent, 364
    classification of, 349–352
    cold-water vs. warm-water, 358
    dissolved gases in seawater and, 360–361
    environmental adaptations
        countershading, 361–362
        dissolved gases, 360–361
        in general, 352–353
        physical support, 353–355
        pressure, 361–363
        salinity, 158, 358–360
        temperature, 356–358
        water's transparency, 361
        water's viscosity, 355–356
    hearing of, 433

macroscopic (large) algae, 380–382
    pelagic. *See* Pelagic organisms
    photosynthetic, 379–382
    physical support of, 353–355
    salinity and, 358–360
    temperature and, 356–358
    water pressure and, 361–363
    water viscosity and, 355–356
    water's high transparency and, 361
Marine provinces, 74–96
Marine science, 1
Marine sediments, 97–129
Marine terrace, 293, 296, 306
Marinelab Undersea Laboratory, 75
Marquesas Islands, 11
Marshes, salt, 296, 323–325
Martha's Vineyard, 342
Martin, John, 190
Massachusetts
    Cape Cod, 439
    Nantucket Island, *Argo Merchant* spill off, 332–334
    West Falmouth Harbor, 331, 333
Matthews, Drummond, 44
Maturity, sediment, 103, 105
Maxim Gorky, 190
Maximum sustainable yield, 396
Maxwellia gemma shell, 445
Mean Lower Low Water (MLLW), 284
Meanders, 214
Mediterranean Intermediate Water, 327
Mediterranean Sea
    circulation, 327
    clues from floor sediment, 115
    early cultures, 7
    early navigators, 12
    map, 115, 328
    Nile sediment and, 297
    Posidonius and, 75
Mega Borg, 335
Meganyctiphanes norvegica, 411
Megaplume, 92
Megaptera novaeangliae, 425
Megapterids, 430
Meiofauna, 448–449, 450
Melanesia, 10, 12
Melon, 427
Melting, latent heat of, 136–137
Melting point, 135
Menhaden, 386
Mercury, minamata disease and, 338–339
Meroplankton, 350, 352
Merry-go-round, coriolis effect example, 169
Mesopelagic zone, 363–364
Mesosaurus, 38–39
Mesosphere, 49
Metal sulfides, 89
Meteor, 76
Meteorite impact event, 118
Meteorite material, 114
Meteorological tides, 284
Methane hydrate, 120–121
Michel, Helen, 118
Micronesia, 10, 12
Microscopic algae, 382
Microscopic biogenous sediment, 105
Mid-Atlantic Ridge

divergent boundary at, 53–54
    fissure in rift valley, 90
    hydrothermal vent near, 462, 464
    "Lucky Strike" vent field, 462
    plate tectonics and, 66
    slow spreading of, 57
    submerging shoreline and, 302
Middle tide zone, 439–441
Mid-latitudes, 185
Mid-ocean ridge
    creation, 90
    defined, 43, 77
    earthquakes along, 47
    features, 87–91
    movement of, 69
    rocks along, 46
    seamounts and tablemounts, 64
Migration, of gray whales, 430–432
Miller, Stanley, 23
Minamata disease, 338–339
Minas Basin, 277
Minerals, 101
Missiles, coriolis effect example, 170–171
Mississippi River Delta, 300, 305
Mixed interference, 247–249
Mixed surface layer, 153
Mixed tidal pattern, 276
Mixotrophs, 456
MLLW. *See* Mean Lower Low Water
Modiolus, 440
Molecules
    defined, 132
    interconnections of, 133
    water, 132–134
Mollusk, 358, 447
Monera, 348
Monodon manaceros, 425
Monsoon, 225
Monthly tidal curves, 279
Monthly tidal cycle, 271
Moon
    declination of sun and, 273
    distance from Earth, 274
    elliptical orbit, 274
    monthly tidal cycle, 271
    phases, 273
    positions, 272
    size, 271
    tidal bulges, 270
Moraine, 320
Moraines, 304
Mountains
    continental drift and, 36–37
    on land, 10
Mousse, 334
Mud flats, 446, 449–450
Munidopsis marianica, 463
Munk, Walter, 192–193
Murray, John (Sir), 406
Museau du singe, 427
Mussels, 354, 440–441, 444, 468
Mutualism, 391–392
Myoblobin, 415
Myomeres, 412
Mysticeti, 428–430
Mytilus, 440, 443

Nadir, 267
Nannoplankton, 106
Nansen, Fridtjof, 202, 206–207
Nantucket Island, *Argo Merchant* spill off, 332–334

Narwhal, 425
National Aeronautics and Space
	Administration (NASA)
	Earth Observing System, 80
	Scatterometer (NSCAT), 80
National Flood Insurance
	Program (NFIP), 288–289,
	310
National Marine Sanctuary, 336
National Oceanic and
	Atmospheric Administration
	(NOAA), 75, 299
National Science Foundation, 97
National Undersea Research
	Program, 75
Natural selection, 24–25
Nautilus, 407
Navigation tools, 14–16
Navigators, early, 10–17
Nazca Plate, 53, 81
Neap tide, 272
Nearshore, 289
Nebula, 19
Nebular hypothesis, 19–20
Negative tide, 284
Nekton, 353, 350. *See also* Fish
	deep-water, 413–414
Nektonbenthos, 350
Nematath, 62
Nematocysts, 441
Nereocystis, 450
Neritic deposits, 103, 107
	distribution of, 116, 119–120
Neritic province, 363
Net primary productivity, 374
Neutral solution, 146–147
Neutrons, 132
New Carissa, 335
New moon, 271
New production, 374
New York, sewage sludge
	disposal, 335
New York City, sea level rise at,
	303
New Zealand, 11
Newton, Isaac (Sir), 265–266
NFIP. *See* National Flood
	Insurance Program
Nile River Delta, 297, 300
Nimbus-7 mission 80
Niphates digitalis, 457
Nitrogen narcosis, 7, 423–424
NOAA. *See* National Oceanic
	and Atmospheric
	Administration
Nodes, 253
Non-point-source pollution, 340
North Atlantic Current, 215
North Atlantic Deep Water, 226
North Atlantic Drift, 215
North Atlantic Gyre, 213–214
North Carolina
	Cape Hatteras, 304, 310, 312
	Outer Banks, 298
North Equatorial Current, 214,
	225
North Pacific Current, 217
North Pacific Gyre, 217
Northeast trade winds, 173
Northern right whale, 424
Norwegian Current, 215
Nova Scotia, 265. *See also* Bay of
	Fundy
NSCAT. *See* NASA
	Scatterometer
Nuclear waste, 125

Nucleus, 132
Nurse shark, 353

Obduction, 93
Observations, 18
Ocean(s)
	beginning of life in, 22–25
	climate patterns in, 182–185
	coastal, 316–345
	comparing to continents, 10
	deep scattering layer (DSL),
		364
	deepest depth in, 10–11
	distribution of life in, 352
	formation of, 22
	four principal, 7–10
	geography of, 7–10
	heat gain and loss from, 167
	light transmission in, 376–379
	origin of, 21–22
	percentage of Earth's water in,
		7
	polar, productivity in, 383–384,
		388
	productivity at margins of,
		375–376
	role in reducing greenhouse
		effect, 189–190
	size and depth of, 9
	surface temperature of world,
		210
	temperate, biological
		productivity in, 388–390
	tides in, 274–280
	tropical, productivity in, 384,
		386–389
	weather, climate and, 175–185
Ocean beach, 296. *See also* Beach
Ocean circulation, 197–235. *See
	also* Ocean currents
Ocean currents. *See also* Currents
	climate and, 208–209
	Coriolis effect on
	deep
		sources of, 226–230
		thermohaline circulation,
			225–226, 228
		worldwide circulation,
			227–239
	defined, 198
	in general, 198
	measuring, 198–199
	sediment transport by, 102
	surface currents. See Surface
		currents
	upwelling and downwelling,
		209–211
Ocean Drilling Program (ODP),
	97, 118
Ocean dynamic topography, 200
Ocean exploration, 10–17
Ocean floor
	age of, 46
	deep. *See* Deep-ocean floor
	human remains on, 460
	mapping of, 93. *See also* Sea
		floor mapping
	paleomagnetism and, 42–43
	pelagic organisms staying
		above, 407–415
	provinces, 77–93
	shallow offshore, 450–458
Ocean salinity. *See also* Salinity
	development of, 22
Ocean surface, satellite
	measurements of, 78

Ocean thermal energy
	conversion (OTEC), 154–155
Ocean trenches
	dimensions of selected, 89
	in general, 43, 85–86
	location of, 88
Ocean waves, 238. *See also* Waves
Oceania, 10. *See also* Pacific
	Ocean
Oceanic common water, 227
Oceanic crust, 21, 50
Oceanic heat flow, 165
Oceanic province, 363
Oceanic ridges, 53, 90
Oceanic rises, 53, 90
Oceanic sediment, 99
Oceanic transform fault, 59
Oceanic-continental
	convergence, 55
Oceanic-Oceanic convergence,
	55
Oceanography
	birth of, 130
	historical notes, 10–17
	interdisciplinary nature of, 2–3
	universities for, 29
	what is?, 1–2
Octacoras, 461
Odontocetes, abilities of, 433
Odontoceti, 426–428
ODP. *See* Ocean Drilling
	Program
Office of Wetlands Protection
	(OWP), 324–325
Offshore drilling rig, 121
Offshore zone, 289
Oil. *See also* Petroleum
	as finite resource, 125
	safe transporting, 340
Oil spills, 329–335
	cleaning, 334–335
Oithona, 356. *See also* Copepod
Oligotrophic, 379
OML. *See* Oxygen minimum
	layer
Ommastrephes, 412
Omnivores, 390
Oncaea conifera, 410
Oolites, 114
Ooze
	calcareous, 106, 110, 114, 115
	coccolith, 110
	diatomaceous, 110
	foraminifer, 110
	globigerina, 110
	in general, 105
	ostracod, 110
	pteropod, 110
	radiolarian, 110
	siliceous, 106–107, 111, 114
	silicoflagellate, 110
Open exposure, 305
Orbital waves, 240
Orbits, elliptical, 274
Orchestoidea, 447, 448
Orcinus orca, 425, 431
Oregon
	Coos Bay, 335
	vent communities off coast of,
		467
Organic carbon, 122
Organic molecules, creation of, 23
Organisms. *See also* Benthic
	organisms; Marine
	organisms; Pelagic organisms
	deep-sea, food sources for, 461

distribution of, continental drift
	and, 38–39
groupings, 349
importance of size, 355
kingdoms of, 348
scientific names for, 367
water content of selected, 355
*Origin of Continents and Oceans,
	The* (Wegener), 34, 39
Origin of Species, The (Darwin),
	346
Orthogonal lines, 252
Osmosis
	in marine organisms, 359–360
	in human body, 157–158
	reverse, 156–157
	wrinkled fingers and, 368
Osmotic pressure, 359
Osprey, 337
Ostracod ooze, 110
OTEC. *See* Ocean thermal
	energy conversion
Out of phase, 246
Outer Banks, 296, 298
Outer core, 49
Outer sublittoral zone, 365–366
Outgassing, 21, 30
Overfishing, 396
OWP. *See* Office of Wetlands
	Protection
Oxygen
	dissolved
		in deep water, 228
		ocean depth and, 366
	importance to life, 22–23
Oxygen minimum layer (OML),
	364, 366
Oyster, 452, 453
Oyster drill snail, 453
Ozone layer, 199

Pachygrapsis crassipes, 446
Pacific Coast, 305
Pacific Decadal Oscillation
	(PDO), 221
Pacific Ocean, 7–8
	blacker smokers, 92
	coral genera, 455
	early navigators, 10–11
	gray whale migration, 430
	gyres, 207
	image of volcano, 77
	island people, 12
	island regions, 10
	Mariana Basin, 461
	Mariana Trench, 131
	ocean trenches along, 86, 88
	pattern of age distribution in,
		46
	pelagic sediment in, 120
	surface currents in, 217–223
	volcanic activity in, 85
	wind-generated waves in, 244
Pacific Plate, 53
Pacific Rim, tsunamis in, 258
Pacific Ring of Fire, 86, 88, 254
Pacific Tsunami Warning Center
	(PTWC), 257–258
Pacific Warm Pool, 218
Padre Island, 296, 326
Pagurus, 441, 445
Paleoceanography, 65–66
Paleogeography, 65–66
Paleomagnetism
	defined, 40
	ocean floor and, 42–43

in rocks, 42
Pangaea, 35–37
Panthalassa, 35
Panuliris, 452
Paralytic shellfish poisoning (PSP), 384
Parasitism, 391–392
Parts per thousand, 142
Passive margins, 77–78, 82, 84
Patchy necrosis, 454
PCBs, 336–338
PDO. *See* Pacific Decadal Oscillation
PDR. *See* Precision depth recorder
Peat bed, 299
Peat deposits, 297, 324
Pectoral fins, 412
Pelagic (open sea) environment, 363–365
Pelagic deposits, 103, 104, 110
distribution of, 116, 119–120
Pelagic environment, animals of, 406–437
Pelagic organisms, 406–437. *See also* Fish
adaptations to avoid being prey, 418–420
adaptations for seeking prey, 415–418
circulatory system modifications, 418
floating, 408–412
marine mammals, 420–432
schooling of, 418–420
staying above ocean floor, 407–415
swimming, 412–415. *See also* Nekton
Pelagic sediment, in each ocean, 120
Pelagophycus, 381
Pelvic fins, 412
Perfect Storm, The (Junger), 236
Peridinium, 351
Perigee, 274
Perihelion, 274
Periwinkles, 440, 443
Persian Gulf War, oil spill during, 329, 331
Persistent organic pollutants (POPs), 336
Peru Current, 217
Peru-Chile Trench, 86, 88, 464
Peruvian anchoveta fishery, 399
Petroleum, 120–121
pollution from, 329
Pfiesteria piscicida, 386–387
pH scale, 146–147
Phaeophyta, 380–381
Philippine Trench, 460
Philosophy of Natural Mathematical Principles (Newton), 265
Phoenicians, 12
Phosphates, 112
Phosphorite, 123
Photophores, 364, 413
Photosynthesis, 24
compensation depth for, 375
defined, 373
productivity and, 373–374
Photosynthetic cells, 355
Photosynthetic marine organisms, 379–382
Phylospadix, 379, 441, 445

Physalia, 412
Physeter macrocephalus, 424
Physical oceanography, 2
Phytoplankton, 349–351, 355, 373, 389, 407
Piccard, Jacques, 10
Picoplankton, 350
Pillow basalts, 87
Pillow lava, 87, 91–92
Ping, 76
Pinnipeds, 420
Pisaster, 440
Plankter, 349
Plankton, 340, 349–350
Plankton nets, 374–375
Planktonic, 106
Planktonic cnidarians, 412
Planktoniella solI, 356
Plantae, 349
Plants
effect on Earth's environment, 25
evolution of, 23–25
multicellular, 348
Plastic
asthenosphere as, 49
in ocean, 340–342
Plate boundaries
convergent boundaries, 55–59
divergent boundaries, 52–55
transform boundaries, 59–60
types of, 52
Plate tectonics
acceptance of theory, 47
applications
coral reef development, 63–65
hotspots, 60–62
seamounts and tablemounts, 62–63
detecting motion, 65
evidence for, 40–47
in general, 35
mantle plumes and, 60–62
predictions of future features with, 66–67, 69
processes of, 43
speed of plates, 67–68
Pleistocene Epoch, 51. *See also* Ice Age
glaciers during, 302
Plunging breaker, 250
Pneumatocysts, 450
Pocillopora verrucosa, 454
Poikilothermic, 416
Polar bears, 420
Polar cell, 172
Polar easterly wind belts, 173
Polar front, 173
Polar highs, 172
Polar regions, 185
Polar wandering, 40–42
Polarity, 42, 133
Pollicipes, 440, 443
Pollution
in coastal waters, 327–336
DDT, 336–338
mercury, 338–339
non-point-source, 340
oil spills, 329–335
PCBs, 336–338
petroleum, 329
sewage sludge, 335–336
what is?, 327–328
World Health Organization on, 328

Polynesia, 10–11, 12
Polyps, 444, 452
POPs. *See* Persistent organic pollutants
Porpoises, 422
dolphins vs., 426
swimming ability of, 432–433
Porter, John, 454
Portuguese man-of-war, 412
Posidonius, 75
Power, tidal, 280–281, 283
Power plants, tidal, 265
Precipitate, 89–90, 112
Precipitation, basis of, 138
Precipitation latitudes, 138
Precision depth recorder (PDR), 76
Prediction, idealized tide, 274–275
Pressure
atmospheric circulation and, 172
on deep-ocean floor, 458
high- and low-, air flow and, 177
water, 361–363
Prevailing westerly wind belts, 173
Primary consumers, 408
Primary productivity, 373–379
availability of nutrients and, 374–375
margins of ocean and, 375–376
photosynthetic productivity, 373–374
solar radiation and, 375
upwelling and, 376
Primary treatment, 335
Prince William Sound, 329–332
Principle of Constant Proportions, 131, 144
Problematum (Aristotle), 6
Producers, 389
Productivity. *See also* Biological productivity
defined, 106
photosynthetic, 373–374
polar oceans, 383–384, 388
primary, 373–379
SeaWiFS image of world, 380
tropical oceans, 398–401
upwelling and, 209, 217
Progressive waves, 238, 240
Protoctista, 349
Protoearth, 20
Protons, 132
Protoperidinium divergens, 383
Protoplanets, 20
Protoplasm, 352–353
Protozoa, 349
Protozoans, in biogenous sediment, 106
Proxigean, 274
Pteropod ooze, 110
Ptolemy, 13
PTWC. *See* Pacific Tsunami Warning Center
Ptychodiscus, 384
Puffer fish, 459
Puget Sound, 305
Purse seine net, 397–398
Pycnocline, 152, 200, 238–239, 323, 382
Pytheas, 12

Quadrature, moon in, 272

Quarter moon, 271
Quartz, 101
Queen angel fish, 414
Queen crab, 446
QuickSCAT/Sea Winds, 80

Radioactivity, 20
Radiolarian ooze, 108
Radiolarians, 106, 107, 351, 408–409
Radiometric age dating, 25–30
Rapid wasting disease, 454
Recreational beach, 290
Red algae, 381–382
Red clays, 105
Red muscle fibers, 415
Red Sea, 327
early navigators on, 12
salinity of, 142–143
Red tides, 382–384, 400
Reflection, wave, 252–253
Refraction, wave, 251–252
Regenerated production, 374
Regional productivity, 382–389
Relict sediments, 103
Relief, 75
Relocation, 310
Remoras, 390–391
Residence time, 145
Resolution, 17, 97–98
Resources
ocean sediments as, 120–124
tidal power, 265, 280–281, 283
wave energy, 260
Respiration, 24
Restoring force, 244
Resultant forces, 268–269
Reversal tidal currents, 281
Reverse osmosis, 156–157
Reversing current, 277
Rhizophora, 379
Rhizosolenia, 351
Rhodophyta, 381–382
Rift valley
East African, 57
fissures, 90
in general, 53, 87
Mid-Atlantic Ridge, 55
Riftia, 461
Rifting, 53
Right whales, 430
Ring of Fire, 86, 88, 254
Rings, 214
Rip currents, 294, 310–311
Rip tides, 310
Ripples, 244
Rip-rap, 307
River valleys, drowned, 306, 320
Rivers, floodplains along, 322
RMS *Titanic,* 34, 162–163
Rock louse, 443
Rocks
in mid-ocean ridge, 46
paleomagnetism in, 42
sedimentary, 99
Rock weed, 443
Rocky bottoms, 450–452
Rocky shores, 439–446
high tide zone, 440
low tide zone, 441–446
marine organisms in, 442–443
middle tide zone, 440–441
spray zones, 439–440
Rogue waves, 249
Rorqual whales, 429
Ross, James Clark (Sir), 438

Ross, John (Sir), 438
Ross Ice Shelf, 190
Rotary current, 277
Rotary drilling, 97
Rough keyhole limpet, 443
Rounded fins, 412
Royal Society of England, 130
Runoff, 144
Ruptert Inlet, 86

Saffir–Simpson Scale, 179
Salinity, 140
 of coastal waters, 318–319
 depth variation, 149–150, 152
 determining, 144
 development of ocean, 22
 expressing, 142
 marine animals and, 358–360
 processes affecting, 147–150
 surface variation, 148–149, 151
 variations of, 142–144
 water conductivity and, 144
Salinometer, 144
Salp, 351
Salt marshes, 296, 323–325, 446
Salt wedge estuary, 322
Saltwater zone, 326
San Andreas Fault, 60, 69, 82, 321
San Clemente island, 306
Sand. See also Beach
 adding to beach, 300
 hard stabilization and, 307
 movement of, on beach,
  290–292
Sand crabs, 448–449
Sand dollar, 354
Sand and gravel industry, 121–122
Sand star, 447, 448
Sandwell, David, 79
Sapphirina auronitens, 410
Sarda, 416
Sardine, 353
Sargasso Sea, 10, 214
Sargasssum, 381
Schooling, 418–420
Scientific method, 17–19
Scientific names, 367
Scomber, 416
Scuba equipment, 7
Scyphozoan, 411. See also
 Jellyfish
"Sea" (wind-generated waves),
  244–245
Sea
 defined, 10
 law of the, 316
 territorial, 316
Sea anemone, 354, 441, 443–444,
  461, 463, 466
Sea arch, 293, 295
Sea area, 244
Sea Around Us, The (Carson), 97
Sea breezes, 175, 177
Sea caves, 293
Sea cucumber, 354
Sea floor. See Ocean floor
Sea floor mapping
 from space, 78–79
 future of, 93
Sea floor maps, 79
Sea floor spreading
 in general, 43–46
 ocean basin formation by, 56
 magnetic evidence of, 45
 plate boundaries and, 53–55
 sea level fluctuation and, 302

sediment accumulation and,
  112
Sea hares, 451
Sea ice, 148
Sea level
 changes in, 300–303
 greenhouse effect and, 303
 movement of continents and,
  311
 rise of, at New York City, 303
Sea lily, 354
Sea lions, 421–422
Sea MARC, 76
Sea otters, 420
Sea salt, mining, 123
Sea squirt, 354
Sea stack, 293, 295
Sea star, 354, 440, 444, 467
Sea turtles, magnetic field and, 44
Sea urchins, 354, 441, 445,
  451–452
Seabeam, 76
Seaknolls, 85
Sealab project, 74
Seals, 421–422
Seamounts, 62–63, 85
Seasat A mission, 80
SeaStar satellite, 379
SeaStar/SeaWiFS, 80, 104
Seawalls, 309–310
Seawater. See also Water; ??
 acidity and alkalinity, 146–147
 comparing with pure water, 155
 density, 150–155
 desalination, 155–157
 dissolved components, 144–146
 dissolved gases in, 360–361
 drinking, 157–158
 freezing, 158
 high transparency, 361
 hydrologic cycle, 148–149
 light transmission, 376–379
 salinity, 140–144, 147–150
 salt in, 134
 water vs., 140
Seaweed, 367–368
SeaWiFS instrument, 374
 image of world productivity,
  379
Secchi disk, 379
Secondary treatment, 335
Sedimentary rock, 40, 99
Sediment-covered shores,
  446–450
 intertidal zonation, 446–447
 life in the sediment, 446–447
 modes of feeding along, 448
 mud flats, 449–450
 sediment on, 446
Sediments
 biogenous, 100
 classification of, 100
 cosmogenous, 100, 114–116
 defined, 99
 evaporative salts, 122
 gas hydrates, 120–122
 hydrogenous, 100, 111–114
 lithogenous, 100–105
 manganese nodules and crusts,
  123–124
 maturity of, 105
 media transporting, 102
 mixtures, 115–116
 pelagic, 120
 petroleum, 120
 phosphorite, 123

relict, 103
 as resource, 120–124
 sand and gravel, 121–122
 submarine canyon, 311
 terrigeneous, 100
 Wentworth scale, 105
Seed-bearing plants, 379–380
Segmented worm, 448
Seismic moment magnitude, 54
Seismic profiling, 81
Seismic reflection profiles, 76
Seismic sea waves, 238, 253
Seismologists, 50
Semidiurnal tidal pattern, 276,
  279
Sepia, 407, 412
Seriola, 416
Sessile animal forms, 440
"Seven Seas," 9–10
Sewage sludge, 335–336
 Boston Harbor project, 335,
  337
 New York's disposal at sea, 335
Sextant, 14–15
Shallow offshore ocean floor,
  450–452
Shallow subtidal zone, 365
Shallow-water waves, 242–244
Sharks, 413, 414
 cancer in, 432
 great white, 417, 432
 myths and facts about, 417
Shelf break, 82
Shellfish, contaminated, 384–385
Shinkai 6500, 34
Shoaling, 248
Shore, 289. See also Coast;
  Coastal regions; Shoreline
 depositional-type, 292, 295–300
 erosional-type, 292–295
 rocky, benthic organisms on,
  439–446
 sediment–covered, benthic
  organisms on, 446–450
Shore crabs, 446
Shoreline, 289, 290. See also
  Coast; Coastal regions;
  Shore
 emerging, 300–301
 evidence of ancient, 302
 movement parallel to, 291
 movement perpendicular to,
  290
 refraction along, 252
 submerging, 300–301
Shrimp, deep-water, 366
Side-scanning sonar, 76–77
Silica, 106, 358
Silicate minerals, 49
Siliceous ooze, 106–107, 111, 114
Silicoflagellate ooze, 108
Sill, 327
Silvertip shark, 414
"Sink level," 302
Siphonophore, 351, 411
Sirenia, 421–422
Slightly stratified estuary, 321
Small, Peter, 7
Small toothed whales, 427
Smith, Walter, 79
Snail, 354
Snails, carnivorous, 440
Snowflakes, 140
Society Islands, 11
SOFAR channel, 192–193
Solar bulges, 270

Solar day, 270
Solar distillation, 155–156
Solar energy, distribution of, 164
Solar heating, uneven, 164–168
Solar humidification, 155–156
Solar radiation, photosynthesis
  and, 375
Solar system, origin of, 19–21
Solid state, 134
Somali Current, 225
Sonar, 76
Sorting, 101
Sounding, 17, 75
South American Plate, 54, 81
South Atlantic Gyre, 213–214
South Equatorial Current, 214
South Pacific Gyre, 217
Southeast trade winds, 173
Southern Ocean, 9. See also
  Antarctic Ocean
Southern Oscillation, 219
Southwest Monsoon Current, 225
Space, sea floor mapping from,
  78–79
Space dust, 114
Spartine, 379
Spawning, 418
Species, 24, 349
 cold-water vs. warm-water, 358
 dearth of, 352
 distribution on Earth, 355
 diversity of, on deep-ocean
  floor, 459
Species diversity, oil spills and,
  332
Sperm whales, 423, 427–428
Spermaceti organ, 427
Spermatophyta, 379–380
Spherules, 114, 116
Spilling breaker, 250
Spirula, 407, 412
Spit, 295
Splash wave, 236
Sponge, 354
 blue-gray, 457
 brown, 457
Sponge weed, 381
Spray zone, 439–440
Spreading center, 43
Spring bloom, 389
Spring tide, 271–272
Squid, 353
 life cycle of, 352
 swimming, 412
SS Marine Sulphur Queen, 29
Standard laboratory bioassay, 329
Standing stock, 396
Standing waves, 253–254
Stationary waves, 253
Steinbeck, John, 372
Stellwagen Bank, 336
Stenella, 416
Stenella attnuata, 397
Stenohaline, 358
Stenothermal, 358
Stick chart, 14
Still water level, 240
Stomatoid, 415
Stommel, Henry, 229
Storm drains, 340–341
Storm surge, 180, 182
Storms. See also Hurricanes;
  Weather
 abyssal, 458
 climate patterns and, 182–185
 in general, 175–178

tropical cyclones, 178–182
Strait of Gibraltar, 115, 327
Strait of Messina, 327
Stranded beach deposits, 300–301
Streamlining, 355–357
Striped bass, 353
Stromatolites, 115
Structure and Distribution of Coral Reefs, The (Darwin), 63
Subduction, 44
Subduction zone, 44
Subduction zone seep biocommunity, 464
Sublittoral zone, 365, 450–452
Submarine, 34
Submarine canyons, 83, 85, 311
Submarine fans, 84
Submerged dune topography, 300–301
Submerging shorelines, 300–301
Submersibles, 34
Subneritic province, 365
Suboceanic province, 365
Subpolar gyres, 202
Subpolar low, 172
Subpolar region, 185
Subtropical convergence, 205
Subtropical gyres, 201–202
Subtropical highs, 172
Subtropical regions, 185
Suez Canal, 327
Summer solstice, 164
Summertime beach, 291–292
Sun
    declination of moon and, 273
    energy radiated by, 186
    heliocentric theory, 19
    photosynthesis and, 375
    size of, 271
    solar bulges, 270
    solar day, 270
    solar distillation, 155–156
    solar energy, 164
    solar heating, 164–168
    solar humidification, 155–156
    solar system origin, 19–21
    tides and, 270
Sunda Strait, 256
Sunmore, 236–237
Superwaves, 249
Supralittoral zone, 365, 439–440
Surf, 247–251
Surf beat, 247
Surf grass, 379–380, 441, 445
Surf zone, 247, 250, 376
Surface area to volume ratio, 356
Surface currents, 198–208
    Antarctic circulation, 211–213
    Atlantic Ocean circulation, 213–217
    boundary, 200
    downwelling and, 211
    Ekman spiral, 202, 204–205
    Ekman transport, 204–205
    equatorial, 200
    equatorial countercurrents, 207
    geostrophic, 205
    gyres, 201–202
    Indian Ocean, 223, 225
    Pacific Ocean, 217–223
    subtropical gyres and, 202
    western intensification, 205–209
    wind-driven, 201

Surface tension, 133
Surface water, converging, 209
Surfing, 250–251, 260
Surging breaker, 250
Suspension feeding, 447
Suspension settling, 84, 87
Susquehanna River, 323
Sverdrup (Sv), 212n8
Swamps, mangrove, 323–325
Swash, 290
Swell, 245–247, 260
Swim bladder, 408
Swimmers, 350, 353
Symbiosis
    of coral and algae, 456–457
    in general, 390–392

Tablemounts, 62–63, 85
TAO. *See* Tropical Atmosphere and Ocean
TAP. *See* Transarctic Acoustic Propagation Experiment
Taxonomy, 349
TDS. *See* Total dissolved solids
Technology
    "liquid air," 28
    use of, oceans and, 3–4
Tectonic estuary, 320–321
Tectonic movements, 301–302
Tektites, 114
Temperate oceans, biological productivity in, 388–390
Temperate regions, 185
Temperature, 134
    of atmosphere, 166–167
    of coastal waters, 318–319
    extremes in ocean and land, 357
    Gulf Stream and surface, 216
    marine organisms and, 356–358
    near Equator, 149
    oceanic heat flow and, 165
    record since 1865, 188
    sea-surface
        anomaly maps, 220
        off southern California, 223
    seawater density and, 152
    surface, of world ocean, 210
    surface salinity and, 151
Temperature–salinity (T–S) diagram, 225, 227
Terra, 80
Terrigeneous sediment, 100
Territorial sea, 316
Testing hypotheses, 18
Tests, 105, 382
Tethys Sea, 35, 327
Texas
    Galveston, hurricane in, 184–185
    Laguna Madre, 326
Texture, 100
Thalassia, 449
Thalassiorsira, 351
Theory, 18–19
Thermal contraction, 139
Thermal expansion, 230
Thermocline, 152, 218, 318, 382
Thermohaline circulation, 225–226, 228–229
Thomson, C. Wyville, 130, 406
Thorvaldson, Erik "the Red," 13
Thunnus, 416
Thunnus albacares, 416
Tidal bore, 278
Tidal bulges

moon and, 270, 273
    sun and, 270
Tidal currents
    coastal, 277–280
    reversal, 281
Tidal cycle, monthly, 271
Tidal patterns, 276–277, 279
Tidal period, 270
Tidal power, 280–281, 283
Tidal power plants, 265
Tidal range, 265, 326
    Bay of Fundy, 280
Tidal waves, 254. *See also* Tsunami
Tide pools, 441
Tide tables, 283–284
Tide zones, 439–446
Tide-generating force, 266–270
    maximum, 284
Tides, 265–287. *See also* Waves
    atmospheric, 281
    defined, 266
    Earth, 281
    Earth's rotation and, 285
    equatorial, 283
    in general, 238, 242, 266
    generating, 266–274
    idealized prediction, 274–275
    low, 275–276
    meteorological, 284
    moon and, 270–271
    negative, 284
    in ocean, 274–280
    power from, 280–281, 283
    sun and, 270
    tropical, 281–282
Tiger grouper, 418
Tintinnid, 351
Titanic, 162–163
TOGA. *See* Tropical Ocean–Global Atmosphere
Tombolo, 296
TOPEX/Poseidon
    mission, 80
    satellite, 199–200
    wave height measurements, 246
Toscanelli, 174
Total dissolved solids (TDS), 144
Trade winds, 173, 201, 219
Transarctic Acoustic Propagation Experiment (TAP), 193
Transfer efficiency, 393
Transform active margins, 81–82
Transform boundaries, 52, 59–60
Transform faulting, 60
Transform faults, 59, 91, 93
Transitional waves, 244
Transitional zone, 326
Transport, longshore, 291
Transport of floating objects, 204
Transporting media
    sediment-, 102
    wind as, 124
Transverse waves, 239–240
Trenches. *See* Ocean trenches
Tridacna gigas, 457
Trieste, 10, 12
Trochoidal waveform, 244
Trophic levels, 392–393
    passage of energy through, 394
Tropic of Cancer, 148–149, 164, 185
Tropic of Capricorn, 148–149, 164, 185
Tropical Atmosphere and Ocean (TAO), 223

Tropical cyclones, 178–182. *See also* Hurricanes
    destruction by, 180, 182
    movement in, 179–180
    origin of, 178–179
    paths of major, 180
    storm surge, 180, 182
    wind speed in, 179
Tropical depression, 179
Tropical Ocean–Global Atmosphere (TOGA), 223
Tropical oceans, productivity in, 384, 386–389
Tropical regions, 164, 185, 379
Tropical storm, 179
Tropical tides, 281–282
Troposphere, 166
Troughs, 240
Truncate fins, 412
Truth, theories and, 18–19
Tsunami
    coastal effects, 254
    historic, 254–256
    occurrence since 1990, 258
    origin of, 255
    record height of, 260
    as shallow-water waves, 242
    warning system, 256–259
Tsunami warning, 257, 261
Tsunami watch, 257
Tube worms, 461, 463
Tuna, 397–398
    body temperatures of, 416
    speed of, 416
Turbidite deposits, 84, 103
Turbidity currents, 83, 103
    evidence for, 86
Tursiops truncatus, 423, 428
Turtle grass, 449
Two Faces of Chemistry, The (Caglioti), 130
Typhoons, 178

U.S. Coast Guard, International Ice Patrol, 162
U.S. coasts
    Atlantic Coast, 304–305
    characteristics of, 304–305
    erosion and deposition, 304
    Gulf Coast, 305
    Pacific Coast, 305
U.S. Environmental Protection Agency (EPA)
    on bioassay use, 329
    on DDT, 336
    Office of Wetlands Protection (OWP), 324–3
U.S. Fish and Wildlife Service, on *Exxon Valdez* spill, 330
U.S. Food and Drug Administration, on mercury concentration in fish, 339
U.S. Navy
    Geosat satellite, 79
    ice patrol, 162
    Sealab project, 74
Uca, 449, 450
Undertow, 311
Underwater habitats, 91
Underwater living, experiments in, 74–96
United Nations Conference on the Law of the Sea, 316–317
United States
    air masses affecting weather, 178

barrier islands along Atlantic
Coast, 296
most destructive hurricane, 183
oceanographic satellite
missions, 80
source of cobalt for, 123
Up-coast, 290
Upper water, 153
Upstream, 291
Upstream side, 307
Upwelling
calcareous ooze and, 111
coastal, 209, 211–212, 376–377,
386
defined, 376
equatorial, 386
nutrient supply and, 376
productivity and, 217
Urchin barren, 466
USS *Ramapo*, 244, 246
Utah, Great Salt Lake, 143

van Bynkershoek, Cornelius, 316
van Sandick, N., 256
van der Waals forces, 134–135
Vapor, 135
Vaporization, latent heat of, 137
Vernal equinox, 164
Vertically mixed estuary, 321
Victoria, 15
Vikings, 13, 174
Vine, Fredrick, 44
Viper fish, 416
Viscosity, 50, 355–356
Visible light, 376
Volcanic arc, 55, 86
Volcano
on abyssal plains, 85
black sand beaches and,
124–125
Krakatau, 256
Loihi, 62
side-scanning image of, 77
Volume ratio, surface area to, 356
Vortex, 214

Walker Circulation Cell, 218
Wallace, Alfred Russel, 346
Walruses, 421
Walsh, Don, 10
Walter, Gilbert T. (Sir), 218
Waning crescent, 273
Waning gibbous, 273
Ward, Barbara, 372
Warm core rings, 214
Warm front, 176, 179

Warm-blooded fish, 416
Warm-water vents, 89
Water
boiling point, 135
brackish, 142, 326
conductivity of, salinity and,
144
deep, sources of, 226–227
density of, 139–144
freezing point, 135
gaseous state of, 135
in general, 131
heat capacity of, 135–136
high transparency of, 361
latent heats in, 136–139
liquid state of, 134–135
properties of, 141
pure
comparing seawater and, 155
as neutral solution, 146
seawater. *See* Seawater
solid state of, 134–135
thermal properties of, 134–139
as universal solvent, 133–134
using electric appliances near,
157
viscosity of, 355–356
Water molecule, 132–134
geometry of, 132–133
interconnections of, 133
polarity in, 133
Water pressure, 361–363
Water spouts, 191
Wave(s)
biggest recorded, 236
capillary, 244–2245
causes of, 238
characteristics of, 240–244
circular orbital motion of, 241
deep-water, 241–243
forced, 275
gravity, 244–245
heavy activity, 291
internal, 238–239, 259
light activity, 291
longitudinal, 238–239
movement of, 238–240
ocean, 238
orbital, 240
power from, 259
progressive, 238, 240
rogue, 249
shallow-water, 242–244
standing, 253
surf, 247–251
in surf zone, 376

swell in, 245–247
tidal, 254. *See also* Tsunami
transitional, 244
transverse, 239–240
wind-generated, 244–253
Wave base, 241
Wave dispersion, 245
Wave height
Wave period, 240
Wave reflection, 252–253
Wave refraction, 251–252
Wave speed, 242
Wave steepness, 240
Wave trains, 245
Wave-cut bench, 289, 296
Wave-cut cliffs, 293
Waveform, 239
trochoidal, 244
Wavelength, 240
Waxing crescent, 272–273
Waxing gibbous, 273
Weather. *See also* Climate
defined, 175
storms and, 175–178
winds and, 175
Weathering, 100–101
"Wedge, The," 253
Wegener, Alfred, 34–37, 39–40
Wentworth scale, 105
Wentworth scale of grain size, 101
Wesly, Claude, 74
West Australian Current, 225
West Falmouth Harbor, oil spill
in, 331, 333
West Wind Drift, 211
Western boundary currents, 200
Western intensification, 205–209,
208
Wetlands
coastal, 323
defined, 323
loss of, 324–325
Whales
baleen, 428–429
blue, 425, 428
bottlenose, 423
finback, 428
gray, 425, 428–434
humpback, 425
humpback, 428, 430
in order Cetacea, 422
intelligence of, 433
killer, 416, 425–426, 431, 432
likelihood of seeing, 399–400
northern right, 424
right, 430

rorqual, 429
small toothed, 427
sperm, 423, 427–428
types of, 424–425
white, 423
White Cliffs of England, 106, 109
White muscle fibers, 415
White plague disease, 454
White pox, 454
White smokers, 89
White whale, 423
Whitecaps, 244
Whittaker, Robert H., 348
Who Speaks for Earth? (Ward),
372
Wind
defined, 168, 175
monsoon, 225
as transporting agent, 124
waves generated by. *See* Wind-
generated waves
Wind belts, 172–173, 176
Wind-driven surface currents, 201
Wind-generated waves, 244–253
breakers, 250
reflection, 252–253
refraction, 251–252
"sea," 245–246
surf, 247–251
swell, 245–247
Winter solstice, 164
Wintertime beach, 291–292
World Health Organization, on
pollution, 328
Worm larva, 351
Worms
annelid, 447–449
segmented, 448
tube, 461, 463

Yellow-band disease, 454
Yellowfin tuna, 416, 418
Yellowstone National Park, 61
Yellowtail, 416
Yucatán peninsula, 118

Zenith, 267
Zero energy level, 240
Zooplankton, 349–351, 392
macroscopic, 408–409
microscopic, 408
in polar oceans, 383–384
Zooxanthellae, 455–457, 468
Zostera, 379, 449–450

Simplified Phylogenetic Tree of Marine Life

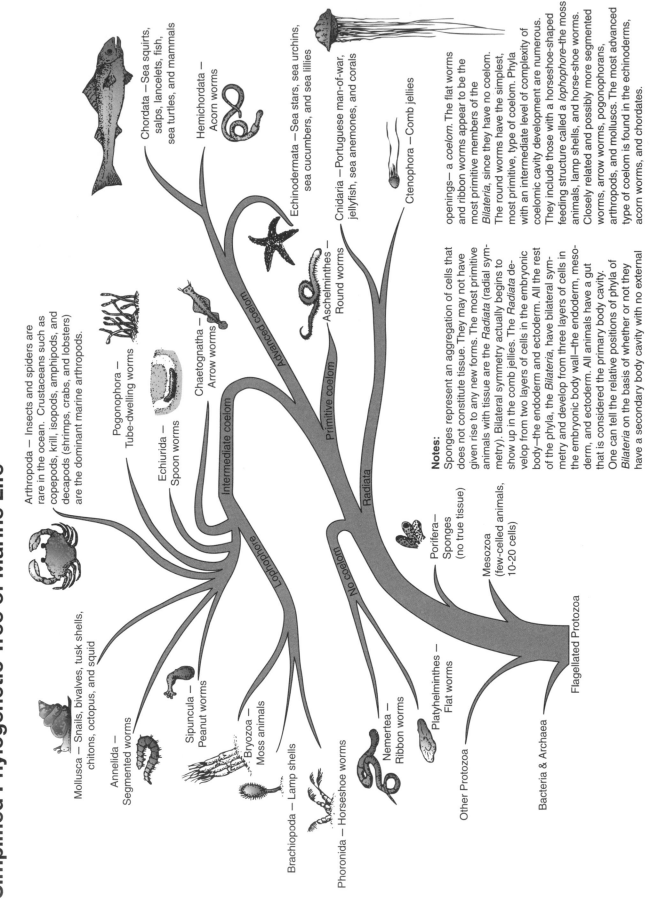

Arthropoda — Insects and spiders are rare in the ocean. Crustaceans such as copepods, krill, isopods, amphipods, and decapods (shrimps, crabs, and lobsters) are the dominant marine arthropods.

Chordata — Sea squirts, salps, lancelets, fish, sea turtles, and mammals

Hemichordata — Acorn worms

Pogonophora — Tube-dwelling worms

Echiurida — Spoon worms

Chaetognatha — Arrow worms

Echinodermata — Sea stars, sea urchins, sea cucumbers, and sea lilies

Cnidaria — Portuguese man-of-war, jellyfish, sea anemones, and corals

Ctenophora — Comb jellies

Aschelminthes — Round worms

Mollusca — Snails, bivalves, tusk shells, chitons, octopus, and squid

Annelida — Segmented worms

Sipuncula — Peanut worms

Bryozoa — Moss animals

Advanced coelom

Intermediate coelom

Primitive coelom

Lophophore

Radiata

No coelom

Porifera — Sponges (no true tissue)

Mesozoa (few-celled animals, 10-20 cells)

Brachiopoda — Lamp shells

Phoronida — Horseshoe worms

Nemertea — Ribbon worms

Platyhelminthes — Flat worms

Other Protozoa

Flagellated Protozoa

Bacteria & Archaea

Notes:

Sponges represent an aggregation of cells that does not constitute tissue. They may not have given rise to any new forms. The most primitive animals with tissue are the *Radiata* (radial symmetry). Bilateral symmetry actually begins to show up in the comb jellies. The *Radiata* develop from two layers of cells in the embryonic body—the endoderm and ectoderm. All the rest of the phyla, the *Bilateria*, have bilateral symmetry and develop from three layers of cells in the embryonic body wall—the endoderm, mesoderm, and ectoderm. All animals have a gut that is considered the primary body cavity. One can tell the relative positions of phyla of *Bilateria* on the basis of whether or not they have a secondary body cavity with no external openings—a *coelom*. The flat worms and ribbon worms appear to be the most primitive members of the *Bilateria*, since they have no coelom. The round worms have the simplest, most primitive, type of coelom. Phyla with an intermediate level of complexity of coelomic cavity development are numerous. They include those with a horseshoe-shaped feeding structure called a *lophophore*—the moss animals, lamp shells, and horse-shoe worms. Closely related and possibly more segmented worms, arrow worms, pogonophorans, arthropods, and molluscs. The most advanced type of coelom is found in the echinoderms, acorn worms, and chordates.

Commonly Used Conversion Factors

Length

1 millimeter (mm)	1,000 micrometers 0.1 centimeter 0.001 meter 0.0394 inch
1 centimeter (cm)	10 millimeters 0.01 meter 0.394 inch
1 meter (m)	100 centimeters 39.4 inches 3.28 feet 1.09 yards 0.547 fathom
1 kilometer (km)	1,000 meters 1,093 yards 3,280 feet 0.62 statute mile 0.54 nautical mile
1 inch (in)	25.4 millimeters 2.54 centimeters
1 foot (ft)	12 inches 30.5 centimeters 0.305 meter
1 fathom (fm)	6 feet 2 yards 1.83 meters
1 statute mile (mi)	5,280 feet 1,760 yards 1,609 meters 1.609 kilometers 0.87 nautical mile

Area

1 square centimeter (cm^2)	0.155 square inch 100 square millimeters
1 square meter (m^2)	10,000 square centimeters 10.8 square feet
1 square kilometer (km^2)	100 hectares 247.1 acres 0.386 square mile 0.292 square nautical mile
1 square inch (in^2)	6.45 square centimeters
1 square foot (ft^2)	144 square inches 929 square centimeters

Temperature

Exact Formula	Approximation (Easy Way)
$= \dfrac{(°F - 32)}{1.8}$	$°C = \dfrac{(°F - 30)}{2}$
$.8 \times °C) + 32$	$°F = (2 \times °C) + 30$

Volume

1 cubic centimeter (cc; cm^3)	1 milliliter 0.061 cubic inch
1 liter (l)	1,000 cubic centimeters 61 cubic inches 1.06 quarts 0.264 gallon
1 cubic meter (m^3)	1,000,000 cubic centimeters 1,000 liters 264.2 gallons 35.3 cubic feet
1 cubic kilometer (km^3)	0.24 cubic mile 0.157 cubic nautical mile
1 cubic inch (in^3)	16.4 cubic centimeters
1 cubic foot (ft^3)	1,728 cubic inches 28.32 liters 7.48 gallons

Mass

1 gram (g)	0.035 ounce
1 kilogram (kg)	2.2 pounds 1,000 grams
1 metric ton (mt)	2,205 pounds 1,000 kilograms 1.1 U.S. short tons
1 pound (lb) (mass)	16 ounces 454 grams 0.454 kilogram
1 U.S. short ton (ton; t)	2,000 pounds 907.2 kilograms 0.91 metric ton

Speed

1 centimeter per second (cm/s)	0.0328 foot per second
1 meter per second (m/s)	2.24 statute miles per hour 1.94 knots 3.28 feet per second 3.60 kilometers per hour
1 kilometer per hour (kph)	27.8 centimeters per second 0.62 mile per hour 0.909 foot per second 0.55 knot
1 statute mile per hour (mph)	1.61 kilometers per hour 0.87 knot
1 knot (kt)	1 nautical mile per hour 51.5 centimeters per second 1.15 miles per hour 1.85 kilometers per hour